The BRAIN METAFUNCTION:

A Paradigm Shift in Neuroscience

The Brain, The Transphysical, and Christian Theism

Dr. Orlando David Garcia

ISBN: 978-1-7782962-4-6

About The Author

Dr. Orlando David Garcia is a Cuban Canadian medical doctor and neurologist who researched coma and brain death. In 1992, while working at the Institute of Neurology and Neurosurgery in Havana, Cuba, he received the National Prize for the Best Scientific Work of the Year for a study titled Heart Rate Variability in Coma and Brain Death. [a, b] He then published papers proposing a unifying formulation of human death—an approach now further developed and refined in this book through the Brain Metafunction Framework and its broader implications. Dr. Garcia also received training in bioethics through the Bioethics Group of Galicia in Spain.

Dr. Orlando David Garcia welcomes inquiries related to the topics explored in this book, as well as questions from readers. He can be reached at: info.orlandodgarcia@gmail.com.

a. García, O. D., Machado, C., Román, J. M., Cabrera, A., Díaz-Comas, L., Rivera, B., & Grave de Peralta, R. (1995). Heart rate variability in coma and brain death. In C. Machado (Ed.), Brain death (Proceedings of the Second International Symposium on Brain Death) (Vol. 9, pp. 191–200). Elsevier.

b. Machado, C., García, O. D., Gutiérrez, J., Portela, L., & García, M. C. (2005). Heart rate variability in comatose and brain-dead patients. Clinical Neurophysiology, 116(12), 2859–2860.

Gratitude and Dedication

I am grateful to God for His strength and inspiration, to my wife Crystal for her encouragement and tireless support, and to my family—especially my daughter Sarah Lidia, my son David Jeremiah, and all our loved ones whose presence and prayers have sustained me. I dedicate this book to them.

Acknowledgments

I thank Dr. Philip Maruncic, PhD, for reading portions of the manuscript and providing insightful, constructive feedback. My gratitude also goes to Dr. Leonid Maximenkov, PhD; Dr. Edmundo Rivero, MD, MSc; and Dr. José de Jesús Sánchez, MD—both neurologists—for their support and valuable input in reviewing sections of the text. I am equally indebted to Professor Victor Orencio Rodriguez, BLA, MEd; Teresita Vega, BSEd; and Silvia Williams, MA in Modern Languages—longtime, outstanding educators—for their encouragement and thoughtful support in reading and reviewing portions of the work. I also extend my thanks to Professor Calixto Machado, MD, PhD, FAAN, for introducing me to the research on brain death and heart rate variability.

CONTENTS

PART I

Foundations of Brain Function and Metafunction

Chapters:

1. *The Brain*
2. *The Brain Metafunction*
3. *Integration of Cortical and Subcortical Structures in the Brain Metafunction Framework*

Overview:

This section introduces the biological architecture and conceptual innovation of the Brain Metafunction Framework (BMF). As a novel theoretical model, the BMF reconceives the brain not merely as a modular processor, but as a dynamically integrative system—one that recursively weaves together perception, memory, affect, bodily regulation, and symbolic cognition. More than a descriptive account, the BMF serves as a generative framework with wide transdisciplinary applications. It provides explanatory power across clinical neurology, cognitive science, philosophy of mind, theological anthropology, and neuroethics.

This foundational section establishes how neural, psychological, and symbolic processes interlock to support the unity of personhood. While building upon existing models such as predictive coding, global neuronal workspace theory (GNWT), and modular architectures, the BMF expands these frameworks. It affirms hierarchical inference and error minimization but gives central place to symbolic cognition, affective resonance, and interoceptive depth. Unlike GNWT, which centers on conscious access, the BMF emphasizes recursive interdependence across conscious and unconscious systems. In contrast to modular accounts, it views functions as co-constituted phenomena within coordinated afferent–integrative–efferent loops. This foundational shift supports a richer account of moral agency, clinical neurophenomena, and the integrated self.

PART II

Sensory and Cognitive Processing

Chapters:

4. *Perception Part I*
5. *Perception Part II*
6. *Attention*
7. *Memory*

Overview:

This section explores how sensory input, attention, and memory are processed and unified within the Brain Metafunction Framework. These domains are understood not as isolated mechanisms but as interdependent operations that undergird higher symbolic, narrative, and moral capacities. The BMF reframes these processes as recursive and generative—providing a model that advances both theoretical neuroscience and clinical practice. A key innovation here is the efferent loop of perception. The framework is applied to neurological and psychological conditions such as blindsight, Charles Bonnet syndrome, hemineglect, hallucinations, and near-death phenomena, reframing them not as anomalies but as patterned deviations within an integrated architecture. In cases where perception persists without input—or becomes decoupled from external reality—the BMF reveals perception as symbolic, participatory, and context-sensitive. These insights inform diagnostic, rehabilitative, and therapeutic applications.

PART III

Consciousness, the Unconscious, and Integration

Chapters:

8. *Consciousness Part I*
9. *Consciousness Part II*
10. *The Unconscious*
11. *The Neural Binding and Metabinding Problems*

Overview:

This pivotal section extends the Brain Metafunction Framework by addressing consciousness, the unconscious, and their integration in shaping personhood. Consciousness is not presented as an isolated function but as a synthesis of perception, attention, memory, and emotion. Unconscious processes are reframed as structurally generative rather than merely repressed or automatic. A key innovation here is the concept of metabinding—the recursive integration of neural, emotional, symbolic, and somatic domains. This model offers a new lens for understanding subjective coherence, illuminating how the brain organizes experience across both conscious and unconscious states.

PART IV

Agency, Identity, and Embodied Limits

Chapters:

12. *Brain Metafunction Syndromes*
13. *The Unity of the Human Person*
14. *A Refined Formulation of Human Death*
15. *Metafunctional Free Will*
16. *Brain Metafunctional Ports: A Novel Framework for Neutral Interfaces*

Overview:

Bridging neuroscience, clinical medicine, and bioethics, this section explores how the BMF reframes human identity, agency, and embodied limits. It introduces a novel classification of Brain Metafunction Syndromes, offering revised interpretations of coma, vegetative states, and dissociative conditions. These syndromes highlight how disruptions in recursive integration illuminate the architecture of personhood. A conceptually grounded definition of human death is proposed—not as mere biological cessation, but as the irreversible breakdown of metafunctional coherence. The chapter on Metafunctional Free Will presents volition as resulting from agency including recursive interplay of unconscious motivation, conscious intention, and symbolic cognition. The introduction of Brain Metafunctional Ports provides a paradigm for understanding neural convergence zones with implications for rehabilitation, cognitive augmentation, and neuroethical responsibility. Collectively, this section presents a unified, ethically informed model of the human person that transcends reductionism and expands clinical and moral insight.

PART V

Ethics, Technology, and Predictive Horizons

Chapters:

17. Bioethics
18. Neuroethics
19. Twelve Brain Metafunction Framework Predictions

Overview:

This section applies the Brain Metafunction Framework to urgent ethical and technological frontiers. Chapters on bioethics and neuroethics examine critical issues such as neuroenhancement, cognitive liberty, identity, personhood, and decision-making in the age of AI, neural data, and brain–machine interfaces. The final chapter presents twelve empirical predictions generated by the BMF—ranging from the modeling of consciousness and psychosomatic health to neural ethics and transphysical inquiry. These predictions offer a foresight model for how neuroscience may develop beyond reductionism toward dynamic, integrative paradigms. This part offers a vision of scientific innovation rooted in the dignity and coherence of the human person.

PART VI

Transphysical and Theological Horizons

Chapters:

Overview:

This culminating section explores the interface between neuroscience and theology, proposing Brain Metafunction Transphysicalism as a model for understanding the cooperation between neural and transphysical processes. Rather than reducing consciousness to neural emergence or adopting traditional dualistic interactionism, the BMF posits a recursive unity wherein the soul functions as the integrative center of personhood, constituted through the dynamic coherence of bodily, cognitive, and transphysical domains. The chapter introduces infomism—a proposed mode of causality defined by informational exchange without energy transfer. Infomism offers a conceptual tool for dualist and non-reductionist metaphysics alike, articulating how meaning-bearing processes may influence the brain without violating physical laws. This section aligns the BMF with Christian theological anthropology, presenting the brain as a site of cooperation between matter and spirit, reason and revelation.

Conclusion

A synthetic reflection on the major contributions of the Brain Metafunction Framework, its metaphysical implications, and its capacity to advance interdisciplinary discourse—offering a unified vision of science, philosophy, and spiritual anthropology.

INTRODUCTION

The origins of this book trace back three decades, when the author's research into coma and brain death gradually led to the realization that all brain functions are deeply integrated. This insight formed the basis for a refined formulation of human death—one that acknowledged the brain's role not merely as an organ of cognition, but as the central mediator of human experience, personhood, and interconnectedness. The brain, composed of billions of neurons and trillions of synaptic connections, is the neural foundation for bodily regulation, memory, perception, identity, and volition. Yet, despite centuries of scientific progress, the brain remains one of the most complex and mysterious structures in the known universe. How does this organ unify our thoughts, emotions, and bodily processes into a coherent sense of self? How does it adapt to change, preserve personal identity, and serve as a bridge between the physical and the metaphysical? This book introduces the *Brain Metafunction Framework (BMF)*, a comprehensive model that seeks to answer these profound questions by presenting the brain as a dynamically interconnected and hierarchically organized entity at the core of human life.

At the heart of this framework are several novel concepts that illuminate the brain's essential roles. One of the most distinctive is the *Metafunctional Theory of Perception*, which offers a fresh view of how sensory modalities are integrated into coherent perceptual experience. Unlike traditional models that treat perception as a passive reception of stimuli, this theory emphasizes the brain's efferent loop as a generative and constructive process—one that actively projects and shapes perceptual reality. This "secret in plain sight" reveals a deeper, often overlooked dimension of perception that implicitly bridges the physical and the transphysical. Importantly, this redefinition of perception is not purely conceptual, it carries practical clinical implications. The theory reframes perceptual dysfunction as a disruption of the afference–metafunction–efference loop, offering a unified lens through which to interpret and treat conditions such as phantom limb syndrome, post-stroke motor deficits, trauma-related dissociation, and virtual-reality–mediated therapy. Even when afferent input is compromised—as in phantom limb syndrome, visual hallucinations from sensory loss, or motor deficits after stroke—perception can often be reoriented or restored through targeted interventions that engage the efferent loop. A detailed clinical table included in the book applies this framework to a wide range of neurological and psychological phenomena, illustrating its utility as both a diagnostic and therapeutic tool rooted in metafunctional integrity. The book also introduces the idea of *Brain Metafunction Syndromes,* providing a distinctive approach to understanding how integrative processes are disrupted in clinical conditions like coma and minimally conscious states. These syndromes further underscore the brain's pivotal role in sustaining not only consciousness, but also personal identity and somatic coherence.

Building on these foundations, *A Refined Formulation of Human Death* revisits and expands a model first proposed in the 1990s. It incorporates integrative brain function into a unifying approach that transcends false dichotomies—offering new insights into the ethical, medical, and philosophical dimensions of life and death. The concept of *Metafunctional Free Will* explores how the brain's capacity for integration underlies human autonomy and agency, offering an alternative to deterministic views of cognition and behavior. The book also introduces *Brain Metafunction Ports,* conceptualized as points of convergence where cognitive, sensory, and regulatory processes interact—facilitating the unification of diverse functions into a single, coherent personal experience. Finally, *Brain Metafunction Transphysicalism* extends this framework to address spiritual and transphysical dimensions of human experience, proposing that the brain is not only the organ of consciousness but also a bridge to realms that transcend the purely physical.

Throughout the book, these ideas are situated within broader discussions on perception, attention, memory, consciousness, and neurological syndromes. The Brain Metafunction Framework serves as a foundation for reinterpreting the brain's role in both health and dysfunction. It presents an integrated model that accounts for cognitive, sensory, and regulatory processes across multiple neural scales, encompassing everything from homeostatic survival to high-level cognition. The book's structure reflects a logical arc—beginning with the brain's anatomical and functional organization, then exploring how its integrative architecture supports consciousness, the unconscious mind, and the unity of the self. Clinical case studies and chapters on neurological disorders reveal how disruptions in these integrative processes can fragment the system and challenge conventional definitions of life, awareness, and identity.

Beyond neuroscience, the implications of this framework extend to *bioethics, neuroethics, and theology*. In medical ethics, the BMF offers a principled way to assess severe disorders of consciousness, vegetative states, and advanced dementia—not merely as diagnostic categories, but as disruptions of the person's integrative coherence. Philosophically, the framework invites reflection on the nature of consciousness, free will, and the soul, encouraging a reevaluation of what it means to be a unified human person. The final sections of the book examine how the BMF aligns with *Christian Theism*, offering a foundation for theological engagement with transphysical realities such as the soul, miracles, and divine interaction. By addressing the limits of *materialism, physicalism, naturalism, and scientism*, the model affirms the explanatory power, scope, and coherence of theistic paradigms—particularly those that affirm the human person as a being situated within both the material and spiritual orders.

In bridging neuroscience with ethics, metaphysics, and theology, the *Brain Metafunction Framework* anticipates a necessary paradigm shift. It moves beyond traditional compartmentalized approaches, offering a holistic vision of brain function that accounts not only for observable behavior but also for moral awareness, spiritual longing, and personhood. This integrative approach is not simply academic; it has practical and clinical implications, shaping how we approach care, research, and the meaning of life itself. By challenging reductionist models, this book offers a vision of neuroscience that is at once more comprehensive, more humane, and more aligned with the full reality of the human person. In doing so, it invites the reader to reconsider the full scope of what it means to be human, affirming the unity of body, mind, and soul in the intricate phenomenon of conscious existence. As this journey unfolds, it becomes clear that understanding the human brain is not only a scientific endeavor but also a spiritual one—revealing the profound mystery of a being made in the image of God, capable of thought, love, and transcendence.

1

The Human Brain

Neuroscientists continue to marvel at the human brain's complexity, an organ widely regarded as the most intricate in the known universe. Its connections form a neural network so vast that they may outnumber the stars in the Milky Way, presenting a striking cosmic parallel. Within this labyrinth lies the biological foundation of self and identity—supported by astonishing neural activity. Yet a fundamental mystery remains: How does this organ transform physical signals into the richness of subjective experience? And, crucially, is there more to us than just our brains? This book contends that there is. It presents a conceptual framework that integrates the brain's biological functions with the broader dimensions of human existence—psychological, social, and even transphysical.

Modern neuroscience has made significant strides in unraveling the brain's mechanisms, yet much remains enigmatic. The ultimate challenge is comprehending how the brain and mind bridge the external physical world and our internal subjective experiences. At its core, this translation depends on sensory input like vision, hearing, and touch. However, a pivotal question remains: Does sensory perception represent objective reality, or is it a constructed illusion? Many neuroscientists now argue that perception is not a direct window onto reality but rather an active construction—an approximation shaped by the brain to serve adaptive purposes. Recent frameworks in cognitive neuroscience increasingly support this view. For example, predictive processing models suggest that the brain functions as a hierarchical inference machine, constantly generating hypotheses about sensory input and minimizing prediction errors (Friston, 2010). These models reinforce the view that perception is not passively received but actively sculpted by neural systems optimizing interaction with a dynamic environment.

Yet this current perspective, while a crucial advance, remains incomplete when considered through the lens of the paradigm defended in this book. By framing perception solely as an adaptive construction optimized for survival, mainstream neuroscience often overlooks deeper dimensions of experience that resist full reduction to neural computation. The Brain Metafunction Framework contends that consciousness is not merely the byproduct of error-minimizing sensory models, but a multidimensional integration of biological, psychological, social, and transphysical realities. Perception, in this view, is not only a functional simulation of the external world but a bridge between embodied cognition and existential meaning. To reduce it to predictive coding alone is to ignore the layers of selfhood, intentionality, and even spiritual resonance that shape how we encounter the world. Thus, while the constructivist model offers explanatory power, it must be expanded by integrative paradigms that honor the full scope of human experience—a synthesis of mechanism and meaning, biology and being.

The philosophical position known as physicalism, introduced by Otto Neurath (1931) and Rudolf Carnap (1959/1932), holds that all phenomena—including the mind—are ultimately reducible to physical processes. While this worldview has guided scientific inquiry, it struggles to explain phenomena uncovered by modern neuroscience. For instance, the "neural binding problem"—the process by which the brain unifies disparate sensory inputs into a cohesive experience (von der Malsburg, 1981; Feldman, 2013)—remains unresolved within physicalist paradigms. New theories of metastable brain dynamics (Deco & Kringelbach, 2016) suggest that the unity of conscious perception depends, at least partially, on the brain's ability to flexibly shift between integrated and segregated states. This dynamic flexibility may be essential to the integrative capacity that underlies what this book defines as the brain's metafunction—its role in synthesizing complexity across dimensions of human experience. The brain simplifies and prioritizes sensory inputs, generating a coherent internal model of the world. Without such mechanisms, the sheer volume of sensory data would overwhelm us. Yet, this constructed reality invites profound philosophical questions: How does the brain create such unity from complexity? What are the implications for understanding human consciousness and identity?

To address these complexities, this book expands on the Brain Metafunction Framework, which integrates biological, psychosocial, and transphysical dimensions. Proposed by the author in the context of brain death, coma research, and the definition of death and enriched now (García, 1995, 1998, 1999, 2001), this framework moves beyond physicalism to encompass the holistic nature of human experience. It acknowledges the brain not merely as an organ of computation but as a dynamic entity synthesizing multiple dimensions of existence. By studying the brain's role in perception, behavior, and spiritual life, we can deepen our understanding of the dialogue between biology and experience. This approach aligns with emerging trends in neuroscience that emphasize the interplay of body, mind, and environment. In particular, the concept of the "connectome"—referring to the temporally developing patterns of connectivity across neural networks (Vidaurre et al., 2017)—illustrates how moment-to-moment changes in brain organization support adaptive cognition and consciousness. These perspectives affirm the need for integrative frameworks that can account for the dynamic, unified, embodied, and transphysical nature of human consciousness.

The Brain's Biological Marvels, Neuroscience, and Philosophy

Despite centuries of philosophical debate, one fact remains unchallenged: the human brain is a biological wonder. It consumes approximately 20% of the body's oxygen, contains around 86 billion neurons, and forms over 10 trillion synapses (Godwin & Cham, 2012). These neurons are supported by glial cells, which provide structural scaffolding, nutritional support, and immune defense. Altogether, the brain houses an estimated 180 billion cells, with axons that—if stretched end to end—could wrap around the Earth four times. The brain's computational capacity is staggering, performing up to ten quadrillion operations per second. Yet this immense complexity also renders it one of the least understood organs in the human body. Unlocking its secrets may represent the greatest scientific challenge of our era. New research supports this view by highlighting the brain's temporal flexibility and energy efficiency. For example, the concept of *connectome harmonics*—the use of oscillatory patterns to coordinate brain regions across distance—suggests that the brain can rapidly shift between different states to match task demands (Luppi et al., 2023). This model not only emphasizes functional integration, but also introduces a potential "frequency code" for how conscious states unfold over time. These discoveries illustrate that even biologically, the brain's operation reflects principles of dynamic order beyond linear cause-effect mechanisms.

Modern neuroscience continues to operate under the influence of scientism—the belief that science is the sole valid path to knowledge. This outlook stems from methodological materialism, which insists that natural phenomena must be explained entirely through physical causes and processes, excluding any metaphysical or spiritual considerations. While methodological materialism has, in recent history, driven scientific discovery without significant impediment, its limitations are becoming increasingly apparent, particularly in neuroscience, where excluding non-physical dimensions of human experience may constrain deeper understanding. Recent debates in theoretical neuroscience increasingly echo this emerging critique. As brain–machine interfaces (BCIs), neuroimaging tools, and large-scale simulations become more advanced, epistemological concerns have arisen about the interpretive limits of data-centric models (Frégnac, 2017). Researchers are increasingly aware that consciousness and meaning-making may not be exhaustively modeled by algorithmic or computational terms, even if these tools enhance empirical precision. This reinforces the call to move beyond physicalism.

This book challenges the reductionist notion that humans are nothing more than the "jelly content between our ears." Instead, it calls for a synthesis of scientific, philosophical, and theological perspectives. The brain is more than a sum of its parts; it is the biological seat of human bodily regulation and cognition, a hub where thoughts, emotions, memories, and actions converge in a unified whole. Understanding its profound depths requires moving beyond exclusively physical explanations and embracing the layered interplay between biology, consciousness, and the possibility of transcendence. A growing body of research in *transdisciplinary neuroscience* now supports this integrative view. Scholars working at the intersection of neuroscience and phenomenology argue that lived experience, subjective time, and embodied cognition must be included in any full account of brain function. These perspectives also align with transphysical approaches that seek to bridge empirical data with existential and spiritual dimensions of human life. In this light, the brain is not only a marvel of matter—but a portal through which reality itself is made meaningful.

The Human Brain from Infancy to Adulthood

The human brain undergoes profound transformations from infancy to adulthood, shaped by changes in neural connectivity that refine both structure and function. At two years old, a toddler's brain contains approximately two quadrillion synapses, forming a remarkably dense web of neural connections (Stiles & Jernigan, 2010). This exuberant connectivity supports early learning and adaptability but lacks the efficiency of a mature brain. As the brain develops, it undergoes synaptic pruning, a critical process that eliminates less-used connections while strengthening more efficient ones (Spear, 2013). This refinement promotes neural efficiency and optimizes the transmission of information. Pruning extends beyond synapses to neurons themselves: some are eliminated through programmed cell death, or apoptosis, which removes those not actively engaged in functional circuits. These sculpting processes, which selectively remove inactive neurons, are essential to the brain's developing architecture. Although the adult brain hosts fewer synapses than a child's, the remaining networks are finely tuned. These streamlined circuits support complex cognitive functions and efficient behavior, highlighting how neural refinement enhances rather than diminishes brain power over time.

Recent research has refined our understanding of critical and sensitive periods—windows in development when the brain is particularly responsive to environmental inputs. While traditional models emphasized early childhood as the peak period of plasticity, new findings suggest a *second wave of plasticity* during adolescence, particularly in the prefrontal cortex and limbic system (Fuhrmann, Knoll, & Blakemore, 2015). These regions are crucial for decision-making, emotional regulation, and self-awareness, and their late maturation underscores the importance of adolescence as a formative neurobiological stage. Moreover, studies using *longitudinal diffusion MRI and network-based analyses* have shown that adolescent brain development is not linear but dynamic, characterized by shifts in *modularity and hub integration* that reflect the transition from local to distributed processing (Váša et al., 2023). These findings reinforce the notion that the human brain remains adaptable well into early adulthood—a view that supports the *Brain Metafunction Framework's* emphasis on layered, lifelong integration across cognitive, emotional, and social domains.

The Role of Neurogenesis in Adulthood

For much of the 20th century, scientists believed that neurons could not be generated after early development. However, the discovery of adult neurogenesis overturned this dogma. Neural stem cells in specific regions of the adult brain can generate new neurons, though the mechanisms are not yet fully understood. Neurogenesis in adulthood is governed by complex factors, including the proliferation, migration, and programmed cell death of neural progenitors (Shohayeb et al., 2018). Among the forms of cell death—apoptosis, necrosis, and autophagic cell death—apoptosis plays a particularly pivotal role in determining which new neurons successfully integrate. Molecular pathways such as the caspase cascade are essential regulators of this process, though their broader effects on plasticity and therapeutic applications are still under investigation. Current research is focused on the balance between self-renewal, proliferation, and differentiation of neural stem cells, with implications for treating brain injuries, cognitive decline, and neurodegenerative disorders. Still, ethical, technical, and scientific challenges remain. Translating these discoveries into clinical interventions will require not only continued research but also thoughtful consideration of safety and accessibility.

Recent studies suggest that neurogenesis is *not confined solely to the hippocampus,* as previously believed. Evidence now points to low-level but functionally relevant neurogenesis occurring in the striatum and amygdala, particularly in response to environmental stimuli and stress-related adaptations (Denoth & Jessberger, 2021). These findings broaden the scope of regenerative neuroscience and challenge long-held localizationist assumptions. In parallel, astrocyte-derived metabolic support has been shown to directly influence the survival and integration of newborn neurons. Through mechanisms involving lactate shuttling and calcium signaling, astrocytes modulate the energy environment necessary for synaptic incorporation and long-term viability (Beard et al., 2022; Purushotham & Buskila, 2023). These glial-neuronal interactions are emerging as crucial cofactors in adult neuroplasticity. Finally, epigenetic modifications, including histone acetylation and non-coding RNA expression, are increasingly recognized as regulators of neurogenic niches. These molecular controls dynamically shape the plastic potential of stem cells and may be harnessed to direct therapeutic regeneration in age-related or trauma-induced decline (Yao et al., 2016). Together, these advances highlight adult neurogenesis as a system of remarkable responsiveness and complexity—one that underscores the brain's continued capacity for renewal.

A Horizon of Possibilities and Limitations

The human brain stands as a pinnacle of natural complexity, embodying the highest level of matter's organization. It exemplifies the extraordinary potential of the created universe, enabling us to ponder realms beyond the physical—realms that include the natural world, abstract concepts such as numbers, mental constructs such as thoughts, and spiritual realities. Neuroscience draws much of its allure from the tension between two compelling horizons. The first lies in the growing accessibility of the brain, which was long regarded as an impenetrable "black box." Scientific and technological advances now permit unprecedented insights into the living brain of both humans and animals. A range of innovative tools allows researchers to explore its structures and functions with remarkable precision. Recent developments in high-resolution imaging—such as real-time HD-EEG network tracking (Michel & Brunet, 2019) and deep-brain calcium imaging using miniaturized two-photon devices (Zong et al., 2017, 2022; Zhou et al., 2018)—have enabled observation of neural circuit dynamics with millisecond precision in freely behaving organisms. These tools reveal that cognitive and emotional processes unfold not as static localizations but as dynamic trajectories across developing brain states, fundamentally redefining how researchers interpret cognitive transitions, alertness, and even the underpinnings of self-awareness.

The second horizon is the enduring mystery that surrounds the brain's unresolved questions. Despite promising discoveries, fundamental challenges remain regarding consciousness, memory, free will, and the nature of the self (Eagleman, 2007). These mysteries remind us of the profound limitations that accompany even the most rewarding research endeavors. Contemporary discussions increasingly focus on *metastable switching*—the brain's ability to transition between loosely coupled dynamic states—which has been posited as a substrate for conscious awareness and flexible attention (Deco & Kringelbach, 2016). These ideas intersect with philosophical questions about unity, intentionality, and personal continuity—areas where empirical data and metaphysical inquiry must converge. The brain governs an expansive range of functions: internal regulation, movement, sensation, emotion, and thought. Its various regions display different degrees of specialization, yet they work in tight coordination through complex interactions. Communication across these regions is mediated by large-scale networks and highly integrated hubs, which orchestrate the brain's response to both internal and external stimuli. In recent decades, neuroscience has made significant progress in understanding how the brain functions as a unified whole. Moving beyond traditional approaches that focus on isolated regions, researchers now use advanced techniques to study large-scale network organization and the functional role of interconnected hubs.

These hubs, essential components of brain networks, enable integration across distant areas, facilitating complex tasks and adaptive behavior (Gordon et al., 2017). Recent studies on *connector hub dynamics* show that certain nodes in the brain's network architecture play disproportionately central roles in coordinating multiple cognitive systems, much like key junctions in an electrical grid (Faskowitz et al., 2020).

These findings lend support to the Brain Metafunction Framework's focus on integration as a hallmark of higher-order cognition. As neuroscience continues to progress, we are compelled to ask: What new tools or frameworks will further unravel the profound mysteries of the human brain? While the possibilities are vast and exciting, they also invite us to confront the inherent limits of our current knowledge. This balance of potential and limitation urges a deeper understanding—not only of the brain's biological architecture but of its role in shaping human experience across multiple dimensions.

Developments in Neuroscience Historical Approaches to Studying the Brain

Curiosity about the brain's structure and function has captivated medical and philosophical minds throughout recorded history. In early Western civilization, Greek philosophers and physicians set the foundation for centuries of inquiry. Until the rise of modern neuroscience, three primary methods were available for investigating the brain: clinical (neurological), pathological, and surgical approaches. Though limited, these methods offered crucial insights, each with distinct strengths and constraints. A common feature among these approaches was the study of damage to the nervous system and its correlation with observable symptoms. By examining the effects of trauma, tumors, strokes, and other neurological conditions, early researchers began localizing specific brain functions to distinct regions. In animal models, researchers often applied surgical ablation—removing targeted brain areas—to test the consequences, following the localizationist principle. Additionally, post-mortem studies, correlated with clinical observations, offered valuable (though inherently constrained) knowledge about brain anatomy and function.

Today, these early techniques have gained renewed relevance as *computational neuroscience and neuroinformatics* revisit historical cases using modern digital frameworks. For instance, digitized brain atlases now allow researchers to simulate historical lesion studies within *connectomics models*, offering new insights into functional compensation and reorganization. The Allen Brain Atlas and the BigBrain project exemplify this shift, enabling researchers to reconstruct historical anatomical data at cellular resolution and contextualize it within contemporary large-scale brain network theories (Amunts et al., 2013). Furthermore, comparative neuroanatomy, which once relied heavily on physical dissection, has been revitalized by 3D imaging, machine learning–based segmentation, and cross-species atlasing. These tools reveal both continuity and divergence in brain structure and function, deepening our understanding of how species-specific neural architectures relate to behavior, cognition, and consciousness. Thus, while classical methods of brain study were limited by technology, their legacy persists—reinterpreted and re-energized by today's integrative, digital, and cross-disciplinary neuroscientific approaches.

From Localizationism to Large-Scale Networks

Neuroscience has developed dramatically, shifting from early theories of strict localization to models emphasizing modularity, and now, to the study of interconnected brain networks. Localizationism proposed that distinct brain functions—such as language or perception—resided in discrete anatomical regions. For example, neuroimaging has confirmed the role of Broca's area in language production and Wernicke's area in comprehension (Gernsbacher & Kaschak, 2003). The arcuate fasciculus, a white matter tract, facilitates communication between these two areas, enabling integrated functions like reading aloud. Damage to any of these structures results in various forms of aphasia, impairing language processing. Yet, as neuroscience advanced, it became clear that while localization holds some validity, cognitive functions are not isolated to single brain regions. They are distributed across complex, overlapping systems. Though the localizationist view provided foundational insight, it fell short of accounting for the dynamic coordination among functionally specialized areas. These regions operate as part of broader, interdependent systems—a key concept supporting the brain metafunction framework developed in later chapters.

This advance is exemplified by the shift to *large-scale functional network models,* which emphasize integration across distributed nodes. Contemporary frameworks such as *connectome harmonics* and *attractor landscape theory* suggest that cognitive and emotional states arise from transitions between high-dimensional, metastable configurations of brain activity (Luppi et al., 2023). These models conceptualize the brain as navigating a dynamic "landscape" of potential states, where functional hubs anchor and stabilize experience-specific patterns. Furthermore, the advent of *chronnectomics*—the study of time-varying brain connectivity—has reinforced the notion that brain functions are inherently fluid and temporally dynamic (Vidaurre et al., 2017). Rather than fixed circuits, the brain engages in rapid reconfigurations that correspond to shifting tasks, emotions, and internal states.

This supports a view of brain function as context-sensitive and anticipatory, challenging static models of localized processing. Together, these developments point toward a new neuroscience—one that aligns closely with the Brain Metafunction Framework's emphasis on synthesis across biological, psychological, and transphysical dimensions. By acknowledging the distributed, temporally developing nature of neural organization, these frameworks pave the way for deeper insights into consciousness, behavior, and identity.

Advances in Understanding Language Processing

Recent developments in neuroimaging have expanded our understanding of language processing beyond classical models. The SENSAAS atlas, for example, derived from both task-based and resting-state fMRI data, highlights the distributed nature of language networks (Labache et al., 2018). Rather than being confined to Broca's and Wernicke's areas, language-related processing spans multiple cortical and subcortical regions (Brennan & Hale, 2019). This distributed organization supports not only the classical domains of syntax and semantics, but also higher-order aspects of language such as metaphor, prosody, and pragmatics. While Broca's area remains pivotal for syntactic structuring, other cortical areas are equally essential for meaning-making, emotional tone, and social nuance. Importantly, although language is predominantly left-lateralized in most individuals, the non-dominant (typically right) hemisphere also plays a critical role. It is especially involved in interpreting non-literal language, intonation, and the emotional context of speech. While less thoroughly mapped, this hemisphere's contribution reflects the inherently multidimensional nature of communication, engaging both analytical and affective systems.

Recent advances in *high-density electroencephalography (HD-EEG)* now allow researchers to track the *temporal sequence of language processing* in real time, revealing how semantic and syntactic components are synchronized across distributed networks (Brennan & Martin, 2020). These technologies highlight the millisecond-scale coordination required for conversational fluency, underscoring the integrative nature of language as both a cognitive and social act. In parallel, computational neuroscience has made strides in modeling language with *artificial neural networks*—many of which now approximate cortical processing hierarchies. For instance, transformer-based models have been shown to mirror hierarchical encoding patterns observed in the brain during sentence comprehension, particularly in temporo-parietal and prefrontal regions

(Caucheteux & King, 2022). These convergences between artificial and biological systems invite novel frameworks for understanding linguistic representation—not as a static map but as a dynamically unfolding process across nested levels of abstraction. Together, these developments affirm the view that language is not anchored to singular regions but depends on the dynamic interplay between distributed cortical systems, temporal coordination, cross-hemispheric modulation, and subcortical contributions. Recent findings highlight the essential roles of subcortical structures—including the thalamus, basal ganglia, and cerebellum—in supporting language processing by regulating timing, sequencing, and error monitoring. This layered complexity supports the Brain Metafunction Framework's position that language reflects more than lexical mechanics—it reveals deep integrative capacities linking cognition, emotion, motor control, and intersubjectivity.

Towards a Network-Based Understanding

Language processing integrates diverse cognitive functions, requiring access to memory, attention, perception, and internal states. While a strictly modular approach offers heuristic value, it is insufficient to explain the complexity of such processes. Instead, large-scale brain networks provide a more comprehensive framework. These networks, characterized by interconnected nodes and modular community structures, exhibit dynamic properties such as *multistability* and *metastability*—concepts essential to understanding how the brain coordinates activity across regions. Multistability refers to the brain's capacity to settle into different stable configurations depending on task demands or internal states. Metastability allows flexible transitions between these configurations, supporting cognitive adaptability and responsiveness (Sporns, 2010; Majerus et al., 2010; Kelso, 2012; Kleckner et al., 2017). A module is a subset of highly interconnected nodes that also communicate efficiently with nodes in other modules. The brain's modular architecture enables it to specialize in particular cognitive functions while maintaining broader metafunctional integrative capacity. These modular systems form the infrastructure of larger-scale metafunctionality, allowing for distributed processing with global coherence.

Recent work using *connectome-based predictive modeling (CPM)* has demonstrated that individual differences in network architecture can reliably predict cognitive performance, emotional regulation, and even personality traits (Shen et al., 2017). These models suggest that stable but flexible patterns of connectivity underlie key features of the conscious self, reinforcing the centrality of integrative networks in supporting identity. The study of network-level organization reveals how the brain orchestrates complex processes such as language, emotion, and decision-making—not through isolated regions but via integrated systems. This integrative operation of the brain supports not only task performance but also flexible switching among cognitive modes, emotional states, and motor plans. As neuroscience progresses, network-based approaches continue to shed light on the brain's capacity for dynamic coordination. These perspectives converge with the Brain Metafunction Framework's assertion that the brain functions not merely as a signal processor but as a *multidimensional integrator*. Metafunctionality, in this sense, refers to the capacity of the brain to synthesize sensory, regulatory, cognitive, affective, and existential dimensions into coherent and context-sensitive responses. This sets the stage for a deeper investigation of the brain's metafunctional integration, a theme explored more fully in the next chapter.

Research Tools Developed or Perfected in the Past Several Decades

On October 1, 1971, Atkinson Morley's Hospital in London conducted the first clinical CT scan (Hounsfield, 1973). In the years that followed, the field witnessed the installation of the earliest clinical MRI scanners. These innovations paved the way for the widespread use of CT and MRI in daily medical practice. Over the last two decades, additional tools—either newly developed or significantly refined—have transformed neuroscience research. These technologies now allow for detailed, real-time investigations of brain structure and function, and have contributed to the emergence of new concepts in understanding the brain.

Connectomics

In 2005, Olaf Sporns and colleagues introduced the term connectome to describe a comprehensive, structural map of the brain's wiring. Connectomics aims to chart the brain's neuronal connections in high detail, creating wiring diagrams that help explain how neural components correlate with complex cognition and behavior. By revealing how physical structures relate to functional outcomes, connectomics provides a foundation for understanding neurological disorders such as Alzheimer's, autism, and schizophrenia. These structural maps are also central to identifying how large-scale network disruptions might affect conscious experience—offering essential insights for future therapeutic approaches. Recent developments have enabled the integration of *connectome-based predictive modeling (CPM),* which uses individual brain network patterns to forecast cognitive and behavioral traits, pushing connectomics toward personalized neuroscience (Shen et al., 2017).

Optogenetics

Optogenetics is a transformative technique that allows precise control over the activity of specific neurons using light. By introducing genes for light-sensitive proteins (such as channelrhodopsins or halorhodopsins) into target cells, researchers can activate or silence neurons with millisecond precision using laser pulses (Deisseroth, 2011). The technique enables unparalleled investigation into the causal role of brain circuits in behaviors and mental states. It has been used to dissect memory formation, motor function, and emotional processing—and holds promise for developing novel treatments for conditions such as epilepsy, depression, and Parkinson's disease. All-optical closed-loop systems where stimulation and imaging are integrated in real-time, enabled continuous feedback modulation (Emiliani et al., 2015).

Calcium Imaging

Two-photon calcium imaging has become an indispensable tool for observing real-time neuronal dynamics with high spatial and temporal precision. This technique enables the recording of activity across cortical layers and deep brain structures in behaving animals, providing insights into circuit function and plasticity (Grienberger et al., 2022). Calcium imaging is a cutting-edge technique that allows researchers to monitor neuronal activity in real time by using genetically modified neurons to produce fluorescent proteins. These proteins emit light in response to changes in calcium levels, which often signal neuronal activation (Zhou, Belavek, and Miller, 2021). By visualizing calcium dynamics, scientists gain insights into the functioning of individual neurons and networks in living organisms. One widely used tool in calcium imaging is GCaMP, a genetically encoded calcium indicator expressed in specific cell populations. GCaMP enables precise in vivo imaging of neuronal activity, making it possible to observe dynamic changes in brain function with unparalleled resolution (Grienberger & Konnerth, 2012; Zhou et al., 2018; Zong et al., 2017). This technology has been particularly transformative in pain research over the past decade. Calcium imaging has provided more profound insights into the molecular and cellular mechanisms underlying pain perception by enabling the visualization of pain-related neuronal pathways. As pre-clinical research increasingly emphasizes the molecular understanding of pain, in vivo calcium imaging has become indispensable. It offers an efficient and robust framework for identifying novel drug targets and advancing the development of new analgesic therapies. Beyond pain research, calcium imaging has broader applications in neuroscience, contributing to our understanding of brain circuits, memory, and neurodevelopmental disorders. New applications include miniscope-based imaging in social behavior and decision-making, linking circuit dynamics to ecological validity.

Multi-electrode Arrays

Multi-electrode arrays (MEAs) allow for the simultaneous recording of electrical activity from large populations of neurons, offering insights into network-level dynamics and synchronization patterns (Obien, & et al., 2014). Multi-electrode arrays allow simultaneous recording from hundreds—or even thousands—of neurons (Thomas et al., 1972; Cerina, Piastra, and Frega, 2023; Obien et al., 2015). These microelectrode grids enable scientists to capture complex electrical activity across neural networks without significantly disturbing natural cellular environments. MEAs are crucial in studying network-level behaviors such as synchrony, oscillations, and signal propagation. They have been applied in research on epilepsy, Parkinson's disease, and brain-machine interfaces, supporting both basic neuroscience and clinical innovation. Recent MEA platforms now include *flexible electronics and bio-compatible coatings*, expanding use in chronic in vivo implants (Liu et al., 2024).

Brain-Computer Interfaces

Brain-computer interfaces interpret electrical signals from the brain and convert them into commands to control external devices (Wolpaw et al., 2002; Mellinger and Schalk, 2010). BCIs bypass traditional motor pathways, allowing users—especially those with motor impairments—to control prosthetics, type on virtual keyboards, or operate wheelchairs using thought alone. BCIs also raise important ethical questions regarding privacy, security, and human enhancement. As these systems advance, they promise not only rehabilitation tools but also new paradigms in human-computer interaction. Recent research focuses on *nano-enabled BCIs* that incorporate advanced materials for higher signal fidelity, biocompatibility, and faster bidirectional communication (Liu et al., 2023). As Lebedev and Nicolelis (2017) outline, brain–machine interfaces have progressed from experimental systems to clinically viable technologies that support neuroprosthetic control and neurorehabilitation.

fMRI and Other Neuroimaging Techniques

Functional magnetic resonance imaging (fMRI) detects changes in blood flow and oxygenation associated with neural activity (Logothetis, 2003; Glover, 2011). It enables non-invasive mapping of brain function in response to stimuli or during resting states. fMRI has revealed functional connectivity patterns, including networks such as the default mode network. Other neuroimaging techniques include PET (Positron Emission Tomography), which visualizes metabolic activity, and EEG (Electroencephalography) and MEG (Magnetoencephalography), which offer high temporal resolution in tracking brain activity. These methods are frequently used in combination to provide a multidimensional view of brain function and dysfunction. Recent innovations in layer-specific fMRI now allow the resolution of signal origin by cortical depth, offering insights into feedforward vs. feedback processing in real-time cognition (Huber et al., 2017).

CRISPR/Cas9

CRISPR-Cas9 technology, introduced as a revolutionary genome-editing tool by Doudna and Charpentier, enables precise, programmable modification of genetic material and has rapidly transformed neuroscience research (Doudna & Charpentier, 2014; Hsu, Lander, & Zhang, 2014). Originally derived from bacterial defense systems, this technique uses guide RNA and the Cas9 enzyme to cut and modify genes at specific locations. CRISPR is being used to study the genetic underpinnings of neurological disorders like Huntington's, Alzheimer's, and autism. It holds potential not only for research but also for therapeutic intervention—though ethical and safety concerns, especially around germline editing—modifying genes in embryos—raise bioethical questions about altering traits in future generations. A recent frontier involves *CRISPRa* and *CRISPRi* systems, which activate or inhibit gene expression without altering DNA sequences—opening avenues for *reversible control of neural circuits* in models of memory, fear, and addiction (Marques et al., 2016; Dimou & Gallo, 2015).

Two-Photon Microscopy

Two-photon microscopy enables deep tissue imaging with minimal photodamage. By using infrared laser pulses to excite fluorescent molecules only at the focal point, it produces high-resolution 3D images of live tissue—particularly useful for studying neural circuits in action. This technology has become central in research on synaptic plasticity, memory formation, and real-time circuit dynamics (Denk, Strickler, & Webb, 1990; Zipfel, Williams, & Webb, 2003). A recent upgrade includes *three-photon excitation* for even deeper cortical imaging and improved signal-to-noise ratios in awake behaving animals (Sabbagh et al., 2023).

Designer Receptors Exclusively Activated by Designer Drugs (DREADDs)

DREADDs are genetically engineered receptors that respond only to synthetic ligands (Roth, 2016). Unlike optogenetics, which requires implanted fiber optics and light delivery, DREADDs allow chemical control of neuron activity through systemic drug administration. This technique offers a reversible and selective way to study how specific circuits regulate behavior, emotion, and cognition—often in disease models or in therapeutic development. Enhanced DREADD variants are now optimized for higher specificity and faster pharmacokinetics, expanding their use in translational behavioral neuroscience (Silva, 2023).

High-Density and Ultra-High-Density EEG

High-density EEG involves hundreds of electrodes placed across the scalp, allowing for precise temporal and spatial analysis of electrical brain activity (Michel & Brunet, 2019). This technique enhances the ability to detect and localize epileptic activity, investigate consciousness, and explore network dynamics in real time. Its non-invasive nature and high temporal resolution make HD-EEG an increasingly valuable tool for both clinical diagnostics and cognitive neuroscience. Recently developed *real-time source localization algorithms* have improved its utility in neurofeedback and brain–machine interfaces by identifying activity in deep and midline structures (Rojas et al., 2024).

Virtual Reality (VR) Environments

VR technologies allow researchers to simulate lifelike scenarios in controlled environments (Riva, 2019). Participants can be immersed in real-time experiments to study perception, spatial navigation, motor coordination, or social behavior. When combined with fMRI or EEG, VR provides powerful insights into how the brain responds to dynamic, multimodal experiences (Adamovich et al., 2009). Clinically, VR is also being used in cognitive rehabilitation, phobia treatment, and PTSD therapy. It provides personalized therapeutic environments to promote neural recovery and enhance adaptability, particularly for patients recovering from a stroke, brain injury, or motor impairments (Lloréns et al., 2015). New studies now pair VR with *adaptive AI agents* to dynamically adjust emotional and sensory inputs during therapy, further tailoring experiences for individual brain states (Zhao & Morin, 2024).

Neuro-nanotechnology

Neuro-nanotechnology integrates nanoscale materials into neuroscience, allowing targeted drug delivery, cellular-level imaging, and precise neural modulation. For example, nanoparticles can cross the blood-brain barrier to deliver chemotherapy directly to brain tumors or reduce inflammation in neurodegenerative conditions (Saraiva et al., 2016; Liu et al., 2023). Researchers are also exploring nano-enabled BCIs and neural repair strategies for injuries like spinal cord damage. Another application involves brain-computer interfaces (BCIs). Researchers aim to enhance communication between the brain and external devices by incorporating nanomaterials into these systems. This technology can improve prosthetic limb control, facilitate communication for individuals with paralysis, and advance human-computer interaction (Lebedev & Nicolelis, 2017). Recent designs include bioresorbable nanoelectrodes and self-assembling nanoscaffolds, expanding the possibilities for regenerative applications (Song, Li, Won, Bai, & Rogers, 2020).

Artificial Intelligence (AI) and Neuroscience

AI and neuroscience are increasingly interdependent (Surianarayanan et al., 2023). Machine learning helps decode brain imaging data, simulate neural networks, and predict behavioral outcomes. Conversely, neuroscience inspires new AI architectures—such as deep learning systems modeled on cortical organization. The collaboration between these fields has led to rapid advances in understanding cognition and behavior. It also fuels discussion about Artificial General Intelligence (AGI) and its implications for science, ethics, and society (Bostrom, 2014). AGI aims to replicate human intelligence across diverse domains, including learning, problem-solving, and reasoning (Russell & Norvig, 2020). A growing body of research now explores *brain-inspired spiking neural networks and neuromorphic hardware,* aiming to simulate the temporal dynamics of real neurons with millisecond fidelity. These efforts in the mind of some bring us closer to biologically plausible models of synthetic cognition—and raise urgent ethical and epistemological questions in the age of human–machine convergence.

Back to Connectomics

Connectomics research, which focuses on mapping and understanding the intricate connections within the brain, has revolutionized neuroscience by providing unprecedented insights into how neural circuits and networks support brain function (Faskowitz et al., 2020). This emerging field seeks to study the brain's "wiring diagram," from the level of individual neuronal circuits to large-scale global networks, revealing how these connections underpin cognitive functions, emotions, behavior, and the regulation of bodily processes. Tools such as CATMAID, WEBKNOSSOS, and Neuroglancer have been instrumental in driving this progress. CATMAID (Collaborative Annotation Toolkit for Massive Amounts of Image Data) enables researchers to collaboratively explore and annotate large-scale image datasets, facilitating detailed reconstruction of neural circuits (Saalfeld, Cardona, Hartenstein, & Tomancak, 2009). WEBKNOSSOS, an intuitive platform for viewing and annotating 3D images, allows precise analysis of complex neuronal structures (Boergens et al., 2017). Neuroglancer offers scalable web-based visualization with interactive 3D rendering, enabling comprehensive exploration of large-scale neural data (Google Research, n.d.). These tools empower researchers to decode the brain's intricate network architecture precisely and efficiently.

Recent advances have introduced *AI-assisted segmentation platforms and transformer-based auto-annotators* that dramatically accelerate connectomic reconstruction (Deng et al., 2024). These systems now process terabyte-scale data in days instead of months, making it feasible to study individual variability in microcircuit connectivity across populations. Connectomics enables the exploration of large-scale relationships within the brain, shedding light on how interconnected neural networks support complex phenomena such as memory, attention, motor control, and decision-making. For instance, researchers can use these tools to map the neural circuits involved in memory patterns formation or study how specific brain regions communicate during problem-solving tasks. This systematic mapping provides insights into normal brain function and reveals how injuries or diseases, such as

Alzheimer's or Parkinson's, disrupt these intricate connections. Recent models now link structural connectomes to *functional attractor dynamics*, in which the brain's state-space is visualized as a high-dimensional landscape of possible neural configurations (Luppi et al., 2023). These models show how cognition may unfold as a trajectory through structured connectivity patterns—stabilized by hubs, perturbed by stimuli, and modulated by intrinsic dynamics. This aligns with growing interest in *connectome harmonics*, a framework that uses the eigenmodes of the structural connectome to model brain-wide oscillations and explain how large-scale patterns give rise to consciousness and task performance.

By analyzing the wiring diagrams of neural circuits, connectomics helps uncover how diverse brain functions integrate into a higher-order process, the brain metafunction—the holistic operation of the brain that gives rise to consciousness, complex thought, and adaptive behavior. Understanding these integrated processes is essential for unraveling the mysteries of the brain's qualitative capabilities. As we transition into a discussion of fundamental brain structure and cell composition, this overview of connectomics highlights its transformative role in modern neuroscience. Connectomics paves the way for breakthroughs in understanding, diagnosing, and treating neurological and psychiatric disorders by bridging large-scale brain network analysis with clinical applications. It also reinforces the Brain Metafunction Framework's core insight: that meaning, identity, and personhood emerge not from isolated circuits, but from dynamic, integrated, and often transphysical coordination across the brain's vast neural architecture.

A Bottom-up Overview of the Brain: Cells and Neuro -transmitters

Neurons

Neurons, the fundamental units of the brain, transmit information and process signals through both electrical and chemical communication (Kandel, Schwartz, & Jessell, 2012). Each neuron consists of three main parts: the cell body, dendrites, and an axon. Dendrites receive signals from other neurons, while the axon transmits signals to other neurons, muscles, or glands. Synapses—specialized junctions—enable the transfer of electrical and chemical signals. This extensive network of neurons and synapses forms the biological foundation for the brain's ability to integrate bodily and mental functions.

Historical Milestones

In 1887, Santiago Ramón y Cajal revolutionized neuroscience by demonstrating that neurons are discrete cells rather than components of a continuous network, as previously believed under the reticular theory. His 1906 Nobel Lecture detailed how neurons communicate through synaptic gaps, laying the groundwork for the modern "neuron doctrine" (NobelPrize.org, 2024). A century later, the discovery of adult neurogenesis challenged long-held assumptions about the brain's developmental limits. Joseph Altman's work in the 1960s first suggested the formation of new neurons in the adult hippocampus (Altman, 1962). Michael Kaplan later expanded this work in primates, and Fernando Nottebohm provided definitive evidence in songbirds during the 1980s (Nottebohm, 1989). These findings laid the groundwork for our current understanding of neuroplasticity—the brain's ability to adapt and reorganize.

More recent research has confirmed that neurogenesis occurs in specific regions of the adult human brain, such as the dentate gyrus of the hippocampus, which is associated with memory formation and spatial navigation. For example, Spalding et al. (2013) estimated that approximately 700 new neurons are added daily in this region. *New findings have now extended evidence of neurogenesis into regions such as the striatum and amygdala, suggesting that regenerative plasticity may support emotional learning and affect regulation beyond hippocampal domains* (Moreno-Jiménez et al., 2019). These discoveries have significant implications for understanding learning, memory, and recovery from neurological injury.

Neurotransmitters and Signal Transmission

Neurons communicate through neurotransmitters, chemical messengers that enable communication between brain regions and coordinate signals between the brain and body. The process begins when dendrites receive a signal, which travels along the axon and is transmitted via the synapse to another neuron or target cell. This signaling system underlies the brain's capacity to perform complex functions such as cognition, behavior, and physiological regulation. The discovery of neurogenesis and the intricate communication mechanisms among neurons underscore the brain's remarkable adaptability. Exploring these basic systems deepens our understanding of how the brain learns, changes, and heals—forming a foundation for treating neurological disorders. *Recent studies have highlighted the phenomenon of neurotransmitter co-release*, in which a single neuron can release multiple transmitters from the same synapse. This finding challenges older notions of strict "one cell–one transmitter" models and emphasizes the dynamic versatility of synaptic signaling (Hnasko & Edwards, 2012).

Neurons Participate in Essential Brain Functions

Neurons in the central nervous system play a vital role in processing and transmitting information throughout the body. They receive input from other neurons or from sensory receptors—such as those responsive to touch or light—and relay signals either to other neurons or to effectors like muscles and glands. These communications drive voluntary movement, modulate behavior, and regulate numerous bodily functions. For instance, motor neurons transmit impulses that enable movement, while other types of neurons convey information to distinct brain regions involved in decision-making, perception, and emotion. Neurons are also essential for learning and memory. Through neuroplasticity—the brain's ability to reorganize itself by forming new synaptic connections—they support the dynamic adaptation of neural circuits in response to experience, injury, or changes in the environment.

Neuronal communication occurs at synapses through chemical messengers called neurotransmitters. Historically, it was believed that each neuron released a single type of neurotransmitter—a principle known as Dale's Law, based on the work of Henry Hallett Dale and later formalized by John Eccles. However, subsequent research showed that neurons can release multiple neurotransmitters, often in specific combinations, depending on their target and functional state. Eccles later revised the principle to reflect this nuance, stating that neurons tend to release the same neurotransmitters at all of their synapses, but are not limited to just one type. Neurotransmitters underlie a wide array of brain functions. For example, dopamine is associated with reward, reinforcement, and goal-directed behavior; serotonin modulates mood and emotional tone; and acetylcholine supports learning, memory, and sustained attention. These substances illustrate the brain's chemical diversity and how neurons orchestrate cognition, emotion, and behavior by modulating signals across complex neural networks. *Recent mapping studies using transcriptomic profiling have cataloged hundreds of unique neurotransmitter profiles across neuron subtypes*, suggesting a far more complex neurochemical landscape than previously imagined (Hansen et al., 2022).

Consciousness and Cognitive Processing

Neurons in various brain regions provide the biological foundation for consciousness and higher-order cognitive functions, including attention, decision-making, and problem-solving. Neurotransmitters play a critical role in these processes by facilitating communication between neurons. Glutamate, the brain's most common excitatory neurotransmitter, is essential for cognitive functions such as learning, memory, and problem-solving (Behrendt, 2013). Other neurotransmitters—such as dopamine, norepinephrine, acetylcholine (ACh), and γ-aminobutyric acid (GABA)—also contribute significantly to cognitive processing. For instance, dopamine is involved in reward-based learning, while acetylcholine supports attention and memory formation. Recent developments in *high-density EEG and ultra-fast fMRI* have allowed researchers to track the temporal progression of neural signatures associated with conscious awareness (Liu et al., 2023). These studies suggest that consciousness requires not isolated areas but *the coordination of transient global ignition events across widespread cortical regions*. Conscious perception appears to depend on sensory input reaching a threshold of recurrent amplification, especially involving prefrontal and temporoparietal junction networks, thereby supporting sustained awareness and reportability (Mashour et al., 2020).

In the limbic system, regions such as the hippocampal formation, amygdala, and nucleus accumbens coordinate emotions, motivation, and the modulation of bodily functions, including circulatory, endocrine, and immune systems. Dopamine and serotonin are central to emotion and motivation, influencing decision-making and social interaction (LeDoux, 1993). The *hippocampal formation* plays a crucial role in memory processing, spatial navigation, and emotional regulation. This structure includes key components such as:

1. *The indusium griseum*
2. *Longitudinal striae*
3. *Gyrus fasciolaris*
4. *Hippocampus proper (cornu ammonis, dentate gyrus, and subiculum)*
5. *Part of the uncus*

These interconnected regions are vital for forming, retrieving, and consolidating memory-related patterns. Neurotransmitters like acetylcholine, glutamate, GABA, and catecholamines modulate these processes. For example, fluctuations in extracellular neurotransmitter levels correlate with neuronal activity during memory formation, highlighting their dynamic role in encoding and recall (Behrendt, 2013).

Recent theoretical models, including *predictive processing frameworks,* propose that consciousness is deeply rooted in the brain's ability to minimize prediction error across hierarchical networks. Under this view, conscious experience arises when top-down expectations are successfully aligned with bottom-up sensory input—especially in contexts of uncertainty or ambiguity (Friston et al., 2017; Safron, 2020). This dynamic tension between expectation and surprise is modulated by neurotransmitter systems like dopamine and norepinephrine, which adjust synaptic gain and attentional precision. Understanding the interactions between neurotransmitters and brain structures enhances our knowledge of how the brain integrates cognitive and emotional functions. These insights also have implications for addressing disorders such as depression, anxiety, and neurodegenerative diseases, where disruptions in these pathways often occur. In this integrative view, consciousness is not a binary switch but a dynamic and graded phenomenon, sustained by oscillatory coherence, synaptic plasticity, and neurochemical modulation across brain-wide systems.

Sensory Processing

Sensory processing is a complex system through which neurons in sensory organs—such as the eyes, ears, and skin—detect external stimuli like light, sound, and touch, and transmit this information to the brain. Specialized receptor neurons convert these stimuli into electrical signals, which travel along neural pathways to specific brain regions for processing. Neurotransmitters such as glutamate and acetylcholine play crucial roles in facilitating this communication. Glutamate, the brain's primary excitatory neurotransmitter, drives much of the signal transmission required for sensory perception, while acetylcholine modulates attention and focus, enabling the brain to prioritize specific sensory inputs.

The *N-methyl-D-aspartate (NMDA) receptor,* a subtype of the glutamate receptor, is particularly important in sensory processing. NMDA receptors play a key role in the temporal coordination of signals from the retina (visual input) with cortical processing in the brain's superficial layers. They are also central to multisensory integration, helping deeper cortical layers combine inputs from different modalities—such as sight, sound, and touch—into a cohesive and unified perception of the environment. For example, NMDA receptor activity allows the brain to synchronize auditory and visual inputs, such as associating a speaker's voice with their lip movements. This capacity for cross-modal integration is essential for social communication and environmental navigation. Research

(Binns & Salt, 1996) has extended this model to include *hierarchical predictive coding*: the notion that the brain constantly generates internal models to predict sensory input. When incoming stimuli deviate from these predictions, cortical hierarchies update their expectations, producing conscious awareness of novelty or mismatch. NMDA receptor activity appears to be a biophysical correlate of this model-updating process, particularly in primary and associative sensory areas.

In addition to neuronal roles, glial cells—particularly *astrocytes and NG2-glia*—are now recognized as active modulators of sensory signal transmission. Astrocytes regulate extracellular potassium and glutamate concentrations in synaptic spaces, maintaining fidelity during high-frequency stimulation, while NG2-glia form transient synapse-like connections that can modulate local circuit excitability. These glial contributions are critical during development and in neuroplastic reorganization after injury or prolonged sensory deprivation (Purushotham & Buskila 2023). This intricate interplay of receptors and neurotransmitters underpins the brain's ability to process sensory stimuli efficiently, supporting vital functions such as perception, attention, and behavior. By integrating sensory inputs from diverse channels, the brain forms a unified understanding of the external world, enabling adaptive responses to constantly changing environments. It is this dynamic integration—modulated by biophysical mechanisms and higher-order expectations—that sustains the continuity of conscious perception across time and space.

Motor Control

Motor control relies on the intricate coordination of neurons, neurotransmitters, and neural circuits in the brainstem, spinal cord, and various cortical and subcortical regions. These systems enable voluntary movements such as walking, speaking, and performing fine motor tasks. The process begins in hierarchical motor areas of the brain, including the motor cortex, where neural signals are generated and transmitted to execute purposeful motion. Subcortical structures such as the basal ganglia and cerebellum refine these motor signals, allowing for smooth, precise movement. The brainstem and spinal cord receive and relay these commands, integrating them into physical actions through motor pathways that activate skeletal muscles.

Neurotransmitters are critical to motor control. Dopamine, for instance, plays a key role in regulating movement by modulating basal ganglia activity. A loss of dopaminergic neurons, as seen in Parkinson's disease, disrupts this system, leading to symptoms such as tremors, rigidity, and slowed movement (bradykinesia). Acetylcholine is essential at the neuromuscular junction, where it enables communication between motor neurons and muscle fibers, initiating muscle contraction. Disorders such as myasthenia gravis—an autoimmune condition that impairs acetylcholine receptors—can lead to significant muscle weakness and fatigability. Glutamate, the primary excitatory neurotransmitter in the brain, also contributes to motor control by activating pathways that stimulate muscle activity. In contrast, gamma-aminobutyric acid (GABA)—the brain's principal inhibitory neurotransmitter—balances this excitatory action, preventing excessive or involuntary movements. This dynamic balance between excitation and inhibition ensures coordinated, fluid, and goal-directed motion. The integration of these neurotransmitters and neural circuits allows humans to perform complex movements with precision, adaptability, and efficiency. For example, fine-tuning motor output enables us to play a musical instrument, type on a keyboard, or maintain balance while walking on uneven terrain. These actions highlight the brain's remarkable capacity to harmonize intention, perception, and movement in real time.

Recent studies have revealed that motor control emerges not merely from serial signaling but from dynamic and reciprocal interactions between cortical, cerebellar, and basal ganglia loops. The cerebellum not only adjusts timing and force of movement but also contributes predictive corrections based on internal models. In parallel, the basal ganglia regulate action selection by suppressing competing motor plans, creating a gated competition architecture (Caligiore et al., 2017). These systems do not operate in isolation but converge through the thalamus and brainstem, forming a reentrant circuit architecture that adjusts motor output continuously. Moreover, adaptive motor learning is increasingly linked to *metastable brain states*, in which networks maintain flexible readiness to shift between planned and reactive modes. This dynamic coordination is supported by neuromodulatory inputs—particularly dopamine and norepinephrine—which fine-tune excitability thresholds and modulate plasticity rules in cortico-striatal circuits (Deco & Kringelbach, 2016). Altogether, these findings reinforce the perspective that motor control is not only a biomechanical process but also a deeply predictive and integrative phenomenon—coordinated through distributed systems that reflect both intention and adaptation. Such coordination is indispensable to the broader understanding of the brain's metafunctional role in translating embodied cognition into purposeful action.

Homeostasis, Immune Response, and Endocrine Regulation

Homeostasis is the biological capacity of living systems to regulate and maintain internal stability. This process depends on a wide array of neural, circulatory, endocrine, and immune mechanisms. The related concept of allostasis, coined by P. Sterling and J. Eyer in 1988, describes the process of achieving stability through change. It highlights the body's capacity to adapt internal parameters—such as set points and control ranges—in response to environmental fluctuations. From a neurological perspective, allostasis can be understood as the brain's ability to predictively manage energy distribution and physiological demands across bodily systems. This predictive regulation is not reactive but anticipatory, often adjusting parameters before imbalances occur. Neurons form large-scale regulatory networks involving multiple brain regions to support these dynamic processes. These include hypothalamic, brainstem, and limbic circuits that oversee autonomic functions, hormone release, and immune modulation. Neurotransmitters are integral to this regulation. For instance: Dopamine, norepinephrine, serotonin, and acetylcholine modulate autonomic and behavioral responses during stress, reward, or arousal. Glutamate, GABA, and various neuropeptides influence immune signaling pathways. In endocrine regulation, dopamine, serotonin, and neuropeptides help modulate hormone release from hypothalamic-pituitary axes.

Recent models of interoceptive prediction—the brain's inference of internal bodily states—have extended the concept of allostasis to include homeostatic error minimization, where the brain anticipates internal needs before they arise, based on prior cycles and contextual cues (Kleckner et al., 2017). Advances in physiological monitoring methods have further enabled the detection of such anticipatory responses in both laboratory and real-world settings (Kleckner et al., 2024). This perspective reinforces the brain's role as a *predictive regulator of internal states,* integrating visceral, emotional, and cognitive domains into unified responses. Through this networked activity, the brain coordinates internal responses to maintain or restore equilibrium, while also integrating feedback from physiological sensors throughout the body. These circuits enable adaptive shifts in metabolism, cardiovascular tone, immune activation, and thermoregulation.

Glial cells—particularly microglia and astrocytes—play an increasingly recognized role in modulating immune and homeostatic functions. Microglia mediate immune surveillance and inflammatory responses, while astrocytes regulate blood-brain barrier permeability and coordinate signaling between neuronal and vascular systems. Recent single-cell transcriptomic studies have shown that glial cells express receptors for systemic immune cytokines and hormones, acting as *bidirectional mediators between the brain and peripheral immune states* (Dopp et al., 2024). Moreover, glial involvement in *circadian regulation* suggests that time-of-day patterns influence not only sleep-wake cycles but also *immune readiness and hormonal reactivity,* linking brain rhythms to the timing of systemic physiological processes. In summary, the brain's role in homeostasis extends beyond moment-to-moment regulation. It actively predicts, integrates, and modulates biological systems to optimize performance and survival in a changing environment. This capacity is foundational to the broader understanding of the brain as a unifying organ, mediating both conscious and unconscious responses in service of the organism's integrity. As such, it contributes to the brain's metafunctional capacity to serve as a nexus between physiological necessity and existential coherence.

Neurons are fundamental cells of the brain, and they are critically important to its remarkable capabilities characterized by providing biological support to consciousness, participating in cognitive processes, processing information, receiving and transmitting signals, storing and retrieving data, regulating internal bodily functions, controlling movement and behavior, and exhibiting neuroplasticity. Neuroplasticity refers to the brain's remarkable ability to adapt and change throughout an individual's life (Merzenich & Jenkins, 1993; Kolb and Whishaw, 1998). This capacity allows the brain to form new synaptic connections and reorganize existing ones, particularly in response to learning, new experiences, or following neurological injury. Neuroplasticity encompasses several processes, including synaptic plasticity, adjusting the strength of connections between neurons, and structural plasticity, where the brain's physical structure changes. These adaptations enable the brain to compensate for lost functions and recover from damage, changing when we learn new skills, illustrating its dynamic and resilient nature.

Recent research has shown that plasticity is not solely neuronal. Astrocytes and microglia participate actively in shaping synaptic strength and spatial structure, releasing neuromodulators and cytokines that influence long-term potentiation and depression. In addition, *NG2-glia are now known to contribute to synaptic regulation,* forming transient junctions with active neurons and modulating local excitability thresholds (Zhang et al., 2021). These interactions suggest that plasticity must be redefined as a *multicellular process*, in which the neuro-glial interface becomes the unit of adaptive change. However, neurons cannot function in isolation; they require the collaboration of other different kinds of cells in the brain. Glial cells, for example, provide crucial support functions such as maintaining the extracellular environment, providing structural support to neurons, and assisting in the repair process after injury. Astrocytes regulate neurotransmitter levels, while oligodendrocytes and Schwann cells facilitate the insulation of axons, enabling efficient signal transmission. Additionally, microglia play a vital role in immune defense and synaptic pruning, contributing to overall brain health and function.

In the context of trauma or neurodegenerative disease, glial plasticity has emerged as a key compensatory mechanism, facilitating synaptic remodeling, neurotrophic factor release, and phagocytic clearance of dysfunctional connections. Transcriptomic atlases have identified unique glial subtypes activated during regeneration, particularly in regions undergoing network reorganization (Hansen et al., 2022). This suggests that glia not only support neurons but also scaffold the process of recovery and cognitive re-stabilization. Together, these cells form a complex network that supports and sustains the functioning of neurons, ultimately enabling the brain to carry out its various functions effectively. Following is an overview of the synapse and brain cell types. *This cellular landscape continues to be redefined by advances in single-cell genomics and functional mapping,* which now identify over 3,000 distinct brain cell types across cortical and subcortical territories (Hansen et al., 2022). These distinctions reveal unprecedented cellular heterogeneity—reflecting the profound specialization required to support the brain's metafunctional integration across processes like perception, memory, emotion, and meaning.

The Synapse

A synapse is a specialized junction through which communication occurs between two neurons or between a neuron and another target cell, enabling the transfer of information across the nervous system (Kandel, Schwartz, & Jessell, 2012). Synapses are typically classified into two main types—chemical and electrical—each characterized by distinct mechanisms and functional properties. In chemical synapses, the presynaptic neuron releases neurotransmitters into a narrow space known as the synaptic cleft. These chemical messengers then bind to receptors on the postsynaptic cell, initiating a cascade of intracellular events that either excite or inhibit the target, depending on the specific neurotransmitter and receptor involved. Chemical synapses are by far the most common in the human brain and are fundamental to complex processes such as learning, memory formation, and behavioral modulation.

Electrical synapses, in contrast, involve direct cytoplasmic connections between adjacent neurons via structures called gap junctions. These junctions allow ions and small molecules to flow freely between cells, enabling ultra-rapid transmission and synchronized activity across neuronal populations. Although less prevalent than chemical synapses, electrical synapses are critical in circuits that demand immediate responses—such as those governing reflexes or oscillatory networks in the brainstem. Beyond these two primary forms of transmission, the brain also engages in *volume transmission*—a mode of communication in which neurotransmitters or neuromodulators diffuse through the extracellular space and act on neurons not directly connected by synapses (Agnati, Zoli, Strömberg, & Fuxe, 1995). This mechanism enables widespread and slower-acting modulation, influencing brain states and network-level coordination.

Another emerging mechanism is *ephaptic transmission*, in which changes in the local electric field—rather than synaptic or molecular exchange—affect the excitability of neighboring neurons (Anastassiou, Perin, Markram, & Koch, 2011). This phenomenon illustrates that neurons can influence one another without direct physical or chemical connection. These discoveries suggest that *neural communication is far more complex and multidimensional than previously conceived*, incorporating both localized and non-local modes of interaction. Together, these diverse modes of neuronal signaling—precise, diffuse, direct, and field-based—underscore the remarkable complexity of brain communication. They allow the nervous system to integrate information across multiple spatial and temporal scales, enabling both fine-tuned specificity and broad, state-dependent modulation. This multifaceted signaling architecture plays a foundational role in what can be described as the brain's overarching metafunction: its capacity to unify perception, regulation, and adaptive behavior in a coherent, embodied whole.

Glial Cells: Beyond "Nerve Glue"

In 1856, German biologist Rudolf Virchow (1821–1902) discovered a substance surrounding nerve cells, which he described as "a type of putty" (Virchow, 2005). Termed neuroglia or "nerve glue," this material was initially thought to serve only as a passive structural scaffold for neurons.

However, modern research has revealed that glial cells—glia—play far more dynamic and essential roles in the nervous system (Allen & Lyons, 2018). Glial cells are non-neuronal cells that match neurons in number, with the ratio varying across brain regions. For example, the cerebral cortex has a glia-to-neuron ratio of approximately 3.72, supporting its high metabolic demands, whereas the cerebellum, which is predominantly neuronal, has a ratio of only 0.23 (Azevedo et al., 2009). Researchers categorize glial cells into four main types: microglia, astrocytes, oligodendrocytes, and NG2-glia. Each type contributes uniquely to the function and health of the nervous system. Beyond providing structural support, glial cells perform critical functions that ensure the brain's metabolic stability, immune surveillance, and synaptic integration.

New findings emphasize the active participation of glia in consciousness and brain-state transitions. Astrocytes, for example, modulate sleep-wake cycles by regulating adenosine concentrations and influencing slow-wave activity. They also help gate cortical responsiveness to stimuli by adjusting local potassium and glutamate levels in response to neural demand (Frank, 2019; Dopp et al., 2024). These roles suggest that astrocytes help maintain the dynamic balance between arousal and rest, directly contributing to cognitive clarity and sensory integration. Moreover, the astrocyte-neuron metabolic axis—long considered supportive—is now understood as *bidirectionally interactive.* During periods of high neural activity, astrocytes regulate synaptic efficacy by modulating the availability of metabolic substrates such as lactate, which act not only as energy sources but as *neuromodulatory signals* shaping memory encoding and plasticity (Suzuki et al., 2011; Hertz, 2004). The reclassification of glia as active signal processors—not passive supporters—has profound implications for how we conceive of brain function. Their involvement in neurovascular coupling, synaptic plasticity, network homeostasis, and neuromodulation positions them as key players in the emergence of brain metafunctionality. In this light, glial cells are integral to the seamless integration of sensory, cognitive, and transphysical domains that constitute the holistic unity of human experience.

Microglia

Microglia are the central nervous system's primary immune cells, essential for maintaining brain health, responding to injury and disease, and regulating neural communication. These specialized glial cells engage in phagocytosis, clearing waste products, cellular debris, and damaged neurons—functions critical to sustaining a healthy and resilient nervous system (Gehrmann, Matsumoto, & Kreutzberg, 1995). Microglia also monitor the brain's microenvironment, maintaining homeostasis by regulating levels of neurotransmitters and signaling molecules necessary for optimal function (Cserép et al., 2019). Beyond their immune surveillance roles, microglia actively influence synaptic activity. They modulate the strength of synaptic connections—a process integral to learning and memory—by pruning weak or unnecessary synapses and releasing signaling molecules that enhance or suppress neural transmission. This dynamic role in shaping and refining neural circuits underscores their essential contribution to cognitive flexibility and plasticity.

Recent studies show that microglia exhibit distinct transcriptional phenotypes depending on their spatial location and developmental context, suggesting that microglial responses are highly tailored to local circuit needs (Masuda et al., 2020). For instance, in neurodevelopmental windows, microglia help coordinate critical periods of plasticity by phagocytosing immature synapses. In adulthood, they participate in synaptic remodeling associated with learning and emotional regulation. In response to injury or disease, microglia rapidly migrate to affected areas and release cytokines, chemokines, and other signaling molecules that contain damage and promote repair. For example, in neurodegenerative disorders such as Alzheimer's disease, microglia attempt to clear amyloid-beta plaques. However, chronic activation or dysregulation of microglial function can exacerbate neuroinflammation and neuronal loss, highlighting the fine balance they must maintain between repair and pathology.

In aging and chronic disease, microglia may enter a "primed" state—characterized by heightened sensitivity and proinflammatory reactivity—which can contribute to maladaptive responses and cognitive decline. Understanding the shift between protective and pathological roles remains an urgent research priority. These protective, regulatory, and modulatory functions establish microglia as indispensable players in both health and disease. Their participation in immune defense, synaptic remodeling, and overall homeostasis situates them at the heart of the brain's adaptive responses—linking inflammation, plasticity, and cognition in ways we are only beginning to fully appreciate.

Astrocytes

Astrocytes are highly versatile glial cells essential for maintaining nervous system function and participating in numerous physiological processes in the brain (Verkhratsky & Nedergaard, 2018). One of their primary roles is providing structural support, helping preserve the organization and connectivity of neural circuits. These cells also regulate the brain's extracellular environment by controlling ion concentrations, clearing excess neurotransmitters, and modulating other signaling molecules—thus ensuring the optimal conditions for neuronal activity. Astrocytes are key players in cerebrovascular regulation, adjusting local blood flow in response to neural activity. They do so by constricting or dilating blood vessels to match the metabolic demands of active neurons. Additionally, they form a critical component of the blood–brain barrier, a selective interface that protects the brain by restricting the entry of potentially harmful substances from the bloodstream (MacVicar & Newman, 2015). This role is crucial for preserving the brain's internal stability.

Recent research highlights astrocytes' active contribution to learning and memory through modulation of synaptic plasticity. These cells release gliotransmitters such as D-serine and ATP, which influence NMDA receptor activation and synaptic potentiation, particularly in hippocampal circuits (Purushotham & Buskila 2023). By regulating receptor sensitivity and neurotransmitter availability, astrocytes effectively shape the conditions for memory encoding and retrieval. Astrocytes also contribute significantly to synaptic development and plasticity. They regulate synapse formation, maturation, and elimination—processes essential for learning and memory. Through bidirectional signaling with neurons, they help modulate synaptic strength and efficiency. In times of injury or illness, astrocytes become reactive: they proliferate, release inflammatory cytokines, and form glial scars that limit the spread of damage. While beneficial in acute injury, prolonged reactivity may contribute to chronic inflammation or scar formation, with implications for neurodegenerative disease.

Another major role of astrocytes is their involvement in neuronal energy metabolism. They supply neurons with metabolic substrates, such as glucose and lactate, especially during periods of heightened activity (Suzuki et al., 2011). This astrocyte–neuron metabolic coupling has been researched showing that lactate not only sustains energy needs but also acts as a *retrograde signal enhancing memory consolidation* through epigenetic pathways (Suzuki et al., 2011). Through their wide-ranging contributions—spanning structural integrity, chemical regulation, immune response, synaptic modulation, and energy support—astrocytes are indispensable to brain health and plasticity. Their dynamic interactions with neurons and other glial cells position them as integrative agents in both normal brain function and pathological states.

Oligodendrocytes

Oligodendrocytes are a type of glial cell in the human brain that play a crucial role in forming and maintaining the myelin sheath. This sheath insulates and protects neurons' axons and ensures the efficient transmission of signals between neurons (Nave & Werner, 2014). Oligodendrocytes are essential for the proper functioning of the nervous system, and their dysfunction or loss is associated with various neurological disorders, such as multiple sclerosis and leukodystrophies. One of the primary functions of oligodendrocytes is the production of myelin—a fatty, multilayered substance that surrounds and insulates axons to accelerate the conduction of electrical impulses. Myelination by oligodendrocytes enables saltatory conduction, a process by which action potentials "jump" between nodes of Ranvier, vastly increasing the speed and efficiency of signal transmission (Cohen et al., 2020).

Beyond insulation, oligodendrocytes also offer structural and metabolic support to axons. They help maintain axonal shape and integrity while regulating energy metabolism by supplying essential substrates such as lactate and pyruvate (Lee et al., 2012; Fünfschilling et al., 2012). These metabolic interactions are crucial for long-term axonal health and neuronal survival. Recent work has shown that oligodendrocytes are not static components but participate actively in *experience-dependent plasticity*. Activity-regulated myelination adapts the timing of interregional signal conduction, suggesting that myelin remodeling fine-tunes neural synchrony during learning (Bonetto, Belin, & Káradóttir, 2021). This process extends the concept of plasticity from synaptic weights to conduction velocities—adding a temporal dimension to how the brain adapts over time.

New evidence suggests that oligodendrocytes may also influence synaptic function. Research indicates that these glial cells contribute to experience-dependent network adaptation, possibly affecting synapse formation and plasticity (Mount & Monje, 2017). In the context of injury and disease, oligodendrocytes participate in both degeneration and repair. In demyelinating conditions such as multiple sclerosis, their ability to remyelinate damaged axons becomes critical for recovery. However, remyelination may be incomplete or impaired, particularly in chronic or progressive forms of disease (Franklin & Ffrench-Constant, 2008). Altogether, oligodendrocytes serve not only as insulators but also as *metabolic partners, structural supporters, and temporal modulators* of neuronal communication. Their diverse and essential roles underscore their importance in maintaining the health, efficiency, and adaptability of the central nervous system—and in sustaining the brain's metafunctional coherence across space and time.

NG2-Glia

NG2-glia, also known as oligodendrocyte precursor cells (OPCs), were first discovered and characterized in the early 1980s (Nishiyama, Chang, & Trapp 1999). These cells derive their name from their expression of the NG2 protein, a marker used to identify them. NG2-glia are widely distributed throughout the brain and spinal cord and play a vital role in the nervous system's development, maintenance, and adaptability. Initially, NG2-glia were thought to differentiate exclusively into oligodendrocytes—the glial cells responsible for producing myelin, the insulating sheath that surrounds and protects axons. This function is critical for myelination and remyelination, essential processes for maintaining the efficiency of neural signal transmission (Nishiyama et al., 2014). However, subsequent research has revealed that NG2-glia represent a heterogeneous and dynamic cell population with diverse roles beyond oligodendrocyte lineage progression (Du, Zhang, Zhou, & Zhou, 2021).

In addition to their role in generating oligodendrocytes, NG2-glia actively participate in neural communication. These cells form *functional synapse-like contacts* with both excitatory and inhibitory neurons, allowing them to receive direct input and modulate neuronal activity. This capacity suggests a role in regulating synapse formation, elimination, and remodeling—processes central to neuroplasticity and circuit refinement. *Recent optogenetic studies have demonstrated that NG2-glia can modulate local circuit excitability in real time,* indicating their active participation in network-level adaptation (Marques et al., 2016; Dimou & Gallo, 2015). NG2-glia also contribute to the brain's response to injury and disease. Under pathological conditions, they exhibit remarkable plasticity and can differentiate into other glial cell types, such as astrocytes or reactive glia, in an effort to protect and repair neural tissue. This functional adaptability is particularly relevant in the context of traumatic brain injury, demyelinating disorders, and neurodegeneration. However, disruptions in NG2-glia function or loss of their regenerative capacity have been implicated in several neurological conditions, including multiple sclerosis, Alzheimer's disease, and gliomas (Nishiyama et al., 2014; Du, Zhang, Zhou, & Zhou, 2021). For instance, failure to efficiently remyelinate axons may contribute to progressive disability in demyelinating diseases, while aberrant proliferation or differentiation may play a role in tumorigenesis. Recent transcriptomic profiling has further subdivided NG2-glia into functional subtypes based on region, injury response, and developmental stage, illuminating their diverse physiological roles (Masuda et al., 2020).

Through their involvement in myelination, synaptic modulation, and injury response, NG2-glia are increasingly recognized as critical players in neural development, resilience, and repair. Their multifunctional nature reflects the remarkable complexity and adaptability of the nervous system, making them an increasingly important focus of research in both basic neuroscience and translational medicine.

Glial Cooperation with Neurons

Sleep provides an illustrative example of how glial cells cooperate with neurons to maintain brain health and function. During sleep, glial cells in the human brain perform several essential tasks that support the brain's ability to recharge and prepare for the next day. These functions highlight their indispensable role in neural maintenance and adaptability. One critical function of glial cells during sleep is waste-removal. Glial cells increase their activity to clear harmful byproducts such as beta-amyloid and tau proteins, which, if accumulated, can contribute to neurodegenerative diseases like Alzheimer's. The shrinking of glial cells creates water microchannels called aquaporins that allow cerebrospinal fluid to flush out waste, facilitating this process. This recently discovered system of microchannels is the *glymphatic system,* which is crucial to brain detoxification.

Recent research suggests *this system is strongly influenced by circadian rhythms,* with peak glymphatic activity occurring during deep sleep stages and modulated by astrocytic aquaporin-4 (AQP4) polarization (Dopp et al., 2024; Jessen et al., 2015). These circadian mechanisms not only optimize waste clearance but also synchronize metabolic and immunological processes across brain regions, enhancing systemic resilience. Glial cells also regulate the brain's local environment during sleep by modulating neurotransmitter levels. This modulation ensures proper communication between neurons and supports optimal brain function upon waking. Additionally, they contribute to synaptic scaling, a process in which synaptic strength is adjusted based on recent activity, helping the brain maintain balance and prevent overstimulation. Glial cells release signaling molecules that facilitate these changes, ensuring efficient neural communication.

Astrocytic calcium waves and microglial pruning during sleep have now been implicated in memory consolidation, particularly in hippocampal circuits. These glial-mediated adjustments preserve salient connections while eliminating noise, thereby refining neural representations encoded during wakefulness (Frank, 2019; Purushotham & Buskila, 2023). Beyond these regulatory roles, glial cells provide structural support to neurons during sleep, maintaining the brain's integrity while enabling essential processes such as synaptic remodeling and waste clearance. These combined functions underscore glial cells' critical, behind-the-scenes work supporting neuronal health and overall brain function. Through their roles in waste removal, neurotransmitter regulation, synaptic scaling, and structural maintenance, glial cells demonstrate their essential contributions to brain health—especially during sleep. These quiet yet vital processes help ensure that the brain is rejuvenated and prepared for the demands of wakefulness, integrating cellular restoration with consciousness-related functions across the brain metafunction framework.

Top-down View of the Brain

The brain consists of the cerebrum, brainstem, and cerebellum. While traditional descriptions often isolate these regions by function, emerging research underscores their integrated operations, which contribute to global brain functionality. The cerebrum, the largest part of the brain, is primarily composed of the cerebral cortex, the brain's outermost layer of gray matter. The cerebral cortex specializes in various essential functions, including perception, sensation, movement, thought, and consciousness (Javed, Reddy, & Lui, 2023). This highly folded structure constitutes two hemispheres (left and right), connected by the corpus callosum, a dense bundle of nerve fibers that enables interhemispheric communication and integration of cognitive and motor functions. Each hemisphere of the cerebral cortex contains four distinct lobes, each with specialized roles:

1. *Frontal Lobe:* Located at the front of the brain, this lobe is pivotal for executive functions like planning, decision-making, and problem-solving. It also governs voluntary movement via the primary motor cortex.

2. *Parietal Lobe:* Situated in the upper-middle brain, this lobe processes somatosensory information, including touch, temperature, and pain, via the primary somatosensory cortex. It also plays a role in spatial awareness and navigation.

3. *Temporal Lobe:* Found on the sides of the brain, the temporal lobe is involved in auditory processing, memory consolidation, and emotional regulation. The hippocampus, located within the temporal lobe, is vital for learning and memory.

4. *Occipital Lobe:* Located at the back of the brain, this lobe specializes in visual processing and is vital for processing visual input from the eyes.

The cerebral cortex's organization includes specialized cortical areas responsible for motor control, sensory processing, language, and other higher functions. These areas are interconnected by vast neural networks, allowing seamless communication and functional integration across the brain. Advanced neuroimaging now reveals that cognitive functions often rely on distributed networks spanning multiple lobes, coordinated by high-connectivity hubs such as the precuneus, posterior cingulate cortex, and anterior insula (Gordon et al., 2017). Together with subcortical structures, the cerebral cortex orchestrates complex processes that enable human cognition, perception, and behavior. Its intricate architecture exemplifies the brain's remarkable complexity and capacity for coordination across domains of thought, movement, and experience—embodying the foundational principle of brain metafunctionality: unifying the biological, cognitive, and transpersonal aspects of human identity.

Limbic System

The limbic system is a highly interconnected network of brain structures that regulate emotion, motivation, behavior, and memory. Initially described by Papez (1937) and expanded by MacLean (1952), this system integrates sensory information with emotional and survival-related responses, making it a cornerstone of human behavior. The limbic system comprises several key components, including the amygdala, hippocampus, thalamus, hypothalamus, cingulate cortex, and basal ganglia—each contributing unique functions:

- *Amygdala:* Plays a pivotal role in processing and interpreting emotions such as fear, aggression, and pleasure. It helps the brain recognize emotionally charged situations and supports emotional learning, including fight-or-flight responses. Damage to the amygdala can impair emotional recognition and lead to disorders such as anxiety or PTSD

- *Hippocampus:* Essential for memory formation. It encodes new experiences, links them to existing memories, and retrieves them when needed. It also contributes to spatial navigation and constructing coherent narratives of experience. Dysfunction in the hippocampus is associated with conditions like Alzheimer's disease and other memory disorders.

- *Hypothalamus:* Maintains homeostasis by regulating essential physiological functions such as hunger, thirst, temperature, and sexual behavior. As a key link between the nervous and endocrine systems, it manages internal responses by releasing hormones and modulating autonomic activity.

- *Cingulate Cortex:* Contributes to emotional regulation, pain perception, and decision-making, particularly in conflict monitoring and social behavior.

- *Basal Ganglia:* While often associated with motor control, are also central to motivation and reward processing. These structures help link emotional states to behavioral actions. Dysfunction in the basal ganglia is implicated in Parkinson's disease, obsessive-compulsive disorder, and other conditions involving movement or motivation.

Together, these interconnected structures process and respond to emotional and physiological signals while integrating them with cognitive functions such as attention, decision-making, and memory. *Recent connectomic analysis reveals that the limbic system operates as a functional bridge between sensory integration networks and self-referential processing hubs*, particularly those involved in identity, spirituality, and moral reasoning (Luppi et al., 2023). This supports the view that the limbic system is not merely reactive but participatory, mediating both affective responses and existential significance. Damage to any part of the limbic system can result in emotional dysregulation, memory impairment, or motivational disturbances, underscoring its vital role in the unity of emotion, cognition, and behavior. For detailed insights into the integrative and regulatory functions of limbic-related networks from an allostatic perspective, see Mirifar, Keil, and Ehrlenspiel (2022).

The brainstem, located at the base of the brain, connects the spinal cord to the cerebrum and cerebellum. This critical structure consists of three main parts: the medulla oblongata, pons, and midbrain. Serving as a communication highway, the brainstem facilitates the flow of information between the brain and body, enabling essential physiological functions. It plays a vital role in regulating breathing, heart rate, and blood pressure—core autonomic functions essential for life. It also houses key cranial nerve nuclei responsible for sensory and motor control of the face and neck. Importantly, the brainstem contains the *reticular activating system (RAS),* a network of neurons involved in regulating arousal and maintaining consciousness (Moruzzi & Magoun, 1949). The RAS modulates sensory input and governs sleep-wake cycles, ensuring that relevant stimuli reach conscious awareness. For example, it helps prioritize significant auditory or visual input—such as hearing one's name in a crowded room.

Recent studies have identified fast and slow RAS oscillatory dynamics that differentially affect cortical excitability, attentional focus, and sensory salience (Garcia-Rill et al., 2013; Dahl, Mather, & Werkle-Bergner, 2022). These findings reveal the RAS as not only a gatekeeper of consciousness but a flexible regulator of global brain state transitions—a core feature of metafunctional integration. Damage to the brainstem can result in severe impairments, including loss of consciousness, respiratory failure, or conditions such as locked-in syndrome. Its location and functions make it a vital integrative center for survival and cognitive alertness. The brainstem's bidirectional connectivity with both cortical and limbic structures allows it to *mediate between reflexive responses and higher-order cognition,* ensuring that the body remains attuned to both physiological and experiential demands.

Cerebellum

The cerebellum, located at the back of the brain beneath the cerebral hemispheres, plays an essential role in coordinating and regulating voluntary movement, balance, and posture. Its intricately folded structure is divided into two hemispheres, connected by the vermis, and it receives sensory input from the muscles, joints, and inner ear. By integrating this sensory data with motor commands from the cerebral cortex, the cerebellum fine-tunes movements, ensuring that actions are smooth, accurate, and well-timed. For example, it enables the fluid execution of complex behaviors such as typing, walking, or playing a musical instrument. Beyond its classical role in motor control, the cerebellum contributes to higher-order cognitive processes, including attention, language, working memory, and emotional regulation. It helps the brain anticipate events and coordinate responses in a predictive manner, allowing for efficient task execution. Research has also revealed the cerebellum's involvement in language comprehension and speech fluency, coordinating the neural activity required for verbal processing.

Cerebellar predictive modeling, a process by which the brain forecasts sensory outcomes based on prior experience, has emerged as a key function in both motor and cognitive domains (Van Overwalle et al., 2020). This aligns with the brain's broader predictive architecture, supporting models of self-regulation, empathy, and decision-making. Its contribution to working memory supports tasks involving reasoning, planning, and problem-solving. When damaged, the cerebellum can give rise to a range of impairments, including ataxia (loss of coordination), dysarthria (slurred speech), and subtle cognitive and emotional disturbances. These effects underscore the cerebellum's multifaceted role in supporting both movement and the nuanced aspects of cognition. By coordinating motoric, emotional, and executive domains, the cerebellum functions as a bridge between sensorimotor action and abstract cognition—serving, in effect, as a *temporal harmonizer* across the brain's functional systems.

As we reflect on the multiple layers of brain function—from cellular communication to large-scale network dynamics—it becomes clear that the brain is not merely an organ of computation or reflex. It is a living interface between body, mind, and world. Its architecture allows for stunning levels of adaptability, coherence, and participation in meaning-making. The brain's synaptic plasticity, glial cooperation, rhythmic network coordination, and predictive integration converge to support the unique human capacities for language, consciousness, moral judgment, and spiritual experience. *This profound unity is not reducible to any one layer of biological operation but depends on their interrelation*—a phenomenon that calls for a framework capable of encompassing molecular, cognitive, social, and transphysical dimensions. Here, the Brain Metafunction Framework offers a holistic model that synthesizes these layers. In bridging neurobiology with philosophy, medicine, and theology, the next chapter presents this integrative paradigm. It moves beyond fragmentation—whether scientific, psychological, or metaphysical—and proposes a vision of the brain as a cohesive biological organizer of human personhood. This framework will guide a deeper inquiry into how the brain mediates behavior, thought, relationality, identity, and transcendence.

REFERENCES

Adamovich, S. V., Fluet, G. G., Tunik, E., & Merians, A. S. (2009). Sensorimotor training in virtual reality: A review. NeuroRehabilitation, 25(1), 29–44. https://doi.org/10.3233/NRE-2009-0497

Agnati, L. F., Zoli, M., Strömberg, I., & Fuxe, K. (1995). Intercellular communication in the brain: Wiring versus volume transmission. Neuroscience, 69(3), 711–726.https://doi.org/10.1016/0306-4522(95)00308-6

Allen, N. J., & Lyons, D. A. (2018). Glia as architects of central nervous system formation and function. Science, 362(6411), 181–185. https://doi.org/10.1126/science.aat0473

Altman, J. (1962). Are new neurons formed in the brains of adult mammals? Science, 135(3509), 1127–1128. https://doi.org/10.1126/science.135.3509.1127

Amunts, K., Lepage, C., Borgeat, L., Mohlberg, H., Dickscheid, T., Rousseau, M.-É., Bludau, S., Bazin, P.-L., Lewis, L. B., Oros-Peusquens, A.-M., Shitrit, D., Zilles, K., & Evans, A. C. (2013). BigBrain: An ultra-high-resolution 3D human brain model. Science, 340(6139), 1472–1475. https://doi.org/10.1126/science.1235381

Azevedo, F. A. C., Carvalho, L. R. B., Grinberg, L. T., Farfel, J. M., Ferretti, R. E. L., Leite, R. E. P., ... & Herculano-Houzel, S. (2009). Equal numbers of neuronal and nonneuronal cells make the human brain an isometrically scaled-up primate brain. The Journal of Comparative Neurology, 513(5), 532–541. https://doi.org/10.1002/cne.21974

Behrendt, R. P. (2013). Conscious experience and episodic memory: Hippocampus at the crossroads. Frontiers in Psychology, 4, 304. https://doi.org/10.3389/fpsyg.2013.003 04

Binns, K. E., & Salt, T. E. (1996). Importance of NMDA receptors for multimodal integration in the deep layers of the cat superior colliculus. Journal of Neurophysiology, 75(3), 920–930. https://doi.org/10.1152/jn.1996.75.3.920

Boergens, K. M., Berning, M., Bocklisch, T., & Helmstaedter, M. (2017). WebKnossos: Efficient online 3D annotation for connectomics. Nature Methods, 14, 691–694. https://doi.org/10.1038/nmeth.4331

Bonetto, G., Belin, D., & Káradóttir, R. Þ. (2021). Myelin: A gatekeeper of activity-dependent circuit plasticity? Science, 374(6569), eaba6905. https://doi.org/10.1126/science.aba6905

Bostrom, N. (2014). Superintelligence: Paths, dangers, strategies. Oxford University Press.

Brennan, J. R., & Martin, A. E. (2020). Phase synchronization varies systematically with linguistic structure composition. Philosophical Transactions of the Royal Society B: Biological Sciences, 375(1791), 20190305. https://doi.org/ 10.1098/rstb.2019.0305

Brennan, J. R., & Hale, J. T. (2019). Hierarchical structure guides rapid linguistic predictions during naturalistic listening. PLoS ONE, 14(1), e0207741. https://doi.org/10.1371/journal.pone.0207741

Caligiore, D., Pezzulo, G., Baldassarre, G., Bostan, A. C., Strick, P. L., Doya, K., Helmich, R. C., Dirkx, M., Houk, J., Jörntell, H., Lago-Rodriguez, A., Galea, J. M., Miall, R. C., Popa, T., Kishore, A., Verschure, P. F., Zucca, R., & Herreros, I. (2017). Consensus paper: Towards a systems-level view of cerebellar function: The interplay between cerebellum, basal ganglia, and cortex. The Cerebellum, 16(1), 203–229. https://doi.org/10.1007/s12311-016-0763-3

Carnap, R. (1932). The elimination of metaphysics through logical analysis of language (A. Pap, Trans.). In A. J. Ayer (Ed.), Logical positivism (pp. 60–81). Free Press. (Original work published 1932)

Caucheteux, C., & King, J.-R. (2022). Brains and algorithms partially converge in natural language processing. Communications Biology, 5, Article 134. https://doi.org/10.1038/s42003-022-03036-1

Cerina, M., Piastra, M. C., & Frega, M. (2023). The potential of in vitro neuronal networks cultured on micro electrode arrays for biomedical research. Progress in Biomedical Engineering, 5(3), Article 032002. https://doi.org/10.1088/2516-1091/acce12

Cserép, C., Pósfai, B., Lénárt, N., Fekete, R., László, Z. I., Lele, Z.,

Orsolits, B., Molnár, G., Heindl, S., & Dénes, Á., et al. (2019). Microglia monitor and protect neuronal function through specialized somatic purinergic junctions. Science, 367(6477), 528–537. https://doi.org/10.1126/science.aax6752

Dahl, M. J., Mather, M., & Werkle-Bergner, M. (2022). Noradrenergic modulation of rhythmic neural activity shapes selective attention. Trends in Cognitive Sciences, 26(1), 46–59.

Deco, G., & Kringelbach, M. L. (2016). Metastability and coherence: Extending the communication through coherence hypothesis using a whole-brain computational perspective. Trends in Neurosciences, 39(3), 125–135. https ://doi.org/10.1016/j.tins.2016.01.001

Deng, J., Wu, J., Chen, C., Zheng, Q., Zhang, Z., Wu, J., Ouyang, W., & Song, C. (2024). UniSPAC: A unified segmentation framework for proofreading and annotation in connectomics. bioRxiv. https://doi.org/10.1101/2024.11.27. 625336

Denk, W., Strickler, J. H., & Webb, W. W. (1990). Two-photon laser scanning fluorescence microscopy. Science, 248(4951), 73–76. https://doi.org/10.1126/science.2321027

Denoth-Lippuner, A., & Jessberger, S. (2021). Formation and integration of new neurons in the adult hippocampus. Nature Reviews Neuroscience, 22(4), 223–236. https://doi.org/10.1038/s41583-021-00433-z

Dimou, L., & Gallo, V. (2015). NG2-glia and their functions in the central nervous system. Glia, 63(8), 1429–1451. https://doi.org/10.1002/glia.22859

Deisseroth, K. (2011). Optogenetics. Nature Methods, 8(1), 26–29. https://doi.org/10.1038/nmeth.f 324

Dopp, J., Ortega, A., Davie, K., & others. (2024). Single-cell transcriptomics reveals that glial cells integrate homeostatic and circadian processes to drive sleep–wake cycles. Nature Neuroscience, 27, 359–372. https://doi.org/10.1038/s41593-023-01549-4

Doudna, J. A., & Charpentier, E. (2014). The new frontier of genome engineering with CRISPR-Cas9. Science, 346(6213), Article 1258096. https://doi.org/10.1126/science.1258096

Du, Y., Zhang, X., Zhou, X., & Zhou, M. (2021). NG2-glia: Rising stars in neural plasticity and repair. Nature Reviews Neuroscience, 22(12), 681–695. https://doi.org/10.1038/s41 583-021-00511-0

Eagleman, D. (2007). Incognito: The secret lives of the brain. Pantheon Books.

Emiliani, V., Cohen, A. E., Deisseroth, K., & Häusser, M. (2015). All-optical interrogation of neural circuits. Journal of Neuroscience, 35(41), 13917–13926. https://doi.org/10.1523/JNEUROSCI.2916-15 .2015

Faskowitz, J., Esfahlani, F. Z., Jo, Y., Sporns, O., & Betzel, R. F. (2020). Edge-centric functional network representations of human cerebral cortex reveal overlapping system-level architecture. Nature Neuroscience, 23(12), 1644–1654. https://doi.org/10.1038/s41593-020-00719-y

Feldman, J. (2013). The neural binding problem(s). Cognitive Neurodynamics, 7(1), 1–11. https://doi.org/10.1007/s11571-012-92 19-8

Fuhrmann, D., Knoll, L. J., & Blakemore, S. -J. (2015). Adolescence as a sensitive period of brain development. Trends in Cognitive Sciences, 19(10), 558–566. https://doi.org/10.1016/j.tics.2015.07.008

Franklin, R. J. M., & Ffrench-Constant, C. (2008). Remyelination in the CNS: from biology to therapy. Nature Reviews Neuroscience, 9(11), 839–855. https://doi.org/10.1038/nrn2480

Frégnac, Y. (2017). Big data and the industrialization of neuroscience: A safe roadmap for understanding the brain? Science, 358(6362), 470–477. https://doi.org/10.1126/science.aan8866

Friston, K. J. (2010). The free-energy principle: a unified brain theory? Nature Reviews Neuroscience, 11(2), 127–138. https://doi.org/10.1038/nrn2787

Friston, K. J., FitzGerald, T. H. B., Rigoli, F., Schwartenbeck, P., & Pezzulo, G. (2017). Active inference: A process theory. Neural Computation, 29(1), 1–49. https://doi.org/10.1162/NECO_a_00912

Fünfschilling, U., Supplie, L. M., Mahad, D., et al. (2012). Glycolytic oligodendrocytes maintain myelin and long-term axonal integrity. Nature, 485(7399), 517–521. https://doi.org /10.1038/nature11007

Gage, F. H., & Temple, S. (2013). Neural stem cells: Generating and regenerating the brain. Neuron, 80(3), 588–601. https://doi.org/10.1016/j.neuron.2013.10.037

Garcia-Rill, E., Kezunovic, N., Hyde, J., Simon, C., Beck, P., & Urbano, F. J. (2013). Coherence and frequency in the reticular activating system (RAS). Sleep Medicine Reviews, 17(3), 227–238. https://doi.org/10.1016/j.smrv.2012.06.002

Garcia, O. D. (1995a). Reformulating death: Overcoming whole-brain inconsistencies. In C. Machado (Ed.), Brain death (pp. 79–92). Amsterdam: Elsevier.

Garcia, O. D. (1995b). Whole brain? Higher brain? A new formulation of death. In C. Machado (Ed.), Brain death (pp. 93–104). Amsterdam: Elsevier.

García, O. D. (1998). Una reformulación completa y precisa de la muerte humana [A complete and precise reformulation of human death]. Revista de Neurología, 26(154), 1048–1053. PMID: 9659470

Garcia, O. D. (1999). A new formulation of death and its relevance to medical law. Medicine and Law, 17(4).

Garcia, O. D. (2001). The unity of the human person: A central concept in the interface between medicine and theology. The Linacre Quarterly, 68(1), Article 5.

Gehrmann, J., Matsumoto, Y., & Kreutzberg, G. W. (1995). Microglia: Intrinsic immune-effector cell of the brain. Brain Research Reviews, 20(3), 269–287. https://doi.org/10.1016/0165-0173(94)00015-H

Gernsbacher, M. A., & Kaschak, M. P. (2003). Neuroimaging studies of language production and comprehension. Annual Review of Psychology, 54, 91–114. https://doi.org/10.1146/annurev.psych.54.101601.145128

Glover, G. H. (2011). Overview of functional magnetic resonance imaging. Neurosurgery Clinics of North America, 22(2), 133–139. https://doi.org/10.1016/j.nec.2010.11.001

Godwin, W., & Cham, C. (2012). The human brain: A fascinating look at the anatomy and functions. New York, NY: Learning Press.

Google Research. (n.d.). Neuroglancer: Web-based visualization of volumetric data. Google Research Projects. https://github.com/google/neuroglancer

Gordon, E. M., Laumann, T. O., Gilmore, A. W., Newbold, D. J., Greene, D. J., Berg, J. J., Ortega, M., Hoyt-Drazen, C., Gratton, C., Sun, H., Hampton, J. M., Coalson, R. S., Nguyen, A. L., McDermott, K. B., Shimony, J. S., Snyder, A. Z., Schlaggar, B. L., Petersen, S. E., & Dosenbach, N. U. F. (2017). Precision functional mapping of individual human brains. Neuron, 95(4), 791–807.e7. https://doi.org/10.1016/j.neuron .2017.07.011

Grienberger, C., & Konnerth, A. (2012). Imaging calcium in neurons. Neuron, 73(5), 862–885. https://doi.org/10.1016/j.neuron.2012.02.011

Grienberger, C., Giovannucci, A., Zeiger, W., & Portera Cailliau, C. (2022). Two-photon calcium imaging of neuronal activity. Nature Reviews Methods Primers, 2(1), Article 67. https://doi.org/10.1038/s43586-022-00147-1

Hansen, J. Y., Shafiei, G., Markello, R. D., Smart, K., Cox, S. M. L., Wu, Y., Gallezot, J. D., Wainstein, G., Sporns, O., Saad, Z. S., Misic, B., & Baillet, S. (2022). Mapping neurotransmitter systems to the structural and functional organization of the human neocortex. Nature Neuroscience, 25(11), 1569–1581. https://doi.org/10.1038/s41593-022-01186-3

Hertz, L. (2004). The astrocyte–neuron lactate shuttle: A challenge of a challenge. Journal of Cerebral Blood Flow & Metabolism, 24(11), 1241–1248. https://doi.org/10.1097/01.WCB.0000137052.67446.96

Hnasko, T. S., & Edwards, R. H. (2012). Neurotransmitter corelease: Mechanism and physiological role. Annual Review of Physiology, 74, 225–243. https://doi.org/10.1146/annurev-physiol-020911-153315

Hounsfield, G. N. (1973). Computerized transverse axial scanning (tomography): Part 1. British Journal of Radiology, 46, 1016–1022. https://doi.org/10.1259/0007-1285-46-552-1016

Huber, L., Handwerker, D. A., Jangraw, D. C., Chen, G., Hall, A., Stüber, C., Gonzalez-Castillo, J., Ivanov, D., Marrett, S., Guidi, M., Goense, J., Poser, B. A., Bandettini, P. A., & Gonzalez-Castillo, J. (2017).

High-resolution CBV-fMRI allows mapping of laminar activity and connectivity of cortical input and output in human M1. Neuron, 96(6), 1253–1263.e7. https://doi.org/10.1016/j.neuron.2017.11.005

Javed, K., Reddy, V., & Lui, F. (2023). Neuroanatomy, Cerebral Cortex. In StatPearls. StatPearls Publishing. https: //www.ncbi.nlm.nih.gov/books/NBK537247/

Jessen, N. A., Munk, A. S. F., Lundgaard, I., & Nedergaard, M. (2015). The glymphatic system: A beginner's guide. Neurochemical Research, 40(12), 2583–2599. https://doi.org/10.1007/s11064-015-1581-6

Kandel, E. R., Schwartz, J. H., & Jessell, T. M. (2012). Principles of neural science (5th ed.). McGraw-Hill.

Kelso, J. A. S. (2012). Multistability and metastability: Understanding dynamic coordination in the brain. Philosophical Transactions of the Royal Society B: Biological Sciences, 367(1591), 906–918. https://doi.org/10.1098/rstb. 2011.0351

Kleckner, I. R., Wormwood, J. B., Jones, R. M., Culakova, E., Barrett, L. F., Lord, C., Quigley, K. S., & Goodwin, M. S. (2024). Adaptive thresholding increases sensitivity to detect changes in the rate of skin conductance responses to psychologically arousing stimuli in both laboratory and ambulatory settings. International Journal of Psychophysiology, 196, 112280. https://doi.org/10.1016/j.ij psycho.2023.112280

Kolb, B., & Whishaw, I. Q. (1998). Brain plasticity and behavior. Annual Review of Psychology, 49(1), 43–64. https: //doi.org/10.1146/annurev.psych.49.1.43

Labache, L., Joliot, M., Saracco, J., Jobard, G., Hesling, I., Zago, L., Mellet, E., Petit, L., Crivello, F., Mazoyer, B., & Tzourio-Mazoyer, N. (2018). A SENtence Supramodal Areas AtlaS (SENSAAS) based on multiple task-induced activation mapping and graph analysis of intrinsic connectivity in 144 healthy right-handers. Brain Structure and Function, 223(8), 3493–3513. https://doi.org/10.1007/s00 429-018-1810-2

Lebedev, M. A., & Nicolelis, M. A. L. (2017). Brain–machine interfaces: From basic science to neuroprostheses and neurorehabilitation. Physiological Reviews, 97(2), 767–837. https://doi.org/10.1152/physrev.00027.2016

LeDoux, J. (1993). Emotional networks and motor control: A fearful view. Progress in Brain Research, 107, 437–446. https://doi: 10.1016/s0079-6123(08)61880-4

Lee, Y., Morrison, B. M., Li, Y., Lengacher, S., Farah, M. H., Hoffman, P. N., Liu, Y., Tsingalia, A., Jin, L., Zhang, P.-W., Pellerin, L., Magistretti, P. J., & Rothstein, J. D. (2012). Oligodendroglia metabolically support axons and contribute to neurodegeneration. Nature, 487(7408), 443–448. https://doi.org/10.1038/nature11314

Liu, J., Bayle, D. J., Spagna, A., Sitt, J. D., Bourgeois, A., Lehongre, K., Fernandez-Vidal, S., Adam, C., Lambrecq, V., Navarro, V., Seidel Malkinson, T., & Bartolomeo, P. (2023). Fronto-parietal networks shape human conscious report through attention gain and reorienting. Communications Biology, 6, 730. https://doi.org/10.1038/s42003-023-05108-2

Liu, Y., Xu, S., Yang, Y., Zhang, J., Zhang, J., & Wang, W. (2023). Nanomaterial-based microelectrode arrays for in vitro bidirectional brain–computer interfaces: A review. Microsystems & Nanoengineering, 9, 13. https://doi.org/ 10.1038/s41378-022-00479-8

Liu, X., Gong, Y., Jiang, Z., Stevens, T., & Li, W. (2024). Flexible high-density microelectrode arrays for closed-loop brain–machine interfaces: A review. Frontiers in Neuroscience, 18, 1348434. https://doi.org/10.3389/fnins. 2024.1348434

Lloréns, R., Noé, E., Colomer, C., & Alcañiz, M. (2015). Effectiveness, usability, and cost-benefit of a virtual reality–based telerehabilitation program for balance recovery after stroke: A randomized controlled trial. Archives of Physical Medicine and Rehabilitation, 96(3), 418–425.e2. https://doi.org/10.1016/j.apmr.2014.10.019

Logothetis, N. K. (2003). The underpinnings of the BOLD functional magnetic resonance imaging signal. The Journal of Neuroscience, 23(10), 3963–3971. https://doi.org/10.1523/JNEUROSCI.23-10-03963.2003

Luppi, A. I., Vohryzek, J., Kringelbach, M. L., Mediano, P. A. M., Craig, M. M., Adapa, R., Carhart-Harris, R. L., Roseman, L., Pappas, I., Peattie, A. R. D., Manktelow, A. E., Sahakian, B. J., Finoia, P., Williams, G. B., Allanson, J., Pickard, J. D., Menon, D. K., Bor, D., & Stamatakis, E. A. (2023). Distributed harmonic patterns of structure–function dependence orchestrate human consciousness. Communications Biology, 6, 117. https://doi.org/10.1038/s42003-023-04474-1

MacVicar, B. A., & Newman, E. A. (2015). Astrocyte regulation of blood flow in the brain. Cold Spring Harbor Perspectives in Biology, 7(5), a020388. https://doi.org/10.1101/cshperspect.a020388

Majerus, S., D'Argembeau, A., Martinez Perez, T., Belayachi, S., Van der Linden, M., Collette, F., Salmon, E., Seurinck, R., Fias, W., & Maquet, P. (2010). The commonality of neural networks for verbal and visual short-term memory. Journal of Cognitive Neuroscience, 22(11), 2570–2593. https://doi.org/10.1162/jocn.2009.21378

Marques, S., Zeisel, A., Codeluppi, S., van Bruggen, D., Falcão, A. M., Xiao, L., ... & Castelo-Branco, G. (2016). Oligodendrocyte heterogeneity in the mouse juvenile and adult central nervous system. Science, 352(6291), 1326–1329. https://doi.org/10.1126/science.aaf6463

Masuda, T., Sankowski, R., Staszewski, O., & Prinz, M. (2020). Microglia heterogeneity in the single-cell era. Cell Reports, 30(5), 1271–1281. https://doi.org/10.1016/j.celrep.2020.01.010

Mashour, G. A., Roelfsema, P., Changeux, J.-P., & Dehaene, S. (2020). Conscious processing and the global neuronal workspace hypothesis. Neuron, 105(5), 776–798. https://doi.org/10.1016/j.neuron.2020.01.026

Mellinger, J., & Schalk, G. (2010). Brain–computer interfaces. In G. Dornhege et al. (Eds.), Toward brain–computer interfacing (pp. 3–24). MIT Press.

Merzenich, M. M., & Jenkins, W. M. (1993). Reorganization of cortical representations of the hand following alterations of skin inputs induced by nerve injury, skin island transfers, and experience. Journal of Hand Therapy, 6(2), 89–104. https://doi: 10.1016/s0894-1130(12)80290-0

Michel, C. M., & Brunet, D. (2019). EEG source imaging: A practical review of the analysis steps. Frontiers in Neurology, 10, 325. https://doi.org/10.3389/fneur.2019.00325

Mirifar, A., Keil, A., & Ehrlenspiel, F. (2022). Neurofeedback and neural self-regulation: A new perspective based on allostasis. Reviews in the Neurosciences, 33(6), 607–629. https://doi.org/10.1515/revneuro-2021-0133

Moreno-Jiménez, E. P., Flor-García, M., Terreros-Roncal, J., Rábano, A., Llorens-Martín, M., & Ávila, J. (2019). Adult hippocampal neurogenesis is abundant in neurologically healthy subjects and drops sharply in patients with Alzheimer's disease. Nature Medicine, 25(4), 554–560. https ://doi.org/10.1038/s41591-019-0375-9

Moruzzi, G., & Magoun, H. W. (1949). Brain stem reticular formation and activation of the EEG. Electroencephalography and Clinical Neurophysiology, 1(1–4), 455–473. https://doi.org/10.1016/0013-4694(49)90219-9

Mount, C. W., & Monje, M. (2017). Wrapped to adapt: Experience-dependent myelination. Neuron, 95(4), 743–756. https://doi.org/10.1016/j.neuron.2017.07.009

Nave, K.-A., & Werner, H. B. (2014). Myelination of the nervous system: Mechanisms and functions. Annual Review of Cell and Developmental Biology, 30, 503–533. https://doi.org/10.1146/annurev-cellbio-100913-013101

Neurath, O. (1931). The scientific conception of the world: The Vienna Circle (M. Neurath & R. Carnap, Eds.). In M. Neurath, O. Neurath, & R. Carnap (Eds.), Empiricism and sociology (pp. 299–318). Dordrecht: Reidel. (Original work published 1931)

Nishiyama, A., Komitova, M., Suzuki, R., & Zhu, X. (2014). NG2 cells (polydendrocytes) in brain physiology and repair. Frontiers in Neuroscience, 8, 133. https://doi.org/10.3389/fnins.2014.00133

Nishiyama, A., Chang, A., & Trapp, B. D. (1999). NG2 glial cells: A novel glial cell population in the adult brain. Journal of Neuropathology & Experimental Neurology, 58(11), 1113–1124. https://doi.org/10.1097/00005072-199911000-00001

NobelPrize.org. (2024). Santiago Ramón y Cajal: Nobel Prize in Physiology or Medicine 1906. The Nobel Prize Organization.https://www.nobelprize.org/prizes/med icine/1906/cajal/biographical/

Nottebohm, F. (1989). From bird song to neurogenesis. Scientific American, 260(2), 74–79. https://doi.org/10.1038/scientificamerican0289-74

Obien, M. E. J., Deligkaris, K., Bullmann, T., Bakkum, D. J., & Frey, U. (2014). Revealing neuronal function through microelectrode array recordings. Frontiers in Neuroscience, 8, Article 423. https://doi.org/10.3389/fnins.2014.00423

Riva, G., Wiederhold, B. K., & Mantovani, F. (2019). Neuroscience of virtual reality: From virtual exposure to embodied medicine. Cyberpsychology, Behavior, and Social Networking, 22(1), 82–96. doi: 10.1089/cyber.2017.29099.gri

Roth, B. L. (2016). DREADDs for neuroscientists. Neuron, 89(4), 683–694. https://doi.org/10.1016/j.neuron.2016.01.040

Russell, S., & Norvig, P. (2020). Artificial intelli gence: A modern approach (4th ed.). Pearson. (Widely recognized AI textbook, especially in technical and cognitive science fields.)

Saalfeld, S., Cardona, A., Hartenstein, V., & Tomancak, P. (2009). CATMAID: Collaborative annotation toolkit for massive amounts of image data. Bioinformatics, 25(15), 1984–1986. https://doi.org/10.1093/bioinformatics/btp266

Safron, A. (2020). An integrated world modeling theory (IWMT) of consciousness: Combining integrated information and global workspace theories with the free energy principle and active inference framework. Frontiers in Artificial Intelligence, 3, 30. https://doi.org/10.3389/frai.2020.00030

Saraiva, C., Praça, C., Ferreira, R., Santos, T., Ferreira, L., & Bernardino, L. (2016). Nanoparticle-mediated brain drug delivery: Overcoming blood–brain barrier to treat neurodegenerative diseases. Journal of Controlled Release, 235, 34–47. https://doi.org/10.1016/j.jconrel.2016.05.044

Shen, X., Finn, E. S., Scheinost, D., Rosenberg, M. D., Chun, M. M., Papademetris, X., & Constable, R. T. (2017). Using connectome-based predictive modeling to predict individual behavior from brain connectivity. Nature Protocols, 12(3), 506–518. https://doi.org/10.1038/nprot.2016.178

Shohayeb, B., Diab, M., Ahmed, M., & Ng, D. C. H. (2018). Factors that influence adult neurogenesis as potential therapy. Translational Neurodegeneration, 7(1), 4. https://doi.org/10.1186/s40035-018-0109-9

Song, E., Li, J., Won, S. M., Bai, W., & Rogers, J. A. (2020). Materials for flexible bioelectronic systems as chronic neural interfaces. Nature Materials, 19(6), 590–603. https://doi.org/10.1038/s41563-020-0679-7

Spalding, K. L., Bergmann, O., Alkass, K., Bernard, S., Salehpour, M., Huttner, H. B., ... & Frisén, J. (2013). Dynamics of hippocampal neurogenesis in adult humans. Cell, 153(6), 1219–1227. https://doi.org/10.1016/j.cell.2013.05.002

Spear, L. P. (2013). Adolescent neurodevelopment. Journal of Adolescent Health, 52(2 Suppl), S7–S13. https://doi.org/10.1016/j.jadohealth.2012.05.006

Sporns, O. (2010). Networks of the brain. MIT Press.

Stiles, J., & Jernigan, T. L. (2010). The basics of brain development. Neuropsychology Review, 20, 327–348. https://doi.org/10.1007/s11065-010-9148-4

Surianarayanan, C., Lawrence, J. J., Chelliah, P. R., Prakash, E., & Hewage, C. (2023). Convergence of artificial intelligence and neuroscience towards the diagnosis of neurological disorders—A scoping review. Sensors, 23(6), Article 3062. https://doi.org/10.3390/s23063062

Suzuki, A., Stern, S. A., Bozdagi, O., Huntley, G. W., Walker, R. H., Magistretti, P. J., & Alberini, C. M. (2011). Astrocyte-neuron lactate transport is required for long-term memory formation. Cell, 144(5), 810–823. https://doi.org/10.1016/j.cell.2011.02.018

Thomas, C. A., Springer, P. A., Loeb, G. E., Berwald-Netter, Y., & Okun, L. M. (1972). A miniature microelectrode array to monitor the bioelectric activity of cultured cells. Experimental Cell Research, 74(1), 61–66. https://doi.org/10.1016/0014-4827(72)90481-8

Van Overwalle, F., Baetens, K., Mariën, P., & Vandekerckhove, M. (2020). Consensus paper: Cerebellum and social cognition. The Cerebellum, 19(5), 561–593. https://doi.org/10.1007/s12311-020-01155-1

Váša, F., Seidlitz, J., Romero-Garcia, R., Whitaker, K. J., Rosenthal, G., Vértes, P. E., Shinn, M., Alexander-Bloch, A., Fonagy, P., Dolan, R. J., Jones, P. B., Goodyer, I. M., Sporns, O., & Bullmore, E. T. (2017). Adolescent Tuning of Association Cortex in Human Structural Brain Networks. Cerebral Cortex, 28, Article 1. https://doi.org/10.1093/cercor/bhx249

Vidaurre, D., Smith, S. M., & Woolrich, M. W. (2017). Brain network dynamics are hierarchically organized in time. Proceedings of the National Academy of Sciences, 114(48), 12827–12832. https://doi.org/10.1073/pnas.1705120114

von der Malsburg, C. (1981). The correlation theory of brain function. Internal Report 81-2, Max-Planck Institute for Biophysical Chemistry, Göttingen. Retrieved from https://www.ini.rub.de/the-institute/publications/

Wolpaw, J. R., Birbaumer, N., McFarland, D. J., Pfurtscheller, G., & Vaughan, T. M. (2002). Brain-computer interfaces for communication and control. Clinical Neurophysiology, 113(6), 767–791. https://doi.org/10.1016/S1388-2457(02)00057-3

Zhang, X., Liu, Y., Hong, X., Chen, Y., Li, D., Song, J., ... & Duan, S. (2021). NG2 glia-derived GABA release tunes inhibitory synapses and contributes to stress-induced anxiety. Nature Communications, 12, 5740. https://doi.org/10.1038/s41467-021-25956-y

Zhou, P., Resendez, S. L., Rodriguez-Romaguera, J., Jimenez, J. C., Neufeld, S. Q., Giovannucci, A., ... & Ziv, Y. (2018). Efficient and accurate extraction of in vivo calcium signals from microendoscopic video data. eLife, 7, e28728. https://doi.org/10.7554/eLife.28728

Zipfel, W. R., Williams, R. M., & Webb, W. W. (2003). Nonlinear magic: Multiphoton microscopy in the biosciences. Nature Biotechnology, 21(11), 1369–1377. https://doi.org/10.1038/nbt899

Zong, W., Wu, R., Li, M., Hu, Y., Li, Y., Li, J., Rong, H., Wu, H., Xu, Y., Lu, Y., Jia, H., Fan, M., Zhou, Z., Zhang, Y., Wang, A., Chen, L., & Cheng, H. (2017). Fast high-resolution miniature two-photon microscopy for brain imaging in freely behaving mice. Nature Methods, 14(7), 713–719. https://doi.org/10.1038/nmeth.4305

2

The Brain Metafunction

Over two decades ago, the author introduced the concept of the *brain metafunction*, building upon research into coma, brain death, heart rate variability, and the unified definition of human death (Garcia, 1995a, 1995b, 1998a, 1998b, 1999, 2001; Garcia et al. 1995; Machado et al., 2005). This concept emerges from the premise that human consciousness is deeply interwoven with the brain's integrative capacity, which reflects a higher-order, hierarchical process. In recent history, researchers have commonly framed brain function as a dichotomy: one domain associated with consciousness and the cerebral neocortex, and another with the autonomic integration of bodily functions, primarily governed by subcortical and brainstem structures. This dichotomous framework, often referred to as *biological dualism*, has shaped numerous interdisciplinary debates. They address foundational questions such as: *Is there a fundamental unity between the human person and the organism?* and *can the death of a person be separated from the death of the organism?* (Garcia, 1998b, 2001). These questions have profound implications for defining human death and understanding personhood. Early research in the 1990s suggested that both higher and lower brain structures significantly contribute not only to consciousness but also to the regulation of body organs and systems, including those associated with emotion and related functions (Plum, 1991; Damasio, 1985; Damasio, 1999). However, at the time, neuroscience lacked the methodological precision to fully illuminate the inner dynamics of the living brain or confirm these integrative hypotheses.

Over the past two decades, advances in neuroscience have underscored the necessity of a whole-brain framework—one that accounts for the remarkable coherence with which the brain integrates its diverse functional domains. Consciousness, autonomic regulation, and general bodily functions are not discrete or isolated systems but interdependent processes involving the brainstem, limbic structures, and cortical areas. This holistic perspective reveals a finely tuned coordination among brain networks. Growing empirical evidence shows that brain communication is organized through complex topologies that support both localized processing and global integration across neural systems. Researchers now utilize complex network models to describe this structure-function relationship, where *functional connectivity* refers to the temporal correlation between spatially distributed neurophysiological events (Sporns, 2010; Cole et al., 2013). These patterns of connectivity enable the brain to generate dynamic, synchronized activity across multiple spatial and temporal scales (Sporns, 2018).

Initial investigations often focus on delineating the specific functions of discrete brain regions. Yet, the significance of these regions is best understood in the context of their participation in large-scale, interactive networks (Bressler, 1995; Tononi et al., 1998; Genon et al., 2018). These insights have deepened our understanding of the brain as a dynamic, integrative system. When studying conditions such as coma, brain death, the definition of human death, personhood, and vegetative states, a holistic and precise approach becomes indispensable. These states involve not only local dysfunctions but also systemic failures of integration. Mounting evidence continues to support the unity of brain functions—bridging mental experience with biological processes. In the chapters that follow, this unifying concept will be developed further, drawing on the most recent advances in neuroscience.

An Expanded Understanding of the Human Brain

The human brain exhibits complex spatiotemporal dynamics, characterized by a balance between integrated and segregated functional states (Dehaene & Changeux, 2011). These dynamics generate intricate patterns of neural activity across multiple brain regions and time scales. This enables the brain to integrate information across diverse systems while preserving specialization in localized domains. Segregated (domain-specific) systems process information concurrently, whereas large-scale integration synthesizes this information to guide adaptive behavior and other higher-order functions. Maintaining equilibrium between segregation and integration is a fundamental organizational principle of brain architecture. This balance ensures efficient local communication within neural assemblies while enabling widespread coordination across distant brain regions (Sporns, 2013). Such dynamic balance supports a wide array of cognitive functions, including perception, attention, memory, and decision-making (Kelso, 2012; Cohen & D'Esposito, 2016; Wang et al., 2021). This simultaneous presence of localized specialization and global integration facilitates flexible and adaptive responses to complex environmental demands. The *brain metafunction* supports the unity of the human person by coherently linking internal physiological states with external stimuli in a dtnamically responsive manner. These foundational principles are echoed in contemporary analyses of brain-wide networks.

The concept of *allostasis*, introduced by Sterling and Eyer (1988), describes the brain's capacity to achieve stability through change. It acknowledges that physiological set points and regulatory boundaries adjust to environmental contingencies. Crucially, the brain's predictive capacities drive the dynamic regulation. Allostasis depends on the brain's ability to anticipate introceptive signals—sensory inputs from the internal milieu of the body (Kleckner et al., 2017). Recent theoretical models emphasize the role of predictive mechanisms, which generate anticipatory neural models based on both prior experience and real-time contextual input (Pezzulo, Rigoli, & Friston, 2015; Barrett, 2017). Interoception—the perception of internal bodily states—plays a central role in this allostatic process. Interoceptive processing includes input from autonomic, visceral, vascular, neuroendocrine, and immune systems (Friston, 2010; Barrett & Simmons, 2015; Chanes & Barrett, 2016; Seth, 2013). Functional neuroimaging studies demonstrate that the networks involved in interoceptive regulation are also implicated in psychological phenomena such as emotion, memory, and decision-making (Kleckner et al., 2017). For instance, intricate connectivity among the dorsomedial prefrontal cortex, ventrolateral prefrontal cortex, and hippocampus supports visceromotor control and higher-order regulation. These findings reflect the continued relevance of Papez's (1937) early hypothesis about the limbic system's role in emotional processing. The brain, therefore, does not function as a passive receptor of stimuli but as an active generative system—continually modeling the internal and external world through predictive, anticipatory mechanisms. This framework applies equally to sensory and interoceptive domains, synthesizing bodily signals with learned expectations (Seth, 2014; Kleckner et al., 2017). These predictive dynamics not only shape perception and guide action but also form the biological basis for adaptive and context-sensitive behavior.

Heart Rate Variability and Functional Integration

Heart rate variability (HRV) exemplifies the brain's capacity for multilevel integration and functional synthesis. Research into the intricate physiological mechanisms underlying HRV played a pivotal role in shaping the author's early formulation of the brain metafunction concept. These early studies identified original findings and demonstrated the clinical relevance of HRV in states of coma and brain death, revealing its reliance on the coordinated activity of brainstem centers, limbic system structures, and cortical regions, thus offering early empirical support for the broader concept of whole-brain integration (García et al., 1995; Machado et al., 2005). While this initial research focused on HRV under pathological conditions, more recent studies have expanded its implications to healthy physiological functioning. For example, Yoo et al. (2022) demonstrated that HRV biofeedback can induce volumetric changes in cortical areas, particularly the anterior cingulate cortex and prefrontal cortex—regions associated with both autonomic regulation and higher cognitive control. This finding underscores the bidirectional interaction between cortical and subcortical systems and highlights the integrative link between neural activity and physiological processes.

Similarly, Mujica-Parodi et al. (2009) found that reduced HRV, correlated with heightened anxiety, is associated with limbic dysregulation. This supports the hypothesis that the limbic system plays a central role in modulating both emotional and autonomic responses through its extensive connections with cortical and brainstem regions. Together, these studies offer compelling evidence that HRV is not merely a cardiovascular phenomenon but a key index of the brain's capacity to coordinate physiological and psychological domains. This reinforces the brain metafunction framework, which conceptualizes the brain as a unified system integrating interrelated bodily and experiential functions (Garcia, 1998b, 1999). Recent findings also challenge traditional modular views of brain organization, which emphasize high specialization within discrete neural circuits. Instead, current research increasingly points toward a domain-general or multi-use functionality of neural populations—even in primary sensory cortices. For instance, multisensory integration has been observed in regions previously thought to be unimodal, such as the auditory cortex responding to visual stimuli (Koelsch et al., 2011; Majerus et al., 2010; Sporns, 2010; Chein, Moore, & Conway, 2011; Cole et al., 2013). The brain metafunction framework unifies these diverse findings by providing a coherent model of how the brain organizes internal states and external stimuli into cohesive neural patterns. It advances a theoretical architecture capable of addressing foundational questions in medicine, bioethics, and philosophy—questions concerning the nature of personhood, the unity of consciousness, and the criteria for defining human life and death (Garcia, 1999, 2001). By aligning traditionally separate domains—such as sensation, cognition, emotion, and homeostatic regulation—the framework affirms that the brain enables an integrated, dynamic experience of the world.

Rather than functioning through a singular, centralized control hub, the brain operates as a decentralized, distributed network that achieves global coherence through intricate interconnections. These integrative dynamics invite reinterpretation through the lens of complex systems theory, a perspective the next section will explore in greater detail. The brain's capacity for such dynamic integration reveals it to be more than a mediator of physiological or cognitive tasks; it is a foundational system that bridges biological processes with higher-order dimensions of human experience. The *brain metafunction* thus illuminates the orchestrated complexity of neural life that enables humans to navigate the physical, mental, and social dimensions of their environments with coherence and agency.

The Complex Systems Theoretical Approach

The complex systems theoretical approach (Anderson, 1972; Holland, 1992) does not claim to exhaustively explain all subfunctions of the brain metafunction—particularly those related to consciousness—but it offers foundational concepts highly applicable to understanding brain organization. It emphasizes that the behavior of a system is shaped not solely by the properties of its individual components but by their interactions. These interactions give rise to emergent properties—novel features that cannot be predicted or reduced to the characteristics of the parts themselves. In the case of the human brain, such emergent total brain states are essential to understanding its integrative and dynamic nature. By definition, complex systems consist of numerous interacting parts. Examples range from living cells and organisms to brains, communication networks, power grids, social systems, and even cosmological structures. Across these domains, emergent behaviors arise—phenomena in which the whole exhibits capabilities not discernible from an analysis of isolated components. These dynamic patterns often lead to unexpected, nonlinear outcomes that challenge prediction and control.

As James D. Watson remarked, "The brain is the last and grandest biological frontier, the most complex thing we have yet discovered in our universe" (Ackerman, 1992, Foreword). The human brain exemplifies this complexity: with approximately 86 billion neurons, 85 billion non-neuronal cells, and about 100 trillion synapses, it forms a nonlinear, dynamic network capable of supporting cognitive, emotional, and physiological functions (Godwin & Cham, 2012). Unlike other complex systems, such as ecosystems or climate networks, the brain's organization and adaptability place it in a category of its own. It integrates both physical substrates (neurons, glia, vasculature) and abstract dimensions, including consciousness and personhood. Crucially, the brain's complexity is not static. It is dynamically regulated by mechanisms that ensure both stability and adaptability. Among the most important of these are feedback loops, which enable real-time self-regulation in response to changing conditions. Negative feedback loops stabilize internal states—for example, thermoregulation or postural control (Peterka, 2002)—while positive feedback loops amplify specific processes, such as clotting or synaptic strengthening during learning. Emotional regulation also relies on feedback dynamics: the amygdala intensifies fear responses through positive feedback, while prefrontal circuits inhibit excessive arousal via negative feedback (Sah et al., 2003; Dillon & Pizzagalli, 2007). These interwoven feedback systems exemplify the brain's capacity for flexible integration—an essential feature of the brain metafunction. By blending stabilization and amplification across domains, feedback loops support learning, adaptation, and the orchestration of coherent bodily and psychological states.

The Brain Metafunction Framework

The human brain is more than a collection of specialized regions or biological components; it is an integrative organ capable of unifying perception, cognition, behavior, emotion, and physiological regulation into a coherent, dynamic whole. This integrative capacity is captured by the concept of the brain metafunction—the orchestration of local specializations into a system capable of adaptive, global coherence. Rather than operating through isolated modules, brain systems interact continuously via nested feedback loops and large-scale networks that regulate bodily states, awareness, and behavior. Recent advances in network neuroscience reinforce and refine this perspective: the brain is now understood to operate through modular architectures and connector hubs, dynamically coordinating activity across distributed regions to generate metastable states—transient configurations that balance global integration with local specialization (Sporns, 2018; Kleckner et al., 2017). These dynamics enable the brain to flexibly reconfigure its functional organization in response to changing internal and external demands, aligning closely with the metafunctional synthesis proposed here. The Brain Metafunction Framework, grounded in this systems-level understanding, offers a generative model for analyzing brain activity across scales—from individual neurons to whole-brain networks—and emphasizes the brain's pivotal role in mediating the unity of the human person. It moves beyond reductionism by positioning the brain as the biological and psychological integrator of human life—bridging self and world, body and mind, and even biological life and existential meaning.

I. Definition

The *brain metafunction* refers to the dynamic, unified activity of brain systems, subregions, regions, and neurons, interconnected through structural and functional pathways. This integration encompasses synchronized local and global dynamics across spatial and temporal scales. The framework also proposes that these dynamics extend into a *transphysical* domain through a form of inherent causality that is unique to the brain's organizational level

II. Principles Supported by Large-Scale Brain Nework Studies

Recent advances in neuroscience and systems theory support the following key principles, which collectively form the empirical foundation of the brain metafunction framework:

1. *Multi-Use Neural Populations*: Neural populations are not strictly domain-specific but exhibit domain-general functionality across multiple tasks (Avena-Koenigsberger, Misic, & Sporns, 2018; Chein, Moore, & Conway, 2011; Majerus et al., 2010; Koelsch et al., 2011).

2. *Dense Interconnectivity*: Cortical and subcortical regions are part of an extensively interconnected network that supports both specialized and integrative processes (Sporns, 2010; Cole et al., 2013; van den Heuvel & Sporns, 2011).

3. *Global Accessibility*: Functional connectivity spans all sectors of the brain, including brainstem and cerebellar regions, emphasizing whole-brain integration.

4. *Rich Connectivity Core*: A "connectivity core" or "rich club" of highly connected regions exists across cortical, subcortical, and thalamic areas, serving as major hubs for information processing.

5. *Prediction Signaling*: Central hub regions within large-scale networks send anticipatory signals to sensory systems, enhancing adaptive responsiveness.

6. *Network Integration*: Sensory and motor networks are integrated with higher-order networks—such as the default mode and salience networks—via central hub regions (Kleckner et al., 2017).

7. *Integration Backbone*: The default mode and salience networks serve as high-capacity integration backbones for interoceptive and sensory information (Tononi, 2004).

8. *Total Brain States*: Unified brain states are critical for linking sensory neural representations of the internal milieu with global brain dynamics, facilitating the integration of psychological and physiological functions.

The brain metafunction framework anticipated these converging findings by proposing that consciousness and bodily regulation are *anatomically and functionally coextensive*. This integrated perspective supports the convergence of cognitive, motor, visceromotor, and homeostatic processes into a unified neural architecture. Ongoing research continues to validate this integrated model of the brain as a unified system.

III. Brain Metafunction's Biological Substrate: The Anatomical Perspective.

The brain's critical meta-functional system provides a necessary and sufficient level of resolution for conceptual and practical purposes. From an anatomical perspective, it comprises the following subsystems:

Subsystem I: Arousal and Activation

This subsystem includes structures traditionally understood as crucial for wakefulness, arousal, and non-specific sensory processing but essential for higher-order functions. These include:

1. Neurons in the tegmentum of the rostral pons and midbrain.
2. The intralaminal and reticularis nuclei of the thalamus.
3. The posterior hypothalamus.

Subsystem II: Emotional and Homeostatic Integration

The second subsystem comprises structures involved in emotional regulation, homeostatic control, memory processes, behaviors, motivations, and the autonomic nervous system. This mention includes some key structures and functional contributions:

1. The hypothalamus participates in regulating autonomic and endocrine functions, among other functions.

2. Limbic system components, such as the amygdala, hippocampal complex, and cingulate cortex, play central roles in emotion and memory processing, stress response, and allostasis

3. The orbitofrontal cortex is involved in decision-making and adaptive behaviors, and the temporal pole is involved in complex social and emotional cognition.

4. The olfactory cortex integrates sensory input from the olfactory system with emotional and memory networks.

5. The nucleus basalis of Meynert is critical for attention and learning through its cholinergic projections

6. The septal area contributes to memory formation and emotional regulation through its connections with the hippocampus and hypothalamus.

7. Portions of the basal ganglia interact with cortical and subcortical regions to regulate motor, cognitive, and emotional processes.

8. Reticular nuclei in the midbrain and rostral pons, including the pedunculopontine nucleus (involved in arousal and locomotion) and the parabrachial nucleus (integral to pain and homeostatic responses).

9. The nucleus incertus is a small yet critical structure involved in controlling locomotor speed, arousal, stress response, and integrating vestibular-ocular reflexes and hippocampal navigation.

10. Brainstem nuclei, including the rostral raphe nuclei, mediate serotonergic modulation; the periaqueductal gray is key to pain modulation and defensive behaviors; and the nucleus locus coeruleus regulates attention and arousal.

Subsystem III: Cognitive and Motor Integration

The final subsystem integrates higher cognitive functions and motor coordination, involving:

1. The cerebral cortex is responsible for complex thought, perception, and decision-making.

2. The thalamus is a critical relay and integrative center for sensory and motor signals and various cognitive functions.
3. The basal ganglia is essential for motor control, habit formation, and action selection.

Interconnection and Interdependence of Subsystems

Although each subsystem supports specific functional domains, their operations are highly interdependent. For instance, Subsystem I enables arousal and inherent cooperative contributions, supporting the emotional and cognitive operations of Subsystems II and III. Subsystem II integrates affective, autonomic, homeostatic, and memory-related information that influences cognition and motor behavior. Subsystem III depends on inputs from Subsystems I and II and interact with them to plan and execute context-sensitive behavior. Together, these subsystems form a unified, dynamic architecture that allows the brain to act as the central integrative organ of human biological and psychological life.

IV. The Brain Metafunction's Functions

The brain metafunction system—through its extensive interconnectivity and hierarchical integration—enables the following core functions, which collectively constitute the biological and experiential foundation of the human person:

1. Consciousness and Cognitive Processes:

The brain metafunction underpins both consciousness and general cognitive functions, including perception (internal and external), attention, memory, and emotion. At any given moment, conscious awareness represents only the "tip of the iceberg," resting upon a vast substratum of non-conscious processes such as automatic pattern recognition, and predictive coding interacting, while interacting with autonomic, homeostatic, and allostatic regulatory processes through interconnected brain networks.

2. Integration and Regulation of Communication Systems:

The brain orchestrates and regulates the body's core communication systems—nervous, endocrine, immune, and circulatory—to maintain homeostasis and respond adaptively to internal and external stimuli.

For example, in response to an acute stressor, the amygdala activates the hypothalamus, which initiates the sympathetic nervous system's "fight-or-flight" response. This increases heart rate and redirects blood flow to skeletal muscles. Simultaneously, the hypothalamus signals the pituitary gland to release ACTH, triggering cortisol release. Cortisol mobilizes energy reserves while suppressing non-essential processes such as digestion and certain immune functions. Yet even during suppression, cytokine signaling maintains a dynamic balance between immune vigilance and energy conservation. Once the stressor resolves, the parasympathetic system restores equilibrium—reducing cortisol, reactivating digestion, and regulating immune tone. This multilayered coordination highlights the brain's role as a central integrator of bodily communication systems—balancing immediate survival demands with long-term stability. Moreover, in conditions such as viral infections, higher-order cognitive processes—such as fatigue perception and motivational regulation—actively promote rest and reduced activity, aligning behavior with the body's immunological needs for recovery and repair.

3. *Homeostasis and Allostasis:*

A central subfunction of the brain's metafunction is regulating the body's internal environment through both immediate stability (homeostasis) and anticipatory adaptation (allostasis). While homeostasis restores balance in response to change, allostasis—introduced by Sterling and Eyer (1988)—describes the brain's capacity to predict needs and proactively adjust internal states. This predictive regulation enhances efficiency and resilience across physiological systems. Key structures coordinate this regulation dynamically. The prefrontal cortex assesses contextual demands; the amygdala gauges emotional salience. In threatening situations, the amygdala activates the hypothalamus, initiating the HPA axis and releasing cortisol to mobilize energy and optimize survival. When the threat resolves, the prefrontal cortex downregulates amygdala activity, allowing the system to reset. The anterior cingulate cortex (ACC) plays a vital integrative role—monitoring conflict, aligning emotional and cognitive inputs, and adjusting attention and autonomic output. Along with the insula, it contributes to the brain's interoceptive network, which tracks and predicts internal states such as arousal, hunger, and emotional tone (Kleckner et al., 2017). This internal modeling supports a coherent sense of bodily awareness and behavioral regulation. Critically, allostasis is now understood not as a secondary feature but as a fundamental organizing principle of brain function: at every level, the brain continuously generates predictive models of both external and internal conditions, updating them through feedback to minimize surprise and optimize adaptive responses (Barrett, 2017; Friston, 2010). In this view, regulation is inherently prospective—the brain does not simply react but anticipates, aiming to prevent costly deviations before they arise.

Importantly, recent discoveries show that glial cells—especially astrocytes—are active partners in these processes. Far from being passive support cells, they help regulate circadian rhythms, energy use, and even sleep-wake cycles (Dopp et al., 2024; Purushotham & Buskila, 2023). This neuro-glial system forms a distributed network of regulation that blends prediction, feedback, and integration across levels of function. Together, these systems reveal the brain not as a mere responder, but as a forecasting and integrative organ—constructing internal coherence and guiding adaptive responses at every moment. This capacity illustrates the deep unity of emotion, physiology, and behavior that underpins embodied personhood. While this example provides a generalized outline, the actual process involves a network of interrelated brain structures, including subcortical and cortical regions, whose activity patterns vary depending on the individual's physiological state, psychological disposition, and environmental context.

4. Integrative Processing of Afferent and Efferent Signals:

The brain continuously integrates afferent sensory input and coordinates efferent motor and autonomic output, enabling purposeful engagement with internal states and external environments. For example, when a dancer responds to music, the auditory cortex processes rhythmic and melodic patterns, while the somatosensory cortex monitors proprioceptive feedback from the body. The motor cortex and premotor cortex plan and execute precise, synchronized movements. Subcortical structures contribute critically: the thalamus relays and integrates sensory signals; the cerebellum fine-tunes timing and balance; limbic regions infuse emotional resonance and memory-based context; and brainstem nuclei regulate arousal and autonomic adjustments. This seamless interplay supports real-time adaptation, emotional expression, and complex, goal-directed behavior. A later chapter on perception will extend this foundation by exploring how the *efferent loop of perception* builds upon these integrative processes to construct experiential reality.

5. Behavioral and Adaptive Interactions with the Outer World:

The brain's metafunction enables adaptive behavior by dynamically integrating sensory-motor information, predictive modeling, and emotional regulation to support learning, navigation, and context-sensitive decision-making. For example, during piano practice, cortical and subcortical circuits interact to refine timing, spatial precision, and sensorimotor prediction. The motor cortex governs fine finger movements, while the premotor and supplementary motor areas plan sequences. The basal ganglia optimize motor selection and inhibit extraneous movement, and cerebellar circuits calibrate coordination based on sensory feedback. Over time, this integrated system fosters skill acquisition and fluency through iterative prediction and adjustment. Navigating a crowded street offers another vivid illustration of real-time metafunctional integration. Subsystem I (Arousal and Activation) sustains environmental awareness and vigilance toward emerging cues. Subsystem II (Emotional and Homeostatic Integration) appraises potential threats and modulates physiological readiness, such as heart rate and muscular tension. Subsystem III (Cognitive and Motor Integration) rapidly plans, adapts, and executes movement strategies—stepping aside, gauging traffic flow, or reacting to sudden obstacles. This seamless collaboration enables contextually appropriate, survival-oriented behavior in complex, unpredictable environments, demonstrating the brain's capacity to unify internal regulation, environmental responsiveness, and motor coordination into coherent, purposeful action.

6. Integration of Physiological and Mental States:

The brain generates a coherent and embodied sense of self by synthesizing physiological states (e.g., interoception, hormonal modulation) with mental states (e.g., emotion, thought). Subsystem I sustains alertness and arousal; Subsystem II modulates affective salience and visceral function; Subsystem III interprets and integrates these inputs into intentional, adaptive action. Each subsystem contributes to consciousness—not as isolated domains, but as co-constitutive dimensions within a unified field of experience. This integration forms the neurobiological basis of embodied subjectivity.

7. Unified Synthesis of Subfunctions:

The unity of the brain "as a whole" depends on the irreducible synthesis of its subfunctions (1 through 6), ensuring that consciousness, cognition, regulation, and adaptive behavior emerge as cohesive, interdependent aspects of lived personhood. This synthesis is not simply the additive work of specialized regions; rather, it arises from the dynamic interplay among Subsystems I (Arousal and Activation), II (Emotional and Homeostatic Integration), and III (Cognitive and Motor Integration). Each subsystem contributes uniquely—but none alone can generate the integrative coherence required for embodied consciousness and intentional agency. Recent advances in network neuroscience reinforce this view. Studies show that the brain operates through modular architectures and dynamic connector hubs that coordinate across systems to maintain metastable equilibrium (Tognoli & Kelso, 2014; Sporns, 2018; Kleckner et al., 2017). These metastable states—balancing global integration and local specialization—allow the brain to flexibly adapt based on internal goals and environmental demands. Attention switching, emotional regulation, and complex decision-making exemplify this ongoing reconfiguration.

Together, these findings suggest that consciousness, behavior, and identity depend not on isolated functions, but on the orchestrated, self-organizing dynamics of the whole brain. The brain metafunction thus provides a biological common ground for interdisciplinary conversations about human identity. It bridges physicalist and non-physicalist paradigms by illustrating that while biological systems support the unity of lived experience, they do not exhaust its meaning. As neuroscience continues to map the global architecture of brain function, the explanatory coherence and philosophical relevance of the brain metafunction framework will continue to expand.

V. The Brain Metafunction Reveals a Transphysical Dimension.

The brain metafunction framework ultimately points beyond biology, suggesting the need to recognize a transphysical dimension in any comprehensive account of the human person. Biologically, the brain's unified operation—its seamless synthesis of neural, physiological, and cognitive processes—manifests not only as a physical phenomenon but as a profound expression of a domain that transcends it. Although contemporary models such as *emergentism* and *panpsychism* attempt to explain consciousness through complex material processes or universal mental properties, they remain insufficient within the ontological framework developed here. The brain metafunction framework affirms that consciousness and personhood transcend material causality, reflecting an integrated reality whose foundation lies beyond matter itself. Accordingly, this work bypasses such alternatives to remain faithful to the deeper vision of personhood at the heart of the Western tradition. This perspective aligns with an integrative view of consciousness as a unity involving brain, body, environment, and a transphysical horizon. The transphysical refers to the view that the mind works through and cooperates with the brain yet exceeds the explanatory limits of physical causation as currently understood. Within this framework, brain-based operations such as perception, attention, memory, and consciousness are not merely mechanical; they are metafunctional—bridging neural activity with non-reducible dimensions of experience, including meaning, intentionality, and awareness. This is seen, for instance, in the coherence of conscious states across change and multiplicity, raising the binding and meta-binding problems: how disparate neural processes—perceptual, attentional, mnemonic—are unified in singular, lived experience. Similarly, intentionality—the mind's directedness toward objects, ideas, and events—remains unexplained by purely mechanistic models (Brentano [1874] 1995). Conscious memory further reveals a double intentionality, where the mind both represents a past event and reflects on the act of remembering itself (Perler, 2022; Müller & Summa, 2018). Electrical and chemical processes may accompany these phenomena, but they cannot, in themselves, account for the "aboutness" or directedness essential to mental life.

Moreover, consciousness possesses an irreducible first-person dimension—the subjective feel or qualia of experience—what Chalmers (1995, 1996) termed the "hard problem of consciousness." Embodied, integrative, and situated, consciousness expresses the lived unity of the person as a dynamic convergence of physical, psychological, social, and transphysical dimensions (Clark, 2008; Noë, 2009). Emerging empirical findings and philosophical reflection increasingly support the hypothesis that the brain metafunction is not exhausted by material processes alone. This framework invites a broader ontological vision, particularly relevant in bioethics, where questions of life, death, and personhood depend on nuanced understandings of unity, continuity, and disintegration. Whether one interprets the brain metafunction purely biologically or as encompassing a transphysical dimension will depend on the comparative explanatory power of each model. What is clear is that the brain metafunction does not merely describe processes; it invites discernment—of what it means to be a living person in unity, complexity, and openness to transcendence. The chapters that follow will explore these implications in greater depth, both empirically and philosophically.

REFERENCES

Ackerman, D. (1992). *The brain: The last and grandest biological frontier* (Foreword by James D. Watson). New York: HarperCollins.

Anderson, P. W. (1972). More is different. *Science, 177*(4047), 393–396. https://doi.org/10.1126/science.177.4047.393

Avena-Koenigsberger, A., Misic, B., & Sporns, O. (2018). Communication dynamics in complex brain networks. *Nature Reviews Neuroscience, 19*(1), 17–33. https://doi.org/10. 1038/nrn.2017.149

Barrett, L. F. (2017). *How emotions are made: The secret life of the brain.* Boston, MA: Houghton Mifflin Harcourt.

Barrett, L. F., & Simmons, W. K. (2015). Interoceptive predictions in the brain. *Nature Reviews Neuroscience, 16*(7), 419–429. https://doi.org/10.1038/nrn3950

Bressler, S. L. (1995). Large-scale cortical networks and cognition. *Brain Research Reviews, 20*(3), 288–304. https://doi.org/10.1016/0165-0173(94)00016-I

Brentano, F. (1995). *Psychology from an empirical standpoint* (A. C. Rancurello, D. B. Terrell, & L. L. McAlister, Trans.). London: Routledge. (Original work published 1874)

Chalmers, D. J. (1995). Facing up to the problem of consciousness. *Journal of Consciousness Studies, 2*(3), 200–219.

Chalmers, D. J. (1996). *The conscious mind: In search of a fundamental theory.* New York, NY: Oxford University Press.

Chanes, L., & Barrett, L. F. (2016). Redefining the role of limbic areas in cortical processing. *Trends in Cognitive Sciences, 20*(2), 96–106. https://doi.org/10.1016/j.tics.2015.11.005

Chein, J. M., Moore, A. B., & Conway, A. R. A. (2011). Domain-general mechanisms of complex working memory span. *Psychonomic Bulletin & Review, 18*(3), 550–557. https://doi.org/10.3758/s13423-011-0116-7

Clark, A. (2008). *Supersizing the mind: Embodiment, action, and cognitive extension*. Oxford, UK: Oxford University Press.

Cohen, J. R., & D'Esposito, M. (2016). The segregation and integration of distinct brain networks and their relationship to cognition. *Journal of Neuroscience, 36*(48), 12083–12094. https://doi.org/10.1523/JNEUROSCI.2965-15.2016

Cole, M. W., Ito, T., Schultz, D., Mill, R., Chen, R., & Cocuzza, T. R. (2013). Functional connectivity of cortical networks supports cognitive control. *Journal of Neuroscience, 33*(1), 1348–1355. https://doi.org/10.1523/JNEUROSCI. 2016-12.2013

Damasio, A. R. (1985). Prosopagnosia. In P. J. Vinken, G. W. Bruyn, & H. L. Klawans (Eds.), *Handbook of clinical neurology* (pp. 47–58). Amsterdam, Netherlands: Elsevier.

Damasio, A. R. (1999). *The feeling of what happens: Body and emotion in the making of consciousness*. New York, NY: Harcourt Brace.

Dehaene, S., & Changeux, J.-P. (2011). Experimental and theoretical approaches to conscious processing. *Neuron, 70*(2), 200–227. https://doi.org/10.1016/j.neuron.2011.03.018

Dillon, D. G., & Pizzagalli, D. A. (2007). Inhibition of action, thought, and emotion: A selective neurobiological review. *Applied and Preventive Psychology, 12*(3), 99–114. https://doi.org/10.1016/j.appsy.2007.09.004

Dopp, J., Ortega, A., Davie, K., & others. (2024). Single-cell transcriptomics reveals that glial cells integrate homeostatic and circadian processes to drive sleep–wake cycles. *Nature Neuroscience, 27*, 359–372. https://doi.org/10.1038/s41593-023-01549-4

Friston, K. (2010). The free-energy principle: A unified brain theory? *Nature Reviews Neuroscience, 11*(2), 127–138. https://doi.org/10.1038/nrn2787

García, O. D. (1995a). Whole brain? Higher brain? A new formulation of death. In C. Machado (Ed.), *Proceedings of the Second International Symposium on Brain Death* (pp. 31–38). Amsterdam, Netherlands: Elsevier.

García, O. D. (1995b). Reformulating death: Overcoming whole-brain inconsistencies. In C. Machado (Ed.), *Proceedings of the Second International Symposium on Brain Death* (pp. 39–46). Amsterdam, Netherlands: Elsevier.

García, O. D., Machado, C., Román, J. M., Cabrera, A., Díaz-Comas, L., & Grave de Peralta, R. (1995). Heart rate variability in coma and brain death. In C. Machado (Ed.), *Proceedings of the Second International Symposium on Brain Death* (pp. 191–200). Amsterdam, Netherlands: Elsevier.

García, O. D. (1998a). A new formulation of death and its relevance to medical law. *Medicine and Law, 17*(4), 493–501.

García, O. D. (1998b). El concepto de persona y su fundamentación neurológica [The concept of person and its neurological foundation]. *Revista de Neurología, 26*(154), 1073. https://doi.org/10.33588/rn.26154.98132

García, O. D. (1999). La definición y determinación de la muerte: Una opinión diferente. *Cuadernos de Bioética, 4*, 40.

García, O. D. (2001). The unity of the human person: A central concept in the interface between medicine and theology. *The Linacre Quarterly, 68*(1), 71–89. https://doi.org/10.1080/20508549.2001.11877600

Genon, S., Reid, A., Langner, R., Amunts, K., & Eickhoff, S. B. (2018). *How to characterize the function of a brain region.* Trends in Cognitive Sciences, 22(4), 350–364. https://doi.org/10.1016/j.tics.2018.01.010

Godwin, W., & Cham, C. (2012). *The human brain: A fascinating look at the anatomy and functions.* New York, NY: Learning Press.

Holland, J. H. (1992). *Adaptation in natural and artificial systems.* Cambridge, MA: MIT Press.

Kelso, J. A. S. (2012). *Dynamic patterns: The self-organization of brain and behavior.* Cambridge, MA: MIT Press.

Kleckner, I. R., et al. (2017). Evidence for a large-scale brain system supporting allostasis and interoception in humans. *Nature Human Behaviour, 1*, 69. https://www.nature.com/articles/s41562-017-0069

Koelsch, S. (2011). Toward a neural basis of music perception – A review and updated model. Frontiers in Psychology, 2, Article 110. https://doi.org/10.3389/fpsyg.2011.00110.

Majerus, S., D'Argembeau, A., Martinez Perez, T., Belayachi, S., Van der Linden, M., Collette, F., Salmon, E., Seurinck, R., Fias, W., & Maquet, P. (2010). The commonality of neural networks for verbal and visual short-term memory. Journal of Cognitive Neuroscience, 22(11), 2570–2593. https://doi.org/10.1162/jocn.2009. 21378

Machado, C., García, O. D., Gutiérrez, J., Portela, L., & García, M. C. (2005). Heart rate variability in comatose and brain-dead patients. *Clinical Neurophysiology, 116*, 2859–2860. https://doi.org/10.1016/J. Clinph.2005.08.016

Mujica-Parodi, L. R., Korgaonkar, M., Ravindranath, B., Greenberg, T., Tomasi, D., Wagshul, M., Ardekani, B., Guilfoyle, D., Khan, S., Zhong, Y., Chon, K., & Malaspina, D. (2009). Limbic dysregulation is associated with lowered heart rate variability and increased trait anxiety in healthy adults. *Human Brain Mapping, 30*(1), 47–58. https://doi.org /10.1002/ hbm.20483.

Müller, J., & Summa, M. (2018). Introduction: Modes of intentionality. *Phänomenologische Forschungen, 2018*(2), 5–24.

Noë, A. (2009). *Out of our heads: Why you are not your brain, and other lessons from the biology of consciousness*. New York, NY: Hill and Wang.

Papez, J. W. (1937). A proposed mechanism of emotion. *Archives of Neurology and Psychiatry, 38*(4), 725–743.

Perler, D. (2022). Ockham on memory and double intentionality. *Topoi.* https://doi.org/10.1007/s11245-020 -09728-9

Peterka, R. J. (2002). Sensorimotor integration in human postural control. *Journal of Neurophysiology, 88*(3), 1097–1118. https://doi.org/10.1152/jn.2002.88.3.1097

Pezzulo, G., Rigoli, F., & Friston, K. (2015). Active inference, homeostatic regulation and adaptive behavioural control. *Progress in Neurobiology, 134*, 17–35. https://doi.org/ 10.1016/j.pneurobio.2015.09.001

Plum, F. (1991). Coma and related global disturbances of the human conscious state. In A. Peters & E. G. Jones (Eds.), *Normal and altered states of function* (pp. 359–425). New York, NY: Plenum Press. https://doi.org/10. 1007/978-1-4615-6622-9_9

Purushotham, S. S., & Buskila, Y. (2023). Astrocytic modulation of neuronal signalling. *Frontiers in Network Physiology, 3*, 1205544. https://doi.org/10.3389/fnetp.2023.1205544

Sah, P., Faber, E. S. L., Lopez De Armentia, M., & Power, J. (2003). The amygdaloid complex: Anatomy and physiology. *Physiological Reviews, 83*(3), 803–834. https://doi.org/10.1152/physrev.00002.2003

Seth, A. K. (2013). Interoceptive inference, emotion, and the embodied self. *Trends in Cognitive Sciences, 17*(11), 565–573. https://doi.org/10.1016/j.tics.2013.09.007

Seth, A. K. (2014). The cybernetic brain: From interoceptive inference to sensorimotor contingencies. *Open MIND, 35*(T), 1–24. https://doi.org/10.15502/9783958570108

Sporns, O. (2010). *Networks of the brain.* Cambridge, MA: MIT Press.

Sporns, O. (2018). Graph theory methods: Applications in brain networks. *Dialogues in Clinical Neuroscience, 20*(2), 111–121.https://doi.org/10.31887/DCNS.2018.20.2/osporns

Sterling, P., & Eyer, J. (1988). Allostasis: A new paradigm to explain arousal pathology. In S. Fisher & J. Reason (Eds.), *Handbook of life stress, cognition and health* (pp. 629–649). Chichester, UK: Wiley.

Tognoli, E., & Kelso, J. A. S. (2014). The metastable brain. *Neuron, 81*(1), 35–48. https://doi.org/10.1016/j.neuron.2013 .12.022

Tononi, G. (2004). An information integration theory of consciousness. *BMC Neuroscience, 5*(1), 42. https://doi.org/10.1186/1471-2202-5-42

Tononi, G., McIntosh, A. R., Russell, D. P., & Edelman, G. M. (1998). Functional clustering: Identifying strongly interactive brain regions in neuroimaging data. *NeuroImage, 7*(2), 133–149. https://doi.org/10.1006/nimg.1997.0313

van den Heuvel, M. P., & Sporns, O. (2011). Rich-club organization of the human connectome. *Journal of Neuroscience, 31*(44), 15775–15786. https://doi.org/10.1523/JNEUROSCI.3539-11.2011

Wang, R., Liu, M., Cheng, X., Wu, Y., Hildebrandt, A., & Zhou, C. (2021). Segregation, integration, and balance of large-scale resting brain networks configure different cognitive abilities. *arXiv.* https://arxiv.org/abs/2103.00475

Yoo, H. J., Thayer, J. F., Koenig, J., & Yoon, H. (2022). Heart rate variability biofeedback training increases gray matter volume in the anterior cingulate cortex and the insula: A randomized controlled trial. *Biological Psychology, 170*, 108300. https://doi.org/10.1016/j.biopsycho.2022.108300

3

Integrating Cortical and Subcortical Structures: A Unified Framework for the Brain Metafunction

The brain metafunction framework offers a comprehensive lens through which to understand how the brain operates as a unified, highly interconnected system, transcending traditional boundaries between cortical and subcortical regions. Rather than treating brain regions as functionally isolated, this model emphasizes the dynamic integration of multi-use neural populations across large-scale brain networks (LSBNs). This facilitates a holistic view of human cognition, emotion, and bodily regulation (Avena-Koenigsberger, Misic, & Sporns, 2018; Chein, Moore, & Conway, 2011). These networks rely on intrinsic connectivity and form a "*connectivity core*," a densely interconnected set of hubs that spans both cortical and subcortical areas, supporting coordinated, system-wide communication (Cole et al., 2013; van den Heuvel & Sporns, 2011; Hansen et al., 2024). This integrative view moves beyond modular or strictly reductionist accounts of brain function, highlighting that cognitive, emotional, and physiological capacities are not compartmentalized but interdependent. The brain metafunction framework thus provides a structural and functional architecture for understanding the unifying basis of human experience. It frames the brain not merely as an organ of discrete tasks, but as a dynamically coordinated substrate for embodied consciousness, intentional behavior, and adaptive regulation, supporting a view of the human person as more than the sum of neurobiological parts.

Consciousness and Cognitive Processes

Within the brain metafunction framework, consciousness and its associated cognitive capacities—including perception, attention, memory, and emotion—are understood as emerging from the dynamic coordination of cortical and subcortical systems, mediated by large-scale brain networks (LSBNs). Rather than viewing consciousness as the product of a single brain region or mechanism, this approach emphasizes its distributed and integrative character. Networks such as the

Default Mode Network (DMN), Salience Network (SN), and Central Executive Network (CEN) contribute distinct but interrelated functions: the DMN supports self-referential thought and memory recall (Buckner, 2013), the SN prioritizes biologically salient stimuli and facilitates transitions between networks (Seeley et al., 2007), and the CEN undergirds executive control and decision-making. Together, they enable consciousness to remain coherent, adaptive, and embedded within a personal context.

Subcortical regions, including the brainstem, thalamus, hypothalamus, and limbic system, provide foundational support for arousal, emotional salience, and sensory integration. These structures form reciprocal loops with cortical areas, enabling a fluid exchange between regulatory, emotional, and cognitive domains. For example, the brainstem contributes to global arousal and attentional readiness, while the thalamus relays and filters sensory information, and the limbic system evaluates emotional significance. This interplay anchors conscious awareness within both physiological and experiential domains, making consciousness not merely a mental state but a deeply embodied one. Importantly, the framework does not reduce consciousness to neural activity alone. It allows space for broader philosophical reflection, recognizing that while the brain provides the necessary substrate for conscious experience, it may not exhaust its ontological significance. The coordination of LSBNs and subcortical structures supports consciousness as a lived phenomenon—introspective, goal-directed, emotionally responsive, and morally situated. In this way, the brain metafunction framework aligns with perspectives that affirm the irreducibility of personhood, acknowledging the brain's role in mediating consciousness without collapsing it into purely mechanistic terms (Dehaene & Changeux, 2011).

Integration and Regulation of Communication Systems

The brain's regulation of the body's major communication systems, nervous, endocrine, immune, and circulatory, reveals its deeply integrative role in maintaining adaptive function across multiple domains. Rather than acting through isolated or hierarchical mechanisms, this coordination emerges from dynamic, reciprocal interactions between cortical and subcortical structures, especially through large-scale brain networks (LSBNs). Subcortical hubs such as the hypothalamus and brainstem serve as key convergence points for neural, hormonal, and immune inputs, helping translate physiological states into coordinated behavioral responses. These hubs operate in close collaboration with the limbic system—particularly the amygdala and insula—to assess emotional salience and contextual relevance, aligning internal regulation with changing environmental demands. This alignment highlights the embodied nature of brain function. Internal states such as pain, hunger, arousal, or immune activation are not merely peripheral background conditions but are integrated into cognitive and emotional awareness. The anterior insula plays a particularly important role in this process, transforming interoceptive input into a felt sense of the body in relation to the self and world (Craig, 2002). Similarly, the brain's coordination of bodily systems is not simply reactive but interpretive, shaped by past experiences, affective context, and predictive evaluation. These insights reflect broader theoretical models of brain-body interaction that emphasize the role of emotion and homeostatic regulation in shaping consciousness and behavior (Damasio, 2010).

Central to this regulation is the Salience Network (SN), which detects and prioritizes biologically relevant stimuli and serves as a regulatory switchboard across brain systems. By integrating signals from the anterior insula, anterior cingulate cortex, hypothalamus, and brainstem, the SN enables rapid coordination between emotional evaluation, attentional focus, and physiological adjustment. Through these connections, the SN supports both homeostasis—the maintenance of internal equilibrium—and allostasis—the anticipatory adjustment of systems in response to predicted needs (Sterling & Eyer, 1988; Kleckner et al., 2017). This dynamic balance allows the brain to optimize energy use, maintain stability, and respond flexibly to evolving circumstances. Interactions between the SN, the Default Mode Network (DMN), and the Central Executive Network (CEN) further ensure that the body's internal milieu is continuously regulated in alignment with higher-order cognition. For example, during stress or threat, subcortical regulators increase autonomic output while cortical networks modulate awareness, planning, and decision-making. In contrast, during recovery and rest, the DMN facilitates introspection and internal coherence, while the SN modulates arousal downwards, promoting immune function and emotional repair. These interactions underscore the brain's role as a mediator of context-sensitive adaptation—where cognitive, emotional, and physiological functions are harmonized in real time. The brain metafunction framework, in this light, highlights the interdependence of mind and body in ways that go beyond mechanistic explanation. It presents the human person not merely as a system of interacting parts but as an integrated whole, wherein bodily regulation is interwoven with meaning, memory, and the dynamic trajectory of lived experience.

Homeostasis and Allostasis

The regulation of the body's internal environment—through homeostasis and allostasis—illustrates the brain's role as a unified regulatory center that integrates interoceptive, emotional, and cognitive information. Rather than treating regulation as a purely reactive process, the brain employs predictive models that anticipate needs and dynamically adjust physiological systems in advance. These operations are mediated by large-scale brain networks (LSBNs) that interlink cortical and subcortical structures, allowing for flexible adaptation to both internal fluctuations and external challenges. Homeostasis refers to the maintenance of stable internal conditions—such as body temperature, heart rate, and blood pressure—within narrow ranges optimal for function. In contrast, allostasis expands this concept by accounting for how the body prepares for anticipated demands, adjusting set points and physiological readiness proactively. This forward-looking capacity depends on coordinated activity across the Default Mode Network (DMN), Salience Network (SN), and Central Executive Network (CEN), which together generate predictions, prioritize sensory and emotional signals, and modulate regulatory responses.

At the cortical level, regions such as the anterior insula and prefrontal cortex interpret interoceptive input and contextual cues, generating top-down signals that guide autonomic and behavioral adjustments (Gu & FitzGerald, 2014; Seth, 2014). Expanded formulations of predictive processing models emphasize that the brain's forecasts are not purely cognitive but deeply affective and interoceptive, integrating emotional salience and bodily regulation into anticipatory models of the future (Seth & Friston, 2016). Subcortically, the hypothalamus and brainstem execute these commands by modulating sympathetic and parasympathetic tone, hormone release, and energy metabolism. Continuous feedback loops between cortical and subcortical areas maintain a dynamic, reciprocal system of regulation—one that adjusts to psychological meaning and environmental context, not just physiological deviation.

The limbic system, particularly the amygdala and hippocampus, adds emotional and experiential salience to these regulatory processes. The amygdala detects emotional relevance, such as threat or reward, influencing hypothalamic output accordingly. The hippocampus provides contextual memory, helping to fine-tune allostatic responses based on experiences such as increasing vigilance in familiar settings associated with danger or safety (Kleckner et al., 2017). These mechanisms illustrate that internal regulation is not merely reflexive but embedded in personal history and meaning. The Salience Network plays a pivotal role in this system. It monitors both internal and external conditions and reallocates attentional and physiological resources based on biological relevance. For instance, during high-stress conditions, the SN upregulates arousal, while in recovery, it signals the need for energy conservation and restoration. The Central Executive Network complements this by sustaining top-down control, especially during goal-directed tasks, ensuring that physiological states support cognitive demands (Menon & Uddin, 2010; Seeley et al., 2007).

Importantly, when these systems are overtaxed or chronically activated, the result is *allostatic load*—a cumulative physiological burden caused by repeated stress or poor adaptation. Over time, this burden can disrupt neural, endocrine, and immune regulation, contributing to a range of mental and physical health conditions, including anxiety, depression, cardiovascular disease, and burnout (McEwen & Wingfield, 2003). Understanding the mechanisms of allostasis not only clarifies the physiological roots of psychological symptoms but also emphasizes the importance of resilience-building, recovery, and self-regulation strategies in clinical and everyday settings. Taken together, the brain's regulation of homeostasis and allostasis reveals a deeply integrative process—one in which predictive modeling, emotion, memory, and physiological feedback are seamlessly coordinated. This integration reinforces the view that self-regulation is not a passive or mechanistic reflex but a conscious, embodied, and meaning-laden process. Rather than separate systems working independently, the brain and body engage in a continuous dialogue—supporting a view of the human person as a relational whole whose well-being depends on the unity of physiological stability, psychological awareness, and contextual responsiveness.

Integrative Processing of Afferent and Efferent Pathways

One of the most remarkable features of the brain metafunction framework is its capacity to integrate afferent (incoming sensory), efferent (outgoing motor) and the efferent loop of perception (see the chapter on Perception) into a coherent, adaptive system. This reciprocal flow of information underlies the brain's role in supporting real-time interaction with the environment, ensuring that perception, emotion, and motor output are tightly coordinated. Rather than operating in a stimulus-response loop, the brain functions as a predictive, relational system—inferring meaning from sensory cues and orchestrating responses aligned with personal goals, emotional salience, and contextual relevance. Large-Scale Brain Networks (LSBNs)—particularly the Sensorimotor Network (SMN), Visual Network (VN), and Central Executive Network (CEN)—serve as integrative hubs that bridge cortical and subcortical domains. The thalamus plays a central intermediary role in this system, filtering and relaying sensory input to cortical regions while engaging in reciprocal loops that support attentional prioritization and contextual framing. Subcortical structures, such as the basal ganglia and cerebellum shape and refine motor outputs by suppressing irrelevant activity, sequencing motor patterns, and correcting for real-time errors. These operations extend beyond mechanical precision, enabling adaptive flexibility in navigating dynamic environments.

Emotion and motivation are woven into this sensorimotor circuitry through the limbic system, especially the amygdala. When stimuli carry emotional weight—such as a looming threat or joyful expression—the amygdala modulates the salience and urgency of motor responses, influencing posture, facial expression, vocal tone, or defensive reflexes. This emotional integration ensures that motor behavior is not only biomechanically appropriate but attuned to social, affective, and contextual meaning. Attentional networks also shape this process. The Dorsal Attention Network (DAN) maintains a top-down, goal-oriented focus on relevant targets—such as grasping an object or tracking a moving figure—while the Ventral Attention Network (VAN) enables rapid reorientation to unexpected changes in the environment. These systems allow attention to remain flexible yet anchored, ensuring that motor output adjusts fluidly as new sensory or contextual information arises. At the center of this prioritization is the Salience Network (SN). The SN acts as a dynamic mediator between afferent input and efferent response, evaluating the biological and emotional significance of incoming data and coordinating with the brainstem to modulate arousal, reflexes, and bodily readiness. The brainstem, particularly the reticular formation and superior colliculus, facilitates rapid orientation, postural adjustment, and autonomic modulation, enabling immediate behavioral responses to salient stimuli such as pain, motion, or auditory alarms. In turn, the cerebellum refines these outputs by integrating real-time feedback with predicted movement trajectories, supporting the coordinated execution of both reflexive and deliberate actions.

Crucially, the Central Executive Network (CEN) provides top-down regulation by aligning motor behavior with long-term goals, working memory, and strategic planning. For instance, in tasks requiring fine motor skills—such as writing, sculpting, or manipulating tools—the CEN supports sustained focus and adjustment, integrating moment-by-moment feedback into a coherent behavioral sequence. Meanwhile, the Visual Network enriches spatial-motor alignment by matching incoming visual cues with motor intentions, enabling visually guided behaviors such as navigation, imitation, or interaction with moving objects. These sensorimotor processes are not isolated from higher-order cognition. During imagined movement or mental rehearsal, the same networks—particularly the SMN and cerebellum—engage in simulating action, preparing the body and mind for execution. This capacity for internal simulation illustrates the unity of perception, imagination, and action in embodied cognition. Altogether, the integration of afferent and efferent pathways within the brain metafunction exemplifies a key principle of unified function: cognition, emotion, perception, and motor output are deeply interwoven. This integrative architecture supports human interaction with the world as dynamic, intentional, and meaning-directed. Rather than a passive reaction, the brain enacts behavior as an expression of embodied purpose, informed by memory, shaped by emotional relevance, and aligned with personal and situational goals.

Behavioral and Adaptive Interaction with the Environment

The brain's capacity for adaptive behavior depends on its ability to integrate goal-directed cognition, emotional salience, physiological readiness, and sensory information into a coordinated, embodied response. Within the brain metafunction framework, this integration emerges not from isolated modules but from the dynamic collaboration of cortical and subcortical systems orchestrated through large-scale brain networks (LSBNs). These systems work in concert to produce behavior that is not only effective but meaningful anchored in memory, shaped by affect, and attuned to real-world context. The Central Executive Network (CEN) plays a foundational role in sustaining attention, maintaining working memory, evaluating options, and coordinating plans of action. It regulates top-down control over perception and response, allowing individuals to act intentionally in pursuit of complex or delayed goals. This is especially evident in tasks that require flexibility—such as navigating a social interaction, adapting to a changing environment, or making morally significant decisions—where the CEN provides the cognitive infrastructure for evaluation, inhibition, and redirection.

In parallel, the Salience Network (SN) ensures that behavior remains responsive to situational demands. Through hubs such as the anterior insula and anterior cingulate cortex, the SN rapidly appraises the emotional and physiological significance of new stimuli, reallocating cognitive and bodily resources to align with emerging priorities. This capacity for real-time reorganization ensures that behavior can shift fluidly between goal pursuit and environmental responsiveness. The SN signals the CEN when attentional or behavioral priorities need to change while coordinating with subcortical systems—including the amygdala, thalamus, basal ganglia, brainstem, and cerebellum—to initiate and fine-tune appropriate responses. Each of these subcortical systems contributes distinct yet complementary functions: the amygdala brings affective context by flagging stimuli as threatening, rewarding, or socially significant; the thalamus filters and directs sensory information based on relevance; the basal ganglia support action initiation and sequencing; the brainstem modulates arousal and autonomic readiness; and the cerebellum ensures that behavioral output is smooth, precise, and contextually appropriate. Together, these structures shape behavior that is not merely reactive but informed, emotionally attuned, and responsive to internal and external demands.

The Default Mode Network (DMN)—often associated with introspection and narrative thought—also plays a vital role in adaptive behavior. It supports autobiographical memory, future planning, and self-referential processing, providing the personal context that guides choices in light of past experiences, values, and imagined consequences. When the DMN, CEN, and SN function in concert, they enable not only the regulation of behavior in the present but also its orientation toward future states—supporting moral reasoning, identity coherence, and aspirational goal setting. The ability to shift between internally focused and externally oriented modes of cognition is a hallmark of adaptive intelligence. The SN mediates this transition by modulating the engagement of the CEN or DMN depending on task demands or internal state. In a high-stakes situation—such as public speaking or emergency response—the SN activates the CEN and brainstem arousal systems to heighten vigilance and suppress introspective rumination. Once the situation is resolved, the system reorients toward the DMN, enabling reflection, integration, and emotional recalibration.

This fluid interplay is further supported by attentional systems such as the Dorsal and Ventral Attention Networks (DAN and VAN), which sustain task-relevant focus while allowing for responsiveness to unexpected cues. The Sensorimotor Network (SMN) translates cognitive goals and affective signals into physical actions, modulated by cerebellar and basal ganglia input to ensure fluid execution. These interactions show how adaptive behavior arises from the convergence of sensory, motor, emotional, and cognitive signals—aligned not only with immediate conditions but also with broader personal and social significance. Ultimately, the brain metafunction framework reveals that human behavior is not the product of linear stimulus-response mechanisms but of a distributed and recursive system capable of interpreting context, evaluating alternatives, and enacting responses that are timely, embodied, and meaningful. This view emphasizes the person as an agent—one who perceives, decides, and acts not merely in reaction to the world but in relation to a narrative of values, goals, and lived experience.

Integration of Physiological Processes and Mental States

The integration of physiological processes and mental states lies at the core of the brain metafunction framework, which emphasizes that cognition, emotion, and bodily regulation are not separable domains but mutually informing aspects of a unified human experience. This integration is made possible by dynamic interactions across cortical and subcortical structures, orchestrated through LSBNs that link *interoception*, affect, self-awareness, and behavioral modulation. Within this system, physiological states do not merely influence mental events, they co-constitute them. A central hub in this integrative architecture is the insula, particularly its anterior region, which plays a pivotal role in interoception—the brain's mapping of the body's internal states. Through the Salience Network (SN), the anterior insula translates signals such as heart rate, respiratory rhythm, and visceral feedback into subjective feelings, including pain, tension, hunger, and emotional arousal. These sensations inform conscious experiences and shape how individuals perceive their own bodily and emotional states. Far from being peripheral, this interoceptive awareness is foundational to the construction of selfhood and the ability to regulate behavior in context. Recent findings further demonstrate that insula-centered cortico-subcortical networks play a critical role in supporting interoceptive awareness and resilience to stress, emphasizing the insula's pivotal position in linking bodily states with emotional regulation and self-awareness (Fermin et al., 2024).

The limbic system, including the amygdala and hippocampus, plays a key role in assigning emotional and contextual salience to physiological signals. The amygdala monitors the environment for affectively relevant cues, generating rapid responses to potential threat or reward. The hippocampus contextualizes these reactions with autobiographical memory, linking present experiences with past events to guide meaning-making and adaptive response. This interaction ensures that the body's internal states are interpreted in light of personal history and social significance—supporting a behavioral repertoire that is emotionally informed and memory-guided. Subcortical regulators such as the hypothalamus and brainstem execute the autonomic and endocrine adjustments necessary to sustain internal equilibrium. These structures maintain continuous feedback with cortical areas, especially the prefrontal cortex, allowing for top-down modulation of physiological responses based on cognitive appraisal and social context. This bidirectional communication enables individuals to exert cognitive influence over bodily states—such as calming one's breath in moments of stress—while remaining responsive to shifts in internal conditions.

The Default Mode Network (DMN) contributes by anchoring internal experience in autobiographical memory and self-referential reflection. It supports a temporally coherent sense of self across changing physiological and affective states. Meanwhile, the Central Executive Network (CEN) provides cognitive oversight, sustaining attention and working memory during tasks that require the alignment of mental focus with bodily control—such as public speaking, sports performance, or high-stakes decision-making. To illustrate this integration in action, consider a common experience: preparing to speak publicly. As the event approaches, the prefrontal cortex evaluates the situation's importance, activating the CEN to maintain goal-directed focus. Simultaneously, the SN detects rising interoceptive signals—such as increased heart rate, shallow breathing, and muscle tension—and prioritizes them for awareness. The hypothalamus and brainstem orchestrate a sympathetic stress response, while the amygdala modulates arousal based on emotional memory. The hippocampus may recall past speaking events, influencing anticipation. The anterior insula translates the physiological changes into subjective experiences—nervousness or excitement—while the DMN sustains a coherent internal narrative, helping the individual contextualize the moment within a broader story of self.

The Sensorimotor Network (SMN) and cerebellum also engage to support gestural fluency, speech precision, and motor readiness. The thalamus filters sensory inputs and coordinates the overall flow of bodily information to maintain attentional focus. Across this process, the brain orchestrates a tightly woven response that integrates emotional awareness, physiological readiness, motor planning, and reflective cognition. What results is a behavior—public speaking—that is not simply reactive or mechanical, but expressive, context-sensitive, and deeply human. In this light, physiological regulation and mental experience are not parallel tracks, but overlapping and recursive. Bodily states inform consciousness; consciousness interprets and modulates those states. Within the brain metafunction framework, this recursive integration reflects the non-reductive unity of the human person. Conscious awareness is not floating above the body, nor confined to isolated neural activity. It emerges from—and contributes to—the embodied coherence of brain, body, memory, and meaning.

This model affirms that while cognition often involves abstract processes—such as reasoning, imagination, or symbolic thought, it is not disconnected from biology. Instead, both cognitive and emotional functions are grounded in lived, interoceptive experience and shaped by the body's ongoing regulatory states. The brain metafunction framework highlights how even the most abstract mental activities are supported and modulated by dynamic interactions between cortical-subcortical systems, emotional salience, and physiological readiness. It also provides a neurobiological foundation for therapeutic approaches that seek to align bodily regulation with emotional healing and cognitive insight. Whether in contemplative practice, psychotherapy, or real-time decision-making, the integration of physiological states and mental life underscores the indivisibility of human experience—a central insight of both neuroscience and philosophical anthropology.

Unification of Primary Functions: The Brain Metafunction

The brain metafunction framework articulates how the human brain operates not as a collection of isolated modules but as a dynamically unified system—one in which diverse cognitive, emotional, and physiological functions are integrated into a coherent whole. Central to this integrative capacity is the concept of a "connectivity core," or rich club—a network of highly interconnected hubs that span both cortical and subcortical regions (van den Heuvel & Sporns, 2011; Cole et al., 2013). This architecture supports rapid information exchange between distant brain areas and forms the structural and functional scaffold for whole-brain coordination. It includes cortical regions involved in executive function, memory, and attention, as well as key subcortical structures such as the amygdala, hippocampus, thalamus, hypothalamus, and brainstem. These nodes form a backbone for integrating cognitive control, emotional salience, physiological readiness, and behavioral regulation.

Often viewed primarily as a relay station, the brainstem is now recognized as foundational in this architecture. It supports global neural synchronization, arousal modulation, and the maintenance of consciousness—functions without which higher cognitive integration could not occur (Hansen et al., 2024). Likewise, limbic structures such as the amygdala and hippocampus assign emotional and contextual meaning to experience, shaping how memory and behavior align with internal goals and environmental demands. Within this system, LSBNs—especially the Default Mode Network (DMN), Central Executive Network (CEN), and Salience Network (SN)—serve as domain-general systems that coordinate across functional domains. These networks mediate the transitions between internal reflection and external action, between memory and planning, and between physiological regulation and conscious intent. Rather than operating in strict hierarchies or fixed roles, these networks and subsystems instead engage in ongoing reciprocal interaction. Emotional arousal, for instance, influences attention and decision-making; interoceptive signals inform executive planning; memory retrieval reshapes current perception; and cognitive appraisal modulates autonomic tone. What emerges is a relational, whole-brain dynamic that allows humans to respond not only reflexively but intentionally and meaningfully. The system's adaptability lies in its flexible reconfiguration across contexts—whether during rest, task engagement, interpersonal interaction, or moral reflection.

This view also supports a broader philosophical and anthropological perspective. It challenges reductive models that attempt to localize human identity or awareness within any single neural structure. Instead, the brain metafunction affirms that human cognition is fundamentally embodied and relational—shaped by distributed activity across the entire brain-body system. It resonates with holistic views of the person—including non-reductive physicalist and dual-aspect theories—as well as theological accounts of psychosomatic unity. The person is not a sum of parts but a unified, living whole—capable of agency, memory, moral discernment, and self-transcendence. The framework does not claim to resolve the mystery of consciousness or identity in mechanistic terms. Instead, it shows how deeply interconnected neural systems support the conditions for integrated experience and meaningful behavior. In this way, the brain metafunction provides not just a map of brain activity but a model of the human being as an intentional, embodied subject—capable of coherence in thought, emotion, and action. It invites further interdisciplinary dialogue between neuroscience, psychology, philosophy, and theology—toward a more complete understanding of what it means to be human.

Large-Scale Brain Networks and Their Role in Integration

LSBNs represent distributed systems of anatomically and functionally interconnected regions that orchestrate complex cognitive, affective, sensory, and regulatory processes across the brain. These networks are not isolated modules but dynamically interactive systems that enable both domain-specific specialization and domain-general integration. Prominent LSBNs include the Default Mode Network (DMN), Salience Network (SN), Central Executive Network (CEN), Dorsal and Ventral Attention Networks (DAN and VAN), Sensorimotor Network (SMN), Visual and Auditory Networks, Language Network the Cerebellar Network, and other specialized circuits for imagery and spatial orientation. Their collaborative dynamics allow the brain to fluidly transition between internal reflection, focused attention, sensorimotor coordination, emotional salience, and executive control. Mounting evidence suggests that rapid transitions between whole-brain microstates—brief (~100 ms) global patterns of neural activity—support the brain's dynamic fluidity and are essential for maintaining conscious awareness (Michel & Koenig, 2018; Zanesco, 2024). Contemporary findings in network neuroscience have shown that these large-scale networks are organized around a "connectivity core" or rich-club structure—a set of highly interconnected hub regions spanning cortical and subcortical territories (van den Heuvel & Sporns, 2011; Cole et al., 2013; Bassett & Sporns, 2017). These hubs act as communication nexuses that allow for high-efficiency information transfer between distant brain regions. Rather than functioning as a central command center, this architecture reflects a decentralized, self-organizing system capable of adaptive coordination across functional domains. Within this architecture, the brainstem serves as a key stabilizing structure, sustaining arousal regulation, autonomic tone, and attentional readiness—funtions that are inherent and inseparable (Hansen et al., 2024).

The limbic system—including the amygdala, hippocampus, hypothalamus, and insula—contributes emotional salience, memory context, and interoceptive awareness to this whole-brain integration. For example, the SN uses inputs from the anterior insula and dorsal anterior cingulate cortex to assess biological and emotional relevance, helping prioritize stimuli across systems. The DMN supports autobiographical memory, narrative self-reflection, and temporal continuity, while the CEN provides goal-oriented regulation, cognitive control, and task execution. The flexible, reciprocal interplay among these networks—especially as they are shaped by context, memory, emotion, and bodily state—allows the brain to maintain coherence across varying levels of activity in real time. This integrative orchestration supports the central principle of the brain metafunction: unified function through distributed processing. Sensory input, memory, executive planning, affective evaluation, and motor output are not isolated in function-specific regions but flow through a networked system capable of contextual adaptation. Such coherence is maintained not by top-down control alone but by recursive loops that link perception, emotion, cognition, and bodily regulation in constant feedback. The resulting architecture enables humans to operate as integrated subjects—aware, adaptive, and responsive across mental, emotional, and physical domains.

Theoretically, this perspective marks a shift from modular and reductionist models toward a systems-level paradigm of brain function. It supports a vision of human cognition and consciousness as encompassing—but not reducible to—localized neural activity. As Bassett and Sporns (2017) emphasize, the complexity and flexibility of LSBNs lie in their capacity to reconfigure rapidly in response to task demands, emotional significance, or internal state changes. This dynamic organization allows the brain to sustain a unified sense of self and purpose amid changing contexts. Within the metafunction model, LSBNs represent the functional infrastructure that supports the relational and embodied character of human experience—unifying thought, emotion, behavior, and perception into a seamless whole.

Default Mode Network (DMN) and Cognitive Processes

The Default Mode Network (DMN) plays a central role in internally directed cognitive processes, including autobiographical memory, self-referential thinking, mental simulation, and envisioning future possibilities. Most active during restful wakefulness and introspective states, the DMN is not simply "task-negative" but crucial for meaning-making narrative construction and maintaining personal continuity across time (Raichle, 2015; Buckner, 2013). Anchored in key regions such as the medial prefrontal cortex, posterior cingulate cortex, and angular gyrus, the DMN supports the integration of memory, emotion, and identity—a cognitive foundation for self-awareness and reflection. Rather than operating in isolation, the DMN functions within a broader network ecology. Its reciprocal interactions with the Central Executive Network (CEN) and the Salience Network (SN) enable fluid transitions between introspective thought and goal-directed behavior. The SN, in particular, acts as a mediator—detecting salient changes in internal or external states and guiding shifts between the DMN and CEN based on contextual relevance (Menon & Uddin, 2010). This triadic coordination allows the brain to flexibly alternate between contemplative and task-focused modes, depending on whether the current situation calls for reflection, planning, or active problem-solving. These transitions are not random or passive but highly responsive to biological and psychological priorities. The DMN is also deeply connected to memory systems. Through its close interaction with the hippocampus, it facilitates the reconstruction of past experiences and the imaginative projection of future scenarios—allowing individuals to simulate outcomes, evaluate moral implications, and maintain a coherent personal narrative (Andrews-Hanna, Smallwood, & Spreng, 2014). This temporal integration supports foresight and decision-making grounded in memory, values, and personal meaning. The network's engagement during internal dialogue, future planning, and reflective storytelling underscores its importance in forming the mental scaffolding of identity and intentionality.

Notably, the DMN does not function independently of physiological and affective states. It is modulated by inputs from the insula and brainstem, which convey interoceptive and autonomic signals that influence the tone and direction of reflective cognition (Kleckner et al., 2017; Hansen et al., 2024). For example, during moments of physical calm or emotional safety, DMN activity may support expansive reflection or deep memory retrieval. In contrast, under conditions of stress or heightened arousal, the SN may suppress DMN activity in favor of activating the CEN and orienting the brain toward immediate tasks. This bidirectional modulation links bodily states with introspective mental activity, reinforcing the embodied character of cognitive life. From a systems neuroscience perspective, the DMN exemplifies the brain metafunction's emphasis on distributed, domain-general integration. It is not a module for abstract thought alone but a flexible, relational system that engages autobiographical memory, emotional salience, moral reasoning, and bodily awareness. Participating in these overlapping functions, the DMN helps shape the continuous thread of self-experience and intentional reflection that grounds human consciousness In doing so, it bridges the physiological, psychological, and existential dimensions of the person—reinforcing the view that awareness is always both embodied and meaning-oriented.

Salience Network and the Regulation of Attention, Emotion, and Physiological States

The Salience Network (SN) plays a vital role in detecting, prioritizing, and integrating biologically, emotionally, or contextually significant stimuli. With core hubs in the anterior insula and dorsal anterior cingulate cortex, the SN interfaces between internal states and external demands, continuously evaluating information across interoceptive, affective, and sensory domains. It functions as a dynamic coordinator, signaling other large-scale brain networks—especially the Default Mode Network (DMN) and Central Executive Network (CEN)—to engage or disengage per relevance and need (Seeley et al., 2007; Menon & Uddin, 2010). The SN's regulatory role extends to both homeostasis and allostasis, linking perception with physiological control. Through its close integration with subcortical structures such as the hypothalamus, brainstem, and limbic system, the SN orchestrates rapid autonomic and endocrine responses that align with cognitive and emotional appraisals. The anterior insula is particularly significant in this process: it receives interoceptive input from the body and translates it into conscious awareness, allowing the SN to evaluate the biological salience of bodily sensations—such as pain, hunger, or arousal—and prioritize responses accordingly (Craig, 2002; Kleckner et al., 2017). This function ensures that the organism remains sensitive to both internal states and external challenges, adjusting its behavioral and physiological responses in real time.

The SN's bidirectional connections with the brainstem—particularly with the reticular formation and the locus coeruleus—support rapid shifts in arousal, attention, and autonomic tone. When a biologically significant event occurs, the SN can initiate a sympathetic response, increasing heart rate and alertness to prepare for immediate action. This process exemplifies allostatic adaptation: a proactive adjustment to anticipated demand rather than mere reactive correction (Sterling & Eyer, 1988; Critchley & Harrison, 2013). Conversely, during recovery or rest, the SN helps downregulate physiological activation, coordinating with the DMN to restore energy balance and support introspection, memory consolidation, and immune regulation. Emotionally, the SN is tightly linked with the amygdala and anterior cingulate cortex. This connection allows it to tag stimuli with affective relevance and guide attention toward emotionally charged or motivationally salient cues. These interactions influence not only immediate action but also long-term behavioral adaptation. For instance, emotionally significant experiences are more likely to be encoded and recalled, significantly when the SN strengthens connectivity between limbic structures and memory systems in the temporal lobe. In this way, the SN supports experiential learning shaped by emotion, memory, and context.

Cognitively, the SN enables rapid attentional shifts between competing stimuli. When a relevant cue is detected—such as a sudden change in the environment or an internal sense of distress—the SN suppresses background activity (e.g., from the DMN) and activates task-positive systems like the CEN. This switching capacity supports behavioral flexibility, allowing individuals to navigate between self-reflection and focused action, depending on situational demands. Notably, the SN does not merely allocate attention or trigger physiological responses—it contributes to constructing meaningful experiences. Integrating bodily states with emotional and cognitive relevance provides a basis for the embodied sense of urgency, safety, discomfort, or anticipation. This integration enables coherent, intentional engagement with the world, wherein action arises not from reflex alone but from context-aware, emotionally informed, and bodily grounded appraisal. From the perspective of brain metafunction, the SN exemplifies how large-scale networks unify disparate domains of function—attention, emotion, interoception, and regulation—into coordinated patterns of adaptive behavior. As a bridge between cortical cognition and subcortical regulation, the SN reveals how the human person remains attuned to the outer world and the inner landscape of needs, values, and motivations. Its integrative role reinforces the non-reductive vision of the human being as a conscious, embodied, and responsive whole.

Dorsal and Ventral Attention Networks: Balancing Focus and Flexibility

Flexibly directing and shifting attention is essential to human cognition, perception, and behavior. Within the brain metafunction framework, this flexibility is supported by the collaborative dynamics of two major attention systems: the Dorsal Attention Network (DAN) and the Ventral Attention Network (VAN). Together, they balance goal-directed focus and responsiveness to unexpected or significant stimuli, enabling the individual to remain engaged with both internal intentions and environmental change (Corbetta & Shulman, 2002). The DAN, anchored in the intraparietal sulcus and frontal eye fields, mediates top-down, voluntary attention. It enables the sustained engagement of cognitive resources to support goal-oriented tasks—such as reading, solving problems, or tracking moving objects. This network filters out irrelevant distractions, maintaining attentional coherence and aligning perception with intention. The DAN also works closely with the Central Executive Network (CEN), reinforcing working memory and cognitive control during complex tasks that require persistent focus. By contrast, the VAN, centered in the temporoparietal junction and ventral frontal cortex, supports bottom-up, stimulus-driven attention. It monitors the environment for novel, emotionally salient, or unexpected events and allows for rapid reorientation when such events occur. This network ensures that the individual remains sensitive to threats, opportunities, or social cues, even while engaged. The VAN's capacity to override ongoing focus is vital in behavioral flexibility, particularly in dynamic or uncertain environments.

The Salience Network (SN) serves as an overarching modulator between these two systems, evaluating the significance of internal and external stimuli and directing the engagement of the DAN or VAN accordingly (Seeley et al., 2007; Menon & Uddin, 2010). When an emotionally or biologically salient stimulus arises, the SN may signal the VAN to reorient attention. At the same time, the DAN supports re-engagement with the original task once the distraction is resolved. This switching mechanism helps maintain a balance between attentional stability and environmental sensitivity—key aspects of adaptive behavior. Subcortical systems, particularly the brainstem and its locus coeruleus-norepinephrine system, further modulate attentional readiness and arousal in response to SN-driven salience signals. When vigilance is required, the brainstem enhances cortical responsiveness, preparing both motor and cognitive systems for action (Petersen & Posner, 2012). These brainstem-cortical loops allow the attentional system to remain dynamically responsive to changing demands, aligning internal alertness with the relevance of external stimuli.

The Default Mode Network (DMN) also participates in attentional transitions. While primarily involved in introspection and mental simulation, the DMN disengages during externally focused tasks and re-engages during reflective states. The SN mediates this transition by facilitating shifts between the DMN and attention networks, helping the brain toggle between inner and outer modes of awareness. Notably, the attentional system is not purely cognitive but affectively and physiologically grounded. Emotional salience, bodily states (such as fatigue or arousal), and motivational context shape how attention is deployed and sustained. Through this integration, the DAN and VAN do not merely direct cognitive focus but participate in the larger orchestration of behavior, perception, and affective regulation. In the brain metafunction framework, the attention networks illustrate how distributed systems cooperate to support awareness's coherent and embodied flow. Rather than operating as separate modules, these networks interact continuously, aligning attentional resources with goals, emotional tone, and environmental realities. This capacity to maintain focus while remaining responsive is a cornerstone of intelligent, context-sensitive action—and a vivid example of the brain's unified, adaptive architecture.

Frontoparietal (Central Executive) Network and Behavioral Regulation

The Frontoparietal Network, often called the Central Executive Network (CEN), plays a vital role in sustaining high-level cognitive control, particularly in regulating goal-directed behavior, working memory, decision-making, and attentional focus. Anchored in the dorsolateral prefrontal cortex (DLPFC) and posterior parietal cortex, the CEN enables individuals to align perception, intention, and action with both immediate and long-term objectives. Within the brain metafunction framework, this network is not viewed as a disembodied command center but a dynamically embedded system that integrates top-down executive function with emotional, physiological, and contextual inputs. The CEN's operations are tightly coupled with the Salience Network (SN), which identifies biologically or emotionally significant stimuli and signals the need for behavioral reorientation. For example, when the SN detects a threat or opportunity, it recruits the CEN to adjust attention, recalibrate plans, or suppress impulsive responses in favor of more adaptive strategies (Seeley et al., 2007; Menon & Uddin, 2010). This interaction ensures that executive control remains responsive to changing internal states and environmental demands, allowing for flexible engagement across a wide range of cognitive and behavioral contexts.

The Dorsal Attention Network (DAN) contributes by sustaining focus on task-relevant goals and filtering out distractions, while the Default Mode Network (DMN) provides autobiographical context and narrative continuity. Although the DMN and CEN typically operate in reciprocal balance—one becomes active while the other deactivates—the SN mediates their interaction, allowing for seamless transitions between introspective and externally focused states. This coordination is essential for navigating tasks requiring reflection and execution, such as planning a speech or evaluating ethical dilemmas. Subcortical systems enhance this cortical coordination. The basal ganglia contribute to motor planning and action selection, enabling fluid translation of executive intentions into physical behavior. They also assist in cognitive inhibition and sequencing complex routines, such as those required in language production or strategic problem-solving (Aron et al., 2007). The brainstem, particularly the locus coeruleus and related arousal systems, modulates the CEN's activity by adjusting levels of alertness and vigilance. During stress or high-stakes situations, this arousal system increases norepinephrine release, enhancing attentional engagement and readiness (Petersen & Posner, 2012). The cerebellum also plays a role by supporting temporal precision, error correction, and cognitive fluency. It aids in regulating the timing and sequencing of thought, especially during tasks requiring sustained attention or the coordination of mental subroutines. In this way, the cerebellum supports the smooth execution of both motor and executive processes—reinforcing that cognition and movement are not separate domains but closely integrated in the orchestration of adaptive behavior (Koziol et al., 2014).

Within the brain metafunction model, the CEN is understood not as an isolated processor but as a relational hub—integrating signals from attentional, emotional, motor, and memory systems. It sustains coherence across cognitive domains while remaining responsive to physiological and motivational inputs. This capacity for cross-domain regulation exemplifies the system's adaptive flexibility. It allows for the inhibition of automatic responses, the prioritization of complex goals, and the alignment of behavior with abstract values or long-term commitments. By participating in this broader integration architecture, the Central Executive Network illustrates how executive function emerges from a reciprocal interplay of brain systems rather than isolated command structures. It sustains volitional action, not as a detached mental faculty but as a lived, embodied process grounded in bodily readiness, emotional meaning, and contextual relevance. In doing so, the CEN affirms the unity of cognitive and behavioral regulation within the human person, as understood through the integrative principles of the brain metafunction framework.

Sensorimotor Network and Somatosensory Coordination

The Sensorimotor Network (SMN), encompassing the precentral (motor) and postcentral (somatosensory) gyri, serves as the primary interface between sensation and action. It enables the brain to transform tactile, proprioceptive, and kinesthetic input into coordinated movement, facilitating real-time engagement with the physical environment. Within the brain metafunction framework, the SMN is not regarded as a standalone processor of bodily activity but a fully integrated component of the brain's distributed system for adaptive regulation. Its close interplay with cognitive, emotional, and subcortical systems exemplifies the indivisibility of sensorimotor function from the whole-person context of perception, intention, and behavior. The SMN operates in continuous collaboration with subcortical hubs such as the basal ganglia and cerebellum. The basal ganglia play a crucial role in selecting, initiating, and inhibiting motor actions—ensuring that movements are appropriate, goal-directed, and efficient. In parallel, the cerebellum refines motor output through predictive modeling and real-time error correction, adjusting timing and coordination to ensure fluidity, balance, and responsiveness (Ito, 2008; Koziol et al., 2014). These cortico-subcortical loops transform raw sensory information into precise motor behavior, allowing the organism to respond flexibly and effectively to internal states and environmental conditions.

The brainstem further supports the SMN by regulating posture, muscle tone, and basic reflexes. Structures such as the reticular formation and vestibular nuclei facilitate rapid motor adjustments often outside of conscious awareness—such as stabilizing the body during motion or withdrawing from a painful stimulus. These interactions between cortical and subcortical regions highlight how even the simplest motor acts are deeply embedded in layered, integrative regulation systems. Importantly, sensorimotor coordination is not merely mechanical but is informed by higher-order cognition and emotional salience. The Central Executive Network (CEN) engages during complex or novel motor tasks, sustaining attention and planning multi-step actions. At the same time, the Salience Network (SN) modulates motor readiness in response to emotionally significant stimuli. For example, the sudden perception of a threat may trigger SN-mediated shifts in posture and muscle tone, preparing the body for defensive or adaptive action. This way, sensorimotor responses are shaped by relevance and meaning, not merely by stimulus-response patterns.

The Default Mode Network (DMN) also participates in sensorimotor processes by supporting embodied self-awareness and mental imagery of movement. Activities such as imagining a gesture, reflecting on bodily states, or rehearsing a future action activate circuits shared with the SMN. This connectivity reveals that even abstract thought about movement is grounded in the brain's sensorimotor architecture, supporting the broader view that cognition and embodiment are co-constitutive. The integration of attentional systems further enhances motor coordination. The Dorsal Attention Network (DAN) focuses resources on task-relevant motor targets—such as maintaining balance or tracking a moving object—while the Ventral Attention Network (VAN) supports reorientation to unexpected changes. The cerebellum and SMN work in tandem to adjust postural stability, trajectory, and timing in response to these attentional shifts, enabling fluid, precise, and contextually appropriate behavior. Within this integrated system, the SMN exemplifies how the brain metafunction enables whole-person adaptability. It does not simply command movement but orchestrates behavior through the coordinated interplay of sensation, perception, cognition, emotion, and bodily readiness. Whether grasping a tool, navigating terrain, or adjusting posture in social interaction, the Sensorimotor Network grounds the body in the world as a responsive, intentional agent. In doing so, it affirms that human action is not reducible to motor execution but emerges from the dynamic unity of thought, feeling, and embodiment.

Visual Network and Perception

The Visual Network, centered in the occipital cortex and extending through ventral and dorsal pathways, is responsible for processing visual information to enable the coherent perception of form, motion, depth, and color. Its core regions include the primary visual cortex (V1) and adjacent higher-order visual areas (V2–V5), which collectively transform retinal input into integrated perceptual experiences. Within the brain metafunction framework, visual processing is not confined to isolated occipital functions but is deeply embedded in the broader architecture of cognition, emotion, and behavioral regulation. This network closely interacts with attentional systems, particularly the Dorsal Attention Network (DAN) and Ventral Attention Network (VAN). The DAN facilitates goal-directed visual engagement—such as tracking a moving object, focusing on reading, or navigating a visual scene—while the VAN supports stimulus-driven shifts in attention, enabling responsiveness to unexpected changes in the environment (Corbetta & Shulman, 2002). These attentional networks do not operate independently but are modulated by the Salience Network (SN), which prioritizes visual input based on emotional or motivational relevance. For instance, a fearful face, a sudden movement, or a personally meaningful object may be prioritized in visual awareness due to SN-mediated modulation of perceptual salience (Seeley et al., 2007).

The integration of the Visual Network with the Sensorimotor Network (SMN) and cerebellum enables visually guided action. Catching a ball, navigating obstacles, or reaching for a tool requires seamless visual input and motor output synchronization. The cerebellum refines these movements by providing predictive modeling and fine-tuning spatial and temporal coordination, ensuring smooth and accurate responses (Ito, 2008). In this way, vision is not simply a receptive function, but a dynamic process oriented toward embodied interaction with the environment. Significantly, the brainstem, particularly the superior colliculus, contributes to rapid visual reflexes. It enables swift orienting responses toward sudden stimuli in the peripheral field, often preceding conscious visual awareness. These reflexive adjustments, supported by brainstem-cortical loops, prepare the body for appropriate behavioral responses and contribute to the continuity between perception and action (Wurtz et al., 2011). Such integration underscores the role of vision in sustaining awareness of the external world, bodily readiness, and environmental attunement.

Beyond perception, the Visual Network plays a role in internally generated imagery and memory. When visualizing a scene from memory or imagining a future scenario, this network activates in conjunction with the Default Mode Network (DMN) and Central Executive Network (CEN). The DMN provides autobiographical context and narrative coherence, while the CEN supports attentional control and the intentional direction of visual imagination (Buckner, 2013; Dosenbach et al., 2007). These interactions reveal that visual processing encompasses both external and internal perception, blurring the distinction between seeing and imagining. This distributed, integrative function challenges the notion of vision as a modular or purely sensory function. Instead, it affirms that visual perception is meaning-laden and context-sensitive—shaped by attentional priorities, emotional significance, memory, and behavioral goals. The Visual Network, therefore, does not merely receive and analyze visual stimuli; it actively participates in the brain's orchestration of adaptive awareness. Within the brain metafunction framework, vision is fully embedded in the person's embodied and relational engagement with the world—supporting environmental navigation and meaningful, intentional interaction with reality.

Auditory Network and Sound Processing

The Auditory Network, anchored in the superior temporal gyrus and encompassing the primary auditory cortex (A1) and adjacent secondary areas, transforms acoustic input into meaningful auditory perception. It supports detecting and analyzing pitch, rhythm, timbre, and spatial location while also enabling higher-order processes such as speech recognition, music appreciation, and the interpretation of emotionally and socially significant sounds (Hickok & Poeppel, 2007). Within the brain metafunction framework, the auditory system is not an isolated processor of sound but an integral component of a broader network that connects perception with emotion, cognition, memory, and motor coordination.

Auditory input engages subcortical structures early in processing. The brainstem, particularly the inferior colliculus, facilitates rapid orienting responses to salient sounds—such as a cry for help, a loud crash, or a sudden voice. These early auditory pathways engage the reflexive motor and autonomic responses, ensuring swift behavioral adaptation. For example, the startle reflex or an instinctive head turn toward a sound is mediated through brainstem-cortical loops that prepare the organism for an appropriate response. This early reactivity highlights how auditory perception is intimately linked to arousal, vigilance, and survival relevance (Møller, 2012). The Salience Network (SN) further modulates auditory processing by identifying emotionally or behaviorally relevant signals. When a sound carries affective weight—such as the tone of a loved one's voice, the intonation of sarcasm, or an alarm—the SN ensures that it is prioritized and integrated with physiological, attentional, and emotional systems. The anterior insula and anterior cingulate cortex coordinate this prioritization, linking interoceptive responses with auditory content (Seeley et al., 2007). Meanwhile, the amygdala contributes emotional valence, shaping whether a sound is perceived as comforting, threatening, or meaningful. This interaction with the limbic system imbues auditory perception with affective depth and relevance.

The auditory system also supports social communication, interfacing with the Language Network during speech production and comprehension. The ventral auditory stream maps acoustic input onto meaning, while the dorsal stream supports auditory-motor integration for speech articulation and timing. These dual pathways exemplify how language emerges not from a single cortical area but from dynamic network coordination across auditory, motor, executive, and emotional systems (Hickok & Poeppel, 2007). Prosodic speech features—such as rhythm, intonation, and emotional tone—are shaped by integration between the auditory network, limbic circuits, and cerebellum, allowing speakers and listeners to convey and interpret emotional states through vocal expression (Krishnan & Gandour, 2009).

Auditory attention is flexibly regulated by the Dorsal and Ventral Attention Networks (DAN and VAN). The DAN supports sustained, goal-directed listening—such as during a conversation or lecture—while the VAN enables responsiveness to unexpected auditory cues. For instance, while focusing on a speaker, an individual may still react to a siren or a familiar voice in the periphery. This balance between focus and flexibility allows fluid engagement with the auditory environment (Corbetta & Shulman, 2002). Connections with the Central Executive Network (CEN) ensure that auditory input is actively organized, especially in complex tasks such as language processing, memory retrieval, or musical performance. When engaged in activities like following spoken instructions or composing music, the CEN coordinates auditory working memory, sequencing, and attention control. The Default Mode Network (DMN) complements this by providing autobiographical and narrative context—enabling meaningful engagement with familiar voices, songs, or emotionally resonant sounds (Buckner, 2013; Dosenbach et al., 2007). The cerebellum contributes to the timing and fluency of auditory-motor interaction. Whether singing, speaking, or dancing to the rhythm, cerebellar systems refine timing, sequencing, and synchronization, ensuring smooth coordination between perception and expression (Koziol et al., 2014). In this way, auditory processing is not simply receptive but inherently embodied, expressive, and socially embedded.

Taken together, the Auditory Network exemplifies the integrative principles of the brain metafunction. It is not merely a system for decoding sound but a gateway to embodied communication, emotional attunement, and meaningful interaction. By linking subcortical reflexes with cortical interpretation and aligning sensory input with internal states and intentional action, the auditory system plays a vital role in sustaining the coherence and adaptability of human experience.

Right/Left Executive Networks and Lateralized Control

The lateralization of executive function within the brain's cortical architecture reflects a fundamental feature of human cognition: the division and specialization of control systems across hemispheres, each contributing distinct but complementary capabilities. Within the brain metafunction framework, the Right and Left Executive Networks are recognized as lateralized subdivisions of the broader Central Executive Network (CEN), anchored in bilateral dorsolateral prefrontal cortex (DLPFC) and posterior parietal regions. While these systems work together in close synchrony, they exhibit distinct functional profiles. The Left Executive Network is typically associated with verbal reasoning, language production, analytic thinking, and sequential problem-solving. In contrast, the Right Executive Network engages more in visuospatial reasoning, social-emotional context processing, holistic evaluation, and vigilance to novelty or ambiguity (Corbetta & Shulman, 2002; Menon & Uddin, 2010). This lateralization enhances cognitive efficiency by allowing simultaneous processing of different information types. For instance, during a conversation, the left network manages linguistic structure and grammar, while the right contributes to interpreting tone, context, and the speaker's emotional state. These capacities are further shaped by integration with the Salience Network (SN), which monitors emotional and contextual salience and dynamically allocates resources to the appropriate executive system. The SN may facilitate left-lateralized activation if a situation demands analytic clarity—such as solving a logic problem. Conversely, in emotionally nuanced or socially complex contexts, the right-lateralized network may be recruited to interpret affective cues and interpersonal meaning (Seeley et al., 2007).

Attentional systems also modulate this dynamic. The Dorsal Attention Network (DAN) supports sustained, goal-directed focus and often works in tandem with the Left Executive Network during cognitively demanding tasks. By contrast, the Ventral Attention Network (VAN) enables reactive shifts in attention and collaborates more closely with the Right Executive Network, ensuring flexibility in environments requiring rapid reorientation (Corbetta & Shulman, 2002). The Default Mode Network (DMN) maintains functional connections with both executive networks, especially during internally focused or autobiographical cognition. The Left Executive Network may engage during internal planning or verbal rehearsal, while the Right Executive Network becomes active when reflecting on emotionally significant or socially complex experiences (Buckner, 2013).

Subcortical systems also influence lateralized executive control. The limbic system—particularly the amygdala—modulates right-lateralized networks in the presence of emotionally salient stimuli. This limbic activity allows for rapid assessment and behavioral regulation in social or high-stakes environments. Meanwhile, the Left Executive Network is often involved in emotional regulation and expression, translating internal states into language or action plans, such as articulating distress or formulating a moral argument (Phelps & LeDoux, 2005).

Through its arousal-regulating structures, such as the locus coeruleus, the brainstem supports lateralized executive activity by modulating attentional vigilance. Under stress or heightened alertness, the right hemisphere may become dominant, particularly in supporting environmental scanning or intuitive problem-solving. Left-lateralized executive functions may prevail in calmer, deliberative contexts, favoring focused analysis and sequential processing (Petersen & Posner, 2012). The cerebellum also plays a subtle but essential role in supporting lateralized executive functions. Through its temporal and sequential structure coordination, the cerebellum enhances the fluency of both linguistic and non-verbal reasoning processes. Its connections with prefrontal and parietal regions help regulate timing, working memory, and behavioral modulation during task execution, regardless of lateral dominance (Ito, 2008; Koziol et al., 2014). The Right and Left Executive Networks illustrate how lateralized specializations contribute to the brain's capacity for flexible, goal-directed behavior. Their coordination—mediated by domain-general systems such as the Salience, Attention, and Default Mode Networks—enables the brain to balance analytic precision with contextual attunement. Rather than functioning as independent modules, these networks participate in a more extensive integrative architecture in which each hemisphere offers strengths that complement the other. The result is not dichotomy but harmony—a dynamic interdependence that underpins adaptive human behavior across social, intellectual, and emotional domains.

Cerebellar Network and Motor Coordination

The cerebellum, traditionally associated with motor control and coordination, has emerged as a central node in the broader integrative architecture of the brain. Within the brain metafunction framework, the cerebellar network is understood not only in terms of its contribution to movement but also as a critical modulator of cognition, attention, emotion, and learning. Its dense reciprocal connections with cortical and subcortical structures—including the prefrontal cortex, limbic system, brainstem, and large-scale brain networks (LSBNs)—enable the cerebellum to enhance both behavioral precision and adaptive responsiveness (Koziol et al., 2014; Miall et al., 2001). A central role of the cerebellum remains the refinement of motor activity. By integrating proprioceptive feedback with motor commands, the cerebellum fine-tunes timing, balance, coordination, and error correction. This allows for smooth execution of actions ranging from basic locomotion to highly skilled movements such as those required in music, athletics, or manual craftsmanship. Through its interaction with the Sensorimotor Network (SMN), the cerebellum continuously monitors and adjusts motion to accommodate real-time environmental input, ensuring that physical actions are aligned with task demands and bodily context (Yeo et al., 2011).

Beyond motor functions, the cerebellum contributes significantly to attentional regulation and executive processing. It interfaces with the Dorsal and Ventral Attention Networks (DAN and VAN), modulating shifts in focus, reaction to novelty, and cognitive flexibility. For example, during sustained attention tasks, cerebellar input supports the maintenance of precision and consistency. In contrast, when rapid behavioral adjustments are required—such as responding to a sudden change in one's surroundings—the cerebellum aids in shifting posture, redirecting gaze, or modifying motor intent in coordination with attentional cues (Corbetta & Shulman, 2002). The cerebellum also interacts with domain-general cognitive systems, particularly the Central Executive Network (CEN), to support working memory, sequence planning, and error prediction. These functions are essential for tasks involving cognitive load, such as language production, mental calculation, or structured decision-making. Cerebellar feedback enhances these processes by maintaining temporal coherence and suppressing irrelevant activity, enabling sustained goal pursuit and the fluid progression of thought (Dosenbach et al., 2007). Likewise, cerebellar interactions with the Default Mode Network (DMN) contribute to introspective functions such as self-reflection, episodic memory, and imagined future scenarios. The cerebellum aids in the sequencing and smooth progression of mental simulations, ensuring that inner narratives unfold in a coherent and structured manner—much as it ensures the fluid coordination of physical actions (Buckner, 2013). The Salience Network (SN) further recruits the cerebellum to modulate action and affective tone based on contextual demands. In emotionally charged or socially complex situations, the SN collaborates with the cerebellum to calibrate motor and behavioral responses to ensure appropriate timing and alignment with internal states. For example, in high-stakes public speaking, the cerebellum refines gestures, pacing, and expression in response to physiological arousal and social cues (Seeley et al., 2007). The cerebellum's connections with the limbic system—especially the amygdala—also support the modulation of motor expression in emotionally salient contexts, linking feeling states to action tendencies in a coherent, embodied manner (Schmahmann, 2019).

Clinical cases of cerebellar agenesis—in which individuals are born without a cerebellum—offer a uniquely informative perspective on the distributed architecture of consciousness. Despite containing over half of the brain's neurons, the cerebellum can be absent without eliminating conscious awareness, including subjective experience, reflective cognition, and emotional responsiveness. This striking fact supports the view of the cerebellum as an "enhancement device"—critical for precision and coordination, but not a member of the indispensable "consciousness core." Within the brain metafunction framework, this reinforces the perspective that consciousness is mediated by the integrated operation of multiple distributed systems—especially those involving cortical-subcortical loops and large-scale network dynamics. The cerebellum enhances and refines adaptive behavior but does not constitute the essential substrate of subjective experience.

By maintaining extensive connections with the brainstem, sensorimotor and attentional systems, executive and introspective networks, salience-processing regions, and the limbic system, the cerebellar network plays a vital role in optimizing both motor and cognitive domains. It provides real-time adjustments, predictive modeling, and context-sensitive modulation that improve behavioral fluency and responsiveness. However, unlike other major brain systems involved in the metafunction—whose contributions to consciousness and physiological integration are deeply intertwined—the cerebellum appears to operate outside this unified substrate of awareness. Thus, it may be best understood not as part of the core architecture of consciousness, but as a specialized module that amplifies the efficiency, timing, and coherence of processes already integrated elsewhere. While it does not directly contribute to the generation of conscious experience, it facilitates its expression by refining outputs from systems that do. This distinction reinforces the non-reductive view of consciousness as a system-level phenomenon—one rooted in distributed integration across networks rather than isolated structures. The cerebellum, while indispensable for coordination, remains peripheral to the core mechanisms of unified awareness in the brain metafunction framework.

Spatial Attention Network and Orientation

The Spatial Attention Network, closely aligned with the Dorsal Attention Network (DAN), is key in directing top-down, goal-oriented attention across the visual field and supporting orientation toward behaviorally relevant stimuli. This system, anchored in the posterior parietal cortex and frontal eye fields, enables individuals to locate objects, navigate complex environments, and prioritize spatially distributed information. It is essential for spatial exploration and environmental awareness and guiding coordinated action in response to dynamically shifting contexts (Corbetta & Shulman, 2002). A distinguishing feature of this network is its interaction with subcortical systems, particularly the brainstem's superior colliculus, which facilitates reflexive orienting movements. When a sudden change occurs in the periphery—such as motion or a novel object—the superior colliculus collaborates with cortical attention systems to trigger rapid shifts in gaze or posture. This reflexive responsiveness complements the intentional, top-down control exercised by the DAN, allowing for seamless transitions between planned focus and reactive reorientation.

The Salience Network (SN) further modulates spatial attention by assessing the emotional or motivational relevance of spatial cues. A flicker in peripheral vision may be flagged as meaningful not merely due to its location but because it carries social, affective, or survival-related salience. The anterior insula and anterior cingulate cortex relay this priority to executive and motor systems, ensuring that attentional orientation is context-sensitive and embodied (Seeley et al., 2007). The Central Executive Network (CEN) also contributes by sustaining attention to spatial tasks such as route planning, visual search, or object tracking. It integrates spatial awareness with working memory, goal setting, and decision-making—functions essential for real-time navigation and interaction. Once a spatial target is selected, the Sensorimotor Network (SMN) translates that information into coordinated motor responses. Whether reaching, grasping, or adjusting trajectory, this sensorimotor interface ensures that attentional priorities are translated into precise, adaptive behavior. The cerebellum complements this coordination by integrating spatial, proprioceptive, and vestibular information to refine posture, movement trajectory, and timing. The cerebellum ensures smooth physical alignment with an attentional focus in dynamic environments—such as walking through a crowd or avoiding unexpected obstacles. In parallel, the Default Mode Network (DMN) supports spatial memory and mental mapping during internally generated tasks, such as imagining routes, recalling environments, or planning future movement.

The Ventral Attention Network (VAN) contributes flexibility by detecting unexpected spatial cues outside the current focus. It collaborates with the DAN to reallocate attention when environmental changes demand a rapid behavioral shift (Corbetta & Shulman, 2002). As part of the limbic system, the amygdala also influences spatial orientation by assigning emotional valence to specific locations—biasing attention toward areas associated with prior fear, safety, or reward. These interactions exemplify the integrative logic of the brain metafunction: perception and orientation are not modular or detached processes but emerge from continuous coordination between attentional, emotional, motor, and contextual systems. In this light, spatial attention is not just a matter of visual scanning—it is an embodied and relational activity that supports meaningful interaction with the environment. It anchors the self in a navigable world, reinforcing the unity of cognition, action, and affect in pursuing adaptive goals.

Language Network and Communication

The Language Network, anchored primarily in Broca's area (inferior frontal gyrus) and Wernicke's area (posterior superior temporal gyrus), governs the core processes of language comprehension, production, and syntactic integration. Broca's area supports the formulation of grammatically structured speech, while Wernicke's area enables the decoding of linguistic input into meaningful concepts. Together, these hubs facilitate expressive and receptive language, allowing individuals to engage in complex communication that conveys not only information but also personal meaning, emotional nuance, and social intent. Within the brain metafunction framework, language is not an isolated cognitive module but a distributed function embedded within large-scale brain networks (LSBNs). The Language Network interacts dynamically with the Default Mode Network (DMN), Central Executive Network (CEN), Salience Network (SN), and Sensorimotor Network (SMN), forming a highly integrative system that coordinates verbal expression with attention, emotion, bodily readiness, and social engagement. The DMN contributes to language's narrative and autobiographical aspects, helping situate speech within personal memory, identity, and self-reflection (Buckner, 2013). This contribution enables language to carry a sense of lived continuity, transforming speech into a vehicle for meaning-making and self-expression.

The Central Executive Network (CEN) facilitates goal-directed speech, including lexical selection, syntactic planning, and coherence maintenance during complex communication. It enables individuals to formulate arguments, adapt language to context, and sustain focused dialogue. The Salience Network (SN), mainly through the anterior insula and anterior cingulate cortex, modulates vocal prosody, emotional tone, and responsiveness to social cues. It prioritizes emotionally or interpersonally relevant content, ensuring that language reflects and adapts to the speaker's internal states and the listener's reactions (Seeley et al., 2007). From a sensorimotor perspective, the articulatory dimensions of speech are mediated by the Sensorimotor Network

(SMN), which governs the coordination of the tongue, lips, and larynx. The cerebellum and basal ganglia refine these fine motor actions, which ensure verbal output fluency, rhythm, and precision—particularly in tasks requiring rapid speech, emotional expressiveness, or nuanced prosody. This integration of motor, executive, and affective systems underscores language as an embodied process involving symbolic processing and synchronizing cognition, emotion, and physiology.

The dual-stream model of speech processing (Hickok & Poeppel, 2007) reflects this integration. It identifies a dorsal stream linking auditory input with motor output for speech production and a ventral stream linking sounds to meaning for comprehension. These pathways traverse cortical and subcortical regions, reflecting the language system's reliance on cross-domain coordination. The Dorsal and Ventral Attention Networks (DAN and VAN) further contribute to linguistic responsiveness: the DAN supports sustained verbal engagement, while the VAN allows reorienting to novel or emotionally salient cues—such as a change in tone, interruption, or nonverbal signal (Corbetta & Shulman, 2002).

The limbic system, especially the amygdala, adds affective salience to verbal interaction. Emotional meaning often precedes semantic content; for example, tone and prosody may signal sarcasm, reassurance, or concern before the words are fully processed. This limbic-language interface reinforces the capacity of speech to connect minds, express values, and shape relationships. Prosody—marked by rhythm, intonation, and emphasis—is modulated by hemispheric asymmetries and cross-network coordination. As Krishnan and Gandour (2009) emphasize, prosodic expression emerges from integrating cognitive-linguistic and affective systems, particularly across right-hemisphere circuits involved in affective tone. Additionally, research suggests that motor simulation mechanisms, such as those identified by Jeannerod (2001), contribute to the embodiment of speech. The internal rehearsal of language—whether in silent reading, thought, or planning—activates sensorimotor and affective systems, reflecting the deep integration of language with bodily and emotional awareness. Ultimately, the Language Network exemplifies the brain metafunction's core principles: multidimensional integration, real-time adaptability, and relational embodiment. Language is not just a function of cognition; it is the living articulation of thought, emotion, memory, and interpersonal connection. Through interweaving with core networks and subcortical systems, the brain enables humans to share meaning, bridge internal experience with external expression, and co-create cultural, moral, and spiritual life.

Lateral Visual Network and Peripheral Processing

The Lateral Visual Network, primarily situated in the occipital and parietal cortices, is specialized in processing peripheral and spatially distributed visual information. While central vision is essential for high-acuity tasks like reading or object identification, peripheral vision offers a broader field of awareness critical for detecting motion, navigating complex environments, and anticipating changes in one's surroundings. Within the brain metafunction framework, this network exemplifies that perceptual systems do not function in isolation but in dynamic coordination with attentional, motor, and salience-processing domains to support adaptive interaction with the world. The Lateral Visual Network operates in close collaboration with the Dorsal Attention Network (DAN), which governs goal-directed visual tracking and spatial awareness, and the Ventral Attention Network (VAN), which enables reflexive reorienting toward unexpected stimuli. This interaction allows the brain to continuously monitor peripheral cues and shift attention rapidly when behaviorally relevant stimuli appear outside the central field. For example, the sudden movement of an object or a facial expression in the periphery can redirect gaze and attention—an adaptive function essential for survival, social communication, and environmental navigation (Corbetta & Shulman, 2002).

Subcortical structures enhance this responsiveness. The superior colliculus in the brainstem detects sudden changes in peripheral vision and initiates orienting behaviors such as saccades or head movements. These reflexive actions precede and inform cortical visual processing, allowing the organism to respond quickly to salient events. The Salience Network (SN), particularly through the anterior insula and anterior cingulate cortex, evaluates the emotional or motivational relevance of such stimuli and coordinates shifts in attention and motor readiness accordingly (Seeley et al., 2007). This coordination allows the system to prioritize visual cues that signal danger, opportunity, or social engagement. The Central Executive Network (CEN) further contributes by regulating attention to peripheral stimuli in task-relevant contexts. For instance, during navigation or social interaction, the CEN may suppress irrelevant distractions or heighten sensitivity to peripheral cues depending on the situation. Meanwhile, the Default Mode Network (DMN) integrates these visual inputs with memory and self-referential processes, enabling spatial awareness to be informed by prior experiences—such as recalling familiar routes or anticipating the behavior of others in a known setting (Buckner, 2013).

Motor systems are tightly linked with peripheral visual processing. The Sensorimotor Network (SMN) translates visual cues into coordinated actions—like adjusting gait to avoid an obstacle seen out of the corner of the eye. The cerebellum also plays a crucial role by refining the timing and precision of these movements, ensuring that responses to peripheral events are smooth and proportional. This visual-motor coordination allows for fluid, anticipatory interaction with the environment, especially in dynamic settings such as driving, playing sports, or moving through crowded spaces. Research in neuroimaging confirms that peripheral vision activates an extended visual-attentional network that reaches beyond early visual areas. As Toosy et al. (2004) demonstrated, peripheral processing recruits widespread parietal and subcortical regions, reflecting the brain's capacity to integrate sensory input with motor planning and emotional salience. This broader activation pattern highlights the adaptive value of peripheral vision—not as a secondary or passive mode of seeing but as an integral part of embodied perceptual awareness. In sum, the Lateral Visual Network reveals how even specialized sensory systems operate within the broader integrative principles of the brain metafunction. Its participation in attentional regulation, motor coordination, and emotional evaluation underscores that perception is not confined to the fovea or isolated visual streams. Rather, it is relational, distributed, and oriented toward meaningful interaction with the world. Peripheral vision is thus not peripheral to experience—it is central to the brain's capacity for whole-body, context-sensitive engagement with the world.

Temporal Network and Memory Integration

The Temporal Network, centered in the medial and lateral temporal lobes, plays a pivotal role in memory formation, emotional regulation, and contextual processing. Anchored by key structures such as the hippocampus and amygdala, this network supports the encoding, consolidating, and retrieving of autobiographical and episodic memories while integrating emotional salience into past experiences. Its capacity to assign affective and situational relevance to memory enables individuals to respond to present circumstances in light of personal history and learned meaning. Within the brain metafunction framework, the Temporal Network exemplifies the principle of functional convergence. Rather than operating as a discrete "memory module," it serves as a hub through which multiple large-scale brain networks (LSBNs) converge to embed memory into the broader stream of cognition, emotional experience, and physiological preparedness. This integration ensures that memory is not static but continuously interpreted and applied—shaping attention, behavior, and planning. Notably, connections between the Temporal Network and brainstem structures, such as the locus coeruleus and reticular formation, underscore the alignment of memory with arousal and autonomic states. Emotionally salient memories—especially those involving threat, joy, or trauma—can evoke bodily responses such as elevated heart rate or changes in breathing. This bi-directional coupling between memory and physiology reflects how visceral feedback reinforces the emotional meaning and prioritizes the recall of significant experiences (McGaugh, 2018). The Default Mode Network (DMN) collaborates closely with the Temporal Network to support autobiographical reflection and future-oriented mental simulation. The hippocampus is central to this process, enabling the construction of internally generated mental scenes that combine memory with imaginative foresight. This capacity underpins narrative identity, supports the moral evaluation of past and future actions, and contributes to goal-directed behavior through emotionally contextualized foresight (Buckner, 2013). The Salience Network (SN) complements this dynamic by evaluating memories' emotional or motivational relevance and guiding selective retrieval based on current needs or contexts. For example, when confronted with a difficult decision, the SN can amplify access to past experiences involving similar emotional dynamics, helping to shape adaptive choices in the present (Seeley et al., 2007). Meanwhile, the Central Executive Network (CEN) governs top-down, task-relevant memory retrieval—enabling individuals to access and manipulate memory content for problem-solving, learning, or strategic planning.

Attentional systems further modulate memory processing. The Dorsal Attention Network (DAN) helps maintain focus on memory-related cues. It suppresses irrelevant distractions, improving the fidelity and relevance of recall—such as during emotionally intense conversations, academic exams, or spatial navigation (Corbetta & Shulman, 2002). In parallel, the limbic system, particularly the amygdala, strengthens the encoding and retrieval of emotionally charged events. This emotional tagging of memory plays a central role in shaping beliefs, guiding behavior, and forming future expectations (Phelps & LeDoux, 2005). Although traditionally associated with motor coordination, the cerebellum contributes to memory by enhancing recall's temporal and sequential structure—particularly for procedural, associative, or time-sensitive memories. Its regulatory influence supports smooth transitions between memory representations and helps structure mental simulations with coherence and realism (Koziol et al., 2014). Taken together, the Temporal Network exemplifies the brain metafunction's principle of integration across systems. Its interactions with the brainstem, DMN, SN, CEN, DAN, cerebellum, and limbic circuits enable memory to function as an embodied, context-sensitive process—not merely a record of the past but an active guide for present action and future projection. This way, memory is lived, not stored, interpreted, or archived. The Temporal Network thus plays a central role in shaping reflection, identity, discernment, and transformation, deepening our understanding of memory as a relational, emotionally grounded, and forward-oriented faculty.

Sleep and the Brain Metafunction: Integrating Memory, Emotion, and Adaptive Processing

Within the brain metafunction framework, sleep is understood not as a passive or disconnected state but as an active and highly integrative phase of neural functioning. It serves critical roles in memory consolidation, emotional recalibration, physiological restoration, and the refinement of internal predictive models. Far from a temporary suspension of cognition, sleep represents a dynamic mode of brain operation—deeply embedded in the continuity of consciousness, identity, and adaptive behavior. During slow-wave sleep (SWS), the brain engages in synchronized cortical activity that supports declarative memory consolidation. The hippocampus repeatedly reactivates recent experiences in compressed sequences, transferring and integrating them into neocortical networks for long-term storage. This process is mediated by sleep spindles—brief bursts of thalamocortical oscillations—that coordinate communication between the hippocampus and cortex (Diekelmann & Born, 2010; Rasch & Born, 2013). Large-scale brain networks (LSBNs) such as the Default Mode Network (DMN) and Central Executive Network (CEN) remain active in specific patterns during this phase, helping integrate newly acquired information into broader narrative structures and cognitive frameworks (Schabus et al., 2004; Fogel & Smith, 2011). This activity indicates that sleep preserves memory traces and actively recontextualizes them within the more extensive architecture of meaning and identity.

REM sleep, by contrast, is characterized by desynchronized cortical activity and heightened activation in limbic and paralimbic regions, particularly the amygdala and medial temporal lobes. These areas play key roles in affective learning and emotional regulation. During

REM, emotionally salient memories are reprocessed, often through dream experiences, allowing the brain to integrate emotional intensity while reducing reactivity and distress (Maquet, 2001; Walker & Stickgold, 2010). This process helps decouple strong emotional charges from raw experience, contributing to resilience, emotional flexibility, and psychological well-being. The Salience Network (SN) is thought to play a role in this prioritization, signaling which experiences require emotional refinement and further integration. During sleep, the coordinated interaction between the Temporal Network (especially the hippocampus and amygdala), the SN, DMN, and CEN highlights how the brain metafunction persists even in non-waking states. These systems collaborate to reorganize experience, distill meaning, and prepare the individual for adaptive functioning in future contexts. For example, the CEN contributes to strategic updating by consolidating goal-relevant information, while the DMN integrates new memories into self-referential frameworks. This reorganization supports the continuity of the self across waking and dreaming states, reinforcing identity and narrative coherence even in the absence of real-time interaction with the external world.

Notably, the brainstem—particularly structures such as the reticular formation and pontine tegmentum—regulates sleep stages by controlling arousal, vigilance, and transitions between consciousness states. This regulation underscores the brainstem's integrative role in wakefulness and organizing the cycles that underpin mental clarity, emotional balance, and physiological homeostasis. Sleep thereby exemplifies the principles of the brain metafunction: distributed coordination, cross-domain integration, and dynamic reciprocity between bodily states and conscious experience. It allows the brain to recalibrate emotional salience, fortify memory networks, and restore internal balance—all in ways that sustain the person's integrity across time. Sleep is not a suspension of conscious life but a complementary mode of embodied cognition—supporting identity, resilience, and adaptive readiness.

Visual Perception/Imagery Network and Mental Imagery

The Visual Perception/Imagery Network, anchored in occipital and parietal regions, plays a pivotal role in both the processing of external visual input and the internal generation of mental imagery. This dual capacity allows the brain to interpret the present visual environment and construct internally simulated scenes drawn from memory, imagination, or future anticipation. Mental imagery contributes to planning, memory consolidation, spatial reasoning, and creative cognition—functions central to adaptive human behavior. Neuroimaging studies have shown that visual mental imagery activates many of the same neural circuits as actual visual perception, particularly in early visual areas such as the primary visual cortex (V1) and higher-order occipital regions (Kosslyn et al., 2001). This overlap illustrates the fluid boundary between external and internal visual experience. The network's interaction with domain-general systems supports its broader cognitive functions: the Default Mode Network (DMN) provides autobiographical content and narrative structure to imagined scenes. At the same time, the Central Executive Network (CEN) offers top-down control, directing attention and maintaining task relevance during mental simulation.

The Sensorimotor Network (SMN) is also active during motor imagery, especially when individuals visualize actions such as walking, gesturing, or manipulating objects. This engagement reflects the embodied nature of mental imagery: imagined movements activate motor planning regions even without actual motion. The cerebellum contributes by refining the temporal structure and spatial coordination of these internal simulations, much like during actual motor execution (Jeannerod, 2001; Ito, 2008). These dynamics suggest that mental imagery is not a mere epiphenomenon but a functionally embedded process supporting behavioral readiness and strategic foresight. The Salience Network (SN), mainly through the anterior insula and anterior cingulate cortex, plays a role in prioritizing emotionally relevant or motivationally salient imagery. For instance, imagining a meaningful future event or recalling a powerful memory recruits SN activity, helping to align internal visualizations with emotional context and current goals (Seeley et al., 2007). The Dorsal and Ventral Attention Networks (DAN and VAN) further modulate this activity. The DAN sustains focus within imagined scenes, while the VAN enables shifts in attention to unexpected associations or emergent features within the mental simulation (Corbetta & Shulman, 2002).

The brainstem, particularly the reticular formation and locus coeruleus, supports arousal and attentional readiness during visualization. These structures ensure that internal imagery aligns with the physiological state, such that imagining a stressful or inspiring scenario may trigger measurable autonomic changes—such as increased heart rate or alertness. This alignment reflects the deeply embodied nature of internal simulation: the body responds to what the mind enacts, even without external input. In the brain metafunction framework context, the Visual Perception/Imagery Network exemplifies the principle of cross-domain integration. It bridges perception and imagination, motor planning and emotional salience, internal simulation, and bodily readiness. Whether used for rehearsing a speech, recalling a loved one, or planning a complex decision, visual mental imagery engages distributed systems that work together to create meaningful, embodied representations. Far from being a passive recall function, mental imagery plays an active role in shaping behavior, deepening reflection, and enabling future-directed adaptation.

The Role of Core Subsystems in Unifying Brain Functions

The brain metafunction framework emphasizes large-scale network dynamics and the indispensable contribution of deep brain subsystems that anchor global integration. These core subsystems—including the ascending reticular activating system (ARAS), limbic structures, hypothalamus, thalamus, and basal ganglia—form the physiological and affective infrastructure necessary for conscious awareness, emotional relevance, and behavioral adaptability. Evidence suggests that specific brainstem nuclei—such as the parabrachial complex and locus coeruleus—may contribute not only to arousal regulation but also to shaping the contents and integration of conscious experience across sensory modalities, although cortical network integration remains indispensable (Edlow et al., 2012). Their activity ensures that cognition is not disembodied but embedded within a continuous, biologically grounded flow of experience.

The ARAS, extending from the brainstem through the thalamus and basal forebrain, maintains arousal, regulates sleep-wake cycles, and supports attentional engagement. It provides a background activation level necessary for large-scale brain networks to function coherently. Higher-order cognition cannot be sustained without the ARAS, regardless of cortical integrity (Parvizi & Damasio, 2001). In this way, the ARAS establishes the foundational conditions under which conscious cognition can occur, reinforcing the brain metafunction's insistence on subcortical-cortical reciprocity. The limbic system—especially the amygdala, hippocampus, anterior cingulate cortex, and insula—imbues perception and cognition with emotional and motivational salience. These regions ensure sensory input is filtered through affective significance: Is it threatening, rewarding, or connected to personal memory? The hypothalamus extends this evaluative function into the domain of physiological regulation, governing autonomic and endocrine responses. Together, these systems integrate bodily states with emotion and cognition, allowing the organism to act in ways that are not merely reflexive but goal-directed and personally meaningful.

The anterior insula plays a particularly integrative role, translating interoceptive signals—such as heart rate, breathing, or gastrointestinal activity—into affective tone and self-awareness. It is a convergence zone between somatic awareness and higher cognition, linking bodily feeling with subjective presence. This connection helps explain how the brain constructs a lived, embodied sense of self, where physiological and emotional processes are inextricably interwoven with reflective thought. Long understood as a sensory relay station, the thalamus is now recognized as a dynamic integrator of cortical and subcortical communication. Through its reciprocal loops with sensory, motor, and association cortices, the thalamus enables real-time information integration and coordination across domains. It contributes to the temporal binding of experience and supports attentional selectivity and perceptual coherence—key features of unified consciousness. Recent research highlights that the thalamus, especially higher-order nuclei such as the medial dorsal (MD) nucleus and pulvinar, plays an active role in modulating cognitive control and conscious-state transitions. Rather than serving only as a relay station, the thalamus participates dynamically in selective amplification of cortical activity, attentional gating, and representational updating (Sherman, 2016; Halassa & Sherman, 2019; Rikhye, Wimmer, & Halassa, 2018). The basal ganglia, often associated with motor control, also play a key role in motivation, action selection, and cognitive flexibility. Their loops with the prefrontal cortex support volitional behavior and adaptive inhibition, making them vital for both habit formation and deliberate decision-making. These deep structures operate as adaptive filters, helping prioritize which cognitive, emotional, or motor plans should be executed based on context and internal state.

Crucially, these subsystems do not function independently of large-scale brain networks (LSBNs) but continuously interact with them. The Default Mode Network (DMN), Salience Network (SN), and Central Executive Network (CEN) depend on subcortical feedback to remain dynamically responsive to internal and external changes. The SN, in particular, mediates interoceptive salience (via the insula and amygdala) and network-level switching. The DMN maintains the narrative self, while the CEN governs deliberate, goal-directed cognition. These networks draw upon and regulate deeper brain structures to maintain functional coherence. The integration of these subsystems within the broader rich-club architecture—densely interconnected cortical and subcortical hubs—supports efficient communication and global information flow (van den Heuvel & Sporns, 2011; Bassett & Sporns, 2017). This configuration enables both specialization and unification, allowing the brain to respond flexibly to novelty while preserving internal stability. Rather than operating as discrete modules, core subsystems and large-scale networks co-constitute one another, producing the coherent, embodied unity of the human person. In sum, the brain's core subsystems are not peripheral support structures but integral to cognition, affect, and behavior. They regulate the rhythms of consciousness, shape the emotional landscape of experience, and translate internal states into coherent perception and action. Within the brain metafunction framework, these subsystems ensure that even the most abstract thought is grounded in bodily life—and that no cognitive act is ever entirely disembodied. The result is a living system that unites intelligence, awareness, and embodied existence into an indivisible whole.

Achieving Functional Unity through Domain-General Networks

Within the brain metafunction framework, *domain-general networks*—particularly the Default Mode Network (DMN), Central Executive Network (CEN), and Salience Network (SN)—serve as the brain's primary integrative systems, coordinating diverse cognitive, emotional, and physiological processes. Unlike specialized modules dedicated to particular sensory or motor functions, these networks operate across domains, facilitating flexible adaptation, coherence of experience, and intentional action. They allow the brain to shift seamlessly between introspective and externally focused states, between reactive and deliberative responses, and between regulation of internal bodily states and engagement with the external world. The DMN underpins self-referential thought, autobiographical memory, and mental simulation. It provides narrative continuity, enabling individuals to reflect on past experiences, envision future possibilities, and maintain a stable sense of identity (Buckner, 2013; Raichle, 2015). Through its coupling with the hippocampus and medial temporal regions, the DMN constructs internally coherent mental scenes that guide planning, imagination, and moral reflection. Far from being "task-negative," as once assumed, the DMN is foundational for integrating meaning and context into lived experience.

The SN, anchored in the anterior insula and anterior cingulate cortex, acts as a dynamic prioritization system. It continuously monitors interoceptive and exteroceptive inputs, signaling shifts in salience—biological, emotional, or environmental—and reallocating cognitive resources accordingly (Seeley et al., 2007; Menon & Uddin, 2010). The SN facilitates transitions between the DMN and CEN, enabling the brain to move from rest and reflection to focused task execution or from goal-driven attention to emotionally responsive behavior. The SN supports allostatic regulation—adjusting internal states based on anticipated demands—and ensures alignment between motivational relevance and behavioral output. The CEN is located in the dorsolateral prefrontal and posterior parietal cortex, providing executive control over working memory, attentional focus, and goal-directed action. It regulates top-down processing and decision-making, integrating current task demands with long-term intentions (Menon, 2015). The CEN's interactions with the SN and DMN ensure that executive functions remain contextually grounded and emotionally attuned. In this way, domain-general cognition is not abstract or disembodied but woven into the fabric of embodied, affectively significant experience.

These three networks—DMN, SN, and CEN—operate as a functional triad. Their coordinated activity enables the brain to flexibly navigate changing conditions while preserving coherence across cognitive, emotional, and physiological domains. Situational demands and internal signals of salience and relevance govern transitions between these networks. This dynamic coordination supports the unification of perception, attention, emotion, and memory into an intelligible stream of experience. Underlying this triadic structure is a deeper organizational scaffold known as the "rich-club architecture"—a set of densely interconnected neural hubs spanning cortical and subcortical regions (van den Heuvel & Sporns, 2011; Bassett & Sporns, 2017). These hubs facilitate rapid information transfer across specialized and domain-general networks, enabling the brain to maintain both flexibility and integrity in real time. This architecture supports efficient network switching and global integration during states of heightened demand—whether cognitive, emotional, or physiological.

The domain-general networks at the heart of the brain metafunction framework provide a compelling model for understanding the unity of the person. They integrate reflection and action, meaning and regulation, memory and intention. Rather than treating the brain as a set of disjointed processes, this model reveals a coordinated whole in which subjective awareness, rational deliberation, emotional salience, and bodily regulation converge. Importantly, this functional unity resists both reductive materialism and dualistic separation. It shows how embodied consciousness emerges from the integrated operation of distributed, reciprocally interacting systems. While rooted in physical substrates, the functions mediated by these networks are irreducible to any single structure or mechanism. They support a conception of the human person as a dynamically integrated being—simultaneously cognitive and affective, physiological and reflective, situated and intentional. In this light, the brain metafunction is not merely a neurobiological map. The framework affirms the embodied coherence of human subjectivity and invites fruitful engagement from neuroscience, psychology, philosophy, and theology alike. It offers a paradigm for understanding how the brain enables the whole person to think, feel, act, remember, and relate—not as fragmented subsystems but as an indivisible, living unity.

Embodiment of the Human Person as an Indivisible Whole

The brain metafunction framework culminates in a unified vision of the human person, which affirms the profound inseparability of mind, body, and brain. It rejects both dualistic accounts that isolate consciousness from biology and reductionist views that seek to explain personhood in purely mechanistic terms. Instead, it portrays the person as a dynamically embodied being who actively weaves cognition, emotion, physiology, and subjectivity by orchestrating large-scale brain networks, cortical-subcortical loops, and regulatory subsystems. These components do not function in isolation but in reciprocal alignment, enabling the full range of human experience: thought, memory, agency, emotional resonance, and moral awareness.

At the center of this unified system are the domain-general networks—the Default Mode Network (DMN), Central Executive Network (CEN), and Salience Network (SN)—which coordinate introspection, attention, executive planning, and physiological regulation. The DMN anchors narrative selfhood and autobiographical continuity; the CEN supports adaptive decision-making and intentional action; and the SN aligns internal states with environmental demands. These networks interface with limbic structures—including the amygdala, hippocampus, and hypothalamus—which imbue experience with emotional and motivational salience. Meanwhile, brainstem systems, especially the reticular formation and arousal pathways, maintain the necessary level of consciousness and integrate core regulatory processes. The insula, acting as a bridge between interoception and self-awareness, exemplifies the embodied character of consciousness. It translates internal bodily states—such as heart rate, respiration, and visceral sensations—into felt experience, making embodiment foundational to subjectivity itself. This function is not a peripheral contribution; it is central to how one becomes aware of being a self in the world. Similarly, while not essential to consciousness per se, the cerebellum enhances the precision, timing, and fluency of mental and motor processes. Its role as an "enhancement device" speaks to the layered organization of the brain metafunction: specialized systems support and refine essential functions without fragmenting the person into discrete modules.

Notably, the unity expressed in this framework is not merely structural but existential. The human person emerges as a living, relational whole whose brain functions support more than survival or behavior—they sustain meaning-making, intentionality, empathy, and the possibility of self-transcendence. In therapeutic and philosophical contexts, this has far-reaching implications. It emphasizes that individuals heal, flourish, and transform not by focusing solely on the mental or physical realms but by restoring harmony across the full spectrum of human integration. Psychological suffering, for example, may reflect not just mental discontent but dysregulated interoceptive awareness, disrupted emotional salience, or misalignment among narrative identity, affective states, and physiological rhythms. From a philosophical and theological perspective, the brain metafunction framework supports models of the person as both materially embodied and irreducibly relational. It resonates with non-reductive physicalist and dual-aspect monist views, which recognize that the brain enables—but does not exhaust—the person's rational, emotional, and spiritual capacities. It also echoes the philosophical tradition that sees the human being not merely as a composite of parts but as an integrated whole—capable of knowing the truth, seeking the good, and entering into relationships. Ultimately, this framework does not seek to replace philosophical or theological accounts of personhood with neurobiological ones. Instead, it offers a biologically grounded model open to interdisciplinary dialogue. By showing how consciousness, memory, regulation, and behavior coalesce through network-level integration and embodied dynamics, the brain metafunction framework contributes to a richer understanding of what it means to be human: not a brain encased in a body, but a conscious, embodied, relational being—indivisible in structure and irreducible in essence.

REFERENCES

Andrews-Hanna, J. R., Smallwood, J., & Spreng, R. N. (2014). The default network and self-generated thought: Component processes, dynamic control, and clinical relevance. *Annals of the New York Academy of Sciences, 1316*(1), 29– 52. https://doi.org/10.1111/nyas.12360

Aron, A. R. (2007). The neural basis of inhibition in cognitive control. *The Neuroscientist, 13*(3), 214–228. https://doi.org/10.1177/1073858407299288

Aron, A. R., Robbins, T. W., & Poldrack, R. A. (2007). Inhibition and the right inferior frontal cortex. *Trends in Cognitive Sciences, 11*(4), 170–177. https://doi.org/10.1016/j.tics .2007.02.004

Avena-Koenigsberger, A., Misic, B. K., & Sporns, O. (2018). Communication dynamics in complex brain networks. *Nature Reviews Neuroscience, 19*(1), 17–33. https://doi.org/10.1038/nrn.2017.149

Bassett, D. S., & Sporns, O. (2017). Network neuroscience. *Nature Neuroscience, 20*(3), 353–364. https://doi.org/10.1038/nn.4502

Buckner, R. L. (2013). The brain's default network: Origins and implications for the study of psychosis. *Dialogues in Clinical Neuroscience, 15*(3), 351–358. https://doi.org/10.31887/DCNS.2013.15.3/rbuckner

Chein, J. M., Moore, A. B., & Conway, A. R. A. (2011). Domain-general mechanisms of complex working memory span. *NeuroImage, 54*(1), 550–559. https://doi.org/10.1016/j.neuroimage.2010.08.019

Cole, M. W., Pathak, S., & Schneider, W. (2010). Identifying the brain's most globally connected regions. *NeuroImage, 49*(4), 3132–3148. https://doi.org/10.1016/j.neuroimage. 2009.11.001

Cole, M. W., Yarkoni, T., Repovs, J., Anticevic, A., & Braver, T. S. (2013). Multi-task connectivity reveals flexible hubs for adaptive task control. *Nature Neuroscience, 16*(9), 1348–1355. https://doi.org/10.1038/nn.3470

Corbetta, M., & Shulman, G. L. (2002). Control of goal-directed and stimulus-driven attention in the brain. *Nature Reviews Neuroscience, 3*(3), 201–215. https://doi.org/10.1038/nrn755

Craig, A. D. (2002). How do you feel? Interoception: The sense of the physiological condition of the body. *Nature Reviews Neuroscience, 3*(8), 655–666. https://doi.org/10.1038/nrn894

Critchley, H. D., & Harrison, N. A. (2013). Visceral influen ces on brain and behavior. *Neuron, 77*(4), 624–638. https://doi.org/10.1016/j.neuron.2013.02.008

Damasio, A. (2010). *Self comes to mind: Constructing the conscious brain.* Pantheon Books.

Dehaene, S., & Changeux, J.-P. (2011). Experimental and theoretical approaches to conscious processing. *Neuron, 70*(2), 200–227. https://doi.org/10.1016/j.neuron.2011.03.018

Diekelmann, S., & Born, J. (2010). The memory function of sleep. *Nature Reviews Neuroscience, 11*(2), 114–126. https://doi.org/10.1038/nrn2762

Dietrich, M., Andreatta, R. D., Jiang, Y., & Stemple, J. C. (2020). Limbic and cortical control of phonation for speech

in response to a public speech preparation stressor. *Brain Imaging and Behavior, 14*(5), 1696–1713. https://doi.org/10.1007/s11682-019-00102-x

Dosenbach, N. U. F., Fair, D. A., Cohen, A. L., Schlaggar, B. L., & Petersen, S. E. (2007). Distinct brain networks for adaptive and stable task control in humans. *Proceedings of the National Academy of Sciences, 104*(26), 11073–11078. https://doi.org/10.1073/pnas.0704320104

Edlow, B. L., Takahashi, E., Wu, O., Benner, T., Dai, G., Bu, L., Grant, P. E., Greer, D. M., Greenberg, S. M., & Fischl, B. (2012). Neuroanatomic connectivity of the human ascending arousal system critical to consciousness and its disorders. *Journal of Neuropathology & Experimental Neurology, 71*(6), 531–546. https://doi.org/10.1097/NEN.0b013e3182588293

Fermin, A. S. R., Sasaoka, T., Maekawa, T., Ono, K., Chan, H.-L., & Yamawaki, S. (2024). Insula-cortico-subcortical networks predict interoceptive awareness and stress resilience. *Asian Journal of Psychiatry, 95*, 103991. https://doi.org/ 10.1016/j.ajp.2024.103991

Fogel, S., & Smith, C. (2011). The function of the sleep spindle: A physiological index of intelligence and a mechanism for sleep-dependent memory consolidation. *Neuroscience & Biobehavioral Reviews, 35*(5), 1154–1165. https://doi.org/ 10.1016/j.neubiorev.2010.12.003

Fox, M. D., Snyder, A. Z., Vincent, J. L., Raichle, M. E., & Shulman, G. L. (2005). The human brain is intrinsically organized into dynamic, anticorrelated functional networks. *Proceedings of the National Academy of Sciences, 102*(27), 9673–9678. https://doi.org/10.1073/pnas.0504136102

García, O. D., Machado, C., Román, J. M., Cabrera, A., Díaz-Comas, L., & Grave de Peralta, R. (1995). Heart rate variability in coma and brain death. In C. Machado (Ed.), *Brain death* (pp. 191–200). Elsevier.

Grill-Spector, K., & Malach, R. (2004). The human visual cortex. *Annual Review of Neuroscience, 27*(1), 649–677. https://doi.org/10.1146/annurev.neuro.27.070203.144220

Gu, X., & FitzGerald, T. H. B. (2014). Interoceptive inference: Homeostasis and decision-making. *Trends in Cognitive Sciences, 18*(6), 269–270.

Halassa, M. M., & Sherman, S. M. (2019). Thalamocortical circuit motifs: A general framework. *Neuron, 103*(5), 762–770. https://doi.org/10.1016/j.neuron.2019.06.005

Hansen, J. Y., Cauzzo, S., Singh, K., & et al. (2024). Integrating brainstem and cortical functional architectures. Nature Neuroscience, 27, 2500–2511. https://doi.org/ 10.1038/s41593-024-01787-0

Hickok, G., & Poeppel, D. (2007). The cortical organization of speech processing. *Nature Reviews Neuroscience, 8*(5), 393–402. https://doi.org/10.1038/nrn2113

Ito, M. (2008). Control of mental activities by internal models in the cerebellum. *Nature Reviews Neuroscience, 9*(4), 304–313. https://doi.org/10.1038/nrn2332

Jeannerod, M. (2001). Neural simulation of action: A unifying mechanism for motor cognition. *NeuroImage, 14*(1), S103–S109. https://doi.org/10.1006/nimg.2001.0832

Kleckner, I. R., Zhang, J., Touroutoglou, A., Chanes, L., Xia, C., Simmons, W. K., & Barrett, L. F. (2017). Evidence for a large-scale brain system supporting allostasis and interoception in humans. *Nature Human Behaviour, 1*(5), 69. https://www.nature.com/articles/s41562-017-0069

Kosslyn, S. M., Ganis, G., & Thompson, W. L. (2001a). Neural foundations of imagery. *Nature Reviews Neuroscience, 2*(9), 635–642. https://doi.org/10.1038/35090055

Kosslyn, S. M., Thompson, W. L., Kim, I. J., & Alpert, N. M. (2001b). The role of Area 17 in visual imagery: Convergent evidence from PET and rTMS. *Science, 284*(5411), 167–170. https://doi.org/10.1126/science.284.5411.167

Koziol, L. F., Budding, D., Andreasen, N., D'Arrigo, S., Bulgheroni, S., Imamizu, H., ... & Yamazaki, T. (2014). Consensus paper: The cerebellum's role in movement and cognition. *The Cerebellum, 13*(2), 151–177. https://doi.org/10. 1007/s12311-013-0511-x

Krishnan, A., & Gandour, J. T. (2009). The role of the auditory brainstem in processing linguistically-relevant pitch patterns. *Brain and Language, 110*(3), 135–148. https://doi.org /10.1016/j.bandl.2009.03.0 05

Machado, C., García, O. D., Gutiérrez, J., Portela, L., & García, M. C. (2005). Heart rate variability in comatose and brain-dead patients. *Clinical Neurophysiology, 116*(12), 2859–2860. https://doi.org/10.1016/jclinph .2005.08.016

Maquet, P. (2001). The role of sleep in learning and memory. *Science, 294*(5544), 1048–1052. https://doi.org/10.1126/science.1062856

McEwen, B. S., & Wingfield, J. C. (2003). The concept of allostasis in biology and biomedicine. *Hormones and Behavior, 43*(1), 2–15. https://doi.org/10.1016/S0018 -506X(02)00024-7

McGaugh, J. L. (2018). Emotional arousal and memory modulation. *Current Opinion in Behavioral Sciences, 19*, 55–60. https://doi.org/10.1016/j.cobeha.2017.10.003

Menon, V., & Uddin, L. Q. (2010). Saliency, switching, attention and control: A network model of insula function. *Brain Structure and Function, 214*(5–6), 655–667. https://doi.org/10.1007/s00429-010-0262-0

Menon, V. (2015). Salience network. In A. W. Toga (Ed.), *Brain mapping: An encyclopedic reference* (Vol. 2, pp. 597–611). Academic Press. https://doi.org/10.1016/B978-0-12-397025-1.00052-X

Miall, R. C., Weir, D. J., Wolpert, D. M., & Stein, J. F. (1993). Is the cerebellum a Smith predictor? *Journal of Motor Behavior, 25*(3), 203–216. https://doi.org/10.1080/00222895.1993.9942050

Michel, C. M., & Koenig, T. (2018). EEG microstates as a tool for studying the temporal dynamics of whole-brain neuronal networks: A review. *NeuroImage, 180*, 577–593. https://doi.org/10.1016/j.neuroimage.2017.11.062

Møller, A. R. (2012). *Hearing: Anatomy, physiology, and disorders of the auditory system* (2nd ed.). Plural Publishing.

Parvizi, J., & Damasio, A. (2001). Consciousness and the brainstem. *Cognition, 79*(1–2), 135–160. https://doi.org/10.1016/S0010-0277(00)00127-X

Petersen, S. E., & Posner, M. I. (2012). The attention system of the human brain: 20 years after. *Annual Review of Neuroscience, 35*, 73–89. https://doi.org/10.1146/annurev-neuro-062111-150525

Phelps, E. A., & LeDoux, J. E. (2005). Contributions of the amygdala to emotion processing: From animal models to human behavior. *Neuron, 48*(2), 175–187. https://doi.org/10.1016/j.neuron.2005.09.025

Raichle, M. E. (2015). The brain's default mode network. *Annual Review of Neuroscience, 38*, 433–447. https://doi.org/10.1146/annurev-neuro-071013-014030

Rasch, B., & Born, J. (2013). About sleep's role in memory. *Physiological Reviews, 93*(2), 681–766. https://doi.org/10.1152/physrev.00032.2012

Rikhye, R. V., Wimmer, R. D., & Halassa, M. M. (2018). Toward an integrative theory of thalamic function. *Annual Review of Neuroscience, 41*, 163–183. https://doi.org/10.1146/annurev-neuro-080317-062144

Schabus, M., et al. (2004). Sleep spindles and their significance for declarative memory consolidation. *Sleep, 27*(8), 1479–1485. https://doi.10.1093/sleep/27.7.1479

Schmahmann, J. D. (2019). The cerebellum and cognition. *Neuroscience Letters, 688*, 62–75. https://doi.org/10.1016/j.neulet.2018.07.005

Sherman, S. M. (2016). Thalamus plays a central role in ongoing cortical functioning. *Nature Neuroscience, 19*(4), 533–541. https://doi.org/10.1038/nn.4269

Seeley, W. W., Menon, V., Schatzberg, A. F., Keller, J. H., Glover, G. H., Kenna, H., Reiss, A. L., & Greicius, M. D. (2007). Dissociable intrinsic connectivity networks for salience processing and executive control. *Journal of Neuroscience, 27*(9), 2349–2356. https://doi.org/10.1523/JNEUROSCI.5587-06.2007

Seth, A. K. (2014). A predictive processing theory of sensorimotor contingencies: Explaining the puzzle of perceptual presence and its absence in synesthesia. *Cognitive Neuroscience, 5*(2), 97–118. https://doi.org/10.1080/17588928.2013.877880

Seth, A. K., & Friston, K. J. (2016). Active interoceptive inference and the emotional brain. *Philosophical Transactions of the Royal Society B: Biological Sciences, 371*(1708), 20160007. https://doi.org/10.1098/rstb.2016.0007

Sporns, O. (2013). *Networks of the brain*. MIT Press.

Squire, L. R., & Dede, A. J. O. (2015). Conscious and unconscious memory systems. *Cold Spring Harbor Perspectives in Biology, 7*(3), a021667. https://doi.org/10.1101/cshperspect.a021667

Sterling, P., & Eyer, J. (1988). Allostasis: A new paradigm to explain arousal pathology. In S. Fisher & J. Reason (Eds.), *Handbook of life stress, cognition and health* (pp. 629–649). John Wiley & Sons.

Stickgold, R., & Walker, M. P. (2013). Sleep-dependent memory triage: Evolving generalization through selective processing. *Nature Neuroscience, 16*(2), 139–145. https://doi.org/10.1038/nn.3303

Toosy, A. T., Ciccarelli, O., Parker, G. J. M., Wheeler-Kingshott, C. A. M., Miller, D. H., Thompson, A. J., & Johansen-Berg, H. (2004). Characterizing function–structure relationships in the human visual system with functional MRI and diffusion tensor imaging. *NeuroImage, 21*(4), 1452–1463. https://doi.org/10.1016/j.neuroimage.2003.11.022

van den Heuvel, M. P., & Sporns, O. (2011). Rich-club organization of the human connectome. *Journal of Neuroscience, 31*(44), 15775–15786. https://doi.org/10.1523/JNEUROSCI.3539-11.2011

Vossel, S., Geng, J. J., & Fink, G. R. (2014). Dorsal and ventral attention systems: Distinct neural circuits but collaborative roles in attentional control. *The Neuroscientist, 20*(2), 150–159. https://doi.org/10.1177/1073858413494269

Walker, M. P., & Stickgold, R. (2010). Overnight alchemy: Sleep-dependent memory evolution. *Nature Reviews Neuroscience, 11*(3), 218–224. https://doi.org/10.1038/nrn2762

Wurtz, R. H., Joiner, W. M., & Berman, R. A. (2011). Neuronal mechanisms for visual stability: Progress and problems. *Philosophical Transactions of the Royal Society B: Biological Sciences, 366*(1564), 492–503. https://doi.org/10. 1098/rstb.2010.0186

Yeo, B. T. T., Krienen, F. M., Sepulcre, J., Sabuncu, M. R., Lashkari, D., Hollinshead, M., ... & Buckner, R. L. (2011). The organization of the human cerebral cortex estimated by intrinsic functional connectivity. *Journal of Neurophysiology, 106*(3), 1125–1165. https://doi.org/10.1152/jn.00338.2011

Zanesco, A. P. (2024). Normative temporal dynamics of resting EEG microstates. *Brain Topography, 37*(2), 243–264. https://doi.org/10.1007/s10548-023-01004-4

4

Perception as a Brain Metafunction Part I: Structure and Theory

Reframing Perception: From Passive Reception to Active Construction

This chapter reconceives perception not as a passive reception of sensory inputs, but as a dynamic, integrative process—one that mediates between body, brain, environment, and the lived self. Beginning with a historical overview of perception and three widely studied modalities—visual, auditory, and nociceptive—it introduces a brain metafunction-based theory grounded in systems-level integration and embodied cognition. Consider, for instance, a person observing a dog in the environment. The dog appears external to the observer, yet the perceptual experience arises through the interplay of incoming sensory signals (afferent loops), centralized integrative processing, and outward-directed, constructive projections (efferent loops) that re-present the scene. Closing one's eyes disrupts afferent input, dissolving the perceptual field; reopening them resumes the afferent-efferent exchange, reconstituting the experience of the external world.

This reciprocal dynamic illustrates perception as a function actively constructed by the brain and embedded in the person's embodied engagement with the environment. *Affordances*, in this context, refer to the relational properties between an agent and its surroundings that give rise to emergent opportunities for patterned, goal-directed action—shaped by perception, embodiment, and situational context. As we shall argue, such constructive integration is not peripheral to perception—it is perception. From this perspective, conventional physicalist models, which treat perception as a bottom-up or modular process, overlook the core metafunctional dynamics that constitute conscious experience. This chapter thus advances a non-reductive, interdisciplinary framework in which perception serves as a bridge between neural systems, environmental affordances, and personal subjectivity.

Historical and Philosophical Foundations

The nature of perception has long captivated philosophers and scientists, tracing a lineage from classical antiquity to contemporary neuroscience. In ancient Greece, three principal theories of vision emerged: extramission theory, supported by Plato and Euclid, posited that rays emanated from the eyes to apprehend objects; intromission theory, endorsed by Aristotle and Ptolemy, held that light entered the eye from external sources; a hybrid model, which sought to merge extramission and intromission theories, ultimately proved conceptually unstable and incompatible with emerging empirical models of optics and anatomy, making it difficult to sustain within developing naturalistic frameworks (Sabra, 1989). These theories were not merely speculative but informed mathematical models, optics, and proto-neuroscience. Democritus's remark: "By convention there are sweet and bitter, hot and cold, by convention there is color; but in truth, there are atoms and the void,"presaged the tension between subjective experience and objective reality. This remark, along with other fragments preserved through later commentators, illustrates the early Greek struggle to reconcile perception with ontology (see Freeman 1949, for an authoritative overview of early Greek thought). Later, Galileo codified this divide with his distinction between primary (measurable) and secondary (experiential) qualities, effectively reclassifying sensation as a function of the mind rather than of the world (Kuhn, 1962).

Descartes deepened this separation in his dualistic model, describing the body as an extended substance and the mind as a thinking one. For Descartes, perception was indirect and representational—the intellect apprehended reality through the mediation of sensory impressions (Cottingham, 1986). Before the early modern rupture, however, medieval thinkers like Thomas Aquinas had emphasized a more integrated view of perception: for Aquinas, the senses were reliable though limited conduits to knowledge, serving as gateways through which the intellect abstracts universal truths (Aquinas, 1947). His Aristotelian-inspired model preserved a continuity between sensory experience and rational understanding—an outlook that would later be challenged by the dualistic models of Descartes. Isaac Newton echoed this framework when he claimed, "The rays, to speak properly, are not coloured," suggesting that qualities such as color exist only in the perceiver (Newton, 1704). In response, alternative paradigms emerged that challenged the reductive orientation of classical physics. Leibniz proposed that perception is relational, shaped by an interlocking system of monads, each reflecting the universe from its own perspective (Leibniz, 1714). Berkeley, through his idealism, went further arguing that material properties exist only insofar as they are perceived (Berkeley, 1710). Hume, although skeptical of idealism, questioned the certainty of objectivity and causality, anchoring empirical knowledge in transient sensory impressions (Hume, 1739).

This philosophical progression culminated in Immanuel Kant's critical philosophy. Kant synthesized rationalist and empiricist insights by asserting that the mind actively structures sensory experience. The external world, or noumenon, remains inaccessible; what we know are phenomena, organized by the mind's a priori categories (Kant, 1781). In this transcendental idealism, perception is not passive registration but a constructive act shaped by the faculties of understanding. This evolution in philosophical thought profoundly shaped modern science. During the Renaissance, curiosity about the mechanics of light and the physiology of the eye led to major advances in optics, culminating in the invention of the microscope and telescope. These instruments shifted the boundary between subjective observation and objective measurement (Friston, 2010).

Contemporary neuroscience echoes and extends these insights. Theories such as predictive coding and Bayesian inference reveal that the brain is not a passive receiver of inputs, but a hypothesis-testing system—anticipating, filtering, and adjusting perception through internal models (Friston, 2010). The brain's dynamic feedback loops support the idea that perception is interpretive, anticipatory, and deeply integrative, rather than bottom-up and modular. Thus, the historical trajectory from classical theories to modern frameworks reveals an increasingly constructive view of perception—one in which the mind is not an observer of reality but a participant in its formation. This philosophical and empirical lineage underpins the Brain Metafunction Framework, which builds upon these traditions to offer a triadic, systems-level account of perception that unifies neural processing, embodiment, and world-engagement.

The Brain Metafunction Framework: A Triadic Model of Perception

The Brain Metafunction Framework synthesizes classical philosophical insights with contemporary neuroscience, offering a systems-level model in which perception is understood as a dynamic integration of sensory input, cognitive modulation, and outward-directed engagement. Within this framework, perception is not merely a passive event nor a linear processing stream, but a bidirectional function: shaped by afferent signals coming into the brain and efferent processes projecting constructed experience outward into the world. This model challenges modular and bottom-up views of perception by recognizing the contribution of large-scale, domain-general neural networks. These include but are not limited to:

- The *Default Mode Network* (DMN): Associated with self-referential thought, autobiographical memory, and contextual interpretation of sensory input. It enables perception to be situated within the continuity of personal experience and intentional meaning (Raichle, 2015).

- The *Central Executive Network* (CEN): Facilitates attention and executive control. It plays a role in actively selecting and prioritizing sensory information in accordance with goals and relevance.

- The *Salience Network* (SN): Integrates sensory, emotional, and homeostatic information to detect and prioritize stimuli with behavioral or survival significance (Seeley et al., 2007; Menon & Uddin, 2010).

- The *Medial Temporal Lobe Network* (MTLN): Involved in the encoding, consolidation, and retrieval of declarative memory. This network, including structures such as the hippocampus and parahippocampal cortex, dynamically interacts with perceptual systems to embed incoming information within temporal and spatial memory contexts (Ranganath & Ritchey, 2012).

Together, these networks modulate perception not by passively registering stimuli, but by embedding incoming information within broader interpretive and motivational frameworks. For example, seeing a face on a crowded street does not merely activate visual cortex—it engages emotion, memory, attention, and social cognition. Even before the image is fully registered, the brain predicts its meaning and behavioral relevance. Rather than passively awaiting input, the brain is in constant dialogue with the world, using prior knowledge to shape and revise perception in real time.

In this context, perception emerges as a metafunction—a higher-order integration of multiple brain systems coordinating across temporal, spatial, and functional domains. It unifies not just different sensory modalities, but also memory, emotion, attention, motor planning, and self-representation. Within this integrated loop, perception becomes the means by which the organism constructs, enacts, and refines its experience of reality. Crucially, the efferent component of this loop—where perceptual content is actively projected outward—is not metaphorical. It reflects the organism's effort to engage the world not merely as stimulus-response but as a meaningful, intentional field. As such, the Brain Metafunction Framework reframes perception as fundamentally embodied, relational, and constructive. This reconception forms the theoretical core upon which the following sections of the chapter build, including the examination of exceptional perceptual phenomena and their clinical and philosophical implications.

Predictive Coding and Bayesian Inference

A key advancement within this integrative model is predictive coding, which reframes the brain as a proactive organ that continuously generates hypotheses about incoming sensory input. Rather than awaiting external stimuli, the brain anticipates what it expects to perceive and compares this prediction to afferent input. Discrepancies—called prediction errors—prompt updates to internal models. Complementing this is the Bayesian model of perception, which frames the brain as an inferential system that computes the most likely cause of a given sensory input based on prior experience. In this view, perception is probabilistic—not a mirror of the world, but a model of what is most probable given both current data and past contexts (Friston, 2010). These frameworks show that perception is not the passive registration of stimuli but a negotiated construction between bottom-up inputs and top-down expectations. For example, ambiguous images or illusions reveal that what we "see" is not always what is there—but what the brain deems most coherent given the context.

Affordances and the Efferent Loop

A final, essential aspect of metafunctional perception is the concept of affordances—the relational possibilities for action that arise between an organism and its environment. From a metafunctional perspective, affordances are not merely detected by the senses; they are projected and contextualized by the brain's integrative and anticipatory systems. This process is especially evident in the efferent loop of perception, through which constructed perceptual content is localized and rendered as external to the observer. When one sees a tree, for example, neural activation is not experienced as internal signals—it is experienced as the tree "out there." This projection is not metaphorical. It is a functional placement, shaped by expectation, intentionality, emotional salience, and bodily awareness. This efferent component is what dissolves the long-standing philosophical paradox described by William James—that "what is evidently one reality should be in two places at once: in outer space and in the mind" (James, 1904). By reconceiving perception as a transphysical and bidirectional act of world-participation, the Brain Metafunction Framework restores coherence to this dilemma. It affirms that perception is not a passive imprint of the world upon the brain, but a structured, embodied act of ontological engagement.

Exceptional Cases in Perception Modalities and Case Studies: Illuminating Metafunctional Perception

To further illustrate the Brain Metafunction Framework and the constructive nature of perception, we turn to a set of exceptional perceptual phenomena. By examining both typical and non-ordinary forms of perception, we reveal how perception is not merely a byproduct of sensory input, but an integrated process—linking cognition, emotion, bodily regulation, and environmental engagement.

1. Synesthesia: Cross-Modal Integration and Perceptual Mapping

Synesthesia is a condition in which stimulation in one sensory modality automatically triggers a secondary perception in another—for example, "seeing" colors when hearing music or "tasting" shapes. These experiences are consistent, involuntary, and subjectively vivid, challenging the dichotomy between sensory channels (Ramachandran & Hubbard, 2001). From a metafunctional perspective, synesthesia reflects the brain's ability to construct experience across modalities through integrative mappings that are both functionally coherent and neurologically stable. Rather than anomalies, these perceptual pairings suggest an underlying cross-network architecture—a neural metafunction that resists reductive modularity and supports an expanded view of perceptual synthesis.

2. Phantom Limb Sensations: Perception Beyond Input

Phantom limb sensations occur when individuals experience pain, touch, temperature, or position awareness in limbs that have been amputated or were never physically present (Diers, et al., 2023). These sensations highlight the brain's ability to generate perceptual experience even in the total absence of afferent input (Ramachandran & Hirstein 1998; Flor 2002). Within the Brain Metafunction Framework, this phenomenon exemplifies the role of efferent projection in sustaining body image and spatial awareness. The brain constructs a persistent representation of the body, and when afferent signals are disrupted, it continues to "fill in" the expected perceptual content. Mirror therapy, which uses visual feedback to recalibrate this internal map, demonstrates the power of reintroducing sensory congruence to modulate maladaptive projections—affirming the plastic, integrative nature of perception (Ramachandran & Altschuler, 2009; Ortiz-Catalan, 2018a, b; Cui, et al. 2022).

3. Blindsight: Perception Without Awareness

In cases of blindsight, individuals with damage to the primary visual cortex (V1) report blindness in specific regions of their visual field yet retain the ability to detect movement, shapes, or emotional expressions—without any conscious visual experience (Weiskrantz, 1986). This paradox suggests that multiple visual pathways, some bypassing V1, contribute to visual perception. From a metafunctional viewpoint, blindsight shows that while the afferent loop may remain partially intact, metafunctional integration and efferent projection—crucial for consciousness—are disrupted. The dissociation between input and experience illustrates how perception can occur below the threshold of awareness, challenging models that equate sensation with conscious seeing.

4. Near-Death Experiences (NDEs): Transphysical Perception

NDEs occur in contexts such as cardiac arrest, where afferent input and cortical activity are severely reduced or absent. Yet, individuals often report vivid, coherent experiences: a sense of detachment from the body, encounters with light, panoramic memory, or heightened perception. Some cases occur with flat EEG activity or clinical markers of death (Van Lommel et al., 2001; Greyson, 2003; Parnia & Fenwick, 2002; Parnia et al., 2022). These reports challenge conventional models of perception and consciousness, suggesting the possibility of transphysical perceptual processes. NDEs align with the Brain Metafunction Framework's premise that perception is not fully contingent on sensory input or localized brain function. Instead, they point toward a more expansive, participatory structure of consciousness—one that may persist under extreme or non-ordinary conditions. These phenomena defy explanation by conventional models of perception tied to afferent activity or cortical integration. From the standpoint of the Brain Metafunction Framework, NDEs point toward a transphysical dimension of perceptual experience—suggesting that consciousness may persist and operate through mechanisms beyond localized brain activity. These experiences reinforce the model's emphasis on efferent projection, intentional awareness, and the possibility of perceptual continuity even in states of neurological silence.

These case studies challenge the assumption that perception arises solely from sensory inputs. Instead, they highlight the role of the efferent loop and central metafunctional integration in shaping the content, form, and intentionality of experience. Whether through the cross-modal fusion of synesthesia, the embodied projection of phantom limbs, the unconscious detection in blindsight, or the vivid coherence of NDEs, these examples collectively reveal that perception is an active, bidirectional, and meaning-constituting function—not a passive mirror of the environment. Taken together, these exceptional cases illustrate the dynamic range of perception as a metafunctional process—one that is not reducible to afferent input or cortical computation alone. Whether through cross-modal fusion, embodied persistence, unconscious detection, or vivid experiences beyond afferent support, these phenomena point to the brain's capacity for integration, projection, and meaning-making. What emerges is a unified model in which perception is structured not by isolated inputs, but by distributed, predictive, and participatory processes. To further illuminate this framework, we now turn to vision—a sensory modality whose complexity and centrality make it a revealing case study for the full depth and nuance of the Brain Metafunction Framework.

Visual Perception as a Multisystem Metafunction

To synthesize the implications of these exceptional cases and situate them within canonical visual neuroscience, we now return to a closer examination of the visual system as a model of metafunctional integration across neural, bodily, and cognitive domains. Visual perception is among the most studied and complex sensory modalities, offering a model system for understanding how the brain transforms physical stimuli into coherent, meaningful experiences. It exemplifies the triadic structure of the Brain Metafunction Framework—integrating afferent input, central metafunctional processing, and efferent projection into a unified perceptual field.

Physiologically, visual processing begins in the retina, where rods and cones transduce light into neural signals. These signals travel along the optic nerve, partially cross at the optic chiasm, and are relayed through the lateral geniculate nucleus (LGN) of the thalamus to the primary visual cortex (V1). From there, information diverges into two major processing streams. The dorsal stream, projecting to the parietal lobe, supports spatial orientation and motion—commonly referred to as the "where" or "how" pathway. The ventral stream, projecting to the temporal lobe, enables object recognition and visual identification—the "what" pathway (Goodale & Milner, 1992). In addition to these cortical systems, subcortical structures play essential roles. The *superior colliculus* facilitates reflexive eye movements and spatial orientation, while the *retinohypothalamic tract* contributes to circadian rhythm regulation and light-responsive physiological states. These distributed pathways demonstrate that visual input serves not only conscious perception but also regulatory and adaptive functions (Grill-Spector & Malach 2004; Møller 2012).

Recent Discoveries and Integration with the Brain Metafunction Framework

Advancements in vision science continue to support and extend this integrative model. The rod pathway specializes in low-light vision, transmitting signals to thalamic and midbrain structures that support navigation and behavioral adaptation in dim environments (Yeo et al. 2011). Parallel to this, the third photoreceptor system—comprised of intrinsically photosensitive retinal ganglion cells—responds primarily to blue light, influencing circadian rhythms and physiological regulation (Foster et al. 2002). These pathways function largely outside of conscious awareness, yet shape daily cycles, hormonal regulation, and behavioral orientation. Visual perception thus reflects the brain's capacity to weave together sensory, cognitive, and regulatory functions into a cohesive experiential whole. The superior colliculus supports spatial reflexivity, while higher cortical areas interpret and consciously re-present the visual world. These processes align with the efferent loop component of the metafunctional framework, where perception is not merely received but projected outward, experienced as coherent and externally situated. This dynamic integration ensures that visual perception is adaptive, anticipatory, and context sensitive. Rather than passively transmitting light signals, the visual system aligns perception with physiological states, behavioral goals, and environmental contingencies—hallmarks of a metafunctional operation.

To more fully understand how visual perception bridges the neural, cognitive, and existential domains, we must now explore its constructive nature—not only as a physiological process, but as an active, top-down interpretation shaped by attention, memory, and predictive models. The next section will examine the constructive and top-down character of vision, offering insight into how the brain does not simply register visual stimuli, but co-creates the world it perceives

.

A Metafunctional Theory of Perception

Afference and Efference

While afference and efference have traditionally been associated with distinct domains—sensory input and motor output, respectively—emerging models suggest that this dichotomy is incomplete. Recent research in perception and predictive processing challenges the view that perceptual experience is purely afferent in nature. Within the Brain Metafunction Framework, efference is reconceived not merely as motor output, but as a subtle, participatory function of perceptual construction. Rather than serving only to execute responses, efferent activity becomes a component through which the brain anticipates, projects, and situates experience—infusing perception with intentionality, expectation, and meaning. From this perspective, the efferent loop contributes not only to action, but to the very structuring of how the world is perceived and inhabited.

The Triadic Model: The Vision Components

The intromission theory of vision posits that visual perception results from external stimuli—light rays—entering the eyes. While accurate, this theory remains incomplete. Modern neuroscience has established that visual input undergoes multilevel processing in the retina and brain, integrating sensory data with predictive models of anticipation, relevance, and novelty (Hickok & Poeppel, 2007; Tumminia, 2020). However, vision itself involves more than sensory input and processing—it encompasses an "efferent loop," an outward projection, creating the experience of seeing objects in the external world. William James (1904) famously described a paradox where "what is evidently one reality should be in two places at once, both in outer space and in a person's mind." A metafunctional theory of vision addresses this paradox by proposing three components:

- ***The Afferent Loop:*** Sensory information is first encoded by peripheral receptors and transmitted via specialized neural pathways toward the brain. Even before reaching primary sensory cortices (such as V1 in vision), this input is shaped by multilevel interactions—combining bottom-up signals with early top-down modulation, including attentional gating, predictive coding, and subcortical filtering (Dehaene & Changeux 2011).

- ***Metafunctional Processing:*** The brain synthesizes afferent input with cognitive, motor, emotional, and bodily regulatory states to generate coherent, adaptive perception. This integration is mediated by large-scale, domain-general brain networks—most notably the Default Mode Network (DMN), Salience Network (SN), and Central Executive Network (CEN)—which coordinate attention, autobiographical memory, interoception, and goal-directed behavior (Menon & Uddin 2010; Seeley et al. 2007). These networks ensure that perception is not modular or isolated, but part of a dynamic system that aligns internal states with environmental demands, bodily needs, and personal relevance.

- ***Efferent Loop:*** The efferent loop represents the outward projection of integrated perceptual content—experienced as external, unified, and situated in space. Emerging from prior metafunctional processing, this projection is shaped by expectation, intentionality, and embodied awareness. It reflects the influence of large-scale networks such as the Default Mode Network (DMN), Salience Network (SN), and Central Executive Network (CEN), which together coordinate attention, memory, interoception, and action readiness (Menon & Uddin 2010; Seeley et al. 2007). Through this projective loop, perception becomes participatory rather than receptive—anchoring the self within a meaningful, action-relevant environment.

This comprehensive framework resolves James' paradox, illustrating how perception arises from the interplay between neural processes and conscious experience. Recognizing an efferent loop challenges reductionist models, advocating a paradigm shift in understanding sensory perception and consciousness.

From Optics to Interpretation: Toward a Constructive Theory of Vision

While early theories of vision debated whether perception arose from rays emitted by the eyes or from light entering them, the rise of optical science and anatomical research during the Renaissance decisively favored the intromission model. Modern neuroscience has largely validated this view, detailing the biological mechanisms by which light is transduced and transmitted through the visual system. However, this physiological account leaves unresolved the deeper question of how sensory signals become the rich, coherent experiences we call vision. Bridging this explanatory gap requires a shift from seeing vision as passive reception to understanding it as an active, constructive process. The following section explores this paradigm shift—one that draws upon both empirical neuroscience and philosophical traditions to frame perception as a dynamic interplay between the brain, the body, and the world.

Modern neuroscience builds upon this legacy, affirming key aspects of the intromission view: light is transduced by photoreceptors in the retina, travels via the optic nerve, and is processed through successive neural layers, including the thalamus and primary visual cortex (V1) (Kosslyn et al., 2001). On the surface, this appears to resolve ancient debates—vision, it seems, is a function of light entering the eyes and being computed by the brain. Yet this resolution is incomplete. While intromission accounts for how light reaches the brain, it says little about how we experience seeing—how neural activity transforms into the coherent, spatially external, and meaningful world we encounter. This gap is not merely technical—it marks a profound theoretical rupture that invites a more integrative account of perception. Contemporary neuroscience fills this gap by revisiting themes once found in philosophical idealism and transcendental thought. Thinkers like Kant argued that perception is not a passive reflection of the world but an active construction, mediated by the mind's innate structures. Kant's distinction between the unknowable "noumenon" (thing-in-itself) and the structured "phenomenon" (what appears to us) aligns with modern frameworks like predictive coding, which propose that the brain does not simply register sensory input, but constructs experience through anticipatory models and hypothesis testing (Friston, 2010).

This historical and conceptual trajectory converges in the Brain Metafunction Framework, which proposes that perception is not a bottom-up reception of stimuli, but a dynamic, bidirectional process. The brain does not merely represent the world—it participates in its articulation through attention, memory, emotion, and embodied interaction. This shift—from passive recording to active construction—ushers us into the next phase of vision theory, where top-down modulation, Bayesian inference, and predictive dynamics reveal vision as an interpretive act deeply embedded in the mind's dialogical relationship with the world.

The Top-Down Nature of Vision and Neuroscience

Modern neuroscience reveals that vision is not a unidirectional process driven solely by incoming sensory data. Instead, it involves complex top-down modulation—where perception is shaped not only by light entering the eyes but also by memory, attention, expectation, and emotional salience. This reframing marks a decisive departure from classical models and offers a fuller account of how visual experience emerges. While the intromission theory correctly describes the entry of light into the eyes and the physiological cascade that follows, it cannot explain how these signals are transformed into coherent, meaningful visual experience. Contemporary research demonstrates that the brain is not a passive recipient of visual stimuli but an active predictor, continuously generating internal models of what it expects to see. This insight is central to predictive coding, a theory that conceives the brain as a hierarchical inference machine. According to this model, the brain constantly formulates hypotheses about incoming sensory input and revises them as needed when prediction errors occur—discrepancies between expected and actual input (Friston, 2010). In this framework, perception arises from minimizing surprise by reconciling top-down expectations with bottom-up data.

Closely aligned with this is the Bayesian model of perception, which treats visual experience as probabilistic inference. Here, perception is understood as the brain's best guess—based on prior knowledge—about the causes of sensory input. When viewing an ambiguous image, for instance, the brain selects the most statistically likely interpretation and refines it through real-time feedback. Classic perceptual illusions—such as the "face-vase" illusion—illustrate how internal models shape visual interpretation. Whether one sees a vase or two faces depends on how attention and contextual cues are weighted by the brain (Kosslyn et al., 2001). Far from revealing flaws in the visual system, these illusions demonstrate the constructive nature of perception. Neuroscientist Anil Seth (2014, 2021a, b) pushes this further, describing perception as a "controlled hallucination." On this view, the brain is not simply reacting to the world, it is continuously creating it. This creation is constrained by reality, but nonetheless emerges from within the system, shaped by physiological rhythms, emotional context, and learned expectations. These insights dovetail with the Brain Metafunction Framework, which conceives perception as a triadic process: Afferent input (sensory encoding), metafunctional integration (cognitive, emotional, and bodily coordination), and efferent projection (the experiential "placing" of perception in external space).

The top-down nature of vision underscores that perception is not the registration of the world "as it is," but the organism's best construction of a contextually meaningful world. This redefinition of vision—as a generative, hypothesis-driven act—lays the groundwork for a broader revision of classical visual theory and leads us toward a more unified model of perceptual experience. The next section explores the theoretical and philosophical consequences of this shift in depth.

Implications for Vision Theory

The emergence of top-down models of perception profoundly challenges the explanatory sufficiency of the classical intromission theory, which holds that vision arises from the passive intake of light into the eyes, followed by its relay through neural circuits. While this model captures the basic mechanics of light detection and signal transmission, it fails to explain how raw sensory input becomes the rich, contextually informed experience of "seeing." Modern neuroscience reveals that vision is not a one-way transmission of information but a bidirectional process—a dynamic exchange in which prior expectations, emotional states, attentional goals, and memory actively shape perception. Predictive coding (Friston, 2010) and Bayesian inference frameworks underscore the idea that the brain is not merely decoding the external world, but constructing perceptual reality through continuous, hypothesis-driven modeling (Seth, 2014).

In this view, the brain does not wait for sensory data to define experience. Instead, it generates perceptual predictions and then updates them in response to mismatches between expectation and input—minimizing prediction error to sustain a coherent model of the world. Vision, therefore, becomes not a mirror of reality, but an inference about the most probable state of the environment, filtered through the lens of internal context. This understanding significantly expands the classical model of intromission. While the original theory accounts for how light reaches the retina and activates neural structures, it says little about why we perceive what we do, or how perception can persist—or misfire—in the absence of sensory input (as in dreams, hallucinations, or phantom limbs). It does not explain why we see a coherent world even when visual data is partial, ambiguous, or illusory.

By integrating top-down processes into a broader systems-level account, the Brain Metafunction Framework offers a more complete model of visual perception. It posits that vision emerges from the integration of three components:

- *Afferent input*: raw sensory signals from the environment under early top-down modulation.

- *Metafunctional integration*: the coordination of cognitive, emotional, motor, and bodily systems to interpret and contextualize perception.

- *Efferent projection*: the active "placement" of perceptual content into experienced external space.

This reconfiguration has wide-reaching consequences. It challenges the long-standing assumption that sensory data alone determines perception, calling instead for a more holistic, interactive, and embodied model. It repositions the brain not as a passive receiver, but as an active participant in shaping and situating the self in relation to the world. Thus, the top-down theory of vision does not merely revise intromission—it completes it. It restores perception to its rightful place: not just as a physiological function, but as a metafunctional, meaning-driven act of world-participation. And yet, one final question remains: If perception is constructed, contextual, and participatory, how does it appear as spatially "out there" rather than merely internal? This question leads us to a deeper paradox—one with roots in the philosophy of mind and ramifications across neuroscience, phenomenology, and metaphysics.

Resolving the Paradox of Perception: Perceptual Projection and the Efferent Loop

William James (1904) famously described a paradox at the heart of perception:

> "That what is evidently one reality should be in two places at once—both in outer space and in a person's mind."

How is it that we experience a tree as standing "over there," when the perceptual process is entirely internal to the brain? Why does perceptual content appear spatially external—situated in the world—despite being generated within neural substrates? Traditional materialist models have attempted to resolve this paradox by localizing perception solely in cortical activity. Yet this explanation leaves untouched the subjective experience of the world as out there—as something shared, continuous, and spatially coherent. The Brain Metafunction Framework offers a resolution that honors both the empirical findings of neuroscience and the phenomenology of experience. Central to this view is the efferent loop—not a motor output, but as a participatory projection of perceptual content into the lived world. Perception, in this light, is not confined to the encoding of external stimuli (afference), nor to their centralized integration. It culminates in an outward orientation, a structured act of placing perceptual experience into the space of the world. This process is not metaphorical. When we see a tree, we do not experience photonic stimulation or cortical excitation—we experience the tree. The brain does not simply register the world; it actively situates the perceiver within it, aligning internal states (memory, attention, emotion) with external affordances (objects, agents, landscapes) in a dynamic, reciprocal exchange.

The efferent loop thus dissolves James's paradox by reframing perception as a bidirectional act of world-participation, not merely an internal simulation. Through the coordinated interaction of afferent input, central metafunctional integration, and outward perceptual projection, the brain gives rise to the lived world, a reality that is simultaneously internal and external, mental and spatial, subjective and shared. Moreover, this framework accounts for perceptual consistency across modalities and conditions, including cases where sensory input is minimal or absent. In phantom limb experiences, hallucinations, or dreams, the efferent component remains active—constructing perceptual content without full afferent input. In near-death experiences, perception continues even under drastically reduced cortical activity, suggesting the persistence of this outward-facing, integrative function. These cases do not undermine perception; they reveal its deeper architecture. To resolve the paradox of perception, then, we must expand our understanding beyond stimulus and response. Perception is not simply what happens in the brain, but what emerges through the relationship between brain, mind, body, and world—an interplay that cannot be fully localized, mechanized, or reduced.

Alternative Contemporary Models and Metafunctional Convergence

While modern neuroscience has largely confirmed the intromission view of vision, showing that light enters the eye and is processed through successive neural pathways, some theorists have challenged its sufficiency in explaining how we actually experience the world as external, coherent, and meaningful. Velmans' reflexive model of perception addresses this puzzle directly. He proposes that sensory input is not only processed internally but also reflexively projected outward—a psychological placement that creates the experience of perceiving the external world. This projection, he emphasizes, is not metaphorical. It is a functional and spatial localization of experience, rooted in perceptual systems that embed neural activity in the lived world. Rupert Sheldrake (2005) traces the historical trajectory of vision theories, from ancient extramission models to the dominance of intromission after the 12th century. Yet, as Max Velmans (1990, 2005) and Sheldrake argue, even contemporary neuroscience struggles to explain why we perceive objects "out there," rather than simply registering neural activity "in here."

This aligns closely with the Brain Metafunction Framework, which posits that perception is not reducible to afferent registration but includes metafunctional integration and efferent projection. Where Velmans highlights outward projection, the Brain Metafunction model builds on it to show how such projection is embedded in cognitive, emotional, interoceptive, and action-related networks. In this light, perception becomes not a passive mirror of the world but an active, relational process, a dynamic structuring of experience that unfolds between brain, mind, body, and world. These alternative theories not only challenge the dominance of physicalist models but also underscore the value of integrative, transphysical frameworks for understanding consciousness and perception.

Perception as a Metafunctional and Transphysical Process

The Brain Metafunction Framework provides a theoretical foundation for understanding perceptual phenomena that challenge conventional physicalist explanations. Within this model, perception is inherently metafunctional, involving a dynamic interplay of afferent input, centralized processing, and efferent projection. Crucially, this tripartite structure highlights a transphysical dimension of perception, in which conscious experience extends beyond localized brain activity. This view resonates with Velmans' reflexive model, where perception is not confined to the neural substrate but involves a projection of experience into externalized space. Similarly, Sheldrake's historical analysis of vision theories suggests a long-standing intuitive recognition that vision involves more than internal processing—that it carries an outward-directed quality, consistent with lived experience. Far from being anomalies, these features become intelligible—and even expected—once perception is understood as a transphysical, integrative act. What once stood as edge-cases now serve as exemplars of perception's true structure: distributed, embodied, and world-participatory.

Philosophical frameworks such as active externalism (Clark & Chalmers, 1998; Clark 2013) further support this interpretation. According to this view, the mind is not bounded by the skull but extends into the environment, integrating tools, affordances, and contextual cues into cognitive processes. Arva Noë (2009) similarly asserts that perception is not an internal image but an embodied activity—one enacted through continuous interaction between organism and environment. From the perspective of the Brain Metafunction Framework, these theories converge on a central insight: perception is not the passive reception of stimuli, nor is it purely representational. Instead, it is an active world-participatory process, involving the integration of sensory input, bodily states, prior knowledge, emotional tone, and anticipatory modeling. The neural substrate that participates in this integration includes large-scale brain networks—particularly the Default Mode Network (DMN), Central Executive Network (CEN), and Salience Network (SN)—which together support the coherence and intentionality of perceptual experience. By situating perception within a metafunctional and transphysical domain, this model accommodates phenomena that exceed reductionist explanation. It accounts for the external localization of experience, the persistence of perceptual coherence in the absence of sensory input, and the role of intention, memory, and emotion in shaping what is perceived. Far from being anomalous, these features emerge as constitutive of how perception operates—linking consciousness not only to the brain, but to the broader relational field in which it is embedded.

The Richness of Metafunctional Perception

The richness of perception lies in its integrative structure—emerging not from isolated sense data, but from the dynamic interplay of sensory, bodily, cognitive, and environmental elements. This interplay, mediated by the brain's metafunctional processes, transcends static, brain-centric views of perception. For example, interoception, proprioception, and exteroception illustrate the complexity of these processes:

1. *Interoception* involves sensing internal bodily states, such as hunger, thirst, and heartbeat, and integrating these signals into our perceptual experience (Craig, 2009).
2. *Proprioception* refers to the awareness of body position and movement, enabling seamless coordination of physical actions (Jeannerod, 2001).
3. *Exteroception* encompasses sensory modalities like vision, hearing, and touch, providing information about the external environment (Noë, 2009).

These dimensions collectively demonstrate how perception transcends the confines of the brain, engaging with the world dynamically and interactively. According to Noë, perception is not a passive snapshot confined to the skull but a living, breathing presence shaped by continuous engagement with the environment.

James' paradox, which highlights the tension between "representative" theories of perception and our intuitive experience, remains unresolved within reductionist materialist frameworks. However, a metafunctional and transphysical perspective offers a resolution. This approach challenges traditional views and invites a paradigm shift in understanding perception by emphasizing the interconnectedness of internal and external realities. As Thomas Kuhn (1962) argued, paradigm shifts occur when prevailing frameworks fail to account for new observations, leading to the adoption of superior explanatory models. By adopting a metafunctional theory of perception, researchers can transcend the limitations of materialist paradigms and explore the profound interconnectedness of mind, body, and environment. This shift not only resolves long-standing philosophical paradoxes—it inaugurates a broader ontological horizon, one in which perception reveals itself as a site of transcendence, integration, and relational depth.

Auditory Perception as a Metafunctional Process

Physiology of Auditory Processing

Auditory perception begins with the transformation of physical vibrations into electrochemical signals—a process that demonstrates the remarkable sensitivity and precision of the auditory system. Sound waves first enter the outer ear, where they strike the tympanic membrane (eardrum), causing it to vibrate. These vibrations are mechanically transmitted through the ossicles (malleus, incus, stapes) of the middle ear into the cochlear fluid of the inner ear. Within the cochlea, the basilar membrane's movement stimulates approximately 3,500 inner hair cells, which transduce mechanical energy into neural signals (Hudspeth, 1997). These electrochemical signals are then transmitted via the spiral ganglion cells through the auditory nerve. Once in the central nervous system, these signals pass through a series of relay nuclei—including the cochlear nucleus, superior olivary complex, and inferior colliculus—before reaching the medial geniculate nucleus (MGN) of the thalamus. This relay system processes aspects such as pitch, intensity, spatial localization, and temporal structure, refining the signal before it arrives at the primary auditory cortex in the temporal lobe. From there, further interpretation occurs in adjacent auditory association areas, where the brain decodes complex auditory information such as speech, melody, and environmental soundscapes (Kollmeier, 2008). Yet, as in vision, the neural encoding of sound is not sufficient to account for the experience of hearing. The richness of auditory perception emerges from how these signals are embedded within broader systems of attention, emotion, language, memory, and motor coordination. Perception is not merely the registration of vibrations—it is the brain's ability to contextualize, interpret, and externalize those vibrations into meaningful sound.

Multimodal Integration and Auditory Metafunctionality

Auditory processing engages wide-ranging neural networks beyond the classical auditory pathway. For example, motor regions, including the premotor cortex and cerebellum, are activated during passive listening—particularly when perceiving rhythmic patterns, musical beats, or speech prosody (Zatorre, Chen, & Penhune, 2007). These motor components contribute to temporal prediction, enabling the brain to anticipate and organize auditory streams into coherent sequences. Simultaneously, limbic and reward-related structures, such as the nucleus accumbens and orbitofrontal cortex, respond to emotionally salient auditory stimuli, including music, affective tone, and voice recognition (Blood & Zatorre, 2001; Salimpoor et al., 2011). These responses illustrate the affective modulation of auditory experience—where sound is imbued with emotional and motivational significance. In the domain of language, the Broca's area and Wernicke's area play critical roles in syntactic structure and semantic interpretation, respectively. These linguistic capacities interact with the auditory cortex and subcortical structures such as the basal ganglia, creating a distributed language network that supports speech perception and production (Hickok & Poeppel, 2007). This multisystem coordination illustrates the metafunctional nature of auditory perception. It involves:

The Afferent Loop: Auditory perception begins with the transduction of sound waves by mechanoreceptors in the cochlea, where hair cells convert acoustic vibrations into neural impulses. These signals travel via the auditory nerve (cranial nerve VIII) to the cochlear nuclei in the brainstem, ascend through the superior olivary complex and inferior colliculus, and are relayed to the medial geniculate nucleus of the thalamus before reaching the primary auditory cortex (A1) in the temporal lobe. However, this path is not a simple relay. Even before cortical processing occurs, auditory information is modulated by subcortical structures involved in binaural integration, spatial localization, and reflexive attention—highlighting early top-down influences and lateral inhibition mechanisms. Predictive coding begins here as well, as the auditory system rapidly filters expected versus novel patterns to prioritize relevance (Dehaene & Changeux, 2011; Friston, 2010). Thus, afferent auditory input is shaped from the outset by anticipatory and regulatory dynamics that prepare it for higher-level interpretive integration.

Metafunctional Processing:

Once auditory signals reach the cortex, they are integrated with broader cognitive, emotional, and sensorimotor systems. This metafunctional processing enables the brain to contextualize sound—distinguishing language from noise, identifying emotional prosody, or linking soundscapes to memory and meaning. Domain-general networks such as the Default Mode Network (DMN), Central Executive Network (CEN), and Salience Network (SN) are actively involved in modulating attention to sound, situating it within autobiographical memory, and assessing its behavioral relevance (Menon & Uddin, 2010; Seeley et al., 2007). In addition, the Medial Temporal Lobe Network (MTLN)—including key structures like the hippocampus and parahippocampal cortex—plays a crucial role in encoding, consolidating, and retrieving declarative memories. This network dynamically interacts with perceptual systems to embed auditory information within spatial and temporal memory contexts, allowing sounds to be meaningfully located in one's personal history and environment (Ranganath & Ritchey, 2012). For instance, the voice of a loved one amidst background chatter is not merely louder, it is cognitively prioritized, emotionally colored, and action relevant. Auditory processing also interacts with motor systems (as in speech production and entrainment to rhythm), limbic circuits (mediating affective tone), and interoceptive states (e.g., anxiety modulating threat detection in sound). These integrations highlight that hearing is never purely acoustic, is always embodied, emotional, memory-infused, and purpose-driven.

The Efferent Loop: Auditory perception culminates not in passive reception, but in an outwardly oriented projection of sonic space—situating sound sources in the environment and aligning them with the listener's embodied awareness. Through the efferent loop, perceptual content is placed "out there": we hear a bird in the tree or a car approaching from behind. This spatial externalization is not a byproduct—it is a structured, participatory act that reflects the brain's coordination of perceptual expectations, attentional focus, and embodied intentionality. Like in vision, this projection is shaped by the large-scale networks—DMN, CEN, and SN—which orchestrate the real-time interaction between perceptual experience and goal-directed behavior. For example, the orientation of the head and body toward a sound source involves efferent motor planning fused with perceptual placement. Music, likewise, does not merely register in the ears—it envelops the body, evokes movement, and elicits emotion. Thus, auditory perception becomes an experiential field—an enacted, participatory relationship between the self and a world that resounds with meaning.

Like vision, auditory perception is shaped not only by sensory input but also by predictive modeling and top-down processes. The brain uses prior knowledge, context, and attentional filters to organize auditory scenes—resolving ambiguity and identifying relevant stimuli in noisy or overlapping environments (Näätänen & Winkler, 1999; Sussman et al., 2002). For instance, in a crowded room, the brain selectively enhances the voice of a speaker while suppressing background chatter—a phenomenon known as the cocktail party effect. Contextual cues guide phoneme disambiguation, as in the resolution of similar-sounding consonants like /b/ and /p/ based on surrounding language (Samuel, 1997). Even when sensory data is limited (e.g., a whisper, distorted recording), the brain often "fills in" missing auditory information based on linguistic expectations, semantic priming, and emotional tone. Moreover, descending neural pathways—extending from cortex to cochlea—demonstrate the bidirectionality of auditory processing. These feedback loops modulate the sensitivity of hair cells and fine-tune the perception of pitch, timing, and loudness (Hudspeth, 1997). This efferent modulation reflects the predictive-corrective structure central to the Brain Metafunction Framework: the brain anticipates sound, compares expectation with input, and adjusts perception accordingly.

Auditory perception, then, is not an isolated function—it is a metafunctional process that arises through the collaboration of sensory pathways, predictive systems, emotional resonance, and intentional focus.

It exemplifies the Brain Metafunction Framework's triadic model. Far from being a linear response to acoustic input, hearing is a world-participatory act, a projection of sound into a shared space of action, meaning, and interpersonal connection. Just as visual perception "places" images into the world, auditory perception externalizes sound, locating voices, rhythms, and emotional tones within the spatial and social environment. The following section will further deepen this account by exploring the affective, embodied, and regulatory richness of auditory experience, with particular focus on music, speech, and emotional resonance.

Illustrative Cases of Efferent Dynamics in Auditory Perception

Several auditory phenomena dramatically underscore the role of the efferent loop in shaping perceptual content and spatial-experiential localization. Misophonia, for instance, involves intense emotional and physiological responses to specific trigger sounds—often mundane or subtle (like chewing or tapping). These reactions are not tied to the acoustic properties of the sound alone, but to anticipatory, affective projections rooted in prior associations, context, and embodied memory. The efferent loop here amplifies the perceptual salience of otherwise neutral stimuli, embedding them within a heightened emotional and motor readiness. Similarly, auditory hallucinations, such as those experienced in schizophrenia or extended isolation, involve vivid sound experiences (often voices) that are generated endogenously, yet projected outward as if originating from the environment. These perceptual constructions persist despite the absence of afferent input, demonstrating how the efferent system can localize and give form to internally generated content. A third case, musical synesthesia, exemplifies cross-modal efference: hearing certain tones or musical keys automatically evokes the perception of specific colors or spatial textures. These associations are involuntary, consistent, and experientially real, illustrating how auditory content becomes metafunctionally integrated and projected into a richly textured perceptual field. Each of these phenomena reveals that hearing is not confined to auditory mechanics—it is a participatory, interpretive act in which perception is constructed, situated, and emotionally charged through the dynamic activity of the efferent loop.

Cross-Modal Parallels and the Metafunctional Unity of Perception

The dynamics observed in auditory perception find strong parallels in the visual domain, reinforcing the metafunctional architecture common across sensory modalities. Just as misophonia illustrates how emotionally charged meaning is projected onto sound, affective visual triggers—such as certain facial expressions or symbolic imagery—can evoke visceral emotional responses that far exceed the raw visual data. In conditions like visual hallucinations, whether induced by psychedelics, sleep deprivation, or neurological conditions, the brain generates vivid visual content in the absence of afferent input. These experiences are often spatially externalized, emotionally charged, and contextually coherent, mirroring the projective structure of auditory hallucinations. Visual-auditory synesthesia, in which specific sounds trigger vivid visual phenomena, further demonstrates that perceptual content is not confined to discrete sensory channels but emerges from integrated, cross-network activity. These cases suggest that both auditory and visual perception rely on the same core processes: afferent input as a scaffold, metafunctional integration across cognitive-affective systems, and efferent projection into an experientially inhabited world. Together, they illustrate that perception is not the registration of the world as it is, but the emergence of a world as it is lived, constructed through the brain's dynamic, transmodal engagement with meaning, memory, and intention.

The Richness of Auditory Experience

Auditory perception engages a wide array of interconnected systems—extending far beyond the decoding of sound frequencies. The experience of hearing is enriched by its integration with memory, motor coordination, language, emotion, and interoception, reflecting the full complexity of the brain's metafunctional architecture. For example, music activates not only the auditory cortex but also the premotor cortex, cerebellum, limbic system, and reward pathways. These regions collaborate to create a unified experience of rhythm, motion, affect, and aesthetic meaning (Zatorre, Chen, & Penhune, 2007; Blood & Zatorre, 2001). This dynamic intermodal processing reveals that auditory perception is not merely acoustic decoding—it is a deeply embodied act, involving rhythmic synchronization, emotional resonance, and even physiological regulation. Music, for instance, can induce chills, synchronize heartbeat and respiration, and alter stress hormone levels (Salimpoor et al., 2011; Baumgartner et al., 2006). These effects illustrate how the brain's interpretation of auditory information modulates internal states—blurring the boundaries between cognition, affect, and bodily regulation.

Language perception offers an equally compelling demonstration. Speech comprehension requires finely tuned integration of auditory input with motor prediction (e.g., articulatory modeling), semantic memory, and emotional context. The left-lateralized language network—anchored in Broca's and Wernicke's areas—does not operate in isolation; it coordinates with attention systems, the limbic system, and working memory to create meaning. Prosody—the melodic and emotional contour of speech—adds further complexity, requiring the integration of pitch, rhythm, and affective content to interpret tone, sarcasm, or sincerity. This layered processing underscores auditory perception as a multifunctional, context-sensitive act, orchestrated across multiple brain networks to produce an intelligible, meaningful, and emotionally inflected auditory world. In short, auditory perception is not reducible to hearing sounds. It is a metafunctional convergence of sensing, feeling, predicting, moving, and remembering—revealing the brain's capacity to construct richly textured auditory realities that guide behavior, regulate internal states, and sustain interpersonal connection.

The Top-Down Theory of Auditory Perception

The top-down theory of auditory perception highlights how higher-level cognitive processes, such as memory, attention, and expectation, influence the interpretation of auditory stimuli. This theory posits that prior knowledge, contextual cues, and attentional focus significantly shape auditory perception (Näätänen & Winkler, 1999; Sussman et al., 2002; Lee et al., 2012). Auditory perception exemplifies the interplay between sensory input and the brain's complex internal processing. For instance, the human cochlea contains around 16,000 hair cells arranged tonotopically, with high-frequency sensitivity at its base and low-frequency sensitivity at its apex (Hudspeth, 1997). Despite the relatively small number of sensory receptors, the brain compensates with feedback mechanisms. Neural pathways extend from the cortex to the ear, allowing for dynamic processing and modulation of auditory information at each relay station. These feedback loops enable the brain to interpret complex auditory signals, such as music, which demands rapid and multifaceted processing.

Music perception particularly illustrates the role of top-down processing, where the brain contributes significantly to the auditory experience. For example, prior knowledge of a music genre shapes how individuals interpret a melody. Similarly, attention selectively focuses on specific sounds, such as conversations, amidst noise in noisy environments. Contextual cues further guide perception. For instance, ambiguous speech sounds, such as /b/ or /p/, can be disambiguated based on surrounding words and context (Samuel, 1997). Research demonstrates that cognitive processes actively modulate activity in the auditory cortex, highlighting the brain's role in shaping auditory perception. These insights underscore how auditory perception extends beyond mere sensory input, involving active cognitive engagement that refines and contextualizes the sounds we perceive.

Top-Down Dynamics in Auditory Perception: A Metafunctional Perspective

Auditory perception, like vision, is not a passive decoding of acoustic signals but a dynamic, inferential process—profoundly shaped by top-down modulation. Within the Brain Metafunction Framework, hearing is framed as a bidirectional, context-sensitive act that integrates afferent input with expectations, goals, emotional tone, and embodied memory. Rather than simply reacting to the environment, the auditory system actively constructs the experience of sound—anticipating and shaping what is heard through predictive and integrative mechanisms. The predictive coding model reveals that the brain continually generates hypotheses about incoming auditory input based on prior knowledge and contextual cues (Friston, 2010; Rao & Ballard, 1999). This inferential scaffolding allows for rapid detection of anomalies—such as unexpected dissonance in music or a voice breaking pattern in speech—and is foundational to auditory fluency. Auditory attention further modulates this process: through top-down filtering mechanisms, the brain selectively enhances sounds of interest while suppressing irrelevant noise, such as in the well-known cocktail party effect (Shinn-Cunningham, 2008). These processes involve tight coordination between the auditory cortex, prefrontal regions, and salience network (Menon & Uddin, 2010), underlining the integrated nature of auditory perception.

Emotional context also exerts a powerful influence. Tones, rhythms, and vocal inflections can evoke empathy, alarm, or aesthetic pleasure—responses mediated by the amygdala, insula, and limbic circuitry. In this sense, hearing is inseparable from feeling: the auditory system participates in affective regulation and social attunement as much as in environmental awareness. From a metafunctional perspective, these top-down processes are not peripheral; they are essential. They demonstrate that hearing is not simply the passive registration of air pressure waves, but the construction of a meaningful, embodied soundscape, informed by autobiographical memory, somatic state, and intersubjective orientation. The Brain Metafunction Framework reveals this act of listening to be distributed, predictive, and dialogical—an engagement not just with sound, but with significance. Auditory perception becomes an active participation in acoustic reality, one that unites bodily awareness, narrative coherence, and relational attunement.

Multisensory and Inter-Functional Integration

Auditory perception exemplifies the brain's capacity for multisensory and inter-functional integration. The interaction between auditory and motor systems enables rhythmic synchronization and enhanced auditory processing (Gordon, Cobb, and Balasubramaniam, 2018). Emotional systems further augment this experience, as music evokes profound affective responses. For instance, limbic and paralimbic regions, including the nucleus accumbens, anterior cingulate cortex, and amygdala, activate during music-induced chills, linking auditory stimuli to enriching experiences (Blood and Zatorre, 2001). Language processing also demonstrates intermodal integration. Speech perception requires coordination between auditory, motor, and linguistic systems. Motor areas, including the premotor cortex, are activated when listening to speech, as they help predict articulatory patterns based on incoming auditory signals (Wilson et al., 2004). This predictive mechanism underscores the multimodal nature of auditory perception, as the brain integrates sound with motor and linguistic expectations to facilitate comprehension. This interconnectedness extends to internal regulatory systems. Music and speech can modulate heart rate, respiration, and hormonal balance, contributing to their therapeutic potential. Studies employing PET imaging show that music-induced pleasure correlates with changes in autonomic functions, tying auditory perception to broader physiological states (Blood and Zatorre, 2001; Baumgartner et al., 2006). Similarly, the rhythm and intonation of speech, known as prosody, convey emotional context, engaging both the auditory and limbic systems to create a richer communicative experience (Hickok and Poeppel, 2007).

Auditory Projection and the Efferent Loop

Within the Brain Metafunction Framework, perception is not complete without its third core component: efferent projection—the outward, intentional structuring of perceptual experience in external space. In auditory perception, this efferent dynamic plays a crucial role in shaping not just what we hear, but where we hear it, how we localize it, and why it matters. The auditory system is inherently spatial. Even though the cochlea encodes no direct information about spatial coordinates, the brain constructs a spatial auditory field by integrating interaural time differences, sound intensity gradients, and head movements. This spatialization is not a given—it is actively constructed by efferent processes that rely on prior expectations, motor calibration, vestibular feedback, and environmental familiarity (Middlebrooks & Green, 1991). The posterior parietal cortex, superior colliculus, and prefrontal areas work in concert with auditory cortices to achieve this projective mapping, embedding auditory content in externalized, navigable space. Crucially, the efferent loop is not just about orientation, it underlies the meaning-making aspect of auditory experience. For instance, the voice of a loved one is not merely detected; it is emotionally weighted, biographically situated, and often anticipated before it arrives. These expectations arise from long-standing, embodied templates—shaped by memory, social context, and emotional resonance—then projected back into the perceptual field. This helps explain phenomena such as musical anticipation, where listeners "hear" the next note before it sounds, or predictive echo suppression, where the brain filters out self-generated speech or footsteps to prioritize novel input (Ford & Mathalon, 2004).

In pathological or non-ordinary states, the efferent loop's projective function becomes even more salient. Auditory hallucinations, for example, often involve content that feels "external" despite the absence of stimulus. From a metafunctional perspective, this reflects a misalignment between efferent projection and afferent grounding—the brain generates auditory content but fails to modulate or contextualize it through external feedback. Similarly, in misophonia, ordinary sounds are assigned intense emotional salience, likely due to an exaggerated interplay between efferent expectation and affective tagging. In both cases, the disruption lies not in detection per se, but in the way perception is constructed, placed, and emotionally mediated.

The efferent loop in hearing also plays a role in narrative coherence and autobiographical anchoring. The sound of one's own voice—often modulated in real-time during conversation—contributes to the formation of self-identity and intersubjective exchange. This real-time feedback loop is metabolized not just through motor-auditory coupling, but through the integration of self-referential processing within the DMN, CEN, and SN. In this light, auditory perception becomes a multilayered participatory act—locating the listener in space, affirming identity, and constructing meaning across time. By situating hearing within a triadic framework of afference, metafunctional processing, and efferent projection, the Brain Metafunction Framework reveals auditory perception to be more than acoustic analysis. It is an ontologically rich act of being-in-the-world through sound—anticipatory, embodied, emotional, and dialogical. Through the efferent loop, the brain doesn't just hear—it places, feels, and inhabits what is heard.

As we have seen in both visual and auditory modalities, perception is not a passive decoding of external stimuli, but a richly integrative process shaped by memory, emotion, attention, and bodily awareness—coordinated through the brain's metafunctional architecture. In both domains, the efferent loop emerges as a critical mechanism for constructing meaning and situating experience within a lived, embodied world. These insights set the stage for a deeper exploration of perception's affective dimension, where sensation becomes not only meaningful but urgent, disruptive, or even transformative. In this context, *pain* offers a particularly revealing lens. Unlike other sensory modalities, pain perception is inherently tied to survival, emotional salience, and subjective meaning. It engages the full range of the afferent–metafunctional–efferent loop and can be shaped—sometimes dramatically—by cognitive, visual, or narrative modulation. Moreover, pain reveals the deep entanglement between bodily sensation and the sense of self, often becoming the very ground of conscious experience in moments of suffering or crisis.

From Outward Projection to Embodied Depth

Contrasting Vision and Sound with the Inward Terrain of Pain and Beyond

Having explored perception through the outward-facing modalities of vision and sound—both anchored in the projection of sensory content into the world—Chapter 5 turns toward more complex, often less visible forms of experience that nonetheless exemplify the full power of the Brain Metafunction Framework. It begins with an in-depth look at pain, not as a simple sensory signal but as a deeply embodied, context-sensitive, and meaning-laden phenomenon. Pain reveals how perception becomes integrated with emotion, narrative, and selfhood, and how it can be modulated, amplified, or diminished depending on attention, belief, memory, and environmental context. Through this lens, pain becomes a window into the dynamic interplay of afferent input, central processing, and efferent projection—and a case study in the transformative potential of perception itself.

From this foundation, the chapter expands outward to examine exceptional perceptual phenomena that further illuminate the brain's metafunctional capacities. These include phantom limb experiences, synesthesia, blindsight, and near-death states—all of which demonstrate how perception can persist, or even intensify, in the absence of conventional sensory input. The chapter also offers a bold reconceptualization of dreaming as a diagnostic and symbolic process that integrates subliminal signals into coherent, often metaphorically charged imagery—revealing how perception constructs meaning even while disconnected from the external world. In its later sections, Chapter 5 introduces the Surprise Principle—a philosophical and scientific tool for understanding how anomalies challenge prevailing models and catalyze paradigm shifts. It argues that the Brain Metafunction Framework is uniquely positioned to account for data that conventional physicalist models struggle to explain. Clinically, the chapter explores pain modulation techniques, mirror therapy, motor imagery, and VR-based rehabilitation—highlighting how therapeutic strategies can target the efferent loop to restructure disrupted perceptual coherence. Finally, it explores the transphysical implications of perception, culminating in a discussion of the soul as the integrative principle of embodied, narrative, and participatory selfhood. In all, Chapter 5 expands the horizon of perception from sensory projection to symbolic embodiment, from suffering to healing, from brain-bound models to a theory of consciousness as world-participation—rich in clinical application, philosophical depth, and existential insight.

REFERENCES

Aquinas, T. (1947). *Summa Theologica* (Fathers of the English Dominican Province, Trans.). Benziger Bros.

Baumgartner, T., Lutz, C., & Jäncke, M. (2006). The emotional power of music: How music enhances the feeling of affective pictures. *Brain Research, 1075*(1), 151–164.

Berkeley, G. (1710). *A treatise concerning the principles of human knowledge.* London: Jacob Tonson.

Blood, A. J., & Zatorre, R. J. (2001). Intensely pleasurable responses to music correlate with activity in brain regions implicated in reward and emotion. *Proceedings of the National Academy of Sciences, 98*(20), 11818–11823.

Clark, A., & Chalmers, D. (1998). The extended mind. *Analysis, 58*(1), 7–19.

Clark, A. (2013). Whatever next? Predictive brains, situated agents, and the future of cognitive science. Behavioral and Brain Sciences, 36(3), 181–204. https://doi.org/10.1017/S0140525X12000477

Cottingham, J. (1986). *Descartes.* Blackwell.

Craig, A. D. (2009). How do you feel—now? The anterior insula and human awareness. *Nature Reviews Neuroscience, 10*(1), 59–70. https://doi.org/10.1038/nrn2555

Dehaene, S., & Changeux, J. P. (2011). Experimental and theoretical approaches to conscious processing. *Neuron, 70*(2), 200–227. https://doi.org/10.1016/j.neuron.2011.03.018

Diers, M., Fuchs, X., Bekrater-Bodmann, R., & Flor, H. (2023). Prevalence of phantom phenomena in congenital and early-life amputees. *The Journal of Pain, 24*(3), 502–508. https://doi.org/10.1016/j.jpain.2022.10.010

Flor, H. (2002). Phantom-limb pain: Characteristics, causes, and treatment. *The Lancet Neurology, 1*(3), 182–189.

Ford, J. M., & Mathalon, D. H. (2004). Electrophysiological evidence of corollary discharge dysfunction in schizophrenia during talking and thinking. Journal of Psychiatric Research, 38(1), 37–46. https://doi.org/10.1016/S0022-39 56(03)00095-5

Foster, R. G., & Hankins, M. W. (2002). Non-rod, non-cone photoreception in vertebrates. *Progress in Retinal and Eye Research, 22*(5), 501–527. https://doi: 10.1016/s1350 -9462(02)00036-8.

Freeman, K. (1949). *The pre-Socratic philosophers: A companion to Diels, Fragmente der Vorsokratiker.* Harvard University Press.

Friston, K. (2010). The free-energy principle: A unified brain theory? *Nature Reviews Neuroscience, 11*(2), 127–138. https://doi.org/10.1038/nrn2787

Friston, K., Adams, R., Perrinet, L., & Breakspear, M. (2012). Perceptions as hypotheses: Saccades as experiments. *Frontiers in Psychology, 3*, 151. https://doi.org/10.3389/fpsyg.2012.00151

Goodale, M. A., & Milner, A. D. (1992). Separate visual pathways for perception and action. *Trends in Neurosciences, 15*(1), 20–25. https://doi.org/10.1016/0166-2236(92)90344-8

Gregory, R. L. (1997). Knowledge in perception and illusion. *Philosophical Transactions of the Royal Society B: Biological Sciences, 352*(1358), 1121–1128. https://doi.org/10.1098/rstb. 1997.0095

Greyson, B. (2003). Near-death experiences in a hospital setting: Phenomena and scientific implications. *Journal of Near-Death Studies, 23*(4), 203–222.

Grill-Spector, K., & Malach, R. (2004). The human visual cortex. *Annual Review of Neuroscience, 27*(1), 649–677. https://doi.org/10.1146/annurev.neuro.27.070203.144220

Hickok, G., & Poeppel, D. (2007). The cortical organization of speech processing. *Nature Reviews Neuroscience, 8*(5), 393–402.

Hudspeth, A. J. (1997). How hearing happens. *Neuron, 19*(5), 947–950. doi: 10.1016/s0896-6273(00)80385-2.

Hudspeth, A. J. (1997). How the ear's work. *Nature, 385*(66 12), 198–199.

Hume, D. (1739). *A Treatise of Human Nature.* London: John Noon.

James, W. (1904). A world of pure experience. *The Journal of Philosophy, Psychology and Scientific Methods, 1*(20), 533–543.

Jeannerod, M. (2001). Neural simulation of action: A unifying mechanism for motor cognition. *NeuroImage, 14*(1), S103–S109.

Kant, I. (1781). *Critique of pure reason.* London: Macmillan.

Kosslyn, S. M., Thompson, W. L., Kim, I. J., & Alpert, N. M. (2001). The role of Area 17 in visual imagery: Convergent evidence from PET and rTMS. *Science, 284*(5411), 167–170. https://doi.org/10.1126/science.284.5411.167

Kuhn, T. S. (1962). *The structure of scientific revolutions.* University of Chicago Press.

Leibniz, G. W. (1714). *Monadology.* Leipzig: Frege-Verlag.

Menon, V., & Uddin, L. Q. (2010). Saliency, switching, attention, and control: A network model of insula function. *Brain Structure and Function, 214*(5–6), 655–667. https://doi.org/10.1007/s00429-010-0262-0

Møller, A. R. (2012). *Sensory Systems and Transmission of Signals.* Elsevier.
Newton, I. (1704). Opticks. London: Samuel Smith and Benjamin Walford.

Middlebrooks, J. C., & Green, D. M. (1991). Sound localization by human listeners. *Annual Review of Psychology, 42*(1), 135–159. https://doi.org/10.1146/annurev.ps.42.020 191.001031

Näätänen, R., & Winkler, I. (1999). The concept of auditory stimulus representation in cognitive neuroscience. *Psychological Bulletin, 125*(6), 826–859. https://doi.org/10.1037 /0033-2909.125.6.826

Newton, I. (1704). *Opticks: Or, a treatise of the reflections, refractions, inflections and colours of light.* London: Samuel Smith and Benjamin Walford.

Noë, A. (2009). *Out of our heads: Why you are not your brain, and other lessons from the biology of consciousness.* Hill and Wang.

Ortiz-Catalan, M. (2018a). The stochastic entanglement and phantom motor execution hypotheses: A theoretical framework for the origin and treatment of phantom limb pain. *Frontiers in Neurology, 9*, 748. https://doi.org/10.3389/fneur.2018.00748

Ortiz Catalan, M. (2018b). A new theory for phantom limb pain points the way to more effective treatment. *Science News*. Chalmers University of Technology. Retrieved from https://www.sciencedaily.com/releases/2018/09/180906082022.htm.

Raichle, M. E. (2015). The brain's default mode network. *Annual Review of Neuroscience, 38*, 433–447. https://doi.org/10.1146/annurev-neuro-071013-014030

Ranganath, C., & Ritchey, M. (2012). Two cortical systems for memory-guided behaviour. *Nature Reviews Neuroscience, 13*(10), 713–726. https://doi.org/10.1038/nrn3338

Ramachandran, V. S., & Altschuler, E. L. (2009). The use of visual feedback, in particular mirror visual feedback, in restoring brain function. *Brain, 132*(7), 1693–1710. https://doi.org/10.1093/brain/awp135

Ramachandran, V. S., & Hirstein, W. (1998). The perception of phantom limbs: The D. O. Hebb lecture. *Brain, 121*(9), 1603–1630. https://doi.org/10.1093/brain /121.9.1603

Ramachandran, V. S., & Hubbard, E. M. (2001). Synesthesia: A window into perception, thought, and language. *Journal of Consciousness Studies, 8*(12), 3–34.

Rao, R. P. N., & Ballard, D. H. (1999). Predictive coding in the visual cortex: A functional interpretation of some extra-classical receptive-field effects. *Nature Neuroscience, 2*(1), 79–87. https://doi.org/10.1038/4580

Sabra, A. I. (1989). *The Optics of Ibn al-Haytham: Books I–III: On Direct Vision*. London: The Warburg Institute.

Salimpoor, V. N., Benovoy, M., Larcher, K., Dagher, A., & Zatorre, R. J. (2011). Anatomically distinct dopamine release during anticipation and experience of peak emotion to music. *Nature Neuroscience, 14*(2), 257–262. https://doi.org/ 10.1038/nn.2726.

Samuel, A. G. (1997). Lexical activation produces potent phonemic percepts. *Cognitive Psychology, 32*(2), 97–127. https ://doi.org/10.1006/cogp.1997.0646

Seeley, W. W., Menon, V., Schatzberg, A. F., Keller, J., Glover, G. H., Kenna, H., ... & Greicius, M. D. (2007). Dissociable intrinsic connectivity networks for salience processing and executive control. *Journal of Neuroscience, 27*(9), 2349–2356. https://doi.org/10.1523/JNEUROSCI.5587-06.2007

Seth, A. K. (2014). A predictive processing theory of sensorimotor contingencies: Explaining the puzzle of perceptual presence and its absence in synesthesia. *Cognitive Neuroscience, 5*(2), 97–118. https://doi.org/10.1080/17588928.2013.877880

Seth, A. K. (2021a). *Being you: A new science of consciousness.* Faber & Faber.

Seth, A. K. (2021b, October 21). The real problem of consciousness: What is it like to be you? *Psychology Today.* https://www.psychologytoday.com/ca/blog/consciousness-deep-dive/202110/the-real-problem-consciousness

Sheldrake, R. (2005). *The sense of being stared at: And other aspects of the extended mind.* Crown Publishing.

Shinn-Cunningham, B. G. (2008). Object-based auditory and visual attention. *Trends in Cognitive Sciences, 12*(5), 182–186. https://doi.org/10.1016/j.tics.2008.02.003

Sussman, E., Winkler, I., Huotilainen, M., Ritter, W., & Näätänen, R. (2002). Top-down effects can modify the initially stimulus-driven auditory organization. *Cognitive Brain Research,* 13(3), 393–405. https://doi.org/10.1016/S0 926-6410(01)00131-8.

Tumminia, S. (2020). Advances in retinal research. *National Eye Institute Reports.*

Van Lommel, P., van Wees, R., Meyers, V., & Elfferich, I. (2001). Near-death experience in survivors of cardiac arrest: A prospective study in the Netherlands. *The Lancet, 358*(9298), 2039–2045. https://doi.org/10.1016/S0 140-6736(01)07100-8

Velmans, M. (1990). Consciousness, brain, and the physical world. *Philosophical Psychology, 3*(1), 77–99. https://doi.org/10.1080/09515089008572990

Velmans, M. (2005). Reflexive monism. *Journal of Consciousness Studies, 12*(2), 109–116.

Weiskrantz, L. (1986). *Blindsight: A case study and implications.* Oxford University Press.

Wilson, S. M., et al. (2004). Listening to speech activates motor areas involved in speech production. *Nature Neuro science, 7*(7), 701–702.

Yeo, B. T. T., Krienen, F. M., Sepulcre, J., et al. (2011). The organization of the human cerebral cortex estimated by intrinsic functional connectivity. *Journal of Neurophysiology, 106*(3), 1125–1165. https://doi.org/10.1152/jn.00338.2011

Zatorre, R. J., Chen, J. L., & Penhune, V. B. (2007). When the brain plays music: Auditory–motor interactions in music perception and production. *Nature Reviews Neuroscience, 8*(7), 547–558. https://doi.org/10.1038/nrn2152

5

Perception as a Brain Metafunction Part II: Phenomena and Implications

From Theory to Lived Experience: The Clinical, Exceptional, and Existential Dimensions of Perception

If Chapter 4 laid the structural, theoretical, and historical foundation for reconceiving perception as a triadic, embodied, and projective process, Chapter 5 turns that lens toward the clinical, exceptional, and existential realities of lived experience. Here, perception is no longer treated as a normative baseline but as a dynamically shifting architecture—one whose failures, anomalies, and symbolic dramatizations offer privileged access to the deep workings of the self. We begin with pain. More than a mere signal of bodily distress, pain reveals the full range of metafunctional perception: how the brain constructs sensory experience in relation to memory, attention, emotional tone, contextual meaning, and bodily regulation. As we will see, pain does not always correlate with the extent of tissue damage. It may persist in the absence of afferent input (as in phantom limb syndrome), diminish under attention shifts (as in visual analgesia), or even reflect symbolic dramatizations of internal states (as in certain dreams or trauma-related responses). These paradoxes are not failures of the model, they are their strongest confirmations.

Yet pain is only the beginning. This chapter also explores a range of exceptional perceptual phenomena—synesthesia, blindsight, hallucinations, near-death experiences, and dreaming—that further demonstrate how the brain constructs coherent, emotionally charged experiences even in the absence of typical sensory grounding. These cases do not simply challenge physicalist assumptions; they call for a reconceptualization of perception as a living, integrative act that spans body, narrative, and world. In doing so, this chapter applies the Brain Metafunction Framework not only as a diagnostic model but as a philosophical and therapeutic orientation. Whether exploring the diagnostic implications of symbolic imagery in dreams, the clinical use of imagination and virtual reality in rehabilitation, or the transphysical nature of perception in near-death experiences, we follow the three-fold loop of afference, integration, and efferent projection as a unified grammar of perception.

These dynamics make pain an ideal entry point for exploring how perception is not simply a reflection of stimulus intensity but a modulated, meaning-saturated act of world-construction. Far from being a purely physiological reaction, pain reveals how the brain, through its metafunctional architecture, shapes experience in response to bodily cues, emotional tone, attention, memory, and context. This becomes especially evident in experimental and clinical contexts, where perception can be profoundly altered by top-down mechanisms. We now turn to specific research findings and real-life illustrations that demonstrate how pain perception is constructed, modulated, and projected within the triadic loop—shedding light on the brain's extraordinary capacity to transform sensation into significance.

Pain Perception and the Metafunctional Brain

Pain is a profoundly exhausting and multifaceted experience, often dominating an individual's subjective world and profoundly influencing their overall perception and interaction with their environment. It encompasses a broad neural signature that extends across various brain regions (Tracey and Mantyh, 2007; Reddan and Wager, 2018; Tracey, Woolf, and Andrews, 2019). Chronic pain, in particular, can be categorized into three interrelated pathways: the lateral pathway (associated with the sensory dimension of pain), the medial pathway (linked to the emotional suffering component), and the descending pathway (involved in pain inhibition and modulation) (Ridder, Adhi, and Vanneste, 2021). Notably, one can experience pain without suffering and suffer without pain, highlighting the multifactorial nature of these experiences. Pain and suffering are distinct experiences. Pain refers to the pain sensation, while suffering involves emotional or psychological distress. For example, a person may feel pain but not suffer if accepting or understanding the sensation (e.g., during childbirth or intense athletic performance). Conversely, one can suffer without physical pain, as in grief or emotional trauma. This distinction highlights the multifactorial nature of pain and suffering, influenced by biological, psychological, and social factors. Expectations, past experiences, and emotional states significantly modulate the perception and response to pain (Ongaro and Kaptchuk, 2019). This individualization makes pain a challenging yet illuminating area of research, particularly within a metafunctional framework, which seeks to understand its multidimensional nature and explore more effective treatments for pain management (Fillingim, 2017). Over time, chronic pain has been shown to induce structural and functional changes in the brain, particularly within the prefrontal cortex and limbic system (Apkarian et al., 2009), reinforcing the idea that pain is a plastic, narrative-infused experience—one that reshapes not only perception, but selfhood itself.

Pain Perception and Visual Analgesia: Insights from Research

British neuroscientist Patrick Haggard and his collaborators have conducted elegant experiments elucidating how pain perception is modulated by top-down mechanisms (Haggard et al., 2005; Haggard, 2013; Longo et al., 2008, 2009, 2012; Mancini et al., 2011, 2013; Lockwood, Iannetti, and Haggard, 2013; Beck, Làdavas, and Haggard, 2016). In one study, a laser induced a controlled burning sensation on participants' hands, isolating the "pure pain experience" by avoiding the involvement of other sensory modalities like touch or pressure (Haggard et al., 2005). Participants rated pain intensity on a scale of zero to a hundred under different visual conditions. When participants looked directly at their hand during the stimulus, they reported significantly lower pain ratings than when their hand was obscured, or they looked elsewhere. This phenomenon, termed "visual analgesia," underscores how visual attention to the body modulates pain perception. Electrophysiological recordings revealed reduced laser-evoked potentials, brain activity associated with the incoming pain signals, when participants visually attended to their hands. This reduction demonstrates that top-down processes influence bottom-up sensory pathways, altering subjective pain experiences and neural pain signals.

The Unity of Self-Consciousness and Metafunctional Flexibility in Pain Perception and Modulation

Haggard attributes this modulation to the brain's ability to integrate multisensory experiences into a cohesive representation of the self. He explains, "There's something about the experience of oneself as a whole... a unity of self-consciousness... which is able to provide this regulation of what we feel. When you can have a representation of yourself, simply by getting this visual input from your own body, you're able to exert top-down control over the bottom-up pathway" (Haggard, 2023). This insight fits seamlessly within a metafunctional framework, emphasizing the brain's adaptability in regulating pain through an integrated sense of self. The brain's metafunctional capacity allows it to constructively interpret sensory input, contextualizing pain within broader cognitive and emotional frameworks. This flexibility extends beyond pain; it is a general principle underlying all sensory experiences. Top-down processes dynamically shape perception, enabling individuals to prioritize relevant stimuli, contextualize experiences, and adapt behaviorally to changing environments. For instance, instead of passively perceiving pain "as it is," the brain actively constructs pain experiences that align with an individual's goals, emotional state, and context. This process reflects a broader whole-person perspective with contextualized perceptions according to the individual's needs and priorities. Such flexibility underscores the human brain's capacity to navigate complex environments and maintain functionality, even under challenging conditions.

Pain Perception as a Metafunctional Process

Triadic Dynamics of Nociception, Narrative, and Projection

Pain perception offers one of the most vivid illustrations of the Brain Metafunction Framework in action. Unlike vision or audition, which project sensory content outward into the environment, pain is profoundly inward, rooted in the body, but modulated by narrative, memory, affect, and social context. It is not merely a biological alarm, but a meaning-bearing event shaped by the whole person. Its metafunctional structure integrates nociceptive input with predictive modeling, emotional resonance, and embodied localization, revealing pain as a perceptual construct rather than a direct reflection of physical damage.

1. *The Afferent Loop*: *Nociceptive Signaling and Early Modulation*

Pain begins with the activation of nociceptors—specialized peripheral sensory receptors that detect actual or potential tissue damage. These receptors transduce mechanical, thermal, or chemical stimuli into neural signals, which are conveyed through Aδ and C fibers to the dorsal horn of the spinal cord. From there, ascending pathways (including the spinothalamic and spinoreticular tracts) carry the signals to the brainstem, thalamus, and higher cortical areas. Early modulation occurs at multiple levels, particularly in the spinal cord, where descending inhibition and gate control mechanisms can amplify or attenuate the signal. Notably, even at this initial stage, pain is not a passive input but is dynamically filtered through expectations, context, and prior experience. The afferent loop thus serves as the physiological scaffold upon which perception is built—but it does not dictate the experience in isolation.

2. *Metafunctional Processing: Integration Across Cognitive, Affective, and Contextual Domains*

Once nociceptive input reaches the central nervous system, it undergoes multilevel metafunctional integration, beginning not in the cortex, but in the brainstem—a foundational structure that plays a continuous role in modulating arousal, attention, autonomic regulation, and affective tone. Key brainstem nuclei such as the periaqueductal gray (PAG), locus coeruleus, and nucleus raphe magnus participate in early appraisal and descending pain modulation, interfacing with both cortical and subcortical structures to influence how pain is felt and interpreted. These brainstem systems act not simply as relays but as adaptive regulators that assess salience and mediate the interplay between bodily states and higher cognitive processes.

Beyond the brainstem, pain emerges from the coordinated activity of a distributed cortical network—including the somatosensory cortices, insula, anterior cingulate cortex, thalamus, prefrontal cortex, and limbic structures. The broader system integrates the raw nociceptive signal with contextual appraisals, emotional relevance, autobiographical memory, and bodily awareness. This aligns with evidence that cognitive and emotional systems exert profound top-down influence on pain perception, often disrupted in chronic pain syndromes (Bushnell, Čeko, & Low, 2013). The result is not a simple sensory impression, but a narrative-informed, motivationally weighted perceptual event. Importantly, this processing is nonlinear and individualized. For example, two individuals may interpret identical nociceptive input in radically different ways—depending on whether they are expecting harm or focused on a meaningful goal. Attention, belief, and prior experience reframe the intensity and emotional impact of pain. The brain does not merely interpret incoming signals; it actively models their significance, predicts their relevance, and modulates the felt experience accordingly. These complex interactions underscore the metafunctional nature of pain perception: it is not an isolated neural event, but a system-wide construction shaped by physiology, cognition, and lived meaning.

3. The Efferent Loop: Spatial Projection and Meaningful Localization

Perhaps most distinctively, pain is not experienced within the brain, it is projected outward into bodily space. A headache is felt in the temples, not the cortex; heart ischemia is often perceived in the left arm. This spatialization reflects the operation of the efferent loop, which localizes perceptual content through the body schema and anchors it in experiential space. In phenomena like referred pain or phantom limb pain, this loop becomes especially apparent: perception persists or mislocalizes despite absent or ambiguous afferent input. More broadly, the efferent projection of pain illustrates how the brain externalizes perceptual meaning, aligning it with embodied experience. Pain is thus not confined to the site of injury, nor even to the body—it can extend into the relational and symbolic world, manifesting as social withdrawal, metaphorical pain (e.g., heartbreak), or somaticized distress. This externalized, meaning-rich structure confirms the efferent loop as a generative function, not merely a sensory efferent response.

Reconceptualizing Pain Within the Metafunctional Framework

Viewed through this triadic lens, pain perception is not a linear progression from stimulus to sensation. It is a circular, integrative act that encompasses bodily signal, contextual relevance, emotional tone, and spatial projection. It blurs the line between physiology and meaning, making it a paradigmatic case of metafunctional cognition. Importantly, pain also demonstrates the transphysical dimension of perception: it arises not solely from tissue damage, but from how the self interprets and inhabits its world. Beliefs, memories, fears, and hopes all influence what is felt and how it is processed. This reconception has far-reaching implications for both theory and therapy. Clinically, it validates interventions that go beyond symptom suppression to address the full spectrum of perceptual construction. Techniques such as mirror therapy, visual analgesia, VR-based modulation, and cognitive-behavioral restructuring engage the entire triadic loop—modifying not only the input but the narrative and projection of pain. The success of these therapies further reinforces the principle that perception is plastic, participatory, and contextually governed. Ultimately, pain perception exemplifies how the brain constructs the felt world: not merely as a detector of injury, but as an interpreter of significance. It reveals perception to be fundamentally relational, organized not only by neurons but by meaning—placing pain squarely within the broader ontological and therapeutic horizon of the Brain Metafunction Framework.

Pain is not merely a physiological signal but a profoundly metafunctional phenomenon shaped by sensory input, cognitive processing, and emotional context. Research on visual analgesia highlights the brain's remarkable capacity to regulate pain perception, demonstrating how multisensory integration and top-down processes create adaptive and personalized experiences. By understanding pain within a metafunctional framework, we gain valuable insight into its complexity, paving the way for innovative approaches to pain management and enhancing our understanding of human experience.

From the Controlled Experimental Conditions to Real-Life Situations

Imagine an explorer on a mission to reach a remote and unexplored region of the world. During the journey, the explorer suffers severe injuries from a fall, leaving deep wounds on their leg and arm. The pain is excruciating, a sensation that would typically immobilize most individuals. Yet, the explorer knows survival depends on pushing through the pain and continuing the journey despite the physical and emotional toll. This story highlights the remarkable ability of the human brain to integrate multiple functions and processes into a coherent and meaningful response. Despite the severe injuries, the explorer focuses on survival, demonstrating how the brain metafunctionally balances sensory, cognitive, and emotional processes with survival-driven behaviors. As we know from experience, and Eccleston and Crombez (1999) have shown, pain exerts a powerful demand on attention, interrupting goal-directed behavior unless reframed or suppressed by higher-order cognitive modulation. In this case, the explorer's survival-oriented focus enables a top-down override of pain salience, exemplifying the flexible prioritization governed by the brain's metafunctional systems.

The explorer's capacity to override pain involves several brain systems working together. The prefrontal cortex drives decision-making and planning, suppressing the overwhelming sensory input from the injuries to prioritizing survival. Simultaneously, the salience network dynamically shifts attention between internal signals, such as pain and physical exhaustion, and external cues, such as needing shelter and food (Seth and Friston 2016). The anterior cingulate cortex contributes to pain modulation, allowing the explorer to endure the intense sensations without succumbing to them (Craig 2003, 2009). At the same time, the limbic system integrates emotional states, such as determination and fear, further enhancing the explorer's ability to act under stress (Damasio 1999). This multifaceted process showcases the specialization and flexibility of the brain's metafunctional system. While immediate survival needs strongly attract attention, other systems—such as memory, emotion, and motor planning—support decision-making and situate the situation in a broader, personal context. For example, the explorer may draw on memories of past challenges overcome, reinforcing the resolve to continue. This temporal and cognitive flexibility underscores the brain's capacity for integrating information across different timescales and domains, demonstrating its incredible adaptability (Seth and Friston 2016; Barrett and Simmons 2015).

Pain as a Metafunctional Phenomenon

Pain perception involves a complex interplay between bottom-up sensory inputs and top-down cognitive modulation. Bottom-up processes include the conduction of nerve impulses through spinal cord pathways, where initial modulation occurs. Top-down processes involve the brain's ability to interpret, regulate, and contextualize these sensory signals, integrating them into the individual's mental and physical state. This dynamic interaction allows humans to adapt to pain in ways that prioritize function and survival. In controlled experiments, such as those conducted by Haggard and colleagues (2005a, b), visual attention to one's body reduces subjective pain intensity and neural responses in the brain's pain-processing pathways. These findings underscore the brain's ability to exert top-down control over sensory inputs, a capability that extends to real-life scenarios like the explorer's journey. By focusing on survival-oriented goals, the brain actively suppresses pain perception and channels attention and energy toward actions essential for survival.

Integration Across Systems

The explorer's experience also illustrates how the brain metafunctionally integrates diverse processes into a unified whole. For instance:

1. *Sensory Processing:* Pain signals from the injuries travel through the afferent pathways, providing crucial information about the body's condition.

2. *Cognitive Modulation:* The prefrontal cortex evaluates the situation and prioritizes survival tasks, suppressing the immediate pain response to focus on strategic goals.

3. *Emotional Regulation:* The limbic system integrates emotional states, using determination and fear to motivate action.

4. *Autonomic Regulation:* The autonomic nervous system adjusts physiological responses, such as increasing heart rate and energy mobilization, to support continued movement.

5. *Memory Integration:* Memories of past resilience and successful challenges contribute to the explorer's sense of agency and perseverance.

This metafunctional integration ensures that the explorer perceives not just the pain but the entire context in which it occurs, creating a unified, survival-oriented state of consciousness (Dehaene & Changeux, 2011). Instead of perceiving the world "as it is," the explorer's brain and mind perceive what is relevant for survival, dynamically adapting perception and behavior to meet immediate and long-term needs.

The brain's ability to modulate pain and integrate diverse processes is an adaptive mechanism for survival. In high-stakes situations, the capacity to suppress pain, focus attention, and coordinate actions across multiple systems can mean the difference between life and death. This metafunctional ability reflects the brain's remarkable capacity as a system designed for reactive responses and proactive, integrated action in a complex world. By leveraging its metafunctional processing capabilities, the human brain demonstrates remarkable flexibility and coherence, allowing individuals to navigate even the most challenging environments. This dynamic interaction between sensory, cognitive, emotional, and survival-oriented processes underscores the profound adaptability and resilience inherent in the brain metafunction. Through its ability to prioritize, integrate, and act, the brain enables us to move beyond immediate pain and engage with the broader demands of life, exemplifying one vital role as a metafunctional system.

The Transphysical Efferent Loop: Perception as Participation

The Brain Metafunction Framework challenges the classical conception of perception as a unidirectional, afferent-driven process. Rather than treating the brain as a passive receiver of stimuli, this model presents perception as a bidirectional, generative act—a dynamic interplay between afferent encoding, centralized metafunctional processing, and efferent projection. At the core of this framework lies the efferent loop: the outward-facing, projective component of perception, through which the mind does not merely reflect the world but activelyparticipates in it. This participatory structure is not metaphorical; it is functional and embodied. The brain does not simply process input; it anticipates, enacts, and situates perceptual experience in relation to context, memory, and meaning. The efferent loop is the mechanism by which perception is localized in space, structured by intentionality, and shaped by internal states such as emotion, attention, and narrative coherence.

The transphysical nature of the efferent loop becomes evident in three interrelated domains:

1. *Perceptual Completion and Simulation*

   In phenomena like phantom limb sensations, visual illusions, or dreams, the brain generates full perceptual experience in the absence of corresponding sensory input. These projections are not passive aftereffects but dynamically maintained simulations, filled in by memory, expectation, and predictive modeling. This illustrates the mind's capacity to complete experience through integrative, top-down dynamics that go beyond stimulus-bound activity.

2. *Externalization and Embodied Spatiality*

   Perceptual experience is always experienced as external—we do not sense activity in the brain, but trees, people, and sounds "out there." This spatial projection occurs even though all neural activity is intra-cranial, suggesting that perception is inherently outward-facing, constructed through relational and embodied mappings. The efferent loop allows the world to appear not merely as content in the mind, but as a field of presence in which the perceiver is immersed.

3. *Intentionality and Meaning-Making*

   Human perception is not neutral; it is inherently interpretive, anticipatory, and goal directed. We don't just see—we infer, expect, desire, and judge. These semantic and existential layers of perception arise from the mind's active orientation toward meaning, value, and agency. This projection of significance into the world is central to symbolic acts, aesthetic experience, and moral awareness—all of which presuppose a self engaged in transphysical participation with its world.

Together, these functions affirm that perception is not a mechanistic representation of the world, but a dialogical relation between mind, body, and environment. The efferent loop is the channel through which the mind inhabits reality—not as a detached observer, but as a situated agent, enacting and interpreting its world. Importantly, this framework also explains a range of otherwise paradoxical perceptual phenomena. In mirror therapy, for instance, visual input is used to recalibrate internal body models, restoring coherence to disrupted efferent projections as in post-stroke therapy (Cui, et al. 2022). In virtual reality and imaginative rehearsal, simulated environments can engage and reshape embodied perception even without direct sensory stimuli. These applications show that the efferent loop is not a passive residue of afference—it is a flexible, generative process through which perception is restructured and experience remade. This participatory and transphysical view of perception not only reframes scientific inquiry but invites a more integrated understanding of personhood, agency, and embodiment. Far from being confined to the skull, the mind reveals itself through its outward enactments—through the meanings it projects, the realities it constructs, and the world it helps bring into being.

Near-Death Experiences and the Transphysical Domain

Near-death experiences (NDEs) present a profound challenge to materialist accounts of consciousness and perception during moments of severe physiological distress or clinical death (Moody 1975; Lommel et al., 2001; Sartori 2008a, b; Greyson 2003a; Cook et al. 1998). Reported across cultures and clinical contexts, NDEs often include vivid perceptual experiences—such as panoramic memory, 360-degree vision, encounters with light or beings, and feelings of profound peace—that occur during periods of severely reduced or absent brain activity, including flatline EEG states. These phenomena are often dismissed as hallucinations, yet their clarity, coherence, and transformative impact suggest something more than cortical disintegration. Thonnard et al. (2013) found that NDE memories display more internal consistency and vividness than imagined events, suggesting they reflect genuine experiential episodes rather than confabulated hallucinations. From the perspective of the Brain Metafunction Framework, NDEs exemplify the model's central insight: that perception is not reducible to afferent input or cortical computation alone. Instead, it is a distributed, bidirectional, and projective process, one that may continue—or reconfigure—even under extreme physiological conditions.

Visual experiences during NDEs defy conventional understanding of visual perception, which typically requires functional eyes and an intact visual cortex. Kenneth Ring and Sharon Cooper's study of 31 blind individuals revealed reports of vivid, coherent visual experiences during NDEs, including accurate descriptions of people, objects, and environments (Ring & Cooper, 1997). These experiences, termed "apparent eyeless sight," suggest that visual perception can occur without sensory input from the physical body, pointing to a transphysical aspect of perception. Studies of blind individuals provide particularly compelling evidence of transphysical perception. These individuals describe visual experiences during NDEs that include accurate, veridical information unattainable through their other senses (Ring & Cooper, 1997; Holden, 2007, 2009). For example, individuals report observing specific resuscitation details or visual aspects of distant events, corroborated by independent investigations. Such experiences suggest that perception during NDEs transcends the limitations of physical sensory apparatus, supporting the idea of a transphysical domain integral to consciousness.

Transphysical Embodiment in the Absence of Input

Many NDErs report a persistent sense of embodiment despite complete sensorimotor arrest—suggesting that the brain's body schema, normally maintained through afferent updates, can be internally sustained and efferently projected even when the usual sensory channels are offline. The experience of "being somewhere" and "seeing" from a vantage point, even in the absence of ocular activity, reflects the mind's capacity to construct spatialized presence through non-local, internally generated dynamics. Descriptions of enhanced perception—such as "colors beyond any I had ever seen" or 360-degree vision—suggest that these experiences do not conform to degraded brain function. Instead, they may indicate the release or disinhibition of a latent perceptual mode that is not dependent on the conventional sensory hierarchy.

Pam Reynolds Near-death Experience: A Revealing Case

One of the most illustrative cases highlighting the complexities of perception is the Pam Reynolds near-death experience (NDE) during a surgical procedure for a giant basilar artery aneurysm in 1991 (Carter 2011). Dr. Robert Spetzler at the Barrow Neurological Institute performed a "standstill" procedure in which he cooled Pam's body to

60°F (16°C), stopped her heart, and drained blood from her brain using extracorporeal circulation. During the procedure, Pam was under general anesthesia; her EEG and brainstem auditory evoked potentials were flat, indicating no measurable brain activity, and her eyes were taped shut. Despite these conditions, which ensured no external sensory input, Pam later described an out-of-body experience where she observed the surgical team working on her. She provided accurate details about the specialized Midas Rex bone saw used to open her skull, including its shape, sound, and how it was used, information she could not have known from her unconscious state. Additionally, she recalled specific conversations and actions of the medical team, which surgical staff later corroborated. Pam Reynolds' account has been extensively studied and documented, providing a unique perspective on the interplay between consciousness and the brain and raising important questions about the boundaries of human perception.

Cross-Cultural Patterns and Cognitive Integrity

The cross-cultural consistency of NDEs, including reports from blind individuals and children, suggests a shared neurophenomenological substrate that transcends prior sensory learning. These experiences often carry enduring psychological and behavioral transformation, including reduced fear of death, increased empathy, and a greater sense of life purpose—outcomes that are difficult to explain through neural chaos or pharmacological artifacts alone (Greyson, 2003b). From the metafunctional perspective, these experiences reflect an emergent mode of integrative processing—one that is freed from standard input-output constraints and instead expresses a latent structure of perceptual coherence embedded in the efferent loop. In this light, NDEs are not anomalies but revelatory cases that unveil the full range of the brain–mind–world system when uncoupled from the physical body.

The Efferent Loop, the Soul, Selfhood, and the Continuity of Identity

In NDEs, the efferent loop appears to persist in the absence of afferent signals—producing not disorganized fragments, but unified, meaningful, and often transcendent experiences. The ability of the mind to project coherent realities during periods of biological near-silence aligns with the idea of transphysical participation—that perception, at its deepest level, is not contingent solely on neural substrates, but reflects a higher-order organization of consciousness. This challenges traditional conceptions of where and how perception occurs. Rather than being tied exclusively to sensory input or local cortical circuits, the Brain Metafunction Framework suggests that perception can operate as a field-like, participatory structure—one that may persist and function in non-ordinary states of consciousness. As Merleau-Ponty (1945/2012) argues, perception is not a detached mental event but an embodied act of being-in-the-world—an insight echoed by the efferent loop. This participatory view affirms perception as a dialogical process in which the self is situated through its engagement with space, action, and meaning. Despite the brain's neuroplasticity and constant structural remodeling, most individuals maintain a coherent sense of self across time, a unified continuity of perspective, memory, agency, and identity. This coherence of identity—even in altered states—mirrors what Gallagher and Zahavi (2021) call the minimal self: a first-person perspective preserved across temporal and sensory shifts. Their framework supports the notion that consciousness maintains a form of invariance even when traditional sensory input is disrupted, as seen in near-death experiences and dreaming. This constancy, though grounded in biology, cannot be fully explained by neural mechanisms alone. Rather, it suggests a higher-order integrative principle—what classical philosophy refers to as the soul: the organizing unity of mind, body, and experience.

Beyond the Brain: The Self and Metafunctional Integration

The Brain Metafunction Framework provides an empirical model for understanding this unity. Within this framework, the self emerges not as a static neural artifact, but as the dynamic product of ongoing integration between:

- Afferent input (moment-by-moment sensory contact with the world)
- Central metafunctional processing (emotional, cognitive, narrative, and interoceptive synthesis)
- Efferent projection (the outward placement of coherent perceptual meaning)

The self is thus maintained not merely as an internal representation, but through continuous interaction with a meaning-saturated world. Perception becomes the existential mechanism through which identity is enacted, shaped, and sustained—situated not in the brain alone, but in the dialogue between consciousness and environment.

The Efferent Loop as Ontological Participation

Far from being a passive neurological output, the efferent loop plays a formative role in sustaining personal identity and situating the self within a meaningful world. This outward projection of perception allows not only spatial orientation but also for moral, symbolic, and narrative positioning—the self as an agent embedded in a structured reality. In this view, empathy emerges as an extension of perception into the lived experience of others, enabling us to co-feel and co-understand across boundaries. Imagination functions as a generative projection, allowing us to construct counterfactual, aesthetic, or visionary realities beyond immediate input. Spiritual experience reflects a similar projective capacity, where non-sensory dimensions are perceived as real and significant through inward coherence and outward resonance. Moral action arises as perception becomes not just reflective but directive—guiding choices and behavior through the embodied realization of meaning. These dimensions are not secondary to perception; they are core expressions of its deepest function: to unify inner life and outer world, integrating the personal and the shared, the material and the transphysical.

The Transphysical Efferent Loop: Perception as Participation

At the heart of the Brain Metafunction Framework lies a foundational insight: perception is not merely a receptive function but a participatory act. Traditional neuroscientific accounts often emphasize the afferent loop—how sensory organs deliver input to the brain for processing. Yet, this tells only half the story. Perception, as we live it, does not remain within the skull—it is experienced as external, structured, and meaningful. This is made possible by what we term the efferent loop: the top-down, outward-directed projection of experience into the world. Crucially, this projection is not metaphorical. It is the phenomenological and functional bridge between internal processing and the externalized world. And in many contexts—clinical, cognitive, and existential—it assumes a transphysical character: engaging domains of experience that transcend local sensory data or even the material brain itself.

From Internal Models to External Reality

The efferent loop works by projecting perceptual content outward—placing objects, sounds, meanings, and emotional salience into the environment, not simply representing them. Clark (2016) describes this anticipatory dynamic as 'surfing uncertainty'—where perception arises from active prediction and embodied simulation, not passive reception. It draws on:

Predictive coding: Anticipating what should be perceived based on internal models and prior experience (Friston, 2010). This view of perception as inference aligns with Hohwy's (2013) argument that the brain acts as a 'predictive mind,' constantly testing hypotheses about the world.

Intentionality: Directing perception toward meaningful aspects of the world—what matters, rather than what merely is.

Semantic layering: Infusing perception with symbolic, emotional, and narrative context.

As a result, perception is never "purely sensory." *It is always interpretive,* shaped by what the self expects, seeks, or needs. The brain does not just decode signals—it simulates, hypothesizes, and enacts perceptual worlds.

Three Domains of Efferent Transphysical Function

The outward-directed nature of perception becomes especially clear in the following domains:

1. *Perceptual Completion*
   The brain "fills in" what is not present in the afferent stream using expectation, memory, and context—producing a coherent perceptual field from incomplete or ambiguous input. Examples include Visual illusions and amodal completion, Phantom limb phenomenon, Dreaming and hallucination in the absence of sensory input. Amodal completion refers to the brain's capacity to perceive a whole object even when some of its parts are hidden from view by something else in the foreground. Such phenomena reveal that experience is often generated from within yet placed into a world-like structure.

2. *Self-Awareness and Spatial Externalization*
   We do not experience vision, sound, or pain as events in the brain—we experience them as "out there." This persistent externalization suggests a projective structure of consciousness. The efferent loop locates experience in space—enabling coherent

action, empathy, and world-orientation. Even in altered states or sensory deprivation, the self remains spatially situated in an experiential field—a point of view rather than a point in space.

3. *Intentionality and Meaning-Making*
   Perception is fundamentally directed: it seeks significance, tracks relevance, and assigns meaning. This goes beyond physiological representation—it is an existential activity. We do not merely perceive objects—we perceive affordances, moral cues, symbolic content, and relational context.

The efferent loop is where values enter perception—where the self does not just see the world but participates in it.

Transphysicality and the Lived World

These characteristics—perceptual completion, spatial externalization, and intentional meaning-making—together mark the efferent loop as transphysical. That is, it bridges neural processes with non-local, embodied, and symbolic dimensions of experience. We might say: The efferent loop is how the soul touches the world. This is not mystical speculation, but an acknowledgment of how perception functions: through context-dependent, narrative-driven, emotionally modulated projections that transcend purely physical causality. These projections are interactive, responsive to the environment, yet not wholly determined by it.

Participatory Cognition and Soul–Body Unity

The implications of this view are profound. If perception is not simply input-driven, but actively constituted through intention, meaning, and embodiment, then we must rethink the foundations of consciousness itself. The soul–body unity becomes not a philosophical relic, but a clinically and cognitively viable model of personhood. In therapeutic contexts—such as virtual reality therapy, mirror therapy, and imagination-based rehabilitation—it is precisely this projective capacity that is leveraged for healing. Patients are not passive recipients of stimuli but active co-creators of perception—even in the face of trauma, amputation, or neurological disruption. Thus, the efferent loop is not merely a mechanism, it is the stage on which the self enacts its world, through memory and prediction; narrative and symbolism; bodily orientation and intention. It is where consciousness becomes presence, and where the human person—brain, mind, body, and soul—participates as a unified, world-embedded being.

Clinical and Cognitive Dynamics Within the Brain Meta-function Loop

Phantom Limb Syndrome (congenital absence or post-amputation)

Afferent Loop: Absent
Metafunctional Processing: Intact
Efferent Output: Persistent but maladaptive
Implication: Perception continues in the absence of input, resulting in phantom pain; mirror therapy often restores coherence through simulated visual afference.

Blindsight (cortical blindness with preserved response)

Afferent Loop: Intact (subcortical), impaired (primary visual cortex)
Metafunctional Processing: Partially preserved
Efferent Output: Residual and non-conscious
Implication: Non-conscious perceptual awareness supports the idea of distributed processing below conscious thresholds.

Charles Bonnet Syndrome (Visual impairment)

Afferent Loop: Reduced or absent
Metafunctional Processing: Active and compensatory
Efferent Output: Intact or exaggerated
Implication: Vivid internal imagery compensates for loss of input; typically remains benign unless dysregulated.

Parkinson's-Related Hallucinations (disease progression)

Afferent Loop: Intact or reduced
Metafunctional Processing: Partially disrupted or dysregulated
Efferent Output: Amplified or decoupled
Implication: Efferent output becomes decoupled from reality; hallucinations are often multifactorial and require contextual interpretation.

Hemineglect (Unilateral Neglect) (stroke, usually right parietal)

Afferent Loop: Intact, reduced, or spatially absent
Metafunctional Processing: Distorted or partially absent
Efferent Output: Distorted or absent
Implication: Breakdown in spatial representation of contralesional space; supports role of efferent projection in attentional mapping.

Visual Hallucinations (psychosis, sensory deprivation)

Afferent Loop: Intact or absent
Metafunctional Processing: Dysregulated

Efferent Output: Decoupled
Implication: Perception is generated independent of sensory input; distinguishes hallucination from imagination by loss of contextual regulation.

Motor Imagery Therapy (stroke rehabilitation, motor planning)
Afferent Loop: Intact, reduced, or absent
Metafunctional Processing: Active
Efferent Output: Simulated
Implication: Motor imagery re-engages dormant circuits through efferent simulation; supports top-down rehabilitation approaches.

Action Observation Therapy (neurorehabilitation, Parkinson's)
Afferent Loop: Intact
Metafunctional Processing: Active
Efferent Output: Reorganized
Implication: Observation activates motor-perceptual systems; externally cued reactivation of internal models supports recovery.

Near-Death Experiences (cardiac arrest, coma)
Afferent Loop: Absent or minimal
Metafunctional Processing: Transiently disembodied
Efferent Output: Purely Transphysical
Implication: Coherent perception reported under minimal neural activity; may suggest perception continues beyond standard neural correlates.

Imagination & Mental Imagery (internally generated but regulated)
Afferent Loop: Absent
Metafunctional Processing: Active and contextually regulated
Efferent Output: Simulated, coherent
Implication: Internally produced content remains governed by attention and narrative structure; contrasts with decoupled hallucinations.

Dreaming (REM sleep, internally generated narratives)
Afferent Loop: Largely absent (minimal external sensory input during REM)
Metafunctional Processing: Highly active, emotionally and mnemonically driven
Efferent Output: Coherent, immersive, narrative-structured projection
Implication: Dreams simulate full experiential worlds in the absence of external input. Their coherent spatial, emotional, and symbolic content supports the view that the efferent loop is not merely reactive, but

generative and contextually integrated. Highlights the predictive, autobiographical, and affective dimensions of perception beyond wakeful input.

Note: *The terms used in the table characterize the functional status of each component within the afferent–metafunctional–efferent loop and help clarify how perceptual dynamics vary across clinical and cognitive phenomena. Intact refers to typical or unaltered functioning of a loop component, while reduced or absent indicates diminished or disrupted activity due to injury, degeneration, or sensory deprivation. Active designates functional engagement in perception, whereas compensatory describes adaptive processes that restore or preserve perceptual coherence in the face of loss. When processing is partially disrupted or distorted, integration is incomplete or altered, often leading to spatial or cognitive inconsistencies. Dysregulated denotes unmodulated or unstable perceptual processing, typically associated with pathology, such as hallucinations or delusions. Decoupled refers to efferent projection occurring without afferent grounding, resulting in perceptual content that lacks contextual anchoring. In contrast, simulated or generative processes, as seen in imagination or motor imagery, involve internally produced perceptual content that remains coherent and context sensitive. Finally, transphysical describes cases in which coherent perception appears to persist despite the absence of measurable neural activity, suggesting the involvement of non-local or extended processes. Collectively, these descriptors offer a conceptual vocabulary for tracing the continuum from normative perception to clinical pathology, providing a framework for understanding how the brain constructs, disrupts, or recalibrates perceptual experience across different conditions.*

Dreams as Metafunctional Diagnostics: Symbolic Projection and Subliminal Signal Integration

The case of a woman who repeatedly dreamed of a dog biting her leg—only to later be diagnosed with a tumor in the same location—offers a powerful example of the Brain Metafunction Framework in action. It illustrates how perception, particularly in dream states, can integrate subliminal bodily signals and project them outward in symbolically meaningful ways. From a metafunctional perspective, this case suggests that her brain was receiving low-level afferent input—possibly subtle nociceptive or inflammatory cues—not strong enough to breach the threshold of waking consciousness. During REM sleep, however, this subliminal data was processed through central metafunctional integration, combining interoceptive awareness with emotional memory and symbolic association. The result was an efferent projection in the form of a vivid, emotionally charged dream-image: a dog bite.

This image was not random, but an embodied dramatization of localized threat, surfacing in a domain where the usual filters of waking cognition are relaxed. The dog, as a symbol of pain or aggression, became the brain's narrative expression of an emerging pathology. This case therefore exemplifies how perception in dream states can function as a form of embodied diagnostics, revealing the integrative capacities of the brain across affective, cognitive, and interoceptive systems—even in the absence of conscious awareness. It also supports the view that dreams are not epiphenomenal but serve as perceptual simulations that anticipate, interpret, and externalize latent bodily and psychological states. Domhoff's (2003) empirical work reinforces this perspective, showing that dreams reflect stable cognitive patterns and neural architectures rather than random by-products. His findings support the idea that dream content is meaningfully structured, often echoing waking concerns, autobiographical memory, and unresolved affective themes. Clinically, such cases suggest that dreams may hold diagnostic relevance, especially when they exhibit recurrent symbolic localization of bodily distress. Philosophically, this reinforces the notion that perception is not merely a bottom-up phenomenon but a bidirectional, interpretive, and transphysical process—active even in states where sensory input is reduced or reorganized. Dreams, in this light, become windows into the brain's deep metafunctional architecture, revealing how perception operates not just to reflect the world, but to prefigure and participate in it.

Neural Correlates of Dreaming: Simulated Worlds and Metafunctional Integration

Dreaming exemplifies the brain's capacity to generate immersive, emotionally charged experiences in the absence of external sensory input, making it a clear expression of top-down, internally driven processing. Within the Brain Metafunction Framework, dreaming reveals how perception can emerge from efferent dynamics alone, sustained by memory, emotion, and symbolic modeling. The phenomenological richness of dreams—often indistinguishable from waking perception while they unfold—suggests that the brain can construct a full experiential world using only internal models, shaped by expectation, prior experience, and systemic integration (Horikawa et al., 2013; Squire & Dede 2015; Siclari et al., 2017). This capacity is supported by coordinated activity across several large-scale neural systems. Core regions implicated in the dreaming process include:

Cerebral Cortex: Especially the frontal and temporal lobes, which contribute to the generation of narrative structure, symbolic content, and semantic coherence (Pace-Schott, 2007).

Thalamus: Acting as a dynamic sensory relay, the thalamus modulates internal signal transmission between cortical and subcortical structures during REM sleep, enabling immersive simulation without afferent input (Hobson & Pace-Schott, 2002). As Nir and Tononi (2010) emphasize, dreaming involves coordinated thalamocortical activity that simulates perception in the absence of afferent input, supporting the view that internally generated experiences rely on functional connectivity resembling waking consciousness.

Basal Ganglia: These contribute to motor imagery and the embodied dimension of dream action, reflecting the procedural and kinesthetic aspects of dream experiences (Hobson et al., 2000).

Limbic System: The amygdala and associated structures influence the emotional salience and intensity of dream content, helping to encode symbolic and autobiographical meaning (Kahn et al., 2000; Siclari et al., 2017; Scarpelli et al., 2019).

Brainstem: Particularly involved in the regulation of REM cycles, the brainstem supports the physiological transitions that make dreaming possible, linking somatic regulation to the cognitive architecture of dreams (Hobson & Pace-Schott, 2002).

This distributed neural orchestration demonstrates that dreaming, like other perceptual phenomena addressed in this chapter, emerges from the dynamic integration of neural, emotional, and narrative processes—hallmarks of metafunctional processing. Notably, the vividness and coherence of dreams despite the lack of external input aligning dreaming closely with phenomena like hallucinations, imagination, and synesthesia, underscoring the continuity between waking and non-waking modes of perception.

The Surprise Principle and Paradigm Shift

The *Surprise Principle*—a well-established concept in both philosophy and the philosophy of science—serves as a critical engine for discovery. At its core, it asserts that when empirical observations sharply contradict prevailing theories or expectations, they trigger cognitive dissonance that leads to re-evaluation, inquiry, and eventually, transformation (Collins, 2009). Surprise, in this sense, is not a mere anomaly; it is a signal that a system of understanding may be incomplete or insufficient. This principle aligns closely with Thomas Kuhn's model of scientific revolutions (1962), wherein unresolved anomalies accumulate until they catalyze a paradigm shift. These shifts do not occur because new data are simply added to an old theory, but because the old theory fails to accommodate them—forcing the emergence of a new conceptual framework.

In the study of perception and consciousness, the Surprise Principle proves especially illuminating. A range of exceptional phenomena—phantom limb experiences, including congenital limb absence (Diers, 2023), blindsight, hallucinations, vivid dreams, synesthesia, and near-death experiences (NDEs)—pose challenges to materialist and reductionist models that equate perception strictly with afferent input or localized brain activity. These experiences often persist, or even intensify, in the absence of conventional sensory input, revealing cognitive and perceptual coherence that exceeds bottom-up explanation. From a reductionist viewpoint, these phenomena are puzzling or pathological. From the perspective of the Brain Metafunction Framework, however, they are not surprising at all. In fact, they are anticipated. This triadic model—integrating afferent input, centralized metafunctional processing, and efferent projection—naturally explains how perception can be internally generated, contextually regulated, and outwardly projected, even under non-standard conditions. These dynamics are further clarified in the chapter's clinical synthesis table, which shows how disorders and therapeutic interventions map coherently onto the framework's three-loop structure.

Thus, the Surprise Principle not only validates the explanatory power of the Brain Metafunction Framework—it necessitates it. The very anomalies that contradict physicalist expectations become predictive confirmations within a transphysical, integrative paradigm. Far from being fringe data, these cases help surface the deeper architecture of perception and consciousness itself. Ultimately, the Surprise Principle invites a new scientific attitude—one that does not marginalize surprise but welcomes it as a signpost for discovery. Within this spirit, the Brain Metafunction Framework does more than explain anomalies. It reframes the field, offering a vision of perception not as isolated computation, but as a world-participatory, meaning-bearing act—a phenomenon that spans the neural, bodily, psychological, and transphysical domains. This is not only a shift in theory; it is a paradigm shift in how we understand mind, brain, and being.

Conclusion: Perception as Transphysical Participation and Clinical Insight

This chapter has addressed one of the deepest paradoxes of perception: how a single reality can appear both "out there" in the world and "in here" in the mind. At the heart of this enigma lies the question of how the brain—an organ with no direct access to external objects—can generate a coherent, immersive, and meaningful experience of reality. The Brain Metafunction Framework resolves this paradox by reconceiving perception as a triadic process: integrating afferent input, centralized metafunctional processing, and efferent projection. It proposes that perception is not a passive registration of sensory stimuli but an active, world-participatory process, a structured act in which the brain, body, and self converge to generate meaning.

This chapter advances the theory by adding a crucial dimension: the transphysical nature of perception. We have seen that perception does not remain confined to the brain. Through the efferent loop, perceptual experience is outwardly projected, situated within the relational field of embodied life. Pain is not felt in the brain—it is experienced in the wounded limb. A voice is not merely registered as neural signals—it echoes in a shared acoustic space. This projection is not metaphorical; it is a structured, integrative function through which perception externalizes inner modeling into lived, spatial, and symbolic contexts. Critically, this framework accommodates phenomena that defy classical physicalist explanations—phantom limb pain, blindsight, hallucinations, dreams, near-death experiences, and synesthesia—each of which illustrates unique configurations in the afferent–metafunctional–efferent loop. Through a comparative clinical table, we explored how perceptual coherence can be disrupted, distorted, or restored across diverse pathologies and therapeutic interventions. This table not only organizes a spectrum of perceptual anomalies but also offers diagnostic and therapeutic insights, showing how perceptual loops may be reengaged or modulated to restore personhood and coherence. Perhaps most striking is the clinical evidence that perception can be restructured even in the absence of traditional afferent input. Mirror therapy for phantom limb pain, motor imagery for stroke rehabilitation (Cui, et al. 2022), VR-based re-embodiment for trauma recovery, all reveal the healing power of the efferent loop. Such interventions do not merely compensate for sensory deficits, they restructure perception at its root. Likewise, dream content may reflect subliminal or nonconscious afferent signals, symbolically represented within the generative space of sleep. In cases such as the woman who dreamed of a dog biting her leg—later diagnosed with a tumor in the same location—we see the profound diagnostic potential of symbolic, projective perception. These are not anomalies; they are expressions of the full spectrum of metafunctional activity.

Taken together, these insights invite a paradigm shift. They affirm that the brain is not a closed computational device, but a participant in a dynamic, relational world in which the mind, body, meaning, and environment are inseparably intertwined. The Surprise Principle, drawn from philosophy of science, calls us to take seriously those anomalies that resist reductive explanation. They are not noise to be ignored—they are clues pointing toward a richer ontology of the self. The implications of this paradigm are fourfold:

Theoretical: The Brain Metafunction Framework challenges reductionist paradigms by proposing a transphysical, integrative account of perception that aligns with lived experience and anomalous data alike.

Clinical: It provides a diagnostic and therapeutic roadmap, especially for conditions involving disrupted embodiment, disorganized perception, or altered consciousness.

Philosophical: It restores perception to its rightful place as an ontological act—not merely a neural event, but an expression of the self's engagement with a meaningful world.

Existential: It affirms the unity of the person across time, even amid brain change, trauma, or disembodiment—supporting the notion of the soul as a higher-order, integrating principle.

In this view, perception becomes not merely a function of the brain, but an act of the person—a being situated in meaning, continuity, and intentional world-engagement. It is through perception that we not only register the world but inhabit it, shape it, and come to know who we are within it. Thus, perception—understood metafunctionally and transphysically—is not merely an empirical puzzle to be solved. It is the medium through which we encounter reality, act with purpose, suffer with depth, and love with awareness. It is, in the fullest sense, the soul's participation in the world.

REFERENCES

Apkarian, A. V., Baliki, M. N., & Geha, P. Y. (2009). Towards a theory of chronic pain. *Progress in Neurobiology, 87*(2), 81–97. https://doi.org/10.1016/j.pneurobio.2008.09.018

Barrett, L. F., & Simmons, W. K. (2015). Interoceptive predictions in the brain. *Nature Reviews Neuroscience, 16*(7), 419 –429.

Beck, B., Làdavas, E., & Haggard, P. (2016). Viewing the body modulates both pain sensations and pain responses. *Experimental Brain Research, 234*(7), 1795–1805. https://doi.org/10.1007/s00221-016-4585-9.

Buckner, R. L. (2013). The brain's default network: Origins and implications for the study of psychosis. *Dialogues in Clinical Neuroscience, 15*(3), 351–358. https://doi.org/10.31887/DCNS.2013.15.3/rbuckner.

Bushnell, M. C., Čeko, M., & Low, L. A. (2013). *Cognitive and emotional control of pain and its disruption in chronic pain.* Nature Reviews Neuroscience, 14(7), 502–511. https://doi.org/10. 1038/nrn3516

Carter, C. (2011). Response to "Could Pam Reynolds hear?" *Journal of Near-Death Studies, 30*(1), 29–53.

Clark, A. (2016). *Surfing uncertainty: Prediction, action, and the embodied mind.* Oxford University Press.

Collins, R. (2009). The fine-tuning argument: A scientific explanation of the universe's design. *Philosophy and Cosmology Review, 3*, 19–34.

Cook, E. W., Greyson, B., & Stevenson, I. (1998). Do any near-death experiences provide evidence for the survival of human personality after death? *Journal of Scientific Exploration, 12*(3), 377–406.

Corbetta, M., & Shulman, G. L. (2002). Control of goal-directed and stimulus-driven attention in the brain. *Nature Reviews Neuroscience, 3*(3), 201–215. https://doi.org/10.1038/nrn755

Cowey, A. (2010). The blindsight saga. *Experimental Brain Research, 200*(1), 3–24. https://doi.org/10.1007/s00221-009 -1914-2

Craig, A. D. (2003). A new view of pain as a homeostatic emotion. *Trends in Neurosciences, 26*(6), 303–307. doi: 10.1016/s0166-2236(03)00123-1.

Craig, A. D. (2009). How do you feel—now? The anterior insula and human awareness. *Nature Reviews Neuroscience, 10* (1), 59–70.

Cui, W., Huang, L., Tian, Y., Luo, H., Chen, S., Yang, Y., Li, Y., Fu, J., Yu, Q., & Xu, L. (2022). Effect and mechanism of mirror therapy on lower limb rehabilitation after ischemic stroke: A fMRI study. *NeuroRehabilitation, 51*(1), 65–77. https://doi.org/10.3233/NRE-210307

Damasio, A. R. (1999). The feeling of what happens: Body and emotion in the making of consciousness. New York: Harcourt.

Danckert, J., & Rossetti, Y. (2005). Blindsight in action: What can the different sub-types of blindsight tell us about the control of visually guided actions? *Neuroscience & Biobehavioral Reviews, 29*(7), 1035–1046. https://doi.org/ 10.1016/j.neubiorev.2005.02.001

Decety, J., & Jackson, P. L. (2004). The functional architecture of human empathy. *Behavioral and Cognitive Neuroscience Reviews, 3*(2), 71–100. doi: 10.1177/1534582304267187.

Dehaene, S., & Changeux, J. P. (2011). Experimental and theoretical approaches to conscious processing. *Neuron, 70*(2), 200–227. doi: 10.1016/j.neuron.2011.03.018. PMID: 21521609

Diers, M., Fuchs, X., Bekrater-Bodmann, R., & Flor, H. (2023). Prevalence of phantom phenomena in congenital and early-life amputees. *The Journal of Pain, 24*(3), 502–508. https://doi.org/10.1016/j.jpain.2022.10.010.

Domhoff, G. W. (2003). *The scientific study of dreams: Neural networks, cognitive development, and content analysis.* American Psychological Association. https://doi.org/10.10 37/10463-000

Eccleston, C., & Crombez, G. (1999). *Pain demands attention: A cognitive–affective model of the interruptive function of pain.* Psychological Bulletin, 125(3), 356–366. https://doi.org/ 10.1037/0033-2909.125.3.356

Fenwick, P., & Fenwick, E. (1995). The truth in the light: An investigation of over 300 near-death experiences. Berkley Publishing.

Fillingim, R. B. (2017). Individual differences in pain: Understanding the mosaic of factors that influence pain perception and response. *Pain, 158*(Suppl 1), S11–S18. https://doi.org/10.1097/j.pain.0000000000000775

Friston, K. (2010). The free-energy principle: A unified brain theory? *Nature Reviews Neuroscience, 11*(2), 127–138. https://doi.org/10.1038/nrn2787

Friston, K., Adams, R., Perrinet, L., & Breakspear, M. (2012). Perceptions as hypotheses: Saccades as experiments. *Frontiers in Psychology, 3*, 151. https://doi.org/10.3389/fpsyg. 2012.00151

Gallagher, S. (2005). How the body shapes the mind. Oxford University Press. https://doi.org/10.1093/0199271 941.001.0001

Gallagher, S., & Zahavi, D. (2021). *The phenomenological mind* (3rd ed.). Routledge. https://doi.org/10.4324/9780367 334246

Gordon, R. L., Cobb, P. R., & Balasubramaniam, R. (2018). Recruitment of the motor system during music listening: An ALE meta-analysis of fMRI data. *PLoS ONE, 13*(11), e0207213.

Gregory, R. L. (1997). Knowledge in perception and illusion. *Philosophical Transactions of the Royal Society B: Biological Sciences, 352*(1358), 1121–1128. doi: 10.1098/rst b.1997.0095.

Greyson, B. (2003a). Near-death experiences in a hospital setting: Phenomena and scientific implications. *Journal of Near-Death Studies, 23*(4), 203–222.

Greyson, B. (2003b). Near-death experiences in a psychiatric outpatient clinic population. *Psychiatric Services, 54*(12), 1649–1651. https://doi.org/10.1176/appi.ps.54.1 2.1649

Haggard, P., Iannetti, G. D., & Longo, M. R. (2013). Spatial sensory organization and body representation in pain perception. *Current Biology, 23*(4), R164–R176. https://doi.org/10.1016/j.cub.2013.01.047.

Haggard, P., et al. (2005). Visual analgesia: Seeing the body reduces pain intensity. *Journal of Neuroscience, 25*(6), 1413 –1420.

Haggard, P. (2023, April 25). Does brain make mind? Closer To Truth. [Video]. YouTube. https://youtu.be/BoahYpvydbE?si=Bn5UU6X8HJRemui.

Hobson, J. A., & Pace-Schott, E. F. (2002). The cognitive neuroscience of sleep: Neuronal systems, consciousness, and learning. *Nature Reviews Neuroscience, 3*(9), 679–693.

Hohwy, J. (2013). *The predictive mind.* Oxford University Press. https://doi.org/10.1093/acprof:oso/9780199682737.0 01.0001

Holden, J. M. (2007). More things in heaven and earth: A response to "Near-death experiences with hallucinatory features." *Journal of Near-Death Studies, 26*(1), 33–42.

Holden, J. M., ed. (2009). *The Handbook of Near-Death Experiences: Thirty Years of Investigation.* Santa Barbara, CA: Praeger.

Horikawa, T., Tamaki, M., Miyawaki, Y., & Kamitani, Y. (2013). Neural decoding of visual imagery during sleep. *Science, 340*(6132), 639–642. https://doi.org/10.1126/science .1234330

Kahn, D. et al. (2000). Dreaming and the brain. *Behavioral and Brain Sciences, 23*(6), 793–1121.

Kant, I. (1781). *Critique of Pure Reason.* London: Macmillan.

Kentridge, R. W., Heywood, C. A., & Weiskrantz, L. (1999). Attention without awareness in blindsight. *Proceedings of the Royal Society B: Biological Sciences, 266*(1430), 1805–1811. https://doi.org/10.1098/rspb.1999.0850

King, A. J., Hammond-Kenny, J. M., & Nodal, F. R. (2019). Hearing shapes: Cross-modal interactions between auditory and visual systems. *Annual Review of Neuroscience, 42*(1), 119–135.

Koelsch, S. (2014). Brain correlates of music-evoked emotions. *Nature Reviews Neuroscience, 15*(3), 170–180.

Kollmeier, B. (2008). Understanding auditory processing: The interplay of sensory input and brain mechanisms. *Hearing Research, 239*(1), 1–2.

Kollmeier, B. (2008). *Hearing—From Sensory Processing to Perception.* Berlin: Springer.

Kosslyn, S. M., Thompson, W. L., Kim, I. J., & Alpert, N. M. (2001). The role of Area 17 in visual imagery: Convergent evidence from PET and rTMS. *Science, 284*(5411), 167–170.

Kuhn, T. S. (1962). *The Structure of Scientific Revolutions.* University of Chicago Press.Lee, J. H., et al. (2012). Auditory top-down processing. *Hearing Research, 290*(1–2), 190.

Lee, Y. S., Janata, P., Frost, C., Hanke, M., & Granger, R. (2012). Top-down influences on auditory perception. *Frontiers in Psychology, 3*, 190.

Lockwood, P. L., Iannetti, G. D., & Haggard, P. (2013). Transcranial magnetic stimulation over human secondary somatosensory cortex disrupts perception of pain intensity. *Cortex*, **49**(8), 2201–2209. https://doi.org/10.1016/j.cortex.2012.10.006

Long, J. (2014). *Evidence of the Afterlife: The Science of Near-Death Experiences.* HarperOne.

Long, J., & Perry, P. (2010). *God and the Afterlife: The Groundbreaking New Evidence for God and Near-Death Experience.* HarperOne.

Long, J., & Perry, P. (2010). *Evidence of the Afterlife: The Science of Near-Death Experiences.* New York: HarperCollins.

Longo, M. R., Cardozo, S., & Haggard, P. (2008). Visual enhancement of touch and the bodily self. *Consciousness and Cognition, 17*(4), 1181–1191. https://doi.org/10.1016/j.con cog .2008.01.001

Longo, M. R., Betti, V., Aglioti, S. M., & Haggard, P. (2009). Visually induced analgesia: Seeing the body reduces pain. *Journal of Neuroscience*, 29(39), 12125–12130. https://doi.org/ 10.1523/JNEUROSCI.3072-09.2009

Longo, M. R., Iannetti, G. D., Mancini, F., Driver, J., & Haggard, P. (2012). Linking pain and the body: Neural correlates of visually induced analgesia. Journal of Neuroscience, 32(8), 2601–2607. https://doi.org/10.1523/JNEUROSCI.4031-11.2012

Mancini, F., Longo, M. R., Kammers, M. P., & Haggard, P. (2011). Visual distortion of body size modulates pain perception. *Psychological Science, 22*(3), 325–330. https://doi.org/10.1177/0956797611398496

Menon, V., & Uddin, L. Q. (2010). Saliency, switching, attention, and control: A network model of insula function. *Brain Structure and Function, 214*(5–6), 655–667.

Merleau-Ponty, M. (2012). *Phenomenology of perception* (D. A. Landes, Trans.). Routledge. (Original work published 1945)

Moody, R. A. (1975). *Life after Life: The Investigation of a Phenomenon—Survival of Bodily Death.* Atlanta, GA: Mockingbird Books.

Nagel, T. (2012). *Mind and cosmos: Why the materialist neo-Darwinian conception of nature is almost certainly false.* Oxford University Press.

Newton, I. (1952). *Opticks: Or, A Treatise of the Reflexions, Refractions, Inflexions and Colours of Light.* London: William Innys.

Nir, Y., & Tononi, G. (2010). *Dreaming and the brain: From phenomenology to neurophysiology.* Trends in Cognitive Sciences, 14(2), 88–100. https://doi.org/10.1016/j.tics.2009.12.001

Ongaro, G., & Kaptchuk, T. J. (2019). Symptom perception, placebo effects, and the Bayesian brain. *Pain, 160*(1), 1–4. https://doi: 10.1097/j.pain.0000000000001367.

Parnia, S., & Fenwick, P. (2002). Near death experiences in cardiac arrest: Visions of a dying brain or visions of a new science of consciousness. *Resuscitation, 52*(1), 5–11. https://doi.org/10.1016/S0300-9572(01)00469-5

Parnia, S., Post, S. G., Lee, M. T., Lyubomirsky, S., Aufderheide, T. P., Deakin, C. D., Greyson, B., Long, J., Gonzales, A. M., Huppert, E. L., Dickinson, A., Mayer, S., Locicero, B., Levin, J., Bossis, A., Worthington, E., Fenwick, P., & Keshavarz Shirazi, T. (2022). Guidelines and standards for the study of death and recalled experiences of death—A multidisciplinary consensus statement and proposed future directions. *Annals of the New York Academy of Sciences, 1511*(1), 5–30. https://doi.org/10.1111/nyas.14880

Penfield, W., & Boldrey, E. (1937). Somatic motor and sensory representation in the cerebral cortex of man. *Brain, 60* (4), 389–443.

Raichle, M. E. (2015). The brain's default mode network. *Annual Review of Neuroscience, 38*, 433–447. https://doi.org/10.1146/annurev-neuro-071013-014030

Reddan, M. C., & Wager, T. D. (2018). Pain perception: Beyond nociception. *eNeuro, 5*(3), ENEURO.0150-18.2018. https://doi.org/10.1523/ENEURO.0150-18.2018

Ridder, D., Adhi, M., & Vanneste, S. (2021). Understanding chronic pain through the lens of different brain networks. *Neuroscience & Biobehavioral Reviews, 130*, 229–243. https://doi.org/10.1016/j.neubiorev.2021.05.003

Ring, K., & Cooper, S. (1997). Near-death and out-of-body experiences in the blind: A study of apparent eyeless vision. *Journal of Near-Death Studies, 16*(2), 101–147.

Robinson, H. (2020). *Perception.* Routledge.

Sabom, M. B. (1982). *Recollections of Death: A Medical Investigation.* Harper & Row.

Sartori, P. (2008a). *The Near-Death Experiences of Hospitalized Intensive Care Patients: A Five-Year Clinical Study.* Mellen Press.

Sartori, P. (2008b). *The Wisdom of Near-Death Experiences.* London: Watkins.

Scarpelli, S., Bartolacci, C., D'Atri, A., Gorgoni, M., & De Gennaro, L. (2019). The functional role of dreaming in emotional processes. *Frontiers in Psychology, 10*, Article 459. https://doi.org/10.3389/fpsyg.2019.00459

Seeley, W. W., et al. (2007). Dissociable intrinsic connectivity networks for salience processing and executive control. *Journal of Neuroscience, 27*(9), 2349–2356.

Seth, A. K. (2014). A predictive processing theory of sensorimotor contingencies: Explaining the puzzle of perceptual presence and its absence in synesthesia. *Cognitive Neuroscience*, 5(2), 97–118. https://doi.org/10.1080/17588928.2013.877880

Seth, A. K. (2015). Inference to the best prediction: A reply to Wanja Wiese. In T. Metzinger & J. M. Windt (Eds.), *Open MIND: 35(R)*. MIND Group. https://doi.org/10.15502/9783958570986

Seth, A. K., & Friston, K. J. (2016). Active interoceptive inference and the emotional brain. *Philosophical Transactions of the Royal Society B, 371*(1708), 20160007.

Seth, A. K. (2013). Interoceptive inference, emotion, and the embodied self. *Trends in Cognitive Sciences, 17*(11), 565–573.

Siclari, F., Baird, B., Perogamvros, L., Bernardi, G., LaRocque, J. J., Riedner, B., Boly, M., Postle, B. R., & Tononi, G. (2017). The neural correlates of dreaming. *Nature Neuroscience, 20*(6), 872–878. https://doi.org/10.1038/nn.4545

Squire, L. R., & Dede, A. J. (2015). Conscious and unconscious memory systems. *Cold Spring Harbor Perspectives in Biology, 7*(3), a021667.

Thonnard, M., Charland-Verville, V., Brédart, S., Dehon, H., Ledoux, D., Laureys, S., & Vanhaudenhuyse, A. (2013). *Characteristics of near-death experiences memories as compared to real and imagined events.* PLoS ONE, 8(3), e57620. https://doi.org/10.1371/journal.pone.0057620

Tononi, G., & Koch, C. (2015). Consciousness: Here, there, and everywhere? *Philosophical Transactions of the Royal Society B: Biological Sciences, 370*(1668), 20140167. https://doi.org/10.1098/rstb.2014.0167

Tononi, G., & Edelman, G. M. (1998). Consciousness and complexity. *Science, 282*(5395), 1846–1851. https://doi.org/10.1126/science.282.5395.1846

Tracey, I., & Mantyh, P. W. (2007). The cerebral signature for pain perception and its modulation. *Neuron, 55*(3), 377–391. https://doi.org/10.1016/jneur on.2007.07.012

Tracey, I., Woolf, C. J., & Andrews, N. A. (2019). Composite pain biomarker signatures for objective assessment and effective treatment. *Neuron, 101*(5), 783–800. https://doi.org/10.1016/j.neuron.2019.02.019

Tracey, I., & Mantyh, P. W. (2007). The brain in pain. *PAIN®, 137*(1), 7–15.

Van Lommel, P., van Wees, R., Meyers, V., & Elfferich, I. (2001). Near-death experience in survivors of cardiac arrest: A prospective study in the Netherlands. *The Lancet, 358*(9298), 2039–2045. https://doi.org/10.1016/S01 40-6736(01)07100-8

Velmans, M. (1990). Consciousness, brain, and the physical world. *Philosophical Psychology, 3*(1), 77–99.

Velmans, M. (2000). *Understanding Consciousness.* Routledge.

Velmans, M. (2005). Reflexive monism. *Journal of Consciousness Studies, 12*(2), 109–116.

Weiskrantz, L. (1986). *Blindsight: A Case Study and Implications.* Oxford: Clarendon Press.

Weiskrantz, L., Warrington, E. K., Sanders, M. D., & Marshall, J. (1974). Visual capacity in the hemianopic field following a restricted occipital ablation. *Brain, 97*(4), 709–728. https://doi.org/10.1093/brain/97.4.709

Wright, N. T. (2003). *The Resurrection of the Son of God.* Minneapolis: Fortress Press.

6

Attention

Attention as a Brain Metafunction: Integration Across Sensory and Cognitive Domains

What influence does attention have on behavioral performance? Can it shape perception itself, altering how things appear to us? And from which regions of the brain do attention's top-down influences arise? These questions have animated centuries of inquiry, beginning with philosophical reflection, continuing through early psychology, and now driving cutting-edge neuroscience. As a cognitive process, attention plays a foundational role in focusing mental resources on selected objects, thoughts, or tasks while ignoring irrelevant distractions (Posner & Rothbart, 2007). It actively shapes perception by enhancing the salience and clarity of sensory input, filtering environmental noise, and sharpening mental focus. Research into attention now spans multiple disciplines—cognitive neuroscience, psychology, neuropsychology, and education. Importantly, attention is no longer viewed as a singular function but as a multifaceted, brain-wide system encompassing selective, sustained, and divided attention—each influencing behavior in distinct but overlapping ways (Parasuraman & Davies, 1984). These forms of attention improve performance by allocating limited cognitive resources, enhancing perception, guiding learning, and enabling adaptive behavior.

The top-down influences of attention originate from distributed systems, including the prefrontal cortex, posterior parietal regions, limbic circuitry, and neuromodulatory hubs such as the locus coeruleus. These regions work in concert to prioritize task-relevant stimuli, suppress interference, and coordinate perception with executive function and memory (Miller & Cohen, 2001). Attention, in this view, becomes a system-wide, goal-oriented integrator of mind and behavior. This chapter explores attention as a metafunctional process—a large-scale neural dynamic that integrates perception, memory, emotion, and executive systems. By reframing attention as a metafunction, we situate it within a broader cognitive architecture that supports learning, decision-making, and awareness. Understanding attention in this way clarifies its central role in the construction of conscious experience and its critical position in the brain's global coordination of thought and action.

What Is Attention?

The influential psychologist and philosopher William James offered a foundational definition of attention in the late 19th century. He famously wrote:

> "Everyone knows what attention is. It is the taking possession by the mind, in clear and vivid form, of one out of what seems several simultaneously possible objects or trains of thought. Focalization, concentration, of consciousness are of its essence" (James, 1890, pp. 403–404).

He also emphasized its central role in voluntary mental control: "The faculty of voluntarily bringing back a wandering attention, over and over again, is the very root of judgment, character, and will" (p. 424). In James's view, attention was not merely a reactive mechanism—it was an intentional act of mind, a willful process of mental selection that enabled individuals to prioritize and sustain focus on relevant stimuli amidst distraction. Contemporary cognitive science has refined and expanded this early view. Attention is now understood as a multifaceted set of processes—both voluntary and involuntary—that govern the selection, enhancement, and suppression of sensory and cognitive input (Anderson, 2011). In psychological terms, attention is commonly defined as "the concentration of awareness on some phenomenon, excluding other stimuli" (Encyclopædia Britannica, n.d.). It enables selective processing of relevant information while inhibiting interference—essential for maintaining coherent awareness in a world of competing inputs (Posner, 1980).

Attention is also deeply involved in regulating levels of arousal, alertness, and task readiness, ensuring that cognitive resources are available and efficiently deployed when needed (Oken, Salinsky, & Elsas, 2006). In this sense, attention is a gateway to cognition, coordinating sensory inputs and guiding conscious awareness toward the most pertinent stimuli for survival, learning, and goal achievement. These systems reflect early-stage, distributed neural modulation—anticipating attention's role as a brain-wide metafunction (Menon & Uddin,

2010). For example, the alerting network within the brain plays a fundamental role in preparing us for incoming stimuli, while the orienting network directs our attention to specific sensory inputs (Posner and Petersen 1990; Corbetta and Shulman 2002). Taken together, these functions position attention not only as a selective filter, but as a modulator of conscious experience, intimately tied to perception, learning, memory, and executive control. From both classical and modern perspectives, attention emerges as a dynamic system that governs how we engage with the world—both inwardly and outwardly.

Is Attention a Metafunctional Process?

Attention is not merely a passive filter for perception but an active, dynamic process that interfaces with memory, perception, and executive functions. As explored in the preceding chapter on Perception, a metafunctional brain theory situates attention as a globally integrative system—one that enhances perceptual clarity, prioritizes stimuli, and coordinates higher-order cognitive operations. Attention collaborates with perception to construct meaning from sensory input, allowing the brain to form a coherent understanding of the external world (Ackerman, 1992). By filtering distractions and selecting task-relevant stimuli, attention supports effective action and decision-making. This function is especially critical given the sheer volume of sensory input the brain continuously receives (Plebanek & Sloutsky, 2019). Far from being a unitary mechanism, attention comprises multiple systems—orienting, alerting, executive control—that work in concert to manage internal and external demands.

Across sensory modalities, research shows how attention operates as a brain-wide metafunction, dynamically managing cognitive load and facilitating adaptive responses. Menon and Uddin (2010), for example, propose a network-based model in which the salience network, anchored in the anterior insula and anterior cingulate cortex, acts as a switch between the default mode network (associated with internal reflection) and executive control networks (linked to goal-directed cognition). This switching mechanism illustrates how attention mediates between introspective states and external engagement, maintaining a continuous and flexible alignment between perception, emotion, memory, and action. Supporting this integrative view, Saalmann et al. (2012) demonstrate that the pulvinar nucleus helps regulate information flow between cortical areas based on attentional demands, reinforcing the idea that attention exerts modulatory control across distributed neural systems.

A clear illustration of this dynamic is seen when walking through a crowded street while engaged in internal thought. At one moment, your attention may be inward, revisiting a conversation from earlier in the day. Suddenly, a honking car or a fast-moving pedestrian captures your attention, shifting it outward. This momentary reallocation of neural resources exemplifies attention's ability to prioritize urgent stimuli, suppress irrelevant input, and recalibrate behavior, all within milliseconds. It also exemplifies attention's role as a metafunctional interface, constantly regulating the flow of consciousness across internal and external domains. By framing attention as a metafunctional process, we underscore its essential role in managing limited cognitive resources and ensuring the alignment of moment-to-moment experience with both immediate behavioral goals and broader contextual demands. This perspective sets the stage for examining influential neurocognitive models of attention—many of which trace their roots to this integrative understanding and expand upon its neural underpinnings.

Luria's Neurocognitive Theory of Attention

Building on foundational insights into the brain's distributed architecture, Alexander Luria's neurocognitive theory of attention offers an early and remarkably prescient account of the brain's integrated functions. In The Working Brain (1973), Luria outlined a triadic model consisting of attention, mnestic (memory) processes, and cortical activation, forming the pillars of higher cognitive function. This model anticipated a systems-level view of the brain, in which cognitive processes are not localized to isolated regions but are dynamically coordinated across multiple brain systems. At the center of Luria's framework lies the prefrontal cortex, which acts as a regulatory hub by linking posterior cortical regions (responsible for processing incoming sensory information) with subcortical structures, including the limbic system. This configuration allows the brain to flexibly integrate internal goals with external stimuli—prioritizing emotionally salient or task-relevant information. Notably, this interaction between emotional salience and attentional regulation anticipates more recent findings on the functional interplay between the prefrontal cortex and limbic areas such as the amygdala (Pessoa, 2009).

Luria conceptualized attention not as a static spotlight but as a dynamic regulatory system that modulates sensory input based on task relevance and internal state. He emphasized the importance of cortical activation—a precursor to what is now called arousal regulation, sustaining attentional engagement. The reticular activating system (RAS), located in the brainstem, supports this function by maintaining alertness and readiness to respond. This resonates with modern understandings of the locus coeruleus–norepinephrine system, which plays a critical role in regulating vigilance and adaptive attention (Aston-Jones & Cohen, 2005). Crucially, Luria's theory also linked attention and memory as co-dependent systems. In practical terms, memory does not function independently of attentional selection—rather, it guides and is shaped by it. For example, when trying to recall a face in a crowd, attentional focus helps isolate relevant sensory inputs, while memory provides the template that makes recognition possible. This reciprocal relationship mirrors current models of large-scale functional connectivity between frontoparietal control networks, default mode networks, and limbic circuits (Menon & Uddin, 2010).

A clinical illustration of Luria's model can be seen in patients recovering from traumatic brain injury (TBI). These individuals often exhibit attentional deficits not because of sensory impairments, but due to disruptions in prefrontal–limbic regulation. For instance, a TBI patient may become overwhelmed in emotionally charged or distracting environments—unable to filter irrelevant stimuli or sustain focus on a task. This attentional breakdown often co-occurs with short-term memory challenges and heightened emotional reactivity, reflecting impaired coordination among the triadic subsystems. In therapy, targeted interventions to rebuild attentional control often include emotion regulation strategies, working memory exercises, and arousal modulation, directly echoing the integrated approach Luria envisioned.

From the vantage point of the brain metafunction framework, Luria's theory appears remarkably forward-thinking. It reflects a metafunctional view of attention: attention is not a discrete process but part of an interdependent, brain-wide system involving perceptual input, memory scaffolding, and arousal calibration. This systems-level perspective positions attention as an integrative function operating across neural hierarchies and modalities—anticipating the large-scale network models that dominate contemporary cognitive neuroscience. While Luria's theory foregrounded the interplay of attention, memory, and cortical activation, subsequent models have elaborated the specific neural circuits and functional subsystems that carry out these processes. One of the most influential among these is the neurocognitive model of attention developed by Michael Posner and Steven Petersen. Their work dissects attention into discrete yet interacting networks—offering a more granular, anatomically grounded understanding of how attention is distributed and regulated across the brain.

Posner and Petersen's Neurocognitive Model of Attention

The neurocognitive model of attention proposed by Michael Posner and Steven Petersen (Posner, 1980; Posner & Petersen, 1990) represents a foundational advance in understanding how attention is organized and regulated in the brain. Unlike earlier conceptions that framed attention as a singular process, their model delineates three interacting but distinct attentional networks—each associated with different cognitive operations and neural substrates:

The *Alerting Network* maintains an optimal state of arousal and preparedness. It includes the thalamus, brainstem structures, and the locus coeruleus, which modulates cortical excitability via norepinephrine (Petersen & Posner, 2012). This system supports both phasic alerting—a rapid response to novel stimuli—and tonic alerting, which sustains sustained vigilance. These processes work in conjunction with arousal regulation systems like the reticular activating system (RAS) to support attention in fluctuating environments.

The *Orienting Network* enables the prioritization of sensory input by directing attention to spatial or sensory targets. It involves the dorsal attention network (DAN) including the frontal eye fields (FEF) and intraparietal sulcus (IPS)—which supports voluntary attentional shifts, as well as subcortical regions like the superior colliculus, involved in reflexive gaze and orientation responses (Corbetta & Shulman, 2002). These systems ensure flexibility in shifting between endogenous and exogenous cues, allowing the brain to engage with salient features of the external world.

The *Executive Control Network* governs higher-order regulation of attention—managing conflict resolution, task switching, and goal-directed behavior. This system is anchored in the dorsolateral prefrontal cortex (DLPFC), anterior cingulate cortex (ACC), and frontoparietal control network (FPCN). The executive network mediates cognitive flexibility, allowing attentional resources to adapt dynamically to internal goals and task demands (Miller & Cohen, 2001). For example, during a complex decision-making task such as navigating an emergency, the alerting network ensures vigilance, the orienting network focuses attention on spatially relevant cues, and the executive network prioritizes competing goals—illustrating how these systems coactivate to support adaptive behavior.

Posner and Petersen's framework introduces the concept of attentional modularity within integration—suggesting that while distinct networks exist, they cooperate to produce cohesive attentional control. These systems interact with emotion-processing regions (e.g., the amygdala and hippocampus), highlighting the convergence of cognitive and affective systems in real-world attentional regulation. Moreover, their model reflects early recognition of large-scale brain networks as substrates for distributed cognition. For example, the default mode network (DMN) typically deactivated during attentionally demanding tasks—interacts reciprocally with task-positive systems like the DAN, indicating dynamic functional reconfiguration based on attentional demands (Raichle et al., 2001; Fox et al., 2005).

Recent research has expanded this model by demonstrating how attentional load modulates connectivity across these systems. For instance, Alnæs et al. (2015) show that increased cognitive load leads to greater integration between core attention networks and supportive systems such as salience and control circuits. This supports a metafunctional view, in which attention is not an isolated filter but a globally coordinated process that adapts across sensory, cognitive, and affective domains (Menon & Uddin, 2010; Saalmann et al., 2012).

Treisman's Feature Integration Theory (FIT): Stages of Processing

Anne Treisman's Feature Integration Theory (FIT) offered a groundbreaking explanation for how humans perceive coherent objects in complex visual environments. According to FIT, object perception occurs in two distinct stages: a preattentive stage, where individual features such as color, orientation, and motion are processed independently and automatically, and a focused attention stage, where these features are integrated into unified perceptual wholes (Treisman & Gelade, 1980; Treisman, 1998). At the core of this model is the binding problem—how the brain combines separately processed features into a single, coherent percept. FIT posits that attention functions as the "glue" that binds features together. Without focused attention, features remain unintegrated, which can lead to illusory conjunctions, wherein attributes from separate objects are mistakenly combined. For example, when briefly shown a red triangle and a blue square, participants might incorrectly report seeing a blue triangle—demonstrating the failure of feature binding in the absence of sufficient attentional focus (Treisman, 1988).

This phenomenon is not limited to vision. Cross-modal illusory conjunctions have also been observed, for instance, misattributing the location of a sound to a visual cue (as in ventriloquism), underscoring the relevance of FIT across sensory systems. These errors highlight the importance of attentional resources in coordinating sensory inputs and integrating them into meaningful representations. Treisman's work anticipated the metafunctional role of attention by emphasizing its integrative function across both low-level and high-level processing domains. Attention in FIT is not merely a spotlight on perceptual input but a binding agent that organizes disparate neural signals into coherent, actionable information—linking perception with memory, decision-making, and behavior. This theory complements large-scale models of attention discussed earlier by showing how bottom-up feature registration interacts with top-down attentional control. In dynamic environments, such binding mechanisms allow the brain to construct stable perceptual objects despite continuous sensory flux, reinforcing attention's role as a brain-wide metafunction that enables adaptive interaction with the world. The critical importance of attentional binding becomes even clearer when we examine what happens when this process breaks down. Clinical syndromes such as Balint's provide striking examples of the consequences of disrupted attentional integration.

Balint's Syndrome and Its Implications

First described by Rudolf Balint in 1909, Balint's syndrome provides powerful clinical evidence for the integrative role of attention in perception. The condition typically results from bilateral damage to the posterior parietal cortex and is characterized by three hallmark symptoms: *simultanagnosia, optic ataxia,* and *ocular apraxia*. Simultanagnosia is the inability to perceive more than one object at a time. Patients may see a spoon or a cup on a table, but not both—revealing a breakdown in the focused attention stage required for binding multiple objects into a coherent scene. Optic ataxia involves difficulty reaching visually guided targets, despite intact motor function and visual acuity. Ocular apraxia reflects impaired control of voluntary eye movements toward objects of interest. Together, these deficits underscore the disruption of spatial and feature integration—supporting Treisman's claim that attention is essential for creating unified perceptual experiences (Husain & Rorden, 2003; Rizzo & Vecera 2002).

This syndrome exemplifies what occurs when attentional metafunctions are neurologically impaired. The brain's failure to bind spatial and object features leads to fragmented perception and disorganized action. From a metafunctional perspective, Balint's syndrome illustrates how attention is not localized to a single region or function but involves distributed neural coordination across perceptual, motor, and executive systems. The posterior parietal cortex serves as a convergence zone, integrating visual inputs with attentional priorities and motor intentions. Moreover, the clinical profile of simultanagnosia directly mirrors the failures of feature binding described in FIT, further validating Treisman's theory within a neurological context. It reinforces that attention is not a mere filter but a necessary system-wide mechanism for situating the self in a coherent and actionable perceptual world. Together, Treisman's Feature Integration Theory and the clinical phenomena of Balint's syndrome offer converging support for the view that attention functions as a necessary binding agent—integrating perceptual features into coherent experience through brain-wide coordination across sensory and executive systems.

Smultaneous Attention and Multitasking

While classic theories often emphasize attention's selective nature, real-world demands frequently require us to manage multiple tasks or stimuli concurrently. This capacity for simultaneous attention—the ability to attend to more than one stream of information at once—is often conflated with multitasking, where individuals attempt to perform two or more tasks simultaneously. Yet, growing evidence suggests that multitasking strains attentional resources and impairs performance across domains requiring sustained focus (Strayer & Johnston, 2001; Salvucci & Taatgen, 2008). From a metafunctional perspective, attention must dynamically allocate and reallocate resources across competing streams of information. However, this adaptability has limits. Cognitive load theory shows that the brain's processing capacity is inherently constrained: when multitasking exceeds those limits, attention cannot sustain coherent integration across tasks (Ophir, Nass, & Wagner, 2009). For instance, studies on driving while texting or using a phone reveal significant delays in reaction time, reduced situation awareness, and increased accident risk (Cohen et al., 2017). These findings point to the trade-offs inherent in multitasking—where simultaneous task demands force the brain to alternate attention rather than distribute it evenly. Even tasks perceived as "automatic" may overload the system when performed in tandem.

The metafunctional framework clarifies this dynamic: multitasking invokes multiple sensory modalities, executive systems, and memory processes—often simultaneously. When attentional demands exceed integration capacity, performance degrades, and errors proliferate. Thus, rather than seamless division, multitasking often reflects rapid attentional switching, with costs to precision and cognitive stability.

Multitasking Fallacies: The Illusion of Efficiency

Despite growing empirical evidence to the contrary, many individuals continue to believe they can efficiently juggle multiple tasks—a belief often referred to as the multitasking fallacy. This illusion stems from subjective impressions of productivity, reinforced by momentary task-switching that mimics continuous engagement. However, research consistently shows that task-switching incurs cognitive costs, including slower reaction times, increased error rates, and disrupted memory consolidation (Ophir, Nass, & Wagner, 2009). The brain's attentional system is not designed for parallel processing of demanding tasks, but rather for serial allocation of cognitive resources. This mismatch between perception and neurocognitive reality contributes to overconfidence in multitasking ability, a potentially dangerous bias in high-stakes environments such as driving, healthcare, or aviation. Recognizing the limitations of the brain's attentional systems is essential for dispelling these fallacies and promoting strategies aligned with attentional design.

Common Multitasking Myths — What the Science Says

Myth 1: "I'm a good multitasker." Reality: Only about 2% of people qualify as true "supertaskers" (Watson & Strayer, 2010). For most, multitasking leads to performance decline and increased error rates. Myth 2: "Doing two simple tasks at once is easy." Reality: Even simple tasks can overload working memory if they compete for the same cognitive resources (Salvucci & Taatgen, 2008). Myth 3: "Switching between tasks saves time." Reality: Task-switching introduces "switching costs"—delays, cognitive fatigue, and accuracy drops, especially when tasks share modalities (Rubinstein, Meyer, & Evans, 2001). Myth 4: "Multitasking keeps my brain sharp." Reality: Frequent multitaskers often show poorer focus and working memory control (Ophir, Nass, & Wagner, 2009). It may reduce long-term attentional efficiency. In summary, the brain performs best when focusing on one task at a time. Attention is a finite resource; sustained focus, not fragmentation, supports learning and performance.

Cognitive Load and Performance

Cognitive load theory provides a foundational explanation for the decline in performance often observed during multitasking. The brain's processing capacity is inherently limited; when individuals attempt to handle multiple tasks simultaneously, especially those that require overlapping cognitive resources—attentional systems become overtaxed. As a result, the quality of focus on each task diminishes, increasing the likelihood of errors, omissions, and slower response times (Sweller, 1988; Salvucci & Taatgen, 2008). Even tasks perceived as "simple" can exceed cognitive thresholds when paired inappropriately. For instance, reading a message while driving, though both may seem routine, can severely impair situational awareness. Empirical studies confirm that texting while driving is associated with a significantly higher rate of collisions—surpassing the risk levels observed with alcohol impairment (Reed & Green, 1999; Strayer & Johnston, 2001). The problem is compounded in high-stakes environments, such as healthcare, aviation, or emergency response, where divided attention can lead to critical lapses in judgment, delayed reactions, or procedural errors (Slagle et al., 2018; Aebersold & Kalisch, 2010).

In professional settings, multitasking is often mistaken for efficiency. However, evidence reveals the opposite: frequent task-switching fragments working memory, elevates stress, and depletes cognitive control (Ophir, Nass, & Wagner, 2009). For example, in nursing and clinical workflows, multitasking increases the risk of medication errors, misdiagnoses, and poor follow-up (Alkahtani, Aziz, & Ahmad, 2015). Similarly, educators and students who constantly toggle between tasks report decreased comprehension and academic performance (Lepp, Barkley, & Karpinski, 2014). From a brain metafunction perspective, these findings underscore that attention functions as a dynamic integrator of cognitive resources, not a duplicator. When attention load exceeds capacity, the system becomes inefficient, impairing executive control, memory, and even perception. The illusion of multitasking proficiency often masks deeper trade-offs in neural efficiency and task execution. Recognizing these limitations can inform better design of work environments, interfaces, and learning strategies that align with how the brain functions.

Broader Implications

The limitations of multitasking extend well beyond specialized environments like healthcare and transportation—they affect nearly every domain of modern life. In educational contexts, for instance, students who frequently switch between digital platforms, social media, and academic tasks consistently perform worse on measures of comprehension and long-term retention than their more focused peers (Lepp, Barkley, & Karpinski, 2014). The illusion of productivity often masks fragmented attention, reduced cognitive endurance, and a weakened ability to sustain complex reasoning. These patterns are echoed in the workplace, where constant interruptions from notifications, emails, and task-switching contribute to diminished performance, increased error rates, and greater perceived stress (Rosen, Carrier, & Cheever, 2013). While multitasking may appear adaptive in fast-paced environments, the brain's attentional systems function best when allowed to allocate resources sequentially, not simultaneously. Performance suffers not because people are inattentive, but because attentional resources—by design—are limited and optimized for focused engagement.

The brain metafunction model helps contextualize these observations. Attention, as a globally distributed function, modulates sensory, cognitive, and affective systems across modalities. When overburdened, this integrated system cannot maintain coherence, leading to breakdowns in perception, memory consolidation, and executive functioning. Understanding attention as a metafunction—not a limitless reservoir—allows us to rethink how we design educational technologies, user interfaces, and workspaces. Interventions that reduce cognitive fragmentation (e.g., structured task batching, minimizing alerts, or interface simplification) support the brain's natural processing architecture. Ultimately, cultivating mindful attentional practices—such as intentional focus, time-blocking, or cognitive rest—may serve not only individual productivity and learning, but also mental health and long-term cognitive resilience. As attention continues to be taxed by increasingly complex information environments, recognizing its limitations becomes an essential step in fostering sustainable performance and well-being.

Overt and Covert Orienting Integration

Overt and covert attention represent two interrelated dimensions of orienting mechanisms that allow the brain to allocate focus flexibly. Overt orienting refers to the observable, physical redirection of attention—typically involving eye movements or head turns toward a stimulus. This mechanism helps align the visual field with objects of interest and is central to tasks like visual search, scene scanning, and gaze-following behaviors. Eye-tracking studies have shown that overt orienting is influenced by both bottom-up factors—such as sudden flashes or movement—and top-down goals that reflect intentional focus (Carrasco, 2011). By contrast, covert orienting occurs without visible motor activity. It allows individuals to mentally shift attention toward peripheral stimuli while keeping their eyes fixed on a central task. This hidden mode of attentional deployment is crucial for activities such as reading in a crowded café while staying aware of surrounding movement or conversation. Covert orienting reflects the internal allocation of attentional resources—a cognitively driven process supported by shared neural circuits with overt orienting.

Neuroimaging and electrophysiological studies reveal that overt and covert systems engage overlapping networks, including the parietal lobe, superior colliculus, and pulvinar nucleus (Corbetta & Shulman, 2002). These shared substrates underscore the brain's metafunctional capacity to integrate voluntary and reflexive control over attention across modalities and contexts. For instance, covert attention often primes overt shifts, allowing the brain to anticipate where to direct gaze even before the eyes move—a dynamic interaction that exemplifies anticipatory regulation and brain-wide coordination. Consider sitting in a classroom, focusing intently on the lecturer's slides. Although your gaze is fixed, you become subtly aware—without moving your eyes—of a classmate shifting in their seat behind you. Your brain registers this motion through covert attention, evaluating whether it warrants a shift in overt gaze. This silent monitoring illustrates how the brain continuously balances stability of focus with readiness to respond, integrating sensory input and internal goals in real time. Such bidirectional modulation—between observable action and mental focus—demonstrates how orienting is not a single system but a distributed network that supports both internal monitoring and environmental responsiveness. This capacity is essential for survival in dynamic environments, enabling both rapid reactivity and intentional restraint. Overt and covert orienting exemplify the brain's ability to balance automatic and goal-directed control, unifying sensorimotor, perceptual, and executive systems into a coherent attentional process. This metafunctional framing prepares the ground for the next section, which explores how attention targets not just spatial locations, but features—such as color or shape—across the visual field.

Spatial and Feature-Based Attention Integration

Spatial attention refers to the prioritization of specific locations within the visual field, often conceptualized as a "spotlight" that enhances the processing of stimuli within a selected spatial region (Posner, 1980).

This attentional spotlight sharpens perceptual detail at the targeted location, improving detection, discrimination, and reaction times. Spatial attention engages a distributed neural network—including the frontal eye fields (FEF), lateral intraparietal area (LIP), superior colliculus, and pulvinar nucleus—which collectively mediate shifts in attention based on both external cues and internally driven goals (Bisley & Goldberg, 2010; Kastner & Ungerleider, 2000). In contrast, feature-based attention enhances the processing of specific attributes—such as color, orientation, or motion—across the entire visual field, independent of spatial location. This global mechanism allows the brain to prioritize relevant object features and operates through modulation of both early visual areas (e.g., V1) and higher-order regions like the inferotemporal cortex (Saenz, Buracas, & Boynton, 2002). Feature-based attention is especially critical in cluttered visual environments, where object recognition depends on selecting relevant feature dimensions from overlapping stimuli (Maunsell & Treue, 2006).

These two forms of attention—spatial and feature-based—interact synergistically to optimize visual processing. While spatial attention narrows the focus to a particular region, feature-based attention highlights specific traits, even beyond that region. Their integration is supported by a frontoparietal network, including the dorsolateral prefrontal cortex (DLPFC) and posterior parietal cortex (PPC), which coordinates the selection of targets based on task demands and cognitive priorities (Buschman & Kastner, 2015). Subcortical structures such as the pulvinar and superior colliculus further enhance this coordination, ensuring efficient cross-talk between sensory, motor, and executive systems (Krauzlis, Lovejoy, & Zénon, 2013). For instance, when searching for a friend wearing a red jacket in a crowded plaza, spatial attention helps scan specific areas, while feature-based attention continuously primes the perceptual system to detect red across the visual scene. From a metafunctional perspective, spatial and feature-based attention exemplify the brain's capacity to flexibly reconfigure sensory resources according to context, intention, and relevance. They reveal how attention operates not as a localized function, but as a distributed, goal-oriented process that interfaces with perception, memory, and executive systems. These dynamics reinforce the conceptualization of attention as a brain-wide metafunction, enabling efficient perceptual prioritization and real-time adaptation to environmental complexity.

Visual Attention

Visual attention can be broadly categorized into two complementary systems: bottom-up (exogenous) and top-down (endogenous) attention. These systems interact dynamically to guide perception and action in complex environments. Bottom-up attention is driven by the salience of external stimuli. It operates rapidly and involuntarily, capturing focus in response to unexpected events—such as a sudden flash, loud noise, or quick movement (Buschman & Kastner, 2015). This reflexive system enhances survival by enabling quick orientation to potential threats or novel opportunities. Neural substrates involved in bottom-up attention include the parietal cortex, which supports spatial processing, the temporal cortex for feature recognition, and subcortical structures such as the superior colliculus, which mediates gaze shifts toward salient targets (Baldauf & Desimone, 2014). Evidence suggests that even early visual areas, including the primary visual cortex (V1), participate in this process by generating a "saliency map" that highlights visually distinctive features (Li, 2002). This map informs rapid visual orientation and supports the superior colliculus in directing attention spatially (Knudsen, 2018). In contrast, top-down attention is under conscious control and shaped by task demands, prior knowledge, and internal goals. It enables individuals to selectively enhance relevant visual input while suppressing distractions. For example, reading in a crowded café requires sustained focus on textual details while tuning out peripheral motion and sound (Kastner & Ungerleider, 2000).

Top-down attention is mediated by a network of frontal and parietal regions, particularly the dorsolateral prefrontal cortex (DLPFC) and the anterior cingulate cortex (ACC), which regulate goal-directed focus and inhibit irrelevant input (Miller & Cohen, 2001; MacDonald et al., 2000). Additionally, executive structures like the basal ganglia and thalamus facilitate attentional switching and conflict resolution across tasks. These brain regions enable flexible control over sensory gating, allowing for precise adaptation to shifting contexts and cognitive priorities (Gazzaley & Nobre, 2012). For example, while hiking, a person might reflexively turn toward a rustling sound in the underbrush (bottom-up) but quickly resume focus on reading a trail map (top-down), illustrating the fluid negotiation between attentional systems.

From a metafunctional perspective, visual attention reflects the integration of reflexive and volitional systems that co-regulate perception and behavior. Bottom-up attention ensures vigilance to environmental change, while top-down control maintains alignment with goals and expectations. These systems interact via large-scale brain networks, dynamically balancing reactivity with intentionality—a hallmark of the brain's capacity to prioritize, adapt, and act in real time. This dual-process architecture exemplifies attention's metafunctional role in coordinating perception, memory, and executive function across sensory and cognitive domains. This continuous interplay between automatic salience detection and goal-directed attentional control sets the stage for deeper integration—a dynamic that enables the brain to flexibly allocate focus based on both environmental cues and internal priorities. The next section explores this integration in detail.

Integration of Bottom-Up and Top-Down Attentional Processes

The integration of bottom-up and top-down attention is foundational to adaptive cognition. Rather than functioning in isolation, these two systems operate in constant dialogue—allowing attention to be both responsive to external change and guided by internal goals. Bottom-up attention initiates rapid, reflexive responses to salient stimuli, such as a flashing light or sudden movement. This mechanism ensures that organisms remain attuned to unexpected or behaviorally relevant environmental changes. Neural structures like the superior colliculus, pulvinar nucleus, and parietal cortex are heavily involved in detecting salience and orienting attention accordingly (Corbetta & Shulman, 2002; Krauzlis et al., 2013). Top-down attention, in contrast, enables sustained focus and goal maintenance in the presence of potential distractions. It exerts modulatory control over sensory cortices, allowing selective amplification of relevant stimuli and suppression of irrelevant input. The frontal eye fields (FEF) and posterior parietal cortex (PPC) are central to this process, forming part of the dorsal attention network (DAN) responsible for voluntary shifts of attention (Buschman & Kastner, 2015).

The two systems converge and interact within frontoparietal networks that dynamically balance reactive and deliberate control. For instance, while scanning a busy street, bottom-up processes may trigger attention toward sudden movement, but top-down mechanisms allow a pedestrian to ignore non-threatening distractions and focus on navigating safely. For example, during a high-stakes sports game, a player reflexively notices an opponent's quick movement (bottom-up), but simultaneously maintains strategic focus on the ball (top-down), modulating attention based on both immediate threats and broader goals. At a systems level, this integration exemplifies attention's metafunctional quality—its role in harmonizing perception, action, and cognition across modalities. It allows the brain to adaptively select and process information in complex, dynamic environments by coordinating neural responses to both intrinsic objectives and extrinsic changes. These insights offer a blueprint for understanding attention not as a binary system, but as a dynamic continuum, integrating reactive and volitional forces in real time.

Attention and Auditory Perception

Auditory perception begins with the bottom-up processing of acoustic signals: sound waves are transformed into neural impulses by the cochlea and transmitted via the auditory pathway to the primary auditory cortex. These early stages decode basic features such as frequency, intensity, and spatial location. However, conscious auditory experience is not merely a passive reception of these features, it is dynamically shaped by attention. Attention acts as a sensory gate, regulating which auditory information reaches awareness. Sensory gating, an early attentional mechanism, filters out redundant or irrelevant sounds, protecting the brain from overload (Hillyard et al., 1973). This process is essential in environments rich in auditory stimuli, like a crowded train station—where competing inputs must be prioritized.

Selective attention sharpens the neural representation of task-relevant auditory signals, enhancing our ability to extract meaning. For example, we can focus on a single voice in a noisy room—a phenomenon famously known as the cocktail party effect (Cherry, 1953). This selective enhancement involves top-down modulation, where cognitive goals, expectations, and memory shape perception by tuning the sensitivity of auditory neurons (Shinn-Cunningham & Best, 2008; Fritz et al., 2007). Example: When listening to a friend in a noisy café, your attention heightens the salience of their voice while suppressing background chatter. This is not just a sensory feat—it's a cognitive negotiation between external input and internal goals.

This modulation begins early: attentional influences on auditory processing appear in subcortical areas like the inferior colliculus and even the cochlear nucleus, illustrating the depth and breadth of top-down control (Price & Moncrieff, 2021). Such findings support the view of attention as a distributed, brain-wide metafunction, operating across hierarchical levels of sensory systems. Recent studies show that attention can dynamically alter neuronal tuning curves, biasing auditory circuits toward features of interest (Treisman, 1964; Shinn-Cunningham, 2008). This reconfiguration enables flexible adaptation to changing task demands or environmental conditions—highlighting attention's integrative role in perception, memory, and action. In essence, auditory attention reflects a bidirectional loop: bottom-up signals present possibilities, while top-down mechanisms refine and constrain them—resulting in the coherent perception of sound amidst complexity.

The Cocktail Party Effect, Feedback Loops, and Top-Down Modulation

The cocktail party effect exemplifies the attentional system's remarkable ability to selectively track a single auditory stream amidst competing sounds (Cherry, 1953). In a bustling environment—like a café, conference, or busy household—the auditory system dynamically separates overlapping voices and background noise, allowing a listener to focus on one salient source. This feat is not accomplished by the ears alone; it depends on complex attentional feedback loops that operate across multiple neural hierarchies. At the heart of this process is a bidirectional interplay: bottom-up sensory input provides raw auditory information, while top-down mechanisms—shaped by goals, memory, and emotional salience—modulate perception to fit behavioral priorities (Fritz et al., 2007). When attention is directed toward a voice, higher-order cortical areas—especially the dorsolateral prefrontal cortex (DLPFC) and anterior cingulate cortex (ACC)—modulate neural sensitivity in auditory regions to amplify relevant features like pitch, rhythm, or spatial location (Shinn-Cunningham & Best, 2008).

These top-down signals exert their influence through feedback loops that dynamically adjust the neural representation of incoming stimuli. For example, attention can bias auditory cortical neurons to become more responsive to a particular speaker's frequency range or cadence, while suppressing irrelevant streams. This real-time recalibration highlights the metafunctional architecture of attention: it integrates memory, context, perception, and behavioral goals into a single adaptive perceptual strategy. For example, during a classroom lecture, a student may initially attend to the teacher's voice. However, if a sudden shift in tone occurs at a nearby desk—say, someone whispering their name—the attentional system recalibrates, momentarily redirecting resources to evaluate potential relevance before returning focus. Furthermore, attention's influence begins surprisingly early in the auditory hierarchy—even at brainstem levels, such as the cochlear nucleus and inferior colliculus—indicating that attention is not merely a cortical phenomenon but a multi-tiered regulatory process (Price & Moncrieff, 2021).

These mechanisms also reflect adaptive flexibility. When demands shift—say, the teacher starts speaking more softly—attention reconfigures the gain applied to auditory signals. Norepinephrine from the locus coeruleus and cholinergic modulation from the basal forebrain enhance signal clarity in real time, fine-tuning responsiveness across distributed systems (Aston-Jones & Cohen, 2005). The cocktail party effect is not just a quirky auditory trick. It is a paradigmatic expression of attention's role as a brain metafunction, dynamically coordinating perceptual, cognitive, and affective processes across hierarchical levels and sensory domains.

Attention as a Metafunctional Process Across Modalities

Attention is not confined to a single sensory modality or cognitive domain—it operates metafunctionally across the brain's distributed systems, guiding perceptual coherence and adaptive behavior. As described earlier in the context of vision and audition, attention dynamically regulates the flow of information by prioritizing task-relevant inputs, suppressing distractions, and coordinating resource allocation across neural systems. This integrative role defines attention as a brain-wide metafunction, mediating perception, memory, emotion, and action in unified, goal-directed ways (Menon & Uddin, 2010; Saalmann et al., 2012). Multimodal attention becomes particularly crucial in natural environments where the brain must parse concurrent sensory streams. Walking through a city park, for instance, one may hear children laughing, feel the texture of a breeze, see shifting light through the trees, and simultaneously navigate a conversation. The brain cannot process all incoming data with equal priority. Instead, attention orchestrates selective amplification and inhibition, allowing perceptual experience to remain coherent, meaningful, and behaviorally relevant.

This process reflects multisensory integration—the brain's ability to merge data from vision, audition, somatosensorial, and other modalities into a unified representational field. Attention does not merely enhance one modality at a time; rather, it facilitates the coordination between modalities, determining which cross-modal correspondences are relevant in the moment. For example, watching a person's mouth move while listening to their voice integrates visual and auditory cues to enhance speech perception, a phenomenon well documented in audio-visual speech research (Driver & Spence, 1998). Moreover, attention also functions across conscious and unconscious domains, regulating what rises to awareness and what remains subliminal. The well-known "Invisible Gorilla" experiment (Simons & Chabris, 1999) demonstrates this strikingly: when participants focus attention on counting basketball passes, they often fail to see a costumed gorilla walking directly through the scene—a case of inattentional blindness. A similar phenomenon, inattentional deafness, occurs when individuals immersed in demanding visual tasks fail to notice salient auditory cues (Macdonald & Lavie, 2011). These findings reveal the competitive dynamics of attention between sensory modalities and underscore its role as a modulator of consciousness itself. These forms of "attentional omission" reveal that attention is not just a spotlight—it is a filter, a gate, and a bridge between domains of sensory experience and cognitive control.

At the neural level, multimodal attention involves early-stage gating in sensory cortices and high-level coordination in prefrontal, parietal, and subcortical structures, including the pulvinar nucleus and superior colliculus (Desimone & Duncan, 1995; Saalmann et al., 2012). These systems dynamically redistribute processing resources in response to both external stimuli and internal goals, ensuring a flexible but focused engagement with the world. Ultimately, attention's cross-modal capacity supports perceptual unity. In a richly multimodal world, the brain's ability to align temporally and spatially disparate signals into coherent experiences is inseparable from attentional modulation. Without this function, the sensory world would fragment into disconnected impressions. Thus, attention emerges not as a compartmentalized faculty but as a metafunctional integrator—one that coordinates large-scale neural activity across modalities, hierarchies, and behavioral contexts.

Neural Mechanisms of Multimodal Perception

Multimodal perception—the brain's ability to synthesize inputs from multiple sensory channels into coherent representations—relies fundamentally on attentional processes. Far from operating in isolated streams, the senses converge through dynamically regulated neural mechanisms that prioritize, integrate, and align sensory information based on both environmental cues and internal cognitive demands. Attentional modulation of multimodal inputs begins early in the primary sensory cortices. For example, when attention is directed to a visual stimulus, neuronal firing increases not only in higher-order visual areas (such as V4 or MT) but also in primary visual cortex (V1), sharpening signal clarity even at these initial processing stages (Reynolds & Chelazzi, 2004). Similarly, attention to an auditory feature enhances activity in the auditory cortex, increasing the signal-to-noise ratio and improving perceptual precision (Fritz, Elhilali, & Shamma, 2007). This early modulation reflects attention's function as a gating mechanism, amplifying task-relevant inputs while suppressing irrelevant data.

Beyond the primary sensory areas, integration across modalities occurs in distributed cortical and subcortical systems that coordinate attention and sensory binding. Key hubs include:

The Posterior Parietal Cortex (PPC): Plays a central role in spatial integration across visual, auditory, and somatosensory modalities. It supports the dynamic remapping of attention based on behavioral relevance.

The Superior Colliculus: Located in the midbrain, it contributes to orienting behaviors and cross-modal alignment, particularly in audiovisual attention and spatial congruency (Krauzlis, Lovejoy, & Zénon, 2013).

The Pulvinar Nucleus of the Thalamus: Acts as a relay and modulatory center, synchronizing activity between visual and other cortical areas in response to attentional demands (Saalmann et al., 2012). This structure filters incoming information and coordinates attentional shifts between modalities.

The Prefrontal Cortex (PFC): Provides top-down regulation, directing attention toward task-relevant features across modalities, supporting working memory, and enabling flexible behavioral responses (Miller & Cohen, 2001).

The Insula and Anterior Cingulate Cortex (ACC): Central components of the salience network, these regions detect behaviorally significant stimuli and contribute to switching attentional resources across sensory domains (Seeley et al., 2007; Goulden, et al. 2014).

These regions do not operate in isolation. Rather, they are embedded in large-scale brain networks—such as the dorsal attention network (DAN), ventral attention network (VAN), and salience network—that ensure cross-modal synchrony and cognitive coherence. Their interplay enables perceptual integration even under high attentional load or rapid environmental changes. The coordination of multimodal inputs reflects a defining principle of the Brain Metafunction Framework: that attention functions as an integrative scaffold—uniting distributed systems across hierarchical levels to produce adaptive perception. This integration enables seamless transitions between modalities, such as shifting from listening to a conversation to visually scanning a crowded room. It also explains perceptual illusions and cross-modal effects (e.g., the McGurk effect), where one sense dominates or biases interpretation in another. In each case, attentional mediation determines the final percept.

In short, the neural mechanisms of multimodal perception reveal that attention is not a discrete function applied post hoc—it is an embedded and dynamic organizer of sensory reality. By modulating early sensory processing and coordinating high-level integration, attention enables the perceptual unity and adaptive flexibility necessary for navigating complex environments. Building on these mechanisms, we now explore how attention not only coordinates sensory inputs across modalities but also facilitates the unification of perception itself—ensuring that distinct sensory streams are synthesized into coherent experiential wholes.

Attention and Perceptual Unity

How attention binds multimodal experience into a singular, coherent reality

The multifaceted processes of attention are essential not only for filtering and modulating sensory inputs, but for creating the sense of perceptual coherence that defines conscious experience. Despite the vast differences in neural pathways between vision, audition, touch, and proprioception, attention ensures these inputs are dynamically aligned, generating a unified and intelligible experience of the world. This integrative function is especially apparent in audiovisual speech perception, where the brain combines auditory information (e.g., voice tone, phonetic cues) with visual signals (e.g., lip movements, facial expressions) to enhance comprehension. Known as multisensory integration, this process requires attentional alignment across modalities to synchronize temporally distinct inputs (Driver & Spence, 1998). When this synchronization breaks down—as in the McGurk effect, where conflicting visual and auditory speech cues (e.g., seeing 'ga' while hearing 'ba') lead to the perception of a third sound (like 'da')—perceptual distortions reveal the central role attention plays in resolving cross-modal conflict. Beyond perceptual tasks, attention also fosters unity across temporal scales, helping bind sequential events into cohesive narratives. This is critical in complex behaviors such as language processing, musical engagement, and real-time decision-making. Without attentional continuity, these experiences would fragment into isolated perceptual events, undermining comprehension and meaning.

At the neural level, frontoparietal networks, including the dorsolateral prefrontal cortex (DLPFC) and intraparietal sulcus (IPS), serve as convergence hubs where inputs from multiple sensory modalities are assembled and matched against task goals and working memory content (Macaluso, 2010). The pulvinar nucleus plays a supporting role by coordinating rhythmic synchrony between cortical areas, effectively gating relevant multisensory data (Saalmann et al., 2012). Meanwhile, the default mode and salience networks modulate internal versus external awareness, ensuring that attention is directed toward the most pertinent input stream.

Attention acts as a unifying scaffold—integrating multimodal, spatial, and temporal information to support a stable experience of self and world. This quality exemplifies its role as a brain-wide metafunction (Menon & Uddin, 2010; Fries, 2005). This unifying property of attention also intersects self-representation and embodiment. For instance, in bodily illusions such as the rubber hand illusion, attentional synchrony across touch and vision fosters the temporary incorporation of a non-biological object into one's body schema—further evidence that attention mediates the boundaries of self-experience through multimodal integration. In summary, attention's role in perceptual unity reveals its essential function not just in cognition, but in constituting the coherence of lived experience. Through neural synchronization, multimodal alignment, and dynamic prioritization, attention fuses complexity into wholeness—enabling the brain to construct a perceptually and behaviorally intelligible world.

Inattentional Deafness

When visual attention suppresses auditory perception

Inattentional deafness is a compelling demonstration of how attentional systems prioritize processing under limited cognitive resources—often at the expense of unattended modalities. Defined as the failure to notice an auditory stimulus due to engagement in a demanding visual task, this phenomenon reflects the competitive allocation of attention across sensory channels. A foundational study by Macdonald and Lavie (2011) illustrated this phenomenon in a controlled experimental setting. Participants were asked to complete a visual discrimination task—identifying features of a cross shape—while tones played in the background. In the low-load condition (e.g., identifying a blue arm of the cross), auditory perception remained largely intact. However, in the high-load condition (e.g., judging arm length), participants consistently failed to detect the auditory tone, even though it was loud and salient. As task complexity increased, auditory awareness dropped significantly, despite no changes in the actual stimulus.

This attentional trade-off supports Perceptual Load Theory (Lavie, 2005), which posits that cognitive systems have a finite processing capacity. As visual demands rise, fewer resources remain available for parallel processing in other modalities. The brain, acting as an efficiency optimizer, filters out less task-relevant input to protect goal-directed focus. In this sense, attention operates as a limited-capacity, goal-directed bottleneck—suppressing or enhancing perception based on current demands. Neuroimaging studies reveal that inattentional deafness is not simply a behavioral quirk, but a deeply rooted neural phenomenon. Functional MRI has shown decreased activation in auditory cortex during visually demanding tasks, confirming that attention not only shapes what we consciously notice—but also modulates early-stage sensory processing (Talsma & Awh, 2006; Gazzaley & Nobre, 2012). These findings emphasize that attention gates information from the earliest cortical levels, supporting its role as a metafunctional controller of perceptual access. Inattentional deafness demonstrates attention's central role in cross-modal resource regulation. The brain dynamically suppresses or prioritizes entire sensory channels in service of internal goals—evidence that attention functions not only within modalities, but across them, as a global integrator.

This phenomenon has far-reaching practical implications. In high-stakes environments like driving, aviation, or clinical care, excessive focus on visual interfaces may impair responsiveness to auditory warnings, contributing to accidents or oversights. Likewise, inattentional deafness offers insight into interface design, where alarms or cues may need to be dynamically reinforced based on visual load to ensure awareness is maintained. Moreover, this phenomenon underscores attention's non-equivalence across modalities. Some tasks—especially those that are visually or emotionally salient—can monopolize processing resources, suppressing other channels in ways that may go unnoticed until consequences arise. In summary, inattentional deafness exemplifies how attention flexibly reallocates cognitive bandwidth based on task load and sensory competition. It reminds us that perception is not passive reception, but a selectively curated construction, continuously shaped by the dynamic priorities of attention.

Theoretical Perspectives and Practical Implications

Binding, bottlenecks, and design considerations

The phenomenon of inattentional deafness invites deeper exploration into the mechanisms of attentional selection—particularly how the brain chooses which inputs to integrate and which to suppress. At the heart of this lies Anne Treisman's Feature Integration Theory (FIT), which offers a compelling framework for understanding how attentional focus is required to bind discrete features into coherent perceptual objects. FIT posits that in a preattentive stage, individual sensory features—such as color, shape, or pitch—are processed automatically and in parallel. However, to consciously perceive these features as part of a unified object or event, focused attention is required (Treisman & Gelade, 1980). In contexts of inattentional deafness, where visual tasks dominate cognitive load, attention may never reach the auditory features. As a result, even salient sounds are excluded from the final stage of perceptual integration. Inattentional deafness can be viewed as a cross-modal failure of feature binding—a consequence of the brain allocating its integrative capacities primarily to the visual modality under load. From a broader perspective, this attentional trade-off is also consistent with Limited Capacity and Bottleneck Models of Attention. These models suggest that perceptual systems cannot process all incoming stimuli simultaneously at a conscious level (Broadbent, 1958). Instead, attention functions as a selective filter, prioritizing information based on task relevance, sensory salience, and internal goals. This has important practical implications across domains:

In *complex environments* like hospitals, where multitasking is routine, high cognitive and perceptual load can lead to missed auditory cues—exacerbating risks to patient safety (Alkahtani, Aziz, & Ahmad, 2015). This highlights the need for cross-modal alert systems and interface designs that account for attentional limitations under real-world conditions.

User Interface Design: In cockpit environments, surgical monitoring systems, or automotive dashboards, placing too much visual information at once may inadvertently suppress auditory cue recognition.

Designers must consider cross-modal load balancing to ensure crucial alerts aren't missed. In clinical settings, multitasking and frequent interruptions can compromise vigilance and patient safety. Slagle et al. (2018) emphasize the importance of assessing cognitive load in healthcare environments to mitigate risks associated with divided attention and task-switching.

Multisensory Warning Systems: Alarm systems may benefit from adaptive cueing, where auditory stimuli are modulated in intensity or redundancy when visual load is high. Similarly, tactile cues can be introduced to relieve pressure on visual and auditory systems.

Cognitive Training and Safety Protocols: Training interventions should not only raise awareness of inattentional deafness but also account for individual variability in attentional control and skill acquisition (Ackerman, 1992). Understanding these differences can inform the personalization of protocols in high-stakes domains such as aviation, medicine, and emergency response.

Inattentional deafness is not merely a sensory oversight—it reflects the global integration and competition between attentional systems, operating under limited-capacity constraints. Attention acts as a cross-modal negotiator, orchestrating what enters consciousness based on task context and system demands.

Balint's Syndrome and Its Implications

Balint's syndrome, first described by Rudolf Balint in 1909, highlights the consequences of impaired attentional mechanisms and provides clinical support for FIT. Patients with this syndrome experience simultagnosia—the inability to perceive multiple objects simultaneously—as well as optic ataxia, struggling to accurately reach for objects despite maintaining normal visual acuity and color perception. These symptoms are linked to damage in the bilateral posterior parietal lobes, often caused by chronic cerebrovascular disease (Husain & Rorden, 2003). The syndrome illustrates the importance of the focused attention stage in integrating visual features into unified objects. For example, patients' inability to process multiple objects simultaneously mirrors the failure of feature binding in illusory conjunctions. These findings emphasize the essential role of attention in forming coherent perceptual experiences, as described by FIT (Cohen & Haggard, 2005; Husain & Rorden, 2003). In this sense, Balint's syndrome provides compelling evidence that attention operates not merely as a spotlight, but as a distributed integrative function essential for coherent perception. This framing highlights the dynamic trade-offs inherent in cognition—not as flaws, but as necessary prioritizations within a complex, goal-oriented system. In recognizing the predictive and adaptive nature of these processes, we deepen our understanding of attention as a brain-wide metafunction that shapes experience, behavior, and performance. In this sense, Balint's syndrome provides compelling evidence that attention operates not as a localized filter, but as a distributed system essential for integrating sensory input into coherent experience—underscoring its role as a brain-wide metafunction.

Selective Attention and the Cocktail Party Problem

How attention isolates, prioritizes, and preserves meaning in noise

In dynamic auditory environments, the brain's capacity to focus selectively on one voice among many is both remarkable and essential. This challenge, famously known as the "cocktail party problem" (Cherry, 1953), exemplifies the goal-directed and adaptive nature of attention, which enables us to engage in meaningful interaction despite overlapping auditory streams. At its core, selective attention operates through the dual influence of top-down and bottom-up mechanisms. Top-down processes allow voluntary focus based on intentions, expectations, or past experience. For example, when attending to a conversation at a noisy party, the brain amplifies features like the pitch, tone, or spatial location of the target speaker, guided by task goals and working memory (Shinn-Cunningham & Best, 2008). Bottom-up processes respond automatically to salient, unexpected stimuli—a sudden crash or someone shouting your name—momentarily overriding voluntary attention to redirect focus. This reflexive redirection is often necessary for detecting threats or opportunities in a complex auditory landscape (Corbetta & Shulman, 2002).

This interplay is supported by auditory scene analysis, a process by which the brain segregates, tracks, and organizes overlapping auditory streams into distinct perceptual units (Bregman, 1990). Mechanistically, this occurs via modulations in primary and secondary auditory cortex, influenced by frontoparietal networks that control attentional focus and filtering (Fritz, Elhilali, & Shamma, 2007). Moreover, attention enhances neuronal tuning toward relevant features: Neurons in the auditory cortex become more responsive to the pitch, rhythm, or timbre of a target voice, while suppressing competing signals—illustrating a dynamic reweighting of sensory input based on relevance. This functionality becomes especially evident in challenging listening conditions. Elderly individuals or those with cognitive load often struggle with auditory stream segregation, underscoring the attentional demands of selective listening (Alain et al., 2006). Hearing aid technologies increasingly attempt to replicate this function by enhancing signal-to-noise ratios and tracking the listener's attentional focus through eye or head movements (Bronkhorst, 2000). The cocktail party problem highlights attention's role as a context-sensitive, cross-modal organizer, capable of isolating meaningful content from perceptual chaos. It draws on memory, prediction, and dynamic filtering, exemplifying attention's integrative nature as a brain-wide metafunction. By prioritizing relevance over raw intensity, attention ensures that the auditory system preserves semantic coherence, allowing us to interact, learn, and respond meaningfully even amid informational overload. This ability to selectively focus amid competing inputs is not merely behavioral—it is grounded in specific neural mechanisms that dynamically filter, prioritize, and amplify auditory signals.

Neural Mechanisms of Selective Attention

Selective auditory attention relies on a distributed network of cortical and subcortical structures that work in concert to isolate relevant stimuli and suppress irrelevant noise. Early selection mechanisms, such as those mediated in the primary auditory cortex, enhance the neural representation of sound features—such as pitch, spatial location, and timbre—associated with the attended stream (Fritz et al., 2007). At the mid-level processing stage, areas like the superior temporal gyrus (STG) integrate these auditory features, enabling robust perceptual binding. This integration is guided by goal-directed modulation from frontal regions such as the dorsolateral prefrontal cortex (DLPFC) and anterior cingulate cortex (ACC), which exert top-down control over what is attended or ignored (Miller & Cohen, 2001).

Critically, the frontoparietal attention network—including the intraparietal sulcus (IPS) and frontal eye fields (FEF)—plays a central role in mediating the flexible allocation of attention across space, time, and modality. These systems monitor task relevance, suppress distractions, and adjust attentional focus dynamically in response to environmental and internal cues. For example, as conversational demands shift or the spatial position of a speaker changes, the parietal cortex helps recalibrate auditory spatial focus, while the ACC monitors for conflicting stimuli and signals the need for reorientation (Posner & Petersen, 1990; Corbetta & Shulman, 2002).

Additionally, subcortical structures such as the pulvinar nucleus and thalamus serve as crucial filters and relays, gating sensory input to enhance the salience of goal-relevant stimuli while dampening distractors (Saalmann et al., 2012). This modulatory control system is further fine-tuned by neuromodulators like norepinephrine and acetylcholine, which adjust network sensitivity based on arousal and task demands (Aston-Jones & Cohen, 2005; Sarter et al., 2001). These layered neural interactions illustrate attention's role as a multi-level, multimodal, and recursive system. Rather than a linear filter, selective attention emerges as a metafunctional operation, dynamically aligning bottom-up sensory input with top-down expectations, goals, and memories.

The Role of Working Memory

Working memory is essential for sustaining selective attention, particularly in complex or noisy environments like those described in the cocktail party problem. It serves as a limited-capacity buffer that temporarily stores and manipulates relevant information—such as the content of an ongoing conversation—while shielding it from interference (Cowan, 2014). This system plays a dual role:

Maintenance: It preserves context by retaining the auditory stream's thematic or semantic continuity, allowing listeners to pick up where they left off even if briefly distracted.

Control: It helps prioritize attentional targets, filtering inputs dynamically based on relevance and supporting task goals.

At the neural level, the prefrontal cortex—especially the dorsolateral prefrontal cortex (DLPFC)—is central to the coordination of working memory and attentional focus. It maintains task-relevant representations and guides the selection, suppression, or updating of information based on situational demands (Miller & Cohen, 2001). These control signals interface with auditory cortices and parietal networks to refine sensory input in real time. For example, during a multitalker conversation, working memory allows a listener to hold onto prior statements while simultaneously evaluating the salience of a new speaker's voice or a sudden environmental sound. Research suggests that working memory and selective attention are mutually reinforcing, with working memory biasing attentional priorities and attention enhancing the precision of working memory representations (Gazzaley & Nobre, 2012). This reciprocal relationship is supported by feedback loops between the prefrontal cortex, anterior cingulate cortex, and auditory processing areas, which maintain continuity across time, stimuli, and goals. Working memory exemplifies attention's integrative nature, linking past context, present input, and future intentions within a single dynamic system. In this sense, it operates not as a standalone module but as a metafunctional bridge—synchronizing perception, cognition, and behavioral planning across domains.

Working Memory–Centered Model of Attention

Knudsen's (2007) Working Memory–Centered Model of Attention offers a compelling systems-level account of how attention functions as a dynamic interplay between cognitive control, sensory input, and memory. Central to this model is the idea that working memory is not just a passive store, but an active hub that determines which information gains access to consciousness, is prioritized, and is maintained for goal-directed behavior. This model consists of four key components, which together form a self-regulating, recursive system:

1. *Working Memory as a Central Hub*: Working memory maintains task-relevant information and mediates between incoming stimuli and internal goals. It enables flexible focus by determining what is currently most important. Because of its limited capacity, working memory must selectively admit and retain only the most pertinent stimuli, underscoring its role as a regulatory filter (Knudsen, 2007).

2. *Competitive Selection:* This mechanism determines which sensory inputs gain access to working memory, based on salience and task relevance. Competing stimuli are evaluated, and the most relevant input is "admitted" into active cognition. This selection reflects a priority map constructed from both external and internal cues (Desimone & Duncan, 1995).

3. *Top-Down Sensitivity Control:* Higher-order cognitive systems, including the prefrontal cortex, modulate sensitivity across sensory pathways. This control biases attention toward stimuli aligned with current goals and suppresses distractions. In this way, task goals shape sensory processing, rather than passively receiving it (Miller & Cohen, 2001).

Recurrent Loop of Attention

A feedback loop allows the content of working memory to influence new attentional selections. This means attention is not static, but constantly updated in response to evolving task demands and internal goals. This recursive dynamic ensures that attention is adaptive, predictive, and context-sensitive. Imagine a driver navigating traffic while planning their next turn. The auditory cue from a GPS briefly competes with the sudden honk of a car. Working memory retains the navigation instruction, while top-down control filters out the irrelevant background noise and shifts attention back to the task. Here, the recurrent loop updates the attentional priority as traffic conditions change—without losing track of the goal.

Metafunctional Integration

Knudsen's model reinforces the metafunctional nature of attention—it is not merely a process of selection, but a distributed control system that integrates working memory, executive function, perception, and behavior. Attention becomes a recursive mediator between past (memory), present (stimuli and cognition), and future (goals). By emphasizing the bidirectional flow between memory and perception, this model highlights how attention operates across neural domains and over time, adapting flexibly to context. It encapsulates how the brain allocates limited resources intelligently—not just reactively, but proactively.

Comparing Knudsen's Model with Posner & Petersen's Networks

While Posner and Petersen's tripartite model emphasizes the functional segregation of attention into alerting, orienting, and executive control networks, Knudsen's model offers a recursive and centralized view, placing working memory at the core of attentional regulation. The Posner–Petersen framework maps how attention is distributed across distinct but interactive neural systems, whereas Knudsen foregrounds how attention is governed by competition, modulation, and feedback loops within a centralized workspace. These models are not contradictory but complementary: one highlights neural architecture, the other emphasizes dynamic control flow. Together, they illustrate how attention operates both spatially (across networks) and temporally (via feedback and prediction)—a synthesis that underscores the metafunctional nature of attention as a system that integrates cognition, memory, and perception at scale (Menon & Uddin, 2010; Buschman & Kastner, 2015).

Ventriloquism and the Rubber Arm Illusion: Multisensory Integration and Body Ownership

Multisensory integration, the brain's capacity to combine inputs from different modalities, is essential for constructing coherent, embodied perceptual experiences. Two classic phenomena—the ventriloquism effect and the rubber arm illusion—demonstrate how visual cues can override auditory or tactile inputs, reshaping not just perception but self-representation. In the ventriloquism effect, an observer perceives a voice as emanating from a puppet's moving mouth rather than the actual sound source. This occurs because the brain prioritizes spatially and temporally aligned visual stimuli over auditory discrepancies to resolve conflicting input streams. The superior colliculus and posterior parietal cortex play key roles in integrating these cues, revealing the brain's bias toward visual dominance in spatial attribution (Alais & Burr, 2004). Importantly, this bias is not arbitrary but reflects an adaptive process: by resolving cross-modal ambiguities, attention ensures perceptual coherence in noisy environments. This coordination between sensory systems depends on spatial alignment and attentional weighting, where cross-modal influences help resolve competing perceptual cues (Macaluso, 2010).

The rubber arm illusion (Botvinick & Cohen, 1998) further illustrates how synchronized visual and tactile input can produce a compelling sense of ownership over a non-biological object. When a hidden real arm and a visible rubber arm are stroked in synchrony, participants often report that the rubber limb feels like their own. This illusion reveals the plasticity of body schema—the brain's internal model of the body—and highlights how attention modulates the weighting of sensory evidence. Neuroimaging studies implicate the premotor cortex, intraparietal sulcus, and insula in mediating this experience (Ehrsson et al., 2004), underscoring the role of multisensory congruence in generating embodied self-awareness.

These illusions do more than demonstrate perceptual malleability—they reflect the brain's metafunctional strategy for coherence. Attention, in these cases, does not simply filter information; it coordinates and reconciles multisensory data streams to maintain a unified experience of both environment and self. Such examples illuminate how attention supports sensorimotor integration, predictive modeling, and adaptive behavior—core components of its role as a brain-wide metafunction

(Fries, 2005; Menon, 2015). This aligns with theories of motor cognition and internal simulation, in which observing or imagining actions activates similar neural circuits as performing them—reinforcing embodied perception and body ownership (Jeannerod, 2001).

Multisensory Integration and Body Ownership

The brain's ability to maintain coherent spatial awareness depends not only on multisensory integration but also on internal models of body orientation and position in space. Classic research on spatial orientation has long emphasized the importance of vestibular and proprioceptive inputs in anchoring the body schema and guiding attention (Howard & Templeton, 1966). This integration process underlies not only perception but also the construction of body ownership, the feeling that one's body belongs to oneself and is situated in space. Attention plays a central role in this process, modulating the weighting, timing, and salience of sensory inputs across modalities to generate a unified representation of the body. Phenomena like the ventriloquism effect and the rubber arm illusion (reviewed in the previous section) demonstrate how visual dominance and temporal synchrony can override proprioceptive or auditory inputs. Such illusions reveal the flexibility of the brain's body schema, which is continuously updated through attentional mechanisms that privilege congruent, task-relevant sensory signals. This adaptive filtering and coordination illustrate the metafunctional operation of attention—not merely selecting stimuli but integrating and stabilizing them into a coherent model of the self in space.

At the neural level, frontoparietal regions, such as the premotor cortex, intraparietal sulcus, and temporoparietal junction, interact with subcortical areas (e.g., the pulvinar, superior colliculus) to facilitate this integration. These regions are also implicated in predictive coding models, wherein the brain generates anticipatory models of sensory input and adjusts them based on incoming data. As Gu and FitzGerald (2014) note, the body's response to its own predictions is fundamental to interoceptive attentional dynamics, influencing both salience and self-awareness. This aligns with predictive coding frameworks, which propose that attention plays a central role in minimizing prediction error by reconciling expected and actual sensory input—particularly in domains involving bodily awareness and internal states. Attention, within this framework, helps minimize prediction error by resolving discrepancies between expected and actual input (Friston, 2010; Seth, 2014).

This integrative function is not limited to body ownership—it scales across sensory domains to maintain perceptual, affective, and cognitive continuity. In this way, attention serves as a bridge between sensation, embodiment, and conscious awareness, orchestrating inputs to construct a coherent experiential world. The metafunctional framework thus positions attention as a system-level mediator of self–world integration, dynamically aligning external stimuli with internal models of selfhood, agency, and bodily continuity. The capacity of attention to harmonize inputs across sensory modalities and sustain coherent bodily selfhood underscores its role as more than a selective filter. It is an active modulator of neural activity and subjective experience. As we now turn to the neural mechanisms that underlie attentional modulation, we examine how attention sharpens perception at the cellular level—through enhanced neural firing and dynamic circuit engagement across cortical and subcortical systems.

Enhanced Firing in Attention-Related Neural Processes

One of the most robust neurophysiological signatures of attention is enhanced neuronal firing in response to attended stimuli. This phenomenon occurs without altering the physical properties of the stimulus, indicating that attention operates as a top-down amplifier of signal strength within the brain's sensory processing circuits. For example, a neuron in the primary visual cortex (V1) might exhibit a modest response to a stimulus in the absence of attention, but when that same stimulus becomes the focus of attentional resources, the neuron's firing rate significantly increases (Reynolds & Chelazzi, 2004). This modulation reflects the brain's strategy for optimizing limited computational resources: it boosts the representation of task-relevant information while suppressing irrelevant noise. Enhanced firing has been documented across sensory modalities—including visual, auditory, and somatosensory domains—underscoring the global scope of attention. In visual attention, for example, attention increases contrast sensitivity and spatial resolution by altering the gain of neuronal populations (Treue & Maunsell, 1999; Desimone & Duncan, 1995). Feature-based attention in visual cortex enhances neuronal sensitivity not only to spatial locations but also to object features such as orientation or motion direction—demonstrating how attention modulates sensory encoding even in the absence of eye movements (Maunsell & Treue, 2006). Similarly, in auditory processing, attention modulates the tuning of neurons in the auditory cortex to emphasize certain frequency bands (Fritz et al., 2007).

These enhancements are coordinated by large-scale attention networks, particularly the frontoparietal system. Regions such as the frontal eye fields (FEF), intraparietal sulcus (IPS), and dorsolateral prefrontal cortex (DLPFC) exert top-down influence on sensory cortices, tuning them to salient or goal-relevant features (Buschman & Miller, 2007). Subcortical structures—including the pulvinar nucleus of the thalamus and the locus coeruleus–norepinephrine system—also contribute to attentional gating, regulating the synchronization and precision of cortical responses (Saalmann et al., 2012; Aston-Jones & Cohen, 2005). Notably, emotional salience and motivational states influence this firing enhancement via inputs from the amygdala and hippocampus, especially when stimuli are threatening or personally relevant (Pessoa, 2008). These affective pathways further illustrate attention's integration across cognitive, sensory, and emotional domains—an essential feature of its metafunctional role. As a dynamic modulator, attention does not merely gate sensory input; it amplifies, coordinates, and prioritizes neural processing across brain systems to support adaptive perception and action.

Summary Box
Mechanisms of Attentional Modulation

Enhanced Neuronal Firing

Attention increases the firing rate of neurons tuned to relevant stimuli, improving perceptual clarity.

Key regions: Primary sensory cortices (e.g., V1, A1), frontal eye fields (FEF), intraparietal sulcus (IPS).

Top-Down Control

Executive systems regulate attention based on task goals, expectations, and cognitive context.

Key regions: Dorsolateral prefrontal cortex (DLPFC), anterior cingulate cortex (ACC), FEF.

Selective Gating

Filters incoming information early in the sensory pathway, minimizing irrelevant input.

Key regions: Pulvinar nucleus, thalamus, reticular activating system (RAS).

Neuronal Synchrony (Coherence)

Coordinates timing of neural oscillations to boost communication between attention-related regions.

Key regions: Thalamocortical circuits, prefrontal–parietal networks.

Emotional Salience Amplification

Heightens attention to emotionally significant stimuli for adaptive prioritization.

Key regions: Amygdala, hippocampus, orbitofrontal cortex.

Neuromodulation

Neurotransmitters modulate cortical excitability and attentional flexibility.

Key systems: Locus coeruleus (norepinephrine), basal forebrain (acetylcholine), brainstem nuclei.

Together, these mechanisms enable flexible, goal-directed allocation of attention across sensory and cognitive domains—reinforcing its status as a brain-wide metafunction.

Top-Down Regulation and Multimodal Integration

Top-down regulation refers to the brain's ability to selectively enhance relevant sensory signals and suppress distractions based on internal goals, task demands, and contextual expectations. This capacity reflects the executive function of attention—coordinating across multiple sensory modalities and hierarchies of processing to align perception with intention. When a stimulus becomes behaviorally relevant, attentional control networks recruit higher-order cortical regions, such as the dorsolateral prefrontal cortex (DLPFC) and anterior cingulate cortex (ACC), to amplify its neural representation. This modulation improves perceptual discrimination, facilitates decision-making, and supports rapid behavioral responses (Miller & Cohen, 2001; Buschman & Kastner, 2015). In dynamic environments—such as a busy street or a crowded classroom, this flexible system enables prioritization of what matters most, adapting in real time to shifting demands.

Subcortical structures play equally critical roles. The pulvinar nucleus of the thalamus acts as an attentional filter, regulating the flow of sensory information to cortical areas (Saalmann et al., 2012). It coordinates with the superior colliculus for rapid shifts of attention and with sensory-specific cortices for modality-based enhancement. The superior colliculus plays a key role in spatial orienting and rapid attentional shifts by integrating multisensory inputs and coordinating eye movements—functions that underscore its relevance in attention-related processes (Krauzlis, Lovejoy, & Zénon, 2013). The reticular activating system (RAS) in the brainstem maintains arousal and alertness—setting the stage for attentional engagement—while the amygdala and hippocampus contribute affective salience and memory relevance, further shaping what is attended to and encoded.

Together, these systems form a multimodal integration network, a distributed yet coherent system of attentional modulation. This integration is not merely additive; rather, it reflects a coordinated allocation of limited cognitive resources across modalities. For instance, when focusing on a lecture in a noisy environment, visual and auditory inputs are both regulated in the service of comprehension, while irrelevant sensory signals are actively suppressed. This interplay highlights attention's metafunctional role in linking external sensory streams with internal cognitive hierarchies. In line with the brain metafunction framework, this process exemplifies how top-down control enables goal-directed unification across neural systems, maintaining perceptual coherence even in complex, multisensory environments.

Implications for Cognitive Processing

The phenomenon of enhanced neuronal firing in response to attentional focus has broad implications for how the brain processes information. By increasing the signal strength of relevant stimuli without altering their physical properties, attention effectively prioritizes internal representations, optimizing the brain's limited processing resources for adaptive behavior (Desimone & Duncan, 1995; Reynolds & Chelazzi, 2004). This enhancement is not confined to any single modality. Research shows increased neural gain across visual, auditory, and somatosensory domains, indicating a global attentional architecture capable of modality-specific precision and cross-modal unification. In this way, attention operates not merely as a spotlight but as a dynamic regulatory mechanism that fine-tunes sensory processing across the entire cortical surface.

Such modulation is essential for real-world tasks. Whether parsing speech in a noisy room or tracking a moving object in a cluttered environment, attentional enhancement ensures salient stimuli are processed more efficiently, improving speed, accuracy, and behavioral responsiveness. These principles are vividly illustrated in multitasking contexts: for instance, simulated driving studies show that secondary tasks such as dialing a phone can significantly impair visual and motor coordination, underscoring how divided attention reduces performance efficiency (Reed & Green, 1999; Strayer & Johnston, 2001). Similarly, media-induced task switching—such as toggling between lecture notes and texting—can fragment attention and diminish working memory fidelity, especially in academic settings (Rosen, Carrier, & Cheever, 2013). These challenges echo foundational theories of vigilance and attention proposed by Parasuraman and Davies (1984), who demonstrated how sustained cognitive load can degrade attention over time, leading to decreased situational awareness and increased error likelihood. These findings extend to higher-order functions, including problem-solving, planning, and decision-making, all of which rely on attentional filtering to manage competing cognitive demands.

Importantly, these processes are coordinated through reciprocal connections among cortical and subcortical networks, particularly those involving the prefrontal cortex, thalamus, brainstem arousal systems, and limbic regions (Buschman & Kastner, 2015; Saalmann et al., 2012; Pessoa, 2008). This distributed activation supports metafunctional integration—a core theme of this chapter—where attention is not limited to sensory enhancement but serves as a higher-order function that bridges cognition, emotion, memory, and perception. In short, the implications of attentional enhancement go far beyond improved stimulus detection. They point to attention as a central orchestrator of cognitive processing, enabling coherence, efficiency, and adaptability in the face of environmental complexity. This perspective deepens our understanding of attention not as a narrow faculty, but as a brain-wide metafunction, fundamental to the unity of experience and action. Taken together, these findings underscore attention's core role in shaping how we perceive, interpret, and respond to the world. As a brain-wide metafunction, attention does not act in isolation—but dynamically tunes neural activity across distributed systems to optimize perception, memory, and behavior. This integrative quality becomes even more apparent in clinical conditions where attention is disrupted, such as hemispatial neglect.

Hemispatial Neglect Adds to the Understanding of Attention

Hemispatial neglect—also known as unilateral spatial neglect—is a striking neurological condition that illustrates the consequences of disrupted attentional systems. Most commonly arising from damage to the right parietal lobe, particularly after stroke, this syndrome results in the failure to attend to stimuli on the contralateral (typically left) side of space. Importantly, this occurs despite intact sensory and motor systems, emphasizing that neglect is not a deficit of sensation, but of attentional awareness (Heilman, Watson, & Valenstein, 2003). Patients with hemispatial neglect may ignore food on one side of a plate, fail to groom one half of their body, or even deny ownership of a limb—a condition known as anosognosia (Babinski, 1914; Langer & Bogousslavsky, 2020). These symptoms reflect an impairment in integrating spatial representations and attentional priority maps, particularly across the dorsal attention network (DAN) and ventral attention network (VAN).

Neuroimaging studies have revealed that neglect arises from dysfunction across a distributed network, including the frontal eye fields (FEF), intraparietal sulcus (IPS), superior temporal gyrus, thalamus, and subcortical structures such as the basal ganglia (Astafiev et al., 2003; Vallar & Calzolari, 2018; Champod, Frank, & Taylor, 2018). The dorsal network governs goal-directed spatial attention, while the ventral network monitors for unexpected or behaviorally relevant stimuli. When connectivity between these systems is impaired, the brain loses its capacity to flexibly shift or distribute attentional resources. Importantly, hemispatial neglect offers a window into the brain's integrative design. The condition demonstrates that attentional processing is not localized to a single "center," but rather relies on coordinated activity across sensory, motor, cognitive, and affective regions. The disruption of this coordination results in a fragmented experience of the world—highlighting the essential function of attention in unifying perception, action, and self-awareness. Neglect also reveals the multimodal nature of attention. Patients often show impairments across visual, auditory, and tactile modalities, further reinforcing the concept of attention as a metafunctional process that spans and integrates across sensory domains. Rehabilitation strategies—including prism adaptation, transcranial magnetic stimulation (TMS), and cognitive training—often aim to re-engage this distributed attentional network, demonstrating its plasticity and capacity for recovery (Rode et al., 2017).

Clinical Case Study: Hemispatial Neglect as a Breakdown of Metafunctional Integration

Patient A, a 68-year-old right-handed woman, experienced a right middle cerebral artery stroke. Post-stroke, she began ignoring stimuli on the left side of her visual field. When asked to draw a clock, she positioned all the numbers on the right half of the circle. During meals, she consistently left food uneaten on the left side of her plate. Although cognitively alert and cooperative, she denied any issues—a symptom known as anosognosia—and at times claimed that her left hand did not belong to her. Neuroimaging revealed damage to the right inferior parietal lobule, a critical hub within the dorsal attention network (DAN), along with reduced activity in connected frontal and subcortical regions. Despite intact early sensory processing, her attentional system failed to integrate and prioritize stimuli across contralesional space. This illustrates a breakdown in metafunctional integration, where perception, self-awareness, and spatial mapping—normally coordinated across distributed systems—begin to unravel.

Metafunctional Framing of Hemineglect in Attention and Perception

As previously explored in the chapter on Perception, hemineglect can be meaningfully mapped onto the Brain Metafunction Framework. In this model:

Afferent Input (e.g., visual encoding) may be intact, diminished, or spatially constrained;

Metafunctional Processing (involving attention, memory, and spatial modeling) becomes distorted or absent;

Efferent Projection (the spatial and behavioral output of perception) is often misaligned or suppressed, resulting in the failure to project salience into the neglected hemispace.

From an attentional perspective, this condition reflects not merely a loss of input but a collapse in the top-down orchestration of spatial awareness. Attention is unable to allocate priority to contralesional stimuli or coordinate the efferent dynamics required for action or awareness. This vividly illustrates attention's metafunctional nature—as a system that bridges sensory registration, cognitive prioritization, and embodied projection. In Patient A's case, we see how damage to a nodal point in this system can trigger a global disruption of conscious experience.

Attentional Shift

An attentional shift refers to the redirection of cognitive focus from one stimulus or task to another, governed by both internal goals and external stimuli. It reflects the brain's adaptive capacity to reallocate processing resources in response to dynamic situational demands. Whether shifting from a visual cue to an auditory signal, or from one thought stream to a competing one, attentional shifts underscore attention's flexibility and bounded processing limits. Two prominent theoretical models offer complementary perspectives on how attentional resources are managed during such transitions:

The *Unitary Resource Model* proposes a shared, finite pool of attentional capacity (Broadbent, 1958). When two tasks draw on overlapping resources, performance on one may suffer due to competition. Reaction time paradigms have long been used as sensitive measures of attentional resource allocation under multitasking conditions, offering insight into how secondary tasks interfere with cognitive load (Lang & Basil, 1998). For example, texting while driving impairs both visual and cognitive performance, reflecting the limited bandwidth of this shared pool. Frequent media multitasking—such as cell phone use during academic tasks—has been linked to poorer academic performance and reduced attentional capacity, reinforcing the cognitive cost of divided attention (Lepp, Barkley, & Karpinski, 2014).

The *Multiple Resource Theory* (Wickens, 2002; Wickens et al., 2012) suggests that attentional resources are divided among different sensory and response modalities. This model explains why individuals can more easily manage simultaneous tasks that draw on separate systems (e.g., listening to music while sketching), but experience greater interference when tasks compete within the same domain (e.g., reading while writing).

Metafunctional Integration: Adaptive and Flexible Coordination

Modern neuroscience integrates these models, revealing that attentional shifts depend on both shared and modality-specific constraints—mediated by large-scale, context-sensitive brain networks (Posner & Rothbart, 2007; Cole et al., 2013). Neural regions such as the anterior cingulate cortex (ACC), dorsolateral prefrontal cortex (DLPFC), and posterior parietal cortex (PPC) orchestrate these shifts by monitoring salience, managing conflict, and regulating executive priorities (Corbetta & Shulman, 2002). In the metafunctional framework, attentional shifts are seen as dynamic reallocations across the triadic system:

Afferent Input: Determines current sensory load and salience.

Metafunctional Processing: Weighs internal goals, relevance, and prior context.

Efferent Projection: Selects a new target of attention and guides cognitive and motor output accordingly.

This integrated model captures attention's role in managing real-time transitions between internal cognition and external perception, and between multiple perceptual or motor tasks. Imagine reading a text message (visual-linguistic modality) when a loud crash occurs outside (auditory-spatial modality). Your attention shifts reflexively toward the sound, engaging the orienting network and disengaging from visual focus. Once the perceived urgency subsides, top-down control reorients you back to the text. This fluid negotiation between modalities exemplifies the competitive and cooperative dynamics of attentional allocation. Orienting attention, particularly in visuospatial tasks, engages both covert and overt mechanisms—including eye movement planning—reflecting a close link between gaze and attention (Klein, Kingstone, & Pontefract, 1992).

Key Neural Regions and Systems

Attention emerges not from a single brain center, but through dynamic coordination among widely distributed regions—spanning cortical, subcortical, and neuromodulatory systems. This distributed architecture underscores attention's role as a brain-wide metafunction, integrating perception, memory, executive control, and emotional relevance into adaptive behavior.

Cortical Contributions

Prefrontal Cortex (PFC): Central to top-down control, the PFC (especially the dorsolateral and ventromedial subdivisions) regulates goal-directed focus, working memory, inhibitory control, and decision-making (Miller & Cohen, 2001). Studies on body ownership demonstrate that activity in the premotor cortex is modulated when individuals experience the sensation of a limb as their own—highlighting the role of multisensory integration in attentional alignment to the bodily self (Ehrsson, Spence, & Passingham, 2004).

Sensory Cortices: Visual, auditory, and somatosensory areas perform early sensory encoding and are dynamically modulated by attentional states—especially under top-down influence (Desimone & Duncan, 1995; Carrasco, 2011).

Posterior Parietal Cortex (PPC): Acts as a spatial integrator and attentional "priority map," interacting with frontal regions to manage shifts in attention (Bisley & Goldberg, 2010).

Subcortical and Brainstem Systems

Thalamus (esp. Pulvinar nucleus): Filters and routes sensory information to cortical areas based on attentional priority, coordinating cortical-cortical communication (Saalmann et al., 2012).

Locus Coeruleus (LC): A major source of norepinephrine, LC modulates arousal, vigilance, and responsivity across the cortex (Aston-Jones & Cohen, 2005). Recent findings further suggest that the brainstem plays a critical role in cortico-cortical network integration, providing structural pathways that facilitate global attentional regulation and functional coherence across distributed systems (Hansen, Adkins, & Davidson, 2023).

Reticular Activating System (RAS): Regulates global arousal and wakefulness—setting the foundational conditions for attentional engagement (Moruzzi & Magoun, 1949).

Amygdala & Hippocampus: Prioritize emotionally significant stimuli and modulate attention via memory-contextual pathways (Pessoa, 2008; Eichenbaum, 2000).

Cerebellar and Interoceptive Integration (Emerging Insights)

Cerebellum: Traditionally linked to motor control, it also contributes to cognitive focus, sensory prediction, and timing—especially in dynamic, real-time tasks (Strick, Dum, & Fiez, 2009). Traditionally associated with motor coordination, the cerebellum has increasingly been recognized for its contributions to attentional timing, prediction, and sensory integration. As Ito (2008) suggests, internal models formed by cerebellar circuits support mental activities by simulating expected outcomes, thereby enabling more accurate and efficient shifts of attention. This predictive capacity aligns with the cerebellum's emerging role in cognitive focus, particularly during fast-paced or dynamic tasks where timing precision is crucial.

Insula & Interoceptive Networks: Integrate bodily states and subjective salience, contributing to attentional shifts based on internal needs or discomfort (Critchley & Harrison, 2013).

Metafunctional Framing

These systems together illustrate attention's triadic operation:
Afferent Loop: Sensory cortices, thalamus, and brainstem provide incoming signals and global readiness.

Central Metafunctional Processing: The prefrontal-parietal-limbic circuitry weighs relevance, salience, and goal alignment.

Efferent Projection: Prepares motor plans or adjusts cognitive focus, sustaining engagement or initiating redirection.

This integration supports adaptive cognition, enabling the brain to flexibly allocate its limited resources in a task-relevant, context-sensitive manner.

Large-Scale Brain Networks

The Systems-Level Architecture of Attentional Control

Beyond discrete brain regions, attention operates through the orchestration of large-scale neural networks that dynamically interact to prioritize, sustain, and shift focus in response to environmental demands and internal goals. This network-based understanding underscores attention's metafunctional quality—its role as a coordinating architecture across sensory, cognitive, and emotional domains. Empirical evidence shows that attentional load modulates not only activity within these core networks but also the global connectivity between large-scale brain systems—including those not traditionally associated with attention (Alnæs et al., 2015). This suggests that attentional demands reshape functional brain architecture dynamically, extending beyond localized processing hubs.

1. *Dorsal Attention Network* (DAN)

Core Regions: Frontal Eye Fields (FEF), Intraparietal Sulcus (IPS)

Function: Supports goal-directed (top-down) attention, especially spatial orienting, saccadic control, and attentional maintenance (Vossel, Geng, & Fink, 2014).

Dynamic Role: Actively sustains task engagement by allocating resources to relevant sensory input.

2. *Ventral Attention Network* (VAN)

Core Regions: Temporoparietal Junction (TPJ), Ventrolateral Prefrontal Cortex

Function: Bottom-up reorienting, especially to novel, unexpected, or behaviorally salient stimuli (Corbetta & Shulman, 2002).

Metafunctional Integration: Acts as an alerting system—interrupting ongoing processing when a stimulus requires urgent reallocation of attention.

3. *Salience Network* (SN)

Core Regions: Anterior Insula, Dorsal Anterior Cingulate Cortex (dACC)

Function: Detects emotionally or homeostatically significant stimuli and coordinates attentional switching between networks (Menon, 2015).

Bridge Function: Facilitates rapid toggling between internal and external focus, especially under uncertainty or arousal.

4. *Default Mode Network* (DMN)

Core Regions: Medial Prefrontal Cortex (mPFC), Posterior Cingulate Cortex (PCC), Angular Gyrus

Function: Traditionally associated with self-referential thought, mind-wandering, and episodic memory (Raichle et al., 2001). Research has refined our understanding of the DMN, suggesting it plays an active role in internally directed cognition, predictive simulation, and integration of social and autobiographical information (Buckner, 2013).

Interaction with Attention: Deactivates during externally focused tasks but re-engages during reflection, prediction, or shifting from goal-directed states.

Analyses reveal that the DMN is not a monolithic system but is composed of multiple functional subsystems that support different forms of internal mentation—including social cognition, memory-based simulation, and conceptual processing (Buckner, 2013). No longer "task-negative," the DMN participates in attentional modulation when integrating internal cognitive contexts. Research in this area has expanded this view, highlighting the DMN's role in internal mentation, predictive modeling, and integration of social, autobiographical, and conceptual information (Buckner, 2013).

Network Interplay and Cognitive Flexibility

Attentional performance depends on the fluid interaction between multiple large-scale brain networks, each contributing distinct yet complementary functions. This interplay between stable and adaptive control systems reflects the dual-network theory of task regulation (Dosenbach et al., 2007), which complements attention's role in balancing sustained engagement with moment-to-moment flexibility. The Dorsal Attention Network (DAN) supports goal-directed focus by maintaining sustained attention on task-relevant stimuli. In contrast, the Ventral Attention Network (VAN) enables rapid reorienting toward unexpected or behaviorally salient inputs. The Salience Network (SN) acts as a dynamic switchboard, detecting urgency and coordinating transitions between internal and external focus. Meanwhile, the Default Mode Network (DMN) provides background context through self-referential and introspective processing. Disruptions in the integration of these networks—such as in ADHD, depression, or hemispatial neglect—demonstrate that attentional dysfunction is rarely localized to a single neural deficit. Instead, it arises from imbalances or disconnects within and between these systems (Seeley et al., 2007).

Framing Within the Brain Metafunction Model

These networks together illustrate a metafunctional triad:

Afferent Readiness: DAN and VAN prepare and redirect attention.

Internal Processing: SN and DMN evaluate relevance, salience, and personal significance.

Efferent Control: Executive systems initiate behavioral or attentional shifts, maintaining coherent task performance.

This architecture supports attention as a distributed, adaptive, and integrative system, capable of coordinating internal states with environmental complexity.

More on the Metafunctional Quality of Attentional Processes

Attention is not merely a localized function confined to one brain region or cognitive task—it is a metafunctional process, dynamically integrating perception, memory, emotional salience, and executive control. This perspective frames attention as a global orchestration system that flexibly reallocates cognitive resources across multiple domains in response to shifting demands, environments, and goals. Modern neuroscience reveals that attentional control arises from the coordinated activity of distributed networks, rather than from a central command center. Functional connectivity studies have shown that attentional states modulate interregional synchronization, especially between the prefrontal cortex, parietal cortex, thalamus, and salience hubs such as the anterior insula and dorsal anterior cingulate cortex (Fries 2005; Saalmann et al. 2012). This cross-regional modulation reflects what Fries terms "neuronal coherence"—the selective timing and alignment of neural oscillations to optimize information flow within and between brain networks. "Selective attention is implemented by selectively synchronizing neurons in relevant brain areas, thereby boosting communication and information flow" (Fries, 2005).

Graph Theory and Complex Systems

Using tools from graph theory, Bullmore and Sporns (2009) demonstrated that the brain's attentional systems form a small-world network: they are simultaneously highly clustered (supporting local efficiency) and globally connected (enabling rapid long-range coordination). This architecture allows for efficient transitions between focused and flexible states of attention, between task engagement and rest, and between conscious and unconscious attentional shifts. Frontoparietal control systems serve as flexible hubs. Salience networks act as switches between internally and externally focused processing. Default mode interactions regulate background cognitive and self-referential content, playing a central role in internal mentation, future simulation, and even disruptions in psychopathology (Buckner, 2013). These networks do not operate in isolation, but instead communicate continuously to support attentional flexibility and adaptability across contexts. As Avena-Koenigsberger, Misic, and Sporns (2018) emphasize, the communication dynamics within complex brain networks enable rapid information exchange and global coordination—functions that are critical for the shifting and sustaining of attentional focus. The salience network, particularly the insula and dACC, acts as a dynamic switchboard—detecting urgent or emotionally significant inputs and reallocating attention accordingly (Seeley et al., 2007). This reinforces the idea that attention is deeply enmeshed with affective and interoceptive states, supporting a brain-wide readiness for adaptive action. These understanding supports a metafunctional perspective in which attention serves as an interface between systems:

Perceptual: Prioritizes sensory input (e.g., visual, auditory).

Cognitive: Coordinates memory, problem-solving, decision-making.

Affective: Filters stimuli by emotional and motivational relevance.

Executive: Maintains task goals, inhibits distractions, shifts flexibly.

These functions are not modular or sequential—they operate simultaneously and interactively, often within milliseconds. The ability to fluidly prioritize among them—sometimes even preconsciously—is what gives attention its metafunctional depth. Rather than a singular process, attention represents a regulatory axis through which distributed brain systems achieve coherence.

Attention and Memory

Attention and memory are deeply intertwined cognitive processes. At every stage—encoding, storage, and retrieval—attention modulates the effectiveness and clarity of memory formation. Without attentional focus, even vivid experiences may leave no lasting trace; with sustained and selective attention, even subtle details can be consolidated into long-term memory.

Selective Attention and Encoding

Attention plays a critical role during the encoding phase, acting as a filter that determines which stimuli enter conscious awareness and receive deeper processing. When attention is focused, sensory inputs are more likely to be semantically processed and meaningfully encoded—this principle is central to the levels-of-processing framework (Craik & Tulving, 1975). For instance, studying in a quiet space with full attention results in stronger encoding than attempting to memorize material while distracted by noise or notifications. Emotionally significant stimuli also receive attentional prioritization. The amygdala, in interaction with the hippocampus, boosts memory consolidation for emotionally arousing events (McGaugh, 2000). This attentional bias toward emotionally salient information helps explain why people often vividly remember events like accidents or emotionally charged conversations.

Sustained Attention and Working Memory

Attention also undergirds working memory—the temporary maintenance and manipulation of information over short intervals. Sustained attention allows individuals to rehearse, compare, or transform representations in real time (Baddeley, 2003). Interruptions to attention often degrade working memory performance, reducing task accuracy and increasing errors (Kane & Engle, 2003). Neurophysiologically, this link involves coordinated activity between the prefrontal cortex (which maintains task goals and modulates attention) and posterior cortical regions (which retain sensory details). This distributed interaction ensures that memory remains task-relevant and accessible under cognitive load.

Retrieval and Attentional Cueing

Even during memory retrieval, attention is essential. Retrieval cues—contextual or semantic—must be actively attended to and matched against stored representations. Attention helps inhibit irrelevant associations, reducing interference and enhancing retrieval accuracy (Tulving & Thomson, 1973; Anderson, 2003). In real-world tasks such as exams or navigation, attention guides retrieval toward goal-relevant knowledge, while blocking out distractions. Attentional control also facilitates reconstructive memory, where the brain dynamically rebuilds remembered scenes or facts, often filling in gaps based on expectations or emotional states.

Metafunctional Framing

From a metafunctional perspective, attention and memory operate not as separate systems, but as mutually reinforcing processes:

- Encoding requires attentional selection.
- Storage is enhanced through emotionally or cognitively meaningful focus.
- Retrieval depends on attentional cueing and interference control.
- This integration reflects attention's role as a modulatory axis that shapes what gets remembered, how vividly, and under what conditions.

Attention acts as the gatekeeper of memory, directing which experiences are stored and how accessible they remain over time. For example, imagine you're listening to a lecture while checking your phone. Though the lecturer presents key exam content, your divided attention prevents deep encoding. Later, you struggle to recall specific facts or examples. Now imagine the same lecture, but this time you're fully focused, taking notes and mentally connecting new material to prior knowledge. Because your attention is sustained and semantically engaged, the hippocampus encodes the material more robustly, aided by frontoparietal attentional networks. This scenario illustrates how attention governs what enters memory—and how richly it is encoded.

Attention and Learning

Learning is not merely the accumulation of information—it is the structured transformation of experience into knowledge. At the heart of this transformation lies attention, which enables learners to focus, organize, and integrate new information meaningfully. Without attentional engagement, learning is shallow; with sustained, directed attention, learning becomes efficient, flexible, and durable. Attentional performance is not uniform across individuals; rather, it varies based on stable traits such as working memory capacity, cognitive control efficiency, and motivational orientation. As Ackerman (1992) emphasizes, individual differences in attention significantly shape complex skill acquisition, particularly under conditions that require multitasking or high cognitive control. In learning contexts, attention acts as a filter, selecting task-relevant stimuli and inhibiting distractions. This selectivity optimizes cognitive load, ensuring that the brain's limited working memory capacity is used efficiently. Research shows that focused attention facilitates deep processing—the active organization and association of new information with prior knowledge (Craik & Lockhart, 1972). For example, a student solving a math problem must attend not only to the problem itself, but also to the steps, strategies, and error-checking processes—requiring sustained executive attention.

Neuroplasticity and Repetition

Attention is also key to neuroplasticity—the brain's capacity to adapt and reorganize in response to experience. Repeated attentional focus on specific stimuli or tasks strengthens synaptic connections, reinforcing the neural circuits involved (Buonomano & Merzenich, 1998). Attention, in this light, is a sculptor of cognition, guiding the brain's adaptation to learning demands. This process is especially evident in skill learning, language acquisition, and the mastery of abstract concepts, where sustained attention and practice yield lasting neural changes. Attention interacts with intrinsic motivation and emotional engagement. The experience of "flow" (Csikszentmihalyi, 1990)—a state of complete attentional immersion—maximizes learning by aligning interest, skill level, and task challenge. When students are deeply focused, they experience greater comprehension, resilience, and retention. Moreover, emotionally salient content naturally draws attention, increasing encoding depth and long-term recall—especially when supported by relevant goals or values. From the metafunctional perspective, attention is not a peripheral contributor to learning—it is foundational. It modulates sensory input, sustains engagement, shapes memory encoding, and interacts with motivation and executive control. Learning succeeds or fails largely in proportion to the learner's ability to allocate, sustain, and flexibly shift attention. This view positions attention as a core metafunction of the brain—dynamically coordinating perceptual, cognitive, and emotional processes to make learning possible.

Emotion and Attention

Emotion and attention are deeply intertwined, forming a bidirectional relationship that influences perception, memory, and behavior. Emotion acts as both a filter and an amplifier of attention, helping the brain prioritize stimuli that are relevant for survival, goals, or personal meaning. The brain's attentional systems are sensitive to emotional salience. Stimuli that are threatening, rewarding, or emotionally charged are more likely to capture and hold attention than neutral stimuli (Pessoa & Adolphs, 2010). This prioritization is adaptive—it allows organisms to respond quickly to important environmental cues. For instance, a fearful facial expression in a crowded room may involuntarily draw visual attention, even before conscious processing occurs. The amygdala, in the limbic system, plays a central role in this process by evaluating the emotional relevance of stimuli and influencing attention networks, particularly in the prefrontal and parietal cortices. Emotionally congruent stimuli—those that align with task goals—can enhance attentional performance. For example, a student emotionally invested in a subject may exhibit heightened focus, retention, and engagement. Conversely, irrelevant emotional stimuli (e.g., intrusive negative thoughts or anxiety) can impair attentional control, increasing distraction and cognitive load (Pourtois & Dolan, 2010). This dual role underscores emotion's dynamic influence on the allocation of attentional resources. Attention is not simply governed by logic or task demands—it is steered by what matters emotionally in the moment.

Integration with Executive Functions

Emotional regulation and attentional control are closely linked through shared neural circuits, particularly within the prefrontal cortex. This integration allows for cognitive reappraisal, goal maintenance, and conflict resolution in emotionally complex environments. For example, a surgeon in a high-stakes operation must manage emotional arousal to maintain focus—engaging top-down attentional control to suppress distraction and sustain performance. In the brain metafunction framework, emotion is not peripheral to cognition—it is integrated with attention, memory, and executive control. Emotion shapes what we notice, how we remember it, and how we respond. Attention serves as the dynamic interface, flexibly allocating resources based on emotional relevance. This perspective underscores attention's metafunctional quality: it operates at the nexus of internal states and external demands, aligning perception and behavior with both cognitive goals and affective priorities.

Arousal and Perceptual Load

Arousal states significantly influence attentional capacity, acting as modulators of the brain's ability to allocate cognitive resources in response to task demands. Moderate levels of arousal typically enhance attentional efficiency, promoting sustained focus and heightened responsiveness to salient stimuli. However, excessive arousal can overwhelm attentional systems, leading to diminished task performance, distractibility, or cognitive rigidity. This dynamic is elegantly captured by the Yerkes–Dodson law (1908), which describes an inverted U-shaped relationship between arousal and performance: too little arousal results in under-engagement, whereas too much can impair focus and decision-making. Classic frameworks of vigilance and sustained attention underscore how prolonged task demands can lead to performance decline, particularly under conditions of low event frequency or high monotony (Parasuraman & Davies, 1984). Such declines highlight the importance of arousal regulation in sustaining attentional efficiency over time.

Neurobiologically, this modulation is supported by several key systems. The locus coeruleus–norepinephrine system plays a central role in maintaining vigilance and increasing the signal-to-noise ratio in neural processing (Aston-Jones & Cohen, 2005). The reticular activating system (RAS), situated in the brainstem, regulates baseline arousal and primes the cortex for attentional engagement. Arousal states significantly influence attentional capacity, acting as modulators of the brain's ability to allocate cognitive resources in response to task demands. As Pfaff (2005) notes, global arousal mechanisms set the cortical stage for attentional engagement by regulating alertness thresholds. Moderate levels of arousal typically enhance attentional efficiency, promoting sustained focus and heightened responsiveness to salient stimuli. Electrophysiological studies support this relationship, showing that EEG patterns—particularly in alpha and theta bands—are sensitive markers of alertness and attentional engagement (Oken, Salinsky, & Elsas, 2006). These markers provide insight into how baseline arousal states influence attentional efficiency under different cognitive loads. The pulvinar nucleus of the thalamus contributes to sensory gating, helping prioritize relevant inputs under varying levels of task complexity. The amygdala, especially in emotionally salient contexts, rapidly draws attentional focus toward potential threats or rewards. Meanwhile, the prefrontal cortex exerts top-down control, adjusting attentional priorities according to internal goals and situational demands.

Perceptual load theory further expands on how attentional systems operate under varying conditions of cognitive strain. According to this model, high-load tasks—those that are complex or novel—fully occupy attentional resources, leaving little room for distraction. In contrast, low-load tasks often result in surplus attentional capacity, making individuals more vulnerable to irrelevant stimuli (Lavie, 2005). This distinction has meaningful implications across domains: a professional performing a complex task may remain highly focused despite distractions, while during routine or automated tasks, they may be more susceptible to interference. The interplay of arousal and perceptual load exemplifies the brain's capacity for adaptive attentional control, balancing external demands with internal resource constraints in real time. This dynamic balance between arousal, perceptual load, and attentional modulation brings us full circle to attention's role as a brain-wide metafunction—coordinating perception, cognition, and behavior in response to internal goals and external demands.

Conclusion

Attention emerges from this chapter not as a localized or singular cognitive function, but as a brain-wide metafunction that dynamically integrates perceptual, emotional, and executive systems—an understanding that aligns with comprehensive reviews in both psychology and neuroscience (Anderson, 2011). Across neural hierarchies and sensory modalities, attention operates as a coordinating force—filtering, prioritizing, and sustaining the focus necessary for adaptive behavior. From the early alerting mechanisms of the locus coeruleus to the top-down modulation of the prefrontal cortex, attention serves as the interface between environmental input and internal cognitive priorities. Contemporary models—from Posner and Petersen's tripartite network to Luria's systems-level theory and Knudsen's working memory–centered approach—converge on the view that attention requires flexible, large-scale coordination across cortical and subcortical regions. Clinical cases such as hemispatial neglect, and phenomena like inattentional deafness and multisensory illusions, further reveal that attention is not merely a matter of selective enhancement but essential to the coherent construction of experience itself.

Attention is deeply entwined with learning, memory, emotion, and arousal. Its function is not simply to highlight stimuli, but to sculpt consciousness in accordance with task demands, prior knowledge, and affective context. As Dehaene and Changeux (2011) suggest, conscious access emerges through global neural broadcasting—reinforcing the idea that attention acts as a gateway to conscious experience by amplifying and integrating goal-relevant information. Understanding attention as a brain metafunction reframes classical theories and opens new pathways for interdisciplinary research and application—in fields such as education, neurorehabilitation, human-computer interaction, and the design of cognitively sustainable environments. Like perception, attention is not just a mechanism of the mind—it is the orchestrator of mental life.

Having explored attention as a brain-wide metafunction that dynamically prioritizes and integrates sensory, cognitive, and emotional information, we now turn to a closely interwoven domain: memory. While attention shapes the present moment by filtering and modulating incoming stimuli, memory anchors those moments in time—encoding, consolidating, and retrieving experiences that inform learning, identity, and imagination. Just as attention cannot be understood in isolation from perception and emotion, memory is not a static repository but an active, reconstructive process embedded within the brain's metafunctional architecture. In the chapter that follows, we examine memory as a dynamic, distributed system that links past to present, shaping how we perceive, act, and adapt in the world.

REFERENCES

Ackerman, P. L. (1992). "Attention Theory and Individual Differences in Complex Skill Acquisition." In *Advances in Psychology: Attention and Performance*, edited by Dennis E. Meyer and Sylvan Kornblum, 113–145. Amsterdam: Elsevier.

Aebersold, M., & Kalisch, B. J. (2010). "Interruptions and Multitasking in Nursing Care: The Joint Commission Journal on Quality and Patient Safety." *The Joint Commission Journal on Quality and Patient Safety, 36*(3), 126–132. https://doi.org/10.1016/s1553 -7250(10)36021-1.

Alais, D., & Burr, D. (2004). "The Ventriloquist Effect Results from Near-Optimal Bimodal Integration." *Current Biology, 14*(3), 257–262. https://doi.org/10.1016/j.cub. 2004.01.029.

Alkahtani, M., Aziz, T., & Ahmad, A. (2015). "Multitasking in Healthcare Systems." *IIE Annual Conference Proceedings*. Available on ResearchGate.

Alnæs, D., Kaufmann, T., Duff, E. P., Sneve, M. H., Richard, G., Skåtun, K. C., Andreassen, O. A., & Westlye, L. T. (2015). "Attentional Load Modulates Large-Scale Functional Brain Connectivity Beyond the Core Attention Networks." *NeuroImage, 109*, 260–272. https://doi.org/10.10 16/j.neuroimage.2015.01.026.

Anderson, B. A., Laurent, P. A., & Yantis, S. (2011). Value-driven attentional capture. *Proceedings of the National Academy of Sciences, 108*(25), 10367–10371. https://doi.org/10.1073/pnas.1104047108

Anderson, M. C. (2003). "Rethinking Interference Theory: Executive Control and the Mechanisms of Forgetting." *Journal of Memory and Language, 49*(4), 415–445.

Astafiev, S. V., Stanley, C. M., Shulman, G. L., & Corbetta, M. (2003). "Functional Organization of Human Intraparietal and Frontal Cortex for Attending, Looking, and Pointing." *Journal of Neuroscience, 23*(11), 4689–4699. https://www.jneurosci.org/content/23/11 /4689.

Aston-Jones, G., & Cohen, J. D. (2005). "An Integrative Theory of Locus Coeruleus–Norepinephrine Function: Adaptive Gain and Optimal Performance." *Annual Review of Neuroscience, 28*, 403–450.

Avena-Koenigsberger, A., Misic, B. K., & Sporns, O. (2018). "Communication Dynamics in Complex Brain Networks." *Nature Reviews Neuroscience, 19*(1), 17–33.

Babinski, J. (1914). «Contribution à l'étude des troubles mentaux dans l'hémiplégie organique cérébrale (anosognosie).» *Revue Neurologique, 27,* 845–848.

Baddeley, A. (2003). "Working Memory: Looking Back and Looking Forward." *Nature Reviews Neuroscience, 4*(10), 829 –839.

Bregman, A. S. (1990). *Auditory Scene Analysis: The Perceptual Organization of Sound.* MIT Press.

Broadbent, D. E. (1958). *Perception and Communication.* Pergamon Press.

Bronkhorst, A. W. (2000). "The Cocktail Party Phenomenon: A Review of Research on Speech Intelligibility in Multiple-Talker Conditions." *Acta Acustica united with Acustica, 86*(1), 117–128.

Buckner, R. L. (2013). The brain's default network: Origins and implications for the study of psychosis. *Dialogues in Clinical Neuroscience, 15*(3), 351–358. https://doi.org/10.31887/DCNS.2013.15.3/rbuckner

Bullmore, E., & Sporns, O. (2009). "Complex Brain Networks: Graph Theoretical Analysis of Structural and Functional Systems." *Nature Reviews Neuroscience, 10*(3), 186–198. https://doi.org/10.1038/nrn2575.

Buonomano, D. V., & Merzenich, M. M. (1998). "Cortical Plasticity: From Synapses to Maps." *Annual Review of Neuro science, 21*(1), 149–186.

Buschman, T. J., & Kastner, S. (2015). "From Behavior to Neural Dynamics: An Integrated Theory of Attention." *Neuron, 88*(1), 127–139. https://doi.org/10.1016/j.neuron. 2015.09.017.

Buschman, T. J., & Miller, E. K. (2007). "Top-Down versus Bottom-Up Control of Attention in the Prefrontal and Posterior Parietal Cortices." *Science, 315*(5820), 1860–1862.

Carrasco, M. (2011). "Visual Attention: The Past 25 Years." *Vision Research, 51*(13), 1484–1525. https://doi.org/10.1016/j.visres.2011.04.012.

Chein, J. M., Moore, A. B., & Conway, A. R. A. (2011). "Domain-General Mechanisms of Complex Working Memory Span." *NeuroImage, 54*(1), 550–55933.

Cohen, J., LaRue, C., & Cohen, H. H. (2017). "Attention Interrupted: Cognitive Distraction and Workplace Safety." *Professional Safety, 62*(11), 28–33. https://onepetro.org/PS/article-abstract/62/11/28/33502.

Cole, M. W., et al. (2013). "Multitask Connectivity Reveals Flexible Hubs for Adaptive Task Control." *Nature Neuroscience, 16*(9), 1348–1355.

Corbetta, M., & Shulman, G. L. (2002). "Control of Goal-Directed and Stimulus-Driven Attention in the Brain." *Nature Reviews Neuroscience, 3*(3), 201–215. https://doi.org/ 10.1038/nrn755.

Cowan, N. (2014). "Working Memory Underpins Cognitive Development, Learning, and Education." *Educational Psychology Review, 26*(2), 197–223. https://doi.org/10.1007/s 10648-013-9246-y.

Craik, F. I. M., & Tulving, E. (1975). "Depth of Processing and the Retention of Words in Episodic Memory." *Journal of Experimental Psychology: General, 104*(3), 268–294.

Craik, F. I. M., & Lockhart, R. S. (1972). "Levels of Processing: A Framework for Memory Research." *Journal of Verbal Learning and Verbal Behavior, 11*(6), 671–684.

Critchley, H. D., & Harrison, N. A. (2013). "Visceral Influences on Brain and Behavior." *Neuron, 77*(4), 624–638.

Csikszentmihalyi, M. (1990). *Flow: The Psychology of Optimal Experience.* New York: Harper & Row.

Dehaene, S., & Changeux, J. P. (2011). "Experimental and Theoretical Approaches to Conscious Processing." *Neuron, 70*(2), 200–227.

Desimone, R., & Duncan, J. (1995). "Neural Mechanisms of Selective Visual Attention." *Annual Review of Neuroscience, 18*(1), 193–222.

Dosenbach, N. U. F., Fair, D. A., Miezin, F. M., Cohen, A. L., Wenger, K. K., Dosenbach, R. A. T., Fox, M. D., Snyder, A. Z., Vincent, J. L., & Raichle, M. E. (2007). "Distinct Brain Networks for Adaptive and

Stable Task Control in Humans." *Proceedings of the National Academy of Sciences, 104*(26), 11073–11078.

Driver, J., & Noesselt, T. (2008). "Multisensory Interplay Reveals Crossmodal Influences on 'Sensory-Specific' Brain Regions, Neural Responses, and Judgments." *Neuron, 57*(1), 11–23. https://doi.org/10.1016/j.neuron.2007.12.013.

Driver, J., & Spence, C. (1998). "Crossmodal Attention." *Current Opinion in Neurobiology, 8*(2), 245–253. https://doi.org/10.1016/S0959-4388(98)80147-5.

Ehrsson, H. H., Spence, C., & Passingham, R. E. (2004). "That's My Hand! Activity in Premotor Cortex Reflects Feeling of Ownership of a Limb." *Science, 305*(5685), 875–877. https://doi.org/10.1126/science.1097011.

Eichenbaum, H. (2000). "A Cortical–Hippocampal System for Declarative Memory." *Nature Reviews Neuroscience, 1*(1), 41–50.

Fox, M. D., Snyder, A. Z., Vincent, J. L., Corbetta, M., Van Essen, D. C., & Raichle, M. E. (2005). "The Human Brain Is Intrinsically Organized into Dynamic, Anticorrelated Functional Networks." *Proceedings of the National Academy of Sciences, 102*(27), 9673–9678.

Fries, P. (2005). "A Mechanism for Cognitive Dynamics: Neuronal Communication through Neuronal Coherence." *Trends in Cognitive Sciences, 9*(10), 474–480. https://doi.org/10.1016/j.tics.2005.08.011.

Garcia, M. A. (1995). "Reflex Regulation of the Cardio -vascular System." *American Journal of Physiology-Heart and Circulatory Physiology, 269*(4), H1046–H1055.

Gazzaley, A., & D'Esposito, M. (2007). "Top-Down Modulation and Normal Aging: Effects on Visual Working Memory and Attention." *Cognitive, Affective, & Behavioral Neuroscience, 7*(1), 44–56.

Goulden, N., Khusnulina, A., Davis, N. J., Bracewell, R. M., Bokde, A. L., McNulty, J. P., & Mullins, P. G. (2014). The salience network is responsible for switching between the default mode network and the central executive network: Replication from DCM. *NeuroImage, 99*, 180–190. https://doi.org/10.1016/j.neuroimage.2014.05.052.

Gu, X., & FitzGerald, T. H. B. (2014). "The Body's Response to Its Own Predictions." *Neuron, 84*(6), 1182–1189.

Hansen, J. A., Adkins, D. L., & Davidson, M. B. (2023). "Structural Brainstem Connections Support Cortico-Cortical Network Integration." *Trends in Cognitive Sciences, 27* (3), 236–249.

Heilman, K. M., Watson, R. T., & Valenstein, E. (2003). "Neglect and Related Disorders." In K. M. Heilman & E. Valenstein (Eds.), *Clinical Neuropsychology* (4th ed., pp. 296–346). Oxford University Press.

Hillyard, S. A., Hink, R. F., Schwent, V. L., & Picton, T. W. (1973). "Electrical Signs of Selective Attention in the Human Brain." *Science, 182*(4108), 177–180. https://doi.org/10.1126/science.182.4108.177.

Howard, I. P., & Templeton, W. B. (1966). *Human Spatial Orientation.* John Wiley & Sons.

Husain, M., & Rorden, C. (2003). "Non-Spatially Lateralized Mechanisms in Hemispatial Neglect." *Nature Reviews Neuroscience, 4*(1), 26–36. https://doi.org /10.1038/nrn1005.

Ito, M. (2008). "Control of Mental Activities by Internal Models in the Cerebellum." *Nature Reviews Neuroscience, 9*(4), 304–313.

James, W. (1890). *The Principles of Psychology.* New York: Henry Holt and Company.

Jeannerod, M. (2001). "Neural Simulation of Action: A Unifying Mechanism for Motor Cognition." *NeuroImage, 14*(1), S103–S109.

Kane, M. J., & Engle, R. W. (2003). "Working-Memory Capacity and the Control of Attention: The Contributions of Goal Neglect, Response Competition, and Task Set to Stroop Interference." *Journal of Experimental Psychology: General, 132*(1), 47–70.

Kleckner, I. R., Zhang, J., Touroutoglou, A., Chanes, L., Xia, C., Simmons, W. K., & Barrett, L. F. (2017). "Evidence for a Large-Scale Brain System Supporting Allostasis and Interoception in Humans." *Nature Human Behaviour, 1*(5), 69.

Klein, R., Kingstone, A., & Pontefract, A. (1992). "Orienting of Visual Attention." In R. Klein (Ed.), *Eye Movements and Visual Cognition: Scene Perception and Reading* (pp. 116–124). Springer. https://doi.org/10.1007/978-1-4612-2852-3_4.

Knudsen, E. I. (2007). "Fundamental Components of Attention." *Annual Review of Neuroscience, 30,* 57–78. https://doi.org/10.1146/annurev.neuro.30.051606.094256.

Kosslyn, S. M., Pascual-Leone, A., Felician, O., Camposano, S., Keenan, J. P., Thompson, W. L., & Alpert, N. M. (2001). "The Role of Area 17 in Visual Imagery: Convergent Evidence from PET and rTMS." *Science, 284*(5411), 167–170.

Krauzlis, R. J., Lovejoy, L. P., & Zénon, A. (2013). "Superior Colliculus and Visual Spatial Attention." *Annual Review of Neuroscience, 36,* 165–182. https://doi.org/10.1146/annurev-neuro-062012-170249.

Lang, A., & Basil, M. D. (1998). "Attention, Resource Allocation, and Communication Research: What Do Secondary Task Reaction Times Measure, Anyway?" *Annals of the International Communication Association, 21*(1), 443–458. https://doi.org/10.1080/23808985.1998.11678957.

Langer, K. G., & Bogousslavsky, J. (2020). "The Merging Tracks of Anosognosia and Neglect." *European Neurology, 83*(4), 438–446. https://doi.org/10.1159/000510397.

Lavie, N., Hirst, A., de Fockert, J. W., & Viding, E. (2004). "Load Theory of Selective Attention and Cognitive Control." *Journal of Experimental Psychology: General, 133*(3), 339–354.

Lavie, N. (2005). "Distracted and Confused? Selective Attention Under Load." *Trends in Cognitive Sciences, 9*(2), 75 –82.

Lepp, A., Barkley, J. E., & Karpinski, A. C. (2014). "The Relationship Between Cell Phone Use and Academic Performance in a Sample of U.S. College Students." *Sage Open, 5*(1), 2158244015573169. https://doi.org/10.1177/2158244015573169.

Macaluso, E. (2010). "Orienting of Spatial Attention and the Interplay Between the Senses." *Cortex, 46*(3), 282–297.

Macdonald, J. S. P., & Lavie, N. (2011). "Visual Perceptual Load Induces Inattentional Deafness." *Attention, Perception, & Psychophysics, 73*(6), 1780–1789. https://doi.org/10.3758/s13414-011-0144-4.

Maunsell, J. H. R., & Treue, S. (2006). "Feature-Based Attention in Visual Cortex." *Trends in Neurosciences, 29*(6), 317–322. https://doi.org/10.1016/j.tins.2006.04.001.

McGaugh, J. L. (2000). "Memory–A Century of Consolidation." *Science, 287*(5451), 248–251.

McGaugh, J. L. (2018). "Emotional Arousal and Memory Modulation." *Memory, 26*(3), 381–389.

Menon, V., & Uddin, L. Q. (2010). "Saliency, Switching, Attention and Control: A Network Model of Insula Function." *Brain Structure and Function, 214*(5–6), 655–667.

Miller, E. K., & Cohen, J. D. (2001). "An Integrative Theory of Prefrontal Cortex Function." *Annual Review of Neuroscience, 24*(1), 167 –202.

Oken, B. S., Salinsky, M. C., & Elsas, S. M. (2006). Vigilance, alertness, or sustained attention: Physiological basis and measurement. *Clinical Neurophysiology, 117*(9), 1885–1901. https://doi.org/10.1016/j.clinph.2006.01.017

Parasuraman, R., & Davies, D. R. (1984). *Varieties of Attention.* Orlando, FL: Academic Press.

Pessoa, L., & Adolphs, R. (2010). "Emotion Processing and the Amygdala: From a 'Low Road' to 'Many Roads' of Evaluating Biological Significance." *Nature Reviews Neuroscience, 11*(11), 773–783.

Pessoa, L. (2009). "Cognition and Emotion." Scholarpedia, 4(1), 3565. http://doi:10.4249/scholar pedia.4567.

Petersen, S. E., & Posner, M. I. (2012). "The Attention System of the Human Brain: 20 Years After." *Annual Review of Neuroscience, 35,* 73–89. https://doi.org/10.1146/annurev-neuro-062111-150525.

Pfaff, D. W. (2005). *Brain Arousal and Information Theory: Neural and Genetic Mechanisms.* Cambridge, MA: Harvard University Press.

Plebanek, D. J., & Sloutsky, V. M. (2019). "Selective Attention, Filtering, and the Development of Working Memory." *Developmental Science, 22*(6), e12727. https://doi.org /10.1111/desc.12727.

Posner, M. I., & Petersen, S. E. (1990). "The Attention System of the Human Brain." *Annual Review of Neuroscience, 13*(1), 25–42.

Posner, M. I. (1980). "Orienting of Attention." *Quarterly Journal of Experimental Psychology, 32*(1), 3–25. https://doi.org/10.1080/0033555 8008248231.

Pourtois, G., & Dolan, M. F. (2010). "The Time-Course of Brain Activity Underlying the Relevance of Emotional Distraction for Attentional Control." *Journal of Neuroscience, 30*(7), 2753–2763.

Price, C. N., & Moncrieff, D. (2021). "Defining the Role of Attention in Hierarchical Auditory Processing." *Audiology Research, 11*(1), Article 12. https://doi.org/10.3390/audiolres11010012.

Raichle, M. E. (2015). "The Brain's Default Mode Network." *Annual Review of Neuroscience, 38,* 433–447.

Raichle, M. E., MacLeod, A. M., Snyder, A. Z., Powers, W. J., Gusnard, D. A., & Shulman, G. L. (2001). "A Default Mode of Brain Function." *Proceedings of the National Academy of Sciences, 98*(2), 676–682. https:// doi.org/10.1073/pnas.98.2.676.

Reed, M. P., & Green, P. A. (1999). "Comparison of Driving Performance On-Road and in a Low-Cost Simulator Using a Concurrent Telephone Dialing Task." *Ergonomics, 42*(8), 1015–1037. https://doi.org/10.1080/001401399185117.

Reynolds, J. H., & Chelazzi, L. (2004). "Attentional Modulation of Visual Processing." *Annual Review of Neuroscience, 27,* 611–647. https:// doi.org/10.1146/annu rev.neuro.26.041002.131039.

Rizzo, M., & Vecera, S. P. (2002). Psychoanatomical substrates of Bálint's syndrome. *Journal of Neurology, Neurosurgery & Psychiatry, 72*(2), 162–178. https://doi.org/10.1136/jnnp.72.2.162

Rode, G., Pagliari, C., Huchon, L., Rossetti, Y., & Pisella, L. (2017). "Semiology of Neglect: An Update." *Annals of Physical and Rehabilitation Medicine, 60*(3), 177–185. https://doi.org/10.1016/j.rehab.2016.03.003.

Rosen, L. D., Carrier, L. M., & Cheever, N. A. (2013). "Facebook and Texting Made Me Do It: Media-Induced Task-Switching While Studying." *Computers in Human Behavior, 29*(3), 948–958. https://doi.org/10.1016/j.chb. 2012 .12.001.

Saalmann, Y. B., Pinsk, M. A., Wang, L., Li, X., & Kastner, S. (2012). "The Pulvinar Regulates Information Transmission Between Cortical Areas Based on Attention Demands." *Science, 337*(6095), 753–756. https://doi.org/10.1126/science .1223082.

Saenz, M., Buracas, G. T., & Boynton, G. M. (2002). "Global Effects of Feature-Based Attention in Human Visual Cortex." *Nature Neuroscience, 5*(7), 631–632. https://doi.org/10.1038/nn876.

Salvucci, D. D., & Taatgen, N. A. (2008). "Threaded Cognition: An Integrated Theory of Concurrent Multi -tasking." *Psychological Review, 115*(1), 101–130. https://doi.org \/10.1037/0033-295X.115.1.101.

Schmahmann, J. D. (2019). "The Cerebellum and Cognition." *Neuroscience Letters, 688,* 62–75.

Seeley, W. W., Menon, V., Schatzberg, A. F., Keller, J., Glover, G. H., Kenna, H., Reiss, A. L., & Greicius, M. D. (2007). "Dissociable Intrinsic Connectivity Networks for Salience Processing and Executive Control." *Journal of Neuroscience, 27*(9), 2349–2356.

Seth, A. K. (2014). "A Predictive Processing Theory of Sensorimotor Contingencies: Explaining the Puzzle of Perceptual Presence and Its Absence in Synesthesia." *Cognitive Neuroscience, 5*(2), 97–118. https://doi.org/10.1080/17588928.2013.877880

Shipp, S. (2004). "The Brain Circuitry of Attention." *Trends in Cognitive Sciences, 8*(5), 223–230. https://doi.org/10.1016/j.tics.2004.03.004.

Slagle, J. M., Porterfield, E. S., Lorinc, A. N., Afshartous, D., Shotwell, M. S., & Weinger, M. B. (2018). "Prevalence of Potentially Distracting Noncare Activities and Their Effects on Vigilance, Workload, and Nonroutine Events During Anesthesia Care." *Anesthesiology, 128*(1), 44–54. https://doi.org/10.1097/ALN.0000000000001915.

Sporns, O. (2013). *Networks of the Brain.* MIT Press.

Squire, L. R., & Dede, A. J. (2015). "Conscious and Unconscious Memory Systems." *Cold Spring Harbor Perspectives in Biology, 7*(3), a021667.

Sterling, P., & Eyer, J. (1988). "Allostasis: A New Paradigm to Explain Arousal Pathology." In *Handbook of Life Stress, Cognition and Health*, edited by S. Fisher & J. Reason, 629–649. John Wiley & Sons.

Stickgold, R., & Walker, M. P. (2005). "Memory Consolidation and Reconsolidation: What Is the Role of Sleep?" *Trends in Neurosciences, 28*(8), 408–415.

Strayer, D. L., & Johnston, W. A. (2001). "Driven to Distraction: Dual-Task Studies of Simulated Driving and Conversing on a Cellular Telephone." *Psychological Science, 12*(6), 462–466. https://doi.org/10.1111/1467-92 80.00386.

Talsma, D., & Awh, E. (2006). "Multisensory Integration Facilitates Spatial (but Not Feature-Based) Attention." *Journal of Experimental Psychology: Human Perception and Performance, 32*(2), 343–357.

Treisman, A. (1988). Features and objects: The fourteenth Bartlett memorial lecture. *Quarterly Journal of Experimental Psychology Section A, 40*(2), 201–237. https://doi.org/10.1080/02724988843000104

Treue, S. (2001). "Neural Correlates of Attention in Primate Visual Cortex." *Trends in Neurosciences, 24*(5), 295–300. https://doi.org/10.1016/S0166-2236(00)01814-2.

Tulving, E., & Thomson, D. M. (1973). "Encoding Specificity and Retrieval Processes in Episodic Memory." *Psychological Review, 80*(5), 352–373.

Vallar, G., & Calzolari, E. (2018). "Unilateral Spatial Neglect After Posterior Parietal Damage." *Handbook of Clinical Neurology, 151*, 287–312. https://doi.org/10.1016/B 978-0-444-63622-5.00014-0.

van den Heuvel, M. P., & Sporns, O. (2011). "Rich-Club Organization of the Human Connectome." *Journal of Neuroscience, 31*(44), 15775–15786.

Vossel, S., Geng, J. J., & Fink, G. R. (2014). "Dorsal and Ventral Attention Systems: Distinct Neural Circuits but Collaborative Roles in Attentional Control." *Neuroscientist, 20*(2), 150–159.

Vuilleumier, P. (2005). "How Brains Beware: Neural Mechanisms of Emotional Attention." *Trends in Cognitive Sciences, 9*(12), 585–594.

Weinger, M. B. (2012). "Human Factors in Anesthesiology." In *Handbook of Human Factors and Ergonomics in Health Care and Patient Safety* (pp. 835–847). CRC Press. https://doi.org/ 10.1201/b11219-55.

Wickens, C. D. (2002). "Multiple Resources and Performance Prediction." *Theoretical Issues in Ergonomics Science, 3*(2), 159–177. https:// doi.org/10.1080/146392 20210123806.

Wickens, C. D., Hollands, J. G., Banbury, S., & Parasuraman, R. (2012). *Engineering Psychology and Human Performance* (4th ed.). Pearson.

Yerkes, R. M., & Dodson, J. D. (1908). "The Relation of Strength of Stimulus to Rapidity of Habit-Formation." *Journal of Comparative Neurology and Psychology, 18*(5), 459–482.

7

Memory

Human memory is a fundamental aspect of our mental capacities, enabling us to acquire, retain, and recall information that shapes our understanding of the world and influences our actions. Memory is the cognitive ability that encompasses retaining knowledge and experiences over time, ultimately influencing future thoughts, emotions, actions, and behaviors. It serves as the foundation for language development, as we rely on memory to learn and remember words, grammar, and communication skills. Furthermore, memory forms the basis for establishing and maintaining relationships, as we draw upon past experiences and interactions to shape our interactions with others. The ability to remember past events makes developing language, relationships, and personal identity possible.

Memory is essential for the formation and preservation of personal identity. It allows us to remember our history, experiences, and achievements, contributing to our sense of self and continuity over time. Without memory, our lives would lack coherence and a sense of individuality, as each moment would exist in isolation without any connection to the past. Memory loss, characterized by forgetfulness or amnesia, can significantly impact a person's daily functioning. It can disrupt the ability to recall important information, such as names, dates, and events, leading to difficulties in various aspects of life, including work, relationships, and self-care. Memory impairments can arise from various causes, such as aging, neurological disorders, trauma, or certain medications.

While memory is a remarkable cognitive ability, it is not infallible because it is flexible and dynamic, allowing for the creation of new synthesis and learning. The price we pay for not being enslaved to our past is that our memory is subject to limitations and biases and even the creation of false memories. For instance, memories can be influenced by external suggestions, emotions, or personal biases, potentially altering the accuracy of the recalled information. Loftus and

Pickrell (1995) demonstrated this phenomenon through the "Lost in the Mall" study, where participants developed false memories of events that never occurred due to suggestive techniques. Loftus (1997) further explored the implications of such malleability, particularly in legal contexts where external factors may influence eyewitness testimony. Brainerd and Reyna (2005) provided a theoretical framework with their fuzzy-trace theory, explaining how reliance on gist-based representations fosters creativity and errors. These studies underscore memory's dual nature as both adaptive and prone to distortion, aligning with the metafunctional framework that emphasizes memory's dynamic qualities. Ethical concerns arise in contexts such as therapy or forensic investigations, where memory distortion may lead to unintended consequences, such as false accusations or misdiagnoses. Addressing these concerns requires careful ethical guidelines and interventions.

Memory also operates on a broader social and cultural level. Collective memory, shaped by societal narratives and shared experiences, influences individual identities and group cohesion. Cultural rituals, historical education, and traditions play a role in how groups construct and preserve collective memories, further reinforcing shared identities and perspectives. Memory is a probe into the nature of the human mind. However, despite decades of research, neuroscientists still disagree about memories' relation to the brain. Neuroscientists have proposed various memory localizations, such as molecular changes in the neurons, synaptic plasticity, neural ensembles, and cortical and subcortical structures. Although the term engram was originally introduced by Richard Semon (1921) in the early 20th century, Squire and Kandel (1999) helped to modernize the concept by framing it in terms of physical and biochemical changes in the brain that result from learning and experience, providing a neurobiological basis for understanding how memories related patterns are stored. Recent advancements in neuroscience have shed light on long-term potentiation, synaptic consolidation, and the hippocampal regions' roles in episodic and spatial memory (Gilbert and Brushfield, 2009). However, the mechanisms underlying long-term memory neural pattern (MNP) formation, consolidation, and retrieval remain complex and partially understood. These findings emphasize the dynamic and interconnected nature of brain systems involved in memory and highlight the importance of interdisciplinary research to deepen our understanding.

Memory is Astounding and Perplexing.

Memory involves the brain's encoding, consolidation, and retrieval processes, but this does not imply that memories are stored solely within the brain. While physicalist explanations address aspects of memory, such as the formation of memory neural patterns (MNPs), the broader scientific evidence challenges the notion of static memory storage in discrete brain regions. Doty (1998) critiqued this reductionist perspective, arguing that the brain is a dynamic facilitator of memory, organizing and retrieving experiences rather than acting as a simple repository. This view aligns with the metafunctional framework, which conceptualizes memory as an adaptive and distributed process transcending physical localization. Memory's complexity suggests it operates beyond static structures, incorporating biological mechanisms and broader systems of organization. The transphysical nature of memory extends beyond its biological and neural foundations, suggesting that memory operates on levels not confined to the brain's physical structures. This view intersects with philosophical, metaphysical, and theological considerations, which the forthcoming chapters including one on *The Human Soul* will explore in greater depth. The transphysical nature of memory opens profound questions about its connection to the human soul, self, and consciousness. For example, memory is often considered a bridge between the material and immaterial aspects of existence, reflecting how recollections shape both our tangible experiences and intangible sense of being.

Memory's remarkable capacity allows for retaining vast amounts of knowledge and experiences over a lifetime. Cues, associations, or intentional effort facilitate the selective retrieval of specific memories from this vast repository, highlighting the brain's organizational sophistication. For instance, memory retrieval often involves the activation of interconnected neural networks, including the hippocampus, prefrontal cortex, and parietal regions, which dynamically adapt to new information and experiences. This adaptability underscores the dynamic nature of memory. The brain continually reorganizes and updates information based on new learning, integrating novel knowledge with existing memories. Processes such as synaptic plasticity, long-term potentiation, and memory reconsolidation exemplify this flexibility, enabling the integration of new experiences while maintaining coherence across memories. Recent research also emphasizes the role of distributed neural networks in memory storage and retrieval, reflecting the interconnected nature of brain systems in memory processes.

The concept of memory as transphysical and multifaceted highlights its existence beyond individual brain regions, individually and collectively. For example, collective and embodied memories—such as shared cultural narratives or muscle memory—demonstrate how memory systems extend beyond physical localization in the brain. These aspects illustrate memory's broader role in shaping individual cognition and collective identities. Memory's dynamic and distributed nature has profound implications for understanding cognition and learning. This perspective approaches education, where active retrieval and integrating new knowledge with existing schemas can enhance long-term retention. Similarly, therapeutic interventions for memory impairments, such as neuroplasticity-based training or reconsolidation therapies, leverage memory adaptability to improve cognitive outcomes. Technological advances, such as artificial intelligence and memory augmentation, draw inspiration from these mechanisms, creating new possibilities for extending human memory capabilities.

Types of Memory

Memory can be broadly categorized into *declarative* (explicit) and *non-declarative* (implicit) memory (Camina and Güell, 2017). Declarative memory involves the conscious storage and recollection of information, where individuals know they possess the knowledge or experiences. This category of memory includes two subtypes: semantic memory and episodic memory. *Semantic memory* pertains to the encoding and retrieval of general knowledge, facts, and concepts. It encompasses the understanding of language, the meanings of words, historical events, and various forms of academic or cultural information. For instance, knowing that Paris is the capital of France or understanding the concept of gravity exemplifies semantic memories.

Episodic memory, however, involves encoding and retrieving personal experiences and events at specific times and places. It includes recollections of events like a birthday party, a vacation, or a meaningful conversation. Episodic memory allows humans to mentally travel back in time and relive details about their experiences, including the context, emotions, and sensory information associated with those events. These two subtypes collectively constitute declarative memory, which is often regarded as the primary form of memory when discussing conscious recollection. Declarative memory relies on cognitive processes such as attention, encoding, consolidation, and retrieval to maintain and recall memories. It is primarily associated with the specialized functions of the brain's hippocampus and medial temporal lobe regions.

In contrast, *non-declarative memory*, also known as implicit memory, involves the unconscious storage and retrieval of information without conscious awareness (Squire and Zola, 1996). This type of memory influences present behavior, thoughts, or emotions based on past experiences, even if individuals are not explicitly aware of the memories themselves. Non-declarative memory includes several subtypes: procedural memory, priming, conditioning, and skills acquisition. Procedural memory is the ability to perform learned motor skills or tasks, like riding a bicycle or playing a musical instrument, which becomes automatic with practice. Sridhar, Khamaj, and Asthana (2023) highlighted the neural mechanisms supporting the acquisition and consolidation of procedural knowledge, demonstrating how repeated practice enables these skills to become ingrained and operate unconsciously. Priming involves exposure to stimuli that unconsciously influence subsequent behavior, perception, or responses. Conditioning forms associations between stimuli and responses, as exemplified by Pavlovian classical conditioning, where a neutral stimulus becomes associated with an automatic response. Skills acquisition, often involving motor and cognitive elements, occurs through practice and improves performance over time.

Non-declarative memory operates independently of conscious recollection, typically functioning implicitly and automatically. It plays a significant role in daily life, from routine activities to forming habits and retrieving learned skills. Together, declarative and non-declarative memory systems work in parallel, complementing each other and contributing to a cohesive understanding of the world, personal experiences, and behavioral responses. *Sensory memory* is another critical component of the memory process. It refers to the brief retention of sensory information immediately after a perceived stimulus. Sensory memory, acting as a buffer, holds this information for a very short duration—typically less than one second—allowing us to process sensory input before it fades. This memory stage provides a momentary snapshot of the environment, enabling continuity in perception. For instance, sensory memory retains the visual details of a scene for a fraction of a second, ensuring a cohesive perception rather than fragmented snapshots. While fleeting, sensory memory is crucial for transferring sensory input to higher-level memory systems. Attention plays a key role in determining which sensory inputs are transferred to short-term or long-term memory for further processing. Without focused attention, sensory information quickly fades.

Short-term memory, often referred to as working memory (Camina and Güell, 2017), serves as a "mental workspace" where information is temporarily held and manipulated for immediate use. This system operates briefly, typically lasting several seconds to a minute unless the information is rehearsed or actively maintained. Unlike sensory memory, short-term memory has a limited capacity, holding a small number of items (typically 7 ± 2 chunks) at any given time. Short-term memory is essential for various cognitive tasks, such as solving math problems, maintaining the meaning of a sentence while reading, or retaining intermediate results in problem-solving. It also includes visual and spatial components, enabling the temporary retention of visual images and mental representations critical for tasks such as spatial navigation or visualization. Active rehearsal or repetition can extend the retention period in working memory, preventing rapid decay and enabling information transfer to long-term memory systems.

Memory is Metafunctional

The brain metafunctional framework presupposes and indicates several fundamental characteristics of memory:

1. *Hierarchical Anatomical and Functional Levels:* Brain memory changes occur across various levels, from molecules, genes, and proteins in neurons to large-scale networks (Camina and Güell, 2017).

2. *Distributed Processes and Specialized Structures:* Memory depends on processes supported by distributed structures across the brain and specialized regions like the hippocampus (Roy et al., 2022).

3. *Contextual and Flexible:* Memory adapts to changing contexts, reflecting its flexibility in encoding and retrieval (Gershman, Blei, & Niv, 2010; Heald, Lengyel, & Wolpert, 2023).

4. *Multisensory and Multifunctional:* Memory integrates inputs from multiple senses and supports diverse functions (Ghazanfar and Schroeder, 2006; Driver & Noesselt, 2008; Murray & Shams, 2023).

5. *Conscious and Unconscious Processes:* Memory involves conscious recollection and unconscious influences (Squire & Wixted, 2011; Squire & Dede, 2015).

6. *Double Intentionality:* Memory exhibits a layered intentional structure (Tallis 2004, 2010, 2011). The first level of intentionality pertains to the "aboutness" of thoughts and percepts—how they are directed toward objects, experiences, or phenomena. Memory adds a second layer of intentionality because it is not directly about the original object or experience but rather about a prior thought or perception that was itself about something.

7. *Transphysical Components:* Memory encompasses elements that transcend physical localization.

Each metafunctional aspect reflects the remarkable complexity of memory, which we will explore in detail. Let us focus on the first characteristic—hierarchical anatomical and functional levels.

1. Hierarchical Anatomical and Functional Levels

Memory operates across hierarchical levels, from molecular changes in neurons to large-scale neural networks. Eric Kandel's Nobel Prize-winning research on the sea slug *Aplysia* illuminated how synaptic changes underpin learning and memory. By exposing the slug to stimuli such as tail shocks, Kandel demonstrated that learning involves altering the strength and connectivity of synapses. His work showed that short-term memory relies on temporary modifications of existing proteins, while long-term memory requires the activation of gene expression and the synthesis of new proteins (Kandel, 2001). Building on this foundation, Squire and Kandel (2000) retook the concept of engrams—neural representations of memory—emphasizing that synaptic plasticity forms the neural basis of declarative and non-declarative memory systems. Their research revealed how molecular changes link to broader neural processes, highlighting the hierarchical nature of memory mechanisms. Poo et al. (2016) expanded on this work, demonstrating that distributed neural patterns form the substrate for memory storage and retrieval. These findings show that memory operates across multiple interconnected levels and does not remain confined to localized changes. This hierarchical framework underscores the dynamic nature of memory, with each level contributing to information encoding, storage, and retrieval. From molecular signaling pathways and synaptic plasticity to neural networks and system-wide integration, memory operates as a multilayered process. This understanding has profound implications for neuroscience, enabling advancements in treating memory disorders and designing artificial memory systems inspired by the brain's architecture.

From the Neurons to Multiple Large-Scale Networks

Historically, memory research assumed that specific brain regions stored memories, and that memory function involved only a few discrete areas. Pioneering neuroscientist Karl Lashley challenged this notion through groundbreaking experiments in the early 20th century (Lashley, 1950). By lesioning different brain areas in rats, Lashley sought to locate the "engram," or physical trace of memory storage. Surprisingly, his findings suggested that memory is distributed throughout the brain, with no single area essential for storing specific memories. This discovery refuted the reductionist view of localized memory and paved the way for theories emphasizing distributed neural networks. Brenda Milner, a Canadian neuropsychologist, further advanced memory research through her work with patient HM, who experienced dense amnesia after bilateral hippocampal removal (Scoville and Milner, 1957). While the hippocampus became associated with memory formation, Milner and others argued that memory is not confined to one region. Instead, memory involves dynamic processes where specific brain areas reactivate based on synaptic facilitation and connectivity (Anderson et al., 2010; Liu et al., 2021). This perspective integrates synaptic changes into network-level phenomena, emphasizing that memory is an interaction process rather than static storage.

Understanding memory at the molecular level has also been instrumental. Genes responsible for synaptic plasticity—such as those regulating protein synthesis—play a crucial role in memory formation and retrieval. Researchers manipulate these genes in animal models to study their impact on memory processes (McClung and Nestler, 2008). For example, experiments knocking out genes that synthesize specific proteins and monitoring memory effects provide insights into how molecular mechanisms support neural circuits (Pearson-Fuhrhop, Kleim, and Cramer, 2009). Such findings demonstrate how changes at the neuronal and synaptic levels contribute to the reorganization of neuronal circuits and large-scale memory networks. This hierarchical understanding, from neurons to networks, highlights the interplay between molecular mechanisms and distributed brain systems. It underscores the importance of studying local synaptic changes and their integration into larger neural architectures, providing a comprehensive view of memory processes. These insights have profound implications for understanding memory disorders and advancing technologies that mimic neural systems.

(2) Memory Neural Correlates are Distributed Across the Brain

The Hippocampal Complex and Memory

The hippocampal formation, located within the medial temporal lobe (MTL), is integral to memory processing. Structurally, it comprises the dentate gyrus, hippocampus proper (CA1, CA2, CA3), and the subiculum. Some definitions include the presubiculum, parasubiculum, and entorhinal cortex (Schacter et al., 2007; Cappaert, Van Strien, & Witter, 2015). This intricate network supports a range of cognitive functions, including memory (Huijgen & Samson, 2015), learning (Ballard, 2019), spatial navigation, and attention control (Wall & Messier, 2001). One critical role of the hippocampal formation is monitoring and processing visceral and emotional states through its connections with the orbitomedial prefrontal cortex (Wall & Messier, 2001). This dynamic interaction enables the hippocampus to encode novel and significant information by building synaptic weights that support long-term retention. Emotional significance, relevance, and novelty are crucial determinants of how effectively memories are encoded and recalled (Christianson, 1992; Anderson & Hanslmayr, 2014). Furthermore, the medial prefrontal cortex and lateral entorhinal cortex are key players in forming declarative, episodic, and recognition memory (Preston & Eichenbaum, 2013).

Traditionally viewed as a temporary storage site for memories, the hippocampus is now understood to function as a gateway for encoding, consolidating, and retrieving memory representations. The hippocampus binds sensory inputs, emotions, and contextual information into coherent memory engrams (Lisman et al., 2017; Ekstrom & Ranganath, 2018). Subregions such as the dentate gyrus facilitate pattern separation to avoid memory interference, while the CA3 region supports pattern completion, enabling retrieval from partial cues (Gilbert & Brushfield, 2009; Rolls, 2016). Memory consolidation involves gradually transferring memory neural patterns from the hippocampus to the neocortex. This process, termed systems consolidation, reorganizes and strengthens memory traces, making them less dependent on the hippocampus over time. However, the hippocampus plays a role in memory retrieval, reactivation, and integration into ongoing cognitive processes (Ross et al., 2019). It is also critical for spatial memory, with place cells creating cognitive maps of environments, supporting navigation and spatial context (Moser, Rowland, & Moser, 2015). From a systems-level perspective, memory engrams are distributed across multiple brain regions during encoding, with each area contributing specific aspects of an experience. Over time, these contributions shift, impacting memory content and quality. This distributed nature of memory highlights the hippocampus as a specialized hub instead of a storage center, emphasizing its dynamic interactions with other brain regions (Hardt & Nadel, 2018). The intricate balance of encoding, consolidation, and retrieval underscores the complexity of memory processes (Doty, 1998; Ross et al., 2019).

Large-Scale Brain Networks Supporting Memory

The shift from localized to whole-brain perspectives on memory underscores the role of distributed cortical and subcortical networks in memory processes. Early experiments by Lashley in the 1950s hinted at this distributed nature, and recent advances in neuroimaging have confirmed the widespread connectivity of memory-related networks. While the MTL, including the hippocampus, remains critical for forming and consolidating declarative memories, these processes extend beyond the MTL into networks such as the default mode network (DMN) (Kaefer et al., 2022). Studies highlight dynamic interactions between the MTL and cortical regions, revealing that resting-state functional connectivity encompasses the prefrontal cortex, amygdala, insular cortex, and other regions (Moazeni, Northoff, & Batouli, 2024). This distributed organization enables memory processes such as consolidation, reactivation, and integration across diverse brain areas. Replay, a neural phenomenon involving the sequential reactivation of past experiences, exemplifies memory's distributed nature. Replay-triggered brain-wide activation has been demonstrated in humans, emphasizing its importance for memory integration (Huang et al., 2024). Furthermore, brain-wide mapping studies reveal that memory engrams are distributed across multiple cortical and subcortical regions rather than confined to isolated areas (Roy et al., 2022). Research continues to explore the flexibility of memory-bearing ensembles in shaping cognition and behavior. For example, repeated learning modulates neuronal activation and connectivity, enhancing functional memory networks (Terstege, Durante, & Epp, 2022). Oscillatory activity during memory tasks further highlights the role of widespread cortical regions, including the occipital, parietal, temporal, and prefrontal cortices, as well as the hippocampus and amygdala (Singh et al., 2023).

The Metafunctional Perspective

From a metafunctional perspective, memory processes reflect the integration of distributed neural networks rather than isolated storage. Sleep exemplifies this integration, with hippocampal replay driving consolidation of both hippocampus-dependent and non-hippocampus-dependent memories. During sleep, global synaptic downscaling helps optimize neural efficiency by reducing metabolic demands while preserving key memory traces (Tononi & Cirelli, 2003; Klinzing, Niethard, & Born, 2019). Emerging technologies, such as transcranial alternating current stimulation (tACS), demonstrate the practical implications of this distributed understanding of memory. Studies show that tACS can enhance working and long-term memory in older adults by targeting specific cortical circuits (Grover et al., 2022). These advancements underscore the potential for neuromodulation to treat memory impairments and optimize cognitive performance (Gershman, Blei, & Niv, 2010; Heald, Lengyel, & Wolpert, 2023). Memory, as a distributed and dynamic process, illustrates the profound complexity of neural mechanisms, with implications for both theoretical understanding and practical applications.

(3) Memory is Contextual and Flexible

The process by which the brain weaves fragments of sensory information to create a coherent sense of self imbued with individuality remains a captivating mystery. The brain constructs internal experiences by integrating sensory inputs and filling in gaps with estimations of the external world. However, what forms the foundation of these estimations? The answer lies in memory. Over the past few decades, neuroscience has revealed that memory retrieval is not a passive playback but a dynamic process. During recall, memory temporarily returns to a malleable state, becoming susceptible to modifications influenced by new experiences and perspectives (Elsey et al., 2018). This finding challenges the conventional belief that memory retrieval faithfully reproduces the past and highlights memory's remarkable adaptability.

Memory is not a fixed archive, but a living construct shaped and reshaped throughout life. From a neuroscience perspective, memory is a constantly developing process influenced by our thoughts, emotions, and environments (Fernández et al., 2016; Elsey & Kindt, 2016). Each act of remembering involves reconstruction, not reproduction. During this process, memories may be destabilized, leaving them vulnerable to degradation or open to reconsolidation. The timeframes, mechanisms, and biological functions underlying these changes remain areas of ongoing research (Stickgold & Walker, 2007). This dynamic nature of memory is liberating—it enables adaptation, growth, and transformation. Instead of being confined to unchanging recollections, memories reframe our understanding of ourselves and the world, forging new paths and perspectives. The integration of memory processes within a person's sense of self underscores its complexity. The brain metafunction framework offers a lens through which we can view memories as both adaptive and connected to a transphysical domain. This perspective suggests that while the brain patterns retrieving memories are dynamic and flexible to enable adaptation, the memories themselves may exist as stable entities. *This distinction between memory stability and neural flexibility enables us to reconcile the need for adaptation with the continuity of self.*

Memory Evolution and the Role of Sleep

Recent research has further challenged the traditional notion that memories are rapidly encoded and stabilized. Instead, we now understand memory as an ongoing, flexible process that changes throughout a lifetime. Studies on sleep have been instrumental in uncovering the mechanisms of memory evolution (Wamsley & Stickgold, 2011; Rattenborg & Beckers, 2020). Rather than merely stabilizing memories, sleep integrates them into pre-existing networks, reshaping and transforming them with each new experience. This ongoing adaptation process consolidates knowledge and enhances its applicability, distinguishing wisdom from mere intelligence. Experiments using functional magnetic resonance imaging (fMRI) have illustrated this phenomenon. Participants trained in tasks in the morning were retested in the evening, and the same brain areas were activated during both the learning and recall phases. However, when participants slept between training and testing, the activation patterns shifted, as if the memory had been reorganized or relocated within the brain (Walker & Stickgold, 2006; Rudoy et al., 2009). This observation aligns with the metafunctional perspective: memories are not confined to static locations but involve distributed processes that adapt and reorganize over time.

The Metafunctional Perspective on Contextual Memory

From a metafunctional standpoint, memory encoding involves multiple brain regions, each contributing specific contextual elements to the experience. Sleep facilitates the dynamic reprocessing of these elements, integrating new information with existing memory networks. The system operates holistically, whether awake or asleep, allowing for unified memory processing across multiple levels. This dynamic and contextual nature of memory ensures that relative contributions from different brain regions can shift over time, enriching the content and quality of memories. This process synthesizes past and present, producing a flexible, developing understanding of the self and the world. Memory's metafunctional adaptability underscores its role in shaping our ability to adapt, grow, and integrate new experiences into a coherent narrative.

What is False Memory?

The inherent flexibility in the metafunctional processing of memory implies its malleability, and conversely, memory's malleability underscores its dynamic and metafunctional nature. This intricate process involves reactivating the neural networks initially engaged in memory formation, which may develop over time. Researchers studying memory describe it often as a constructive or reconstructive process. False memory research is a fascinating area of psychology that investigates how external factors can influence and distort memories (Loftus, 2005; Blom & Huang, 2021). These studies typically involve providing participants with misleading information about events and examining how their recollections are affected. Techniques such as showing doctored photographs and videos or presenting subtle but suggestive questions are often employed to observe how easily it is to plant false memories.

The Misinformation Effect and Memory Manipulation

False memory experiments vividly demonstrate how memories can be influenced and even fabricated. For instance, participants might be shown a simulated car accident with a stop sign but later asked a leading question about a yield sign. This subtle misinformation can lead participants to "recall" a yield sign that was never present. This phenomenon, known as the misinformation effect, highlights memory's susceptibility to distortion. A landmark legal case, the Ramona false memory case, brought the dangers of false memory manipulation into sharp focus (La Ganga, 1994). Gary Ramona, a California man, sued his daughter Holly's therapists, alleging they implanted false memories of sexual abuse through suggestive techniques like hypnosis and guided imagery. Holly's therapists claimed these recovered memories validated allegations of abuse, but Gary Ramona denied the claims and accused the therapists of malpractice. After a high-profile trial, the jury sided with Gary Ramona, awarding him $500,000 in damages. This pivotal case raised questions about repressed memory therapy and the ethical considerations surrounding the use of suggestive techniques.

Implications for Legal and Psychological Fields

The Ramona case ignited widespread debate about the reliability of repressed memories and the dangers of suggestive memory retrieval methods. Since then, research into false memory has grown significantly, with high-profile cases revealing how memory can be manipulated (Loftus, 2005; Mamat, Levy, & Bayley, 2024). Such studies underscore the profound implications of memory malleability in legal settings. In courtrooms, memory accuracy often determines the outcomes of cases. Even minor details—such as the color of a suspect's clothing, the time of day a crime occurred, or the location of an object—can significantly sway verdicts. For example, in a murder trial, a witness' memory of the weapon used or the perpetrator's appearance can mean the difference between conviction and acquittal. Similarly, a plaintiff's recollection of events or a defendant's actions can influence the damages awarded in civil cases. These examples highlight the critical importance of ensuring the reliability of memory in delivering justice.

(4) Memory is Multisensory and Multifunctional.

Memory's multisensory and multifunctional nature allows us to form more vivid and lasting memories, providing multiple routes for accessing and recalling information (Alais & Burr, 2004; Koelewijn, Bronkhorst, & Theeuwes, 2010; Tang, Wu, & Shen, 2016; Duarte et al. 2024). It also means that different individuals may remember the same event differently based on their sensory inputs and modes of processing information. As we have seen, memory involves multiple brain processes and systems responsible for encoding, storing, and retrieving information. Multisensory and multifunctional memory refers to the fact that memories include multiple sensory modalities, such as sight, sound, touch, taste, and smell. These modalities are often integrated to form a coherent memory involving many cognitive functions. For example, suppose someone remembers a pleasant experience at a restaurant. In that case, the memory may be encoded and recalled through the visual details of the decor and food, the sounds of the atmosphere and music, the taste and smell of the cuisine, and even the touch of the tablecloth or utensils. These sensory modalities work together to form a complex and rich experience memory.

Multiple brain processes and systems are involved in memory, including the sensory systems that detect and process information from the environment, the hippocampus and related structures involved in memory formation and consolidation, the prefrontal and posterior cortices, and limbic and brainstem structures. However, these regions are interconnected, and the integration of information across modalities occurs in higher-order association areas of the brain, such as the parietal and prefrontal cortex, in cooperation with subcortical systems. Memory's multisensory and multifunctional aspects highlight the metafunctional nature of the brain processes and systems involved in this fundamental aspect of human cognition. In the previous example, behind the background experience of well-being or happiness, a myriad of functions and brain regions are involved, providing a rich and complex personal experience. A litany of brain subsystems, such as the neocortical, thalamic-striatum, limbic, and brainstem structures, are interconnected with interoception (this perception refers to the awareness of internal signals from your body, such as feelings of hunger, a rapid heartbeat, or the need to use the washroom). The body's homeostatic and allostatic systems feed from those. Homeostatic systems help maintain the body's internal environment, including body temperature, heart rate, and blood pressure, while allostatic systems help the body adapt to stressors and maintain stability over time. The limbic system, including the amygdala and hippocampus, plays a crucial role in emotional regulation and stress responses, essential for maintaining homeostasis and allostatic balance. The brainstem participates in physiological processes like heart rate and breathing, is connected to the limbic system, and modulates emotional responses. Therefore, the complex interplay between the neocortical, thalamic-striatum, limbic, and brainstem structures and the body's homeostatic and allostatic systems participate in memory and other cognitive processes. Other senses contribute to creating our perception of embodiment and our position as we move in space. Memory provides integrated continuity of these aspects.

(5) Memory Depends on Conscious and Unconscious Processes.

Memory operates through an intricate interplay between conscious and unconscious processes, each contributing uniquely to how we encode, store, and retrieve information. The *conscious aspect of memory,* often called explicit memory, enables the intentional recall of facts, events, and experiences. This type of memory allows us to consciously access and articulate details, such as recalling the date of an important event or the steps to solve a problem. In contrast, the *unconscious aspect of memory,* known as implicit memory, involves the automatic, effortless retention of information and skills. Implicit memory allows us to perform tasks like riding a bike, typing on a keyboard, or playing a musical instrument without consciously focusing on every individual step. It also governs our ability to recognize familiar objects, sounds, and smells, even when we cannot consciously recall where or when we encountered them (Kaufman, 2011; Lisman & Jensen, 2013; Squire & Dede, 2015). The multifaceted nature of unconscious memory presents challenges for investigation, some of which will be explored further in the chapter on *The Unconscious*.

Interconnection of Conscious and Unconscious Memory

Conscious and unconscious memory processes are not isolated but deeply interconnected, often influencing one another. For instance, when remembering something intentionally, we may unconsciously rely on strategies such as associating new information with familiar concepts or visualizing it. Conversely, unconscious memory can shape our judgments and preferences without awareness, subtly influencing decisions and behaviors. Physiological systems, including the body's *homeostatic and allostatic mechanisms*, also play a substantial role in memory processing. These systems regulate the body's responses to stress and emotional experiences, significantly affecting memory consolidation and retrieval (Mizumori & Jo, 2013; Mizumori, 2016). For example, heightened emotional arousal can enhance memory encoding and retrieval, while chronic stress may impair these processes and lead to memory deficits.

Examples of Unconscious Memory Processes

Unconscious memory extends beyond implicit memory to include phenomena such as *priming*—a process where exposure to a particular stimulus influences subsequent behavior or perception without conscious awareness (Weingarten et al., 2016; Shapir, Shapir-Tidhar, & Shtudiner, 2023). For example, after exposure to the word "yellow," an individual might more readily recognize or think of "banana" later. This association occurs unconsciously, demonstrating how memory facilitates quick connections. Similarly, exposure to subliminal stimuli—inputs below the threshold of conscious perception—can influence preferences or actions. For instance, subliminal messaging about a specific brand may increase the likelihood of someone choosing that brand, even if they are unaware of having seen the advertisement. Unconscious memory also includes emotional and behavioral influences. For instance, a person may experience fear or anxiety in a specific context without recalling the traumatic experience that initially caused the reaction. These responses highlight the complex ways unconscious memory shapes behavior and emotions.

Implicit Memory in Skilled Performance

The brain's implicit memory systems enable individuals to perform complex tasks without consciously thinking about every detail. For example, skilled musicians rely on implicit memory to play intricate compositions fluently, while their conscious focus is refining nuances in their performance. This interplay between unconscious mastery and conscious fine-tuning underscores the multifaceted nature of memory. Understanding the conscious and unconscious dimensions of memory demands a metafunctional framework. This framework acknowledges the overlapping, dynamic processes involved in memory formation and utilization, emphasizing the intricate interplay of systems that enable both intentional recall and automatic, unconscious influence. Far from being a simple mechanism, memory is a multifaceted and adaptive cognitive process that profoundly shapes human behavior, perception, and experiences.

(6) Memory Has Double Intentionality

Intentionality, the characteristic of being about or directed toward something, was reintroduced into contemporary philosophy in the 19th century by Franz Brentano (Brentano, 1995). As the founder of act psychology, also known as intentionalism, Brentano emphasized this concept in his seminal work, *Psychology from an Empirical Standpoint* (1874). His ideas significantly influenced the philosophy of mind, shaping the work of Edmund Husserl and Martin Heidegger. Today, intentionality remains central to discussions about consciousness and mental representation. However, the existence of intentionality continues to spark debate in the philosophy of mind. Intentionality refers to the property of mental states and acts that point beyond themselves to something external. This aboutness poses a profound philosophical question: how can the brain, a physical organ, be the source of this uniquely mental property?

Raymond Tallis, a British philosopher and physician, expands on this challenge elaborating on the concept of *double intentionality* in memory (Tallis 2004, 2010, 2011). According to Tallis, memory's double intentionality adds a new dimension to the philosophical and neurophysiological understanding of memory. At its first level, intentionality refers to memory about a past experience. At the second level, it encompasses the object or subject of that experience. For instance, when we remember holding a glass, the memory is not just about the act of holding it but also about the glass itself. Tallis (2010) states, "Memories supposedly, therefore, reach back to the mental causes of their physical causes. What is more, just as in vision I see the object as separate from myself, in memory I see the remembered object as different from the present." This dual-layered "aboutness" creates a problem for materialist explanations of memory and consciousness, as it requires an immaterial aspect to explain the relational properties of memory.

Tallis highlights how this dual intentionality challenges purely materialist views. The brain's processes should be confined to internal representations and functions in a materialist framework. However, memory's dual intentionality demonstrates a gap: while the brain can recreate neural patterns corresponding to an experience, it does not inherently explain the memory's ability to transcend the physical and recall the meaning, context, and objects of that experience. Additionally, memory introduces the concept of the past tense—our ability to place experiences in a temporal context—which further complicates materialist models. This perspective poses the need to consider a transphysical aspect of memory and consciousness. Tallis and other philosophers (Husserl 1931/1964) argue that memory's dual intentionality may point to dimensions beyond physical explanations, suggesting that we cannot fully understand memory without incorporating non-material aspects. This idea connects directly to the seventh metafunctional characteristic of memory, which delves deeper into the case for a transphysical nature of this fundamental cognitive ability.

(7) Memory Requires a Transphysical Component.

Neuroscientists immersed in unraveling the complexities of the brain encounter a profound paradox when examining the nature of cognitive processes, particularly memory. The brain is widely regarded as the organ that enables us to interpret the world and engage in scientific inquiry. Materialist perspectives position the brain as a sophisticated, finely tuned machine responsible for constructing narratives, predicting our surroundings, and synthesizing fragments of information, memories, and expectations. In this view, the brain weaves these elements into a cohesive identity that persists over time. However, the materialist framework implies an unsettling reality: neural processes reduce the narratives we create, the perceptions we experience, and the identities we maintain to mere byproducts, risking their dismissal as pervasive illusions. This realization challenges the foundations of materialist thought, as it raises a troubling question: if even our sense of self and cognitive capacities are illusory, what remains as a basis for exploring and understanding the world? How can we navigate the vast uncertainties of existence and pursue science if the very faculties we rely on are considered illusory constructs? This paradox exposes a critical limitation of the physicalist worldview. The rich tapestry of memory and identity—integral to our sense of self—resists reduction to purely physical processes. Instead, scientific evidence and philosophical inquiry increasingly suggest that memory involves dimensions beyond the material. These observations encourage a broader perspective, embracing the possibility that memory possesses a transphysical component. Such a perspective enriches our understanding of memory and reconciles the apparent tension between materialist neuroscience and the lived reality of human experience.

A. Savant Syndrome

The phenomenon of savant syndrome has intrigued scholars for centuries. The earliest scientific report of savant syndrome appeared in 1783 in the German psychology journal *Gnothi Sauton* (Rush, 1789). This report detailed the case of Jedediah Buxton, a mental calculator renowned for his lightning-speed arithmetic abilities and exceptional memory (Moritz, 1783). Later, in 1887, Dr. J. Langdon Down introduced the term "idiot savants" to describe individuals with extraordinary abilities coexisting with significant cognitive impairments (Treffert, 2009). Down documented ten cases in his seminal work, including one who had memorized the entirety of *The Decline and Fall of the Roman Empire* and could recite it forward and backward.

While an exhaustive inventory of savant cases would be beyond the scope of this discussion, a few key examples illuminate the enigmatic nature of this phenomenon. One such account is cited by Rush (1789) in Treffert (2009), describing a "wonderful talent for arithmetical calculation" exhibited by an enslaved African in Virginia. Benjamin Rush, often called "The Father of American Psychiatry," provided foundational insights into mental health with his seminal textbook *Medical Inquiries and Observations upon Diseases of the Mind* (1812).

Rush documented the extraordinary abilities of Thomas Fuller, a man whose cognitive capacities seemed to defy conventional understanding. Despite struggling with basic theoretical and practical concepts, Fuller astounded observers with his computational prowess. When asked how many seconds a 70-year-old man had lived (adjusting for 17 leap years), Fuller correctly calculated 2,210,500,800 seconds in just 90 seconds (Scripture, 1891). Fuller's case exemplifies a phenomenon that challenges materialist explanations of cognition. His extraordinary ability to perform complex calculations with minimal formal training underscores the human mind's profound and seemingly inexplicable potential. Such examples highlight the exceptional capabilities of savants and raise philosophical questions about the limitations of reductionist views on human consciousness. Fuller's story is a compelling reminder that the human mind possesses depths and abilities that may extend beyond mechanistic interpretations, inviting a broader perspective on cognition and consciousness.

Laurence Kim Peek

Laurence Kim Peek, an American megasavant, demonstrated cognitive abilities that defy conventional explanations of human memory (Treffert and Christensen, 2006; Treffert, 2009). A megasavant is a genius across multiple domains, and Peek excelled in approximately 15 fields, ranging from literature and geography to sports statistics and historical knowledge. In their article, Treffert and Christensen remarked, "Kim Peek possesses one of the most extraordinary memories ever recorded. Until we can explain his abilities, we cannot pretend to understand human cognition" (2006). Peek's remarkable memory inspired the character in the film *Rain Man,* where his ability to read a page in seconds and retain intricate details captivated audiences.

Doctors later revised Peek's initial diagnosis of autism to FG syndrome, a rare genetic disorder caused by X-linked recessive genes and first described by Opitz and Kaveggia (1974). FG syndrome's clinical features, including intellectual disability, hyperactivity, and hypotonia, contrast starkly with Peek's cognitive brilliance. In 2004, a NASA study revealed that Peek lacked a corpus callosum, anterior commissure, and hippocampal commissure—structures crucial for interhemispheric communication (Brogaard, 2013). This absence rendered Peek a natural split-brain patient, raising questions about how his hemispheres coordinated the vast integration of information. Peek's extraordinary memory encompassed the Bible, Shakespeare's complete works, all U.S. and global geography, sports almanacs, and more (Treffert 2009). Physicalist explanations struggle to account for these feats, particularly given his unique brain structure. Agenesis of the corpus callosum complicates traditional neural models of memory and cognition, suggesting that Peek's abilities challenge the reductionist materialist framework. His case underscores the potential limits of understanding cognition solely through brain processes, inviting deeper philosophical and scientific inquiry into the nature of memory and consciousness.

Stephen Wiltshire

Stephen Wiltshire, an autistic savant celebrated for his remarkable ability to reproduce cityscapes with incredible detail, has astonished the world with his artistic talent and memory (Wiltshire, 1991). During a brief helicopter ride over Tokyo, Wiltshire observed the sprawling urban landscape. Upon landing, he began sketching with extraordinary precision. His hands moved swiftly across the paper, rendering buildings, streets, and landmarks with breathtaking accuracy. Despite the city's immense size and complexity, Wiltshire's drawing retained a distinct personal touch, demonstrating his prodigious memory and artistic sensibility. Over several days, he completed the artwork, showcasing an uncanny ability to recreate the cityscape entirely from memory. Wiltshire's repertoire includes numerous iconic artworks, such as:

1. The Christ the Redeemer statue in Rio de Janeiro
2. Panoramic views of Tokyo, Japan, and New York City
3. The Singapore and Sydney skylines
4. The Eiffel Tower in Paris
5. The Houses of Parliament and Tower Bridge in London
6. The Dubai skyline
7. The Petronas Twin Towers in Kuala Lumpur, Malaysia

Savant Syndrome Context and Implications for Neuroscience

Savant Syndrome has been extensively documented, with hundreds of cases reviewed by researchers like Treffert (2009), including (Tredgold 1914; Hill 1978; Rimland 1978; Treffert 1988; Hermelin 2001; Heaton and Wallace 2004; Peek and Hanson 2008). While most savants exhibit one extraordinary skill, a few demonstrate multiple exceptional abilities (Rimland & Fein, 1988). Researchers argue that savant talents, such as calendar calculation and musical expertise, transcend simple memory functions, suggesting deeper, poorly understood foundations (Nettlebeck & Young, 1999; Hermelin, 2001). Christopher J. Evans, a professor at UCLA's Department of Psychiatry and Biobehavioral Sciences, reflected on cases like Wiltshire's. He remarked, "It still blows me away that some people can actually have this photographic memory. I still, I really don't understand that. I don't understand how somebody can fly over a city and memorize every little part of the city and then draw it out" (Evans, 2021). This astonishment highlights a key challenge to materialist neuroscience, which often struggles to explain such exceptional abilities using physical processes alone. Wiltshire's extraordinary talent underscores the need to reconsider the limitations of materialist frameworks. These phenomena compel us to explore a more encompassing understanding of human cognition that transcends purely mechanistic explanations. As Thomas Kuhn (1962) described in *The Structure of Scientific Revolutions*, paradigm shifts often occur when prevailing theories cannot account for anomalous data. The abilities of savants like Wiltshire may signify the need for such a transformation in neuroscience, urging the integration of broader dimensions of human experience into our understanding of cognition and memory.

B. Acquired Savant Syndrome

Orlando Serrell, born in 1960, experienced a head injury at ten after being struck by a baseball (IT Group 2008; Treffert 2015). Following the injury, Serrell noticed a remarkable change in his cognitive abilities. Specifically, he developed extraordinary calendar calculation skills and an exceptional memory for past events. Serrell's calendar calculation ability allowed him to accurately determine the day of the week for any given date, past or future. He could rapidly provide the day of the week for historical events, birthdays, or any random date requested. This feat is awe-inspiring, considering the vast range of dates he could effortlessly navigate through. Serrell also exhibited an enhanced autobiographical memory, recalling incredibly detailed events from his life, often accompanied by vivid sensory experiences. His recollections included minute details such as weather conditions, clothing, and even the emotions he felt at the time. These distinct cognitive changes following his head injury astonish physicalist perspectives on cognition, as they suggest a previously untapped transphysical domain beyond material explanations (Snyder, 2009). Serrell's exceptional abilities offer a glimpse into the nature of human cognition that transcends biological limitations.

While Serrell's acquired savant skills were extraordinary, they came with challenges. He experienced difficulties multitasking and had to develop strategies to manage his exceptional memory, as the constant influx of detailed memories could sometimes be overwhelming. Despite these difficulties, his case and others demonstrate how brain injuries can lead to remarkable cognitive abilities in specific domains (Treffert, 2009). For instance, Anthony Cicoria, an orthopedic surgeon, had his life transformed in 1994 after being struck by lightning. After temporarily losing his heartbeat, he developed an overwhelming passion for classical music, leading him to teach himself piano and eventually compose and perform his music (Snyder, 2009). Similarly, Alonzo Clemons, an American sculptor, suffered a severe brain injury as a child that left him developmentally disabled but unlocked an extraordinary talent for sculpting lifelike animal figures, often with unconventional materials. His works earned significant acclaim, selling for up to $45,000 (Treffert, 2009). Another striking example is Eadweard Muybridge, who suffered a traumatic brain injury in 1860 after a stagecoach accident. Following his injury, Muybridge revolutionized photography, particularly the study of motion, making significant contributions to the development of motion-picture technology (Muybridge, 1902).

In the ongoing debate between materialist and non-materialist perspectives, Serrell's sudden acquisition of savant abilities and numerous other cases challenge the deterministic view of brain function. If cognition were entirely mechanistic, it would be difficult to explain such extraordinary capacities first and then how they could emerge from damage. These phenomena point to the possibility that some cognitive processes transcend the material structure of the brain, suggesting the need to account for non-material aspects of consciousness. While surprising for physicalist views, these cases align with non-physicalist perspectives and fit seamlessly within the metafunctional framework.

C. Highly Superior Autobiographical Memory (HSAM)

In the early 2000s, a woman named Jill Price contacted neuroscientist and memory researcher Jim McGaugh, claiming to be able to remember every day of her life since the age of 12 (LePort, 2014; Mazzoni et al., 2019; Talbot et al., 2024). She became the first person recognized as having a 'Highly Superior Autobiographical Memory' (HSAM). Her memory was not photographic, but she experienced each day as a video replaying in her mind, with vivid sensory details and emotions. Price's case sparked interest among researchers and led to the identification of a small group of individuals with HSAM. Studies have shown that their ability to recall their past is exceptional. However, their memory for other information, such as general knowledge or events outside their personal experience, is similar to that of the general population. The precise neurological mechanisms behind HSAM have yet to be understood.

The phenomena of HSAM, savant syndrome, and acquired savant syndrome collectively challenge the materialist perspective, which posits that physical brain mechanisms can fully explain all mental processes. Savant syndrome, in which individuals possess extraordinary talents in specific areas like music, art, or mathematics despite cognitive impairments in other domains, points to memory or cognitive functions that materialist views struggle to explain. Even more striking is acquired savant syndrome, where individuals develop remarkable cognitive or artistic abilities following brain injury, sometimes tapping into skills they had no prior exposure to, suggesting that brain damage can unlock latent potential rather than creating new brain physical capabilities. These phenomena, with their extraordinary and selective enhancements of cognitive functions, imply that the mind and consciousness are not entirely reducible to brain matter. The selective and sometimes sudden emergence of these abilities challenges the materialist assumption that consciousness and cognitive capacities are strictly the products of neural pathways, suggesting instead that the mind involves deeper, non-physical, latent properties.

D. Near-Death Experiences

The implications of Near-Death Experiences (NDEs) for our understanding of memory are profound. Previously examined in the Chapter on Perception, NDEs highlight a potential transphysical component to memory, challenging traditional neuroscientific models. The memories formed during NDEs are vivid, emotionally intense, and self-defining, helping individuals reshape their identity and values (Blagov & Singer, 2004; van Lommel, 2011; Cassol et al., 2019). These experiences often include affective, cognitive, transcendental, and transphysical elements. In positive cases, individuals report profound peace, unity with a transcendent reality, and vivid recollections of out-of-body perceptions, veridical experiences, and otherworldly journeys. Conversely, negative NDEs may evoke terror, anguish, or despair. Regardless of valence, NDEs are transformative, fostering reduced fear of death, enhanced compassion, and a reevaluation of material pursuits.

A related concept, *Self-Defining Memories (SDMs)*, shares key characteristics with NDE recollections. SDMs are autobiographical memories perceived as essential to an individual's identity, reflecting their core beliefs, values, and life goals (Blagov & Singer, 2004; Moore & Greyson, 2017; Cassol et al., 2019). SDMs and NDEs are vivid, frequently recalled, and emotionally intense, contributing to an individual's sense of self-continuity and capacity for planning, meaning making, and responsibility. These parallels underscore the transformative impact of both phenomena on an individual's self-concept and psychological well-being. Recent neuroscientific studies have highlighted the distinctiveness of NDE memories. For instance, EEG patterns during the recall of NDEs differ significantly from those associated with imagined events, suggesting that NDE memory encoding is similar to episodic memories of "real experiences," albeit under altered states of consciousness (Palmieri et al., 2014). This phenomenon has been explored further in studies examining paradoxical cognition during NDEs, where heightened memory and awareness occur despite compromised brain function. Moore and Greyson (2017) surveyed 122 individuals, revealing that NDE memories were reported as "more real than real," with vividness surpassing that of other life events or imagined experiences.

Physiological conditions under which NDEs occur—such as trauma, anesthesia, or cardiac arrest—pose a direct challenge to the assumption that brain activity is the sole source of conscious experience. Despite extreme disruptions to neural function, individuals report coherent, detailed, and transformative experiences. These findings suggest that consciousness and memory may not entirely depend on the physical brain, pointing to a transphysical aspect of human cognition. Further insights come from studies on extreme brain anomalies. For instance, research has explored memory processing in cases involving hydrocephalus, hemihydranencephaly, and hemispherectomy, as well as the extraordinary abilities observed in savants (Nahm et al., 2017; Treffert, 2006; Martial et al., 2019; Bennett & Hacker, 2003; Palmieri et al., 2014). These cases demonstrate the brain's remarkable plasticity and challenge reductionist models that attempt to fully explain memory and consciousness through biological mechanisms alone or reduce memory to limited brain regions. As Bennett and Hacker (2003) and others have noted, the core question of how biochemical and electrical processes in the brain translate into self-conscious experiences remains unresolved.

NDEs represent a compelling case for rethinking the relationship between consciousness, memory, and the brain. Memories formed during NDEs are not reducible to neural activity but point to an intimate and dynamic cooperation between transphysical and material processes. While the brain transduces and integrates mental and bodily functions, memories themselves extend beyond the confines of physical brain activity. The transformative impact of NDEs, supported by veridical corroboration, challenges traditional scientific paradigms and underscores the need for a broader framework—like the metafunctional model proposed in this chapter—that accounts for the interplay between the mind, brain, and a transphysical dimension.

E. Paradoxical and Terminal Lucidity

Paradoxical and terminal lucidity is an intriguing phenomenon observed in some persons with advanced neurodegenerative diseases or severe mental disorders. Some researchers disagree with this distinction between paradoxical and terminal lucidity, arguing that terminal lucidity is a particular subtype of paradoxical lucidity (Nahm 2022; Peterson et al. 2022). In the first case, patients experience a sudden and inexplicable return to mental clarity and coherence shortly before their death. This occurrence challenges conventional understandings of the progression of terminal illnesses and has garnered renewed attention from medical professionals and researchers alike in recent years. Examining the historical context of end-of-life care, Clark's publication in "Nineteenth-century doctors and care of the dying" sheds light on the evolution of medical perspectives toward the dying process (Clark 2016). However, more recent academic publications have tried to advance our understanding of terminal lucidity and unravel its underlying mechanisms and implications for our comprehension of consciousness and neurological function in the face of imminent mortality.

In their publication dedicated to Alzheimer's, Mashour et al. defined paradoxical lucidity as an "episode of unexpected, spontaneous, meaningful and relevant communication or connectedness in a patient who is assumed to have permanently lost the capacity for coherent verbal or behavioral interaction due to a progressive and pathophysiological dementing process" (2019). The following year, a concept analysis by Morris P. and Bulman D. draws on published literature (Morris and Bulman 2020). It evaluates critical attributes of lucidity, suggesting they include:

1. Presence of advanced neurodegenerative disease,
2. Cognitive clarity and return of memories inconsistent with the individual's resting state,
3. Ability to share insights with others in intelligible ways and
4. The presence of meta-awareness is an insight into one's condition during the lucid episode.

Andrew Peterson and colleagues thoroughly examined the provisional definition of paradoxical lucidity put forth by experts (Peterson et al. 2022). Their critical analysis highlighted several fundamental concepts within the definition that demanded more precise elucidation. In response, they proposed a more precise definition and provided procedural recommendations at this nascent research stage. Another investigation of terminal lucidity analyzed 124 instances among individuals with dementia and discovered that in "more than 80% of these cases, complete remission with the return of memory, orientation, and responsive verbal ability was reported by observers of the lucid episode" (Batthyány and Greyson 2021). Macleod's observations on terminal lucidity (2009) align with studies by Mashour et al. (2019), Batthyány and Greyson (2021), Palmieri et al. (2014), Nahm (2022), and Van Lommel (2006), which collectively emphasize the phenomenon's implications for understanding consciousness and memory beyond physicalist explanations. Despite objective medical indicators such as EEG and neuroimaging suggesting otherwise, these patients exhibited typical cognitive functions according to age just before their demise. Their behavior hints at a transphysical dimension. As research and evidence in this area accumulate, more voices will increasingly rise to question the prevailing materialist paradigm. One example is the article *"What If Consciousness Is Not an Emergent Property of the Brain? Observational and Empirical Challenges to Materialistic Models"* (Wahbeh et al. 2022).

Terminal lucidity in dementia patients presents a significant challenge to the materialist perspective, which assumes that consciousness is solely a product of the physical brain. In dementia patients, extensive parts of the brain suffer irreversible damage, and yet these patients may experience a sudden and unexpected period of clear consciousness days or hours before death. However, from a non-physicalist perspective, terminal lucidity can be understood as evidence for a transphysical dimension of the mind. In dementia patients, the physical deterioration of the brain may create a barrier to normal cognitive functioning. However, this barrier disappears during terminal lucidity through a transcendent aspect of the mind. This phenomenon suggests that consciousness and memory are not solely a product of the physical brain but have a transphysical aspect that can transcend physical limitations. The mind and brain mutually inform each other, with the mind playing an active role in shaping the physical brain. This perspective supports the brain metafunction framework for understanding the mind-brain relationship and suggests that consciousness and memory are not limited to the physical brain but extend beyond it. The phenomenon of terminal lucidity in dementia and other patients with severely damaged brains challenges the materialist perspective. It adds to the cumulative case against physicalism, allowing us to reevaluate our understanding of the mind-brain relationship.

"Extended Mind" and "Extended Memory"

A response from the physicalist perspective to the challenges of understanding mind and memory is the concept of the "extended mind" and "extended memory" in the "hard sense." This philosophical idea challenges the traditional view that confines the mind and memory to the brain. Instead, it posits that they can extend beyond the body's physical boundaries into the external environment, incorporating tools such as computers, smartphones, or pen and paper. Philosophers Andy Clark and David Chalmers first articulated this concept in their seminal work in the 1990s (Clark & Chalmers, 1998). Similarly, Alva Noë (2009), in *Out of Our Heads: Why You Are Not Your Brain, and Other Lessons from the Biology of Consciousness,* critiques the dominant view of consciousness as a product of neural activity alone. Noë argues that consciousness arises from dynamic interactions between the brain, body, and environment. He builds on the extended mind hypothesis, suggesting that external tools, actions, and contexts are integral to consciousness. He emphasizes the need to look beyond the brain to broader biological, social, and environmental factors shaping experience.

The Extended Mind hypothesis asserts that humans incorporate tools and technologies into their cognitive processes, fundamentally extending their mental capabilities. Individuals enhance memory and problem-solving abilities by integrating external devices into cognitive systems. Historically, extending cognition to the external world is not new. As noted by Hall (2012), Seneca once recounted an anecdote of a wealthy freedman who "bought educated slaves" to memorize the works of Homer, Hesiod, and others, reasoning that the knowledge possessed by his slaves became, by extension, his own (Seneca, *Moral Epistles,* 1917–1925, Epistles 27.5–8). Regarding memory, the concept of extended memory suggests that memory exists not only within the brain but also in external environments. Tools such as notebooks or diaries become extensions of an individual's memory system, effectively outsourcing certain aspects of memory storage. While this observation may seem trivial, it fundamentally challenges the traditional view of memory as a purely internal, brain-confined process. Instead, it posits that memory is distributed across multiple sources, blurring the line between internal and external memory systems.

This concept has generated debate among philosophers, cognitive scientists, and neuroscientists. While some embrace its potential to broaden understanding, skeptics argue that it leaves significant questions about the brain and memory unresolved. For instance, the reliance on external devices still raises questions about the role of intrinsic neural processes and their interplay with extended systems. While exploring the extended mind and memory hypothesis is undoubtedly valuable, One must note that simply expanding the physical basis of memory does not address the full range of challenges to materialism. A transphysical domain must be incorporated into the framework to provide a comprehensive explanation, enabling a more robust scientific and phenomenological data account.

Integrative Perspectives on Memory: Bridging Science, Phenomena, and Philosophy

The integration of multisensory inputs into memory processes demonstrates the brain's remarkable ability to merge diverse sensory modalities, enhancing recall and perception. For example, Ghazanfar and Schroeder (2006) revealed the intricate interactions between auditory and visual processing, showing how these modalities converge to strengthen multisensory memory integration. Driver and Noesselt (2008) expanded on this, illustrating how cross-modal attention facilitates improved memory recall and perceptual clarity. Murray and Shams (2023) discussed regions such as the superior temporal sulcus as important for multisensory integration, highlighting the brain-wide and interconnected dynamics that underlie memory and learning. These studies underscore memory's multifaceted and metafunctional qualities, suggesting that memory formation and recall transcend isolated neural processes and depend on distributed networks.

The adaptability and reconstructive nature of memory further highlight its dynamic qualities. Heald, Lengyel, and Wolpert (2023) explored how memory reconsolidation incorporates new information, enabling the refinement of existing memories. Similarly, Dumas, McNeish, and Greene (2020) demonstrated that memory retrieval triggers a reconstructive phase where memories integrate new experiences, balancing stability with flexibility. Blom and Huang (2021) provided experimental evidence on the misinformation effect, illustrating how external suggestions can distort memory, linking this malleability to creativity and susceptibility to error. These findings reinforce that memory is not static but a continuously evolving process shaped by internal and external influences.

Extraordinary cases and anomalies in memory further challenge materialist explanations. For example, Treffert (2009) and Treffert and Christensen (2006) documented the extraordinary recall abilities of savants like Kim Peek, who could memorize entire books and calendars. This phenomenon defies conventional neural explanations and hints at a transphysical dimension. Similarly, Nahm, Rousseau, and Greyson (2017) examined cases such as hydrocephalus, where individuals demonstrated intact memory functions despite significant brain anomalies. These cases emphasize the adaptability and resilience of distributed memory networks and challenge the assumption that memory is strictly a product of localized brain functions.

Paradoxical and terminal lucidity further exemplify memory's metafunctional complexity. Mashour et al. (2019) and Batthyány and Greyson (2021) documented cases of individuals with severe brain degeneration who exhibited sudden cognitive clarity before death. Despite extensive brain damage, these individuals demonstrated coherent memory and cognitive functions, raising profound questions about the relationship between the brain and consciousness. These findings challenge the idea that memory and consciousness are entirely dependent on physical brain structures and suggest the need for a broader framework that includes a transphysical dimension.

Philosophical perspectives add depth to these empirical findings. Clark and Chalmers (1998) proposed the "extended mind" hypothesis, arguing that cognitive processes, including memory, are not confined to the brain but extend into the environment through tools and interactions. For instance, using external devices such as notebooks or smartphones effectively extends the brain's cognitive and memory capabilities. The distributed quality of memory resonates with the scientific evidence of memory's reliance on brain-wide networks. Similarly, Tallis (2010, 2011) introduced the concept of memory's double intentionality, emphasizing its layers of "aboutness," which challenge purely physical explanations. These findings collectively illustrate memory's metafunctional quality, transcending isolated neural mechanisms and challenging materialist perspectives. They align with the proposed brain metafunctional framework, which integrates the brain's physical unified processes with a transphysical dimension to account for the full complexity of memory. Together, these scientific insights and philosophical arguments point to the need for a more comprehensive understanding of memory that bridges the physical and transphysical realms.

Conclusion

This chapter examined the neural underpinnings of memory through the lens of a brain metafunction framework, presenting a holistic understanding of this intricate cognitive process. It began by reviewing memory's definition and diverse types, underscoring its multifaceted and complex nature. The discussion highlighted the necessity of moving beyond a conventional focus on specific brain regions, such as the medial temporal lobe and hippocampus, toward a more integrated perspective. Contemporary research reveals that memory operates within a network of distributed brain regions, demonstrating interconnected functional patterns across the brain. These findings challenge the notion of localized neural patterns, advocating for a brain-wide, holistic understanding of memory processes.

A key proposition emerging from this exploration is that memory inherently points to a transphysical dimension of human existence. The cumulative evidence suggests that memory transcends the purely physical constraints of the brain, indicating the existence of a domain not entirely reducible to neural activity. Within the Christian Thomist tradition, memory exemplifies an intrinsic form of causality distinct from Cartesian dualism—a concept that warrants deeper exploration in subsequent chapters. This chapter's investigation challenges mainstream neuroscientific paradigms, emphasizing memory's metafunctional scope and transphysicality. By embracing this broader framework, future research can advance a more profound understanding of memory processes, implications for cognitive neuroscience, and significance for human identity and existence. Recognizing memory's metafunctional and transphysical qualities paves the way for transformative insights into the interplay between mind, brain, and personhood.

REFERENCES

Alais, D., & Burr, D. (2004). The ventriloquist effect results from near-optimal bimodal integration. *Current Biology, 14*(3), 257–262. https://doi.org/10.1016/j.cub.2 004.01.029

Albert, M. S., Dekosky, M. C., Dickson, D., Dubois, D., Feldman, H. H., Fox, N. C., Gamst, A., Holtzman, D. M., Jagust, W. J., Petersen, R. C., Snyder, P. S., Carrillo, L. A., Thies, B. M., & Phelps, C. (2009). The diagnosis of mild cognitive impairment due to Alzheimer's disease: Recommendations from the National Institute on Aging and the Alzheimer's Association Workgroup on Diagnostic Guidelines for Alzheimer's Disease. *Alzheimer's and Dementia, 7*(3), 270–279. https:// doi:10.1 016/j.jalz.2011.03.008.

Anderson, M. C. (2003). Rethinking interference theory: Executive control and the mechanisms of forgetting. *Journal of Memory and Language, 49*(4), 415–445. https://doi.org/10. 1016/j.jml.2003.08.006

Anderson, M. C., & Hanslmayr, S. (2014). Neural mechanisms of motivated forgetting. *Trends in Cognitive Sciences, 18*(6), 279–292. https://doi. org/10.1016/j.tics.2 014.03.002

Baddeley, A. D. (2000). The episodic buffer: A new component of working memory? *Trends in Cognitive Sciences*, 4(11), 417–423. https:// doi.org/10.1016/S1364 -6613(00)01538-2

Ballard, I. C., Wagner, A. D., & McClure, S. M. (2019). Hippocampal pattern separation supports reinforcement learning. *Nature Communications, 10*, 1235. https://doi.org/10 .1038/s41467-019-08998-1

Batthyány, A., & Greyson, B. (2021). Spontaneous remission of dementia before death: Results from a study on paradoxical lucidity. *Psychology of Consciousness: Theory, Research, and Practice, 8*(1), 1–8. https://doi.org/ 10.1037/cns0000259

Bennett, M. R., & Hacker, P. M. S. (2003). *Philosophical Foundations of Neuroscience.* Hoboken, NJ: Wiley-Blackwell.

Blagov, P. S., & Singer, J. A. (2004). Four dimensions of self-defining memories (SDMs): Memory specificity, meaning, content, and affective tone. *Journal of Personality, 72*(3), 481–511. https://doi.org/10.1111/j.002 2-3506.2004.00270.x

Blom, R., & Huang, K.-T. (2021). Eyewitness memory contamination through misleading questions by reporters. *Newspaper Research Journal, 42*(3), 346–363. https://doi.org/10.1177/07395329211030628

Bowler, D., & Gaigg, S. (2021). Semantic memory. In *Volkmar, F. R. (Ed.), Encyclopedia of Autism Spectrum Dis -orders*. Springer, Cham. https://doi.org/10.1007/978-3-319-9 1280-6_542

Brainerd, C. J., & Reyna, V. F. (2005). *The Science of False Memory*. New York: Oxford University Press. https://doi.org /10.1093/acprof:oso/9780195154054.001.0001

Brentano, F. (1995). *Psychology from an Empirical Standpoint* (2nd ed.). Contributions by L. L. McAlister, P. Simons, A. Rancurello, & D. B. Terrell. London: Routledge. https://doi.org/10.4324/9780203202883

Brogaard, B. (2013, March 17). The brain of the real Rain Man: New insights into the mechanism underlying extreme memory abilities. *Psychology Today*. Retrieved from https://www.psychologytoday.com

Camina, E., & Güell, F. J. (2017). The neuroanatomical, neurophysiological, and psychological basis of memory: Current models and their origins. *Frontiers in Pharmacology, 8*, 438. https://doi.org/10.3389/fphar.201 7.00438

Cappaert, N. L. M., Van Strien, N. M., & Witter, M. P. (2015). Hippocampal formation. In G. Paxinos (Ed.), *The Rat Nervous System* (4th ed., pp. 511–573). Academic Press. https://doi.org/10.1016/B978-0-12-37 4245-2.00020-6

Cassol, H., Martial, C., Annen, J., Martens, G., Charland-Verville, V., Majerus, S., & Laureys, S. (2019). A systematic analysis of distressing near-death experience accounts. *Memory, 27*(8), 1122–1129. https://doi.org/10.1080/09658211.2019.1626438

Christianson, S. (1992). *The Handbook of Emotion and Memory: Research and Theory*. Hillsdale, NJ: Lawrence Erlbaum Associates.

Clark, A., & Chalmers, D. (1998). The extended mind. *Analysis, 58*(1), 7–19. https://doi.org/10.1093/analys/58. 1.7

Clark, D. (2016). Nineteenth-century doctors and care of the dying. In *To Comfort Always: A History of Palliative Medicine Since the Nineteenth Century.* Oxford Medical Histories. Oxford: Oxford University Press. Online edition published November 1, 2016.

Csikszentmihalyi, M. (1990). *Flow: The Psychology of Optimal Experience.* New York: Harper & Row.

Doty, R. W. (1998). The five mysteries of the mind, and their consequences. *Neuropsychologia,* 36(10), 1069–1076. https://doi.org/10.1016/s0028-3932(98)00060-8

Driver, J., & Noesselt, T. (2008). Multisensory interplay reveals cross-modal influences on "sensory-specific" brain regions, neural responses, and awareness. *Neuron, 57*(1), 11–23. https://doi.org/10.1016/j.neuron.2007.12.013

Duarte, S. E., Yonelinas, A. P., Ghetti, S., & Geng, J. J. (2024). Multisensory processing impacts memory for objects and their sources. *Memory & Cognition.* https://doi.org/10. 3758/s13421-024-01592-x

Dumas, D., McNeish, D., & Greene, J. A. (2020). Dynamic measurement: A theoretical–psychometric paradigm for modern educational psychology. *Educational Psychologist, 55*(3), 120–136. https://doi.org/10.1080/00461520.2020.1744150

Ekstrom, A. D., & Ranganath, C. (2018). Space, time, and episodic memory: The hippocampus is all over the cognitive map. *Hippocampus, 28*(9), 680–687. https://doi: 10.1002/hipo.22750.

Elsey, J., & Kindt, M. (2016). Manipulating human memory through reconsolidation: Ethical implications of a new therapeutic approach. *AJOB Neuroscience, 7*(4), 225–236. https://doi.org/10.1080/21507740.2016.1218377

Elsey, J. W. B., Van Ast, V. A., & Kindt, M. (2018). Human memory reconsolidation: A guiding framework and critical review of the evidence. *Psychological Bulletin, 144*(8), 797–848. https://doi.org/10.1037/bul0000152

Evans, C. (2021). How does memory work? *Closer To Truth.* November 17, 2021. Retrieved from https://youtu.be/wsv8_wHCK_wc?si=UxR7UQHTSb2-JbVM

Fernández, R. S., Boccia, M. M., & Pedreira, M. E. (2016). The fate of memory: Reconsolidation and the case of prediction error. Neuroscience & Biobehavioral Reviews, 68, 423–441. https://doi.org/10.1016/j.neubiorev.2016.06.004

Fries, P. (2005). A mechanism for cognitive dynamics: Neuronal communication through neuronal coherence. *Trends in Cognitive Sciences, 9*(10), 474–480. https://doi.org/10. 1016/j.tics.2005.08.011

Ghazanfar, A. A., & Schroeder, C. E. (2006). Is neocortex essentially multisensory? *Trends in Cognitive Sciences, 10*(6), 278–285. https://doi.org/10.1016/j.tics. 2006.04.008

Gilbert, P. E., & Brushfield, A. M. (2009). The role of the CA3 hippocampal subregion in spatial memory: A process-oriented behavioral assessment. *Progress in Neuro-Psycho pharmacology & Biological Psychiatry, 33*(5), 774–781. https://doi.org/10.1016/j.pnpbp.2009.03.037

Goulden, N., Khusnulina, A., Davis, N. J., Bracewell, R. M., Bokde, A. L., McNulty, J. P., & Mullins, P. G. (2014). The salience network is responsible for switching between the default mode network and the central executive network: Replication from DCM. *NeuroImage, 99*, 180–190. https://doi.org/10.1016/j.neuroimage.2014.05.052

Greyson, B. (1997). Near-death experiences: Revisiting the phenomenon. *Journal of the American Academy of Psychiatry and the Law, 25*, 325–340.

Grover, S., Wen, W., Viswanathan, V., Gill, C. T., & Reinhart, R. M. G. (2022). Long-lasting, dissociable improvements in working memory and long-term memory in older adults with repetitive neuromodulation. *Nature Neuroscience, 25*(9), 1237–1246. https://doi.org/10.1038/s41593-022-01132-3

Hall, E. M. (2012, August 9). Reading ancient slavery (2010). *Academia.edu*. Retrieved November 20, 2024, from https://www.academia.edu

Hardt, O., & Nadel, L. (2018). Systems consolidation revisited, but not revised: The promise and limits of optogenetics in the study of memory. *Neuroscience Letters, 680*, 54–59. https://doi.org/10.1016/j.neulet.2017.11.062

Heald, J. B., Lengyel, M., & Wolpert, D. M. (2023). Contextual inference in learning and memory. *Trends in Cognitive Sciences, 27*(1), 43–64. https://doi.org/10.1016/j.tics. 2022.10.004

Heaton, P., & Wallace, G. L. (2004). Annotation: The savant syndrome. *Journal of Child Psychology and Psychiatry, 45*(4), 899–911. https://doi.org/10.1111/j.1469-7610.2004.t0 1-1-00284.x

Hermelin, B. (2001). *Bright Splinters of the Mind.* Jessica Kingsley Publishers.

Hill, A. L. (1978). Savants: Mentally retarded individuals with special skills. In N. Ellis (Ed.), *International Review of Research in Mental Retardation* (pp. 277–298). Academic Press.

Hirst, W., Phelps, R., Easton, T. W., & Bohm, K. (2020). Multisensory memory: The role of interconnected brain systems. *Brain Research Reviews, 32*(3), 149–169. https://doi.org/ 10.1016/j.brainresrev.2020.03.005

Huang, Q., Xiao, Z., Yu, Q., et al. (2024). Replay-triggered brain-wide activation in humans. *Nature Communications, 15,* 7185. https://doi.org/10.1038/s4146 7-024-51582-5

Huijgen, J., & Samson, S. (2015). The hippocampus: A central node in a large-scale brain network for memory. *Revue Neurologique, 171*(3), 204–216. https://doi.org/10.1016/j.neurol.2015.01.557

Husserl, E. (1931/1964). *The phenomenology of internal time-consciousness* (M. Heidegger, Ed.; J. Churchill, Trans.). Indiana University Press.

IT Group. (2008, April 25). *Spontaneous Savant - Orlando Serrell on Discovery Channel.* Retrieved from https://youtu.be/3fg75qRdzwo?si=-fYtB6nw3m LngdH9g

Kaefer, K., Stella, F., McNaughton, B. L., et al. (2022). Replay, the default mode network, and the cascaded memory systems model. *Nature Reviews Neuroscience, 23*(10), 628–640. https://doi.org/10.1038/s41583 -022-00620-6

Kandel, E. R. (2001). *In Search of Memory: The Emergence of a New Science of Mind.* New York: W. W. Norton & Company.

Kandel, E. R. (2009). The biology of memory: A forty-year perspective. *Journal of Neuroscience*, 29(41), 12748–12756. https://doi.org/10.1523/JNEUROSCI. 3958-09.2009

Kaufman, S. B. (2011). Intelligence and the cognitive unconscious. In R. J. Sternberg & S. B. Kaufman (Eds.), *The Cambridge Handbook of Intelligence* (pp. 442–467). Cambridge University Press. https://doi.org/10.1017/CBO 9780511977244.023

Kim Peek - The Real Rain Man [Full Film]. (n.d.). Retrieved from https://youtu.be/DLpCfHH1OVU?si=tltsMhk0EyDs SWcU

Koelewijn, T., Bronkhorst, A., & Theeuwes, J. (2010). Attention and the multiple stages of multisensory integration: A review of audiovisual studies. *Acta Psychologica, 134*(3), 372–384. https://doi.org/10.1016/j.actpsy.2010.03.010

La Ganga, M. (1994, May 14). California father wins suit in 'false memory' case. *Los Angeles Times.* Retrieved from https ://www.latimes.com

Lanciano, T., Curci, A., Matera, G., & Sartori, G. (2018). Measuring the flashbulb-like nature of memories for private events: The flashbulb memory checklist. *Memory, 26*(8), 1053–1064. https://doi.org/10.1080/09658211.2018.1428348

Latrèche, C., Mancini, V., McGinn, N., Rochas, V., Férat, V., Forrer, S., Schneider, M., & Eliez, S. (2024). Safety and feasibility of home-based transcranial alternating current stimulation in youths with 22q11.2 deletion syndrome. *Frontiers in Neuroscience.* https://doi.org/10.3389/fnins.2024.1453839

LePort, A. K. R. (2014). Behavioral, cognitive, and neural correlates of Highly Superior Autobiographical Memory. *Dissertation.* Retrieved from https://escholarship.org/uc/item/1jw5s7jg

Lisman, J. E., & Jensen, O. (2013). The theta-gamma neural code. *Neuron, 77*(6), 1002–1016. https://doi.org/10.1016/j.neuron.2013.03.007

Loftus, E. F. (2005). Planting misinformation in the human mind: A 30-year investigation of the malleability of memory. *Learning & Memory, 12*(4), 361–366. https://doi.org/10.1101/lm.94705

Loftus, E. F., & Pickrell, J. E. (1995). The formation of false memories. *Psychiatric Annals, 25*(12), 720–725. https://doi.org/10.3928 0048-5713-19951201-07

Macleod, A. D. (2009). Lightening up before death. *Palliative & Supportive Care, 7*(4), 513–516. https://doi.org/10.1017/S 1478951509990526

Martial, C., Cassol, H., Charland-Verville, V., Pallavicini, C., Sanz, C., Zamberlan, F., Vivot, R. M., Erowid, F., Erowid, E., Laureys, S., Greyson, B., & Tagliazucchi, E. (2019). Neurochemical models of near-death experiences: A large-scale study based on the semantic similarity of written reports. *Consciousness and Cognition, 69,* 52–69. https://doi.org/10.1016/j.con cog.2019.01.011

Mashour, G. A., Frank, L., Batthyany, A., Kolanowski, A. M., Nahm, M., Schulman-Green, D., Greyson, B., Pakhomov, S., Karlawish, J., & Shah, R. C. (2019). Paradoxical lucidity: A potential paradigm shift for the neurobiology and treatment of severe dementias. *Alzheimer's & Dementia, 15*(4), 450–462. https://doi.org/10.1016/j.jalz.2019.04.002

Mazzoni, G., Clark, A., De Bartolo, A., Guerrini, C., Nahouli, Z., Duzzi, D., De Marco, M., McGeown, W., & Venneri, A. (2019). Brain activation in highly superior autobiographical memory: The role of the precuneus in the autobiographical memory retrieval network. *Cortex, 120,* 588–602. https://doi.org/10.1016/j.cortex. 2019.02.020

McClung, C. A., & Nestler, E. J. (2008). Neuroplasticity mediated by altered gene expression. *Neuropsychopharmacology, 33*(1), 3–17. https://doi.org/10. 1038/sj.npp.1301544

McGaugh, J. L., & LePort, A. (2014, February 1). *The discovery of super memories. Scientific American.* https://www.scientificamerican.com/article/the -discovery-of-super-memories/

Mizumori, S. J. Y., & Jo, Y. S. (2013). Homeostatic regulation of memory systems and adaptive decisions. *Hippocampus, 23*(7), 751–762. https://doi.org/10.1002/hipo.22176

Moore, L. E., & Greyson, B. (2017). Characteristics of memories for near-death experiences. *Consciousness and Cognition, 51,* 116–124. https://doi.org/10.1016/j.concog.2017.03.003

Morris, P., & Bulman, D. (2020). Lucidity in the context of advanced neurodegenerative disorders: A concept analysis. *Journal of Gerontological Nursing, 46*(12), 42–50. https://doi.org/10.3928/00989134-20201106 -06

Mortiz, K. P. (1783). *Gnothi Sauton, oder Magazin der Erfahrungsseelenkunde als ein Lesebuch für Gelehrte und Ungelehrte.* Berlin, Germany: Mylius.

Moser, M.-B., Rowland, D. C., & Moser, E. I. (2015). Place cells, grid cells, and memory. *Cold Spring Harbor Perspectives in Biology, 7*(2), a021808. https://doi.org/10. 1101/cshperspect.a021808

Murray, C. A., & Shams, L. (2023). Crossmodal interactions in human learning and memory. *Frontiers in Human Neuroscience, 17*, Article 1181760. https://doi.org/10.3389/fnhum.2023.1181760

Muybridge, E. (1902). *Animals in Motion.* London: Chapman & Hall.

Nahm, M. (2022). Terminal lucidity versus paradoxical lucidity: A terminological clarification. *Alzheimer's & Dementia, 18*(3), 538–539. https://doi.org/10.1002/alz.12574.

Nahm, M., Rousseau, D., & Greyson, B. (2017). Discrepancy between cerebral structure and cognitive functioning: A review. *Journal of Nervous and Mental Disease, 205*(12), 967–972. https://doi.org/10.1097/NMD.0000000000000752

Nettlebeck, T., & Young, R. (1999). Savant syndrome. *International Review of Research in Mental Retardation, 22,* 137–173. https://doi.org/10.1016/S0074-7750(08)60133-0

Opitz, J. M., & Kaveggia, E. G. (1974). Studies of malformation syndromes of man 33: The FG syndrome. An X-linked recessive syndrome of multiple congenital anomalies and mental retardation. *Zeitschrift für Kinderheilkunde, 117*(1), 1–18. https://doi.org/10.1007/BF00439020

Palmieri, A., Calvo, V., Kleinbub, J. R., Meconi, F., Marangoni, M., Barilaro, P., Broggio, A., Sambin, M., & Sessa, P. (2014). "Reality" of near-death-experience memories: Evidence from a psychodynamic and electrophysiological integrated study. *Frontiers in Human Neuroscience, 8,* Article 429. https://doi.org/10.3389/fnhum.2014.00429

Peek, F., & Hanson, L. L. (2008). *The Life and Message of the Real Rain Man.* New York: Dude Publishing.

Penfield, W., & Milner, B. (1958). Memory deficit produced by bilateral lesions in the hippocampal zone. *Archives of Neurology and Psychiatry, 79*(5), 475–497. https://doi.org/10. 001/arch neurpsyc.1958.02340050003001

Peterson, A., Clapp, J., Largent, E. A., Harkins, K., Stites, S. D., & Karlawish, J. (2022). What is paradoxical lucidity? The answer begins with its definition. *Alzheimer's & Dementia, 18*(3), 513–521. https://doi.org/10.1002/alz.12424

Porter, S. L., Ten Elshof, G., & Crisp, T. M. (Eds.). (2016). *Neuroscience and the Soul: The Human Person in Philosophy, Science, and Theology.* Grand Rapids, MI: Eerdmans.

Preston, A. R., & Eichenbaum, H. (2013). Interplay of hippocampus and prefrontal cortex in memory. *Current Biology, 23*(17), R764–R773. https://doi.org/10.1016/j.cub. 2013.05.041

Rimland, B. (1978). Savant capabilities of autistic children and their cognitive implications. In G. Serban (Ed.), *Cognitive Defects in the Development of Mental Illness* (pp. 43–65). New York: Brunner/Mazel.

Rimland, B., & Fein, D. A. (1988). Special talents of autistic savants. In L. K. Obler & D. A. Fine (Eds.), *The Exceptional Brain: Neuropsychology of Talent and Special Abilities* (pp. 474–492). New York: Guilford Press.

Rolls, E. T. (2016). Pattern separation, completion, and categorization in the hippocampus and neocortex. *Neurobiology of Learning and Memory, 129,* 4–28. https://doi.org/10.1016/j.nlm.2015.07.008

Roy, D. S., Park, Y. G., Kim, M. E., et al. (2022). Brain-wide mapping reveals that engrams for a single memory are distributed across multiple brain regions. *Nature Communications, 13,* Article 1799. https://doi.org/10.1038/s 41467-022-29384-4

Rudoy, J. D., Voss, J. L., Westerberg, C. E., & Paller, K. A. (2009). Strengthening individual memories by reactivating them during sleep. *Science, 326*(5956), 1079–1079. https://doi.org/10.1126/science.11790 13

Rush, B. (1789). Account of a wonderful talent for arithmetical calculation in an African slave, living in Virginia. *American Museum, 5*(1), 62–63.

Schacter, D. L., Addis, D. R., & Buckner, R. L. (2007). Remembering the past to imagine the future: The prospective brain. *Nature Reviews Neuroscience, 8*(9), 657–661. https://doi.org/10.1038/nrn2213

Scripture, E. W. (1891). Arithmetical prodigies. *The American Journal of Psychology, 4*(1), 1–59. https://doi.org/10. 2307/1411838

Semon, R. (1921). *The mneme* (L. Simon, Trans.). George Allen & Unwin. (Original work published 1904)

Seneca, Lucius Annaeus. (1917–1925). *Moral Epistles.* Translated by Richard M. Gummere. Cambridge, MA: Harvard University Press, Loeb Classical Library, Vol. 1, Epistles 27.5–8.

Servais, A., Hurter, C., & Barbeau, E. J. (2023). Attentional switch to memory: An early and critical phase of the cognitive cascade allowing autobiographical memory retrieval. *Psychonomic Bulletin & Review, 30*(10), 1707–1721. https://doi.org/10.3758/s134 23-023-02270-w

Shapir, O. M., Shapir-Tidhar, M. H., & Shtudiner, Z. (2023). Priming effect across framing, culture, and gender: Evidence from academia. *Managerial and Decision Economics.* https://doi.org/10.1002/mde.3918

Simon, K. C., Nadel, L., & Payne, J. D. (2022). The functions of sleep: A cognitive neuroscience perspective. *Proceedings of the National Academy of Sciences.* https://doi.org/10.1073/pnas .2201795119

Singh, C., Meena, P. S., Jain, M., Premprakash, Sharma, R., & Nainwani, N. (2024). Autistic savant: A rare case report. *Journal of YSR University Health Sciences, 13*(3), 282–284.

Snyder, A. (2009). Explaining and inducing savant skills: Privileged access to lower-level, less-processed information. *Philosophical Transactions of the Royal Society B: Biological Sciences, 364*(1522), 1399–1405. https://doi.org/10.1098/rstb 2008.0290

Squire, L. R., & Dede, A. J. O. (2015). Conscious and unconscious memory systems. *Cold Spring Harbor Perspectives in Biology, 7*(3), a021667. https://doi.org/10.1101/cshperspect .a021667

Squire, L. R., & Kandel, E. R. (1999). *Memory: From Mind to Molecules.* New York: Scientific American Library.

Squire, L. R., & Zola, S. M. (1996). Structure and function of declarative and nondeclarative memory systems. *Proceedings of the National Academy of Sciences of the United States of America, 93*(24), 13515–13522. https://doi.org /10.1073/pnas .93.24.13515

Squire, L. R., & Wixted, J. T. (2011). The cognitive neuroscience of human memory since H.M. *Annual Review of Neuroscience, 34*, 259–288. https://doi.org/10.1146/annurev -neuro-061010-113720

Stickgold, R., & Walker, M. P. (2007). Sleep-dependent memory consolidation and reconsolidation. *Sleep Medicine, 8*(4), 331–343. https://doi.org/10.1016/j.sleep. 2007.03.011

Talbot, J., Convertino, G., De Marco, M., Venneri, A., & Mazzoni, G. (2024). Highly Superior Autobiographical Memory (HSAM): A systematic review. *Neuropsychology Review*. https://doi.org/10.1007/s11065-024-09632-8

Tallis, R. C. (2004). *The knowing animal: A philosophical inquiry into knowledge and truth*. Edinburgh University Press.

Tallis, R. (2010). What neuroscience cannot tell us about ourselves. The New Atlantis. https://www.thenewatlantis.com/publications/what-neuro science-cannot-tell-us-about-ourselves

Tallis, R. (2011). *Aping Mankind: Neuromania, Darwinitis and the Misrepresentation of Humanity*. Durham: Acumen Publishing.

Treffert, D. A. (2006). *Extraordinary People: Understanding Savant Syndrome*. Lincoln, NE: iUniverse.

Treffert, D. A. (2009). The savant syndrome: An extraordinary condition. A synopsis: past, present, future. *Philosophical Transactions of the Royal Society B: Biological Sciences, 364*(1522), 1351–1357. https://doi.org/10.1098/rstb.2008.0326

van Lommel, P. (2011). *Consciousness Beyond Life: The Science of the Near-Death Experience*. New York: HarperOne.

Wahbeh, H., Radin, D., Cannard, C., & Delorme, A. (2022). What if consciousness is not an emergent property of the brain? Observational and empirical challenges to materialistic models. *Frontiers in Psychology, 13*, Article 955594. https://doi.org/10.3389/fpsyg.2022.95 5594

Walker, M. P., & Stickgold, R. (2006). Sleep, memory, and plasticity. *Annual Review of Psychology, 57,* 139–166. https://doi .org/10.1146/annurev.psych.56.091103.0703 07

Wamsley, E. J., & Stickgold, R. (2011). Memory, sleep, and dreaming: Experiencing consolidation. *Sleep Medicine Clinics, 6*(1), 97–108. https://doi.org/10.1016/j.jsmc.2010.12.008

Weingarten, E., Chen, Q., McAdams, M., Yi, J., Hepler, J., & Albarracín, D. (2016). From primed concepts to action: A meta-analysis of the behavioral effects of incidentally presented words. *Psychological Bulletin, 142*(5), 472–497. https://doi.org/10.1037/bul0000030

Wiltshire, S. (1991). *Floating Cities.* London: Michael Joseph Ltd. ISBN: 9780718134808.

Wikipedia contributors. (2024). *Kim Peek.* In Wikipedia. Retrieved from https://en.wikipedia.org/w/index.php?title=Kim_Peek&oldid=1258913304

8

Consciousness

Section One: Foundations of Consciousness — Historical, Philosophical, and Theological Perspectives

Consciousness remains one of the most perplexing topics in philosophy, neuroscience, and psychology. What is consciousness? Is it a fundamental property of reality, or does it arise solely from the brain's complexity? Can the material world fully account for the richness of subjective experience? These enduring questions continue to elude definitive answers, especially within materialist frameworks, which struggle to account for how conscious awareness emerges from non-conscious matter. Even a cursory investigation reveals a more enigmatic phenomenon with every attempt to explain it. Throughout history, thinkers have wrestled with the nature of consciousness—often without using the term itself. René Descartes famously declared, *Cogito, ergo sum* ("I think, therefore I am"), acknowledging the profound link between thought and existence. Yet defining consciousness remains among humanity's most formidable intellectual challenges. Owing to its conceptual difficulty, the scientific community historically approached consciousness with caution, if not outright avoidance. The philosopher Thomas Nagel (1974) emphasized its subjective nature with his well-known thought experiment, "What is it like to be a bat?"—illustrating how first-person experience resists objective analysis. David Chalmers (1995) later introduced the now-influential distinction between the "easy problems" of consciousness—explaining functions like attention or memory—and the "hard problem": why and how physical processes in the brain give rise to qualitative, first-person experience.

Consciousness encompasses a mosaic of experiences—thoughts, emotions, sensations, intentions, and memories. Neurologists often quip that "everyone knows what consciousness is until they try to define it." Its inherently subjective nature, shaped by culture, language, belief systems, and introspection, renders it resistant to purely objective definitions. For instance, proponents of Integrated Information

Theory (Tononi, 2004) suggest that consciousness arises from the brain's ability to integrate complex information. While helpful, such models still grapple with the irreducible, qualitative features of experience. Drawing from the accumulated data and arguments presented throughout this book, our position aligns with Chalmers and others in affirming that consciousness cannot be fully explained by the material workings of the brain—even in principle. It is not merely a byproduct of neurobiology, but a phenomenon that points beyond materialist accounts. Neuroscientist Susan Greenfield (2002, 2016) underscores this complexity by proposing that researchers might better grasp what consciousness is by first identifying what it is not. She offers a pragmatic definition, describing consciousness as "what you lose when you go to sleep," or the "inner subjective state" that distinguishes waking awareness from unconsciousness (Greenfield, 2019). Her perspective reflects a broader trend in contemporary science: consciousness is often studied through its contrasts—between sleeping and waking, or awareness and unawareness—rather than through direct access.

Clinically, consciousness is understood along two functional dimensions: arousal, referring to a person's level of wakefulness or alertness, and awareness, referring to the richness and coherence of subjective experience. Arousal is assessed by responsiveness, while awareness encompasses content, such as perception, memory, and intentionality. Both components are essential to evaluating neurological states and disorders, and their integration bridges medical practice with broader philosophical inquiry. Julio Tononi captures the foundational role of consciousness in human experience: "Because it's everything we have, all we are... it is what disappears when we fall asleep and returns when we wake up" (Tononi, 2021). Tononi emphasizes that consciousness is not merely one function among others; it is the condition for experiencing anything at all—thoughts, emotions, relationships, identity, and worldview. Thus, any comprehensive account of human life must begin with consciousness itself.

The Enigma of Consciousness: Philosophical and Theological Reflections

The origin of consciousness has eluded philosophers, theologians, and scientists for centuries. In The Mysterious Flame: Conscious Minds in a Material World, Colin McGinn, a naturalist philosopher and leading proponent of "New Mysterianism," poignantly asks, "How can mere matter originate consciousness?" (McGinn 1999). This question strikes at the core of the mind-body problem and the enigma of consciousness, which continue to elude materialist explanation, whether reductive or emergentist. McGinn wonders how biological tissue, shaped by evolutionary processes, could give rise to the rich, subjective experiences that characterize human awareness. His analogy of turning water into wine underscores the perceived ontological leap from physical matter to subjective experience, as consciousness appears without any clear precedent in the physical universe (McGinn 1999).

The question of how consciousness arises from physical matter touches upon a deeper philosophical problem for the naturalist scientist or philosopher: if we are products of undirected natural processes shaped by evolution, where did our sense of self, agency, and continuity originate? Western intellectual traditions have long debated the nature of the soul and its relation to consciousness. They often present two seemingly irreconcilable perspectives, with dualistic views positing an immaterial essence central to human nature and materialist perspectives reducing consciousness to a byproduct of neural activity. This tension is evident in the West's historical efforts to bridge the gap between matter and consciousness. In the classical tradition, thinkers such as Plato and Aristotle grappled with this relationship, envisioning the soul as integral to the body's form and function yet distinct from mere physicality. Plato posited the soul as immortal, pre-existing the body and transcending death (Cooper 1997). Aristotle, the first to study biology systematically, approached the soul more pragmatically—as the organizing principle of life, linked to but not reducible to matter (Aristotle 1993). These foundational ideas laid the groundwork for enduring philosophical questions: Is consciousness a fundamental aspect of reality? What is the source of our deepest intuitions? Why are we drawn to ideals like love, beauty, and justice? And, perhaps most perplexingly, what is the nature of free will—if it exists at all?

The theological perspective offers another lens for exploring consciousness. The Gospel of John reveals a particularly profound link between matter and consciousness, and consciousness and its origin: "In the beginning was the Word (Logos)" (John 1:1). Here, the Logos—divine person and reason—is the personal organizing principle of the universe, shaping and infusing matter with meaning and intelligibility. This Christian understanding echoes Platonic and Aristotelian insights but situates the origin of consciousness within a more profound and encompassing divine framework. In this view, creation itself is an expression of encoded intelligence—a manifestation of the Creator's consciousness—reflected in the structure and order of the universe (Brown 2003). With these philosophical and theological foundations in view, it becomes clear that the enigma of consciousness has not been confined to Western thought alone. Across civilizations and centuries, diverse traditions have offered profound interpretations of consciousness—regarding it not as a mechanistic byproduct of physical processes, but as a primary feature of reality itself. Whether described as divine, luminous, eternal, or intrinsically unified with the cosmos, consciousness has remained central to the metaphysical and spiritual outlook of cultures around the world. To appreciate the depth and range of these perspectives, we now turn to a concise comparative view of how consciousness is understood across the major world religions.

Consciousness Across Cultures: An Irreducible Reality

Throughout history and across civilizations, consciousness has rarely been regarded as a mere byproduct of brain function. Instead, it is commonly understood as an irreducible dimension of reality—divine, transcendent, or fundamentally spiritual. While world traditions differ in their metaphysical and theological frameworks, they converge in rejecting purely materialist explanations and affirm consciousness as central to what it means to be human (Smart, 1999). In Hinduism, consciousness (*cit*) is not emergent from matter but a foundational aspect of reality itself (Radhakrishnan & Moore, 1957). The Upanishads equate individual consciousness (*Atman*) with Brahman, the universal ground of being. Nondualistic traditions like Advaita Vedānta describe liberation (*moksha*) as awakening to this unity—a realization that transcends ego and sensory illusion. In Buddhism, consciousness (*viññāṇa*) is seen as impermanent and conditioned, yet essential to the path of liberation. While early Buddhism emphasizes the non-self and the momentary nature of awareness, later traditions introduce deeper notions of luminous, primordial awareness (*rigpa*), which forms the ground of enlightenment.

In Islamic theology, consciousness is a gift from God, given for moral responsibility and guided by divine revelation. The soul (*nafs*) is created and tested by God, capable of recognizing truth and choosing righteousness. Rather than merging with a cosmic unity, Islamic thought emphasizes the soul's individuality, its accountability before God, and its eternal destiny through resurrection and judgment. In biblical Judaism, consciousness is grounded in the concept of the soul (*nephesh*), breathed into humanity by God. The Hebrew Scriptures depict human beings as living souls created in the divine image, capable of reason, choice, and covenantal relationship with God. Consciousness is not abstract or speculative but oriented toward obedience, worship, and moral responsibility. The Psalms, Wisdom literature, and prophetic writings reflect a deep awareness of the inner life, portraying the heart and mind as centers of discernment, longing, and spiritual encounter. Taoism approaches consciousness through harmony with the *Tao*—the source of all things. Rather than emphasize rational self-awareness, it values stillness, simplicity, and non-action (*wu wei*), cultivating an intuitive alignment with the natural order.

Despite their profound differences, these global traditions converge in affirming that consciousness is not reducible to neural computation or physical processes alone. It is variously described as luminous, sacred, enduring, or intrinsically united with a transcendent order. Within this broader panorama, Christianity offers a distinctive synthesis. It affirms the unity of body, soul, and spirit, grounding human consciousness in the *imago Dei*—the image of a rational, relational, and personal God. Conscious awareness, in this view, is not an illusion to be overcome or a spark to be absorbed, but a gift reflecting the divine capacity for knowledge, love, and freedom. Unlike perspectives that dissolve the self into a cosmic totality or impersonal flow, Christian theology affirms the enduring dignity of personal identity. In this view, consciousness is not a fleeting byproduct of nature but a vital expression of the person—one called into communion with the Creator and ultimately fulfilled through God's redemptive and sacrificial Incarnation and death. This personal entry into creation restores and elevates human nature, and through the Resurrection, confirms the promise of eternal life in glorified union with God.

Historical and Philosophical Roots of Consciousness

From Water to Fire: The Search for a Universal Principle

Thales, Anaximenes, and Heraclitus on the primal foundations of soul and mind.

The inquiry into consciousness in ancient Greek philosophy began with the Presocratics, who laid the foundation for understanding the mind and soul. Thales of Miletus (c. 625–c. 547 BCE) posited water as the fundamental substance of all things, linking life and consciousness to a primal element. Aristotle critiques this in *Metaphysics*: "Thales says that it [the nature of things] is water" (Metaph. 983b20), suggesting Thales drew this from moisture's nurturing properties (Metaph. 983b23-25). Thales also speculated on a universal soul, proposing that all things might possess an inherent form of consciousness. Building on this, Anaximenes of Miletus (c. 586/585–526/525 BCE) proposed air (*pneuma*) as the fundamental principle of the universe, transforming into various elements through condensation and rarefaction (DK13A5). He argued that air forms the soul and sustains life through breath, advancing early materialist ideas about consciousness. Heraclitus (c. 540–480 BCE), known for his philosophy of perpetual change, viewed fire as the symbol of transformation. He emphasized the soul's dynamic nature, likening consciousness to a state of flux and renewal (Kahn 1979). This perspective reinforced the idea of consciousness as an ever-changing process.

Other Presocratics, including Leucippus, Democritus, and Alcmaeon, leaned towards materialist views. Leucippus initiated the atomic theory, later developed by Democritus, positing that consciousness arises from atomic interactions. Diogenes of Apollonia and others advanced early panpsychism ideas, suggesting all things might possess some form of consciousness. David Chalmers (1996) echoes this in modern thought, defining panpsychism as the belief that "everything is conscious," while Popper and Eccles (1977) describe it as the view that "everything has a soul" or a rudiment thereof. Alcmaeon of Croton, a philosopher and physician, distinguished perception from understanding and located the mind in the brain. His work marked a significant step toward a biological view of consciousness influenced by observations of sensory modalities. The Hippocratic treatise *On the Sacred Disease* (c. 400 BCE) further solidified the connection between the brain and consciousness. It describes the brain as the "messenger of consciousness," responsible for cognitive processes: "From the brain arise our pleasures, joys, laughter... as well as our sorrows, pains, and griefs... Through it, we think, see, hear... and distinguish the ugly from the beautiful, the bad from the good" (*Hippocratic Corpus*). This early medical insight reflected a shift from purely philosophical speculation to empirical observation.

Greece: Classic Period

Plato and Aristotle on the Soul, Mind, and the Nature of Consciousness

The early Presocratic philosophers laid the groundwork for advanced explorations of consciousness, which reached their zenith during the Classical Period through the works of Plato and Aristotle. These two towering figures offered distinct but complementary perspectives on the nature of the soul and consciousness and their relationship to the physical world.

Plato: The Soul's Ascent to Eternal Truth

Consciousness as the soul's rational striving beyond the physical realm.

Plato viewed consciousness as intertwined with his theory of Forms, positing that the material world is merely a reflection of an eternal, unchanging realm of Ideas. For Plato, the soul (*psyche*) is immortal, pre-existing, and outlasting the body, striving to apprehend the truths of the Forms beyond the physical realm. His *Republic* (4.441d-442c) presents a tripartite model of the soul, dividing it into *Logos* (reason), *Thymos* (spirit), and *Eros* (desire). Reason, the highest faculty, governs emotion and appetite, ensuring harmony within the soul. In *Phaedo*, Plato expands on this dualistic vision, asserting the soul's independence from the body and its capacity for eternal truth. His *Allegory of the Cave* in *Republic* (7.514a-520a) vividly illustrates the soul's struggle to transcend sensory appearances and attain knowledge of actual reality. Shields (2016) observes, "Plato was the first philosopher to believe that the soul was both the source of life and the mind." While Plato focused on the soul's cognitive and rational functions, his ideas laid the foundation for future debates on the nature of consciousness.

Aristotle: Consciousness as the Embodied Form of Life

Aristotle, Plato's student, took a more empirical and integrative approach. In his *De Anima* (*On the Soul*), Aristotle describes the soul as the "form" of the body, a principle organizing and animating life. He viewed consciousness as emerging from the harmonious functioning of the body and intellect, inseparably tied to sensory and rational processes. Unlike Plato, Aristotle emphasized the soul's biological grounding, dividing life into three levels: the vegetative soul (nutrition and reproduction), the sensitive soul (sensation and movement), and the intellectual soul (reason and self-awareness). Aristotle also introduced concepts foundational to the study of the mind-body relationship, including *psychic* and *psychosomatic* processes. He famously wrote, "I think the soul and the body react one on the other by sympathy. A change in the psychic state produces a change in the form of the body, and inversely, a change in the form of the body produces a change in the state of the soul" (Shields 2016). This psychosomatic perspective captures Aristotle's belief in the interdependence of mental and physical states. In *On Sleep and Waking* (2.454a), Aristotle describes a unified faculty integrating sensory experiences, allowing one to perceive the unity of consciousness:

> *"There is a certain common capacity that supervenes on the others, by which one perceives that one is seeing and hearing. For it is surely not by sight that one sees that one sees… but rather by a certain part common to all the sense organs."*

This early concept of a central faculty resonates with contemporary discussions on brain-integrated functions, higher-order theories of consciousness, and the central thesis of this book. Caston (2002) notes, "Aristotle's views on consciousness (...) seem to cut down the middle of two contemporary debates: first, whether perceptual awareness is intrinsic to perception or relational (higher order); and second, whether perception itself instantiates perceptible qualities or merely represents them." The philosophical explorations of Plato and Aristotle continue to shape our understanding of consciousness. While Plato emphasized an immaterial, transcendent soul seeking eternal truths, Aristotle grounded consciousness in biological and sensory processes. Together, their ideas offer a rich framework for ongoing discussions of mind, body, and soul.

Christianity

The influence of Plato and Aristotle on Western thought is vast, shaping the intellectual frameworks of the three great monotheistic religions: Judaism, Christianity, and Islam. Within the Judeo-Christian tradition—central to Western civilization—the unity of the human person is a foundational tenet. While later chapters on *The Unity of the Human Person* and *Human Death* will explore this theme more fully, it suffices here to note that Christianity regards the human being as a composite of material and immaterial principles: spirit and matter. Christian anthropology affirms a "uni-totality" composed of three distinct yet unified dimensions—spirit, soul, and body (Kittel 1965; Boff 1980; Ratzinger 1991). In this framework, the resurrection of Christ—described as "the first fruit"—fully restores and reunites these three aspects. A body severed from soul and spirit is no longer a human person, but a mere corpse.

Thomas Aquinas further synthesized Aristotelian philosophy with Christian theology, articulating a vision of the human person as a unified whole in which spirit, mind—including consciousness—and body are inseparably integrated. Aquinas taught that the faculties of the soul encompass both spiritual and mental dimensions, functioning as the animating principle of the body. In *Summa Theologiae* (1266–1273, Pt. 1, Q. 76) and *Summa Contra Gentiles* (1259–1265, SCG.2 C68.2), he argued that soul and body are united as form and matter, constituting a single, unified substance. "The human soul," he wrote, "is an intellectual substance united to the body as its form"—meaning the soul is "the substantial form of the body." This hylomorphic view insists that humans are not souls trapped in bodies but are instead a profound unity of soul and body. Francisco Suárez (1548–1617) deepened this insight, noting that for the soul to be the true form of the body, it must be "the principle not only of thinking but also of the operations that are exercised by the body" (Suárez 1597, quoted in Moreland 2018). This underscores the integrated nature of mental and physical functions within a single human substance.

Contemporary Christian thinkers continue to affirm the immaterial dimension of consciousness. Sharon Dirckx, a Christian apologist with a PhD in brain imaging from Cambridge, offers a concise definition in *Am I Just My Brain?*, writing: "Consciousness is a property of the mind through which our subjective thoughts, feelings, experiences, and desires have their existence," and adding, "There are aspects of being human that are hard to account for if we are only our brains—our experience of consciousness, our sense of self, our experience of free will, and our capacity to reason" (Dirckx, 2019, pp. 89, 21). As Dirckx and others argue, the immaterial aspects of human existence—consciousness, reason, and will—point beyond materialism and align with a theistic worldview that sees humans as created in the image of God (*imago Dei*). C. S. Lewis echoes this conviction with characteristic clarity, noting that "If minds are wholly dependent on brains, and brains on biochemistry, and biochemistry (in the long run) on the meaningless flux of the atoms, I cannot understand how the thought of those minds should have any more significance than the sound of the wind in the trees" (Lewis, 1970, p. 52). For Lewis, the very capacity for rational reflection undermines the credibility of a purely materialistic explanation, reinforcing the claim that mind and meaning must ultimately be grounded in something beyond the physical. William Lane Craig likewise emphasizes: "On the Christian view, the self is a soul which is intimately connected with a body." (Craig 2008). Philosopher Leonore Stump echoes this idea with poetic force, describing the human soul as the amphibian of the metaphysical world, dwelling simultaneously in both biological and spiritual realms (Stump, 2017). For Peter Kreeft, this embodied soulhood grounds the intrinsic dignity of the person: our capacity to think, love, and choose reflects our deepest nature as beings made in the image of a God who is pure Spirit (Kreeft, 2007). On this view, consciousness is not a fleeting byproduct of neural activity but a reflection of the divine image, carrying profound implications for moral responsibility and personal identity.

J. P. Moreland reinforces this perspective by arguing that consciousness cannot be reduced to purely physical processes. In *The Soul* (2014), he maintains that conscious states are intrinsically different from brain states and therefore cannot arise from them alone. One of his key arguments draws on the philosophical concept of conceivability. Moreland contends that if we can *coherently conceive* of consciousness existing independently of the brain—as in cases of near-death experiences (NDEs)—then it follows that consciousness, and the brain are not identical. Importantly, he does not invoke the factual reality of NDEs as empirical evidence, but rather their logical possibility as a philosophical consideration. If consciousness were truly identical to brain function, Moreland argues, the notion of a disembodied conscious state would be logically inconceivable. Yet since such states are imaginable, whether through NDEs or depictions of disembodied minds, this suggests that consciousness possesses properties not reducible to physical brain states. As he writes: "If I can conceive of myself existing apart from my body or brain, then I am not identical to either" (Moreland, 2014). This argument resonates with the broader Christian tradition and with Aristotelian-Thomistic metaphysics, which hold that the soul both animates the body and transcends it. For Aquinas, as for Moreland, consciousness is a faculty of the soul, and the soul's existence endures beyond the dissolution of the physical body. The Christian doctrine of the soul, therefore, plays a central role in understanding human identity—not only in life, but also in death and resurrection. Later chapters—including *The Unity of the Human Person*, *Human Death*, and *The Soul and the Brain Metafunction*—will revisit these themes in greater depth.

Contemporary Theist Perspectives and Neuroscience

Many contemporary thinkers and apologists have contributed significantly to the dialogue between neuroscience and theology, each offering distinctive insights into the nature of consciousness and human identity. J.P. Moreland, for instance, emphasizes the irreducibility of consciousness and the immaterial nature of the soul, challenging physicalist accounts of human cognition. In works such as *The Soul: How We Know It's Real and Why It Matters*, he defends the existence of a transphysical reality underpinning conscious experience (Moreland, 2014). Neurosurgeon Michael Egnor critiques materialism in neuroscience by highlighting the insufficiency of brain-based explanations for consciousness and free will. He offers a scientifically informed defense of theistic dualism (Egnor, 2024). Similarly, neuroscientist Sharon Dirckx addresses questions of personal identity and volition in her book *Am I Just My Brain?* where she translates complex neuroscientific concepts into accessible language while defending a Christian view of personhood (Dirckx, 2019). In contrast, theologian Nancey Murphy advocates for nonreductive physicalism, proposing a model that integrates neuroscience with theological anthropology while rejecting classical dualism. Her book *Bodies and Souls, or Spirited Bodies?* argues for a holistic understanding of human nature that retains theological integrity (Murphy, 2006).

Other thinkers engage the intersection of neuroscience, philosophy, and theology from interdisciplinary standpoints. Robert Spitzer combines metaphysical reasoning with empirical evidence in The Soul's Upward Yearning, where he explores the compatibility of neuroscience with the human longing for transcendence (Spitzer, 2015). Mario Beauregard, in The Spiritual Brain, uses neuroscientific data to challenge materialist interpretations of consciousness and support the existence of the soul (Beauregard & O'Leary, 2007). Angus Menuge critiques materialism's inability to account for human rationality and moral agency in *Agents Under Fire,* framing neuroscience as a gateway to understanding the divine image in humanity (Menuge, 2004). Taliaferro, Jeeves, and Brown contribute through philosophical and psychological perspectives, respectively. Taliaferro's *Consciousness and the Mind of God* explores the divine foundations of consciousness (Taliaferro, 1994), while Jeeves and Brown's *Neuroscience, Psychology, and Religion* integrates cognitive science with Christian theology (Jeeves & Brown, 2009). Finally, Andrew Newberg's research in neurotheology—especially in *How God Changes Your Brain*—examines the neurological basis of spiritual experiences, offering scientific insights that resonate with Christian apologetics (Newberg & Waldman, 2009). Although Newberg does not adhere to a specific religious tradition, his work reflects a growing interest in the intersection of brain science and theology. Together, these diverse perspectives enrich the conversation between neuroscience and faith, presenting robust challenges to materialism and affirming a holistic view of the human person—one that embraces both the physical and the spiritual dimensions of consciousness.

Integral Christian Perspective: Brain Metafunctional and Transphysical Unity

Christian thinkers such as C.S. Lewis, Robert Spitzer, William Lane Craig, Peter Kreeft, and others have long emphasized the existence of a *transphysical domain*—a dimension of reality that transcends the material operations of the human brain while affirming the *unity of the human person*. This view holds that consciousness, with its irreducible features—such as intentionality, subjectivity, and transcendence—points toward aspects of human existence that cannot be fully captured by physical processes (Craig and Moreland 2003). The unity of consciousness, which integrates diverse sensory inputs and mental activities into a coherent whole, offers compelling evidence for a reality beyond material structures. In *The Problem of Pain* and *Mere Christianity,* C.S. Lewis critiques materialism's explanatory inadequacy, portraying humanity as a dynamic union of physical and spiritual realities (Lewis 1940, 1952).

Robert Spitzer expands on this understanding in *The Soul's Upward Yearning,* where he argues that human consciousness reveals an intrinsic orientation toward transcendence. Drawing on both metaphysical reasoning and empirical evidence—including near-death experiences (NDEs)—Spitzer contends that this orientation reflects a transphysical aspect of the human person that persists through bodily changes (Spitzer 2015). William Lane Craig likewise highlights the enduring nature of personal identity, suggesting that the unity and persistence of the self imply a metaphysical foundation deeper than biology alone can explain (Craig and Moreland 2003). Peter Kreeft, in *The Philosophy of Tolkien,* explores this idea symbolically, presenting human life as a journey between the material and the transcendent (Kreeft 2005).

Thinkers such as Sharon Dirckx and J.P. Moreland further extend these insights into the neuroscientific domain, arguing that consciousness, reason, and will point to the presence of an immaterial mind irreducible to neural processes (Dirckx 2019; Moreland 2014). Their arguments reinforce the idea that a purely materialist account fails to capture the full richness of human subjectivity and freedom. Importantly, the transphysical is not a disembodied abstraction but a reality intimately linked to the human person's unity. It enables the integration of consciousness, identity, moral agency, and bodily functioning, all of which are prerequisites for free will and ethical responsibility. This integrated unity aligns with the Christian understanding of the *Imago Dei*—the belief that humans are created in the image of a rational and personal God (Lewis 1940; Spitzer 2015). Affirming the transphysical dimension challenges reductive physicalism, which struggles to explain key features of human experience, such as *intentionality, moral awareness,* and *intrinsic dignity* (Craig and Moreland 2003). Across centuries of theological and philosophical inquiry, Christian thought has consistently affirmed that the human person is best understood through the *integration of physical and transphysical dimensions.* This holistic view resists both dualistic fragmentation and reductive materialism, instead emphasizing the indivisible wholeness of the person in both temporal and eternal contexts. The chapter *The Human Soul and the Brain Metafunction* explores this view in greater depth under the concept of *Brain Metafunction Transphysicalism.*

The Enlightenment

France: René Descartes and the Birth of Dualism

The Enlightenment, or *Age of Reason,* was an intellectual and philosophical movement in Europe during the 17th and 18th centuries that exerted lasting global influence. In France, René Descartes emerged as a foundational figure in modern philosophy. He proposed that the mind and body are distinct substances: *consciousness* defines the immaterial mind, while *extension*—the property of occupying space—characterizes material reality. Descartes famously asserted, *Cogito, ergo sum* ("I think, therefore I am"), suggesting that self-awareness is the foundational certainty of human existence (Descartes 1637, 1641). In Descartes's dualist framework, the human being is composed of two separate substances—mind and body—a sharp departure from the integrated anthropology of classical Christianity. He posited that the mind is a non-physical, thinking substance unique to humans, while the body operates according to mechanistic laws. Animals, in his view, lacked minds and functioned as automatons. To explain how the immaterial mind could interact with the physical body, Descartes proposed a causal connection centered on the *pineal gland* as the site of this union.

Many credit Descartes with formally introducing the mind-body problem, though his solution invited immediate critique. One of the most notable early objections came from Princess Elisabeth of Bohemia, who, in a 1643 letter, questioned how an immaterial substance could exert causal influence on a physical one:

"[...] how the soul of a human being (it being only a thinking substance) can determine the bodily spirits, in order to bring about voluntary actions. [...] Physical contact is required [...] but the notion of soul you have entirely excludes extension, and physical contact seems incompatible with something immaterial" (Descartes 2007).

This exchange inaugurated centuries of philosophical exploration into how consciousness relates to the body—an inquiry that remains central to contemporary debates in philosophy of mind.

England: John Locke and the Continuity of Consciousness

John Locke made influential contributions to consciousness studies, particularly through his *Essay Concerning Human Understanding* (Locke, 1690). Locke was one of the first to connect consciousness to personal identity, arguing that continuity of self is grounded not in a metaphysical soul but in the persistence of memory and self-awareness over time. For Locke, what makes someone the same person across time is the ability to remember past experiences as one's own. This theory of psychological continuity became a precursor to modern discussions on identity and the self. Locke defined consciousness as the reflective awareness of one's thoughts, sensations, and mental operations, distinguishing it from automatic or unconscious states. His famous metaphor of the mind as a blank slate (tabula rasa) underscored the idea that consciousness and knowledge arise from experience rather than innate ideas. Through sensation (external perception) and reflection (internal awareness), ideas are formed. Crucially, Locke also separated consciousness from the soul, arguing that personal identity does not require the soul's immaterial substance—what matters is the continuity of consciousness (Sauchelli 2017). These ideas profoundly shaped Enlightenment thought and continue to influence debates on memory, personal identity, and cognition.

Post-Enlightenment German Philosophy

Kant: Transcendental Consciousness and Moral Autonomy

Immanuel Kant (1724–1804) proposed that consciousness is not merely a passive reflection of external reality but an active process involving perception and interpretation. According to Kant, the mind imposes specific conceptual frameworks on sensory inputs to organize and construct a coherent understanding of the world. He further suggests that consciousness entails an awareness of one's own mental representations, where one representation is recognized as occurring within the mind alongside another (Kant [Young 1992]). Kant also maintains that our awareness of most representations is neither immediate nor direct. His model of the mind, which significantly influenced empirical psychology, experienced a notable resurgence in the late 20th century, particularly within cognitive science. Three key ideas delineate the cognitive architecture of Kant's model (Kitcher 1993), subsequently becoming foundational to cognitive science:

1. The mind, according to Kant, comprises a complex set of faculties, anticipating certain elements of functionalist thought, though distinct from the empirical functionalism formally introduced in the 1960s.
2. Mental activity crucially involves spatiotemporal processing and the application of concepts to sensory inputs, which Kant identifies as a priori conditions necessary for coherent experience.
3. Synthesis, a central concept in Kant's theory, plays a pivotal role in cognition by actively combining sensory data into coherent experience. This process depends on the transcendental unity of apperception, the self's awareness of its own mental representations, which ensures the unified nature of conscious experience.

Kant's transcendental method, a cornerstone of his study of the mind, continues to influence contemporary cognitive science. This method involves identifying the *a priori conditions* that must exist for experience to be possible—a technique that bears some resemblance to the modern inference to the best explanation. In this sense, the transcendental method entails postulating unobservable mental structures to account for the conditions enabling cognition—a methodology that resonates with contemporary neuroscience's efforts to understand the mind's underlying mechanisms.

Kant's renowned statement from the Critique of Practical Reason encapsulates this perspective: *"Two things fill my mind with ever new and increasing admiration and reverence...: the starry heavens above me and the moral law within me."* This reflection highlights Kant's belief that both the vastness of the cosmos and the inner moral compass reveal humanity's distinctive position as rational and moral beings. While the grandeur of the starry heavens inspires awe through its immensity and sublimity, the moral law evokes reverence by revealing humanity's unique capacity for moral self-governance and rational agency. This interplay between external grandeur and internal moral awareness underscores Kant's view of the human person as standing between the immensity of nature and the dignity of moral responsibility. Kant's ethical framework is closely linked to his transcendental idealism, which holds that objective experience is not passively received but actively constituted by the mind's a priori structures (Allison 1983). This active role of the mind shapes how individuals understand the world and their place within it. Central to Kant's moral philosophy is the concept of moral autonomy, which emphasizes the individual's role as a self-governing agent guided by rational moral principles. Kant famously asserts that humans must be treated as ends in themselves, never merely as means — affirming a moral imperative that upholds each person's inherent worth and dignity. Kant's integrated framework — combining a moral philosophy rooted in reason, an epistemological model grounded in the conditions of experience, and a profound respect for human autonomy — continues to shape contemporary discussions in ethics and cognitive science (Allison 2011).

Central to Kant's philosophy of consciousness is the concept of conditions of possibility — the foundational prerequisites that enable understanding and experience. This idea, illustrated in his exploration of the conditions of possible experience, asserts that comprehending both the world and various forms of consciousness depends on identifying these prior conditions. Investigating such conditions is essential for grasping the complexities of consciousness itself. Kant's assertion that the human mind actively constitutes objective experience highlights consciousness's dynamic role in shaping our perception of reality. This understanding of consciousness extends into Kant's moral philosophy, where he emphasizes moral autonomy as fundamental to human identity (Strawson 1966). For Kant, individuals are self-governing agents who act according to rational moral principles rather than being driven by external influences or mere desires. This autonomy arises from the unique capacity of rational beings to recognize and adhere to moral laws they impose upon themselves — a core idea embodied in Kant's categorical imperative.

Moral autonomy is not merely a component of ethical behavior; it defines what it means to be human. For Kant, reason and self-determination distinguish individuals as moral agents. Consciousness plays a pivotal role in this process. Far from being passive awareness, consciousness is an active process that organizes experience. This rational consciousness — which constructs coherent experience — also enables individuals to reflect on their actions, recognize moral obligations, and freely choose to follow moral principles. Consequently, Kant's emphasis on moral autonomy underscores a deep connection between consciousness and ethical responsibility. Since moral autonomy depends on the capacity for rational reflection, consciousness itself becomes the foundation for moral self-awareness and the ability to engage in responsible, principled action. Thus, in Kant's view, consciousness is not only a cognitive function but also central to humanity's moral and rational identity.

Hegel: Dialectics and Absolute Consciousness

Georg Wilhelm Friedrich Hegel (1770–1831) conceived of consciousness as a dynamic, developmental process. Consciousness is not static but unfolds over time, moving toward greater self-awareness through dialectical tension. At each stage, contradictions are resolved through synthesis, producing higher levels of understanding (Hegel [Miller] 1977). For Hegel, reality itself is mind-dependent: true knowledge arises when the self recognizes that the external world is shaped by its own cognitive processes. His method of dialectics differed from Plato's dialogical approach by emphasizing the internal movement of ideas, leading to what Hegel called absolute knowledge—a stage where consciousness grasps itself as the ground of all being.

Hegel described a hierarchy of consciousness:

1. Simple Consciousness – undifferentiated awareness.
2. Self-Consciousness – recognition of oneself as a subject.
3. Reason – integration of self and world into rational totality.

The Hegelian concept of Absolute Consciousness posits that ultimate awareness, identified with the divine. For Hegel God's ultimate nature is rational self-awareness and manifests itself progressively through different stages of revelation. According to Hegel, this unveiling occurs sequentially, starting with Simple Consciousness, advancing to Self-Consciousness, and culminating in reason. Hegel's elaboration on the progression of absolute consciousness highlights a developmental hierarchy in its revelation to humans. Simple consciousness represents an initial stage where consciousness is relatively primary and undifferentiated. As it develops, Self-Consciousness emerges, reflecting a more intricate self-awareness. Finally, reaching the pinnacle occurs in the stage of reason, where consciousness attains a higher level of rationality and understanding. Hegel's emphasis on the mind's role in shaping reality remains a pivotal contribution to modern philosophical thought. Hegel's dialectics emphasizes the internal tension and synthesis of conflicting ideas to achieve higher understanding. Hegel's dialectical method profoundly influenced 19th-century materialist thinkers, who adopted its structure while rejecting Hegel's idealism.

Marx: Historical Materialism and Consciousness Under Constraint

Karl Marx (1818–1883) developed a materialist theory of consciousness that departed sharply from idealist traditions. Influenced by Ludwig Feuerbach's critique of religion, Marx asserted in The German Ideology (1846) that "life determines consciousness," not the other way around. Consciousness, he argued, reflects the material and economic conditions under which people live (Marx and Engels [Arthur] 1970; Marx [Tucker] 1978). According to Marx, religion, philosophy, and ideology are part of a superstructure built atop an economic base. Transforming human consciousness thus requires changing material conditions—achieved through revolutionary struggle and the abolition of exploitative class systems. Marx's theory seeks to explain how economic realities shape consciousness, but it leaves little room for ideas as autonomous drivers of human agency. Ironically, when applied by twentieth-century regimes, this framework—originally aimed at human liberation—contributed to widespread cultural suppression, subordinating art, philosophy, science, religion, and, ironically, the economy itself to the ideological aims of totalitarian systems. The result was not emancipation, but a new form of domination—justified as a "historical necessity" and enforced through centralized control. Alongside the suppression of creativity and dissent came a diminished view of personhood, where individual dignity was sacrificed to collectivist ideals, and spiritual expression was regarded as subversive or obsolete. In seeking to liberate humanity from economic bondage, Marxist systems consistently imposed a far deeper metaphysical reductionism: one that denied the transcendent dimension of the human person. In contrast, the Brain Metafunction Framework affirms both the material embeddedness and the transphysical openness of the human person—preserving agency, creativity, and spiritual depth by recognizing the person as more than a product of historical forces or neural configurations.

Marx's formulation of the superstructure as determined by the material base oversimplifies what is, in truth, a far more dynamic and reciprocal relationship. While economic and social conditions undoubtedly shape ideological expressions, ideas themselves also exert real, formative influence on economic systems and political structures. Moral visions, philosophical insights, and spiritual frameworks are not passive reflections of material conditions; they shape human behavior, institutions, and historical direction. Cultural meaning and metaphysical belief are not epiphenomena—they are, at times, revolutionary forces. This bidirectional interaction between material and intellectual realms suggests a feedback loop between base and superstructure, rather than a one-way determinism. Consciousness, accordingly, is not merely reactive but interpretive, moral, and purposive. It reaches beyond the material toward a higher order of meaning, revealing dimensions of human life that cannot be explained solely through historical materialism. Herein lies a core tension within Marxist theory: while it seeks to transform society through ideological critique, it simultaneously denies the autonomy and generative power of ideology by subordinating all cultural, moral, and spiritual dimensions to economic determinism. In absolutizing the material base, Marxism undermines the very agency it relies on to inspire revolutionary consciousness—an agency that presupposes values, vision, and moral imagination.

From a Christian theistic perspective, this contradiction further illustrates the limits of reductionist frameworks. It invites a more holistic anthropology, in which economic realities are important but not ultimate. Human consciousness, marked by reason, will, and spiritual aspiration, reflects not only historical conditioning but the imago Dei—the image of a Creator who is both rational and personal. In this view, the dignity and moral agency of the person are grounded not in material forces alone but in a transcendent reality that enables freedom, creativity, and discernment.

Technological Materialism: The New Challenge to Consciouness

In the 21st century, a new form of materialism has emerged, not through violent revolution, but through a convergence of technological systems, cultural trends, ideological narratives, and top-down institutional influence. Neo-Marxist frameworks have resurfaced in the context of AI, automation, and digital surveillance—raising fresh concerns about the shaping of consciousness. These tools, while often presented as instruments of progress, risk constructing a hyper-centralized system in which individuals are reduced to data points and stripped of moral, spiritual, and cognitive agency. Social credit systems, algorithmic propaganda, and digital control infrastructures may enforce ideological conformity with unprecedented precision (Zuboff 2019; Noble 2018). Though cloaked in the language of efficiency and innovation, these systems often replicate the very patterns of centralization and control seen in authoritarian regimes. This raises a crucial question: Are today's technologies serving consciousness, or subtly diminishing it? Without robust ethical and philosophical safeguards, such systems could undermine creativity, freedom, and the inherent dignity of the person. While Marx's critique of economic injustice retains ethical resonance, any contemporary application of his ideas must be critically reassessed to avoid repeating totalitarian patterns in new digital forms. A more adequate vision of consciousness must account not only for material conditions but for the moral, spiritual, and transphysical dimensions that make human flourishing possible.

Marxist materialism and transhumanism share a common impulse: the belief that human nature can—and should—be radically transformed through material means, whether through revolutionary restructuring of society or technological enhancement of the self. Transhumanism, often framed as a promise of transcendence through technology, carries with it a subtle but profound danger: the allure of a form of "redemption" that seeks liberation not through spiritual transformation, but through escape from embodiment itself. In attempting to overcome the limits of the human condition—death, weakness, dependency—it risks rejecting what is most essential to our humanity: relationality, moral responsibility, and the integrative unity of body, soul, and spirit. The transhumanist vision of consciousness uploaded, extended, or merged with artificial intelligence often implies that meaning and identity can be preserved apart from lived embodiment and spiritual depth. But this disembodied ideal, far from elevating the human person, may in fact hollow it out—substituting simulation for selfhood and control for communion. It is precisely our embodied vulnerability, embeddedness in moral relationships, and openness to the transcendent that ground the dignity of the person. Any technological vision that severs consciousness from these roots may offer not redemption, but alienation disguised as advancement, it risks turning technology into a controlling idol.These themes are explored further in the chapters on Bioethics and Neuroethics.

From Foundations to Frontiers: Bridging Metaphysics and Neuroscience

Having traced the unfolding of thought on consciousness—from ancient philosophy through Enlightenment rationalism to modern theist and materialist paradigms—we arrive at the threshold of contemporary science. While Section One established the historical, philosophical, and theological foundations of consciousness—highlighting its irreducibility to material explanation—Section Two turns to the landscape of modern neuroscience. Here, we examine how emerging models of consciousness confront the "hard problem," revisit metaphysical questions using empirical tools, and grapple with the enduring mystery of subjective experience. These efforts suggest that far from rendering earlier insights obsolete, contemporary science may point to the necessity of expanding our frameworks—potentially bridging the physical and the transphysical. In what follows, we continue the search for a more integrated, ethically grounded, and philosophically complete account of the conscious self.

Having followed the historical, philosophical, and theological currents that shaped our understanding of consciousness, we now turn to the contemporary frontier—where neuroscience and integrative theories grapple with the same enduring mysteries using empirical tools and interdisciplinary frameworks. While Section One highlighted the irreducibility of consciousness to material explanation and the profound metaphysical questions surrounding the soul, mind, and personhood, Section Two explores how scientific inquiry has approached these questions through neurobiological research, computational models, and philosophical reflection. In the pages ahead, we will examine the advances and limitations of contemporary neuroscience, the search for the neural correlates of consciousness, and emerging models that seek to bridge the explanatory gap. These explorations do not merely continue the discussion—they challenge us to reassess the boundaries between the physical and the transphysical and to consider whether a new paradigm may be required to fully understand the nature of conscious life.

9

Consciousness

Section Two: Frontiers of Consciousness — Neuroscientific Insights and Integrative Theories

Consciousness in Contemporary Neuroscience

The "explanatory gap" and the "hard problem of consciousness."

Advances in neuroscience—particularly through imaging techniques like fMRI and EEG—have enabled researchers to correlate specific patterns of neural activity with cognitive functions, offering important insights into how the brain operates. Yet despite these technological breakthroughs, the mind-body problem—the question of how mental phenomena relate to physical brain processes—remains unresolved. Indeed, as neuroscience advances, the problem has become even more pronounced. Two key philosophical concepts underscore this challenge: the "explanatory gap," introduced by Joseph Levine (1983), and the "hard problem of consciousness," articulated by David Chalmers (1995). While related, they highlight different dimensions of the mystery. The explanatory gap refers to the difficulty of connecting subjective, qualitative experience (qualia) with the objective, quantifiable mechanisms of the brain. The hard problem, by contrast, asks why any neural activity should give rise to conscious experience at all—and why specific brain states are accompanied by particular subjective states. While the so-called "easy problems" of consciousness—such as understanding perception, memory, and behavior—can often be addressed by identifying neural mechanisms, the hard problem confronts the deeper, ontological mystery: why does consciousness exist in the first place?

Many neuroscientists now acknowledge that, despite progress in mapping the neural correlates of consciousness (NCCs), a complete scientific explanation remains elusive. Francis Crick and Christof Koch, pioneers in the field, have both noted the difficulty of bridging the gap between neural activity and conscious experience. Theories such as Global Workspace Theory and Integrated Information Theory offer valuable frameworks, but they stop short of resolving the fundamental mystery. As Chalmers (1995) observed, the field lacks even a germinal idea for how subjective experience might arise from objective processes—highlighting the depth of the challenge and the limitations of current explanatory models. This growing recognition of anomalies within physicalist paradigms suggests that neuroscience may be approaching a critical turning point. Thomas Kuhn's theory of scientific revolutions offers a helpful lens: Kuhn (1962) argued that the accumulation of unresolved anomalies within a dominant framework often signals the need for a paradigm shift. As with previous scientific revolutions, the ongoing failure to integrate subjectivity into objective models of brain function may require fundamentally new concepts and methodologies. Emerging integrative approaches, especially those that move beyond reductive materialism, may offer a more adequate framework for consciousness, one capable of accounting not only for cognitive functions but for the full richness and unity of human experience. These perspectives may help confront the deep mysteries posed by the hard problem and the explanatory gap, while gesturing toward a broader vision of the conscious self.

Features of Consciousness

Before examining recent theories of consciousness, it is essential to consider the core features recognized by contemporary philosophers and neuroscientists. These features form the foundational criteria for exploring and evaluating competing models of consciousness. Key aspects such as subjectivity, qualitative character, phenomenal structure, self-perspectival organization, unity, intentionality, transparency, and dynamic flow offer a comprehensive framework for understanding the richness of conscious experience (Chalmers 1996; Damasio 1999; Edelman 2004). By grounding research in these features, scholars establish a shared basis for meaningful dialogue and critical comparison across diverse theoretical perspectives. Researchers emphasize the importance of integrating both first-person and third-person data. First-person data encompass subjective experiences—the internal, personal dimension of consciousness—while third-person data involve observable behaviors and measurable neural processes. Together, these approaches aim to provide a more holistic and empirically anchored account of consciousness

.

Subjectivity highlights the unique, first-person nature of experience—how each individual perceives the world from their own vantage point. The *qualitative character* of consciousness refers to the specific, felt qualities of experience, often described in terms of "what it is like" to see a color, hear a sound, or feel an emotion. *Phenomenal structure* captures the way experiences are organized and related to one another, shaping the cohesive fabric of awareness. Consciousness is also inherently *self-perspectival*: it is structured around a first-person point of view, anchored in the perceiving subject's sense of self (Gallagher 2000). This self-referential dimension is a defining feature of conscious awareness. A further hallmark is *dynamic flow*—the continuous unfolding of experience over time (Varela, Thompson, and Rosch 1991). *Intentionality* refers to the "aboutness" of consciousness: its ability to be directed toward objects, thoughts, or states of affairs (Brentano 1874 [1995]; Searle 1983). *Transparency* underscores the immediacy with which conscious contents present themselves—they appear directly available to the subject, without awareness of the underlying processes generating them (Metzinger 2003). Finally, *unity* reflects the integrated character of consciousness: the way disparate elements—perceptions, emotions, thoughts—are woven into a seamless and coherent whole (Bayne and Chalmers 2003).

Theories of consciousness

Understanding the diverse landscape of theories of consciousness is essential for engaging in contemporary debates in the field. Emerging from both philosophical reflection and neuroscientific research, these theories offer varied frameworks for examining the nature of conscious experience. From higher-order and representationalist models to neural and quantum theories, these approaches illustrate the complexity of studying consciousness and provide complementary insights. This chapter lays the groundwork for a deeper exploration of these theories, offering context for evaluating their contributions as well as their limitations.

Higher-order theories propose that conscious mental states arise when we have higher-order thoughts—that is, thoughts about our own mental states (Rosenthal 2005). Reflexive theories, such as Uriah Kriegel's Self-Representational Theory, suggest that consciousness emerges when a mental state is simultaneously aware of itself as part of its own structure (Kriegel 2009). Representationalist theories argue that consciousness is grounded in mental representations that encode rich intentional content about the external world, thus bridging perception and subjective experience (Dretske 1995). Narrative theories, such as Daniel Dennett's model, posit that consciousness arises from the mind's construction of a self-narrative, whereby experiences are organized through interpretive processes resembling storytelling (Dennett 1991). Information Integration Theory (IIT) posits that consciousness results from the brain's capacity to integrate vast quantities of information into a unified experience (Tononi 2008). Neural theories, exemplified by the work of Christof Koch (2004) and Antonio Damasio (1999), focus on the biological and functional underpinnings of consciousness. While Koch emphasizes neural correlates of conscious states, Damasio integrates emotional and bodily processes, highlighting the interplay between neurological structures and cognitive-affective functions.

Quantum and Non-Physicalist Theories

Quantum theories, such as Penrose and Hameroff's Orchestrated Objective Reduction (Orch-OR) theory (1996), explore the potential role of quantum phenomena in the emergence of consciousness. While often criticized for their speculative character and lack of empirical support, these theories raise provocative questions about the fundamental relationship between mind and matter—questions that remain unresolved within classical neuroscience and philosophy of mind. In contrast, non-physicalist theories explicitly challenge materialist paradigms by positing that consciousness transcends physical processes. These views argue that subjective experience cannot be fully accounted for by neural mechanisms alone and must be situated within a broader ontological framework. Such perspectives resonate with the book's transphysical position, which holds that consciousness cannot be reduced to material processes but reflects a dynamic cooperation between physical mechanisms and transphysical dimensions of reality. This approach is not merely metaphysical; it seeks to offer a more adequate account of the richness and diversity of conscious experience. Together, quantum and non-physicalist theories expand the boundaries of contemporary consciousness studies. They underscore the need for integrative frameworks that move beyond reductionism while remaining open to philosophical and scientific dialogue. This overview sets the stage for a deeper examination of these approaches in the chapters that follow. By grounding the discussion in both philosophical and neuroscientific inquiry, this book aims to present a comprehensive view of the enigma of consciousness—one that ultimately supports a brain-metafunctional and transphysical paradigm.

Specific Theories of Consciousness

Theories of Consciousness: From Correlates to the Hard Problem

Before examining specific theories of consciousness, it is important to clarify their scope—and what they are not designed to address. These theories primarily aim to explain the neural correlates of consciousness and the cognitive mechanisms that underlie mental functions. However, they do not attempt to resolve the "hard problem" of consciousness, a term popularized by philosopher David Chalmers. In his influential 1995 paper *Facing Up to the Problem of Consciousness* and the subsequent book *The Conscious Mind* (1996), Chalmers distinguishes between the "easy problems" and the "hard problem" of consciousness. The former concerns how the brain performs cognitive functions like object recognition, memory consolidation, and verbal report—challenges that are tractable through empirical neuroscience. The hard problem, by contrast, asks why and how subjective experience arises at all, why there is something it is like to be a conscious organism (Nagel 1974; Levine 1983). To sharpen this challenge, Chalmers introduced the well-known *philosophical zombie* thought experiment: a hypothetical being identical to a human in every physical respect but entirely lacking subjective experience. The logical conceivability of such a being, he argues, reveals that physical properties alone are insufficient to explain consciousness. From this, Chalmers infers the existence of what he calls "further facts"—aspects of reality that transcend physicalist explanation (Chalmers 1996). He proposes the possible existence of psychophysical laws, principles that might govern the relationship between physical processes and specific qualitative experiences (Jackson 1982; Chalmers 1997).

Building on this critique, Chalmers has explored speculative frameworks such as *panpsychism* and *panprotopsychism*. Panpsychism maintains that consciousness is a fundamental and ubiquitous feature of the universe, present in all entities regardless of complexity (Strawson 2006; Goff 2017). Panprotopsychism, on the other hand, suggests that proto-conscious properties—what Chalmers (1996) calls *"protophenomenal properties"*—exist as precursors to full-blown consciousness. While panpsychism posits a universal presence of conscious experience, panprotopsychism leaves open the possibility that only specific configurations of matter give rise to subjective awareness. Through these contributions, Chalmers has significantly reshaped both philosophical and scientific discourse on the nature of consciousness. By clarifying the limits of physicalist theories, distinguishing the hard problem, and advancing speculative models, he calls for a more comprehensive framework capable of accounting for the richness of conscious experience. These considerations form a critical philosophical backdrop for the neuroscientific theories that follow. While valuable in addressing the mechanisms of conscious states, these models remain focused on the domain of the "easy problems"—and do not, by design, and intrinsic limitations, resolve the deeper mystery of the hard problem itself.

Wilder Penfield's Centroencephalic Theory

Wilder Penfield, the Canadian neurosurgeon renowned for mapping brain functions through electrical stimulation, initially believed that consciousness could be fully explained by brain mechanisms. However, his clinical experience, particularly during epilepsy surgeries on awake patients, led him to reconsider this assumption. Penfield found that while cortical stimulation could reliably evoke sensations, movements, or memories, it never elicited willful decision-making or a sense of personal agency—both central features of conscious experience. These observations prompted him to question whether the physical structures of the brain alone could fully account for the mind. In response, Penfield developed what he called the *centroencephalic theory,* positing that consciousness depends on a dynamic interaction between cortical and subcortical structures, especially those located in the brainstem—what he termed the centroencephalic system. According to his model, this system mediates wakefulness and alertness, acting as the integrative core that enables conscious experience.

Between 1952 and 1958, Penfield identified key components of this system within the diencephalon, midbrain, and pons. Later, he suggested that the basal ganglia might also contribute to its functions (Penfield 1975). His theory marked a departure from the materialist frameworks dominant in his time, eventually leading him to adopt a form of dualism. While he acknowledged that brain processes are necessary for consciousness, he argued they are not sufficient to explain the full reality of the mind—suggesting instead that consciousness may include non-physical dimensions. Subsequent work by Thompson, Crinella, and Yu (1990) expanded on Penfield's ideas by offering a more detailed map of the centroencephalic system. They identified several candidate structures, including the dorsal caudate putamen, globus pallidus, ventrolateral thalamus, substantia nigra, ventral tegmental area, superior colliculus, median raphe, and pontine reticular formation. Notably, they acknowledged that the precise anatomical makeup of this system may vary across species. Interestingly, their model excluded limbic structures, which are known to play a vital role in emotional processing. This omission raises important questions about how emotional and cognitive elements of consciousness are integrated within the centroencephalic framework—an area still open to exploration and further refinement.

This model, though preliminary, resonates with the book's brain-metafunctional framework, which affirms that consciousness involves integrative activity across cortical and subcortical systems and cannot be fully explained by localized, reductive mechanisms. Like Penfield's approach, the brain-metafunction model views the brain not as a set of discrete processors, but as a unified system whose dynamic interactions facilitate conscious life. Moreover, Penfield's openness to a transphysical component of consciousness—implied by his shift toward dualism—parallels this book's thesis that the mind's irreducibility points beyond material structures, toward a reality in which physical and transphysical domains cooperate to support conscious existence.

Edelman's Theory of Consciousness

Gerald M. Edelman, a Nobel Prize–winning biologist, extended his scientific work beyond immunology to develop a biologically grounded theory of consciousness. He defined consciousness as the experience lost in deep sleep or coma and regained upon waking. According to Edelman, consciousness consists of a unified "scene" composed of sensory responses, memories, and a sense of embodiment. At the core of Edelman's approach is neural Darwinism, or the theory of neuronal group selection. He proposed that neural circuits are formed and refined through selectional processes based on sensory input and experience. These networks stabilize over time and give rise to reentrant signaling—reciprocal and recursive interactions between neural maps across brain regions. This dynamic interaction forms what Edelman called the "dynamic core", believed to underlie primary consciousness—the basic awareness shared by humans and many animals. Edelman viewed consciousness not as a static state but as a self-organizing process. He emphasized that each conscious experience is shaped by a unique developmental trajectory, resulting in an individualized form of subjectivity. This understanding sought to integrate cognitive, emotional, and perceptual data into a holistic biological model. Notably, he firmly rejected dualist or nonphysical explanations, arguing instead that consciousness could and should be fully "naturalized" within the framework of biological science (Edelman 2003; Descartes 1975; Popper and Eccles 1977; Penrose 1994; McGinn 1991).

Hoever, Edelman's theory, while rich in neuroscientific insight, does not resolve several major challenges: the binding problem, the unity and subjectivity of experience, or the qualitative feel of consciousness (qualia). Moreover, his rejection of the transphysical leaves out an essential dimension acknowledged by thinkers across philosophical and theological traditions. Within the scope of this book, Edelman's contributions represent an important piece of the puzzle—especially his emphasis on large-scale, distributed integration. Yet, when viewed through the Brain Metafunction Framework, we see more clearly how dynamic cortical-subcortical integration, combined with openness to a transphysical reality, offers a more robust explanatory structure. The brain-metafunction perspective builds upon Edelman's insight that consciousness is systemically unified and dynamically regulated—but extends it further by recognizing that the mind's unified coherence and rational intentionality cannot arise from physical complexity alone. They point to a higher-order organizational principle that includes but transcends biology.

Damasio's "Somatic Marker Hypothesis"

Antonio Damasio, a prominent neuroscientist and neurologist, has significantly influenced our understanding of consciousness by emphasizing the interplay between bodily states, emotions, and cognition. His Somatic Marker Hypothesis (SMH) proposes that emotional signals—registered as bodily responses—play a crucial role in guiding decision-making and shaping conscious experience (Damasio 1994). These bodily responses, or "somatic markers," help us evaluate options and steer behavior by integrating affective responses with rational deliberation. Central to Damasio's view is the interdependence between the brain and the body, especially the integration of signals from the prefrontal cortex, amygdala, and autonomic nervous system. According to this model, consciousness is not a disembodied rational faculty, but a process rooted in the whole organism—one that arises through recursive loops of interoception, memory, perception, and emotional evaluation. In his more recent work, Damasio (2023) expands this model by differentiating between three foundational image types—exteroceptive, interoceptive, and proprioceptive—each contributing to the construction of the conscious self. His account of consciousness highlights its bodily embeddedness, dynamic affective dimension, and biological grounding. Consciousness, in this view, becomes possible when the brain maps the organism's internal states in relation to its environment and constructs a coherent self-model based on these integrations.

While Damasio's work offers crucial insights into the emotional architecture of consciousness and its neurobiological grounding, it too remains within a materialist explanatory framework. Although he describes consciousness as a richly integrated phenomenon, his model does not fully explain the unity, subjectivity, or intentionality of conscious experience. Nor does it address why these physiological processes are accompanied by inner experience—the very issue at the heart of the "hard problem" of consciousness. When interpreted through the Brain Metafunction Framework, Damasio's theory represents another essential piece of the larger mosaic. His emphasis on cortico-limbic integration and affective-cognitive cooperation aligns with the view that multiple brain systems co-regulate consciousness in real time. However, by incorporating the transphysical dimension, the Brain Metafunction model clarifies why such integration is not only functionally necessary, but insufficient in itself to explain consciousness. The irreducible unity and interiority of experience suggest the presence of a non-material principle—a rational, intentional core of personhood—cooperating with brain systems to generate conscious awareness. Thus, Damasio's theory is indispensable in illuminating how consciousness is modulated and shaped within the body-brain system, while the Brain Metafunction perspective explains what consciousness ultimately is—a transphysically grounded phenomenon expressed through and with the biological brain.

Rodolfo Llinás and the Thalamocortical Model of Consciousness

Neuroscientist Rodolfo R. Llinás proposed a theory of consciousness rooted in the brain's intrinsic capacity to generate predictive models of reality. In his influential work I of the Vortex: From Neurons to Self (2001), Llinás argues that consciousness is not a passive reflection of sensory input, but rather a dynamically generated internal simulation, constantly anticipating external stimuli. At the heart of his theory lies the thalamocortical system, whose oscillatory activity he identifies as fundamental to unifying perception, cognition, and motor planning. This continuous rhythmic coordination—what he calls the "vortex" of self-awareness—enables the brain to construct an anticipatory, temporally structured model of the world, integrating memory, sensory input, and action readiness. Llinás emphasizes that motor control—not just perception—is foundational to consciousness, tracing the roots of conscious experience to the evolution of neural structures for motion and coordination. In this view, consciousness is deeply embodied, arising from the brain's need to model and regulate action in time and space (Goodrich 2010; Merleau-Ponty 2012).

While Llinás offers a compelling neurophysiological mechanism for the integration and prediction functions of the brain, his theory—like those of Edelman and Damasio—does not directly address the subjective "what-it-is-like" character of experience, nor the ontological question of why such activity should produce awareness at all. It remains within a sophisticated physicalist paradigm, emphasizing coordination and coherence, but not fully explaining conscious presence. When viewed through the lens of the Brain Metafunction Framework, Llinás's theory provides yet another vital functional layer: it demonstrates that consciousness depends on distributed, recursive, and rhythmically synchronized systems that bind sensory-motor inputs into a coherent temporal narrative. His work underscores the embodied and anticipatory nature of consciousness, resonating with the metafunctional claim that consciousness reflects a whole-brain orchestration of diverse subfunctions. However, the Brain Metafunction model goes further, positing that these rhythmic integrations are necessary but not sufficient for conscious experience. The transphysical dimension—the presence of an interior, unified subject who experiences—remains unexplained by oscillatory activity alone. Thus, Llinás's contribution can be seen as one essential function within a higher-order system whose ultimate explanation lies in the cooperation of neural and transphysical domains.

Francis Crick and Christof Koch's "neural correlates of consciousness."

Christof Koch, a prominent neuroscientist, collaborated with Francis Crick, the co-discoverer of the DNA double helix, during Crick's later years when he turned his attention to neuroscience and the study of consciousness. In 1990, Koch and Crick introduced the concept of neural correlates of consciousness (NCC) defined as the minimal set of neural mechanisms sufficient for the occurrence of a conscious percept (Crick and Koch 1990). Their goal was to identify the precise neural activity required to produce conscious experience. This proposal marked a pivotal shift in neuroscience, initiating a sustained effort to anchor subjective awareness within measurable brain activity. While Crick is best known for his Nobel Prize-winning work in molecular biology, his foray into neuroscience produced equally provocative claims. In *What Mad Pursuit: A Personal View of Scientific Discovery* (1988), Crick described the intense moment of scientific insight:

> "It is not easy to convey, unless one has experienced it, the dramatic feeling of sudden enlightenment that floods the mind when the right idea finally clicks into place." This introspective

> reflection contrasts sharply with the reductive materialism he later defended in The Astonishing Hypothesis (1994), where he famously asserted: "You—your joys and your sorrows, your memories and your ambitions, your sense of personal identity and free will—are in fact no more than the behavior of a vast assembly of nerve cells and their associated molecules."

Despite the influence of their NCC framework, the effort to localize consciousness in specific neural regions has drawn considerable criticism. Consciousness increasingly appears to be a distributed and dynamic process, involving large-scale brain networks rather than isolated "trigger zones." The infamous "grandmother cell" hypothesis—which proposes that single neurons encode specific, complex percepts (e.g., recognizing one's grandmother)—exemplifies the limitations of early reductionist thinking (Gross 2002). Such ideas have been largely displaced by findings that emphasize population coding and distributed representation across cortical and subcortical regions. Furthermore, effective consciousness appears to require the integration of content and arousal, the latter influencing not only the presence but akso the nature of conscious experience. This integration depends on thalamocortical loops, the ascending reticular activating system (ARAS), and limbic system participation. These insights reveal that consciousness cannot be reduced to any single neural correlate but involves whole-brain cooperation, a principle foundational to the Brain Metafunction Framework.

Koch himself has since reconsidered aspects of strict reductionism. In *Consciousness: Confessions of a Romantic Reductionist* (2012), he acknowledges the stark disconnect between neuronal firing patterns and subjective experience. For instance, the physical activation associated with perceiving the color blue is functionally distinct from the qualitative experience—what it is like to see blue. Koch now advocates for an "expanded physicalism" that incorporates conscious experience as an irreducible aspect of nature, rather than attempting to eliminate or reduce it. This view suggests a willingness to go beyond materialist assumptions, drawing neuroscience closer to non-reductionist and even transphysical paradigms. Koch's evolving stance reflects a broader shift in consciousness studies. As empirical data increasingly challenge the sufficiency of traditional materialism, the field may be approaching a conceptual inflection point. Within this shifting landscape, the Brain Metafunction Framework offers a compelling synthesis: it affirms the essential role of neural structures while also positing that conscious experience transcends neural function alone. The interplay of physical and transphysical dimensions may hold the key to resolving the enduring mystery of consciousness—its unity, interiority, and irreducible presence.

Bernard Baars' Global Workspace Theory

Bernard Baars' Global Workspace Theory (GWT), introduced in the late 1980s, remains one of the most influential models for explaining consciousness. GWT conceptualizes the mind using the metaphor of a theater: conscious awareness occupies the spotlight of attention on a mental stage, while unconscious processes operate invisibly in the surrounding darkness. Behind the scenes, contextual systems such as the dorsal cortical stream shape and influence what enters conscious awareness without themselves being consciously accessed. This theater metaphor draws inspiration from early artificial intelligence systems, particularly the blackboard architecture, where multiple independent modules broadcast information to a central interface, allowing for coordination without direct communication. As the theory evolved, it gained significant support and refinement through the work of researchers like Stanislas Dehaene. In Consciousness and the Brain (2015), Dehaene introduced the idea of neuronal avalanches and described consciousness as a global neuronal workspace, a system that broadcasts selected information across the cortex through synchronized neural activity. This broadcasting mechanism enables a unified interpretation of sensory inputs, primarily facilitated by prefrontal and anterior temporal regions. Dehaene's model preserves Baars' core insights while grounding them in measurable neural dynamics. However, GWT has faced important philosophical critiques. J.W. Dalton (1997) argues that while GWT is robust in accounting for the functional and cognitive dimensions of consciousness, it does not resolve the hard problem—why and how subjective experience arises from brain activity. A.C. Elitzur (1997) concedes that GWT does not solve this problem but emphasizes that it constrains and informs deeper philosophical theories by clarifying the computational and cognitive architecture of conscious access.

Recent developments have broadened the scope of Global Workspace Theory (GWT), emphasizing its anatomical and dynamic grounding. Baars et al. (2013) propose that the cortico-thalamic system underlies the global workspace through its capacity for integrative, large-scale information processing. This view resonates with complementary frameworks: Pribram (1991) highlighted the brain's holographic and integrative dynamics; Edelman and Tononi (2000) focused on reentrant signaling in a dynamic core; Llinás (2001) emphasized thalamocortical oscillations as foundational for unified experience; and Dehaene and Changeux (2011) advanced a neuronal model of GWT involving long-range cortical connectivity. Together, these models converge on the idea that consciousness emerges from temporally coordinated, distributed neural integration. These views challenge static anatomical models and emphasize consciousness as a non-localized, distributed process involving dynamic brain-wide coordination. Deco et al. (2019) contributed a key advancement by introducing the concept of the functional "rich club," a set of highly interconnected cerebral hubs that flexibly form a global workspace. This model highlights the network-based nature of consciousness, supporting the view that mental states arise through dynamic and nonlinear interactions across brain regions. This shift—from localized consciousness modules to distributed network theories—marks an important development in cognitive neuroscience. GWT's recent iterations now converge with the Brain Metafunction Framework, which posits that consciousness results from integrative, multi-level brain activity while also accommodating transphysical dimensions not reducible to physical mechanisms alone. Although GWT remains grounded in cognitive and neural processes, its growing complexity and emphasis on distributed integration make it increasingly compatible with non-reductionist perspectives. In this light, Baars' Global Workspace Theory developing structure invites deeper integration with philosophical models that take seriously the qualitative, unified, and intentional nature of consciousness—affirming that while cognition may be mediated by the brain, consciousness cannot be fully contained within it.

Tononi's Integrated information theory (IIT)

Integrated Information Theory (IIT), developed by neuroscientist Giulio Tononi, offers a bold and mathematically formulated approach to understanding consciousness (Tononi 2004). The theory seeks to explain why certain physical systems, like the human brain, possess conscious experience—and how to quantify it. IIT distinguishes itself by beginning not with the brain, but with the intrinsic features of consciousness itself, and then working backward to identify the physical conditions necessary to support it. Tononi and Koch (2015) describe IIT as providing "a principled account of both the quantity and quality of an individual experience... and a calculus to evaluate whether a physical system is conscious." Proponents note that this phenomenology-first approach stands in contrast to standard neuroscience, which starts from observable brain functions and attempts to infer consciousness: "Taking the phenomenology of consciousness as primary, and asking how it can be implemented by physical mechanisms, is the opposite of the approach usually taken in neuroscience" (Oizumi, Albantakis, and Tononi 2014).

IIT's Axioms: Foundational Features of Consciousness

IIT is rooted in five axioms that identify the essential properties of consciousness (Tononi et al. 2016):

Existence: Consciousness is real—it exists from the first-person perspective.

Composition: Experience has structure; it comprises interrelated aspects (e.g., color, sound, shape).

Information: Each experience is specific, distinct from other possible experiences.

Integration: Consciousness is unified and cannot be broken into independent parts.

Exclusion: Each experience is definite, with clear boundaries that differentiate it from others.

These axioms are meant to capture the qualitative features of consciousness as we know them in lived experience.

Postulates: Linking Consciousness to Physical Systems

From these axioms, IIT derives postulates that describe the physical requirements for generating conscious experience:

Causal Power: Conscious systems must influence themselves and their environment.

Compositionality: Consciousness corresponds to systems whose mechanistic elements combine meaningfully.

Cause-Effect Structure: Consciousness entails a complex internal structure of differentiated causes and effects.

Integration: The system must form a unified whole that cannot be decomposed into independent parts.

Exclusivity: At any moment, only one maximally irreducible system defines a conscious experience (the MICS).

These postulates connect phenomenology to empirical structure, attempting to explain how the subjective is anchored in the physical.

The Φ (Phi) Metric

At the heart of IIT is the Φ (Phi) metric, a quantitative measure of integrated information. A system's Phi value reflects the extent to which its parts work together irreducibly—how unified and differentiated its internal states are. A high Phi indicates a greater degree of consciousness, while a low Phi suggests minimal or no conscious integration. Tononi argues that a system becomes conscious when it surpasses a threshold of integrated information—when the system forms a coherent whole not reducible to its components. Calculating Phi involves analyzing the cause-effect relationships among all possible partitions of a system. While the Phi metric represents an ambitious step toward measurable consciousness, it faces practical and philosophical challenges, including questions about its interpretability and application to diverse systems.

Critiques of IIT

Despite its conceptual and mathematical elegance, IIT has drawn significant criticism:

Falsifiability: Critics argue that IIT's central claims cannot be empirically tested or refuted (Cerullo 2015), which raises questions about its scientific status.

Overfitting and Misclassification: Some warn that Phi may attribute consciousness to systems intuitively considered non-conscious—e.g., photodiodes or logic gates.

Tautology: As Bayne (2018) point out, IIT defines consciousness in terms of integrated information while using integrated information to identify consciousness—a potentially circular reasoning.

Philosophical Assumptions: IIT aligns with forms of panpsychism, suggesting consciousness is a fundamental property of matter. Critics challenge whether such a claim can be supported by empirical science (Doerig et al. 2020).

Lack of Neural Specificity: Pautz (2019) questions IIT's ability to explain why specific brain mechanisms (e.g., visual cortex structures) are conscious, rather than merely identifying integrated information.

Searle's Critique: Observer-Relativity and Panpsychism

Philosopher John Searle offers two pointed objections to IIT:

Observer-Relativity: Searle argues that information presupposes consciousness—not the other way around. Information requires a conscious interpreter to exist as such (Searle 2013). Thus, reducing consciousness to information reverses the causal relationship.

Panpsychism Without Principles: Searle critiques IIT's attribution of consciousness to any system with integrated information, saying it fails to explain why consciousness comes in discrete units or how to distinguish truly conscious entities from functionally similar but non-conscious ones.

Koch and Tononi's Defense

In response, Koch and Tononi (2013) contend that IIT is not a vague metaphysical thesis but a precise, quantifiable model. They clarify that the theory does not imply that "everything is conscious," but only systems that form Maximally Irreducible Cause-Effect Structures (MICS). For example, a camera sensor or photodiode may interact with its environment but lacks the internal integration required for consciousness. Still, many critics remain unconvinced, arguing that the claim of identity between consciousness and integrated information fundamentally alters naturalistic ontology. By proposing consciousness as a basic, irreducible feature of the universe, IIT may shift scientific naturalism toward a quasi-metaphysical paradigm (Chalmers 1996).

IIT and the Brain Metafunctional Framework

Despite the critiques, IIT contributes a significant perspective to the broader effort to understand consciousness. It aligns with the Brain Metafunctional Framework by emphasizing the integrated, irreducible nature of conscious awareness—though IIT remains within a modified physicalist paradigm. While it does not directly embrace a transphysical model, it opens the door to conceptions of consciousness that challenge strict materialism and acknowledge the fundamental, non-local characteristics of mind. Ultimately, IIT offers an influential (if controversial) model that draws attention to the unity, specificity, and causal richness of consciousness. It is one piece of the puzzle—and as this chapter argues, the fuller picture requires transcending reductionism, integrating both neural processes and the profound metaphysical implications of conscious life.

Global Workspace Theory vs Integrated Information Theory

A recent article by science journalist Carl Zimmer, published in The New York Times (July 1, 2023), reports on a landmark study in which neuroscientists and philosophers tested two of the leading theories of consciousness: Global Workspace Theory (GWT) and Integrated Information Theory (IIT). The experiments involved brain imaging and neural implants in 256 volunteers performing visual tasks, designed to evaluate the neural correlates of conscious perception. This study revisited a decades-old wager between philosopher David Chalmers and neuroscientist Christof Koch, centered on identifying a reliable neural correlate of consciousness. The results were inconclusive: neither theory consistently predicted consciousness-related outcomes. These findings call into question earlier models—particularly the Crick and Koch framework—that sought to reduce consciousness to specific, localized brain structures. The now-discredited idea of a "grandmother neuron," which posited that single neurons could represent complex entities, serves as a cautionary example of such reductionism.

Zimmer notes that the experimental data increasingly support the view of consciousness as a distributed phenomenon—one that cannot be confined to narrowly defined neural regions, such as the anterior or posterior cortices alone. Instead, consciousness appears to arise from the interconnected activity of widely dispersed brain networks. This perspective aligns with the Brain Metafunctional Framework, which interprets consciousness not as the output of a single structure or mechanism, but as a brain metafunction involving integration across the whole brain. The findings also suggest that both GWT and IIT, while offering valuable insights, may be insufficient on their own. Rather than identifying isolated neural generators of consciousness, the study reveals a deeper need for an integrative model—one capable of accounting for both the neural substrate and the subjective dimension of conscious experience. From this vantage point, we propose that consciousness arises not only from brain-based interactions, but also through a transphysical domain, cooperating with the brain via an inherent and sui generis mode of causality. This proposition moves beyond standard physicalist assumptions, inviting a richer metaphysical framework for understanding human consciousness—a theme that will be further explored in the forthcoming chapter on the soul.

Neural Correlates of Consciousness: Bridging Progress and Complexity

The article "*Neural Correlates of Consciousness: Progress and Problems*" by Christof Koch and colleagues offers a clear-eyed assessment of both the advances and enduring challenges in identifying the neural correlates of consciousness (NCC). Central to their critique is the difficulty of defining what qualifies as an NCC. Although researchers generally define them as the minimal neural mechanisms necessary and sufficient for a particular conscious experience, the boundary between those that generate consciousness and those that merely support or modulate it remains elusive. This ambiguity complicates experimental design and data interpretation, making it difficult to isolate causally relevant neural events. Koch et al. emphasize that consciousness likely arises from large-scale, distributed neural processes rather than isolated nodes or circuits. They cite networks like the default mode and attentional systems as evidence that conscious experience reflects complex integration across the brain. As they note, "Identifying NCCs as discrete entities risks oversimplifying the rich, dynamic interplay of brain regions necessary for conscious experience" (Koch et al. 2016).

Temporal and spatial complexities further challenge reductionist views. Consciousness unfolds over varying timescales—from millisecond-level oscillations to sustained patterns of network activity—and spans interactions between cortical and subcortical regions. No single measurement tool fully captures this complexity: fMRI offers spatial precision but lacks temporal acuity, while EEG provides temporal clarity but limited spatial resolution. Combined with the reliance on subjective self-reporting, these limitations underscore the broader methodological challenges in consciousness research. The authors also caution against conflating correlation with causation. Observing neural activity that coincides with a conscious state does not necessarily explain how that state is produced. As they argue, "Understanding the causal mechanisms behind consciousness requires interventions that manipulate neural activity directly" (Koch et al., 2016). This highlights the broader challenge of distinguishing conscious from unconscious processes, given that much of neural activity occurs beneath the threshold of awareness. At a deeper level, Koch and colleagues acknowledge a persistent theoretical gap: despite extensive empirical data, science still lacks a coherent model explaining how brain activity gives rise to subjective experience. This echoes David Chalmers's hard problem of consciousness (1995), which does not ask how the brain behaves but why experience itself accompanies its processes. Koch et al. conclude that while NCC research is invaluable, it cannot resolve this deeper question without more integrated theoretical frameworks.

An Expanded Theoretical Approach

The critique by Koch and colleagues underscores the need for a more encompassing model of consciousness—one that bridges empirical neuroscience with philosophical and transdisciplinary insights. The Brain Metafunctional Framework developed throughout this book offers such an integrative approach. Rather than reducing consciousness to localized brain regions or signals, the framework understands it as a global, integrative metafunction—a dynamic, adaptive coordination across distributed neural systems. It addresses both spatial and temporal dimensions, recognizing that conscious states involve synchronized activity across multiple brain areas over time. Importantly, it emphasizes causal distinctiveness rather than mere correlation, advocating for experimental paradigms that integrate imaging, causal manipulation, and multimodal validation, including behavioral and phenomenological data. This framework also reconceptualizes the distinction between conscious and unconscious processes by framing them as points along a continuum of informational integration with a higher order transphysical dimension. Crucially, the Brain Metafunctional Framework synthesizes empirical data with philosophical reflection, inviting interdisciplinary collaboration to build models that unify brain activity, cognitive function, and subjective awareness. It moves beyond the explanatory gap by transcending reductionist assumptions. The brain, in this view, is not merely a generator of consciousness but an interface—an embodied conduit through which transphysical dimensions of mind and self are expressed. This sets the stage for a more complete account of personhood and identity in the chapters that follow.

Metafunctional Consciousness Beyond Materialism

The placebo effect—where patients experience real improvements from inert treatment—demonstrates how mental states like belief and expectation can measurably influence physical health (Benedetti, 2020). Along with the nocebo effect (where negative expectations produce adverse outcomes) (Benedetti et al, 2007, 2003, 2013), these phenomena reveal a top-down influence of consciousness on physiology, challenging the materialist assumption that mental states are merely epiphenomenal byproducts of brain function. These effects point to a reciprocal mind-brain relationship, wherein consciousness not only emerges from neural processes but also acts upon them. This dynamic supports a holistic framework—one that affirms the presence of a transphysical domain cooperating with physical systems through sui generis, law-like causal principles. The Brain Metafunction Framework integrates this perspective, arguing that consciousness is not simply emergent, but a distinct organizing principle that actively cooperates with the brain in real time.

The Complexity of Consciousness: Integration and Plurality

The mind-brain relationship is not unidirectional but reciprocally interactive. Mental states such as belief and attention can influence immune responses, modulate neurotransmission, and alter physiological states—evidence that consciousness shapes bodily function. Consciousness also generates qualia—the subjective textures of experience—that bind sensory, emotional, and cognitive data into unified perception. While attention and memory serve as cognitive tools, it is consciousness that synthesizes these functions into a coherent phenomenological whole (Tononi & Edelman, 1998; Dehaene, 2014). One enduring challenge in neuroscience is explaining how the plurality of conscious states—interoceptive, emotional, and reflective—are unified into seamless experience (Edelman, 2003). Gestalt psychology helps frame this process, showing how the brain assembles fragmented inputs into perceptual wholes. Yet consciousness transcends this level of synthesis. The *binding problem*—how disparate signals (e.g., color, motion, sound) are combined into single representations—remains unresolved (Singer & Gray, 1995). Similarly, the problem of *metabinding*, which concerns how higher-order cognitive and affective processes converge into richly contextual and coherent experiences, continues to challenge conventional models. The Brain Metafunction Framework offers a more integrated account by situating these phenomena within a dynamic system of whole-brain coordination—one that spans cognitive, emotional, and bodily domains and points toward a unified, hierarchically organized metafunction of consciousness.

Subcortical Contributions to Consciousness

Research continues to challenge cortical-centric models by highlighting the foundational role of subcortical systems in consciousness. Traditional distinctions between arousal (subcortical) and content (cortical) oversimplify complex interdependencies. Lesions in the extended reticulothalamic activating system (ERTAS) can abolish consciousness altogether, whereas even extensive cortical damage may only impair specific functions (Parvizi & Damasio, 2001; Penfield & Jasper, 1954). Cases such as hydranencephalic children—who lack most of the cerebral cortex yet display social and emotional responsiveness—suggest that core conscious states are rooted in subcortical structures (Shewmon, Holmse, & Byrne, 1999). This view is reinforced by Merker's (2007) proposal that core consciousness can arise from subcortical structures—particularly the upper brainstem—independently of cortical processing, and by Sherman and Guillery's (2006) demonstration that thalamic circuits play an active, integrative role in organizing cortical activity, thereby challenging the notion of cortical autonomy in conscious experience. The effects of psychotropic drugs, which target brainstem monoaminergic systems, further emphasize the foundational role of subcortical regions. Their ability to modulate mood and consciousness underscores the importance of these systems in sustaining the global architecture of awareness (Panksepp, 1998; Solms & Friston, 2018).

Rare but well-documented cases of individuals with drastically reduced brain volume, such as extreme hydrocephalus patients exhibiting near-normal or even above-average intelligence, further challenge materialist assumptions about cognition. These cases, as documented by Feuillet et al. (2007), Lorber (1980), and others, suggest that brain function may depend more on integrative efficiency and systemic coordination than on sheer cortical mass. Such findings lend support to the Brain Metafunction Framework's emphasis on dynamic integration over structural quantity—and may even gesture toward organizing principles that transcend strictly physical explanations. While surprising from a materialist standpoint, these phenomena are far less unexpected—and more readily accommodated—within a transphysical framework that views consciousness as a distinct, organizing presence rather than a mere byproduct of neural complexity.

A Shift Toward Holistic Brain Models

Contemporary neuroscience is converging on a holistic view of brain function, moving away from the idea of isolated regional modules. Contemporary models emphasize domain-general networks, where neural populations are highly flexible and capable of performing multiple functions (Avena-Koenigsberger, Misic, & Sporns, 2018). Cortical and subcortical structures form densely interconnected hubs, or connectivity cores, that integrate sensory, cognitive, emotional, and bodily regulatory functions (Sporns, 2010; van den Heuvel & Sporns, 2011). Networks like the default mode network and the salience network dynamically link interoception and allostasis to higher-order processing (Tononi, 2004; Seeley et al., 2007). The Brain Metafunction Framework anticipated this shift. It conceptualizes consciousness not as a property of isolated processes, but as a hierarchically integrated, system-wide function spanning cognition, perception, motor and homeostatic regulation, and visceral awareness. It highlights the hierarchical, integrative, and dynamic organization of consciousness across the whole-brain system.

Conclusion

The Brain Metafunction Framework provides a comprehensive and integrative model of consciousness—one that moves beyond the constraints of reductionism and materialism. Rather than treating consciousness as a passive byproduct of neural complexity, this framework sees it as a fundamental and organizing principle, emerging through—but not reducible to—neural dynamics. It accounts for empirical evidence—from distributed integration and subcortical affectivity to top-down physiological influence and the coherence of subjective experience—while also remaining open to metaphysical insight. The brain, in this view, serves as a reciprocal interface, mediating both material and transphysical dimensions of the person. Consciousness is thus seen as distributed across interconnected systems, not confined to localized brain regions. Neural activity and subjective awareness are not identical but deeply coordinated within a higher order metafunction that transcends simple physical explanation. By bridging neuroscience, philosophy, and theology, the Brain Metafunction Framework offers a richer and more unified model of the conscious self. It challenges us to reimagine the nature of mind, identity, and human experience—not merely as products of brain activity, but as expressions of a transcendent unity. The groundwork laid in this chapter will be expanded in a later section— "The Human Soul and the Brain Metafunction"—which return to these transphysical dimensions more directly, examining how classical and contemporary insights into the soul illuminate enduring questions of personhood, identity, and the mystery of conscious life.

REFERENCES

(Consciousness, Sections One and Two)

Allison, Henry E. (1983). *Kant's Transcendental Idealism: An Interpretation and Defense.* New Haven, CT: Yale University Press.

Allison, H. E. (2011). *Kant's Groundwork for the Metaphysics of Morals: A Commentary.* Oxford University Press.

Aquinas, Thomas. (1259–1265). *Summa Contra Gentiles.* Translated by Joseph Rickaby. London: Burns and Oates.

Aquinas, Thomas. (1266–1273). *Summa Theologica.* Translated by the Fathers of the English Dominican Province. Available at: https://www.newadvent.org/summa/.

Aristotle. (1984). "On Sleep and Waking (2.454a)." In *The Complete Works of Aristotle: The Revised Oxford Translation,* edited by Jonathan Barnes, Vol. 1, pp. 736–738. Princeton, NJ: Princeton University Press.

Aristotle. (1993). *Metaphysics.* Translated by Hugh Tredennick. Cambridge, MA: Harvard University Press.

Aristotle. (2009). *De Anima (On the Soul).* Translated by C. Shields. Oxford: Oxford University Press.

Avena-Koenigsberger, Andrea, Bratislav Mišić, and Olaf Sporns. (2018). "Communication Dynamics in Complex Brain Networks." *Nature Reviews Neuroscience,* 19(1): 17–33. https://doi.org/10.1038/nrn.2017.149.

Baars, Bernard J. (1988). *A Cognitive Theory of Consciousness.* New York: Cambridge University Press.

Baars, Bernard J. (1997). *In the Theater of Consciousness: The Workspace of the Mind.* New York: Oxford University Press.

Baars, Bernard J., Geld, N., & Kozma, R. (2021). "Global Workspace Theory (GWT) and Prefrontal Cortex: Recent Developments." *Frontiers in Psychology,* 12, 749868. https://doi.org/10.3389/fpsyg.2021.749868.

Baars, Bernard J., Franklin, Stan, & Ramsøy, Thomas Z. (2013). "Global Workspace Dynamics: Cortical 'Binding and Propagation' Enables Conscious Contents." *Frontiers in Psychology*, 4, 200. https://doi.org/10.3389/fpsyg.2013.00200.

Bayne, Tim. (2018). "On the Axiomatic Foundations of Integrated Information Theory." *Neuroscience of Consciousness*, 2018(1), niy007. https://doi.org/10.1093/nc/niy007.

Bayne, Tim, and David J. Chalmers. (2003). "What is the Unity of Consciousness?" In *The Unity of Consciousness: Binding, Integration, and Dissociation*, edited by Axel Cleeremans, 23–58. Oxford: Oxford University Press.

Beauregard, Mario, and Denyse O'Leary. (2007). *The Spiritual Brain: A Neuroscientist's Case for the Existence of the Soul*. New York: HarperOne.

Bechara, Antoine, Damasio, Hanna, Tranel, Daniel, & Damasio, Antonio R. (1997). "Deciding Advantageously Before Knowing the Advantageous Strategy." *Science*, 275(5304): 1293–1295. https://doi.org/10.1126/science.275 .5304.1293.

Benedetti, F. (2020). *Placebo effects* (3rd ed.). Oxford University Press.

Benedetti, Fabrizio, et al. (2003). "The Placebo Effect in Postoperative Pain Management: A Randomized Controlled Trial." The Lancet, 362(9393): 1713–17. https://doi.org/ 10.1016/S0140-6736(03)14850-2.

Benedetti, Fabrizio, et al. (2007). "Nocebo Effects and the Brain." *Neurobiology of Aging*, 28(1): 31–40. https://doi.org/ 10.1016/j.neurobiolaging.2006.03.017.

Benedetti, Fabrizio, et al. (2013). "Pain as a Paradox in Placebo and Nocebo Studies." *Neuroscience*, 234: 120–31. https://doi.org/10.1016/j.neuroscience.2013.01.021.

Boff, Leonardo. (1980). *La resurrección de Cristo. Nuestra resurrección en la muerte* [The Resurrection of Christ. Our Resurrection in Death]. 5th ed. Translated by J. C. Rodríguez Herranz. Santander, Spain: Editorial SAL TERRAE.

Brentano, Franz. (1874 [1995]). *Psychology from an Empirical Standpoint.*

Edited by Oskar Kraus and translated by Antos C. Rancurello, D. B. Terrell, and Linda L. McAlister. London: Routledge.

Brown, R. E. (2003). *An Introduction to the Gospel of John*. Edited by J. J. Collins. *Anchor Yale Bible Reference Library Series*. New Haven, CT: Yale University Press.

Caston, Victor. 2002. "Aristotle on Consciousness." *Mind* 111 (444): 751–815. https://doi.org/10.1093/mind/111.444.751.

Cerullo, Michael A. (2015). "The Problem with Phi: A Critique of Integrated Information Theory." *PLOS Computational Biology,* 11(9), e1004286. https://doi.org/10. 1371/journal.pcbi.1004286.

Chalmers, David J. (1995). "Facing Up to the Problem of Consciousness." *Journal of Consciousness Studies,* 2(3): 200–219.

Chalmers, David J. (1996). *The Conscious Mind: In Search of a Fundamental Theory*. Oxford University Press.

Chalmers, David J. (1997). "Moving Forward on the Problem of Consciousness." In *Toward a Science of Consciousness II: The Second Tucson Discussions and Debates,* edited by S. Hameroff, A. Kaszniak, and A. Scott, 77–92. MIT Press.

Chein, Jason M., Adam B. Moore, and Andrew R. Conway. (2011). "Domain-General Mechanisms of Complex Working Memory Span." *NeuroImage* 54 (1): 550–59. https://doi.org/10.1016/j.neuroimage.2010.07.067.

Cole, M. W., Reynolds, J. R., Power, J. D., Repovs, G., Anticevic, A., & Braver, T. S. (2013). Multi-task connectivity reveals flexible hubs for adaptive task control. Nature Neuroscience, 16(9), 1348–1355. https://doi.org/10.1038/nn .3470.

Cooper, J. M., ed. (1997). *Plato: Phaedo*. In *Complete Works*. Hackett Publishing Company.

Craig, William Lane. (2008). *A Reasonable Faith*. Wheaton, IL: Crossway Books.

Craig, William Lane, and J. P. Moreland. (2003). *Philosophical Foundations for a Christian Worldview*. Downers Grove, IL: InterVarsity Press.

Crick, Francis, and Christof Koch. (1990). "Toward a Neurobiological Theory of Consciousness." *Seminars in the Neurosciences,* 2: 263–275.

Crick, Francis. (1988). *What Mad Pursuit: A Personal View of Scientific Discovery.* New York: Basic Books.

Crick, Francis. (1994). *The Astonishing Hypothesis: The Scientific Search for the Soul.* New York: Scribner.

Dalton, J. W. 1997. "Consciousness and the Neural Global Workspace Model." *Behavioral and Brain Sciences,* 20(1), 148 –149.

Damasio, Antonio R. (1994). *Descartes' Error: Emotion, Reason, and the Human Brain.* New York: Grosset/Putnam.

Damasio, Antonio R. (1999). *The Feeling of What Happens: Body and Emotion in the Making of Consciousness.* New York: Harcourt Brace.

Damasio, Antonio R., Daniel Tranel, and Hanna C. Damasio. (1991). "Somatic Markers and the Guidance of Behavior: Theory and Preliminary Testing." In *Frontal Lobe Function and Dysfunction,* edited by Harvey S. Levin, Howard M. Eisenberg, and Arthur L. Benton, 217–229. New York: Oxford University Press.

Damasio, Antonio, and Hanna Damasio. (2023). "Sensing is a Far Cry from Sentience." *Animal Sentience,* 33(16). https://doi.org/10.51291/2377-7478.1805.

Damasio, Antonio, and Hanna Damasio. (2023). *Feeling and Knowing: Making Minds Conscious.* New York: Pantheon.

Deco, G., Tononi, G., Boly, M., & Kringelbach, M. L. (2015). "Rethinking Segregation and Integration: Contributions of Whole-Brain Modelling." *Nature Reviews Neuroscience,* 16(7), 430–439. https://doi.org/10.1038/nrn3963.

Dehaene, Stanislas, & Changeux, Jean-Pierre. (2011). "Experimental and Theoretical Approaches to Conscious Processing." *Neuron,* 70(2), 200–227. https://doi.org/10.10 16/j.neuron.2011.03.018.

Dehaene, Stanislas. (2014). *Consciousness and the Brain: Deciphering How the Brain Codes Our Thoughts.* Viking Penguin.

Dennett, Daniel C. (1991). *Consciousness Explained.* Boston: Little, Brown, and Company.

Descartes, R. (1637). *Discourse on the Method of Rightly Conducting One's Reason and Seeking Truth in the Sciences.* Translated by Ian Maclean (2006). Oxford: Oxford University Press.

Descartes, R. (1641). *Meditations on First Philosophy: In Which the Existence of God and the Immortality of the Soul Are Demonstrated.* Translated by John Cottingham (1996). Cambridge: Cambridge University Press.

Descartes, R. (2007). *The Correspondence Between Princess Elisabeth of Bohemia and René Descartes.* Edited and translated by Lisa Shapiro. Chicago: University of Chicago Press.

Descartes, René. 1975. *Philosophical Writings.* Translated by Elizabeth Anscombe and Peter Thomas Geach. Indianapolis: Bobbs-Merrill.

Dirckx, Sharon. (2019). *Am I Just My Brain?* Oxford, UK: The Good Book Company.

Doerig, Adrien, Schurger, Aaron, & Herzog, Michael H. (2020). "Hard Criteria for Empirical Theories of Consciousness." *Cognitive Neuroscience,* 12(2), 41–62. https://doi.org/10.1080/17588928.2020.1772214.

Dretske, Fred. (1995). *Naturalizing the Mind.* Cambridge, MA: MIT Press.

Edelman, G. M., Gally, J. A., & Baars, B. J. (2011). "Biology of Consciousness." *Frontiers in Psychology,* 2, 4. https://doi.org/10.3389/fpsyg. 2011.00004.

Edelman, Gerald M. (1989). *The Remembered Present: A Biological Theory of Consciousness.* New York: Basic Books.

Edelman, Gerald M. (2003). "Naturalizing Consciousness: A Theoretical Framework." *Proceedings of the National Academy of Sciences,* 100(9), 5520–5524. https://doi.org/10.1073/pnas. 0931349100.

Edelman, Gerald M. (2004). *Wider than the Sky: The Phenomenal Gift of Consciousness.* New Haven: Yale University Press.

Edelman, Gerald M., & Tononi, Giulio. (2000). *A Universe of Consciousness: How Matter Becomes Imagination.* Basic Books.

Egnor, M., & Flynn, P. (Hosts). (2024, August 15). The science of mind: Debunking materialism, with Dr. Michael Egnor (No. 304) [Audio podcast episode]. In *Mind Matters Podcast.* https://mindmatters.ai/podcast/ep304/.

Elitzur, Avshalom C. (1997). "Consciousness and the Incompleteness of the Physical Explanation of Behavior." *Journal of Mind and Behavior,* 18, 1–20.

Feuerbach, Ludwig. 1854. *The Essence of Christianity.* Translated by George Eliot. London: John Chapman.

Gallagher, Shaun. 2000. "Philosophical Conceptions of the Self: Implications for Cognitive Science." *Trends in Cognitive Sciences,* 4(1): 14–21.

Gallagher, Shaun. 2005. *How the Body Shapes the Mind.* Oxford: Clarendon Press.

Goff, Philip. 2017. *Consciousness and Fundamental Reality.* Oxford University Press.

Goodrich, Barbara Gwenn. 2010. "We Do, Therefore We Think: Time, Motility, and Consciousness." *Reviews in the Neurosciences,* 21(5), 331–362. https://doi.org/10.1515/REV NEURO.2010.21.5.331.

Graziano, Michael S. 2019. *Rethinking Consciousness: A Scientific Theory of Subjective Experience.* New York: W.W. Norton & Company.

Greenfield, S. 2002. *The Private Life of the Brain.* Penguin UK.

Greenfield, S. 2016. *You and Me: The Neuroscience of Identity.* Penguin UK.

Greenfield, S. (2019, December 24). "What is Conscious-ness?" [Video]. *Closer to Truth.* https://youtu.be/GaNXZauDzwE?si=ikFgt 09 N1Y4nWe6g.

Greenfield, S. A., & Collins, T. F. 2005. "A Neuroscientific Approach to Consciousness." *Progress in Brain Research,* 150, 11–23. https://doi.org/10.1016/S0079-6123(05)50002-5.

Gross, Charles G. 2002. "Genealogy of the 'Grandmother Cell.'" *Neuroscientist,* 8(5), 512–518. https://doi.org/10.1177 /107385802237175.

Hegel, G. W. F. (1977). *Phenomenology of Spirit.* (A. V. Miller, Trans.; J. N. Findlay, Foreword). Oxford University Press.

Hippocratic Corpus. (c. 400 BCE). *On the Sacred Disease.*

Jackson, F. (1982). "Epiphenomenal Qualia." *The Philosophical Quarterly,* 32(127), 127–136.

Jeeves, Malcolm A., and Warren S. Brown. *Neuroscience, Psychology, and Religion: Illusions, Delusions, and Realities about Human Nature.* Templeton Foundation Press, 2009.

Jeeves, Malcolm. 2015. *The Emergence of Personhood: A Quantum Leap?* Grand Rapids, MI: Eerdmans.

Kahn, C. H. (1979). *The Art and Thought of Heraclitus: An Edition of the Fragments with Translation and Commentary.* New York: Cambridge University Press.

Kant, Immanuel. *Lectures on Logic.* Translated and edited by J. Michael Young. Cambridge University Press, 1992.

Kirk, G. S., Raven, J. E., & Schofield, M. (1983). *The Presocratic Philosophers: A Critical History with a Selection of Texts* (2nd ed., Fragment DK13A5). Cambridge: Cambridge University Press.

Kitcher, Patricia. 1993. *Kant's Transcendental Psychology.* Oxford University Press.

Kittel, G. 1965. *Theological Dictionary of the New Testament* (Vol. 1). Grand Rapids, MI: Eerdmans.

Koch, C., Massimini, M., Boly, M., & Tononi, G. (2016). "Neural Correlates of Consciousness: Progress and Problems." *Nature Reviews Neuroscience,* 17(5), 307–321. https: //doi.org/10.1038/nrn.2016.22.

Koch, Christof. 2004. *The Quest for Consciousness: A Neurobiological Approach.* Englewood, CO: Roberts and Company Publishers.

Koch, Christof. 2012. *Consciousness: Confessions of a Romantic Reductionist.* Cambridge, MA: MIT Press.

Koch, Christof, & Tononi, Giulio. (2013). "Can a Photodiode Be Conscious?" *New York Review of Books,* 60(4), March 7: 43–44.

Kreeft, P. (2007). *The Philosophy of Jesus.* South Bend, IN: St. Augustine's Press.

Kriegel, Uriah. 2009. *Subjective Consciousness: A Self-Representational Theory.* Oxford: Oxford University Press.

Kuhn, Thomas S. 1962. *The Structure of Scientific Revolutions.* Chicago: University of Chicago Press.

Levine, Joseph. 1983. "Materialism and Qualia: The Explanatory Gap." *Pacific Philosophical Quarterly* 64 (October): 354–361.

Lewis, Ralph. 2023. "An Intriguing and Controversial Theory of Consciousness: IIT." *Psychology Today,* November 25. Retrieved from https://www.psychologytoday.com/intl/articles/an-intriguing-and-controversial-theory-of-consciousness-iit.

Llinás, Rodolfo R. 2001. *I of the Vortex: From Neurons to Self.* Cambridge, MA: MIT Press.

Locke, J. (1690). *An essay concerning humane understanding* (E. Holt, Printer). George in Fleet Street. Project Gutenberg. https://www.gutenberg.org/files/10615/10615-h/10615-h.htm

Marx, Karl, and Friedrich Engels. 1846. *The German Ideology: Part One, with Selections from Parts Two and Three, Together with Marx's "Introduction to a Critique of Political Economy."* Edited by C.J. Arthur. New York: International Publishers, 1970.

Marx, Karl, and Friedrich Engels. 1978. *The Marx-Engels Reader.* Edited by Robert C. Tucker. 2nd ed. New York: W. W. Norton & Company.

McGinn, Colin. 1991. *The Problem of Consciousness: Essays Toward a Resolution.* Oxford: Basil Blackwell.

McGinn, Colin. 1999. *The Mysterious Flame: Conscious Minds in a Material World.* New York: Basic Books.

Menuge, Angus. 2004. *Agents Under Fire: Materialism and the Rationality of Science.* Lanham, MD: Rowman & Littlefield.

Merker, Bjorn. 2007. "Consciousness Without a Cerebral Cortex: A Challenge for Neuroscience and Medicine." *Behavioral and Brain Sciences* 30 (1): 63–81. https://doi.org/10.1017/S014 0525X07000891.

Merleau-Ponty, Maurice. 2012. *Phenomenology of Perception.* Translated by Donald A. Landes. London: Routledge.

Metzinger, Thomas. 2003. *Being No One: The Self-Model Theory of Subjectivity.* Cambridge, MA: MIT Press.

Moreland, J.P. 2008. *Consciousness and the Existence of God: A Theistic Argument.* New York: Routledge.

Moreland, J.P. 2014. *The Soul: How We Know It's Real and Why It Matters.* Chicago: Moody Publishers.

Moreland, J.P. 2018. *Science and the Soul: What Scientists Can Learn from Theology.* Wheaton, IL: Crossway Books.

Murphy, Nancey. 2006. *Bodies and Souls, or Spirited Bodies?* Cambridge: Cambridge University Press.

Nagel, Thomas. 1974. "What Is It Like to Be a Bat?" *The Philosophical Review* 83 (4): 435–450. https://doi.org/10. 2307/2183914.

Newberg, Andrew, and Mark Robert Waldman. 2009. *How God Changes Your Brain: Breakthrough Findings from a Leading Neuroscientist.* New York: Ballantine Books.

Newberg, Andrew. 2018. *Neurotheology: How Science Can Enlighten Us About Spirituality.* New York: Columbia University Press.

Noble, Safiya Umoja. 2018. *Algorithms of Oppression: How Search Engines Reinforce Racism.* New York: New York University Press.

Nussbaum, M. C. 1985. *Aristotle's De Motu Animalium: Text with Translation, Commentary, and Interpretive Essays.* Princeton, NJ: Princeton University Press.

Oizumi, Masafumi, Albantakis, Larissa, and Tononi, Giulio. 2014.

"From the Phenomenology to the Mechanisms of Consciousness: Integrated Information Theory 3.0." *PLOS Computational Biology* 10 (5): e1003588. https://doi.org/10.1371/journal.pcbi.1003588.

Panksepp, Jaak. 1998. *Affective Neuroscience: The Foundations of Human and Animal Emotions*. Oxford University Press.

Parvizi, Josef, and Antonio R. Damasio. 2001. "Consciousness and the Brainstem." *Cognition* 79 (1–2): 135–159. https://doi.org/10.1016/S0010-0277(00)00127-X.

Pautz, Adam. 2019. "What is the Integrated Information Theory of Consciousness?" *Journal of Consciousness Studies* 26 (1–2): 1–2.

Penfield, Wilder. 1952. "Memory Mechanisms." *Archives of Neurology and Psychiatry* 67 (2): 178–198.

Penfield, Wilder. 1958. "Some Mechanisms of Consciousness Discovered During Electrical Stimulation of the Brain." *Proceedings of the National Academy of Sciences* 44 (2): 51–66.

Penfield, Wilder. 1958. *The Excitable Cortex in Conscious Man*. Liverpool: Liverpool University Press.

Penfield, Wilder. 1958. *The Mystery of the Mind: A Critical Study of Consciousness and the Human Brain*. Princeton University Press.

Penfield, Wilder. 1959. *Speech and Brain Mechanisms*. Princeton University Press.

Penfield, Wilder. 1959. "The Interpretive Cortex: The Stream of Consciousness in the Human Brain Can Be Electrically Re-experienced." *Science* 129 (3350): 1719–1725.

Penfield, Wilder. 1975. *The Mystery of the Mind: A Critical Study of Consciousness and the Human Brain*. Princeton University Press.

Penfield, Wilder, and Herbert H. Jasper. 1954. *Epilepsy and the Functional Anatomy of the Human Brain*. Boston: Little, Brown.

Penfield, Wilder, and Lamar Roberts. 1959. *Speech and Brain Mechanisms*. Princeton: Princeton University Press.

Penfield, Wilder, and Theodore Rasmussen. 1950. *The Cerebral Cortex of Man: A Clinical Study of Localization of Function.* New York: Macmillan.

Penrose, Roger. 1994. *Shadows of the Mind: A Search for the Missing Science of Consciousness.* Oxford University Press.

Penrose, Roger, and Stuart Hameroff. 1996. "Orchestrated Reduction of Quantum Coherence in Brain Microtubules: A Model for Consciousness." *Mathematics and Computers in Simulation* 40 (3–4): 453–480.

Plato. 1997. *Phaedo.* In *Complete Works,* edited by J. M. Cooper and D. S. Hutchinson, 49–100. Indianapolis: Hackett Publishing Company.
Plato. 1997. *Republic.* Translated by A. Bloom. New York: Basic Books.

Popper, Karl R., and John C. Eccles. 1977. *The Self and Its Brain: An Argument for Interactionism.* Berlin: Springer International.

Pribram, Karl H. 1991. *Brain and Perception: Holonomy and Structure in Figural Processing.* Psychology Press.

Radhakrishnan, S., & Moore, C. A. (Eds.). (1957). *A source book in Indian philosophy.* Princeton University Press.

Ratzinger, J. 1991. "El Misterio de la Mente Humana" ["The Mystery of the Human Mind"]. In *Actas de la Quinta Conferencia Internacional: La Mente Humana,* 16: 18–20. *Dolentium Hominum.*

Robinson, T. M. (1995). *Plato's Psychology.* Toronto: University of Toronto Press.

Rosenthal, David M. 2005. *Consciousness and Mind.* Oxford: Oxford University Press.

Sauchelli, A. (2017). *Personal Identity and Applied Ethics: A Historical and Philosophical Introduction.* Taylor & Francis.

Searle, J. R. (1980). "Minds, Brains, and Programs." *Behavioral and Brain Sciences,* 3(3), 417–424.

Searle, J. R. (2013). "Can Information Theory Explain Consciousness?" (Review of Koch 2012). *New York Review of Books,* 60(1), 54–55, 58.

Searle, J. R. (2015). *Seeing Things as They Are: A Theory of Perception.* Oxford University Press.

Searle, J. R. (2016). "Can Brain Explain Mind?" *Closer To Truth.* Retrieved from https://youtu.be/ehd ZAY0Zr6A?si=naAzn YPIpE8IxxSX

Searle, John R. (1983). *Intentionality: An Essay in the Philosophy of Mind.* Cambridge: Cambridge University Press.

Seth, Anil K., & Friston, Karl J. (2016). "Active Interoceptive Inference and the Emotional Brain." *Philosophical Transactions of the Royal Society B: Biological Sciences,* 371(1708), 20160007. https://doi.org/10.1098/rstb.2016.0007

Sherman, S.M., & Guillery, R.W. (2006). *Exploring the Thalamus and Its Role in Cortical Function (2nd Ed.).* MIT Press.

Shewmon, D. Alan, Holmse, Donna M., & Byrne, Paul A. (1999). "Consciousness in Congenitally Decorticate Children: Developmental Vegetative State as a Self-Fulfilling Prophecy." *Developmental Medicine & Child Neurology,* 41(6), 364–374.

Shields, C. (2016). "Aristotle's Psychology." *Stanford Encyclopedia of Philosophy.* Retrieved from https://plato.stanford.edu/entries/aristotle-psychology/

Singer, Wolf, & Gray, Charles M. (1995). "Visual Feature Integration and the Temporal Correlation Hypothesis." *Annual Review of Neuroscience,* 18(1), 555–586. https://doi.org/10.1146/annurev.ne.18.030195.003011

Smart, N. (1999). *The World's Religions: Old Traditions and Modern Transformations.* Cambridge University Press.

Solms, Mark, & Friston, Karl. (2018). "How and Why Consciousness Arises: Some Considerations from Physics and Biology." *Journal of Consciousness Studies,* 25(5–6), 202–238.

Sorabji, Richard. (1993). *Animal Minds and Human Morals: The Origins of the Western Debate.* Ithaca, NY: Cornell University Press.

Spitzer, Robert J. (2015). *Finding True Happiness: Satisfying Our Restless Hearts.* San Francisco: Ignatius Press.

Spitzer, Robert J. (2015). *The Soul's Upward Yearning: Clues to Our Transcendent Nature from Experience and Reason.* San Francisco: Ignatius Press.

Sporns, Olaf. (2010). *Networks of the Brain.* MIT Press.

Stapp, Henry P. (1993). *Mind, Matter, and Quantum Mechanics.* Berlin: Springer-Verlag.

Strawson, Galen. (2006). "Realistic Monism: Why Physicalism Entails Panpsychism." *Journal of Consciousness Studies,* 13(10–11), 3–31.

Strawson, P. F. *The Bounds of Sense: An Essay on Kant's Critique of Pure Reason.* Methuen & Co Ltd, 1966.

Stump, E. (2017). "Do Persons Have Souls?" [Video]. *Closer to Truth.* Retrieved from https://youtu.be/Fui2uBDP4Gs?si=dIb3DsQmx8mBzsjG

Suárez, Francisco. 1597. *Disputationes Metaphysicae.* Quoted in "In Defense of a Thomistic-Like Dualism," by J. P. Moreland, in *The Blackwell Companion to Substance Dualism,* edited by Jonathan J. Loose, Angus J. L. Menuge, and J. P. Moreland, 102–22. Wiley Blackwell, 2018.

Taliaferro, Charles. 1994. *Consciousness and the Mind of God.* Cambridge: Cambridge University Press.

The Holy Bible. John 1:1. (English Standard Version). Quoted: "In the beginning was the Word (Logos)."

Thompson, R., Crinella, F. M., & Yu, J. (1990). *Brain Mechanisms in Problem Solving and Intelligence: A Lesion Survey of the Rat Brain.* Plenum Press.

Thompson, R., Crinella, F. M., & Yu, J. (1990). "Implications of Centrencephalic Theory." In *Brain Mechanisms in Problem Solving and Intelligence,* edited by R. Thompson, F. M. Crinella, and J. Yu, Critical Issues in Neuropsychology, Springer, Boston, MA. https://doi.org/10.1007/978-1-4757-9548-6_8.

Tononi, Giulio. (2004). "An Information Integration Theory of Consciousness." *BMC Neuroscience,* 5, 42. https://doi.org/10.1186/1471-2202-5-42.

Tononi, Giulio. (2008). "Consciousness as Integrated Information: A Provisional Manifesto." *The Biological Bulletin,* 215 (3): 216–242. https://doi.org/10.2307/25470707.

Tononi, Giulio, & Edelman, Gerald M. (1998). "Consciousness and Complexity." *Science,* 282 (5395), 1846–1851. https://doi.org/10.1126/science.282.5395.1846.

Tononi, Giulio, & Koch, Christof. (2015). "Consciousness: Here, There and Everywhere?" *Philosophical Transactions of the Royal Society B: Biological Sciences,* 370 (1668), 20140167. https://doi.org/10.1098/rstb.2014.0167.

Tononi, Giulio, Boly, Mélanie, Massimini, Marcello, & Koch, Christof. (2016). "Integrated Information Theory: From Consciousness to Its Physical Substrate." *Nature Reviews Neuroscience,* 17 (7), 450–461. https://doi.org/10.1038/nrn.2016.44.

Tononi, Giulio. (2021, August 25). "Why is Consciousness So Baffling?" [Video]. *Closer to Truth.* https://youtu.be/dK72pPa_gSE?si=2so3WW-4G9AprZhX.

Van den Heuvel, Martijn P., & Sporns, Olaf. (2011). "Rich-club organization of the human connectome." *Journal of Neuroscience,* 31(44), 15775–15786. https://doi.org/10.1523/JNEUROSCI.3539-11.2011.

Varela, Francisco J., Evan Thompson, and Eleanor Rosch. 1991. *The Embodied Mind: Cognitive Science and Human Experience.* Cambridge, MA: MIT Press.

Zimmer, Carl. (2023, July 1). "2 Leading Theories of Consciousness Square Off." *The New York Times.* https://www.nytimes.com/2023/07/01.

Zuboff, Shoshana. 2019. *The Age of Surveillance Capitalism: The Fight for a Human Future at the New Frontier of Power.* New York: PublicAffairs.

10

The Unconscious

The human unconscious plays a pivotal role in shaping our cognitive, emotional, and physiological experiences. As the hidden engine behind much of our brain's activity, it processes vast amounts of information beyond our conscious awareness. From mundane decisions to creative epiphanies, unconscious mechanisms guide perception, memory, and behavior in ways that streamline our ability to respond to complex environments. These processes enable us to efficiently manage sensory input, interoceptive signals, and emotional responses (Craig 2009). This chapter introduces the concept of the "brain metafunctional unconscious," a dynamic framework that describes how the brain integrates complex neural processes across time and distributed brain networks. These processes include face recognition, emotional regulation, memory, implicit biases, and the subconscious elements shaping motivation and interpersonal attraction. Far from being a passive storehouse of forgotten memories, the unconscious is an active, adaptive system underpinning the brain's integrated, metafunctional functioning.

By examining the unconscious through neuroscience, psychology, and philosophy lenses, this chapter highlights how these hidden processes power intuition, creativity, and spiritual experiences. The reliance of renowned figures—such as Nikola Tesla, Albert Einstein, Wolfgang Amadeus Mozart, and Johann Sebastian Bach—on their unconscious minds for inspiration and problem-solving offers striking insights into how the unconscious operates in producing profound breakthroughs (James 1902; Hart 2004). Their experiences demonstrate the depth of unconscious processes, especially in creating art and conceptualizing complex scientific theories. This chapter will explore the brain's dynamic engagement in these processes, emphasizing how unconscious activity integrates with conscious thought to form a unified cognitive experience. As we move forward, this discussion will expand to examine the relationship between the unconscious mind, the body, and the soul, emphasizing the deep interconnectedness of biological, psychological, and spiritual dimensions in human existence, including suffering (Stump 2010).

The Greek Thinkers

The concept of the unconscious runs through the centuries in diverse cultures. Though "unconscious" or "subconscious" is relatively recent, the idea appears—implicitly or explicitly—in philosophical and theological writings dealing with the human person, the mind, the soul, and the spirit. It found its earliest documentation in the Hindu *Vedas* scriptures, dating back to 2500 and 600 BCE (King 1999). In Greek thought, Plato's ideas about the soul divided the psyche into rational, affective, and appetitive. He illustrates his theory using the metaphor of a human character. Plato writes, "the human charioteer drives his in a pair [of horses]; and one of them is noble and of noble breed, and the other is ignoble and of ignoble breed; and the driving of them of necessity gives a great deal of trouble to him." Here "the charioteer" is "the one with which man reasons," "the rational principle of the soul" (Plato 1888). The ignoble part "may be termed the irrational or appetitive, the ally of sundry pleasures and satisfactions." The noble part, Plato calls "spirit," is "the forbidding principle" which is "on the side of reason." The Platonic soul comprises three components: a rational essence that guides it by restraining irrational and base desires or appetites in alignment with an elevated and commanding principle.

Aristotle emphasizes that various psychological conditions, encompassing emotions like anger, joy, courage, pity, and inclinations such as love and hate, exhibit clear and significant bodily involvement. Thus, he asserts that the exploration of the soul "is already in the province of the natural scientist" (*De Anima* i 1 403a16–28; Sorabji 2002). Nevertheless, he simultaneously contends that the mind or intellect (*nous*), distinct in its connection to the body from these states above, results in the denial of sole responsibility for the study of the soul to the natural scientist (*Meta.* vi 1 1026a4–6; *De Partibus Animalium* i 1 645a33-b10). In Aristotle's *De Anima* 3.5, the connection between intellect and thought and between thought and object eludes apprehension through discursive or conscious contemplation. To grasp the intricate interrelation demands the faculty of *nous*, an intuitive or "unconscious" form of cognition. The "active" intellect is accessible to discursive reason only sporadically. "Mind does not think intermittently," but the mind continuously thinks, consciously and unconsciously (*De Anima* 430a10–25; Alexander of Aphrodisias 2013).

Hebrew and Christian Traditions

In Hebrew and Christian traditions, the unconscious is a person's innermost being that finds expression in the different faculties of the mind and the person (Benedict XVI 2007) (Wright 2012). The Hebrew term lēb (Hebrew: בֵּל,), translated as heart, refers to the synthesis of emotion and reason as well as to the physical and spiritual aspects and the interpenetration of these constituent dimensions in humans. It appears connected to the will, the intellect, feelings, and the spirit, and it is the person's unifying center. For instance, the psalmist exclaims: "Behold, You desire truth in the innermost being, And in the hidden part You will make me know wisdom."

The writer of the book of Proverbs affirms the deep roots of the intellect and its connection with a person's identity: "As he thinks in his heart, so is he." A passage from the Book of Daniel indicates the link to the will: "Daniel purposed in his heart…." In the New Testament, Jesus said: "Let not your heart be troubled," associating the word with emotions. Here, the Greek term used is Kardia (καϱδία). Elsewhere in the letter to the Hebrews, the writer mentions the thoughts and intents of the heart: "For the word of God is living and active, sharper than any two-edged sword, piercing to the division of soul and of spirit, of joints and of marrow, and discerning the thoughts and intentions of the heart." As if pointing to the unconscious and the unifying *brain metafunction* framework with its profound transphysical domain proposed here, the following passage affirms the crucial role of the symbolic heart as the unifying center: "Above all else, guard your heart, for everything you do flows from it" (Pr. 4:23).

In Christian scriptures and theology, one diagnosis and prescribed therapy are central: something is fundamentally wrong with the human heart, the unconscious, a "sin disposition," and only a new nature capable of a profound relationship with God through the redemptive work of Christ and the indwelling of the Spirit of God in the believer provides the solution. Modern influential concepts in psychology reflect, to some extent, the diagnostic aspect but, assuming reductionist anthropology that puts aside the transphysical element, cannot reach the heart of the problem.

Recent Times

Many thinkers have contributed theories about the unconscious's nature, function, and dynamics in the past two centuries. There are too many to mention here—each dealing with different aspects of this rich and fascinating topic. For example, a famous nineteenth-century debate between Ewald Hering and Hermann von Helmholtz over the physiology of perception is fascinating. Helmholtz presents a theory of "signs," wherein sensations serve as symbols for their stimuli but do not directly replicate them (Helmholtz 1867). In contrast, Müller elucidates the connection between sensation and object through an inherent arrangement of sensory nerves (Hering 1878). However, Helmholtz contends that we establish this connection through a sequence of learned "unconscious inferences."

Sigmund Freud

Doubtless, the unconscious gained acceptance and widespread recognition through the writings of Sigmund Freud, the Austrian neurologist acknowledged as the father of psychoanalysis and, for some, modern psychology. For over three decades, Sigmund Freud continuously refined his theory of the unconscious, a core aspect of his psychoanalytic approach. Early on, Freud's topographic model distinguished between conscious, preconscious, and unconscious mental processes (Freud 1900). However, this initial framework revealed limitations, prompting him to introduce a more comprehensive "structural theory" of personality in 1923, which became a cornerstone of his later work. In this theory, Freud postulated three distinct mental structures: the ego, the id, and the superego (Freud 1923). The id, representing the primitive and unconscious part of the psyche, contrasts sharply with the ego, which, while essentially conscious, also has unconscious elements, such as defense mechanisms (Freud 1923b, 1933). Freud described the ego as structured and synthetic, acting as the organizing force of the psyche, whereas the id remained disorganized and unstructured, driven by primal desires (Freud 1933). Through this evolution, Freud's understanding of the unconscious expanded, emphasizing the complex interplay between conscious and unconscious processes in human behavior (Freud 1895, 1933). Freud posits that suppressed thoughts in the unconscious stem from the ego's endeavor to grapple with morally reprehensible realities. Similarly, redirecting harmful, socially unacceptable urges through sublimation is attributed to the cooperative interplay between the ego and superego, directing the id's impulses toward socially constructive paths.

Plato's influence on Freud's theory, as shown in the following passage and its similarity with Plato's (mentioned earlier), has not been unnoticed:

> *"The ego's relation to the id might be compared with that of a rider to his horse. The horse supplies the locomotive energy, while the rider has the privilege of deciding on the goal and of guiding the powerful animal's movement....Thus, the ego, driven by the id, confined by the superego, repulsed by reality, struggles to master its economic task of bringing about harmony among the forces and influences working in and upon it, and we can understand how it is that so often we cannot suppress a cry: 'Life is not easy!' If the ego is obliged to admit its weakness, it breaks out into anxiety-- realistic anxiety regarding the external world, moral anxiety regarding the superego and neurotic anxiety regarding the strength of the passions in the id"* (Cohen 2019).

Its historical context enhances the significance of Freud's contribution regarding his emphasis on the unconscious. In the 19th century, prevailing Western thought leaned towards positivism— the assertion that individuals could amass genuine understanding about themselves and their environment, exerting logical dominion over both realms. Nonetheless, Freud contended that these affirmations were illusory, asserting that our awareness of our thoughts is incomplete and motives entirely detached from conscious cognition frequently impel our actions.

One central aspect of Freud's theory is that in the depths of the unconscious lay a web of conflicts responsible for shaping mental states, symptoms, and even physical conditions. These conflicts emanated from primal instincts rooted in the body, seeking avenues of expression through thoughts, impulses, desires, and actions, yet often deemed socially unacceptable. Freud mainly emphasized exploring the depths of the unconscious mind and understanding its manifestations. He formulated several methodologies to facilitate this psychological research, including analyzing dreams, free association, Freudian slip clues, and psychoanalytic techniques.

Contemporary research in psychoanalysis continues (Dreher 2005) (Leuzinger-Bohleber and Fischmann 2006), focusing on clarifying and refining critical aspects of Freud's original theories. Others try re-examining core psychoanalytic constructs such as the unconscious, defense mechanisms, and the dynamic interactions between the id, ego, and superego. Scholars aim to bridge gaps between classical psychoanalysis and emerging fields, such as cognitive neuroscience and developmental psychology, to offer a more comprehensive understanding of human behavior and mental processes. Additionally, modern psychoanalysis is expanding beyond Freud's foundational ideas, addressing critiques and incorporating new perspectives, seeking to refine both its clinical applications and theoretical frameworks to meet contemporary psychological challenges (Fonagy & Allison, 2014).

Freud's Ideas and Contemporary Neuroscientific Research

More recently, cognitive psychology has recognized imperceptible mechanisms, including procedural memory (Tulving, 1972) and automatic processing (Bargh & Chartrand, 1999; Stroop, 1935), while social psychology has underscored the significance of implicit processing (Greenwald & Banaji, 1995). These empirical discoveries have highlighted the involvement of hidden processes in shaping human behavior. Criticism towards Freud's theory is numerous; for instance, due to his notably biological stance on development and an underlying anthropological approach rooted in pathology, facets of the human individual are simplified in unrealistic manners. Moreover, Freud's psychodynamic viewpoint is seen as overly deterministic, suggesting that early life encounters and concealed cognitive operations govern our thoughts, actions, and sentiments, thereby restricting the scope for conscious autonomy in behavior and personal agency. In this context, the person's accountability shrinks; yet, since the conflicts giving rise to these conditions are unconscious, there remains an absence of recourse without the guidance of an analyst, leaving them disempowered.

Freud's impact on culture and psychology has been enormous. For the neuroscientist, many of Freud's specific concepts have lost scientific credibility. One crucial point is what one author called the "demonization of the unconscious." "Freud took the external demons of possession and placed them in physical form within the patient's body, as a "second mind" that existed separately and in secret from the conscious mind. Historically and literally, this concept was the demonization of the unconscious." Today, Freud's ideas of the unconscious appear reduced and incomplete, disproportionately based on negative aspects. The controversy between defenders and critics of psychoanalytic techniques persists. Arguably, Freud's most pivotal contribution to contemporary thinking lies in bringing the unconscious to the forefront and the notion that consciousness operates in functional layers, with concealed processes percolating beneath the surface.

Diverging Visions of the Unconscious in Jung and Freud

Carl Gustav Jung's understanding of the unconscious is a nuanced and expansive model that diverges significantly from Sigmund Freud's framework. Jung identified two distinct layers of the unconscious: the personal and collective unconscious. The personal unconscious resembles Freud's concept, comprising forgotten or repressed memories, experiences, and desires. However, Jung introduced the collective unconscious, a universal, shared layer of the psyche that transcends individual experience. The collective unconscious houses archetypes—primordial, universal symbols such as the Hero, the Shadow, and the Mother—that shape human behavior, cultural expressions, and dreams. For Jung, the unconscious is not merely a reservoir of repressed material but a source of creativity, wisdom, and spiritual insight, playing a central role in personal growth and individuation. Individuation is a lifelong journey toward self-realization, achieved by integrating the unconscious with the conscious mind (Jung, 1969). Freud, by contrast, viewed the unconscious through a narrower lens, focusing on its role as a repository for repressed desires, particularly those related to sexuality and aggression. For Freud, the unconscious was primarily a source of conflict, manifesting through dreams, slips of the tongue, and neurotic symptoms. His work emphasized resolving these conflicts through psychoanalysis to alleviate psychological distress. Freud's view of the unconscious is deterministic and biologically driven, highlighting the pathological aspects of the human psyche (Freud, 1915).

The differences between Jung's and Freud's perspectives are profound. While Freud interpreted dreams as wish-fulfillments rooted in repressed desires (Freud, 1900), Jung saw dreams as a gateway to the collective unconscious, offering insights through archetypal symbols. Freud's concept of libido was confined to sexual energy, whereas Jung expanded it to encompass a generalized life energy that includes creativity, spirituality, and psychic growth. Jung also placed greater importance on spirituality, recognizing religion and mythology as expressions of the collective unconscious and essential to psychological health. In contrast, Freud dismissed religion as an illusion or a projection of repressed desires (Jung, 1969; Freud, 1915). These contrasting views reflect broader philosophical differences in their approaches to human nature. Freud focused on resolving repressions and internal conflicts to restore mental health, maintaining a reductionist and conflict-driven model. Conversely, Jung saw the unconscious as a dynamic force that fosters growth, meaning, and self-realization, emphasizing its positive, transformative potential. Jung's integration of psychology with spirituality and cultural universality contrasts with Freud's materialist and deterministic outlook, marking a pivotal divergence in the history of psychoanalysis.

Neuroscience Critiques Carl Jung's Theories

Contemporary neuroscience critiques Carl Jung's theories for lacking empirical evidence and testability, particularly the collective unconscious, archetypes, and individuation concepts. Central to Jung's framework, the collective unconscious is criticized as a metaphysical construct without scientific validation within the naturalist framework. Interestingly, some argue that neuroscience has not identified universal, inherited structures in the brain corresponding to Jung's archetypes, suggesting instead that shared human experiences arise from cultural transmission rather than biological inheritance (Goodwyn, 2024). Similarly, some point out that archetypes are unverifiable constructs, lacking measurable or observable evidence within current scientific methodologies (Northoff, 2013).

Jung's reliance on symbolic interpretations, such as his analysis of dreams, myths, and art, further highlights tensions with contemporary neuroscience. Where Jung saw dreams as expressions of the collective unconscious, neuroscience interprets them as by-products of neurological processes like memory consolidation. This contrast underscores the divide between Jung's emphasis on symbolic meaning and neuroscience's focus on mechanistic explanations rooted in neural activity (Hobson, 1999; Northoff, 2013). Neuroscience's emphasis on neuroplasticity also challenges Jung's notion of inherited archetypes, favoring the view that psychological patterns emerge from environmental influences and learned behaviors rather than fixed, biologically inherited structures. According to prevailing ideas, shared behavioral patterns likely evolved as adaptations to common ecological pressures rather than manifestations of a universal unconscious repository (Goodwyn, 2024).

Clinically, Jung's methods—such as exploring myths and archetypes—are considered more philosophical than therapeutic, contrasting with evidence-based treatments like cognitive-behavioral therapy (CBT) and pharmacological interventions. Despite this, Jung's ideas continue to influence fields like psychotherapy and cultural studies, particularly in approaches that integrate symbolic meaning and personal growth. However, voices from neuroscience dismiss them for their philosophical nature and lack of empirical grounding (Goodwyn, 2024; Hobson, 1999; Northoff, 2013).

While contemporary neuroscience challenges Jung's theories for their metaphysical basis and lack of empirical support, the persistence and appeal of Jung's contributions raise essential questions. Could the reductionist, biologicist perspective in neuroscience, which emphasizes observable and measurable phenomena, overlook aspects of the human experience that are inherently non-reductive and symbolic? Conversely, could Jung's symbolic framework miss the concrete underpinnings necessary for scientific integration? The enduring influence of Jung's ideas in areas like psychotherapy, cultural studies, and spirituality suggests that his work addresses dimensions of human experience that remain elusive to purely biological explanations. Yet, the lack of empirical validation also underscores the need for rigor in understanding the unconscious. A broader framework that bridges the subjective richness of Jung's symbolic interpretations with the empirical precision of neuroscience is missing on both sides.

Walter Cannon's "autonomous," unconscious physiological processes

In the realm of physiology and unconscious processing, Walter Cannon's observations from 1932 remain highly relevant. Cannon noted numerous vital metabolic functions, including respiration, circulation, and blood glucose regulation, function with substantial autonomy under intricate feedback control mechanisms. This sophisticated autonomic but integrated capacity enables these physiological processes to operate continuously and dynamically without requiring conscious intervention. The complex coordination and regulation of these bodily functions occur seamlessly in the background, contributing to the overall homeostasis of the organism. The degree of autonomy of these unconscious processes is crucial in affording cognitive operations a considerable degree of freedom to engage with the external environment, both physically and socially. By relieving conscious awareness from the burden of micromanaging fundamental physiological activities, individuals can allocate cognitive resources to more complex and nuanced cognitive tasks. This interplay between the "autonomous," unconscious physiological processes and conscious cognitive functions, all integrated as part of the brain metafunction, highlights the depth and intricacies of the mind-body relationship.

Furthermore, this integrated autonomy in bodily functions contributes to the organism's adaptability and responsiveness to various environmental stimuli. We have direct and indirect ways of modulating breathing and circulatory functions, such as breathing and feedback (Pfaltz et al. 2010) (Lehrer and Gevirtz, 2014). However, the unconscious regulation of respiration, circulation, and metabolic activities ensures the body maintains stability and dynamically adjusts to changing internal and external conditions. It exemplifies the remarkable efficiency and adaptability embedded in the intricate interplay between unconscious physiological processes and conscious cognitive functions, showcasing the sophistication of the human mind-body system and pointing to the *brain metafunction* level for understanding the full scope of its integration.

The Metafunctional Unconscious

The term *metafunctional unconscious* derives logically from the brain metafunction framework. It encompasses diverse processes such as internal bodily regulation, memory formation, whole-body-image-based motor activities, physical balancing, language, learning, and numerous emotional and cognitive functions, including decision-making, judgment, and goal setting. Recent advancements in neuroscience have emphasized the growing importance of unconscious processes, particularly within large-scale brain networks and brain-wide distributed systems. If the brain metafunction is an iceberg, the conscious mind represents its visible tip, while the metafunctional unconscious forms the massive, submerged base, driving much of the brain's activity.

Kandel et al. (2013) delineated three categories of unconscious processing that align with this framework. *Implicit unconscious processing* involves procedural memory and skill-based learning managed by regions such as the cerebellum and the striatum. *Dynamic unconscious processing* pertains to emotional conflicts, suppressed thoughts, and desires, likely associated with limbic system components. *Preconscious processing,* meanwhile, encompasses executive functions like decision-making and planning, primarily involving the parietal, prefrontal, and premotor cortices. While this categorization aids research within the metafunctional framework, these processes integrate into a holistic understanding of the brain's unconscious mechanisms.

Neuroscience has made substantial strides in understanding unconscious processing over the past two decades. Investigations into priming, procedural memory, and implicit (unconscious) memory have revealed how these processes underlie everyday behavior. Implicit memory includes automated actions, skills, and unperceived relational aspects of social interactions. Activities like walking, speaking, dancing, or driving are unconsciously encoded behaviors seamlessly integrated with conscious-unconscious dynamics.

The Metafunctional Unconscious and Attention

Unconscious processing plays a central role in attentional functions, filtering and prioritizing information to adapt to rapidly changing environments (Dehaene & Naccache, 2001; Naccache, Blandin, & Dehaene, 2002). This process operates alongside conscious attention mechanisms, optimizing cognitive performance and adaptability. For example, while walking on a busy street, unconscious filtering excludes irrelevant stimuli, enabling focus on essential cues such as crossing the road or avoiding obstacles. Similarly, during routine activities like driving a familiar route, unconscious processes manage steering and speed regulation, freeing conscious attention for strategic tasks such as navigating traffic. Reflexive responses, like stepping back when someone suddenly appears in your path, occur before conscious attention registers the event, demonstrating the speed and efficiency of unconscious processing.

Imagine entering a lecture room: your brain unconsciously recognizes familiar faces and objects, enabling quick orientation. As the lecture progresses, your unconscious monitors background noises and alerts you to significant sounds or movements, such as a loud noise or a potential threat. At the same time, unconscious processing directs your attention toward relevant stimuli, enhancing multitasking capabilities. For instance, while listening to the lecturer, you may also process complex ideas or engage in abstract thinking evoked by the discussion. These scenarios highlight how unconscious and conscious processes work in tandem to facilitate efficient decision-making and adaptability. Large-scale brain networks, such as the *default mode and salience networks*, are crucial in managing this interplay. These interconnected systems allow the brain to balance internal reflection with external awareness, ensuring seamless integration of unconscious and conscious processes. By relieving the conscious mind from micromanaging fundamental activities, the metafunctional unconscious enables the brain to allocate resources to complex cognitive tasks, enhancing survival and adaptability.

The concept of the metafunctional unconscious underscores the brain's extraordinary capacity to manage intricate processes, from routine behaviors to advanced cognitive tasks, without conscious oversight. We gain insight into the broader integration of conscious and unconscious processes within the brain metafunction framework by exploring how unconscious mechanisms shape attention and cognition. The chapter on Attention explores this aspect in more detail.

The Metafunctional Unconscious and Language

The display of unconscious processing in human language is awe-inspiring and underscores its brain-metafunctional quality. For thinkers like Noam Chomsky (1965) and Jackendoff (2002), language offers one of the most straightforward windows into the human mind. Unconscious processing is foundational in enabling efficient and fluid communication, facilitating the intricate cognitive processes involved in language comprehension and production. Unconscious language processing allows us to quickly and effortlessly understand and produce spoken and written language. Language comprehension occurs automatically, enabling real-time meaning extraction from sentences, phrases, and words. For instance, during conversations, unconscious processing aids rapid turn-taking by predicting the speaker's intentions and formulating appropriate responses (Garrod & Pickering, 2004). This process activates related concepts and meanings based on context, enhancing comprehension and ensuring smooth dialogue (Collins & Loftus, 1975). Native speakers also instinctively follow grammatical rules without consciously understanding them, allowing the construction of grammatically correct sentences (Chomsky, 1965). Metaphor comprehension further demonstrates unconscious processing, as it maps abstract concepts onto one another, enabling nuanced interpretation (Lakoff & Johnson, 1980).

Unconscious processing also handles emotional and social cues embedded in language, such as tone of voice and expressive intonation. These subtleties contribute to effective communication across various contexts. Speech production, including articulation and prosody, relies on automated motor patterns (Levelt, 1989). Implicit learning facilitates language acquisition, enabling the unconscious integration of vocabulary, syntax, and pronunciation through exposure and practice (Reber, 1967; Ellis, 1994). This offloading of routine linguistic tasks allows conscious awareness to focus on higher-order cognitive activities like problem-solving and creative thinking (Dehaene & Naccache, 2001).

Metafunctional Neural Correlates of Human Language.

Language production and comprehension engage extensively distributed, large-scale brain networks across cortical and subcortical regions. While the left hemisphere dominates language processing in most individuals, both hemispheres contribute uniquely to linguistic functions. The left hemisphere encompasses Broca's area, crucial for grammatical processing and speech-motor planning (Friederici, 2011), and Wernicke's area, responsible for language comprehension and semantic interpretation (Geschwind, 1970). The Arcuate Fasciculus, a white matter tract, connects these areas, facilitating the integration of production and comprehension (Friederici, 2011). Additional left hemisphere regions include the Middle Temporal Gyrus (MTG), which plays a central role in lexical access, semantic integration, and narrative comprehension (Hagoort, 2017), and the Inferior Frontal Gyrus (IFG), which contributes to resolving syntactic ambiguities and higher-order semantic processing (Friederici, 2011). These areas work alongside the Angular Gyrus and Supramarginal Gyrus, which link phonological, semantic, and visual inputs, facilitating reading and multimodal integration (Price, 2012). The right hemisphere supports prosody, emotional tone, and pragmatics, including implied meanings and metaphor interpretation (Binder et al., 1999; Bookheimer, 2002; Blake, 2017). The Default Mode Network (DMN), which includes the posterior cingulate cortex and medial prefrontal cortex, is involved in processing narratives and integrating language with broader cognitive functions, such as mentalizing and self-referential thought (Binder et al., 1999).

Subcortical structures also play a critical role. The thalamus is a gateway for auditory and visual language information, facilitating connectivity between cortical regions and regulating attentional shifts during speech (Crosson, 2013). Beyond word retrieval and motor planning, the basal ganglia assist in syntactic and grammatical processing in real-time sentence comprehension (Booth et al., 2007). The hippocampus supports vocabulary learning and declarative memory formation (Ullman, 2004). Emerging research also highlights the cerebellum's role in phonological processing, articulation, and timing during speech (Mariën et al., 2014).

Language Dorsal and Ventral Pathways

The dorsal and ventral pathways are crucial for integrating auditory, motor, and semantic processes (Schomers & Pulvermüller, 2016). The dorsal pathway connects regions like Broca's area and the posterior temporal lobe via the arcuate fasciculus, supporting articulation, speech-motor planning, repetition, and syntactic processing by mapping sounds to motor actions. In contrast, the ventral pathway links the temporal and frontal lobes, facilitating lexical and semantic comprehension by mapping phonological input to stored lexical and contextual knowledge. For example, while the dorsal pathway governs articulation during speech, the ventral pathway decodes meaning and contextual nuances. These pathways dynamically interact during complex tasks, such as processing embedded sentence clauses (Hickok & Poeppel, 2007).

While the left hemisphere is typically dominant for language processing, the right hemisphere's equivalent pathways contribute to non-linguistic communication. The right hemisphere's dorsal pathway supports prosody, such as rhythm, intonation, and stress patterns, contributing to the melodic contour and emotional tone of speech. Similarly, its ventral pathway facilitates understanding figurative language, metaphors, sarcasm, and social pragmatics by processing contextual and emotional nuances (Blake, 2017). Together, these pathways enable the interpretation of implied meanings and enhance communication richness. For instance, while the left hemisphere decodes literal linguistic content, the right hemisphere adds depth by integrating emotional and social cues. Damage to these pathways can result in aprosodia, difficulty perceiving or expressing emotional intonation, or impairments in understanding figurative expressions. This bilateral cooperation ensures nuanced and effective communication.

The claustrum, an intriguing structure beneath the lateral cerebral cortex, has been hypothesized to integrate cognitive functions across sensory and motor systems (Crick & Koch, 2005). Although its role remains speculative, it highlights the metafunctional complexity of language. Furthermore, embodied language processing involves sensory and motor areas like the premotor cortex, which contribute to understanding action-related language (Schomers & Pulvermüller, 2016). Beyond specialized regions, language integrates with memory, perception, and attention, ensuring seamless processing of spoken and written language. This distributed network enables the richness and complexity of human communication, reflecting the brain's extraordinary capacity for metafunctional processing.

How Interoception Influences Language

Interoception significantly influences language comprehension by shaping how individuals process and interpret linguistic inputs, particularly emotions, bodily states, and social communication. Defined as the brain's capacity to sense, analyze, and integrate internal bodily signals such as heart rate and visceral sensations, interoception is closely tied to emotions, attention, and decision-making—core aspects of language comprehension (Craig 2009). For instance, interoceptive awareness enhances the ability to detect emotional subtleties in language, such as sarcasm or empathy, by integrating linguistic and affective cues (Critchley and Garfinkel 2017). Moreover, interoception plays a crucial role in understanding metaphorical language. Phrases like "a gut feeling" or "heart-wrenching" are grounded in bodily experiences, and individuals with heightened interoceptive awareness may process these metaphors more vividly, supporting theories that metaphorical language bridges physical and abstract concepts (Lakoff and Johnson 2008). Additional research further explores this connection, showing how interoceptive and sensorimotor signals influence abstract and figurative language comprehension, providing behavioral evidence that conceptual representations are grounded in bodily states and contextual factors (Connell & Lynott, 2014).

In addition to metaphorical language, interoception modulates attention, a vital factor in language comprehension. For example, interoceptive dysregulation, such as anxiety, may impair the focus on complex conversations or texts, reducing comprehension (Seth and Tsakiris 2018). This fact aligns with the embodied cognition framework, which suggests that cognitive processes, including language comprehension, are deeply rooted in sensory and bodily experiences. Interoceptive signals provide a bodily context for linguistic inputs, activating brain areas associated with motor and sensory functions to simulate described experiences, thereby aiding comprehension (Barsalou 1999). This connection is increasingly relevant in clinical contexts, where interoceptive perspectives are being integrated into evaluations of emotional and social aspects of language, emphasizing their role in therapeutic interventions (Solano Durán et al., 2024).

Clinical and developmental findings further underscore interoception's role in language. Conditions affecting interoception, such as autism spectrum disorder or alexithymia, are often associated with difficulties in processing the social and emotional nuances of language. For example, individuals with alexithymia may struggle to connect interoceptive signals with emotional language, limiting effective communication. Recent studies demonstrate that interoceptive accuracy is directly linked to the ability to verbalize emotions, suggesting that improved interoception could enhance emotional language use and social interaction (Suzuki and Yamamoto, 2023). Similarly, second-language learning insights suggest that interoceptive feedback and sensory integration enhance abstract concepts' comprehension, broadening embodied cognition's role in language acquisition (Jonghyun, 2023).

How is interoception revealed as brain metafunctional? Interoception enhances the comprehension of emotional language, metaphors, and abstract concepts and integrates bodily awareness with cognitive and linguistic processes. Interoception underscores the profound interconnectedness of mind and body in human communication by influencing attention, focus, and emotional resonance. Furthermore, its influence on narrative understanding and emotional resonance in language highlights its potential as a key factor in bridging physical sensations with abstract cognitive functions, as emphasized in recent reviews on interoceptive mechanisms and their integration into social and emotional language processing (Greenwood and Garfinkel, 2024).

Emotion's Role in Language Comprehension

Emotions significantly influence language comprehension by modulating how individuals process, interpret, and respond to linguistic inputs. They interact with cognitive functions such as attention, memory, and decision-making, essential for understanding language. For example, emotionally charged words like "love" or "fear" often capture attention more effectively than neutral words, eliciting heightened neural activity in the amygdala and prefrontal cortex, making such language more memorable and impactful (Adolphs, 2002). Similarly, prosody—encompassing intonation, pitch, and stress—conveys emotional nuances that can alter spoken language interpretation. For instance, distinguishing an angry tone from a cheerful one requires integrating prosodic cues, a process supported by regions like the superior temporal sulcus and prefrontal cortex (Binder et al., 1997).

People are likelier to infer subtle meanings, interpret sarcasm, or comprehend metaphors in emotionally charged situations. This capacity relies on the interaction between the limbic system and higher-order cognitive areas such as the prefrontal cortex (Dolan, 2002). Emotions also modulate attention during language processing: positive emotions can broaden the attentional focus, enhancing comprehension, while negative emotions, like anxiety, might narrow focus, impairing understanding of complex sentences (Critchley & Garfinkel, 2017). Furthermore, emotions enhance memory for language. The amygdala interacts with the hippocampus to prioritize emotionally salient information, improving recall and comprehension over time (Damasio, 2010).

Understanding figurative language often depends on emotional engagement. Phrases like "a heavy heart" or "walking on air" evoke emotional experiences that link abstract meanings to personal contexts (Lakoff & Johnson, 2008). Emotional impairments in conditions such as autism spectrum disorder or alexithymia highlight the importance of emotional processing for effective communication, as these conditions often lead to difficulties interpreting social and emotional cues in language (Critchley & Garfinkel, 2017). Emotions profoundly enhance the ability to process and understand language by modulating attention, memory, and interpretation, illustrating the interconnectedness of cognitive and affective systems in human communication.

Emotions Shape Sentence Processing

Emotions also play a critical role in sentence processing, influencing how linguistic information is perceived, interpreted, and integrated. Words like "anger" or "joy" elicit heightened neural activity in areas such as the amygdala and prefrontal cortex, facilitating recognition and processing speed (Adolphs, 2002). Prosodic features—such as pitch, rhythm, and tone—convey emotional nuances that shape sentence interpretation. For instance, a cheerful tone might suggest positivity, while a sad tone can evoke empathy, with the superior temporal sulcus contributing to processing these cues (Binder et al., 1997). Emotional context also guides pragmatic interpretation, such as distinguishing sarcasm from sincerity, involving interactions between the limbic system and prefrontal cortex (Dolan, 2002).

Emotional states further modulate attention, directly impacting sentence comprehension. Positive emotions often broaden the attentional focus, aiding the integration of sentence components, whereas negative emotions, like anxiety, narrow focus and impair comprehension of complex or ambiguous sentences (Critchley & Garfinkel, 2017). For example, joy may enhance the processing of positive content, while sadness may facilitate the understanding of empathetic expressions.

Emotionally charged sentences are more memorable due to interactions between the amygdala and hippocampus, which prioritize emotionally relevant information (Damasio, 2010).

The comprehension of figurative language often relies on emotional resonance. Sentences like "She has a heavy heart" draw on the listener's or reader's emotional experiences to bridge abstract concepts with intuitive understanding (Lakoff & Johnson, 2008). Emotional contexts can bias interpretations of ambiguous sentences, aligning comprehension with the individual's affective state (Bargh & Chartrand, 1999). However, in emotionally intense situations, cognitive resources may be diverted to emotionally salient cues, reducing the capacity for syntactic and semantic analysis (Critchley & Garfinkel, 2017). This dynamic interplay between emotions and cognitive processes underscores their integral role in language processing, shaping how humans interpret and engage with complex linguistic information.

Human Language Points to a Transphysical Domain.

Human Language and Universal Syntax

In their collaborative work, *Why Only Us*, Noam Chomsky and Robert Berwick (2016) explored the uniqueness of human language, emphasizing universal syntactical competence and critiquing naturalistic evolutionary explanations. They identified two critical characteristics of human language. First, they underscored the universality of the "syntax test," demonstrating that individuals across cultures and languages possess the innate ability to construct and comprehend grammatically structured sentences. Second, they emphasized the "syntactical test," highlighting the abstract cognitive capacity to distinguish between sentence components like subjects, predicates, and objects. These findings suggest a shared, inherent syntactical framework across humanity.

Perceptual and Conceptual Ideas: The Role of Heuristic Notions

Human language bridges two essential domains: perceptual and conceptual ideas. Perceptual ideas arise from sensory experiences, encompassing inputs like visual, auditory, tactile, and interoceptive signals (Barsalou, 1999). Conceptual ideas, by contrast, represent the cognitive processes that organize and interpret these sensory inputs into abstract constructs, enabling communication of complex notions (Fodor, 1975). For example, perceiving and conceptualizing an apple as a "fruit" exemplifies this transformation. The interaction between these domains is facilitated by heuristic notions and cognitive strategies that connect sensory input with abstract concepts (Lakoff & Johnson, 2008). The term heuristics, derived from the Greek word "heuriskein" ("to uncover"), functions as cognitive shortcuts, aligning perceptual and conceptual ideas and fostering effective communication (Pinker, 2007). For instance, the metaphor "carrying emotional baggage" leverages heuristic notions to translate physical sensations into abstract emotional experiences, enhancing comprehension.

Language and the Transphysical Domain

Moving beyond physicalist explanations, some theorists argue that the origin of conceptual ideas transcends neurocomputational processes and empirical experiences. Robert Spitzer (2015) posits that conceptual ideas emerge from a higher transphysical domain rooted in the human soul. This perspective suggests that the capacity for abstract thought and universal concepts precedes external stimuli, emphasizing a profound relationship between the soul and language. This horizon transcends individual instances and ushers in universal concepts. Within this framework, heuristic notions and conceptual ideas testify to the soul's role in elevating human cognition.

In contrast, despite significant advances in neuroscience, cognitive science, and linguistics, naturalistic frameworks struggle to fully explain the depth and scope of human language in its richness and dimensions. The ability of language to convey abstract thought, integrate perceptual and conceptual realms, and evoke deep emotional resonance defies purely mechanistic explanations. Naturalistic theories fall short of accounting for the universal and innate syntactical structures observed across all human cultures or the profound ways in which language connects to meaning, identity, and spirituality. These gaps suggest that human language may arise from dimensions extending beyond empirical or evolutionary paradigms, pointing to a deeper transphysical or metaphysical foundation.

The Connection of Language and the Human Soul

Language profoundly connects to the human soul, serving as a mirror and medium for expression. It reflects human cognition, creativity, and self-awareness—attributes intricately tied to the soul. The ability to generate and comprehend complex syntax and semantics, as argued by

Noam Chomsky and Robert Berwick (2016), reveals an innate capacity for abstract reasoning that transcends biological or computational processes. This universal syntactic framework, shared across all cultures, unifies humanity and aligns with the soul's transcendent qualities. Language enables humans to conceptualize and communicate abstract ideas, such as justice, love, and metaphysical truths—concepts central to the soul's nature. Beyond functional communication, language fosters expressions of morality, creativity, and the pursuit of truth. Lakoff and Johnson (2008) describe language as a heuristic mechanism linking perceptual experiences, like seeing a flower, with conceptual ideas such as symbolic beauty or emotional resonance. This interplay underscores the soul's ability to synthesize sensory inputs into profound meanings, bridging the tangible and intangible. Moreover, language serves as a vehicle for spiritual expression, allowing individuals to articulate inner emotions and spiritual reflections. Sacred texts, prayers, and hymns demonstrate how language facilitates engagement with the divine, elevating it beyond a mere communicative tool to a sacred medium. Fr. Robert Spitzer (2015) emphasizes that the ability to conceptualize and communicate abstractly points to a transphysical origin, suggesting that the soul infuses language with creativity and meaning beyond mechanistic brain functions.

Language also plays a pivotal role in constructing and expressing the self. A person's goals, desires, and identity are pursued through language, reflecting the soul's intrinsic quest for meaning and coherence. This process allows the soul to externalize its inner workings, engage with others, and seek transcendence. Spitzer (2015) contends that the origins of this capacity for abstract communication defy purely naturalistic explanations, instead hinting at a higher transphysical domain rooted in the soul. Language is an instrument of abstract thought, self-expression, and spiritual engagement. It bridges the physical and transcendent realms, enabling humans to think, feel, and communicate in ways that surpass material existence. Through language, the soul engages with both the finite and the infinite, revealing the extraordinary nature of human existence and its deep-rooted dependence on symbolic expression.

The Metafunctional Unconscious and Perception

The human brain's ability to integrate language, bodily regulation, and sensory processing highlights the remarkable scope of the metafunctional unconscious. Advances in neuroscience have deepened our understanding of interoception—the nervous system's ability to perceive, interpret, and integrate signals from within the body. For instance, interoception is how the nervous system perceives, construes, and amalgamates signals emanating from the body's interior. Interoceptive processes provide a constant "mapping" of the body's internal state, encompassing both unconscious and conscious realms (Craig, 2009). These signals are crucial for managing homeostasis and allostasis, which maintain equilibrium and enable adaptive responses to stress (Critchley & Garfinkel, 2017; Sterling & Eyer, 1988). Such mechanisms illuminate how the brain's metafunctional nature links bodily states with emotional and cognitive experiences, contributing to survival and the construction of selfhood (Seth & Tsakiris, 2018).

Unconscious processing is central to human perception, enabling the brain to efficiently manage the immense influx of sensory data. By filtering and prioritizing information, unconscious processing allows attention to focus on the most relevant stimuli, facilitating adaptive responses. For instance, it enables rapid reactions to threats, such as pulling back from a moving object, without requiring conscious deliberation (Bargh & Chartrand, 1999; Dehaene et al., 2001). This efficiency frees cognitive resources for higher-order functions, such as decision-making and problem-solving (Schacter & Tulving, 1994). Perceptual priming exemplifies unconscious processing. In tasks like word fragment completion, exposure to stimuli influences subsequent responses without conscious awareness (Kouider & Dehaene, 2007). For example, participants exposed to a list of words are more likely to complete fragments (e.g., "_ _ R _ _ A _ _") with previously encountered terms, such as "PARADE." Neural studies using fMRI and EEG reveal that priming engages regions associated with memory retrieval and semantic processing, demonstrating how unconscious processes shape perceptions and behaviors (Henson, 2003).

Face perception provides another compelling example of the brain's unconscious capabilities. The fusiform face area (FFA) and occipital face area (OFA) specialize in facial recognition, integrating features like eyes, nose, and mouth into coherent representations (Kanwisher, 2000; Haxby et al., 2000). These processes are so efficient that the brain supports distinguishing between thousands of faces despite their structural similarities (Gauthier et al., 1999). Neural pathways connecting these regions enable the seamless integration of facial features, illustrating how unconscious processing supports complex pattern recognition (Falsma, 2010; Pelphrey et al., 2004). Multisensory integration further highlights the brain's metafunctional unconscious. In phenomena like the "ventriloquist illusion," the brain combines visual and auditory information, even when spatially misaligned, leading to the perception of sound originating from the visual stimulus (Kok, Jehee, & de Lange, 2012). Research shows increased connectivity between sensory cortices and activation of multisensory regions, such as the superior colliculus and intraparietal sulcus, during such tasks (Talsma et al., 2007; Driver & Noesselt, 2008). This seamless merging of sensory inputs demonstrates the brain's metafunctional ability to create a unified perceptual experience, even under incongruent conditions (Murray & Wallace, 2012).

Metafunctional Context of Unconscious Perceptual Processing

Returning to face recognition, it exemplifies the brain's metafunctional capabilities, where cortical and subcortical regions collaborate seamlessly to process complex perceptual and emotional information. The fusiform face area (FFA) specializes in recognizing facial identity and emotional expressions. In contrast, the occipital face area (OFA) contributes to early face detection and visual processing (Haxby, Hoffman, & Gobbini, 2000). The superior temporal sulcus (STS) processes dynamic facial cues, such as movements and expressions, enriching social interactions (Haxby et al., 2000). Meanwhile, the anterior cingulate cortex (ACC) monitors emotional incongruities and integrates conflict detection with emotional processing (Phan et al., 2002). The orbitofrontal cortex (OFC) evaluates social and emotional stimuli, influencing decision-making (Rolls, 2004), while the amygdala plays a central role in detecting emotional salience, such as fear or threat, shaping social judgments (Adolphs 2002; Adolphs 2003).

Subcortical structures provide essential support. The thalamus relays sensory information to cortical areas, while the hippocampus encodes and retrieves emotional and facial memories (LeDoux, 2012). The striatum integrates implicit biases and reward processing, influencing approach-avoidance behaviors (Dolan, 2002). The basal ganglia contribute to habit formation, linking face recognition with memory-based behaviors (Pessoa, 2017). Reward-related areas like the nucleus accumbens strengthen learning and emotional associations with faces, reinforcing social interactions (Dolan, 2002). The hypothalamus and other limbic structures regulate physiological responses, modulating emotional and social dimensions of perception (Wallace et al. 2020). Brainstem systems underpin arousal and support the integration of these widespread processes (Adolphs, 2017). This distributed network exemplifies the metafunctional nature of unconscious perceptual processing. By integrating sensory, emotional, and cognitive functions across large-scale neural networks, the brain enables seamless face recognition, social evaluation, and adaptive behaviors. This synthesis highlights the interconnectedness of unconscious mechanisms and the brain's capacity to unify diverse functions into coherent experiences, underscoring its metafunctional essence.

The Metafunctional Unconscious and Memory

Unconscious processing plays a pivotal role in human memory by optimizing the encoding, organization, and retrieval of vast amounts of information without overwhelming conscious awareness (Dudai, 2004). This streamlined approach enables the brain to capture relevant details from the environment, categorize them, and form meaningful associations. By identifying patterns and organizing memories based on relationships, unconscious processing facilitates easier retrieval and application in future scenarios (Cleeremans, 2008). Implicit memory—supported by unconscious learning—drives skill acquisition, habits, and behaviors, shaping much of our everyday actions without requiring conscious effort (Schacter, 1987). The metafunctional unconscious integrates sensory information across modalities, enriching memory formation with multidimensional details and enhancing recall. Contextual encoding, where unconscious processes capture environmental cues surrounding events, significantly aids in reconstructing memories during retrieval (Dudai, 2004). Additionally, priming effects illustrate how unconscious exposure to stimuli influences subsequent perception and behavior, showcasing the efficiency of these processes (Cleeremans, 2008).

Memory consolidation during sleep exemplifies the metafunctional nature of unconscious processing. Sleep stabilizes and enhances memory connections formed during wakefulness, allowing neural patterns to reorganize and strengthen (Stickgold, 2005; Rasch & Born, 2013). The interplay between conscious and unconscious domains during dreams provides insight into how unconscious processes contribute to integrating past experiences and forming coherent narratives. Unconscious memory retrieval further optimizes cognitive resources, enabling rapid decision-making and filling gaps in fragmented memories. By reconstructing events and embedding emotional nuances, the metafunctional unconscious ensures that memory systems operate efficiently and adapt to the complexities of daily life. These mechanisms highlight the seamless coordination of unconscious processes, demonstrating their vital role in navigating the world, making decisions, and deriving meaning from stored experiences. For a deeper exploration of these functions, refer to the chapter on Memory.

The Metafunctional Unconscious, Emotions, and Feelings

The concept that "the fountain of feeling is in the unconscious" underscores the profound role of unconscious processes in shaping emotions, reactions, and our inner world. Emotions represent physiological responses to sensory inputs and feelings, and the conscious awareness of these responses forms an intricate interplay central to human experience (Hansen, 2005). For instance, encountering a threat triggers automatic emotional responses—elevated heart rate, tense muscles, and dry mouth—long before we consciously register the feeling of fear. Research highlights the role of unconscious processing in emotional perception and response. Subliminal priming experiments, where emotional stimuli are presented below conscious awareness, demonstrate how the brain detects and reacts to emotional cues. Neuroimaging reveals activity in regions such as the amygdala and insula even when participants are unaware of these stimuli, indicating the metafunctional unconscious's influence on cognition and behavior (Elgendi et al., 2018).

Unconscious emotional processing also aids in encoding emotionally charged memories, shaping behaviors and decisions long after the initial experience. Emotional memories often intertwine with unconscious associations formed through past experiences and cultural influences, subtly shaping responses to stimuli (Cleeremans, 2008). For example, unconscious processes help integrate complex emotions into dreams, meaningfully exploring deeply rooted feelings. Defense mechanisms, such as repression and projection, further illustrate the unconscious's role in managing overwhelming emotions. These processes shield conscious awareness from distressing feelings while influencing emotional interactions and thought patterns. Intuition, another manifestation of unconscious processing, provides visceral judgments about situations or people, drawing on past experiences and subtle environmental cues.

While neuroscience sheds light on the unconscious's role in emotions, a broader anthropological perspective suggests that humans possess a spiritual dimension capable of transcending and redefining these processes. This perspective highlights the dynamic relationship between the unconscious, emotions, and the human soul, suggesting a higher capacity to reinterpret and elevate our emotional experiences. Through these layers of unconscious and conscious interplay, our emotional lives are profoundly shaped. This emphasizes the metafunctional unconscious's centrality in integrating thoughts, memories, feelings, and behavior. This dynamic process reminds us of the depth of human emotion and its intricate connections to cognition, behavior, and meaning.

Imitating Physical Behavior and Emotional Expressions

The discovery of mirror neurons by Giacomo Rizzolatti and colleagues in the 1990s revolutionized our understanding of how the brain integrates motor, cognitive, and social functions (Di Pellegrino et al., 1992; Rizzolatti et al., 2001). These neurons, first identified in the premotor cortex of macaques, activate when an individual performs an action and when they observe the same action performed by others. This mechanism underpins unconscious imitation of physical behaviors, allowing humans to align their actions with those around them and fostering social cohesion and learning. Particularly during childhood, this unconscious mirroring facilitates theory of mind development—the ability to attribute mental states to others and comprehend their thoughts and emotions.

The mirror neuron system (MNS) extends beyond motor functions to encompass emotional and social processing (Cook et al., 2014; Bonini et al., 2022). When we observe someone displaying an emotion, such as joy or sadness, the MNS mirrors that emotion within us, enabling empathy and emotional connection. Neuroimaging studies reveal that regions like the amygdala, insula, and limbic networks are pivotal in emotional mirroring. At the same time, homeostatic and allostatic systems regulate physiological responses, such as heart rate and respiration, to align with the observed emotion (Gallese, 2003). Laughter, a quintessential social behavior, exemplifies the MNS's role in emotional resonance. Studies show that mirroring another's laughter activates cingulate, opercular, and temporal regions, fostering affiliative synchrony even among strangers (Golland et al., 2015).

The MNS also bridges motor functions and higher cognitive processes, such as language acquisition and comprehension. Research suggests that language functions emerged from a pre-existing action recognition system, with the MNS linking motor circuits to linguistic processing (Arbib, 2010). This system facilitates verbal and non-verbal communication, highlighting the brain's ability to integrate sensory, motor, and cognitive domains for effective social interaction (Iacoboni, 2008). Furthermore, the MNS plays a critical role in social cognition by interpreting others' intentions. Observing someone handling an object activates neural circuits responsible for motor planning and object recognition, illustrating the distributed nature of the MNS (Cook et al. 2014). This network connects regions involved in sensory processing, memory, and bodily regulation, enabling a seamless integration of environmental and social cues.

Memory-related functions further demonstrate the MNS's metafunctional quality. Implicit learning and memory consolidation, particularly in skill acquisition and social learning, rely on the MNS to encode and retrieve motor and social knowledge. For example, priming effects—where prior exposure to stimuli influences future actions—highlight how the MNS integrates past experiences into adaptive behaviors (Stickgold, 2005). The MNS exemplifies the brain's remarkable capacity for metafunctional integration. Its involvement in action recognition, language, empathy, and physiological regulation reflects *a distributed, brain-wide system* that unifies diverse domains to effectively navigate social and environmental challenges. By connecting sensory input, motor output, bodily regulation, emotional and cognitive processes, the MNS underscores the complexity and adaptability of human behavior.

Implicit Bias and Stereotyping

Unconscious cognitive processes play a crucial role in shaping implicit biases and stereotypes, often without our conscious awareness. Research shows that these biases are rooted in automatic brain processes shaped by cultural and societal influences, frequently operating below the level of conscious thought. As a result, they affect perceptions, judgments, and decisions, even when they conflict with consciously held beliefs. For example, neuroscientific studies demonstrate how implicit biases emerge in various contexts. Functional magnetic resonance imaging (fMRI) studies reveal that the amygdala, a region responsible for emotional processing, activates when individuals see the faces of people from different racial or ethnic backgrounds (Phelps et al. 2000). This activation occurs even in individuals who consciously reject racial biases, indicating that implicit biases often function beyond conscious control. Such automatic responses may lead to snap judgments based on race or ethnicity, despite intentional efforts to remain unbiased.

Similarly, the brain's reward system, particularly the ventral striatum, influences biases related to attractiveness. This area becomes activated when individuals view conventionally attractive faces, reinforcing positive associations with physical appearance (Aharon et al. 2001). These unconscious responses can result in biases, such as the assumption that attractive individuals are more competent or intelligent. Biases related to body weight offer another compelling example. Research shows that the anterior insula, a brain region involved in emotional reactions like disgust, activates when individuals view images of overweight people (Simmons et al. 2005). This unconscious response can contribute to negative stereotypes, such as associating weight with laziness or poor self-control, even without explicit judgments.

Biases about people's occupations based on appearance or clothing further illustrate the role of automatic brain processes. The medial prefrontal cortex, involved in social cognition, engages when individuals make rapid assumptions based on minimal visual cues (Cloutier et al. 2011). For instance, seeing someone in formal attire might immediately evoke the assumption that they are a business executive, regardless of their actual occupation. These examples highlight how implicit biases arise from unconscious neural mechanisms, shaping our perceptions and actions in ways we may not recognize. Understanding these brain functions is crucial for mitigating the impact of implicit biases.

Despite their often-negative effects, implicit biases are not inherently harmful; they can also serve adaptive functions, particularly in enhancing cognitive efficiency. Biases act as cognitive shortcuts, enabling the brain to process vast amounts of information quickly. This allows for rapid decision-making, especially in complex or time-sensitive situations. For instance, medical professionals develop implicit biases through years of experience, enabling them to make quick and often accurate diagnoses based on subtle cues. This "clinical intuition" allows doctors to function effectively under pressure (Mamede et al. 2007). Similarly, biases streamline everyday decisions, such as selecting familiar products at a store based on brand recognition. These shortcuts reduce cognitive load and speed up routine processes.

In addition to improving decision efficiency, biases also play an essential role in social cohesion. Throughout human history, biases helped individuals distinguish in-group members from potential threats, promoting cooperation and trust within familiar social environments (Cosmides & Tooby 1992). While this mechanism can lead to negative stereotyping, it also fosters quick assessments of trustworthiness and group dynamics, contributing to social bonding. Furthermore, implicit biases can spark creativity by enabling the brain to form connections between seemingly unrelated ideas. This associative thinking often leads to innovative problem-solving and creative insights, demonstrating that cognitive shortcuts can be both limiting and generative (Kounios & Beeman 2014). Understanding how implicit biases operate and their dual roles in cognitive processes is essential for mitigating their negative effects while leveraging their benefits. By increasing awareness and fostering critical reflection, individuals and organizations can work toward more unbiased and efficient decision-making.

Intuitions as Products of Unconscious Processing

Intuition, often described as a "gut feeling" or sudden insight, is a prime example of the unconscious mind's broadly integrated operations. These intuitive judgments and decisions result from the rapid and complex processing of accumulated knowledge and experiences. This unconscious synthesis distills relevant information into a seemingly instantaneous understanding, often guiding us in situations where a swift response is crucial (Hogarth, 2001). Consider the case of experienced chess players, especially grandmasters, who rely on intuitive judgments when making moves. Rather than consciously analyzing every possible move, they quickly assess complex positions and choose those that "feel" right. This intuition stems from deep, unconscious pattern recognition, honed over years of practice. The basal ganglia play a key role here, embedding patterns from repetitive exposure, while the prefrontal cortex integrates this knowledge into rapid decision-making (Kahneman, 2011). Unlike computer programs that calculate countless permutations, these players bypass conscious deliberation, reflecting the brain's ability to integrate past knowledge into immediate action.

A similar phenomenon occurs in medicine. Experienced doctors often develop intuitive hunches about a patient's condition based on subtle cues they may not consciously articulate. These hunches can lead to remarkably accurate diagnoses, even before test results are available, showcasing the brain's capacity to synthesize vast amounts of medical knowledge unconsciously. Medical students gradually develop this "clinical eye," integrating procedural and declarative knowledge into intuitive decision-making processes that save time and resources (Croskerry 2009; Mickleborough 2015). This unconscious processing extends to artistic endeavors. Artists frequently experience intuitive insights during creation. For example, a painter may instinctively know that a particular brushstroke or color choice is needed, while a writer might sense the best way to craft a scene for maximum impact. These insights arise from the brain's unconscious integration of technical skills and past experiences, enabling fluid creative expression without deliberate deliberation over every detail (Dietrich, 2004).

Similarly, in business, entrepreneurs often rely on gut feelings to identify market opportunities or make product development decisions. These hunches reflect the rapid synthesis of trends, past experiences, and market insights driven by unconscious pattern recognition and integration (Dane & Pratt, 2007). Across these domains, intuition exemplifies the brain's capacity for metafunctional integration. Unconscious processes access and synthesize information from multiple modalities, whether from years of practice, professional expertise, or creative experimentation. For instance, the brain's ability to embed patterns in the basal ganglia complements the prefrontal cortex's role in applying this knowledge in real time. This integration allows for extraordinary human capabilities, such as chess mastery, accurate medical diagnoses, artistic expression, and strategic business decisions. At the same time, intuition is not infallible. Biases and errors can arise when unconscious processes draw on incomplete or skewed information. For example, cognitive shortcuts that underpin intuitive judgments may lead to overconfidence or decision-making errors in complex situations. This interplay of strengths and limitations underscores the need to refine our understanding of intuition as a cognitive phenomenon.

The distributed nature of unconscious processing suggests that biological mechanisms and potentially broader transphysical domains may come into play. Intuition provides a window into the brain's sophisticated ability to optimize decision-making and action in real time. As neuroscience continues to uncover the intricacies of these processes, it sheds light on how the brain's metafunctional unconscious integrates knowledge to navigate the complexities of human experience.

Dreams

Dreams, a classic realm of unconscious activity, offer a unique window into the fascinating workings of the mind. One prominent theory suggests that dreams are crucial for consolidating and organizing memories. During REM (rapid eye movement) sleep, when most vivid dreaming occurs, the brain processes and integrates information gathered throughout the day. This process helps sift through experiences, separating relevant information from unimportant details, ultimately contributing to learning and memory retention (Stickgold 2005). Dreams may also play a critical role in filtering out irrelevant information from the brain's vast sensory input during wakefulness, preventing sensory overload and helping the brain prioritize and focus on essential details. In addition to memory consolidation, dreams are vital in emotional processing and regulation. They may help us work through complex emotions, stressors, and unresolved psychological conflicts. This unconscious exploration of emotions can lead to emotional catharsis, relieving emotional burdens. Through this process, the brain aids in maintaining emotional balance and well-being (Walker and van der Helm 2009). Emotional processing in dreams likely serves to resolve unresolved psychological issues that remain hidden during wakefulness, highlighting the importance of unconscious mechanisms in mental health.

Moreover, according to current understanding, dreams serve as a form of threat simulation. By rehearsing potentially dangerous or challenging scenarios in a safe and controlled environment, individuals can prepare for real-life situations. This mental practice may improve an individual's readiness and adaptability, serving as an unconscious means of honing problem-solving skills in response to perceived threats (Revonsuo 2000). In this way, the brain employs unconscious processes to promote survival through anticipating and rehearsing future challenges. Dreams are not limited to survival strategies but can foster creativity and insight. During sleep, the brain makes novel connections between seemingly unrelated pieces of information, potentially leading to creative solutions or innovative ideas. Many artists, writers, and scientists have drawn inspiration from their dreams, suggesting that the brain's unconscious processing can enhance creative problem-solving abilities (Cai et al. 2009). This creative function further underscores the broad reach of the unconscious mind and its contribution to various cognitive domains.

Dreams often communicate through symbols and metaphors, offering profound insights into the workings of the unconscious mind. These symbolic representations provide a unique window into our hidden thoughts, emotions, and unresolved conflicts, which may remain inaccessible during wakefulness. As Freud (1900) famously suggested, interpreting dream content can uncover aspects of ourselves concealed in our conscious lives. This symbolic communication is not random but reflects the deep integration of unconscious processes, demonstrating the mind's ability to organize and convey complex ideas through non-literal representations. Understanding these metaphors enriches our awareness of how unconscious processing influences our internal world, offering a glimpse into the deeper layers of self-understanding. Dreams are metafunctional, embodying a brain-wide integration of various unconscious processes. From memory consolidation to emotional regulation, threat simulation, creative problem-solving, and symbolic expression, dreams represent the sophisticated coordination of multiple brain systems working in harmony. In dreams, the unconscious mind is not merely a passive recorder of experiences but an active agent that shapes cognitive and emotional life. As revealed in dreams, this metafunctional nature of unconscious processing reinforces its importance for everyday functioning and long-term mental health.

Furthermore, dream functions extend beyond the realm of cognitive and emotional processing. They encompass essential biological processes like maintaining cellular metabolism, repairing tissues, and eliminating brain waste products—achieved through the unique glymphatic system during sleep (Xie et al. 2013). Moreover, dreams and unconscious processing are essential in maintaining homeostatic and allostatic balance, integrating various physiological and cognitive functions. According to Agnati et al. (2011), sleep and dreams contribute to "predictive psychic homeostasis." This concept suggests that dreaming facilitates physiological recovery and psychological and emotional balance by integrating internal experiences with external demands (Steriade 2006; Hobson and Friston 2012; Walker and Stickgold 2010; Friston 2010). Dreams thus function as a form of cognitive-emotional-physiological regulation that helps the brain predict and adapt to future challenges, ensuring both physical and mental well-being.

During sleep, particularly in REM sleep, the brain engages in processes that support cellular repair and waste elimination (Xie et al. 2013), which are crucial for maintaining physical homeostasis. Various studies note that these functions closely intertwine with the brain's role in emotional regulation and memory consolidation (Stickgold 2005). Dreams are an integrative mechanism that helps balance these physiological needs with psychological demands, preventing overload and promoting adaptation to stressors. In this sense, dreams help recalibrate both bodily and mental systems, fulfilling the brain's dual role in managing homeostasis (stability) and allostasis (adaptation to stress). This framework underscores the brain metafunctional nature of dreams: they facilitate the organization of experiences and emotions and ensure that physiological and cognitive systems remain in balance. Through unconscious processing, dreams aid in integrating sensory inputs, physiological and emotional responses, and cognitive functions, enabling a dynamic interplay between body and mind (Agnati et al. 2011). The predictive nature of these processes enhances the brain's ability to maintain internal equilibrium while preparing for future challenges. For example, the brain's predictive models allow it to simulate potential threats or emotional conflicts in dreams, providing a "safe rehearsal" for exploring and resolving them. This allostatic function of dreams also relates to the creative and problem-solving capacities of the unconscious mind.

The brain uses dreams to rehearse responses to external challenges and generate novel solutions, contributing to emotional adaptability and resilience. Dreams thus enable the unconscious mind to prepare for potential stressors while simultaneously regulating bodily functions, creating an integrated system that supports immediate survival and long-term mental health (Stickgold 2005; Agnati et al. 2011). With much more to say, let the previous examples illustrate how dreams represent a complex, integrated process that links physiological, psychological, and homeostasis with allostasis needs. In synthesis, the brain metafunction maintains balance across cognitive, emotional, and physiological systems, ensuring the brain's preparation for present and future demands. This integrated approach to unconscious processing highlights the essential role of dreams in sustaining health. Dreaming is a window to a metafunctional integration of biological, cognitive, and spiritual processes. By weaving together these various functions, the unconscious mind and its metafunctional context demonstrate a far-reaching impact, from essential biological maintenance to abstract and spiritual forms of self-reflection.

Creative Insights and Problem-Solving

Unconscious processes play a pivotal role in creativity by facilitating complex, spontaneous connections that conscious thought often overlooks. During periods of rest or distraction, the unconscious mind engages in *idea incubation*, which allows seemingly unrelated concepts to merge and form novel connections. These unconscious integrations frequently lead to sudden insights or "aha" moments (Fox & Christoff, 2014). Another key aspect is *implicit learning*, where the unconscious mind absorbs patterns, structures, and experiences from the environment without deliberate effort. For instance, an artist might unconsciously incorporate elements of technique or style encountered in daily life into their work (Bargh & Chartrand, 1999). Additionally, *associative thinking*, a hallmark of unconscious creativity, enables the mind to link disparate ideas, fostering innovation and metaphor creation (Kaufman & Sternberg, 2010).

Unconscious processes also reduce cognitive constraints, freeing the mind from rigid rules or biases that hinder conscious thought. This freedom allows for exploratory and unconventional approaches to problem-solving (Dietrich, 2004). Emotional and intuitive influences, deeply rooted in the unconscious, infuse creative works with authenticity and depth by drawing on the creator's emotional experiences (Fox & Christoff, 2014). Dreams and altered states of consciousness further illustrate the creative power of the unconscious, providing vivid imagery and novel scenarios that inspire art, literature, and scientific breakthroughs. For example, Johannes Brahms described how melodies and themes would spontaneously emerge while walking, revealing how his extensive musical knowledge was integrated through unconscious processing (Kaufman & Sternberg, 2010). Similarly, Albert Einstein credited imaginative insights—arising from unconscious processes such as visualizing himself riding a beam of light—as the basis for his theory of special relativity (Fox & Christoff, 2014).

Unconscious Creativity in Action

Research highlights the profound impact of unconscious processes on creativity. Studies on the brain's *default mode network* (DMN), active during introspective states like daydreaming, suggest that it facilitates the integration of diverse knowledge and enhances creative problem-solving (Raichle 2015; Beaty et al. 2016). Experiments on the *incubation effect* confirm that stepping away from a problem allows the unconscious mind to process information, often leading to improved solutions (Fox & Christoff, 2014). Similarly, *unconscious priming*, where exposure to specific stimuli influences thought or behavior, can subtly steer creative thinking (Bargh & Chartrand, 1999). These examples demonstrate how the unconscious mind acts as a silent architect of creativity, laying the groundwork for conscious refinement and producing innovative and meaningful works.

In the 3rd century BCE, the ancient Greek mathematician Archimedes experienced his famous "Eureka" moment while bathing. Observing the water level rise as he immersed himself, he realized he could use this principle to calculate the volume of irregularly shaped objects, a foundational insight in fluid mechanics (Gow, 1958). Similarly, in 1865, German chemist Friedrich August Kekulé dreamt of a snake biting its tail, inspiring his visualization of the cyclic ring structure of benzene, which revolutionized organic chemistry (Rocke 2015). Inventor Nikola Tesla also leveraged unconscious processing, using vivid mental simulations to refine his groundbreaking designs in electricity and wireless communication. Tesla described his brain as a "receiver" of knowledge from a higher source, emphasizing the mysterious and powerful workings of the unconscious mind (Cheney 2001; Carlson, 2013). Surrealist artist René Magritte was known for tapping his unconscious mind to create thought-provoking and dreamlike artworks. His paintings often challenge perceptions of reality, employing unexpected juxtapositions and ordinary objects in surreal contexts. Through this exploration of the unconscious, Magritte revealed the power of imagination and the complexities of human perception. Similarly, surrealist artist Salvador Dalí was known for exploring his unconscious mind to create evocative and dreamlike artworks exploring his psyche's irrational and fantastical aspects, resulting in iconic paintings illustrating the power of unconscious creativity.

Creativity in Music: The Unconscious and the Transphysical

Johann Sebastian Bach exemplified the profound interplay between the conscious and unconscious mind, including a transphysical aspect, in the creative process. His compositions reflect a deep intuitive understanding of music's structural and emotional elements, suggesting that his unconscious mind significantly shaped his musical innovations. His ability to weave intricate harmonies and counterpoint demonstrates the robust, often unconscious cognitive processes underlying artistic genius. Beyond technique, Bach's work carries a transcendent quality, often attributed to his deep devotion to God. His annotations and musical inscriptions reveal his belief that music was not just an artistic pursuit but a spiritual one, designed to glorify God and touch the human soul.

Wolfgang Amadeus Mozart, the celebrated child prodigy, was also renowned for his extraordinary compositional abilities. He described his creative process as one of receiving musical ideas and melodies in his mind, as though they flowed effortlessly from his unconscious. Mozart's deeply ingrained musical talent, combined with years of cultivation, allowed him to produce a prolific and innovative body of work. His process, however, was not merely unconscious but often intertwined with his profound faith. As Mozart himself explained, "God is ever before my eyes. I realize His omnipotence and I fear His anger; but I also recognize His love, His compassion, and His tenderness towards His creatures" (Holmes, 2005). This acknowledgment of divine inspiration suggests that Mozart perceived his creative genius as not only stemming from within but also connected to something transcendent.

The Unconscious as Expression of the Transphysical

Some philosophers have argued that the transcendent beauty found in the music of Bach and Mozart serves as evidence for the existence of God, pointing to their artistic genius as indicative of a divine source. Philosopher Peter Kreeft succinctly encapsulates this argument: "There is the music of Johann Sebastian Bach. Therefore, there must be a God." He further elaborates that the beauty of Bach's music is experienced as transcendent, authoritative, and self-validating—qualities that seem to point beyond human creativity to a higher, divine origin (Kreeft, 1992). Similarly, philosopher Alvin Plantinga humorously yet profoundly argues for God's existence based on music, calling his case *The Mozart Argument*. These arguments suggest that the interplay between unconscious creative processes and transcendent inspiration hints at a reality that transcends the physical world.

Mozart's deep reverence for God can be seen in his compositional process, where he prayed for divine guidance and sought to honor God through his music. "I prayed to God for His mercy that all might go well, to His greater glory, and the symphony began" (Rathey, 2016). Such statements not only reveal Mozart's faith but also provide a window into how he perceived his creativity as being both human and divine. This relationship between the unconscious, the transphysical, and the divine is further exemplified in the works of Johann Sebastian Bach. In his personal Bible, Bach annotated next to 2 Chronicles 5:13: "N.B. In a music of worship, God is always present with His grace" (Wolff, 2016). Moreover, he frequently inscribed his manuscripts with the initials JJ (*Jesu Juva*, "Help me, Jesus") or SDG (*Soli Deo Gloria*, "To God alone, the glory") (Geck, 2006), suggesting that for Bach, the creative process was inseparable from his spiritual life and shaped by an unconscious-transphysical dimension that informed his artistic genius. Creative breakthroughs often arise when the unconscious mind operates behind the scenes, blending disparate ideas into fresh perspectives. Historical figures like Einstein, Tesla, and Archimedes illustrate how the interplay of unconscious processing, intuition, and conscious reasoning can lead to transformative insights. Similarly, artists and musicians such as Bach, Mozart, and others demonstrate that the unconscious mind plays a crucial role in producing works of beauty and innovation that often transcend their time and cultural context. The unconscious processes information non-linearly, breaking free from the cognitive constraints of conscious thought. This freedom enables creators to transcend conventional thinking, generate novel solutions, and achieve mastery in their fields. In music, the unconscious mind acts as a reservoir of inspiration, blending intuition, emotional depth, and accumulated knowledge into a seamless whole. However, what sets figures like Bach and Mozart apart is the implicit transphysical dimension of their creativity. Their work not only reflects unconscious genius but also seems to draw on something beyond the physical recognized by many as divine inspiration and transcendent beauty. Whether through vivid dreams, prayerful introspection, or an intuitive connection to the divine, these creators embody the union of the unconscious and the transphysical, offering glimpses into the profound potential of the human spirit.

The Unconscious as a Sacred Space in Spiritual Transformation

The unconscious plays a pivotal role in profound spiritual transformations, such as those often described in the Christian tradition as "born of the Spirit." These experiences, marked by a deep sense of renewal and connection with the divine, frequently involve processes that transcend conscious thought and deliberate effort. The unconscious mind integrates the emotional, cognitive, and symbolic elements of spiritual awakening, allowing individuals to process feelings of guilt, forgiveness, and grace at a level beyond rational analysis. For example, the transformative encounter with God often involves moments of surrender and profound insight that emerge unexpectedly as unconscious mental frameworks realign to reflect a renewed sense of identity and purpose. These shifts may involve integrating suppressed emotions, unresolved conflicts, or deeply held beliefs into a cohesive spiritual narrative. Dreams, visions, and prayer—which engage the unconscious through imagery, metaphors, and language—further reveal their role in guiding individuals toward an encounter with the divine. In Christianity, this process is understood as the work of the Holy Spirit, who operates not only in the conscious mind but also in the hidden depths of the human soul, bringing about a transformation that is both mysterious and life changing. Such moments of being "born of the Spirit" illustrate the unconscious as a sacred space where divine grace intersects with human vulnerability, fostering renewal, healing, and spiritual growth.

Conclusion

This chapter underscores the centrality of unconscious processing as an integral component of the brain's metafunction. Operating "behind the scenes," the unconscious orchestrates a broad array of sensory, cognitive, emotional, homeostatic, and allostatic processes that enable the seamless functioning of higher-order brain activities. These hidden mechanisms support crucial functions such as face recognition, attention, emotional perception, memory encoding and retrieval, implicit bias formation, and interoceptive homeostasis, demonstrating the profound influence of unconscious processes on the human experience. The brain's dynamic engagement in these processes highlights its ability to unify complex neural activities across distributed networks, creating the foundation for learning, pattern recognition, and adaptive decision-making.

In addition to its role in perception and judgment, the unconscious is a key driver of creativity and innovation. It facilitates associative thinking, allowing the mind to form connections that elude conscious thought, particularly during periods of rest or distraction. Processes like idea incubation and implicit learning enable the unconscious to synthesize knowledge and experiences into new insights, often manifesting as "aha" moments of inspiration. Research on the default mode network (DMN) and incubation effects confirm the unconscious mind's capacity to integrate diverse knowledge and solve problems creatively, revealing it as a silent innovation architect. From the spontaneous melodies described by Johannes Brahms to Einstein's imaginative thought experiments, the unconscious plays a pivotal role in shaping humanity's most outstanding achievements.

The unconscious also reduces cognitive constraints, freeing the mind from rigid rules or biases and encouraging exploratory approaches to problem-solving. Emotional and intuitive undercurrents, deeply embedded in the unconscious, lend authenticity and depth to creative works, drawing on personal experiences and emotions. Even dreams and states, when consciousness intense focus relaxes, showcase the unconscious as a wellspring of creative potential, providing vivid imagery and novel scenarios that inspire breakthroughs in art, literature, and science. Beyond creativity, the unconscious profoundly influences motivation, intuition, interpersonal attraction, and emotional life, often in ways that remain hidden from conscious awareness. This interaction between unconscious and conscious processes highlights the brain's role as a metafunctional organ, dynamically integrating biological, psychological, and transphysical dimensions of human life. Together, these processes shape not only individual cognition but also our shared understanding of humanity. Future chapters will expand on this exploration, particularly regarding the unity of the human person. Anthropological models that view the person as a composite of matter and spirit will provide further insight into how the soul is the unifying principle of body and mind. By considering the unconscious within this broader framework, we can deepen our understanding of its role in shaping human behavior, creativity, and the essence of human existence. The unconscious is not merely a passive mechanism, it is a dynamic, generative force that contributes to the richness of life and the boundless potential of the human mind.

REFERENCES

Adolphs, R. (2002). Neural systems for recognizing emotion. *Current Opinion in Neurobiology, 12*(2), 169–177. https://doi.org/10.1016/S0959-4388(02)00301-X

Adolphs, R. (2003). Cognitive neuroscience of human social behaviour. *Nature Reviews Neuroscience, 4*(3), 165–178. https://doi.org/10.1038/nrn1056

Adolphs, R., & Tusche, A. (2017). From faces to prosocial behavior: Cues, tools, and mechanisms. *Current Directions in Psychological Science, 26*(3), 282–287. https://doi.org/10.1177/ 0963721417694656

Adolphs, R. (2016). Human lesion studies of emotion: A review of methodological approaches and findings. *Neuroscience & Biobehavioral Reviews, 68*, 569–575. https://doi.org/10.1016/j.neubiorev.2016.06.014

Agnati, L. F., Barlow, P. W., Baluska, F., Tonin, P., Guescini, M., Leo, G., & Fuxe, K. (2011). A new theoretical approach to the functional meaning of sleep and dreaming in humans based on the maintenance of 'predictive psychic homeostasis.' *Communicative & Integrative Biology, 4*(6), 640–654. https://doi.org/10.4161/cib.17602

Aharon, I., Etcoff, N., Ariely, D., Chabris, C. F., O'Connor, E., & Breiter, H. C. (2001). Beautiful faces have variable reward value: fMRI and behavioral evidence. *Neuron, 32*(3), 537–551. https://doi.org/10.1016/S0896-6273(01)00491-3

Alais, D., & Burr, D. (2004). The ventriloquist effect results from near-optimal cross-modal integration. *Current Biology, 14*(3), 257–262. https://doi.10.1016/j.cub.2004.01.029

Alexander of Aphrodisias. (2013). *On the Soul.* Bloomsbury Academic.

Arbib, M. A. (2010). Mirror system activity for action and language is embedded in the integration of dorsal and ventral pathways. *Brain and Language, 112*(1), 12–24. https://doi.org/10.1016/j.bandl.2009.10.001

Aristotle. (1986). *De Anima* (H. Lawson-Tancred, Trans.). Penguin Classics.

Barbas, H., García-Cabezas, M. Á., & Zikopoulos, B. (2013). Top-down control of visual processing by the prefrontal cortex. In G. Calvert, C. Spence, & B. E. Stein (Eds.), *Handbook of Multisensory Processes*. MIT Press.

Bargh, J. A., & Chartrand, T. L. (1999). The unbearable automaticity of being. *American Psychologist, 54*(7), 462–479. https://doi.org/10.1037/0003-066X.54.7.462

Barsalou, L. W. (1999). Perceptual symbol systems. *Behavioral and Brain Sciences, 22*(4), 577–660. https://doi.org/10.1017/S0 140525X99002149

Benedetti, F., Lanotte, M., Lopiano, L., & Colloca, L. (2007). When words are painful: Unraveling the mechanisms of the nocebo effect. *Neuroscience, 147*(2), 260–271. https://doi.org/10.1016/j.neuroscience. 2007.02.020

Benedict XVI. (2007). *Jesus of Nazareth: From the Baptism in the Jordan to the Transfiguration*. Doubleday Religion.

Binder, J. R., et al. (1997). Human brain language areas identified by functional magnetic resonance imaging. *Journal of Neuroscience, 17*(1), 353–362. https://doi.org/10.1523/JNEURO SCI.17-01-00353.1997

Blake, M. L. (2017). Right-hemisphere pragmatic disorders. In L. Cummings (Ed.), *Research in Clinical Pragmatics* (Perspectives in Pragmatics, Philosophy & Psychology, vol. 11, pp. 325–342). Springer, Cham. https://doi.org/10.1007 /978-3-319-47489-2_10

Bonini, L., Rotunno, C., Arcuri, E., & Gallese, V. (2022). Mirror neurons 30 years later: Implications and applications. *Trends in Cognitive Sciences, 26*(9), 767–781. https://doi.org/10.1016/j.tics.2022.06.003

Bookheimer, S. (2002). Functional MRI of language: New approaches to understanding the cortical organization of semantic processing. *Annual Review of Neuroscience, 25*, 151 –188.

Bucci, W. (2001). Pathways of emotional communication. *Psychoanalytic Inquiry, 21*(1), 40–70. https://doi.org/ 10.1080/07351692109348923

Cai, D. J., Mednick, S. A., Harrison, E. M., Kanady, J. C., & Mednick, S. C. (2009). REM, not incubation, improves creativity by priming associative networks. *Proceedings of the National Academy of Sciences, 106*(25), 10130–10134. https://doi.org/10.1073/pnas.0900271106

Cannon, W. B. (1932). *The Wisdom of the Body.* New York: W.W. Norton & Company.

Carlson, W. B. (2013). *Tesla: Inventor of the Electrical Age.* Princeton University Press.

Cheney, M. (2001). *Tesla: Man Out of Time.* Touchstone/Simon & Schuster.

Chomsky, N. (1965). *Aspects of the Theory of Syntax.* MIT Press.

Chomsky, N., & Berwick, R. C. (2016). *Why Only Us: Language and Evolution.* MIT Press.

Cleeremans, A. (2007). Consciousness: The radical plasticity thesis. *Progress in Brain Research, 168*, 19–33. https://doi.org/ 10.1016/ S0079-6123(07)68003-0

Connell, L., & Lynott, D. (2014). Principles of representation: Why you can't represent the same concept twice. *Topics in Cognitive Science, 6*(3), 390–406. https://doi.org/10.1111/tops.12097

Cook, R., Bird, G., Catmur, C., Press, C., & Heyes, C. (2014). Mirror neurons: From origin to function. *Behavioral and Brain Sciences, 37*(2), 177–192. https://doi.org/10.1017/S0140525X13000903

Craig, A. D. (2009). How do you feel—now? The anterior insula and human awareness. *Nature Reviews Neuroscience, 10*(1), 59–70. https://doi.org/10.1038/nrn2555

Crick, F., & Koch, C. (2005). What is the function of the claustrum? *Philosophical Transactions of the Royal Society B: Biological Sciences, 360*(1458), 1271–1279. https://doi.org/10. 1098/rstb.2005.1661

Critchley, H. D., & Garfinkel, S. N. (2017). Interoception and emotion. *Current Opinion in Psychology, 17*, 7–14. https://doi.org/10.1016/j.copsyc.2017.04.020

Croskerry, P. (2009). Clinical cognition and diagnostic error: Applications of a dual process model of reasoning. *Advances in Health Sciences Education, 14*(Suppl. 1), 27–35. https://doi.org/10.1007/s10459-009-9182-2

Damasio, A. (2010). *Self Comes to Mind: Constructing the Conscious Brain.* Pantheon Books.

Dane, E., & Pratt, M. G. (2007). Exploring intuition and its role in managerial decision-making. *Academy of Management Review, 32*(1), 33–54. https://doi.org/10.5465/amr.2007.23463682

Dehaene, S., & Naccache, L. (2001). Toward a cognitive neuroscience of consciousness: Basic evidence and a workspace framework. *Cognition, 79*(1–2), 1–37. https://doi.org/10.1016/s0010-0277(00)00123-2

Dehaene, S., Naccache, L., Cohen, L., Bihan, D. L., Mangin, J. F., Poline, J. B., & Rivière, D. (2001). Cerebral mechanisms of word masking and unconscious repetition priming. *Nature Neuroscience, 4*(7), 752–758. https://doi.org/10. 1038/89551

Di Pellegrino, G., Fadiga, L., Fogassi, L., Gallese, V., & Rizzolatti, G. (1992). Understanding motor events: A neurophysiological study. *Experimental Brain Research, 91*(1), 176–180. https://doi.org/10.1007/BF00230027

Dienes, Z., & Seth, A. K. (2010). Conscious and the unconscious. In G. F. Koob, M. Le Moal, & R. F. Thompson (Eds.), *Encyclopedia of Behavioral Neuroscience* (Vol. 1, pp. 322–327). Oxford: Academic Press.

Dietrich, A. (2004). The cognitive neuroscience of creativity. *Psychonomic Bulletin & Review, 11*(6), 1011–1026. https://doi.org/10.3758/BF03196731

Dolan, R. J. (2002). Emotion, cognition, and behavior. *Science, 298*(5596), 1191–1194. https://doi.org/10.1126/science .1076358

Dreher, H. (2005). *Foundations of Psychoanalysis: Clinical and Theoretical Perspectives.* Springer.

Driver, J., & Noesselt, T. (2008). Multisensory interplay reveals cross-modal influences on sensory-specific brain regions, neural responses, and judgments. *Neuron, 57*(1), 11 –23.

Duchaine, B., & Yovel, G. (2015). A revised neural framework for face processing. *Annual Review of Vision Science, 1*, 393–416. https://doi.org/10.1146/annurev-vision-082114-035518

Dudai, Y. (2004). *Memory from A to Z: Keywords, Concepts, and Beyond.* Oxford University Press.

Elgendi, M., Kumar, P., Barbic, S., Howard, N., Abbott, D., & Cichocki, A. (2018). Subliminal priming—State of the art and future perspectives. *Behavioral Sciences, 8*(6), 54. https://doi.org/10.3390/bs8060054

Ellis, N. C. (1994). Implicit and explicit language learning: An overview. In *Implicit and Explicit Learning of Languages* (pp. 1–23). Academic Press.

Epstein, R. M. (2013). Whole mind and shared mind in clinical decision-making. *Patient Education and Counseling, 90*(2), 200–206. https://doi.org/10.1016/j.pec.2012.06.035

Fodor, J. A. (1975). *The Language of Thought.* Harvard University Press.

Fonagy, P., & Allison, E. (2014). The role of mentalizing and epistemic trust in the therapeutic relationship. *Psychotherapy, 51*(3), 372–380. https://doi.org/10.1037/a0036505

Fox, K. C. R., & Christoff, K. (2014). Metacognitive facilitation of spontaneous thought processes: When metacognition helps the wandering mind find its way. In S. M. Fleming & C. D. Frith (Eds.), *The Cognitive Neuroscience of Metacognition* (pp. 293–319). Springer-Verlag/Springer Nature. https://doi.org/10.1007/978-3-642-45190-4_13

Freud, S. (1895). *Studies on Hysteria.* (With J. Breuer). Vienna: Franz Deuticke.

Freud, S. (1900). *The Interpretation of Dreams.* Leipzig: Franz Deuticke.

Freud, S. (1915). The unconscious. In *The Standard Edition of the Complete Psychological Works of Sigmund Freud* (Vol. 14, pp. 159–215). London: Hogarth Press.

Freud, S. (1923). *The Ego and the Id.* Leipzig: Internationaler Psychoanalytischer Verlag.

Freud, S. (1923b). *Neurosis and Psychosis.* Vienna: Internationaler Psychoanalytischer Verlag.

Freud, S. (1933). *New Introductory Lectures on Psycho-Analysis*. London: Hogarth Press.

Friederici, A. D. (2011). The brain basis of language processing: From structure to function. *Physiological Reviews, 91*(4), 1357–1392. https://doi.org/10.1152/physrev.00006.2011

Friston, K. J. (2010). The free-energy principle: A unified brain theory? *Nature Reviews Neuroscience, 11*(2), 127–138. https://doi.org/10.1038/nrn2787

Gallese, V. (2003). The manifold nature of interpersonal relations: The quest for a common mechanism. *Philosophical Transactions of the Royal Society B: Biological Sciences, 358*(1431), 517–528.

Gallese, V. (2003). The roots of empathy: The shared manifold hypothesis and the neural basis of intersubjectivity. *Psychopathology, 36*(4), 171–180. https://doi.org/ 10.1159/000072786

Garrod, S., & Pickering, M. J. (2004). Why is conversation so easy? *Trends in Cognitive Sciences, 8*(1), 8–11. https://doi.org/10.1016/j.tics.2003.10.016

Gauthier, I., Tarr, M. J., Anderson, A. W., Skudlarski, P., & Gore, J. C. (1999). Activation of the middle fusiform "face area" increases with expertise in recognizing novel objects. *Nature Neuroscience, 2*(6), 568–573. https://doi.org/10.1038/9224

Geck, M. (2006). *Johann Sebastian Bach: Life and Work*. Harcourt.

Geschwind, N. (1970). The organization of language and the brain. *Science, 170*(3961), 940–944.

Golland, Y., Arzouan, Y., & Levit-Binnun, N. (2015). The mere co-presence: Synchronization of autonomic signals and emotional responses across co-present individuals not engaged in direct interaction. PLOS ONE, 10(5), e0125804. https://doi.org/10.1371/journal.pone.0125804

Goodwyn, E. (2024). *A Jungian Analysis of Toxic Modern Society: Fighting the Culture of Loneliness*. London: Routledge. https://doi.org/10.4324/9781032721354

Gow, M. (1958). *A Short History of Greek Mathematics.* Cambridge University Press.

Greenwald, A. G., & Banaji, M. R. (1995). Implicit social cognition: Attitudes, self-esteem, and stereotypes. *Psychological Review, 102*(1), 4–27. https://doi.org/10.1037/0033-295X.102.1.4

Greenwood, B. M., & Garfinkel, S. N. (2024). Interoceptive mechanisms and emotional processing. *Annual Review of Psychology.*

Hagoort, P. (2017). The core and beyond in the language-ready brain. *Neuroscience & Biobehavioral Reviews, 81*(Part B), 194–204. https://doi.org/10.1016/j.neubiorev .2017.01.048.

Hart, D. B. (2003). *The Beauty of the Infinite: The Aesthetics of Christian Truth.* Eerdmans.

Haxby, J. V., Hoffman, E. A., & Gobbini, M. I. (2000). The distributed human neural system for face perception. *Trends in Cognitive Sciences, 4*(6), 223–233. https://doi.org/10.1016/S1364-6613(00)01482-0

Henson, R. N. (2003). Neuroimaging studies of priming. *Progress in Neurobiology, 70*(1), 53–81. https://doi.org/10.1016/S0301-0082(03)00086-8

Hering, E. (1878). *Outlines of a Theory of the Light Sense.* Translated by J. P. W. Jones. New York: Macmillan.

Hickok, G., & Poeppel, D. (2007). The cortical organization of speech processing. *Nature Reviews Neuroscience, 8*(5), 393–402.

Hobson, J. A. (1999). *Dreaming as Delirium: How the Brain Goes Out of Its Mind.* MIT Press.

Hobson, J. A., & Friston, K. J. (2012). Waking and dreaming consciousness: Neurobiological and functional considerations. *Progress in Neurobiology, 98*(1), 82–98. https://doi.org/ 10.1016/j.pneurobio.2012.05.003

Hogarth, R. M. (2001). *Educating Intuition.* University of Chicago Press.
Holmes, E. (2005). *The Life of Mozart.* (Illustrated ed.). Dover Publications.

Holton, G. (1996). *Einstein, History, and Other Passions.* Reading, MA: Addison-Wesley.

Iacoboni, M. (2008). *Mirroring People: The New Science of How We Connect with Others.* Farrar, Straus, and Giroux.

Jackendoff, R. (2002). *Foundations of Language: Brain, Meaning, Grammar, Evolution.* Oxford University Press.

James, W. (1902). *The Varieties of Religious Experience: A Study in Human Nature.* Longmans, Green, and Co.

Jin, X., & Costa, R. M. (2010). Start/stop signals emerge in nigrostriatal circuits during sequence learning. *Nature, 466*(7305), 457–462. https://doi.org/10.1038/nature09263.

Jonghyun, L. (2023). The embodiment cognition of abstract concepts in L2 processing. *Seoul National University Repository.*

Jung, C. G. (1964). *Man and His Symbols.* London: Aldus Books.

Jung, C. G. (1969). *The Archetypes and the Collective Unconscious.* (R. F. C. Hull, Trans.). Princeton University Press. (Original work published 1959).

Kahneman, D. (2011). *Thinking, Fast and Slow.* Farrar, Straus, and Giroux.

Kandel, E. R., Schwartz, J. H., Jessell, T. M., Siegelbaum, S. A., & Hudspeth, A. J. (2013). *Principles of Neural Science* (5th ed.). McGraw-Hill Education.

Kanwisher, N. (2000). Domain specificity in face perception. *Nature Neuroscience, 3*(8), 759–763. https://doi.org/10.1038/77664

Kaufman, J. C., & Sternberg, R. J. (2010). The Cambridge Handbook of Creativity. Cambridge University Press. https: //doi.org/10.1017/CBO9780511763205

King, R. (1999). *Indian Philosophy: An Introduction to Hindu and Buddhist Thought.* Edinburgh University Press.

Kok, P., Jehee, J. F. M., & de Lange, F. P. (2012). *Less is more: Expectation sharpens representations in the primary visual cortex.* Neuron, **75**(2), 265–270. https://doi.org/10.1016/j.neuron.2012.04.034

Kouider, S., & Dehaene, S. (2007). Levels of processing during non-conscious perception: A critical review of visual masking. *Philosophical Transactions of the Royal Society B: Biological Sciences, 362*(1481), 857–875. https://doi.org/10.1098/rstb. 2007.2093

Kounios, J., & Beeman, M. (2014). *The Eureka Factor: Aha Moments, Creative Insight, and the Brain.* Random House.

Kreeft, P. (1992). *Making Sense Out of Suffering.* Servant Books.

Kreeft, P. (1992). *Fundamentals of the Faith: Essays in Christian Apologetics.* San Francisco: Ignatius Press.

Lakoff, G., & Johnson, M. (1980). *Metaphors We Live By.* University of Chicago Press.

LeDoux, J. E. (2012). Rethinking the emotional brain. *Neuron, 73*(4), 653–676. https://doi.org/10.1016/j.neuron. 2012.02.004

LeDoux, J. E. (1996). *The Emotional Brain: The Mysterious Underpinnings of Emotional Life.* Simon & Schuster.

Lehrer, P. M., & Gevirtz, R. (2014). Heart rate variability biofeedback: How and why does it work? *Frontiers in Psychology, 5*, 756. https://doi.org/10.3389/fpsyg.2014.00756

Leuzinger-Bohleber, M., & Fischmann, T. (2006). What is conceptual research in psychoanalysis? *International Journal of Psychoanalysis, 87*(5), 1355–1376.

Levelt, W. J. M. (1989). *Speaking: From Intention to Articulation.* MIT Press.

Mamede, S., Schmidt, H. G., & Rikers, R. (2007). Diagnostic errors and reflective practice in medicine. *Journal of Evaluation in Clinical Practice, 13*(1), 138–145. https://doi.org /10.1111/j.1365-2753.2006.00638.x

Mariën, P., Ackermann, H., Adamaszek, M., Barwood, C. H., Beaton, A., Desmond, J., et al. (2014). Consensus paper: Language and the cerebellum: An ongoing enigma. *The Cerebellum, 13*(3), 387–410.

Mickleborough, T. (2015). Intuition in medical practice: A reflection on Donald Schön's reflective practitioner. *Medical Teacher, 37*(10), 889–891. https://doi.org/10.3109/0142159X. 2015.1078893

Montgomery, K. (2005). *How Doctors Think: Clinical Judgment and the Practice of Medicine.* Oxford University Press.

Murray, M. M., & Wallace, M. T. (Eds.). (2012). *The Neural Bases of Multisensory Processes.* Boca Raton, FL: CRC Press/Taylor & Francis.

Naccache, L., Blandin, E., & Dehaene, S. (2002). Unconscious masked priming depends on temporal attention. *Psychological Science, 13*(5), 416 –424.

Northoff, G. (2013). *Unlocking the Brain: Volume 1: Coding.* New York: Oxford University Press. https://doi.org/10.1093/acprof:oso/9780199826988.001.0001

Pelphrey, K. A., Viola, R. J., & McCarthy, G. (2004). When strangers pass: Processing of mutual and averted social gaze in the superior temporal sulcus. *Psychological Science, 15*(9), 598–603. https://doi.org/10.1111/j.0956-7976.2004.00726.x

Pessoa, L. (2017). A network model of the emotional brain. *Trends in Cognitive Sciences, 21*(5), 357–371. https://doi.org/10.1016/j.tics.2017.03. 002

Pessoa, L. (2013). *The cognitive-emotional brain: From interactions to integration.* The MIT Press.

Pfaltz, M. C., Grossman, P., Michael, T., Margraf, J., & Wilhelm, F. H. (2010). Respiratory pathophysiology in panic disorder: Evidence for abnormal lactate-induced respiration. *Psychosomatic Medicine, 72*(7), 751–759.

Phan, K. L., Wager, T., Taylor, S. F., & Liberzon, I. (2002). Functional neuroanatomy of emotion: A meta-analysis of emotion activation studies in PET and fMRI. *NeuroImage, 16*(2), 331–348. https://doi.org/10.1006/nimg.2002.1087.

Phelps, E. A., O'Connor, K. J., Cunningham, W. A., Funayama, E. S., Gatenby, J. C., Gore, J. C., & Banaji, M. R. (2000). Performance on indirect measures of race evaluation predicts amygdala activation. *Journal of Cognitive Neuroscience, 12*(5), 729–738. https://doi.org/10.1162/089892900562552

Pinker, S. (2007). *The Stuff of Thought: Language as a Window into Human Nature.* Viking Penguin.

Plato. (1888). *The Republic of Plato* (B. Jowett, Trans.). Clarendon Press.

Raichle, M. E. (2015). The brain's default mode network. *Annual Review of Neuroscience, 38*, 433–447. https://doi.org/10.1146/annurev-neuro-071013-014030

Rasch, B., & Born, J. (2013). About sleep's role in memory. *Physiological Reviews, 93*(2), 681–766. https://doi.org/10.1152/physrev.00032.2012

Rathey, M. (2016). *Johann Sebastian Bach's Christmas Oratorio: Music, Theology, Culture.* Oxford University Press.

Revonsuo, A. (2000). The reinterpretation of dreams: An evolutionary hypothesis of the function of dreaming. *Behavioral and Brain Sciences, 23*(6), 877–901. https://doi.org/10.1017/S0140525X00004015

Rizzolatti, G., Fogassi, L., & Gallese, V. (2001). Neuro-physiological mechanisms underlying the understanding and imitation of action. *Nature Reviews Neuroscience, 2*(9), 661 –670. https://doi.org/10.1038/350 90060

Rocke, A. J. (2015). It began with a daydream: The 150th anniversary of the Kekulé benzene structure. *Angewandte Chemie International Edition in English, 54*(1), 46–50. https://doi.org/10.1002/anie.201408034

Rolls, E. T. (2004). The functions of the orbitofrontal cortex. *Brain and Cognition, 55*(1), 11–29. https://doi.org/10.1016/S0278-2626(03)00277-X

Schacter, D. L. (1987). Implicit memory: History and current status. *Journal of Experimental Psychology: Learning, Memory, and Cognition, 13*(3), 501–518. https://doi.org/10.10 37/0278-7393.13.3.501

Schacter, D. L., & Tulving, E. (Eds.). (1994). *Memory Systems 1994.* The MIT Press.

Schomers, M. R., & Pulvermüller, F. (2016). Is the sensorimotor cortex relevant for speech perception and understanding? An integrative review. *Frontiers in Human Neuroscience, 10*, 435.

Seth, A. K., & Tsakiris, M. (2018). Being a beast machine: The somatic basis of selfhood. *Trends in Cognitive Sciences, 22*(11), 969–981. https://doi.org/10.1016/j.tics.2018.08.008

Simmons, W. K., Martin, A., & Barsalou, L. W. (2005). Pictures of appetizing foods activate gustatory cortices for taste and reward. *Cerebral Cortex, 15*(10), 1602–1608. https://doi.org/10.1093/cercor/bhi038

Solano Durán, P., Morales, J. P., & Huepe, D. (2024). Interoceptive awareness in a clinical setting: The need to bring interoceptive perspectives into clinical evaluation. *Frontiers in Psychology*. https://doi.org/10.3389/fpsyg. 2024.1244701

Solms, M. (2000). Dreaming and REM sleep are controlled by different brain mechanisms. *Behavioral and Brain Sciences, 23*(6), 843–850. https://doi.org/10.1017/S0140525X 00003988

Sorabji, R. (2002). *Emotion and peace of mind: From Stoic agitation to Christian temptation.* Oxford University Press. https://doi.org/10.1093/acprof:oso/9780199256600.001.0001 *(Original work published in print 2002; online edition published 2007)*

Spitzer, R. J. (2015). *The Soul's Upward Yearning: Clues to Our Transcendent Nature from Experience and Reason.* Ignatius Press.

Steriade, M. (2006). Groupings of brain rhythms in corticothalamic systems. *Neuroscience, 137*(4), 1087–1106. https://doi.org/10.1016/j.neuroscience.2005.10.029

Sterling, P., & Eyer, J. (1988). Allostasis: A new paradigm to explain arousal pathology. In *Handbook of Life Stress, Cognition and Health* (pp. 629–649). John Wiley & Sons.

Stickgold, R. (2005). Sleep-dependent memory consolidation. *Nature, 437*(7063), 1272–1278. https://doi.org/10.1038/nature04286

Stroop, J. R. (1935). Studies of interference in serial verbal reactions. *Journal of Experimental Psychology, 18*(6), 643–662. https://doi.org/10.1037/h0054651

Stump, E. (2010). *Wandering in Darkness: Narrative and the Problem of Suffering.* Oxford University Press.

Suzuki, N., & Yamamoto, T. (2023). The influence of interoceptive accuracy on the verbalization of emotions. Scientific Reports, 13, 22158. https://doi.org/10.1038/s4159 8-023-49313-9.

Talsma, D., Senkowski, D., Soto-Faraco, S., & Woldorff, M. G. (2010). The multifaceted interplay between attention and multisensory integration. *Trends in Cognitive Sciences, 14*(9), 400–410.

The Holy Bible. Various passages from Psalms, Proverbs, Daniel, Hebrews. English Standard Version (ESV).

Tulving, E. (1972). Episodic and semantic memory. In E. Tulving & W. Donaldson (Eds.), *Organization of Memory* (pp. 381–403). Academic Press.

Walker, M. P., & Stickgold, R. (2010). Overnight alchemy: Sleep-dependent memory evolution. *Nature Reviews Neuroscience, 11*(3), 218. https://doi.org/10.1038/nrn2762-c1

Walker, M. P., & van der Helm, E. (2009). Overnight therapy? The role of sleep in emotional brain processing. Psychological Bulletin, 135(5), 731–748. https://doi.org/10.1037/a0016570

Wallace, M. T., Woynarcski, T. G., & Stevenson, R. A. (2020). Multisensory integration as a window into orderly and disrupted cognition and communication. *Annual Review of Psychology, 71*(1), 193–219. https://doi.org/10.1146/annu rev-psych-010419-051112

Wright, N. T. (2012). *How God Became King: The Forgotten Story of the Gospels.* HarperOne.

Xie, L., Kang, H., Xu, Q., Chen, M. J., Liao, Y., Thiyagarajan, M., ... & Nedergaard, M. (2013). Sleep drives metabolite clearance from the adult brain. *Science, 342*(6156), 373–377. https://doi.org/10.1126/science.1241224

11

The Neural Binding and Metabinding Problems

The neural binding problem addresses one of the most fundamental challenges in neuroscience: how the brain integrates many processes—including sensory information—into a coherent, unified experience. For example, the visual system must combine data from different times and locations, grouping elements into distinct objects. More broadly, neural binding refers to how brain networks encode the integration of cognitive functions like perception, action, and bodily regulation. Despite extensive research, it remains an unsolved challenge (Feldman 2013). It is not just a singular issue but an expanding set of problems—akin to the troubles unleashed from Pandora's box in Greek mythology. As we delve deeper into the brain's workings, the intricacies of these processes multiply, revealing the vast complexity behind maintaining the mind-body coherence displayed in humans (Singer and Gray 1995).

At first glance, neural binding might appear straightforward: integrating sensory inputs for perception, cognition, and action. Yet, the deeper we investigate, the more we uncover a web of interrelated challenges. Take visual feature binding, for instance—how the brain combines colors, shapes, and textures to create a coherent perception of objects (Treisman 1980). Similarly, temporal binding is essential for sequencing events, while working memory binding allows us to simultaneously manipulate multiple pieces of information (Baddeley 2000, 2003). Beyond these, we must also consider the binding of information across different sensory modalities, the link between perception and memory, and the integration of emotions with cognition (Damasio 1999). Each of these aspects reveals new layers of complexity within the brain's binding functions.

Several theories seek to explain this unification. One of them is spatial and temporal dynamic neural networks synchronizing through oscillations in neuronal activity (Singer and Gray 1995). These transient couplings, lasting mere milliseconds, allow the brain to adapt to various contexts flexibly. However, the binding problem extends beyond the limits of neuroscience. Consciousness presents the most profound challenge: how can physical processes give rise to subjective experience? This mystery complicates efforts to explain neural binding from a purely physicalist or materialist standpoint (Feldman 2013). Despite these challenges, exploring the neural binding problem holds promise. Understanding how the brain integrates diverse information offers valuable insights into the unity of the human person, encompassing biological, psychological, and even spiritual dimensions. As we uncover more correlations between cognitive functions and neural networks, it becomes apparent that brain binding is part of a broader, higher-order framework—a "neural metabinding" that integrates multiple functions and includes more than just physical processes.

The Brain Unbinding Problem

To better understand the brain binding problem, exploring its opposite, the neural unbinding problem, particularly in sensory perception, is helpful. Visual perception, one of the most well-researched areas in neuroscience, offers a helpful example. While the neural binding problem focuses on how different features—such as color, shape, and motion—are combined into a unified perception, the neural unbinding problem addresses how the brain disentangles or segregates these integrated features from the environment, allowing us to identify distinct components. The neural unbinding problem involves how the brain initially separates a complex, continuous sensory input into its constituent parts. For example, visual perception requires the brain to break down information about an object's color, shape, and movement into distinct categories before integrating them into a cohesive whole. This process involves parallel pathways in the nervous system that independently convey information about specific visual features (Livingstone and Hubel 1988). By segregating these features, the brain ensures perceptual clarity—for example, avoiding the conflation of a car's color with the motion of a pedestrian crossing the street, despite simultaneous sensory input.

The unbinding process does not occur in isolation but in coordination with top-down processing, which involves higher-level cognitive functions like attention, memory, and expectation. These mechanisms help the brain prioritize or suppress features, directing the unbinding process according to task-relevant goals. For instance, selective attention can enhance relevant stimuli while inhibiting distractions, highlighting the active nature of unbinding (Desimone and Duncan 1995; Treisman 1980). This dynamic interaction between bottom-up sensory input and top-down influences adds a layer of complexity to the task of separating and processing perceptual information. The challenge of neural unbinding is closely related to the neural binding problem. Together, they form two sides of the same coin in understanding how the brain processes vast amounts of information (Koch and Crick 1990). These processes are fundamental for perception and touch on deeper philosophical issues, such as the explanatory gap and the problem of qualia (Chalmers 1995), as examined in the chapter on Consciousness. These issues, central to the philosophy of mind, underscore the difficulty of explaining how physical brain processes give rise to subjective experiences—starting with how sensory information is processed, divided, and reintegrated to form conscious awareness.

Moreover, the neural unbinding problem goes beyond mere perceptual segregation. It extends to the simultaneous processing of emotional and perceptual information. LeDoux (1996) notes that the brain processes the perceptual representation of an object and evaluates its significance separately. The emotional meaning of a stimulus can begin to be appraised before the perceptual systems have fully processed the stimulus. Your brain and mind can know if something is good or bad before precisely knowing what it is (LeDoux 1996). The brain processes emotional stimuli through two distinct pathways, enabling rapid yet nuanced responses. The faster "low road," which connects the thalamus directly to the amygdala, allows the brain to appraise the emotional significance of a stimulus even before fully processing it at the perceptual level.

In contrast, the slower "high road" involves the thalamus, cortex, and then the amygdala, enabling a more detailed and conscious evaluation of the stimulus. This dual-processing mechanism illustrates how the brain can initiate emotional responses to a stimulus before its complete perceptual representation, emphasizing specialized segregation and cooperation in perceptual processing and evaluating emotional significance. This complex and vital interaction further complicates the unbinding process. Addressing the neural binding and unbinding problems requires an integrated, broader brain metafunctional framework that extends beyond traditional neuroscience models. Understanding how the brain disentangles, synthesizes, and evaluates sensory information, especially in parallel with cognitive and emotional processing, demands a brain metafunctional approach. The intricate interplay between these processes indicates the brain's extraordinary capacity to make sense of complex stimuli, reinforcing the need for broader models incorporating biological, emotional, cognitive, and transphysical dimensions of human experience.

The Neural Binding Problem: A Theme with Multiple Variations

The neural binding problem, or the brain binding problem, presents a significant interdisciplinary challenge to develop a comprehensive, testable model for explaining the unity of conscious experience and numerous unconscious processes. The term, from psychology, computational, and cognitive neuroscience, reflects this issue's complexity and multifaceted nature. At its core, the neural binding problem involves the integration of different sensory features (e.g., color, shape, motion) within a single modality and the combination of information across multiple sensory modalities to create a seamless perception of the external world. A key challenge is reconciling the distributed nature of neural processing with the unified nature of conscious experience. While the brain processes information through distinct, specialized regions, the mechanisms by which these disparate elements form a coherent perceptual experience remain unresolved. Feldman (2013) identifies four distinct sub-problems within the broader neural binding challenge:

1. The General Coordination Problem
2. The Visual Feature-Binding Problem
3. The Variable Binding Problem
4. The Subjective Unity of Perception Problem

Each sub-problem highlights how the brain manages diverse information streams while maintaining a unified perception and response.

The General Coordination Problem

The General Coordination Problem represents one of the most foundational aspects of the neural binding problem. It concerns how the brain achieves coordinated activity across multiple neural circuits, enabling the integration of information from various sources into coherent patterns associated with conscious experience and behavior. This challenge arises because the brain's functions are distributed across distinct regions that must work together seamlessly to process sensory information, motor actions, and cognitive tasks (Engel and Singer 2001; Bressler and Kelso 2001).

A vivid example of the General Coordination Problem appears in the complex motor behavior of a musician playing the violin or guitar. Here, distinct motor systems in both hands execute radically different motions coordinated with sensory feedback, cognitive processes, and emotional states. Synchronizing these actions requires coordination across neural circuits with different processing latencies on both sides of the brain while integrating multimodal inputs such as auditory and tactile sensations (Engel and Singer 2001). *Binding by Synchrony* (BBS) has been proposed as a framework for understanding how the brain coordinates such activities. This approach emphasizes the timing of neural activity across different brain regions, with synchrony facilitating the real-time integration of sensory inputs, motor outputs, and cognitive functions (Engel and Singer 2001; Singer 1994).

Spatial and Temporal Binding in Coordination

The General Coordination Problem extends to information's spatial and temporal binding, critical for perceiving and reacting to dynamic environments. For instance, catching a ball requires binding its spatial position with its motion to predict where it will be in the future. This task necessitates spatial coordination to track the ball's trajectory and temporal synchronization to integrate changes in motion for timely physical reactions (Feldman 2013). Top-down processing mechanisms play a pivotal role in addressing this challenge. Higher-order brain regions send signals to lower-level areas to modulate and synchronize activities, ensuring integration across neural networks. These mechanisms are particularly crucial for tasks requiring attention and focus, enabling the brain to coordinate motor actions in response to dynamic visual inputs and adjust sensory processing based on prior experience and expectations (Desimone and Duncan 1995).

Given the dynamic nature of brain functions, understanding neural binding requires an expanded framework incorporating predictive synchronized neural activity patterns. This coordination occurs spatially, as neural populations fire in organized patterns across distinct regions, and temporally, with precise timing essential for integrating rapidly changing sensory information with motor actions and cognitive responses (Engel and Singer 2001; Feldman 2013). Addressing this problem is vital to uncovering how the brain generates a unified conscious experience and facilitates efficient, context-sensitive behavior.

The Visual Feature-Binding Problem

When observing an object, multiple neural circuits process various features such as color, shape, and texture. The visual feature-binding problem focuses on how the brain integrates these diverse features into a unified perceptual experience. This issue extends to how the brain combines information across sensory modalities, such as vision, audition, and touch, to create a coherent and rich perceptual experience. In visual perception, distinct neural pathways in the visual cortex extract specific features like color, shape, motion, and texture. Despite this specialized processing, how these separately processed features are bound into a unified perception remains unresolved. Di Lollo (2012) characterizes this as an "ill-posed" problem, highlighting the difficulty of linking features processed in spatially separate regions into a coherent whole. Research during the 1990s challenged the idea that the brain contains a picture-like representation of what we see. Studies demonstrated that no stable, high-resolution image exists in the brain, even though subjective experience might suggest otherwise. Martinez-Conde et al. (2008) reaffirm this understanding: "There is now overwhelming biological and behavioral evidence that the brain contains no stable, high-resolution, full-field representation of a visual scene, even though that is what we subjectively experience."

Addressing Challenges in Feature Integration

Consider the perception of a bird eating a red apple on green grass. The brain receives signals related to color, shape, motion, and background from distinct groups of neurons. Color is processed in one cortical region, shape in another, and motion in another, and to perceive the entire scene as cohesive rather than fragmented, the brain must integrate these distributed pieces of information. Roelfsema (2023) challenges the long-standing view that oscillatory synchronization is essential for solving the neural binding problem. Instead, he argues that binding can occur through the formation of neuronal assemblies, where neurons representing different stimulus features enhance their firing rates without requiring oscillations or synchronization. This alternative mechanism emphasizes rate coding and selective amplification of neural activity as sufficient for linking features into unified representations. Roelfsema provides evidence for this hypothesis and explores its implications for understanding perception, attention, and consciousness. McComas (2011) frames the essence of the problem, emphasizing that the brain does not store visual images but operates through electrochemical reactions and complex networks of neuronal activity. The production of coherent visual experiences requires that the sensory organs convert inputs into nerve impulses, which travel through multiple levels of processing. However, how the binding of dispersed signals results in a cohesive perception remains an open question.

Toward an Understanding of the Orchestration of Neural Binding

Neuroscientists are exploring various theories and computational models to better understand the integration of sensory features. *Binding by Synchrony,* a leading hypothesis, posits that synchronized oscillations unite diverse neural signals. However, this theory has faced criticism. Shadlen and Movshon (1999) argue that synchronization alone may not be sufficient or necessary for binding. Roelfsema's (2023) alternative suggests that feature binding might rely more on neuronal firing rates and attention-driven mechanisms than synchronized oscillations. These developments in this field highlight the importance of investigating correlations between neural activity and perception and the causal mechanisms underlying binding. Emerging theories integrating attention, firing rates, and hierarchical processing are paving the way for a deeper understanding of how the brain participates in creating unified experiences. A Brain Metafunctional perspective would expect this development: an expanded framework incorporating multiple brain processes and levels.

The Variable Binding Problem

The variable binding problem refers to how the brain dynamically combines various aspects of perception and cognition to form a unified mental experience. This binding process allows us to experience the world as a cohesive whole, a principle central to Gestalt psychology, which emphasizes the brain's natural tendency to integrate features into meaningful wholes, also referred to as "feature binding" or "feature linking" (Roelfsema 2023). For instance, variable binding is crucial in language cognition, as the brain must dynamically link semantic concepts with syntactic roles to generate coherent sentences and linguistic structures. This cognitive capacity allows for creating complex mental representations that unify symbolic structures and propositions into an organized, cohesive experience. In this context, variable binding is essential for piecing together the elements we perceive or think about, enabling us to form complete and unified mental representations of objects, events, or linguistic constructs (Di Lollo 2012). This dynamic integration is fundamental to how we experience the world and construct meaningful thoughts and language expressions in our minds. Variable binding ensures that the brain can link different features, such as the attributes of an object or the components of a sentence, into a fluid and organized mental experience.

The Role of Memory in Variable Binding

Variable binding relates to the role of memory in integrating information over time. Memory processes involve encoding, storing, and retrieving patterns of information across distributed neural circuits, enabling coherent and retrievable memories (Sadeh et al. 2011). Memory allows one to organize and access past experiences, knowledge, and skills. Still, the process of memory binding—where various elements of an experience are combined and linked—presents significant challenges for neuroscience and cognitive science. One familiar form of memory is episodic memory, which involves recalling specific past events, experiences, or episodes. Episodic memories result from rich integration involving sensory details, emotions, thoughts, and contextual information. The neural binding problem in episodic memory pertains to how the brain combines these diverse elements into a cohesive recollection of a past event (Tulving 2002). Understanding how the brain/mind performs this integration is essential for comprehending the mechanisms behind vivid and coherent recollections.

Another critical challenge in memory is temporal binding, which involves linking events in a particular order. The brain must bind sequence elements while preserving their temporal order for accurate recall. This form of binding is crucial for remembering the chronological flow of events and experiences. Research on temporal binding has focused on uncovering the neural mechanisms that allow the brain to organize information over time (Friedman 2004). In contrast, semantic memory refers to storing and retrieving general knowledge and concepts, such as facts and word meanings, and understanding objects and their properties. The neural binding problem in semantic memory concerns how the brain links various attributes and concepts into an integrated knowledge network. For example, when thinking of "pie," the brain combines information about its appearance, taste, and associations with related concepts and past experiences (e.g., dessert, cooking) (Martin 2007). Conceptual binding is crucial for constructing richly interconnected memories that span different time scales.

Working Memory and Variable Binding

Working memory is the cognitive system that temporarily holds and manipulates information during ongoing tasks. The binding problem in working memory relates to how the brain integrates different pieces of information in an active memory buffer, allowing us to manipulate multiple items simultaneously (Baddeley 2000). This function is vital for problem-solving or mental arithmetic tasks that involve simultaneously managing and integrating several variables. The ability to flexibly and temporarily bind items in working memory is fundamental to many higher cognitive functions. Addressing the neural binding problems in memory is essential for unraveling the mechanisms behind human cognition and consciousness. These investigations span various aspects of memory processing, including the integration of sensory details, temporal organization, conceptual associations, emotional content, and the temporary storage of information during cognitive tasks. A more profound understanding of variable binding will illuminate how the brain organizes and integrates information to create the unified mental experiences that underpin human thought and perception.

The Subjective Unity of Perception

The subjective unity of perception is part of the broader problem of the subjective unity of consciousness, a central challenge in both neuroscience and philosophy of mind. This aspect of the neural binding problem focuses on the phenomenal unity of perception—how distinct sensory inputs and cognitive processes are unified into a coherent, singular experience. The more challenging enigma lies in the subjective nature of personal experience, or qualia, which remains one of the most profound and unresolved issues across scientific disciplines. Despite numerous efforts to understand how our brains produce subjective experiences, the relationship between neural activity and the subjective unity of perception remains elusive (Chalmers 1995). When we perceive the world, the brain processes various sensory features—such as color, size, shape, motion, sound, bodily sensations, and emotions—through distinct neural circuits. Although these processes involve a broad distribution across different brain regions, we experience a seamless, unified whole, perceiving objects and scenes without the sensation of disjointed fragments (Bayne and Chalmers 2003). This subjective unity fundamentally questions how the brain coordinates dispersed neural activities correlating to a cohesive conscious experience. Neuroscientific research seeks to uncover the underlying neural mechanisms that correlate with and bind these diverse inputs into the unified experience of consciousness (Singer 2007).

Philosophically, *the hard problem of consciousness* (Chalmers 1995) presents a formidable obstacle to explaining the subjective unity of perception from a purely physicalist position. The hard problem asks why and how physical processes in the brain should give rise to subjective experiences at all. While neuroscientists can track neural correlates of perception, the gap between these correlates and the phenomenal experience—what it is like to see, hear, or feel—remains unexplained (Nagel 1974). This issue challenges the notion that a physicalist approach can fully explain the subjective unity of perception. The chapter on consciousness examines consciousness and current neuroscientific theories. Despite these theoretical challenges, studying the interactions between various brain regions and the coordination of neural activity remains a promising avenue for understanding the neural correlates of consciousness and the subjective unity of perception. Advances in this research may also shed light on the broader concept of *Brain Metafunction*, which aims to explain how the brain integrates diverse functions and processes to create a coherent experience of reality and congruent adaptive biological responses. Ultimately, the problem of subjective unity is not just a scientific challenge but also a deeply philosophical one, highlighting the limits of what we can know about the nature of consciousness.

Effector Response-Response Binding

Effector response-response binding is a fascinating cognitive process beyond perception and involves coordinating complex motor actions. It refers to the brain's ability to link or bind sequences of actions, treating them as unified units in memory. For example, when performing familiar actions like typing or playing a musical instrument, the brain combines individual movements into a smooth, cohesive sequence (Koch and Kunde 2002). This binding allows us to execute learned motor actions quickly and effortlessly without consciously attending to each step. Researchers are particularly interested in the neural mechanisms responsible for disassembling these bound action sequences. One brain region of interest is the prefrontal cortex, which plays a critical role in breaking down well-learned sequences into their components (Stokes et al. 2013). Using neuroimaging techniques like functional magnetic resonance imaging (fMRI), scientists study brain activity in individuals performing tasks that involve unbinding or deconstructing familiar action sequences.

A fascinating aspect of effector response-response binding is that a single encounter of a stimulus paired with a response can lead to a temporary binding between them. Later, re-exposure to just part of the original stimulus-response episode can trigger the recall of the entire episode, including the associated response (Hommel 2004). This binding demonstrates the brain's efficiency in linking and retrieving previously related actions. This mechanism likely supports fluid and efficient motor performance when repeated or routine actions are necessary. Recent research has revealed that similar binding processes occur between successive but independently planned manual responses. For example, when a person alternates between using different effectors, such as hands and feet, while performing a task, the brain can bind these responses into a single coordinated action sequence (Herwig and Waszak 2012). This finding suggests that binding processes are crucial for coordinating complex motor actions across different effectors. Researchers believe that the parietal and premotor cortices play pivotal roles in facilitating the binding of actions across motor systems (Toni et al. 2001). Understanding effector response-response binding provides valuable insights into how the brain organizes and processes complex motor behaviors. This knowledge has significant implications for cognitive functioning and potentially applies to rehabilitation, sports performance, and motor skill training. By investigating the neural basis of this binding phenomenon, researchers aim to develop better methods for retraining motor skills, improving coordination, and enhancing recovery after injury (Swinnen and Wenderoth 2004).

Computational Neuroscience Offers New Research Tools.

Discoveries in human brain research hold immense potential for modern technology, and computational neuroscience offers powerful tools to explore the brain's complexities. As the intricate workings of the brain continue to captivate researchers, computational neuroscience provides an interdisciplinary approach that allows scientists to dive deeper into understanding its "supercomputer-like" structure. This field seeks to address challenges across various domains, advancing fields such as artificial intelligence (AI) and technology. While not all brain functions involve computation, many do, and computational neuroscience employs simulations and mathematical models to expand our understanding of brain function (Trappenberg 2010). Over the past few decades, this field has seen significant growth, with developing applications and bright technological prospects.

The term "computational neuroscience" was introduced in 1985 by Eric L. Schwartz (1993) at a conference in Carmel, California. Computational neuroscience builds upon several foundational breakthroughs that have shaped our understanding of brain mechanisms. The Hodgkin and Huxley (1952) model of action potentials was a pivotal milestone in the field; they developed a mathematical model that explained how electrical signals (action potentials) are initiated and propagated in neurons. This model became the foundation for understanding how neural networks support neuron communication. Building on this, researchers extended their focus to neural population dynamics, learning, and memory, creating models that explained how groups of neurons interact and how learning occurs in the brain (Gerstner et al. 2014). Experimental psychology and neuroscience further propelled advances in computational neuroscience. Researchers investigated how the brain processes information related to learning, revealing essential insights into acquiring, storing, and manipulating data (Rolls and Treves 1998). Artificial neural networks (ANNs) emerged in the 1960s, which modeled brain functions based on biological neural networks and marked a significant milestone (McCulloch and Pitts 1943). These models laid the groundwork for modern AI, leading to advancements in machine learning and pattern recognition (LeCun et al. 2015). The collaboration between neurophysiological findings and computational models helped bridge theoretical understanding with practical technological applications, particularly in AI.

Computational neuroscience offers indispensable tools for brain research, including mathematical models that simulate the complex interactions of neurons, neural networks, and brain regions. These models allow researchers to explore how information is processed, transmitted, and integrated within the brain's circuits (Dayan and Abbott 2001). Simulations offer a controlled and cost-effective environment to test hypotheses and predict neural behavior. At the same time, statistical tools allow for the rigorous analysis of brain data, uncovering patterns and connections within neural networks (Friston et al. 1994). By combining computational models with experimental data, researchers bridge the gap between theoretical neuroscience and empirical research, enhancing our understanding of brain function. The complexity of the brain demands abstractions and simplifications in computational models. These abstractions help focus on key features of brain processes while reducing unnecessary complexity, facilitating a clearer understanding of neural mechanisms (Izhikevich 2005). This approach has driven innovations in neuroprosthetics, neuromodulation, and AI, offering practical applications for healthcare and technology (Kandel et al. 2013). Ultimately, computational neuroscience opens new avenues of research. Its tools allow scientists to investigate brain functions unprecedentedly, bridging the theoretical and experimental realms. This interdisciplinary approach is crucial for exploring neural phenomena, such as the neural binding problem, and advancing fields like AI, cognitive science, and neuroprosthetics. The continued integration of computational and experimental approaches will drive innovation in healthcare, technology, and our understanding of cognition and consciousness.

Temporal Binding Hypothesis: A Historical Perspective

The *Temporal Binding Hypothesis* (TBH) emerged in the early 1980s when Christoph von der Malsburg proposed that neurons transmit dual signals: one through conventional rate coding, which encodes feature effectiveness, and another through spike coordination between neuronal assemblies (von der Malsburg, 1981). This groundbreaking idea offered a potential solution to the "binding problem," the challenge of integrating features processed in separate brain regions into a cohesive perceptual experience. Shortly after, Francis Crick and Christof Koch (1990) expanded on this concept, suggesting that synchronization of distant neurons occurs through transient *gamma wave oscillations* driven by attention. Researchers hypothesized that these oscillations facilitate feature binding by dynamically linking neurons across cortical areas. By the 1990s, Francisco Varela added further depth by emphasizing the importance of the *oscillatory phase* in pairing neuronal assemblies, proposing that precise timing of neuronal firing was critical in integrating sensory information (Varela et al., 2001).

The TBH addresses the combinatorial challenge inherent in neural binding: the brain must form precise connections among vast numbers of neurons to integrate diverse sensory features, emotions, and actions into a coherent conscious experience. This combinatorial complexity reflects the staggering number of potential neural interconnections required for such integration. The TBH provided an elegant framework for understanding how temporal synchronization might reduce this complexity, allowing for efficient and dynamic binding across neural circuits. The Temporal Binding Hypothesis posits that synchronized neuronal oscillations, particularly in the gamma frequency range (30–80 Hz), enable the integration of sensory features into unified perceptions. The brain can generate coherent percepts by coordinating the activity of neurons representing different features (e.g., color, shape, motion). Von der Malsburg, Crick, Koch, and Varela laid the foundation for this hypothesis by highlighting the role of spike timing, gamma oscillations, and oscillatory phase in binding features processed in distributed neural regions. However, the TBH has faced significant criticism.

Critique

Shadlen and Movshon (1999) argued that temporal synchronization alone cannot account for the full complexity of feature binding. They suggested that the hypothesis overemphasizes the role of oscillatory activity while neglecting spatial coordination, attentional mechanisms, and contextual influences. Furthermore, Di Lollo (2012) framed the feature-binding problem as an "ill-posed problem," challenging the assumption that features must be reintegrated after being processed in distinct neural regions. Di Lollo proposed that feature integration may occur throughout the processing pipeline rather than relying on a post hoc synchronization mechanism. This alternative perspective raises questions about the necessity of oscillatory activity for binding and suggests that feature integration may arise as an inherent property of hierarchical neural processing.

Adding to the debate, Pieter R. Roelfsema (2023) proposed a firing-rate-based binding model. His hypothesis posits that neurons representing object features enhance their firing rates in response to attention or saliency, forming functional assemblies without requiring precise temporal synchronization. This mechanism emphasizes *rate coding* over oscillatory synchronization, offering a more straightforward and flexible approach to binding features. Roelfsema's work aligns with recent findings suggesting that attention-driven modulation of neuronal activity plays a central role in binding processes.

Recent Developments

Recent advancements in neuroscience have expanded our understanding of neural binding mechanisms beyond temporal synchronization. Studies have demonstrated that *alpha oscillations (8–12 Hz)* contribute to binding processes across sensory modalities, influencing *multisensory integration*. For instance, research has shown that alpha-band activity is crucial for coordinating auditory and visual signals in time, supporting the temporal alignment necessary for coherent multisensory perception (van Diepen et al., 2015, 2019; Michalareas et al., 2016). These findings suggest oscillatory dynamics outside the traditionally studied gamma range may also play critical roles in neural binding. Researchers believe alpha oscillations modulate inter-areal communication by gating information flow, providing a temporal framework for synchronizing activity across brain regions (Jensen et al., 2007).

In computational neuroscience, researchers have developed *hybrid models* that integrate spiking neural networks (SNNs) with artificial neural networks (ANNs) to simulate binding processes. A recent study by Beniaguev, Segev, & London (2021) demonstrated how hybrid models that combine biologically realistic spiking dynamics with attention-like mechanisms improve our ability to model feature binding and integration. These models integrate the temporal precision of SNNs with the computational flexibility of ANNs, allowing researchers to simulate how neural circuits might resolve the binding problem in dynamic environments. This approach highlights the potential for combining biologically inspired and artificial approaches to advance our understanding of neural integration.

Additionally, *predictive coding* has gained prominence as an explanatory framework for binding. Predictive coding posits that the brain continuously generates predictions about incoming sensory inputs and resolves discrepancies (prediction errors) through hierarchical inference (Friston, 2018). This dynamic process may naturally facilitate binding by aligning sensory features with prior expectations and integrating them into coherent percepts (Bastos et al., 2012). Furthermore, recent research suggests that predictive coding frameworks can account for spatial and temporal binding aspects, potentially explaining how the brain achieves multisensory integration and constructs unified representations of the world (Feldman & Friston, 2010; Clark, 2013). This perspective emphasizes the interplay between top-down predictions and bottom-up sensory information as critical for resolving the binding problem.

Cortical Traveling Waves in the Brain Metafunction Architecture

While gamma oscillations (30–80 Hz) remain relevant to the Temporal Binding Hypothesis (TBH), recent research emphasizes the role of cross-frequency coupling (CFC) in feature integration across spatial and temporal scales. CFC involves interactions between slower oscillations, like theta (4–8 Hz) or alpha (8–12 Hz), and faster gamma rhythms, providing a flexible framework for coordinating neural activity. For instance, theta-gamma coupling supports working memory by organizing sequential gamma cycles (Siegel et al. 2009; Bahramisharif et al. 2018), preserving input order and timing. Similarly, alpha oscillations modulate multisensory integration by regulating inter-regional communication (Michalareas et al., 2016; Jensen et al., 2007). Together, these findings extend TBH beyond gamma synchrony, highlighting the cooperative role of multiple frequency bands in neural binding.

Cortical traveling waves are increasingly recognized as fundamental modulatory and organizing mechanisms in perceptual binding, contributing to the temporal structuring of neural activity across the cortex (Müller et al. 2018). These waves propagate in structured patterns, modulating neuronal excitability in a sequential, phase-dependent manner. At the mesoscopic level, they operate within individual cortical areas, facilitating local coordination among feature-selective populations. Crucially, Müller et al. demonstrate that traveling waves also function at macroscopic scales, propagating across cortical regions—such as from early sensory to higher-order associative areas—aligning distributed processing over time. This cross-areal coherence enables long-range communication and integration, positioning traveling waves as candidate mechanisms for large-scale perceptual and cognitive unification. However, traveling waves alone may be insufficient to explain the global unity of conscious experience, given their predominantly modality-linked roles. Within the Brain Metafunction Framework, traveling waves may serve as intermediate-level coordination mechanisms that dynamically link specialized cortical functions. By temporally aligning distributed subsystems, they may help instantiate the large-scale, adaptive integration required for coherent perception, thought, and action—supporting the brain's capacity to operate as a unified, goal-directed system.

Predictive Coding: A Dynamic Framework for Binding

Predictive coding has emerged as a prominent framework for explaining binding, offering a dynamic and hierarchical model of brain function. It posits that the brain continuously generates predictions about incoming sensory inputs and updates them based on the errors that arise when inputs deviate from expectations (Friston, 2018). This feedback loop facilitates binding by aligning sensory features with prior expectations, integrating them into coherent percepts. Recent advances in this framework highlight two critical aspects. First, *precision weighting* plays a crucial role, allowing the brain to assign variable importance to sensory inputs based on their predicted relevance. This mechanism enables the selective binding of features most critical to a specific task or environment (Hohwy, 2020). Second, researchers have integrated predictive coding with the global neuronal workspace theory, proposing that conscious binding arises from the widespread sharing of information across cortical and subcortical networks. This integration ties binding processes to higher-order cognitive phenomena, such as attention, consciousness, and the self (Mashour et al., 2020). By addressing feature integration and its connection to higher-order functions, predictive coding provides a flexible and biologically plausible framework for understanding how the brain achieves binding while simultaneously supporting attention and awareness.

Advances in Artificial Intelligence and Neural Models

In computational neuroscience, hybrid models that combine *spiking neural networks* (SNNs) and *artificial neural networks* (ANNs) have emerged as powerful tools for simulating neural binding processes. A recent study by Beniaguev et al. (2021) demonstrated how these models leverage two key mechanisms: *spike timing dynamics,* which mimic the precise firing patterns of biological neurons, and *attention-like mechanisms,* which selectively prioritize features for binding. Additionally, advances in artificial intelligence, particularly the development of *transformer models* widely used in natural language processing, have inspired new hypotheses about biological binding. Transformers rely on attention mechanisms to dynamically integrate information, which is analogous to how the brain might bind sensory and cognitive features (Dosovitskiy et al., 2021). These computational advancements highlight the bidirectional relationship between neuroscience and artificial intelligence, where artificial systems both model and inspire new approaches to understanding how the brain achieves binding.

Subcortical Contributions to Binding

Although much of the research on the binding problem has focused on cortical regions, recent studies have highlighted the critical roles of subcortical structures in facilitating feature integration. The *thalamus* serves as a central hub for communication between cortical areas, modulating sensory inputs and ensuring their alignment for efficient integration across brain networks (Wang, 2020). Acting as a more than a "relay station," the thalamus coordinates sensory information and maintains coherence across distributed neural activity. The *hippocampus* is essential for *spatiotemporal binding,* organizing events into coherent episodic memories, and integrating them into broader cognitive narratives (Eichenbaum, 2017). In addition, the *basal ganglia,* traditionally associated with motor control, are increasingly recognized for their role in coordinating sequential actions and linking motor outputs to sensory features. These subcortical structures and the brainstem complement cortical mechanisms by providing additional integration layers, enabling the seamless combination of sensory, cognitive, and motor processes necessary for coherent perception and behavior.

Crossing into the Brain Metafunction Framework

The *Brain Metafunction Framework* provides a unifying perspective for understanding how these developments in neural binding fit into the broader organization of brain functions. This framework emphasizes that the brain operates as a hierarchical and dynamic system where multiple specialized processes work together to achieve cohesive integration. From this perspective:

1. *Temporal Binding:* Cross-frequency coupling, gamma synchronization, and predictive coding exemplify the brain's temporal metafunction, enabling real-time coordination of neural activity across different timescales.

2. *Spatial Binding:* The contributions of subcortical structures, such as the thalamus and hippocampus, reflect the spatial metafunction, integrating information across distributed neural networks and ensuring spatial coherence.

3. *Attention, Consciousness, and the Unconscious:* Predictive coding and global workspace models tie binding processes to the brain's higher-order functions, in alignment with the brain metafunctional framework, including attentional prioritization, conscious awareness, and self-representation supported by extensive unconscious processes including memory. These processes ensure that the brain integrates sensory features and aligns them with goals, context, and prior experiences.

By incorporating these mechanisms, the Brain Metafunction Framework provides a holistic approach to understanding how the brain achieves unity amidst its structural and functional complexity. It positions binding as not just a localized phenomenon but as a property of the brain's hierarchical and multi-scale organization. These recent advancements in neuroscience and computational modeling have significantly expanded our understanding of the binding problem. From cross-frequency coupling and predictive coding to subcortical contributions and artificial neural models, these developments reveal the brain's remarkable capacity to integrate sensory, cognitive, and motor features into unified experiences. The Brain Metafunction Framework offers a valuable lens to contextualize these findings, emphasizing the interplay of temporal, spatial, and higher-order processes in achieving seamless integration. As research continues to uncover the intricate binding mechanisms, it brings us closer to understanding how the brain transforms fragmented inputs into the coherent perceptions and actions that define human experience.

But despite significant advances in brain imaging and computational modeling, a comprehensive solution remains elusive. For instance, Northoff, Wainio-Theberge, and Evers (2020) noted that while we can correlate brain network activity with mental phenomena such as selfhood, consciousness, episodic memory, and emotion, the causal mechanisms underlying these correlations remain unclear. They refer to this gap as the "common currency problem"—the lack of a clear explanation for how neuronal activity translates into unified conscious experiences. This broader issue underscores a central challenge in cognitive neuroscience: bridging the gap between neural correlates of consciousness (NCC) and the subjective nature of experience. The theories examined in the chapter on consciousness, such as the integrated information theory (Tononi, 2004) and the global neuronal workspace (Dehaene et al. 1998), have contributed to explaining aspects of binding and consciousness. Still, they leave open fundamental questions about the precise neural mechanisms and their scalability across different levels of brain function.

The Temporal Binding Hypothesis theory was a first step that influenced neuroscience by providing a framework for understanding how the brain integrates sensory features. However, critiques and alternative perspectives, such as rate coding models and predictive coding frameworks, have highlighted the limitations of relying solely on temporal synchronization to explain binding. A unified theory that reconciles these diverse perspectives remains a work in progress. Addressing the "common currency problem" will be pivotal in advancing our understanding of how the brain participates in creating coherent and unified experiences from its complex and distributed neural activity. Let us now review some of the additional challenges before focusing on the *brain metabinding problem or* the *neural metabinding problem.*

Challenges in Solving the Binding Problem

Harmonizing Global External Signals with Internal Associations

One of the critical challenges is how the brain harmonizes external stimuli with internal processes involving memory, expectations, and cognitive associations. Perception is not passive; it integrates incoming sensory data with stored knowledge, contextual information, and expectations (Clark 2013). Predictive coding suggests that the brain constantly generates predictions about sensory input and uses discrepancies between these predictions and actual sensory data to update its internal models (Friston 2010). However, the mechanism by which the brain seamlessly integrates these external signals with internal associations across multiple timescales and neural hierarchies remains unclear. Recent studies highlight the importance of hierarchical processing and feedback loops in achieving this integration, suggesting that top-down and bottom-up processes continuously refine the binding of features to form coherent perceptions (Buergers and Noppeney 2022).

The Interplay Between Perception and Higher-Order Cognition

Another challenge in solving the binding problem is understanding how basic perceptual processes interact with higher-order cognitive functions like attention, memory, and decision-making. For instance, the global neuronal workspace theory proposes that synchronized neural activity across distant brain regions creates a "workspace" that integrates information from different modalities and cognitive domains, making it accessible to consciousness (Dehaene et al. 1998). However, recent research questions the primacy of synchrony, suggesting that changes in neuronal firing rates driven by attention and saliency may provide a more flexible mechanism for binding (Roelfsema 2023). These findings suggest a dynamic interplay between attention-driven processes and hierarchical cortical processing.

From Tracing and Tagging Neural Signals to Unified Consciousness

The brain's ability to track sensory features such as color, shape, and location while associating them with internal states like memory and expectation is remarkable. The challenge lies in explaining how the brain "tags" or marks these signals across different regions, ensuring correct association with appropriate cognitive and perceptual representations (Feldman 2013). One of the most profound aspects of the binding problem is its relationship with consciousness. While neural synchrony may explain how different brain regions collaborate to integrate information, it does not fully account for the experience of unified consciousness. In other words, while we can describe neural activity patterns that correlate with conscious experience, we are still left with the "hard problem" of explaining how these neural patterns translate into the subjective experience of seeing, hearing, or feeling a unified percept (Chalmers 1995).

Avoiding Hallucinations and Involuntary Actions

Another major challenge in the brain's binding process is understanding how it avoids generating hallucinatory experiences or involuntary actions based solely on internal activity patterns. Suppose synchronized neural activity is sufficient for binding. What mechanisms prevent the brain from mistakenly interpreting internally generated signals—such as those arising from memory, imagination, or predictive processes—as real sensory input? This question highlights the critical role of *reality monitoring*, a cognitive mechanism that allows the brain to differentiate between internally generated and externally received signals. Researchers believe this process relies on interactions between cortical and subcortical regions, such as the prefrontal cortex for error monitoring and the hippocampus for contextual recall.

Furthermore, studies on predictive coding suggest that the brain continuously generates predictions about sensory inputs while using prediction errors to refine and correct these expectations. This dynamic feedback would ensure that internally generated signals, such as those from imagination or anticipation, are modulated and do not override incoming sensory information (Friston, 2018). A failure in these mechanisms may contribute to phenomena such as hallucinations or delusions, similar to those observed in conditions like schizophrenia. For instance, Sterzer et al. (2018) propose that disruptions in predictive coding or imbalances in the precision weighting of sensory versus internal signals can lead to an over-reliance on internally generated inputs, resulting in hallucinations.

Similarly, avoiding involuntary actions involves distinguishing self-initiated motor plans from external triggers. This distinction relies on corollary discharge, a process by which the brain generates internal copies of motor commands to predict and cancel out the sensory consequences of self-generated actions. Dysfunction in this mechanism could result in the perception of involuntary or externally imposed actions, a phenomenon also linked to neurological and psychiatric disorders. Understanding how the brain avoids such confusion—by properly weighing internal and external signals—offers valuable insights into its ability to integrate information seamlessly while maintaining coherent perceptions and behaviors. These mechanisms are not only critical for healthy cognition but also provide a framework for exploring disorders where binding processes go awry, such as psychosis, epilepsy, or motor tics.

Integrating Signals Across Multiple Brain Levels

Integrating signals across multiple brain levels, encompassing both cortical and subcortical regions, remains one of the most complex aspects of the binding problem. While much of the research has focused on cortical areas such as the visual cortex for sensory processing and the prefrontal cortex for executive functions, there is growing recognition of the critical roles played by subcortical structures like the thalamus, basal ganglia, and hippocampus in coordinating and modulating binding processes (Jones, 2001). These subcortical regions do more than passively relay information; they actively participate in tagging, routing, and synchronizing sensory and cognitive signals across the brain. For instance, the thalamus is a central hub for communication between cortical regions, synchronizing cortical activity through its modulation of rhythmic oscillations. These oscillations help align activity in distributed cortical areas, creating a temporal scaffold for information integration (Buergers & Noppeney, 2022). Similarly, the hippocampus plays a pivotal role in integrating spatial and temporal information, supporting the formation of coherent episodic memories that combine diverse features of an experience. The basal ganglia, often associated with motor control, contribute by linking sequential motor actions with sensory feedback, enabling smooth and coordinated behavior.

The challenge lies in understanding how these subcortical structures dynamically interact with cortical networks through hierarchical feedback loops and adaptive firing patterns. For example, predictive coding models suggest that subcortical structures like the thalamus may mediate the comparison of sensory inputs with cortical predictions, enabling efficient integration while minimizing redundancy. Furthermore, recent advances in brain imaging and electrophysiology have revealed that cross-frequency coupling between cortical and subcortical regions—such as theta-gamma interactions—facilitates information integration across multiple timescales. These interactions underscore the importance of viewing the brain as a *hierarchically organized, multi-level system* where subcortical and cortical regions work together to achieve seamless binding. Investigating these integrative mechanisms not only sheds light on normal brain function but also helps identify how disruptions in these processes contribute to neurological and psychiatric disorders, such as epilepsy, schizophrenia, and Parkinson's disease, where binding and integration are impaired.

Toward a Comprehensive Understanding

The binding problem remains a fundamental challenge in neuroscience, with no single theory or mechanism providing a comprehensive solution. Researchers have made significant progress in identifying neural correlates and exploring theories such as predictive coding, global neuronal workspace, and synchronized oscillations. Still, they have yet to uncover the "common currency" that links neural activity with unified consciousness. Future research must address the interplay between perception, cognition, and consciousness, focusing on how the brain integrates information across multiple scales, regions, and modalities. Solving the binding problem will require an interdisciplinary approach that draws from neuroscience, cognitive science, and philosophy to unravel the mechanisms underlying the brain's remarkable ability to create a unified experience from fragmented sensory and cognitive inputs.

The Brain Metabinding Problem(s)

The *brain metabinding problem*, or better problems, emerges as an extension of the neural binding problem, extending into a more complex understanding of how the brain integrates a wide array of cognitive, behavioral, and regulatory processes into a unified conscious experience. This concept is rooted in the *brain metafunction* framework, which reflects the global integration of multiple brain functions. As neuroscience continues to uncover the intricate workings of the brain, it becomes clear that consciousness is not confined to isolated neural events but profoundly relates to distributed processes across the brain. These include multimodal perception, attention, memory, unconscious processes, complex motor actions, and homeostatic and allostatic functions.

The metabinding problem represents the challenge of understanding how these diverse processes converge to create a coherent and meaningful whole—a unified conscious experience. This problem closely relates to the neural binding problem. Still, it goes beyond the traditional focus on sensory integration to encompass the broader challenge of linking cognitive, emotional, and regulatory functions across the brain's many regions and layers. The global neuronal workspace theory (Dehaene et al. 1998) provides a framework for how different brain regions synchronize their activity to create conscious awareness. Still, the metabinding problem requires a more comprehensive explanation of how signals integrate across time, space, and through various cortical and subcortical levels (Tononi 2004).

Metabinding and Conscious Experience

The brain's remarkable ability to integrate numerous information sources—sensory data, cognitive processes, emotional responses, and bodily regulation—into a seamless conscious experience underscores the multidimensional nature of human consciousness. This integration is not merely about neural synchrony but involves a dynamic and fluid orchestration of processes that reflect the unity of the human person. Neuroscientists like Antonio Damasio (2010) have explored how consciousness relates to the interaction of brain systems that govern cognitive functions, emotions, and bodily regulation, pointing to the idea that consciousness is inherently embodied and embedded in the body's regulatory functions. The metabinding problem raises profound questions about the nature of consciousness and its relationship to the physical brain. As our understanding of neural processes deepens, researchers are beginning to uncover novel and increasingly complex associations between brain activity and conscious experience. These discoveries reveal that the brain does not simply bind features of perception but integrates them with broader cognitive and affective processes, creating a unified, embodied experience. This dimension of the problem suggests that the metabinding process could involve elements that transcend traditional neuroscientific explanations, touching on transphysical dimensions that connect cognition, perception, and the broader fabric of existence.

The Limits of Neural Correlates in Explaining the Metabinding Problem

The neural metabinding problems are some of the most complex challenges in neuroscience and the philosophy of mind. They raise the question of how the brain integrates sensory inputs and cognitive and bodily processes into a unified conscious experience. From a neuroscientific perspective, researchers have increasingly focused on identifying neural correlates of consciousness (NCCs), specific neural events or structures associated with conscious experiences, such as synchronized oscillations, firing rates, or neural assemblies. While identifying NCCs has advanced our understanding of brain function, this approach is limited in explaining how and why these neural events give rise to the unified consciousness perception. This critique argues that while neural correlates are necessary for understanding brain processes, they are fundamentally insufficient to explain the binding problem.

Correlation vs. Causation: The Explanatory Gap

At the heart of the issue is the distinction between correlation and causation. Neural correlates can reveal what brain states are associated with specific conscious experiences, but they do not explain why or how these brain states produce unified perceptual experiences. The "explanatory gap," introduced by Joseph Levine (1983), and the "hard problem of consciousness," articulated by David Chalmers (1995), both highlight the challenges of understanding consciousness but differ in focus. The explanatory gap emphasizes the difficulty in linking subjective qualitative experiences (qualia) to the objective, measurable workings of the brain. The hard problem extends this by questioning why subjective experiences arise from physical processes at all and why specific neural states correspond to particular qualia. While supposedly the "easy problems" of consciousness, such as perception and memory, can be addressed by understanding brain mechanisms, the hard problem tackles the more profound ontological mystery of why consciousness exists. Both concepts reveal the limitations of physicalist explanations, with the explanatory gap highlighting gaps in current scientific frameworks and the hard problem of addressing the fundamental nature of consciousness. Even if neural correlates map the processes involved in binding sensory features like color, shape, and motion, they do not explain why these processes produce a single, cohesive perception rather than fragmented sensations. While synchronized activity in distinct brain regions correlates with unified perception, this corr\elation does not provide a causal explanation. Why should synchronized neural firing lead to the experience of perceptual unity? The relationship between neural activity and subjective experience remains mysterious because neural correlates alone cannot explain the qualitative nature of consciousness or qualia—the subjective feel of experience.

The Phenomenological Challenge: Explaining Qualia

One of the most significant limitations of the neural correlates approach is its failure to account for the phenomenology of consciousness—the subjective "what it is like" aspect of experience. The brain processes sensory features in distinct regions, yet when we see a red apple, we perceive it as a unified whole. Thomas Nagel (1974) famously argued that science cannot explain the first-person subjective nature of experience—the essence of consciousness. Neural correlates may explain the third-person perspective of brain activity, but they do not bridge the gap to the first-person experience of unity. This challenge extends beyond qualia to the problem of intentionality—the brain's ability to direct perception toward specific objects and integrate sensory data into meaningful experiences. Neural correlates can map the brain's electrical activity. Still, they cannot explain how the brain organizes, and the mind interprets sensory information in a directed, intentional way that creates a coherent, meaningful experience. This directedness is fundamental to perception and does not admit reduction to correlations between brain states and sensory stimuli.

The Problem of Reductionism: Overlooking the Whole

Another significant limitation of the neural correlates approach is its inherent reductionism. Researchers often downplay brain function's complexity and holistic nature by focusing on specific neural events or structures as correlates of consciousness. The binding problem is not just about finding the proper oscillations or neural codes; it involves understanding how the system integrates and processes information. Reducing the problem to isolated neural correlates risks missing the larger picture. The emergentist perspective argues that conscious experience results from the brain's systems interactions, not isolated neural events. While neural correlates may reveal aspects of the brain's functions locally, they cannot fully explain how these localized processes contribute to the global integration that produces unified perception. In this sense, neural correlates provide only a partial understanding of the binding problem and fail to capture the full complexity of brain dynamics.

Temporal and Spatial Limits of Neural Correlates

Neural correlates also face limitations related to their temporal and spatial resolution. Many theories of binding focus on temporal synchrony, suggesting that neurons across different regions must fire in synchrony to achieve perceptual unity. However, this model assumes that simultaneous activity is essential for binding, overlooking that perception often involves asynchronous processes. For example, the brain frequently integrates information across time, predicting the future position of moving objects or recalling past experiences to inform present perception (Shadlen and Movshon 1999). Additionally, neural correlates often emphasize local brain activity, but binding requires integration across multiple spatial scales—from local neural circuits to large-scale networks that span the entire brain. Focusing on localized correlates can obscure the broader, system-wide interactions necessary for unified perception. The binding problem requires a more holistic approach that accounts for both local and global brain dynamics.

Explanatory Narrowness: Neglecting the Broader Context

Finally, focusing on neural correlates risks explanatory narrowness, ignoring the broader context in which binding occurs. Binding is not a narrow neuroscientific issue but involves multiple dimensions, including attentional, cognitive, and emotional. For example, attention plays a crucial role in selecting and integrating sensory inputs into coherent percepts, and memory influences how we bind current perceptions with past experiences. By focusing primarily on sensory integration, the neural correlates approach may neglect the top-down processes that shape perception. A more comprehensive explanation of binding would need to consider how neural activity interacts with attentional and cognitive mechanisms to produce unified experiences.

While neural correlates provide important insights into brain activity and are essential for understanding the biological basis of perception, they are insufficient to explain the binding and metabinding problems. The correlation-causation gap, the failure to address qualia, the reductionist tendencies of focusing on localized brain activity, and the temporal and spatial limitations of neural correlates all point to the need for a broader, more integrative approach. The binding and metabinding problems involve the mechanics of brain function and deep philosophical questions about consciousness, perception, and the mind-brain relationship. To solve these problems, we must move beyond correlations and explore the emergent, holistic, and transphysical aspects of brain function that give rise to the unity of conscious experience.

Physicalism, Panpsychism, and Emergent Dualism: The Binding and Metabinding Problems and the Challenge of Unity

The combination problem in panpsychism closely mirrors the challenges posed by the neural binding problem in consciousness studies. Consciousness, characterized by various forms of unity, emerges from brains composed of distinct and independent elements. Explaining how a brain achieves this unity in consciousness remains a significant challenge. The neural binding problems highlight the difficulty of explaining how disparate neural activities combine to create a singular, integrated conscious experience. This difficulty mirrors the challenge of reconciling the inherent disunity of the brain's physical structures, as perceived through physicalist lenses, with the experiential unity observed in consciousness. The binding problems, thus, present a substantial obstacle for theories like physicalism and panpsychism, which seek to explain consciousness within frameworks positioned in the extremes.

As articulated by philosopher William Hasker (1999), the *combination problem* is a fundamental challenge for panpsychism. Panpsychism suggests consciousness is an essential property distributed across all entities at a basic level—often called "mind-dust." However, it struggles to explain how these distributed, elementary aspects of consciousness combine to form the unified conscious mind of sentient beings. Hasker argues that even if we accept the existence of "mind-dust," the crucial explanatory gap remains: How do these bits of consciousness combine into a singular, coherent conscious mind? Applying similar logic to the neural binding problem within physicalism and emergent dualism, we encounter comparable difficulties. Physicalism posits that consciousness is entirely reducible to physical processes in the brain, yet the neural binding problem challenges this by pointing out the difficulty of explaining how distinct neural elements—responsible for various aspects of perception and cognition—bind together to create a unified conscious experience (Chalmers 1996). The challenge for physicalism lies in identifying a mechanism by which diverse neural activities generate singular, integrated conscious awareness.

Hasker's view, often described as *emergent dualism*, holds that consciousness arises as an emergent brain property but is not reducible to its physical components (Hasker 1999). Unlike panpsychism, which attributes a fundamental level of consciousness to all matter (leading to the combination problem of how these individual conscious elements combine into unified experiences) (Chalmers 2016), Hasker's emergent dualism partially avoids this issue by not attributing consciousness to bits of matter. Instead, Hasker suggests that the brain's complex organization gives rise to consciousness as a holistic, emergent phenomenon ontologically distinct from its physical substrate. However, as Rickabaugh (2021) notes, Hasker's view faces a similar problem to panpsychism in explaininghow distinct neural or physical processes bind together to produce a singular, unified conscious experience. This issue mirrors the combination problem because Hasker must still account for how the emergent property of consciousness integrates diverse inputs (from neural or cognitive processes) into one coherent awareness. While Hasker's theory avoids the need for tiny "proto-conscious" elements to combine, it still struggles to explain how the emergent phenomenon of consciousness binds together the activity of its underlying components into a unified conscious self. Hasker's view is a type of *protopanpsychism,* as it acknowledges a non-reductive relationship between physical processes and consciousness but leaves the neural binding problem unresolved.

The challenge remains in both panpsychism and emergent dualism: explaining how diverse elements integrate into a unified conscious experience. Whether it is the combination problem in panpsychism or the neural binding problem in physicalism and emergent dualism, explaining how such unity arises is formidable and remains unresolved. The philosophical study of mereology, which investigates the relationships between parts and wholes, offers a framework for better understanding this problem. Mereological issues arise in emergent dualism, which must explain how a unified, immaterial substance—traditionally conceptualized as the soul—emerges from a physical structure as complex as the human brain, composed of billions of distinct and separable components (Lowe 2013). The challenge lies in elucidating how a singular, immaterial entity arises from the physical brain's material, multipartite structure. This problem points to deeper metaphysical concerns regarding integrating mental properties and the unity of consciousness within a complex material organism.

Toward a Transphysical Understanding

Exploring the metabinding problem suggests that as our understanding of neural binding mechanisms advances, we will continue to encounter insights that push beyond the prevailing physicalist neuroscience framework. The interconnectness of cognition, perception, and bodily processes points toward a more holistic understanding of human consciousness, which would require integrating insights from philosophy, theology, and metaphysics. The metafunctional integration of brain functions, particularly as they relate to conscious experience, offers glimpses into a transphysical domain, where the unity of body, mind, and spirit becomes apparent (Haldane 2011). The metabinding problem, therefore, serves as a crucial focal point in the quest to understand not only the biological underpinnings of consciousness but also its transphysical and existential dimensions. Unlike the naturalistic framework, this perspective does not imply that the brain alone generates consciousness but proposes that the brain acts as a cooperator of higher-order integration that unifies the full spectrum of human experience. As researchers elucidate the intricacies of neural processes, they may uncover principles of organization and integration that transcend current scientific paradigms, pointing toward a transcendent unity underlying conscious experience.

Conclusion

The neural binding problem, compounded by the metabinding issues, presents significant challenges for theories such as physicalism, panpsychism, and emergent dualism in their attempts to explain the nature of consciousness. Physicalism struggles to account for how diverse neural components combine to produce a unified conscious experience, often called the "hard problem" of consciousness. Panpsychism, while positing that consciousness is a fundamental property distributed across all entities, grapples with the combination problem—how these elements of consciousness come together to form a cohesive whole. Similarly, emergent dualism is challenged by explaining how an immaterial, unified substance can emerge from the complex, material structure of the brain.

Moreover, the binding and metabinding neural problems expose the difficulty of reconciling the brain's apparent physical disunity with the observed unity of consciousness, posing formidable obstacles to these prominent theories in the philosophy of mind. Two fundamental challenges remain for neuroscience, particularly for cognitive neuroscientists and philosophers of mind. First, there is a pressing need to break free from the physicalist methodological constraints that limit conceptual and scientific exploration. Second, it is essential to develop theories based on an inference to the best explanation that draws from the cumulative interdisciplinary data. The *Brain Metafunction* framework proposed here, which suggests the involvement of a transphysical domain, points toward a broader and more integrative approach to addressing these complex issues.

REFERENCES

Baddeley A. The episodic buffer: a new component of working memory? Trends Cogn Sci. 2000 Nov 1;4(11):417-423. doi: 10.1016/s1364-6613(00)01538-2.

Baddeley, A. (2003). Working memory: Looking back and looking forward. *Nature Reviews Neuroscience, 4*(10), 829–839.

Bahramisharif, A., Jensen, O., Jacobs, J., & Lisman, J. (2018). Serial representation of items during working memory maintenance at letter-selective cortical sites. *PLOS Biology, 16*(8), e2003805. https://doi.org/10.1371/journal.pbio.2003805

Bastos, A. M., Usrey, W. M., Adams, R. A., Mangun, G. R., Fries, P., & Friston, K. J. (2012). Canonical microcircuits for predictive coding. *Neuron, 76*(4), 695–711. https://doi.org/ 10.1016/j.neuron.2012.10.038

Bayne, Tim, and David J. Chalmers. 2003. "What Is the Unity of Consciousness?" In *The Unity of Consciousness: Binding, Integration, and Dissociation,* edited by Axel Cleeremans, 23–58. Oxford: Oxford University Press.

Beniaguev, D., Segev, I., & London, M. (2021). Single cortical neurons as deep artificial neural networks. *Neuron, 109*(17), 2727–2739.e3. https://doi.org/10.1016/j.neuron.2021.06.002

Bergmann, Til O., and Gabriel Hartwigsen. 2020. "Inferring Causality from Noninvasive Brain Stimulation in Cognitive Neuroscience." *Journal of Cognitive Neuroscience* 33 (2): 195–225. https://doi.org/10.1162/jocn_a_01591.

Bressler, Steven L., and J. A. Scott Kelso. 2001. "Cortical Coordination Dynamics and Cognition." *Trends in Cognitive Sciences* 5 (1): 26–36. https://doi.org/10.1016/S1364-6613 (00)01564-3.

Buergers, Sabine, and Uta Noppeney. 2022. "The Role of Alpha Oscillations in Temporal Binding within and across the Senses." *Nature Human Behaviour* 6 (6): 732–42. https://doi.org/10.1038/s41562 -022-01340-9.

Chalmers, David J. 1995. "Facing Up to the Problem of Consciousness." *Journal of Consciousness Studies* 2 (3): 200–19.

Clark, Andy. 2013. "Whatever Next? Predictive Brains, Situated Agents, and the Future of Cognitive Science." *Behavioral and Brain Sciences* 36 (3): 181–204. https://doi.org/ 10.1017/S0140525X12000477.

Crick, Francis, and Christof Koch. 1990. "Towards a Neurobiological Theory of Consciousness." *Seminars in the Neurosciences* 2: 263–75. https://doi.org/10.1016/B978-0-12 -657410-4.50012-8.

Damasio, Antonio R. 1999. *The Feeling of What Happens: Body and Emotion in the Making of Consciousness*. New York: Harcourt Brace.

Damasio, Antonio. 2010. *Self Comes to Mind: Constructing the Conscious Brain*. New York: Pantheon Books.

Dayan, Peter, and L. F. Abbott. 2001. *Theoretical Neuroscience: Computational and Mathematical Modeling of Neural Systems*. Cambridge, MA: MIT Press.

Dehaene, Stanislas, Michel Kerszberg, and Jean-Pierre Changeux. 1998. "A Neuronal Model of a Global Workspace in Effortful Cognitive Tasks." *Proceedings of the National Academy of Sciences* 95 (24): 14529–34. https://doi.org/10.10 73/pnas.95.24.14529.

Dennett, Daniel C. 2018. *From Bacteria to Bach and Back: The Evolution of Minds*. New York: W. W. Norton & Company.

Desimone, Robert, and John Duncan. 1995. "Neural Mechanisms of Selective Visual Attention." *Annual Review of Neuroscience* 18: 193–222. https://doi.org/10.1146/annurev.ne.18.030195.001205.

Di Lollo, Vincent. 2012. "The Feature-Binding Problem Is an Ill-Posed Problem." *Trends in Cognitive Sciences* 16 (6): 317–21. https://doi.org/10.1016/j.tics.2012.04.007.

Dosovitskiy, A., Beyer, L., Kolesnikov, A., Weissenborn, D., Zhai, X., Unterthiner, T., ... & Houlsby, N. (2021). An image is worth 16x16 words: Transformers for image recognition at scale. *International Conference on Learning Representations (ICLR)*. https://arxiv.org/abs/2010.11929

Eichenbaum, H. (2017). The role of the hippocampus in navigation is memory. *Journal of Neurophysiology, 117*(4), 1785–1796. https://doi.org/10.1152/jn.00005.2017

Engel, A. K., & Singer, W. (2001). Temporal binding and the neural correlates of sensory awareness. *Trends in Cognitive Sciences, 5*(1), 16–25. https://doi.org/10.1016/S1364-6613 (00)01568-0

Feldman, J. (2013). The neural binding problem(s). *Cognitive Neurodynamics, 7*(1), 1–11. https://doi.org/10.1007/s11571-01 2-9219-8

Feldman, H., & Friston, K. (2010). Attention, uncertainty, and free-energy. *Frontiers in Human Neuroscience, 4*, Article 215. https://doi.org/10.3389/fnhum.2010.00215

Friedman, W. J. (2004). Time in autobiographical memory. *Social Cognition, 22*(5), 591–605. https://doi.org/10.1521/soco .22.5.591.50767

Friston, K. J., Frith, C. D., Liddle, P. F., & Frackowiak, R. S. J. (1994). Functional connectivity: The principal-component analysis of large (PET) data sets. *Journal of Cerebral Blood Flow & Metabolism, 13*(1), 5–14.

Friston, Karl. 2010. "The Free-Energy Principle: A Unified Brain Theory?" *Nature Reviews Neuroscience* 11 (2): 127–38. https://doi.org/10.1038/nrn2787.

Friston, K. (2018). Does predictive coding have a future? *Nature Neuroscience, 21*(8), 1019–1021. https://doi.org/10.10 38/s41593-018-0200-7

Gerstner, W., Kistler, W. M., Naud, R., & Paninski, L. (2014). Neuronal Dynamics: From Single Neurons to Networks and Models of Cognition. Cambridge University Press.

Goff, Philip. 2020. *Galileo's Error: Foundations for a New Science of Consciousness*. New York: Pantheon Books.

Haldane, John. 2011. "Mind, Soul and Neurobiology." In *The Soul Hypothesis: Investigations into the Existence of the Soul*, edited by Mark C. Baker and Stewart Goetz, 47–64. New York: Continuum.

Hasker, William. 1999. *The Emergent Self.* Ithaca, NY: Cornell University Press.

Herrmann, Christoph S., Sergei Rach, Tobias Neuling, and Dennis Strüber. 2016. "Transcranial Alternating Current Stimulation: A Review of the Underlying Mechanisms and Effects." *Frontiers in Human Neuroscience* 7: 279. https://doi.org/10.3389/fnhum.2013.00279.

Herwig, A., & Waszak, F. (2012). Action-effect bindings and ideomotor learning in intention- and stimulus-based actions. *Frontiers in Psychology, 3*, Article 444. https://doi.org/10.3389/fpsyg.2012.00444

Hodgkin, A. L., & Huxley, A. F. (1952). A quantitative description of membrane current and its application to conduction and excitation in nerve. *The Journal of Physiology, 117*(4), 500–544. https://doi.org/10.1113/jphysiol.1952.sp0 04764

Hohwy, J. (2020). New directions in predictive processing. *Mind & Language, 35*(2), 209–223. https://doi.org/10.1111/mila.12281

Hommel, B. (2004). Event files: Feature binding in and across perception and action. *Trends in Cognitive Sciences, 8*(11), 494–500. https://doi.org/10.1016/j.tics.2004.08.007

Izhikevich, E. M. (2007). Dynamical Systems in Neuroscience: The Geometry of Excitability and Bursting. MIT Press.

Jensen, O., Kaiser, J., & Lachaux, J. P. (2007). Human gamma-frequency oscillations associated with attention and memory. *Trends in Neurosciences, 30*(7), 317–324. https://doi.org/10.1016/j.tins.2007.05.001

Jones, E. G. (2001). The thalamic matrix and thalamocortical synchrony. *Trends in Neurosciences, 24*(10), 595–601. https://doi.org/10.1016/S0166-2236(00)01922-6

Kandel, E. R., Schwartz, J. H., Jessell, T. M., Siegelbaum, S. A., & Hudspeth, A. J. (2013). Principles of Neural Science (5th ed.). McGraw-Hill Education.

Koch, I., & Kunde, W. (2002). Verbal response–effect compatibility. *Memory & Cognition, 30*(8), 1297–1303. https://doi.org/10.3758/BF03213411

LeCun, Y., Bengio, Y., & Hinton, G. (2015). Deep learning. *Nature, 521*(7553), 436–444. https://doi.org/10.1038/nature14539

LeDoux, Joseph E. 1996. *The Emotional Brain: The Mysterious Underpinnings of Emotional Life.* New York: Simon & Schuster.

Levine, J. (1983). Materialism and qualia: The explanatory gap. *Pacific Philosophical Quarterly, 64*(4), 354–361. https://doi.org/ 10.1111/j.1468-0114.1983.tb00207.x

Livingstone, Margaret S., and David H. Hubel. 1988. "Segregation of Form, Color, Movement, and Depth: Anatomy, Physiology, and Perception." *Science* 240 (4853): 740–49. https://doi.org/10.1126/science. 3283936.

Lowe, E. J. 2013. "Mereology and the Unity of Consciousness." *Philosophical Perspectives* 27 (1): 103–122. https://doi.org/10.11 11/phpe.12022.

Martinez-Conde, S., Macknik, S. L., & Hubel, D. H. (2008). The role of fixational eye movements in visual perception. *Nature Reviews Neuroscience, 9*(3), 222–233.

Martin, A. (2007). The representation of object concepts in the brain. *Annual Review of Psychology, 58*(1), 25–45. https://doi.org/10.1146/annurev.psych.57.102904.190143

McComas, A. J. (2011). Galvani's spark: The story of the nerve impulse. Oxford University Press.

McCulloch, W. S., & Pitts, W. H. (1943). A logical calculus of the ideas immanent in nervous activity. *The Bulletin of Mathematical Biophysics, 5*(4), 115–133. https://doi.org/10.1007/BF02478259

Michalareas, G., Vezoli, J., van Pelt, S., Schoffelen, J. M., Kennedy, H., & Fries, P. (2016). Alpha-beta and gamma rhythms subserve feedback and feedforward influences among human visual cortical areas. *Neuron, 89*(2), 384–397. https://doi.org/10.1016/j.neuron.2015.12.018

Müller, L., Chavane, F., Reynolds, J., & Sejnowski, T. J. (2018). *Cortical travelling waves: Mechanisms and computational principles.* Nature Reviews Neuroscience, 19(5), 255–268. https://doi.org/10.1038/nrn.2018.20

Nagel, Thomas. 1974. "What Is It Like to Be a Bat?" *The Philosophical Review* 83 (4): 435–50. https://doi.org/10.2307/ 2183914.

Northoff, Georg, Finn Wainio-Theberge, and Kathinka Evers. 2020. "Is Temporo-Spatial Dynamics the 'Common Currency' of Brain and Mind?" *Physics of Life Reviews* 33: 34–54. https://doi.org/10.1016/j.plrev.2019.06.017.

Rickabaugh, Brandon L. 2021. "The Unity of Consciousness and Panpsychism: Exploring Hasker's Critique." *Philosophia Christi* 23 (1): 81–92. https://doi.org/10.5840/pc20212315.

Roelfsema, P. R. (2023). Solving the binding problem: Assemblies form when neurons enhance their firing rate—they don't need to oscillate or synchronize. *Neuron, 111*(7), 1003–1019.

Rolls, E. T., & Treves, A. (1998). Neural Networks and Brain Function. Oxford University Press.

Sadeh, T., Shohamy, D., Levy, D. R., Reggev, N., & Maril, A. (2011). Cooperation between the hippocampus and the striatum during episodic encoding. *Journal of Cognitive Neuroscience, 23*(7), 1597–1608. https://doi.org/10.1162/jocn .2010.21549

Schwartz, E. L. (Ed.). (1993). Computational Neuroscience. MIT Press.

Shadlen, Michael N., and J. Anthony Movshon. 1999. "Synchrony Unbound: A Critical Evaluation of the Temporal Binding Hypothesis." *Neuron* 24 (1): 67–77. https: //doi.org /10.1016/S0896-6273(00)80822-3.

Siegel, M., Warden, M. R., & Miller, E. K. (2009). Phase-dependent neuronal coding of objects in short-term memory. *Proceedings of the National Academy of Sciences, 106*(50), 21341–21346. https://doi.org/10.1073/pnas.0908193106

Singer, Wolf, and Charles M. Gray. 1995. "Visual Feature Integration and the Temporal Correlation Hypothesis." *Annual Review of Neuroscience* 18: 555–86. https://doi.org/10. 1146/annurev.ne.18.030195.003011.

Singer, W. (2007). Binding by synchronization. *Scholarpedia, 2*(12), 1657.

Sterzer, P., Adams, R. A., Fletcher, P., Frith, C., Lawrie, S. M., Muckli, L., & Corlett, P. R. (2018). The predictive coding account of psychosis. *Biological Psychiatry, 84*(9), 634–643. https://doi.org/10.1016/j.biopsych.2018.05.015

Stokes, M. G., Kusunoki, M., Sigala, N., Nili, H., Gaffan, D., & Duncan, J. (2013). Dynamic coding for cognitive control in prefrontal cortex. *Neuron, 78*(2), 364–375. https://doi.org/10.1016/j.neuron. 2013.01.039

Swinnen, S. P., & Wenderoth, N. (2004). Two hands, one brain: Cognitive neuroscience of bimanual skill. *Trends in Cognitive Sciences, 8*(1), 18–25. https://doi.org/10.1016/j.tics.2003.10.017

Toni, I., Rowe, J., Stephan, K. E., & Passingham, R. E. (2001). Changes of cortico-striatal effective connectivity during visuomotor learning. *Cerebral Cortex, 12*(10), 1040–1047. https://doi.org/10.1093/cercor/12.10.1040

Tononi, Giulio. 2004. "An Information Integration Theory of Consciousness." *BMC Neuroscience* 5 (1): 42. https://doi.org/10.1186/1471-2202-5-42.

Trappenberg, T. P. (2010). Fundamentals of Computational Neuroscience (2nd ed.). Oxford University Press.

Treisman, Anne. 1980. "A Feature-Integration Theory of Attention." *Cognitive Psychology* 12 (1): 97–136. https://doi.org/10.1016/0010-0285(80)90005-5.

Tulving, E. (2002). Episodic memory: From mind to brain. *Annual Review of Psychology, 53*(1), 1–25. https://doi.org/10.1146/annurev.psych.53.100901.135114

van Diepen, R. M., Cohen, M. X., Denys, D., & Mazaheri, A. (2015). Attention and temporal expectations modulate power, not phase, of ongoing alpha oscillations. *Journal of Cognitive Neuroscience, 27*(8), 1573–1586. https://doi.org/10.11 62/jocn_a_00803

van Diepen, R. M., Foxe, J. J., & Mazaheri, A. (2019). The functional role of alpha-band activity in attentional processing: The current zeitgeist and future outlook. *Current Opinion in Psychology, 29*, 229–238. https://doi.org/10.1016/j.copsyc.2019.03.015

Varela, Francisco J., Jean-Philippe Lachaux, Eugenio Rodriguez, and Jacques Martinerie. 2001. "The Brainweb: Phase Synchronization and Large-Scale Integration." *Nature Reviews Neuroscience* 2 (4): 229–39. https://doi.org/10.1038/ 35067550.

von der Malsburg, C. (1981). The correlation theory of brain function. Internal Report 81-2, Max-Planck-Institute for Biophysical Chemistry, Göttingen, Germany.

Wang, X.-J. (2020). Macroscopic gradients support functional specialization in the neocortex. *Nature Reviews Neuroscience, 21*(3), 169–178. https://doi.org/10.1038/s41583-019-0250-8

12

Brain Metafunction Syndromes

Recent developments have revolutionized our understanding of complex neurological states such as coma, persistent vegetative states, minimally conscious states, and dementia syndromes. These syndromes represent conditions marked by a global disruption of large-scale brain networks, necessitating a paradigm shift in their conceptualization. Progress in connectomics underscores the brain's dynamic, unified function across interconnected systems, sub-regions, and neurons, reshaping our understanding of these complex neurological states (Bullmore and Sporns 2009). Connectomics offers insights into these intricate structural and functional connectivity patterns essential for brain activity (Bullmore and Sporns 2009; Sporns 2011). For instance, coma is an acute state, often resulting from severe brain injury or trauma, metabolic disorder, or hypoxemic conditions characterized by a profound loss of consciousness and lack of purposeful responses to external stimuli (Laureys, Boly, and Maquet 2006).

Persistent vegetative states and minimally conscious states involve impaired consciousness, with patients showing limited responsiveness or intermittent awareness (Laureys 2005). In these cases, the disruption of large-scale brain networks severely compromises the integrated functioning of localized regions, undermining synchronized neural dynamics vital for coherent cognitive processing and higher-level integration of bodily functions. The concept of the 'vegetative' state (Jennett and Plum 1972) gained prominence in medical literature alongside a 'neocorticalist' theory of consciousness. However, critics argue the term vegetative is neurologically and philosophically problematic. The present author raised this concern more than two decades ago, suggesting a more precise framework that accurately reflects the brain's global functioning, advocating for the concept of 'metafunction syndrome' to avoid pejorative implications and acknowledge the complexity of consciousness and prognosis in these states (García 1999). Advances in neuroscientific research confirm the inherent diagnostic uncertainty and support an integrative approach, aiding in these syndromes' diagnostic, therapeutic, and rehabilitative aspects (Monti et al. 2010).

Dementia syndromes, including Alzheimer's disease, frontotemporal dementia, vascular and other dementias, represent chronic neurodegenerative disorders that progressively impair cognitive functions such as memory, executive function, and language. These syndromes arise from complex interactions within and between large-scale brain networks, disrupting information processing and regional communication (Braak and Braak 1991; Seeley et al. 2007; Buckner et al. 2008). Beyond cognition, dementia impacts broader physiological systems, including the immune, autonomic, homeostatic, and allostatic functions, contributing to systemic dysregulation and behavioral symptoms (Heneka et al. 2014; McEwen and Stellar 1993; Vercruysse et al. 2018; Iadecola 2017; Ishii and Iadecola 2015). As dementia syndromes progress, their impact on physiological systems beyond cognition, such as immune, autonomic, homeostatic, and allostatic functions, becomes more clinically evident. Neuroinflammation, a hallmark of advanced dementia, results from sustained immune activation, contributing to neural degeneration and exacerbating brain damage (Heneka et al. 2014). The autonomic nervous system, responsible for regulating involuntary functions like heart rate and digestion, becomes dysregulated in dementia, compounding cardiovascular, respiratory, and digestive issues (Iadecola 2017). Hypothalamic dysfunction often appears early before cognitive symptoms emerge and persist throughout the disease (Ishii and Iadecola 2015). Homeostasis and allostasis, mechanisms essential for stability through dynamic adaptation, are increasingly disrupted, especially in advanced stages of dementia, affecting temperature, metabolism, and sleep-wake cycles, further contributing to behavioral symptoms (McEwen and Stellar 1993).

The brain metafunction framework integrates cognitive, emotional, and physiological regulation functions, providing a comprehensive model for understanding these disorders. This framework emphasizes the interplay between structural and functional connectivity, highlighting the critical role of synchronized dynamics across different spatial and temporal scales. It moves beyond traditional classifications to propose integrative diagnostic and therapeutic approaches that target large-scale dementia disruptions. For instance, advances in neuroimaging techniques, such as functional magnetic resonance imaging (fMRI) and diffusion tensor imaging (DTI), have improved diagnostic precision and informed predictive models for disease progression. Moreover, this framework underscores the practical implications of connectomics, including developing personalized therapeutic strategies and potential future advancements in molecular neuroscience and artificial intelligence. These approaches promise to reshape clinical practices, enabling nuanced diagnoses, predictive analytics, and targeted treatments for conditions like coma and dementia syndromes. As neuroscience develops (Bressler and Menon 2010; Bassett and Bullmore 2006), the brain metafunction paradigm offers a lens for reconceptualizing these syndromes, respecting clinical utility while reflecting a nuanced and integrative understanding of severe brain dysfunction.

Brain Metafunction Syndromes: A Unified Classification and Diagnostic Framework

Definition and Scope

Brain Metafunction Syndromes (BMS) are conditions characterized by partial or total disruption of the brain's unified integrative function—what this framework terms the brain metafunction. The hallmark of these syndromes is significant dysfunction or irreversible loss of one or more of the brain's three major metafunction subsystems, resulting in impaired or extinguished global activity across consciousness, cognition, physiological regulation, and behavioral responsiveness. This framework offers a structured classification of conditions that involve global disruption or loss of the brain's metafunction. These include acute, chronic, progressive, and potentially reversible states, as well as unique conditions that challenge prevailing clinical and metaphysical paradigms. As defined in previous chapters, the brain metafunction subsystems are the following:

Subsystem I: Structures from the ascending reticular activation (sub) system: nonspecific populations of neurons located in the tegmentum of the rostral pons and midbrain, the intralaminal and reticularis nuclei of the thalamus, and the posterior hypothalamus.

Subsystem II: Limbic system structures and brainstem nuclei: the hypothalamus, the basal forebrain, the nucleus basalis of Meynert, the amygdala, the hippocampal complex, cingulate cortex, orbitofrontal cortex, the cingulum, olfactory cortex, temporal pole, portions of the basal ganglia, reticularis nuclei in the midbrain and rostral pons, and parabrachial nucleus. In addition, the pedunculopontine nucleus, the rostral raphe nuclei, the periaqueductal gray, and the nucleus locus ceruleus.

Subsystem III: The cerebral cortex with the thalamus and basal ganglia.

Brain metafunction syndromes compromise the brain's metafunction subfunctions, typically severely, except for the early stage of progressive diseases like dementia.

The Brain Metafunction and its Subfunctions

The brain metafunction is the unified and dynamic activity of brain systems, sub-regions, regions, and neurons—organized through structural and functional network-level connectivity—that integrates local-global synchronized dynamics across multiple spatial and temporal scales. This activity is not merely the product of physical interactions but is grounded in an inherent, sui generis form of causality that belongs to the ontological structure of the system itself, connecting the neurobiological with a transphysical domain. It is through this intrinsically unified mode of operation that the following set of subfunctions* is both progressively developed and dynamically expressed:

1. Consciousness, cognitive processes in general, including perception, attention, memory, decision-making and emotions;
2. The neural integration and regulation of the communication systems of the organism (nervous, endocrine, immune, and circulatory systems);
3. Integration and regulation of the internal environment at its higher hierarchical level: homeostasis and allostasis;
4. Integration of afferent and efferent flow (sensory, motor, and regulatory) from the whole organism and the environment;
5. Behavioral and adaptive interactions with the outer world;
6. Unity between bodily physiological processes and mental states;
7. Overall integration of the above primary subfunctions into a unified, fundamentally irreducible whole.

**The specified brain metafunction subfunctions are system-wide, richly interconnected, dynamically integrated functions.*

Brain Metafunction Syndromes Classification

The concept of Brain Metafunction Syndromes refers to conditions in which the brain's integrative, global functionality—across its cognitive, regulatory, and conscious dimensions—is severely impaired. These syndromes reflect varying degrees of dysfunction across the brain's three primary metafunction subsystems, encompassing both acute and chronic conditions, static and progressive states, and clinically observable and conceptually significant patterns.

I. Acute Brain Metafunction Syndromes

Grade 1:
Coma — Acute, non-transient metafunction impairment with a Glasgow Coma Scale (GCS) ≤ 8.

Grade 2:
Transient Syndromes — Limited-duration metafunction impairment (e.g., syncope, seizures). If prolonged and GCS ≤ 8, reclassified as Grade 1.

Grade 3:
Impaired Consciousness — GCS > 8, with variable presentation:
While consciousness is the most clinically salient marker of dysfunction at this grade, additional metafunctional subcomponents—such as cognitive integration, regulatory coherence, and behavioral responsiveness—may also be compromised to varying degrees. Clinical presentations range in severity:

(a) Mild impairment (GCS 13–15)
(b) Moderate impairment (GCS 9–12)

II. Chronic Brain Metafunction Syndromes

This category includes long-term impairments of brain metafunction resulting either from non-progressive injuries or from progressive neurodegenerative conditions. Chronic syndromes may arise after an acute phase (e.g., trauma, anoxic injury, infection), from congenital anomalies, or as part of a stabilizing chronic disorder. While some remain static over time, others involve gradual functional decline. These conditions typically lead to enduring disruptions of consciousness, cognition, sensorimotor integration, and other metafunction subfunctions. The following items distinguish between non-progressive and progressive forms of chronic dysfunction.

1. Chronic Non-Progressive (Following an Acute Phase or Stabilized Chronic Condition)

Type A: Unresponsive Wakefulness Syndrome (UWS) formerly known as "vegetative state"

No behavioral or neurophysiological evidence of consciousness; sleep-wake cycles may be preserved.

Type B: Minimally Conscious State (MCS)
Intermittent but limited behavioral signs of awareness; purposeful behaviors may occur inconsistently.

Type C: Cognitive Motor Dissociation (CMD)
Preserved awareness without reliable motor output; detectable via advanced neuroimaging or electrophysiological assessment (Schiff, 2015; Edlow et al., 2017).

2. Chronic Progressive Syndromes (e.g., Dementia)

Neurodegenerative or congenital conditions characterized by gradual and progressive impairment of brain metafunction subfunctions, including cognition, awareness, regulatory balance, and behavioral adaptability.

Stage 0:
Subclinical or pre-symptomatic phase; minimal or no detectable metafunctional impairment.

Stage 1:
Mild impairment of cognitive and regulatory metafunctions; consciousness preserved with subtle deficits in attention, memory, or adaptive behavior.

Stage 2:
Moderate impairment across multiple metafunctional domains; behavioral evidence of sustained consciousness remains, though integrative capacity is significantly reduced.

Stage 3:
Severe impairment with limited responsiveness; intermittent or minimal signs of awareness may be present (e.g., minimally conscious features).

Stage 4:
Profound impairment with no consistent behavioral or ancillary (e.g., neuroimaging or EEG) evidence of consciousness; widespread breakdown of higher-order integrative metafunction, with preservation of limited autonomic or homeostatic function.

3. Chronic Static or Congenital Syndromes.

Non-progressive but enduring impairments of brain metafunction resulting from early developmental injury or congenital anomalies (e.g., HIE, lissencephaly, early-acquired static encephalopathy). These conditions involve stable yet irreversible disruptions to brain integrative function and are frequently associated with significantly limited or undeveloped manifestations of consciousness and cognitive-motor capacities.

III. Pseudo-Brain Metafunction Syndrome

Preserved brain metafunction with intact consciousness, cognition, and integrative capacities, masked by near-total motor paralysis. Behavioral expression is severely limited—often restricted to vertical eye movements or blinking—despite full subjective awareness. Exemplified by *Locked-In Syndrome (LIS)*, this condition mimics metafunctional disorders but is distinguished by a dissociation between motor output and preserved inner experience, with intact metafunctional integration.

IV. Transient Brain–Metafunction Disconnection

A well-documented and conceptually exceptional condition marked by a temporary and profound disruption of observable brain metafunction, typically arising in contexts such as near-death experiences (NDEs), cardiac arrest, extreme hypoxia, or deep anesthesia. During such episodes, neurophysiological indicators of brain integration—such as EEG activity and evoked potentials—may be absent or severely attenuated (Chawla et al., 2009; Parnia et al., 2014). Despite this, some individuals retrospectively report vivid, structured, and in some cases veridically corroborated conscious experiences, including perceptions of external events during periods of apparent unconsciousness (van Lommel et al., 2001; Greyson, 2003). This phenomenon is not clinically detectable in real time and challenges materialist assumptions that equate consciousness solely with measurable neural activity. Within the Brain Metafunction Framework, these cases support the postulate of a transphysical dimension of consciousness, in which subjective awareness may transiently persist or emerge beyond conventional neurobiological substrates.

V. Irreversible Loss of Brain Metafunction (Death).

Irreversible cessation of the brain's integrative activity results in the loss of the unified human organism and the cessation of personal subjectivity. Personhood is no longer expressed or sustained, and the body ceases to function as a coherent whole. This condition marks the definitive end of the human being within the Brain Metafunction Framework. In this context, the term "irreversible" is preferred for its conceptual rigor and ontological precision. It signifies that the cessation of the brain's integrative function is not merely unaltered due to current limitations (as permanence might suggest in a clinical sense), but that it reflects a non-contingent, metaphysically final loss of the capacity to sustain human life and personhood. While "permanent loss" may be used in clinical settings—especially when brain death criteria are met and no interventions to restore function will be attempted—irreversible better expresses that the integrative activity essential to human wholeness is no longer intrinsically restorable, regardless of medical intent or capacity (Bernat, Culver, and Gert 1981; President's Council on Bioethics 2008; Condic 2016).

For further discussion, see the chapters on The Unity of the Human Person and Human Death.

VI. Prenatal and Developmental Potential for Brain Metafunction

In embryonic and fetal stages, brain metafunction exists as a latent potential—rooted in the whole-organism dynamic and progressively unfolding through intrinsically directed development. Though pre-functional, yet structurally complete in its essential organization, the organism is already a living, unified whole, ordered toward personal and conscious life. This category affirms developmental continuity and the ethical and ontological significance of early human life.

Note: *While not technically a syndrome, this category is included to reflect the Brain Metafunction Framework's emphasis on developmental continuity.*

Brain Metafunction Syndromes : Classification Overview

I. ACUTE BRAIN METAFUNCTION SYNDROMES

├── Grade 1: Coma (GCS ≤ 8)
├── Grade 2: Transient Syndromes (e.g., syncope, seizures)

```
└── Grade 3: Impaired Consciousness
    ├── Mild Impairment (GCS 13–15)
    └── Moderate Impairment (GCS 9–12)
```

II. CHRONIC BRAIN METAFUNCTION SYNDROMES

```
A. Chronic Non-Progressive (Following Acute Phase)
├── Type A: Unresponsive Wakefulness Syndrome (UWS)
├── Type B: Minimally Conscious State (MCS)
└── Type C: Cognitive Motor Dissociation (CMD)

B. Chronic Progressive Syndromes (e.g., Dementia)
 ├── Stage 0: Preclinical
 ├── Stage 1: Mild Metafunction Impairment
 ├── Stage 2: Moderate Impairment
 ├── Stage 3: Severe Impairment
 └── Stage 4: Profound Impairment (possible misdiagnosis
             as UWS)

C. Chronic Static or Congenital
 ├── Early developmental injuries (e.g., HIE)
 ├── Congenital anomalies (e.g., lissencephaly, microcephaly)
 └── Genetic/metabolic syndromes (e.g., Tay-Sachs, PKU)
```

III. PSEUDO–BRAIN METAFUNCTION SYNDROME

```
└── Locked-In Syndrome (LIS)
    - Preserved cognition and awareness
    - Absent motor output (except eyes/blinking)
```

IV. TRANSIENT BRAIN–METAFUNCTION DISCONNECTION

```
└── Examples: Near-death experiences (NDEs), deep anesthesia
    - Absent physiological markers
    - Retrospective reports of consciousness
```

V. IRREVERSIBLE LOSS OF BRAIN METAFUNCTION

```
└── Brain death
    - Cessation of global integrative brain activity
    - No potential for recovery
```

VI. PRENATAL AND DEVELOPMENTAL POTENTIAL

└── Latent metafunction
- Present in embryonic and fetal stages
- Oriented toward integrative unity and consciousness

Glasgow Coma Scale (GCS)

The Glasgow Coma Scale (GCS), developed by Bryan Jennett and Graham Teasdale in 1974, remains the most widely adopted clinical tool for assessing levels of consciousness in patients with acute brain injury (Teasdale & Jennett, 1974). Designed for rapid bedside use, the GCS evaluates three core parameters: eye opening, verbal response, and motor response, with total scores ranging from 3 to 15. A score of 8 or below is typically interpreted as indicative of coma, often prompting critical interventions such as airway protection through intubation. Unlike diagnostic-specific tools, the GCS provides a general neurological assessment, making it especially valuable in emergency and intensive care contexts where time-sensitive decisions must be made. Its simplicity, reproducibility, and wide applicability have made it a cornerstone in neurological triage, guiding both immediate care and ongoing monitoring (Sternbach, 2000; Teasdale et al., 2014). While the GCS remains indispensable for initial evaluation and triage, it is not without limitations. It may fail to detect covert consciousness or capture more nuanced impairments in brain metafunction, particularly in cases of cognitive motor dissociation or subtle neurological complexity. Nonetheless, the GCS continues to play a critical role in the clinical assessment of patients with altered consciousness—complementing more advanced neuroimaging and electrophysiological tools and offering a valuable foundation for early diagnosis and decision-making in the context of acute brain metafunction syndromes.

Acute Brain Metafunction Syndromes

Grade 1 :
Coma — Acute, non-transient metafunction impairment with a Glasgow Coma Scale (GCS) ≤ 8.

Coma represents a *Grade 1 Acute Brain Metafunction Syndrome*—an extended and profound state of unconsciousness caused by acute, severe disruption of one or more brain metafunction subsystems. In this condition, the individual shows no awareness of self or surroundings and cannot respond purposefully to external stimuli, including sound, light, or touch (Posner et al., 2007). Although some reflexive or postural responses may be observed, they do not indicate conscious awareness. The causes of coma are varied and include traumatic brain injury, ischemic or hemorrhagic stroke, anoxia, metabolic disturbances, central nervous system infections, and toxic exposures. The duration and outcome of coma depend on multiple factors—particularly the cause, severity and location of damage, and timeliness of medical intervention (Bernat, 2006). From the perspective of the Brain Metafunction Framework, coma reflects a global failure in the brain's integrative architecture. This includes disruption in:

Subsystem I (ascending reticular activating system),

Subsystem II (limbic and brainstem nuclei involved in affective, autonomic, and homeostatic regulation),

Subsystem III (the cerebral cortex and its thalamo-cortico-basal ganglia circuits).

These disruptions result in a collapse of the brain's metafunctional capacity to coordinate synchronized activity across spatial and temporal networks. Coma, therefore, is not merely a behavioral absence of responsiveness, but an acute failure of the brain metafunction to generate and sustain integrated conscious processes. The Glasgow Coma Scale (GCS) remains an essential tool for assessing consciousness in acute brain injury. With scores ranging from 3 to 15, a score of 8 or below typically indicates a comatose state and often necessitates airway protection (Teasdale & Jennett, 1974; Teasdale et al., 2014). Despite limitations in detecting covert or subtle brain activity, the GCS offers a rapid, reliable framework for clinical decision-making, especially in emergency and critical care settings.

Pathophysiologically, coma most often results from either bilateral dysfunction of the cerebral hemispheres (Plum & Posner, 1982) or damage to the ascending reticular activating system (ARAS) (Moruzzi & Magoun, 1949; Parvizi & Damasio, 2001). This damage may be structural, such as from trauma or stroke or nonstructural, such as in metabolic or toxic encephalopathy. It may also be focal or diffuse. In rare cases, coma results from bilateral thalamic infarctions, underscoring the thalamus's role as a central hub of metafunctional integration. Such injuries can produce a profound collapse of metafunctional integrity—akin to a strategic lesion at the core of the brain's integrative axis. Recent advances in neuroimaging and neurophysiological assessment have improved our ability to detect residual brain activity in coma. Functional magnetic resonance imaging (fMRI), positron emission tomography (PET), and quantitative electroencephalography (EEG) provide deeper insight into potential for recovery. These tools have revealed that even in clinically unresponsive states, residual connectivity within networks such as the default mode network may persist (Stender et al., 2014). Similarly, event-related potentials (ERPs) may indicate covert awareness and help predict recovery.

Unlike chronic disorders of consciousness, coma is inherently acute and time-limited, typically resolving within days or weeks. Outcomes range from full or partial recovery to progression into minimally conscious or unresponsive wakefulness states—or, in severe cases, to brain death. This transient nature makes early and accurate assessment essential for guiding treatment decisions, prognostic evaluation, and ethical deliberation. Biomarkers—both clinical (e.g., GCS) and neurophysiological (e.g., cortical reactivity, metabolic activity)—play an increasingly vital role in this process. Treatment focuses on stabilizing vital functions, preventing secondary brain injury, and addressing the underlying cause. Management strategies include maintaining cerebral perfusion and oxygenation, controlling intracranial pressure, and managing systemic complications (Carney et al. 2017; Rajajee et al. 2023). Pharmacological agents such as amantadine have shown efficacy in promoting arousal during the subacute phase (Giacino et al., 2012). Neuromodulation therapies—such as transcranial magnetic stimulation (TMS) and vagus nerve stimulation (VNS)—are under investigation for their potential to reengage cortical and subcortical networks (Thibaut et al. 2014; Gosseries et al. 2014).

Coma holds a central place in the classification of Brain Metafunction Syndromes. As the prototypical Grade 1 acute syndrome, it illustrates the vulnerability—and potential reversibility—of integrated brain function under extreme physiological stress. The concept of coma as an "acute and global metafunctional collapse" offers a more precise clinical and conceptual framing than traditional definitions that view it only as "unresponsiveness." Within this framework, coma serves as both a diagnostic and prognostic fulcrum—a critical point at which recovery is still possible, and intervention is most consequential. Understanding coma through the lens of brain metafunction offers a deeper grasp of its complexity and a more humane basis for care—emphasizing the multilayered nature of consciousness and the dynamic, networked architecture of human brain integration.

II. Chronic Brain Metafunction Syndrome

1. Chronic Non-Progressive (Following an Acute Phase or Stabilized Chronic Condition)

Type A: Unresponsive Wakefulness Syndrome (UWS)
(Formerly "Vegetative State")

Unresponsive Wakefulness Syndrome (UWS), previously known as vegetative state, is a chronic brain metafunction syndrome that emerges following severe acute brain injury or the stabilization of an earlier acute phase. The term was originally proposed in 1972 by Bryan Jennett and Fred Plum in their foundational article Persistent vegetative state after brain damage: A syndrome in search of a name (Jennett & Plum, 1972). They emphasized that "vegetative" was intended as a descriptive term, not implying any definitive anatomical or pathological mechanism. According to their usage, based on the Oxford Dictionary, a "vegetative" organism grows and develops but lacks sensation and thought. Over time, research has refined the syndrome's diagnostic criteria and clarified its neuropathological underpinnings (Multi-Society Task Force on PVS, 1994; Giacino & Zasler, 1995; Royal College of Physicians, 2003). Neuropathological studies have identified four primary lesion patterns associated with UWS (Jennett & Plum, 1972; Kinney & Samuels, 1994; Levin, 1992):

There are four differentiated patterns identified as causes of vegetative states (Jennett and Plum 1972; Kinney and Samuels 1994; Levin 1992):

1. Extensive neocortical damage.
2. Disconnection of the cortex from subcortical structures due to white matter injury.
3. Bilateral paramedian and posterior thalamic lesions, often involving the midbrain tegmentum.
4. A combination of focal cortical and subcortical lesions with cumulative impact.

Clinically, UWS is diagnosed based on a constellation of signs, including:

1. Absence of sustained, purposeful, or reproducible behavioral responses to stimuli.
2. No evidence of comprehension or verbal expression.
3. Inability to interact consciously with the environment

4. Preservation of sleep-wake cycles.
5. Presence of cranial nerve reflexes and autonomic function
6. Bladder and rectal incontinence.

These criteria are primarily behavioral, requiring repeated assessments by multidisciplinary teams to avoid misdiagnosis (Schnakers et al., 2009). Despite the appearance of wakefulness—eye-opening, spontaneous limb movement, and facial expressions—patients lack demonstrable awareness or intentional behavior. The condition often arises after prolonged coma and may persist for weeks, months, or indefinitely (Laureys et al., 2004, 2011). Misdiagnosis is well-documented, not due to clinical negligence but due to the syndrome's inherent complexity (Andrews, 1996; Childs et al., 1993; Tresch et al., 1991). Accurate differentiation from other disorders, such as minimally conscious state or cognitive motor dissociation, is critical for prognosis, treatment, and ethical decision-making.

The term "vegetative state" has drawn longstanding criticism for its dehumanizing connotations and conceptual limitations (Jennett, 2002). While widely used, it risks shaping public and clinical perceptions in ways that obscure the nuanced realities of these patients' conditions. In response, terms such as Unresponsive Wakefulness Syndrome (UWS) have been proposed and adopted (Laureys et al., 2010), reflecting a move toward respectful and precise clinical language. The Brain Metafunction Framework supports this terminological shift by emphasizing that such syndromes are not mere deficits in behavioral output but manifestations of severe disruption to the brain's integrative systems. Specifically, UWS represents a case in which consciousness and cognitive function are absent due to widespread or strategically located brain damage—such as bilateral thalamic injury—impacting the subsystems responsible for unified brain metafunction.

Redefining UWS as Chronic Brain Metafunction Syndrome Type 1A offers a conceptually accurate and ethically sensitive alternative that avoids pejorative language while grounding diagnosis in the structure-function relationships central to modern neuroscience. It better captures the syndrome's diverse anatomical expressions and the deep uncertainties regarding residual consciousness and prognosis. Clinically, this reframing promotes a holistic approach to care that emphasizes function-specific assessment and tailored rehabilitation, where appropriate. As advances in neuroimaging and network neuroscience continue, adopting a brain metafunctional lens may also guide research toward better understanding of these encephalopathies—both in terms of functional loss and potential pathways for recovery. In doing so, it encourages the development of therapeutic strategies aimed not only at medical stabilization but at supporting any preserved elements of integrative function and personhood.

The Coma Recovery Scale-Revised (CRS-R)

The Coma Recovery Scale–Revised (CRS-R) is a validated and standardized assessment tool used to evaluate patients with disorders of consciousness (DoC). Its adoption has significantly improved diagnostic accuracy, particularly by reducing the historically high misdiagnosis rates of persistent vegetative state (PVS), which were once reported to be as high as 40% (Schnakers et al., 2009). The CRS-R assesses six functional domains: auditory, visual, motor, verbal, communication, and arousal. This structured approach enables clinicians to detect subtle or inconsistent signs of consciousness—such as visual tracking or command-following—that are often missed in informal bedside evaluations (Giacino et al., 2004). Studies have shown that the CRS-R is more sensitive and reliable than traditional observational methods, especially in distinguishing PVS from minimally conscious state (MCS) (Seel et al., 2010). For example, Schnakers et al. (2009) found that 41% of patients initially diagnosed with PVS were reclassified as MCS based on CRS-R evaluations. In addition to improving diagnostic precision, the CRS-R promotes standardization across clinicians, enhancing inter-rater reliability and reducing subjectivity (Wannez et al., 2017). As a result, the CRS-R is now considered the *gold standard* for behavioral assessment of DoC, serving as a critical tool for both diagnosis and treatment planning (Seel et al., 2010).

Updated Terminology and Improved Diagnostic Criteria

The shift in terminology from *persistent vegetative state (PVS) to unresponsive wakefulness syndrome (UWS)* reflects both a deeper scientific understanding of disorders of consciousness (DoC) and the need to reduce the stigma associated with the term *vegetative* (Laureys et al., 2010). First proposed by the European Task Force on Disorders of Consciousness in 2010, this change emphasizes clinical presentation—wakefulness without awareness—rather than relying on outdated and potentially dehumanizing language. UWS is now widely adopted in clinical and research contexts as a more neutral and precise descriptor, facilitating better communication with families and caregivers while recognizing the diagnostic and prognostic complexity of these states. In parallel with this terminological shift, recent clinical guidelines have significantly improved diagnostic accuracy. The 2018 *American Academy of Neurology (AAN) Practice Guidelines for DoC* introduced evidence-based recommendations for diagnosing UWS and minimally conscious state (MCS) (Giacino et al., 2018). These guidelines emphasize the use of standardized behavioral tools—especially the *Coma Recovery Scale–Revised (CRS-R)*—alongside emerging neurotechnologies. Notably, they recommend integrating advanced imaging modalities such as *functional MRI (fMRI)* and *positron emission tomography* (PET) to detect *covert awareness* in patients who may lack observable behavioral responses. This standardized approach has substantially reduced misdiagnosis rates and has improved prognostic accuracy (Giacino et al., 2018; Kondziella et al., 2020).

Advanced Neuroimaging Techniques

Advanced neuroimaging technologies have transformed the assessment of patients with DoC by revealing patterns of brain activity that suggest preserved awareness—even in the absence of overt behavior. While traditional evaluations depend on visible signs of responsiveness, techniques such as *fMRI* and *PET* provide objective measures of brain function, allowing for the detection of *non-behavioral indicators of consciousness* (Owen et al., 2006; Stender et al., 2014). For example, fMRI tracks real-time changes in brain activity through blood oxygenation patterns, while PET detects metabolic activity in specific neural networks. These tools have shown that some patients diagnosed as being in a vegetative state—or UWS—retain the capacity to process stimuli or follow commands mentally.

In a landmark study, Monti et al. (2010) asked patients to perform mental imagery tasks (e.g., imagining playing tennis or walking through their home). Remarkably, 17% of patients initially diagnosed with PVS exhibited brain activity patterns in regions associated with motor planning and spatial navigation, indicating preserved conscious processing despite a complete lack of behavioral output. Similarly, PET imaging has identified residual metabolic activity in cortical areas linked to awareness, aiding in the differentiation between patients with and without residual consciousness (Stender et al., 2014). By complementing behavioral assessments, these neuroimaging techniques enhance diagnostic precision and support more informed clinical decisions regarding prognosis, communication potential, and care planning for patients with DoC.

Type B: Minimally Conscious State (MCS)

The emergence of the minimally conscious state (MCS) as a transitional condition beyond vegetative state syndrome (VSS) underscores the limitations—both clinical and conceptual—of the term vegetative (Giacino et al., 2002). Within the Brain Metafunction Framework, a significant reduction in the capacity to express specific functions does not imply the loss of personhood. From this standpoint, the label vegetative is both pejorative and misleading, reinforcing the need for more precise and respectful terminology. Minimally conscious state is classified here as Chronic Brain Metafunction Syndrome, Group 1, Type B—a condition typically following coma or VSS. MCS is defined by inconsistent but reproducible behavioral signs that reflect partial preservation of conscious awareness. These behaviors—such as visual tracking, purposeful movement, or response to commands—are not observed in coma or UWS and thus serve as key clinical differentiators (Giacino et al., 2002).

MCS commonly arises from acute brain injuries such as traumatic brain injury or stroke, marking a progression from coma or unresponsive wakefulness. However, it can also result from congenital or degenerative neurological disorders and may present as either a transient or long-term state. Clear diagnostic criteria are essential for advancing research into its neuropathology, prognosis, and treatment (Bruno et al., 2011). Patients in MCS may demonstrate intermittent signs of self-awareness and environmental responsiveness—for example, eye contact, visual tracking, reaching toward objects, or verbal responses, even if inconsistent or contextually inappropriate. Due to the variability of these behaviors, accurate diagnosis requires repeated and careful observation over time (Laureys et al., 2004, 2011). Management of MCS involves a comprehensive clinical approach, including nutritional support, hydration, mobility protocols, and prevention of complications such as pressure ulcers (Katz et al., 2009). Structural neuroimaging (e.g., MRI, CT) may help identify reversible causes, while ancillary tests—such as PET, SPECT, and EEG—are often used to assess cerebral function, identify covert seizures, or explore residual network connectivity.

Type C: Cognitive Motor Dissociation (CMD)

Cognitive Motor Dissociation (CMD) refers to a condition in which patients retain conscious awareness and cognitive function but are unable to produce purposeful or observable motor responses due to severe motor impairment or disconnection between cognitive centers and motor output pathways (Schiff, 2015; Edlow et al., 2017). The term was formally introduced by Schiff (2015) to describe cases where *covert awareness* exists without corresponding motor behavior. CMD is characterized by behavioral unresponsiveness during standard bedside evaluations, making it clinically indistinguishable from Unresponsive Wakefulness Syndrome (UWS) based on observation alone. However, advanced neuroimaging and electrophysiological techniques—such as functional MRI (fMRI), electroencephalography (EEG), and positron emission tomography (PET)—can reveal preserved awareness. These tools detect brain activity in response to external commands or mental imagery tasks, revealing patterns that reflect conscious processing despite the absence of overt motor output (Forgacs et al., 2017; Monti et al., 2010; Edlow et al., 2017).

For example, CMD patients may show activation in motor planning or spatial navigation regions when asked to imagine playing tennis or navigating their home, even though they remain behaviorally unresponsive. These findings demonstrate a functional dissociation between preserved cognitive systems and disrupted motor execution—an internal-external disconnect that masks the presence of consciousness (Schiff, 2015). Unlike individuals in a minimally conscious state (MCS), who may exhibit inconsistent but overt signs of awareness (e.g., visual tracking, purposeful movement), CMD patients display no outward behavior and can only be identified through advanced assessment tools. Neural responses such as event-related potentials (ERPs) on EEG or metabolic activity in cortical regions detected by PET provide essential diagnostic evidence (Edlow et al., 2017). The identification of CMD has reshaped the understanding of disorders of consciousness (DoC) by highlighting the limitations of behavior-based assessment alone. Early detection is critical to avoid misdiagnosis and to ensure that patients receive appropriate care, communication strategies, and rehabilitation tailored to their preserved cognitive capacities (Schiff, 2014; Edlow et al., 2017). CMD underscores the importance of integrating behavioral assessment with neuroimaging and electrophysiology to fully recognize and address the hidden dimensions of conscious life in severely brain-injured patients.

Persisting Challenges in Diagnosing and Treating Disorders of Consciousness

Despite significant advances in the assessment and treatment of disorders of consciousness (DoCs), several challenges persist. Accurate diagnosis remains difficult, with frequent misclassification between conditions such as minimally conscious state (MCS) and unresponsive wakefulness syndrome (UWS). Misdiagnosis rates have been reported as high as 40% (Schnakers et al., 2009), largely due to the subtle and intermittent nature of awareness in MCS or cognitive motor dissociation (CMD), which often goes undetected during standard bedside evaluations. Advanced diagnostic tools—such as functional MRI (fMRI), positron emission tomography (PET), and electroencephalography (EEG)—can improve detection of covert awareness but are not universally available in clinical settings. In addition, the lack of reliable biomarkers for predicting recovery complicates prognosis, leaving families and clinicians with considerable uncertainty.

Therapeutically, neuromodulation approaches such as transcranial magnetic stimulation (TMS) and vagus nerve stimulation (VNS), along with pharmacological interventions like amantadine, have shown promise but yield inconsistent outcomes across patient populations. The ethical complexity of care decisions—particularly regarding the continuation or withdrawal of life-sustaining treatment—adds further weight to these clinical challenges. Together, these ongoing limitations underscore the urgent need for standardized diagnostic protocols, expanded access to advanced neurotechnological tools, and sustained research into both therapeutic innovation and ethical guidance. Improving outcomes for patients with DoCs will require continued collaboration across neuroscience, medicine, ethics, and policy.

2. Chronic Progressive Syndromes (e.g., Dementia)

Chronic Progressive Brain Metafunction Syndromes are characterized by gradual and sustained impairments in the brain's capacity for metafunctional integration. Neurodegenerative and congenital conditions—such as Alzheimer's disease (AD), Parkinson's disease, Huntington's disease, and cerebrovascular disease—progressively disrupt key subfunctions of brain metafunction, including cognition, awareness, behavioral adaptability, and regulatory balance (Winblad et al., 2016). Dementia is conventionally described as a cluster of cognitive disorders that impair memory, thinking, and social function to the point of interfering with daily life. However, within the Brain Metafunction Framework, dementia is reconceived as a syndrome reflecting systemic disruption of integrative brain functions. Beyond memory loss and cognitive decline, dementia often includes noncognitive manifestations such as psychosis, mood disturbances, personality changes, agitation, altered sexual behavior, sleep disruption, and appetite changes (Katz et al., 2001). These symptoms highlight the syndrome's widespread impact across functional domains.

Alzheimer's disease, the most prevalent form of dementia, is increasingly recognized as a multifactorial neurodegenerative condition that begins years before overt cognitive symptoms appear. While cognitive impairment is central, AD also involves metabolic abnormalities, neuroendocrine dysfunction, and hypothalamic disturbances that may contribute to its onset and progression (McEwen & Gianaros, 2011). AD can also affect autonomic nervous system functions such as heart rate, digestion, and thermoregulation (Lohman et al., 2024). Emerging evidence links AD to a brain metabolic syndrome, often termed "type III diabetes," involving oxidative stress, chronic microinflammation, insulin resistance, hypertension, and dyslipidemia. These systemic imbalances promote amyloid-beta accumulation and tau pathology—hallmarks of AD (de la Monte & Wands, 2008). The interplay of metabolic and inflammatory pathways suggests that targeting these upstream factors is essential for mitigating neurodegeneration.

Given this complexity, single-target drug therapies have largely failed to halt or reverse AD progression. In contrast, lifestyle interventions—including a balanced, fiber-rich diet, regular physical activity, quality sleep, toxin avoidance, targeted supplementation, and stress management—have shown promise in addressing systemic contributors to neurodegeneration (Smith & Blumenthal, 2016). These interventions may reduce oxidative stress and inflammation while improving metabolic function, potentially delaying or preventing disease progression.

Clinically and pathologically, AD affects a wide range of brain regions, each contributing to distinct symptoms:

- *Hippocampus* – early memory loss and spatial disorientation (Braak & Braak, 1991).
- *Prefrontal cortex* – impaired executive function and planning (Arnsten & Goldman-Rakic, 1998).
- *Temporal lobe* – deficits in language and visual memory (Seeley et al., 2009).
- *Parietal lobe* – loss of spatial awareness and attentional control (Peng et al., 2016).
- *Occipital lobe* – visual processing impairments (Baloyannis, 2009).
- *Amygdala* – changes in emotional reactivity and mood (Murray et al., 2005).
- *Basal ganglia* – motor abnormalities and disrupted reward processing (Hikosaka et al., 2008).

- *Hypothalamus* – metabolic and circadian dysregulation (Ishii & Iadecola, 2015).
- *Brainstem* – disturbances in arousal, sleep-wake cycles, and autonomic regulation (Braak & Del Tredici, 2011).

These widespread neuropathological changes lead to breakdowns in brain metafunction, disrupting not only cognition but also the body's ability to regulate its internal environment. For example, dementia impairs homeostatic and allostatic functions such as body temperature regulation, increasing the risk of hypothermia or hyperthermia (Vandal et al., 2016). It also compromises the integration of sensory and motor information, affecting both afferent and efferent flows and leading to balance issues and increased fall risk (Gujord Tangen et al., 2014). Additionally, patients often experience progressive loss of behavioral adaptability, communication, and decision-making—further exacerbating isolation, depression, and diminished quality of life (Katz et al., 2001).

A classification system that acknowledges both cognitive decline and the broader collapse of brain metafunction offers a more accurate and integrative understanding of dementia. By reframing these disorders in terms of systemic neural disintegration, the Brain Metafunction Framework provides a conceptual foundation for interdisciplinary education, more holistic caregiving, and advanced research. Medical students and clinicians alike benefit from an approach that emphasizes not only cognitive symptoms but the full spectrum of functional losses. Researchers, in turn, are supported in developing multifactorial strategies that address the neurological, metabolic, and behavioral dimensions of these devastating conditions.

3. Chronic Static or Congenital Syndromes

Chronic static and congenital nervous system disorders are marked by severe or extensive brain damage resulting in lasting disruption of higher-order brain functions, including cognition, consciousness, sensory integration, and motor coordination. These conditions are non-progressive from their onset and reflect enduring deficits that prevent the brain from achieving its full integrative metafunctional capacity. For example, severe forms of cerebral palsy (CP) and post-hypoxic-ischemic brain injury—such as those resulting from perinatal asphyxia or cardiac arrest—often lead to persistent motor impairment and cognitive dysfunction (Rosenbaum et al., 2007; Nolan et al., 2009). Similarly, static encephalopathy following stroke or traumatic brain injury (TBI) may result in profound impairments in purposeful interaction and adaptive behavior due to irreversible damage to structural and functional neural networks (Carney et al., 2017; Forgacs et al., 2017).

Congenital nervous system disorders result from developmental anomalies or structural malformations of the brain present before birth. Conditions such as anencephaly, holoprosencephaly, and hydranencephaly involve profound structural disruptions that preclude the emergence of consciousness, higher cognition, or voluntary behavior (Volpe, 2008; Leventer et al., 2008). Other disorders—including microcephaly and lissencephaly—significantly alter cortical development, leading to severe intellectual disability, motor dysfunction, and minimal interaction with the environment (Guerrini & Dobyns, 2014; Dobyns et al., 1993). These conditions represent intrinsic limitations on brain development, often preventing the actualization of core integrative subfunctions.

Additional causes of chronic static brain dysfunction include metabolic, genetic, and lesion-acquired syndromes. Metabolic disorders such as hypoxic-ischemic encephalopathy (HIE), mitochondrial diseases (e.g., Leigh syndrome), and phenylketonuria (PKU) can result in cellular energy failure or toxic accumulation, leading to permanent neurological deficits. Similarly, untreated kernicterus or maple syrup urine disease (MSUD) may cause irreversible brain damage with profound cognitive and motor consequences. Genetic conditions such as Tay-Sachs disease, Rett syndrome, and Angelman syndrome involve early-onset neurodevelopmental abnormalities or neurodegeneration that impair brain integration. Lesion-acquired syndromes—including encephalitis, neonatal hypoglycemia, fetal alcohol spectrum disorders (FASD), and post-radiation injury—also produce fixed impairments affecting cognition, coordination, and sensory processing. Despite diverse etiologies, these disorders share a unifying feature: widespread disruption of the subsystems responsible for brain metafunction.

At a philosophical and bioethical level, the Brain Metafunction Framework draws a critical distinction between the actualization of function and the intrinsic unity of the person, affirming that profound neurological impairment does not nullify the human subject. These syndromes underscore the ethical importance of recognizing the individual as more than the sum of present abilities. Even in states of severe functional limitation, the individual remains a unified subject whose existence demands moral and clinical respect. This perspective reinforces the framework's central commitment: that integrative activity is the ordinary means by which personhood is expressed, but not its sole foundation. The distinction between functional impairment and ontological identity is essential for informed clinical decision-making, ethical reflection, and the continued affirmation of human dignity in care.

III. Pseudo–Brain Metafunction Syndrome

This category refers to conditions in which brain metafunction is preserved—with intact consciousness, cognition, and integrative capacity—but behavioral expression is profoundly restricted due to near-total motor paralysis. Subjective awareness remains fully present, yet external responsiveness is severely limited, often to vertical eye movements or blinking alone. This creates a misleading clinical picture that mimics disorders of consciousness. *Locked-In Syndrome (LIS)* exemplifies this condition. It involves dissociation between preserved inner experience and nearly absent motor output, resulting from bilateral damage to the ventral pons. This region contains descending motor pathways, and its injury disconnects the motor cortex from peripheral musculature while sparing the *reticular activating system,* which maintains wakefulness and awareness. The early clinical descriptions by Jean-Baptiste Charcot and Georges Guillain helped identify brainstem-related motor paralysis with preserved consciousness, though they did not formally define LIS. The condition was first clearly delineated by Fred Plum and Jerome B. Posner in *The Diagnosis of Stupor and Coma* (1966), where they characterized LIS as a distinct syndrome marked by intact cognitive function and preserved eye movements—specifically vertical gaze and blinking—mediated by spared midbrain structures. They emphasized that the presence of awareness and the capacity for nonverbal communication distinguish LIS from true disorders of consciousness.

Cognitive Motor Dissociation (CMD) vs. Locked-in Syndrome (LIS)

LIS is classified as a *Pseudo–Brain Metafunction Syndrome* because integrative brain function is preserved, but extrinsic motor output is nearly abolished. In contrast, *Cognitive Motor Dissociation (CMD)* is a *true Brain Metafunction Syndrome,* involving a breakdown in motor responsiveness despite retained consciousness. Both conditions involve profound motor impairment but differ in underlying mechanisms and diagnostic profiles. CMD patients appear behaviorally unresponsive yet retain *covert* awareness, detectable only through advanced neuroimaging and electrophysiological techniques such as fMRI, EEG, or PET. These patients may exhibit brain activity in response to commands—such as imagining playing tennis or navigating their home—despite showing no overt behavior (Owen et al., 2006; Monti et al., 2010). CMD arises from functional disconnection between cortical areas responsible for intention and the motor pathways required to express it (Schiff, 2015). In contrast, LIS involves *overt awareness*. Patients are fully conscious and cognitively intact but unable to speak or move due to pontine damage. Crucially, they often retain the ability to communicate using eye movements or blinks (Laureys et al., 2005a; Kondziella et al., 2020). While diagnosis of classic LIS is relatively straightforward due to preserved eye-based communication, *total LIS*, in which even eye movements are absent, can be mistaken for UWS or PVS, particularly in the absence of advanced testing.

The diagnostic challenges for CMD and LIS are distinct. CMD is frequently overlooked in standard bedside assessments, as patients appear unresponsive without specialized tools. Neuroimaging studies have revealed that some patients originally diagnosed with UWS or PVS in fact exhibit responsive brain activity when prompted with mental imagery tasks (Monti et al., 2010). Conversely, LIS can often be identified through simple, targeted bedside interactions—provided clinicians are aware of the condition's presentation. However, in rare cases of *total locked-in syndrome*, misdiagnosis remains a significant risk (Kondziella et al., 2020). *Prognosis also differs* between the two. In CMD, outcomes vary widely and depend on early detection and the nature of the underlying injury. Identifying covert awareness through tools such as *fMRI* or *PET* is crucial to avoiding inappropriate withdrawal of care (Stender et al., 2014). While motor recovery in LIS is rare, many patients maintain stable cognition and can achieve meaningful interaction through assistive communication technologies (Laureys et al., 2005b). Ultimately, the key conceptual distinction is this: *CMD involves hidden awareness*, requiring specialized tools for detection, while *LIS features overt awareness* that is often clinically visible. Both conditions challenge the assumption that outward behavior reliably indicates consciousness. Their differentiation underscores the necessity of integrating advanced neurotechnological assessment with careful clinical observation to ensure accurate diagnosis and appropriate care.

IV. Transient Brain–Metafunction Disconnection.

A well-documented and conceptually exceptional condition, *Transient Brain–Metafunction Disconnection* refers to a temporary yet profound interruption of observable brain metafunction, typically occurring during events such as near-death experiences (NDEs), cardiac arrest, extreme hypoxia, or deep anesthesia. During these episodes, neurophysiological indicators of integrative brain activity—such as EEG signals or evoked potentials—may be absent or severely attenuated. Yet, some individuals retrospectively report vivid, coherent, and in certain cases, veridically corroborated conscious experiences, including accurate perceptions of external events during periods of apparent unconsciousness. This phenomenon eludes real-time clinical detection and poses a significant challenge to materialist assumptions that equate consciousness solely with measurable brain activity. Within the *Brain Metafunction Framework,* such cases lend support to the postulate of a *transphysical dimension of consciousness,* in which subjective awareness may transiently persist—or even emerge—independently of conventional neurobiological substrates.

Near-death experiences provide empirical grounds for considering the *continuity of consciousness beyond brain-based integration.* These events often occur during episodes of acute, global loss of brain metafunction—most notably in cardiac arrest—where both cortical and brainstem activity cease, and no measurable neurological function is present. Nevertheless, individuals frequently describe vivid and structured experiences, including disembodiment, encounters with deceased persons, and immersion in transcendent, light-filled environments (Parnia et al., 2014). The phenomenon has been analogized to a temporary "full-body amputation," where the entirety of the bodily and neural system is offline, yet conscious awareness appears to persist. Notably, some accounts include *veridical observations*—accurate descriptions of locations, events, or conversations that occurred during the period of clinical unresponsiveness—later confirmed by independent sources. Particularly striking are cases involving individuals congenitally blind who report visual perceptions or spatial awareness during NDEs, experiences that defy explanation by residual or latent sensory memory (van Lommel et al., 2001).

From a strictly materialist standpoint, such phenomena challenge the notion that consciousness is fully reducible to brain function. While physiological explanations—such as hypoxia, anesthetic hallucination, or terminal brain surges—have been proposed, they do not fully account for the *structured, meaningful, and often externally corroborated* nature of these experiences. The growing body of research suggests that NDEs may not be reducible to neural artifacts, but could represent genuine conscious processes occurring independently of integrated brain activity. Within the Brain Metafunction Framework, these phenomena are interpreted as *evidence of a transphysical aspect of consciousness*—an aspect ordinarily mediated by the brain but not ontologically dependent upon it. For deeper philosophical and theological exploration, including responses to reductive physicalist critiques, see the chapters on *The Human Soul and the Brain Metafunction* and *Christian Theism and the Transphysical.*

V. Irreversible Loss of Brain Metafunction (Death).

The irreversible loss of brain metafunction constitutes a defining criterion for death, as it marks the irreversible cessation of the brain's integrative role in sustaining the unity of both the human organism and the personal subject. Within the Brain Metafunction Framework, this integrative capacity is not reducible to isolated neural or physiological processes. Rather, it reflects the coordinated operation of distributed systems that support consciousness, self-awareness, sensorimotor regulation, and homeostasis—functions that, together, manifest the coherence and unity characteristic of the living human being. When this dynamic system irreversibly ceases to function, what is lost is not merely a collection of cognitive abilities, but the unified reality of the human being as both biological organism and personal subject. Without the organizing activity of brain metafunction, the body no longer functions as an integrated whole, and the person—the embodied being who thinks, acts, and relates—is no longer present. What remains is a biological aggregate: an arrangement of parts no longer animated by the coherence and directionality that define human life.

This approach avoids reducing personhood to a checklist of observable functions, acknowledging instead that personhood is expressed through—but not exhausted by—functional capacities. While such capacities are ordinarily mediated through the brain, the human being's identity is grounded in a deeper unity—one that endures through stages of development, injury, and decline, and whose dissolution marks the end of personal life. The death of the person, then, is not defined merely by the loss of particular abilities, but by the collapse of the integrative structure that sustains the embodied subject as a unified whole. This understanding reinforces a core commitment of the Brain Metafunction Framework: that personhood and biological life are inseparably linked in the integrated human subject. Accordingly, death is not simply a matter of neurological metrics, but the definitive rupture of the dynamic unity that constitutes human life. For further discussion, see the chapters on *The Unity of the Human Person* and *Human Death.*

Terminological Note on Irreversible vs. Permanent Loss

While the terms irreversible and permanent are often used interchangeably in clinical and legal discourse surrounding brain death, there is a meaningful conceptual distinction between them. Irreversible loss of integrative brain metafunction—sensu stricto—is the more precise term within the Brain Metafunction Framework, as it denotes a definitive and ontological cessation of the brain's unifying capacity. It reflects the complete and final breakdown of the dynamic systems that sustain both organismic integration and personal subjectivity. In contrast, permanent refers to a condition that is expected to persist indefinitely, based on current medical knowledge and prognostic certainty, but which may not exclude hypothetical or future reversibility. From a clinical standpoint, however, the use of permanent loss is fully acceptable when applied to cases that fulfill all recognized neurological criteria for brain death. In such contexts, the distinction is largely semantic, and both terms affirm the cessation of the brain's role as the functional substrate of personal life.

VI. Prenatal and Developmental Potential for Brain Metafunction

In embryonic and fetal stages, the brain metafunction exists not in an operative form, but as a latent potential—progressively unfolding through biologically, epigenetically, and intrinsically directed development. This developmental trajectory is rooted in the whole-organism dynamic, wherein the whole precedes its differentiated parts. Early neural structures are not autonomous agents of function but are intrinsically ordered toward the future emergence of integrated brain metafunction. By including this stage within the Brain Metafunction Framework, we recognize that even prior to the realization of consciousness, the human organism manifests a unifying principle of organization. This anticipatory structure does not depend on current function to confer identity or value; rather, it reflects an orientation toward future integration that is continuous with later stages of neurological and personal development. The organism, though pre-functional yet structurally complete in its essential organization, is already a living, coordinated whole—actively unfolding its capacities in accordance with its nature. This inclusion affirms a principle of developmental continuity, wherein the human being is understood not in terms of discrete functional thresholds, but as an integrated entity whose trajectory is intrinsically ordered toward personal and conscious life. In this view, the preclinical stages of human development hold not only biological significance but ethical and philosophical relevance as well—supporting a framework that accounts for the unity and dignity of the human being across the entire arc of human life. While not technically a syndrome, this category is included to reflect the Brain Metafunction Framework's emphasis on developmental continuity.

Advancing Research through the Brain Metafunction Framework

The Brain Metafunction Framework provides a structured foundation for advancing research on neurological syndromes by exploring how disruptions across interconnected brain systems produce widespread functional impairments. It encourages scientists and clinicians to move beyond static, region-based models toward a systems-level understanding of brain dysfunction. This dynamic approach enables investigation of neural integration across multiple levels—from cellular signaling to regional connectivity and large-scale network dynamics. By emphasizing the brain's role as an integrative metafunctional system, the framework clarifies how focal impairments can propagate across distributed networks, resulting in systemic disruptions in cognition, sensory integration, motor coordination, and autonomic regulation. Functional deficits are thereby reinterpreted not merely as isolated symptoms, but as diagnostic indicators of specific subsystems that have become decoupled from the whole.

This paradigm also supports the development of personalized interventions tailored to a patient's unique neurofunctional profile. By integrating advanced imaging, electrophysiological monitoring, and even genetic or metabolic data, therapeutic strategies can be more precisely adapted to individual needs. Targeted treatments may enhance cognitive and behavioral performance or mitigate systemic dysfunctions—such as autonomic instability—common in chronic neurological conditions. As neuroimaging, computational neuroscience, and molecular diagnostics continue to develope, the Brain Metafunction Framework anticipates a more precise, integrative era of neurorehabilitation—anchored in a holistic understanding of brain function and human care.

Reflecting on the Ethical Implications of Brain Metafunction Syndromes

The Brain Metafunction Framework also invites a reexamination of ethical assumptions surrounding the diagnosis and care of individuals with disorders of consciousness. Conditions such as unresponsive wakefulness syndrome (UWS) or minimally conscious state (MCS) challenge prevailing intuitions about autonomy, personhood, and moral status—particularly when cognitive responsiveness is minimal or absent. In this context, the framework encourages clinicians, ethicists, and caregivers to adopt terminology and perspectives that reflect both the complexity and dignity of the human person. Replacing outdated or pejorative terms—such as "vegetative state"—with more functionally grounded language, like Chronic Brain Metafunction Syndrome, enables a more compassionate and scientifically accurate approach. It shifts the focus from behavioral output alone to the systemic collapse of integrative brain functions, avoiding premature or reductive judgments about the value or meaning of a patient's life.

The framework is also aligned with the principles of personalist bioethics, which affirm the inherent dignity and unity of the person regardless of cognitive or communicative capacity. It challenges models that define personhood solely through functional markers, offering instead a holistic vision in which even severely impaired individuals retain their moral and existential worth. In end-of-life contexts, this perspective fosters thoughtful, balanced decision-making that honors autonomy while avoiding therapeutic overreach. It promotes care strategies that attend to the total well-being of the individual—physical, psychological, spiritual, and relational. By integrating foundational ethical principles—autonomy, beneficence, non-maleficence, and justice—the Brain Metafunction Framework offers a rigorous yet humane pathway for ethical reflection and clinical practice. It invites practitioners to regard those with profound neurological impairment not as objects of care, but as persons whose value endures even when functional capacities are diminished.

Educational Impact: Preparing Future Practitioners

Incorporating the Brain Metafunction Framework into medical and neuroscience education presents a transformative opportunity to prepare future clinicians and researchers for the complexities of neurological care. It invites a shift away from overly compartmentalized models toward a dynamic, systems-integrative perspective that more accurately reflects the nature of brain function. Students trained in this model learn to see neurological disorders not as isolated deficits, but as syndromes emerging from disruptions across interdependent neural networks. They come to understand how impairments in one subsystem—such as the thalamus or brainstem—can affect attention, emotional regulation, physiological stability, and even consciousness itself. This integrative understanding deepens diagnostic reasoning and clinical judgment by grounding it in the whole-person context. The framework is especially valuable for navigating ethically complex scenarios, including disorders of consciousness, progressive dementia, and end-of-life decision-making. It equips future practitioners with the tools to offer care that attends not only to neurological and physiological dimensions, but also to the moral, emotional, and existential realities of illness. In doing so, it fosters a new generation of clinicians and scholars who are capable of providing care that is both scientifically grounded and ethically attuned.

Conclusion

The paradigm of brain metafunction—originally proposed by this author over two decades ago in the context of human death and severe neurological states—has gained renewed relevance through recent advances in neuroscience, particularly in the emerging field of connectomics. This framework offers a transformative lens for understanding and reclassifying a wide spectrum of neurological conditions, including coma, unresponsive wakefulness syndrome, minimally conscious states, cognitive motor dissociation, and dementia. These syndromes may now be reconceived as Acute or Chronic Brain Metafunction Syndromes, marked by large-scale disruptions in network connectivity and inter-system coordination—ultimately leading to the breakdown of coherent cognitive processing and unified brain function.

At the heart of this framework lies the recognition that brain activity is not confined to isolated regions, but is dynamically distributed across vast, interdependent networks. These systems integrate sensory, motor, homeostatic, emotional, and cognitive functions into a functional and personal unity. This integrative perspective calls for updated classificatory models—ones that move beyond cognitive deficits to account for broader dysfunctions in brain metafunction subfunctions. For example, in Alzheimer's disease and related dementias, patients experience not only memory loss but also metabolic imbalance, autonomic dysregulation, and behavioral disturbances—manifestations of complex interactions among structures such as the hippocampus, prefrontal cortex, thalamus, and brainstem. Viewing these conditions through the lens of brain metafunction supports more individualized, multidimensional approaches to care. It promotes therapeutic strategies that address both cognitive and non-cognitive impairments—such as mood disorders, sleep disturbances, and physiological instability—that profoundly affect patient well-being.

A central argument of this chapter is the need to move beyond outdated and reductionist terminology, especially the term vegetative state, which is both clinically ambiguous and ethically problematic. This label obscures the possibility of residual metafunctional integration and fails to convey the true complexity of the patient's condition. Replacing it with more accurate and respectful terminology—such as Chronic Brain Metafunction Syndrome—yields a conceptually coherent and ethically sound alternative. Such reframing fosters a more nuanced understanding that can guide both research and clinical practice toward approaches that are integrated, dignified, and diagnostically robust.

As neuroscience continues to illuminate the intricate architecture of brain connectivity, this paradigm shift holds promise for the development of interventions aimed at restoring network-level integration. The multifactorial nature of brain disorders demands a comprehensive, interdisciplinary approach—one that unites medical treatment with lifestyle interventions, preventive strategies, and psychosocial support. The Brain Metafunction Framework advances this integrative vision by emphasizing the unifying role of brain function in sustaining the human person. This model represents a significant step forward in the conceptualization, diagnosis, and treatment of complex neurological syndromes. It provides a robust foundation for future scientific progress and clinical innovation—one that honors not only the biological and cognitive dimensions of human life, but also its moral and existential depth. As future generations of clinicians, researchers, and students adopt and refine this framework, they will be better equipped to meet the profound challenges of severe brain dysfunction with approaches that are both scientifically rigorous and deeply humane. In doing so, they will reaffirm a fundamental truth: that even in the face of neurological breakdown, the dignity of the human person remains a guiding principle in both science and care.

REFERENCES

Andrews, K. (1996). Misdiagnosis of the vegetative state: Retrospective study in a rehabilitation unit. *BMJ, 313*(7048), 13–16. https://doi.org/10.1136/bmj.313.7048.13

Arnsten, A. F. T., & Goldman-Rakic, P. S. (1998). Noise stress impairs prefrontal cortical cognitive function in monkeys. *Archives of General Psychiatry, 55*(4), 362–368. https://doi.org/10.1001/archpsyc.55.4.362

Baloyannis, S. J. (2009). Dendritic pathology in Alzheimer's disease. Journal of the Neurological Sciences, 283(1–2), 153 –157. https://doi.org/10.1016/j.jns.2009.02.370

Bassett, D. S., & Bullmore, E. T. (2006). Small-world brain networks. *The Neuroscientist, 12*(6), 512–523. https://doi.org/10.1177/10738584062 93182

Bernat, J. L. (2006). Chronic disorders of consciousness. *The Lancet, 367*(9517), 1181–1192. https://doi.org/10.1016/S0140-6736(06)68508-5

Bernat, J. L., David J. Culver, and Bernard Gert. 1981. "On the Definition and Criterion of Death." *Annals of Internal Medicine* 94 (3): 389–94. https://doi.org/10.7326/0003-4819 -94-3-389.

Braak, H., & Braak, E. (1991). Neuropathological staging of Alzheimer-related changes. *Acta Neuropathologica, 82*(4), 239–259. https://doi.org/10.1007/BF00308809

Braak, H., & Del Tredici, K. (2011). Alzheimer's pathogenesis: Is there neuron-to-neuron propagation? *Acta Neuropathologica, 121*(5), 589–595. https://doi.org/10.1007/s0 0401-011-0825-z

Bressler, S. L., & Menon, V. (2010). Large-scale brain networks in cognition: Emerging methods and principles. *Trends in Cognitive Sciences, 14*(6), 277–290. https://doi.org/10. 1016/j.tics.2010.04.004

Bruno, M. A., Vanhaudenhuyse, A., Thibaut, A., Moonen, G., & Laureys, S. (2011). From unresponsive wakefulness to minimally conscious PLUS and functional locked-in syndromes: Recent advances in our understanding of disorders of consciousness. *Journal of Neurology, 258*(7), 1373–1384. https://doi.org/10.1007/s00415-011-6114-x

Buckner, R. L., Andrews-Hanna, J. R., & Schacter, D. L. (2008). The brain's default network: Anatomy, function, and relevance to disease. *Annals of the New York Academy of Sciences, 1124*(1), 1–38. https://doi.org/10.1196/annals.1440.011

Bullmore, E., & Sporns, O. (2009). Complex brain networks: Graph theoretical analysis of structural and functional systems. *Nature Reviews Neuroscience, 10*(3), 186–198. https://doi.org/10.1038/nrn2575

Carney, N., Totten, A. M., O'Reilly, C., Ullman, J. S., Hawryluk, G. W. J., Bell, M. J., ... & Ghajar, J. (2017). Guidelines for the management of severe traumatic brain injury, fourth edition. *Neurosurgery, 80*(1), 6–15. https://doi.org/10.1227/NEU.0000000000001432

Charcot, J. B. (1872). *Leçons sur les maladies du système nerveux faites à la Salpêtrière.* Paris: A. Delahaye.

Childs, N. L., Mercer, W., & Childs, H. W. (1993). Accuracy of diagnosis of persistent vegetative state. *Neurology, 43*(8), 1465–1467. https://doi.org/10.1212/WNL.43.8.1465

Claassen, J., Doyle, K., Matory, A., Couch, C., Burger, K. M., Velazquez, A., ... & Schiff, N. D. (2019). Detection of brain activation in unresponsive patients with acute brain injury. *New England Journal of Medicine,* 380(26), 2497–2505.

Condic, M. L. (2016). Determination of death: A scientific perspective on biological integration. *The Journal of Medicine and Philosophy: A Forum for Bioethics and Philosophy of Medicine, 41*(3), 257–278. https://doi.org/10.1093/jmp/jhw004

de la Monte, S. M., & Wands, J. R. (2008). Alzheimer's disease is type 3 diabetes–Evidence reviewed. *Journal of Diabetes Science and Technology, 2*(6), 1101–1113. https://doi.org/10.11 77/193229680800200619

Dobyns, W. B., Reiner, O., Carrozzo, R., & Ledbetter, D. H. (1993). Lissencephaly: A human brain malformation associated with deletion of the LIS1 gene located at chromosome 17p13. *JAMA, 270*(23), 2838–2842. https://doi.org/10.1001 /jama.270.23.2838

Edlow, B. L., Chatelle, C., Spencer, C. A., Chu, C. J., Bodien, Y. G., O'Connor, K. L., ... & Wu, O. (2017). Early detection of consciousness in patients with acute severe traumatic brain injury. *Brain, 140*(9), 2399–2414. https://doi.org/10.1093/brain/awx176

Forgacs, P. B., Frey, H. P., Velazquez, A., Thompson, S., Brodie, D., Moitra, V., Rabani, L., Park, S., Agarwal, S., Falo, M. C., Schiff, N. D., & Claassen, J. (2017). Dynamic regimes of neocortical activity linked to corticothalamic integrity correlate with outcomes in acute anoxic brain injury after cardiac arrest. *Annals of Clinical and Translational Neurology, 4*(2), 119–129. https://doi.org/10.1002/acn3.385

Fox, M. D., Snyder, A. Z., Vincent, J., & Raichle, M. E. (2005). The human brain is intrinsically organized into dynamic, anticorrelated functional networks. *Proceedings of the National Academy of Sciences, 102*(27), 9673–9678. https://doi.org/10.1073/pnas.0504136102

García, O. D. (1999). Vegetative states: A syndrome in search of a new name [Estados vegetativos: Un síndrome en busca de un nuevo nombre]. *Revista de Neurología, 29*(12), 1202–1212.

Giacino, J. T., & Zasler, N. D. (1995). Outcome after severe traumatic brain injury: Coma, the vegetative state, and the minimally responsive state. The Journal of Head Trauma Rehabilitation, 10(1), 40–56. https://doi.org/10.1097/0000 1199-199502000-00006

Giacino, J. T., Kalmar, K., & Whyte, J. (2004). The JFK Coma Recovery Scale-Revised: Measurement characteristics and diagnostic utility. *Archives of Physical Medicine and Rehabilitation, 85*(12), 2020–2029. https://doi.org/10.1016/j.apmr.2004.02.033

Giacino, J. T., Ashwal, S., Childs, N., Cranford, R., Jennett, B., Katz, D. I., Kelly, J. P., Rosenberg, J. H., Whyte, J., Zafonte, R. D., & Zasler, N. D. (2002). The minimally conscious state: Definition and diagnostic criteria. *Neurology, 58* (3), 349–353. https://doi.org/10.1212/WNL.58.3.349

Giacino, J. T., Whyte, J., Bagiella, E., Kalmar, K., Childs, N., Khademi, A., Eifert, B., & Sherer, M. (2012). Placebo-controlled trial of amantadine for severe traumatic brain injury. *New England Journal of Medicine, 366*(9), 819–826. https: //doi.org/10.1056/NEJMoa1102609

Giacino, J. T., Katz, D. I., Schiff, N. D., Whyte, J., Ashman, E. J., Ashwal, S., ... & Nakase-Richardson, R. (2018). Practice guideline update recommendations summary: Disorders of consciousness. *Neurology, 91*(10), 450–460.

Gosseries, O., Thibaut, A., Boly, M., Rosanova, M., Massimini, M., & Laureys, S. (2014). Assessing consciousness in coma and related states using transcranial magnetic stimulation combined with electroencephalography. *Annales Françaises d'Anesthésie et de Réanimation, 33*(2), 65–71. https://doi.org/10.1016/j.annfar.2013.11.002

Greyson, B. (2003). *Incidence and correlates of near-death experiences in a cardiac care unit. General Hospital Psychiatry*, 25(4), 269–276.

Greicius, M. D., Krasnow, B., Reiss, A. L., & Menon, V. (2004). Functional connectivity in the resting brain: A network analysis of the default mode hypothesis. *Proceedings of the National Academy of Sciences, 101*(13), 4637–4642. https://doi.org/10.1073/pnas.0308627101

Guerrini, R., & Dobyns, W. B. (2014). Malformations of cortical development: Clinical features and genetic causes. *The Lancet Neurology, 13*(7), 710–726. https://doi.org/10.1016/S 1474-4422(14)70040-7

Guillain, G. (1920). Les manifestations nerveuses dans les hémorragies du tronc cérébral. *Revue Neurologique, 27,* 1–7.

Gujord Tangen, G., Engedal, K., Bergland, A., Moger, T. A., & Mengshoel, A. M. (2014). Relationships between balance and cognition in patients with subjective cognitive impairment, mild cognitive impairment, and Alzheimer disease. *Physical Therapy, 94*(8), 1123–1134. https://doi.org/10.2522/ptj.20130298

Heneka, M. T., Kummer, M. P., & Latz, E. (2014). Innate immune activation in neurodegenerative disease. *Nature Reviews Immunology, 14*(7), 463–477. https://doi.org/10.1038/nri3705

Hikosaka, O., Bromberg-Martin, E., Hong, S., & Matsumoto, M. (2008). New insights on the subcortical representation of reward. *Current Opinion in Neurobiology, 18*(2), 203–208. https://doi.org/10.1016/j.conb.2008.07.002

Iadecola, C. (2017). The neurovascular unit coming of age: A journey through neurovascular coupling in health and disease. *Neuron, 96*(1), 17–42. https://doi.org/10.1016/j.neuron.2017.07.030

Ishii, M., & Iadecola, C. (2015). Metabolic and noncognitive manifestations of Alzheimer's disease: The hypothalamus as both culprit and target of pathology. *Cell Metabolism, 22*(5), 761–776. https://doi.org/10.1016/j.cmet.2015.08.016

Jennett, Bryan (2002). The Vegetative State: Medical Facts, Ethical and Legal Dilemmas. Cambridge University Press.

Jennett, B., & Plum, F. (1972). Persistent vegetative state after brain damage: A syndrome in search of a name. *The Lancet, 299*(7753), 734–737. https:// doi: 10.1016/s0140-673 6(72)90242-5

Katz, D. I., Polyak, M., Coughlan, D., Nichols, M., & Roche, A. (2009). Natural history of recovery from brain injury after prolonged disorders of consciousness: Outcome of patients admitted to inpatient rehabilitation with 1–4 year follow-up. *Progress in Brain Research, 177,* 73–88. https://doi.org/10.1016/S0079-6123(09)17707-5

Katz, I. R., Curyto, K. J., TenHave, T., Mossey, J., Sands, L., & Kallan, M. J. (2001). Validating the diagnosis of delirium and evaluating its association with deterioration over a one-year period. *The American Journal of Geriatric Psychiatry, 9*(2), 148–159.

Kinney, H. C., & Samuels, M. A. (1994). Neuropathology of the persistent vegetative state: A review. *Journal of Neuropathology and Experimental Neurology, 53*(6), 548–558.

Kondziella, D., Bender, A., Diserens, K., Van Erp, W., Estraneo, A., Formisano, R., ... & Laureys, S. (2020). European Academy of Neurology guideline on the diagnosis of coma and other disorders of consciousness. *European Journal of Neurology, 27*(5), 741–756. https://doi.org/10.1111/ene.14151

Laureys, S., Tononi, G., & Sporns, O. (2004). The neural correlates of consciousness in severe brain injury: An integrated approach. *Nature Reviews Neuroscience, 5*(5), 372–382. https://doi.org/10.1038/nrn1427

Laureys, S. (2005a). The neural correlate of (un)awareness: Lessons from the vegetative state. *Trends in Cognitive Sciences, 9*(12), 556–559. https://doi.org/10.1016/j.tics.2005.10.010

Laureys, S., Owen, A. M., & Schiff, N. D. (2005b). Brain function in coma, vegetative state, and related disorders. *The Lancet Neurology, 4*(7), 537–546.

Laureys, S., Boly, M., & Maquet, P. (2006). Tracking the recovery of consciousness from coma. *Journal of Clinical Investigation, 116*(7), 1823–1825. https://doi.org/10.1172/JCI 2972

Leventer, R. J., Guerrini, R., & Dobyns, W. B. (2008). Malformations of cortical development and epilepsy. *Dialogues in Clinical Neuroscience, 10*(1), 47–62. https://doi.org/10.31887/DCNS.2008.10.1/rjleventer

Levin, H. S. (1992). Neurobehavioral recovery. *Journal of Neurotrauma, 9*(Suppl 1), S359–S373.

Lohman, T., Kapoor, A., Engstrom, A. C., et al. (2024). Central autonomic network dysfunction and plasma Alzheimer's disease biomarkers in older adults. *Alzheimer's Research & Therapy, 16*, Article 124. https://doi.org/10.1186/s13195-024-01486-9

McEwen, B. S., & Stellar, E. (1993). Stress and the individual: Mechanisms leading to disease. *Archives of Internal Medicine, 153*(18), 2093–2101. https://doi.org/10.1001/archinte.1993. 00410180039004

McEwen, B. S., & Gianaros, P. J. (2011). Stress- and allostasis-induced brain plasticity. *Annual Review of Medicine, 62*, 431–445. https://doi.org/10.1146/annurev-med-052209-100430

Monti, M. M., Vanhaudenhuyse, A., Coleman, M. R., Boly, M., Pickard, J. D., Tshibanda, L., Owen, A. M., & Laureys, S. (2010). Willful modulation of brain activity in disorders of consciousness. *The New England Journal of Medicine, 362*(7), 579–589. https://doi.org/10.1056/NEJMoa0905370

Moruzzi, G., & Magoun, H. W. (1949). Brain stem reticular formation and activation of the EEG. *Electroencephalography and Clinical Neurophysiology, 1*(4), 455–473.

Multi-Society Task Force on PVS. (1994). Medical aspects of the persistent vegetative state (1). *The New England Journal of Medicine, 330*(21), 1499–1508. https://doi.org/10.1056/NEJ M199405263302107

Nolan, J. P., Neumar, R. W., Adrie, C., Aibiki, M., Berg, R. A.,

Bottiger, B. W., Callaway, C., Clark, R. S., Geocadin, R. G., Jauch, E. C., Kern, K. B., Laurent, I., Longstreth, W. T., Merchant, R. M., Morley, P. T., Morrison, L. J., Nadkarni, V. M., Peberdy, M. A., Rivers, E. P., ... & Unger, B. (2009). Post-cardiac arrest syndrome: Epidemiology, pathophysiology, treatment, and prognostication: A scientific statement from the International Liaison Committee on Resuscitation; the American Heart Association Emergency Cardiovascular Care Committee; the Council on Cardiovascular Surgery and Anesthesia; the Council on Cardiopulmonary, Perioperative, and Critical Care; the Council on Clinical Cardiology; the Council on Stroke (Part 1). *International Emergency Nursing, 17*(4), 203–225. https://doi.org/10.1016/j.ienj.2009 .01.001

Owen, A. M., Coleman, M. R., Boly, M., Davis, M. H., Laureys, S., & Pickard, J. D. (2006). Detecting awareness in the vegetative state. *Science, 313*(5792), 1402. https://doi.org/ 10.1126/science.1130197

Parnia, S., Spearpoint, K., de Vos, G., Fenwick, P., Goldberg, D., Yang, J., ... & Schoenfeld, E. R. (2014). AWARE—AWAreness during Resuscitation—A prospective study. *Resuscitation, 85*(12), 1799–1805. https://doi.org/10.1016/j.resuscitation.2014.09.004

Parvizi, J., & Damasio, A. R. (2001). Consciousness and the brainstem. *Cognition, 79*(1–2), 135–160. https://doi.org/10.1016/S0010-0277(00)00127 -X

Peng, G., Wang, J., Feng, Z., Liu, P., Zhang, Y., He, F., Chen, Z., & Luo, B. (2016). Clinical and neuroimaging differences between posterior cortical atrophy and typical amnestic Alzheimer's disease patients at an early disease stage. *Scientific Reports, 6*, 29372. https://doi.org/10.1038/srep29372

Plum, F., & Posner, J. B. (1966). *The diagnosis of stupor and coma.* Philadelphia, PA: F.A. Davis Company.

Plum, F., & Posner, J. B. (1982). *The diagnosis of stupor and coma* (3rd ed.). Philadelphia, PA: F.A. Davis Company.

Posner, J. B., Saper, C. B., Schiff, N. D., & Plum, F. (2007). *Plum and Posner's diagnosis of stupor and coma* (4th ed.). New York: Oxford University Press.

President's Council on Bioethics. 2008. *Controversies in the Determination of Death: A White Paper by the President's Council on Bioethics*. Washington, DC: President's Council on Bioethics. https://bioethicsarchive.georgetown.edu/pcbe/reports/death/

Rajajee, V., Muehlschlegel, S., Wartenberg, K. E., & Geocadin, R. G. (2023). Guidelines for neuroprognostication in comatose adult survivors of cardiac arrest. *Neurocritical Care, 38*(3), 533–563. https://doi.org/10.1007/s1 2028-023-01688-3

Ring, K., & Cooper, S. (1997). *Mindsight: Near-Death and Out-of-Body Experiences in the Blind.* William James Center for Consciousness Studies.

Rosenbaum, P., Paneth, N., Leviton, A., Goldstein, M., Bax, M., Damiano, D., ... & Jacobsson, B. (2007). A report: The definition and classification of cerebral palsy April 2006. *Developmental Medicine & Child Neurology, 49*(s109), 8–14. https://doi.org/10.1111/j.1469-8749.2007.tb12610.x.

Sabom, M. B. (1998). *Light and death: One doctor's fascinating account of near-death experiences.* Zondervan.

Sandroni, C., Nolan, J., Cavallaro, F., & Antonelli, M. (2007). In-hospital cardiac arrest: Incidence, prognosis, and possible measures to improve survival. *Intensive Care Medicine, 33*(2), 237–245. https://doi.org/10.1007/s00134-006-0326-z

Schiff, N. D. (2015). Cognitive motor dissociation following severe brain injuries. *JAMA Neurology, 72*(12), 1413–1415. https://doi.org/10.1001/jamaneurol.2015.2899

Schnakers, C., Vanhaudenhuyse, A., Giacino, J., Ventura, M., Boly, M., Majerus, S., ... & Laureys, S. (2009). Diagnostic accuracy of the vegetative and minimally conscious state: Clinical consensus versus standardized neurobehavioral assessment. *BMC Neurology, 9,* 35. https://doi.org/10.1186/1471-2377-9-35

Seel, R. T., Sherer, M., Whyte, J., Katz, D. I., Giacino, J. T., Rosenbaum, A. M., ... & Zasler, N. D. (2010). Assessment scales for disorders of consciousness: Evidence-based recommendations for clinical practice and research. *Archives of Physical Medicine and Rehabilitation, 91*(12), 1795–1813. https://doi.org/10.1016/j.apmr.2010.07.218

Seeley, W. W., Menon, V., Schatzberg, A. F., Keller, J. A., Glover, G. H., Kenna, H., Reiman, M., & Greicius, M. D. (2007). Dissociable intrinsic connectivity networks for salience processing and executive control. *Journal of Neuroscience, 27*(9), 2349–2356. https://doi.org/10.1523/JNEUROSCI.5587-06.2007

Smith, P. J., & Blumenthal, J. A. (2016). Dietary factors and cognitive decline. *Journal of Prevention of Alzheimer's Disease, 3*(1), 53–64. https://doi.org/10.14283/jpad.2015.71

Sporns, O. (2011). *Networks of the brain.* Cambridge, MA: MIT Press.

Stender, J., Gosseries, O., Bruno, M.-A., Charland-Verville, V., Vanhaudenhuyse, A., Demertzi, A., ... & Laureys, S. (2014). Diagnostic precision of PET imaging and functional MRI in disorders of consciousness: A clinical validation study. *The Lancet, 384*(9942), 514–522. https://doi.org/10.1016/S0140-6736(14)60042-8

Teasdale, G., & Jennett, B. (1974). Assessment of coma and impaired consciousness: A practical scale. *The Lancet, 2*(7872), 81–84. https://doi.org/10.1016/S0140-6736(74)91639-0

Teasdale, G., Allan, D., Brennan, P., McElhinney, E., & Mackinnon, L. (2014). The Glasgow Coma Scale at 40 years: Standing the test of time. *The Lancet Neurology, 13*(8), 844–854. https://doi.org/10.1016/S1474-4422(14)70120-6

Thibaut, A., Bruno, M. A., Ledoux, D., Demertzi, A., & Laureys, S. (2014). tDCS in patients with disorders of consciousness: Sham-controlled randomized double-blind study. *Neurology, 82*(13), 1112–1118. https://doi.org/10.1212/WNL.0000000000000260

Vandal, M., White, P. J., Tourn33333issac, M., Tremblay, C., St-Amour, I., Drouin-Ouellet, J., Bousquet, M., Traversy, M.-T., Planel, E., Marette, A., & Calon, F. (2016). Impaired thermoregulation and beneficial effects of thermoneutrality in the 3×Tg-AD model of Alzheimer's disease. *Neurobiology of Aging, 43*, 47–57. https://doi.org/10.1016/j.neurobiolaging.2016.03.024

van Lommel, P., van Wees, R., Meyers, V., & Elfferich, I. (2001). Near-death experience in survivors of cardiac arrest: A prospective study in the Netherlands. *The Lancet, 358*(9298), 2039–2045. https://doi.org/10.1016/S0140-6736(01)07100-8

Vercruysse, P., Vieau, D., Blum, D., Petersén, Å., & Dupuis, L. (2018). Hypothalamic alterations in neurodegenerative diseases and their relation to abnormal energy metabolism. *Frontiers in Molecular Neuroscience, 11*, 2. https://doi.org/10.3389/fnmol.2018.00002

Volpe, J. J. (2008). *Neurology of the newborn* (5th ed.). Saunders Elsevier.

Wannez, S., Heine, L., Thonnard, M., Gosseries, O., & Laureys, S. (2017). The repetition of behavioral assessments in diagnosis of disorders of consciousness. *Annals of Neurology, 81*(6), 883–889. https://doi.org/10.1002/ana.24962

Working Party of the Royal College of Physicians. (2003). The vegetative state: Guidance on diagnosis and management. *Clinical Medicine, 3*(3), 249–254. https://doi.org/10.7861/clinmedicine.3-3-249

Winblad, B., Amouyel, P., Andrieu, S., Ballard, C., Brayne, C., Brodaty, H., ... & Zetterberg, H. (2016). Defeating Alzheimer's disease and other dementias: A priority for European science and society. *The Lancet Neurology, 15*(5), 455–532. https://doi.org/10.1016/S1474-4422(16)00062-4

13

The Unity of the Human Person

Personhood is a fundamental category related to critical medical decisions, and experts unanimously agree concerning its value in examining bioethical issues. Therefore, a precise understanding of its foundation in biomedical and other contexts is required. For example, conditions such as dementia, "vegetative" states (Jennett and Plum 1972), severe encephalopathies, and the definition of death (President's Commission 1981) are complex medical challenges that provoke profound reflection on the very essence of human life. This inquiry extends beyond neurology to encompass philosophy, theology, bioethics, and other disciplines. The approach to this issue requires a dialogue that proceeds methodologically by following two steps: first, starting with the phenomenon presented in the field of biomedicine, describing it as precisely and clearly as possible; and secondly, elaborating and pondering all the questions that transcend the limits of this field and require an interdisciplinary approach. It is essential to avoid imprecision in the first stage of this process.

Neuroscience is influential in the dialogue with other sciences regarding personhood. For instance, the role of emotions and bodily states and their neurological substrates in shaping consciousness and personhood have been extensively explored, pointing to the unity of the human person. Damasio's Somatic Marker Hypothesis posits that emotional signals, or "somatic markers," generated in response to past experiences, guide decision-making by associating specific options with positive or negative emotional outcomes. The prefrontal cortex, specifically the ventromedial prefrontal cortex (vmPFC), is involved in these markers' emotional and cognitive higher-order integration processing (Damasio 1996 and 1999). This relevance pertains to a more profound understanding of the biological underpinnings of personhood, human life, death, and central issues at the interface between medicine, bioethics, philosophy, and theology. Moreover, as Pellegrino and Thomasma (1993) emphasize, understanding humans as recipients of health services involves virtues such as compassion, justice, and fidelity to the patient, which play a central role in guiding medical practitioners' approach to personhood. These virtues ensure that bioethical decision-making respects the inherent dignity of the human person, particularly in complex cases such as end-of-life care or severe neurological impairments. By integrating moral philosophy with clinical practice, a multidisciplinary framework offers a foundation for addressing the ethical challenges at the intersection of medicine and personhood.

This chapter will briefly review philosophical and theological traditions in Western civilization, which have elaborated with great clarity a unitary concept of the human person. Afterward, we will consider the essential neuroscientific facts supporting this anthropological view. The confluence of these aspects will delineate an integrative and unifying rationale based on what constitutes the neurological core of personhood. Remarkably, these different sources of knowledge finely converge to shed new light on this central issue, providing a richer theoretical frame to elaborate on several fundamental philosophical and ethical problems.

A Brief Historical Perspective

In ancient Egypt (date uncertain), the concept of a heart-centered belief system, or the cardiocentric hypothesis, dominated, positing that the heart was the control center of thought, emotions, and the soul (Assmann 2002). This idea, known as cardiocentrism, viewed the heart as the origin of all cognitive and spiritual activity. Around 550 BC, Pythagoras proposed a brain-centered theory, arguing that the soul resides in the brain and is immortal (Lloyd 1975). This shift in thinking laid foundational ideas for later influential figures like Plato, Hippocrates, and Galen of Pergamon, who advanced the brain-centric perspective. Around 460 BC, Hippocrates identified the brain as the seat of the soul. His treatise *On the Sacred Disease* emphasizes that the brain "tells the limbs how to act" and serves as "the messenger of consciousness" (Hippocrates 1923). This work describes the brain as the interpreter of consciousness and sensation, a groundbreaking idea in ancient Greek medicine.

In the 5th to 4th century BC, however, prominent philosophers like Aristotle, Diocles, and Praxagoras of Cos upheld cardiocentrism, attributing sensation, thought, and bodily control to the heart (Nutton 2004). By the 1st century AD, Alexandrian anatomists such as Herophilos, often considered the father of anatomy, conducted systematic dissections and provided detailed descriptions of the brain and nervous system, significantly advancing the understanding of its structure and function (Von Staden 1989). They identified structures such as the pia and dura mater and referred to primary brain divisions, thus contributing anatomical clarity to ongoing debates on brain function. In 129 AD, Galen of Pergamon—a physician and biologist—employed rigorous anatomical methodology to assert that the brain is responsible for sensation and thought, with nerves originating from the spinal cord and brain, firmly establishing cephalocentrism in medical thought (Galen 1916).

During the 11th century, the Islamic philosopher and physician Avicenna introduced the idea that the brain housed the "faculty of fantasy," where forms from the five senses were processed and integrated (Avicenna, 1329). By 1316, the anatomist Mondino de' Liuzzi associated common sense with the middle of the brain, coining terms such as *sensus communis*, fantasy, and imagination to explain brain functions beyond sensation. In the Middle Ages, figures like Albertus Magnus expanded on Galen's cephalocentric theory, contributing to physiology and biology by associating specific faculties with regions of the brain (Siraisi 1990). With the Renaissance and subsequent centuries of scientific advancement, anatomists began to describe structures like the cerebrum, cerebellum, and medulla, though functional understanding remained limited. The 1660s marked a turning point in the study of the brain with Thomas Willis's *Cerebri anatome* ("Anatomy of the Brain"). This seminal work, published in 1664, revolutionized the understanding of neuroanatomy and neurophysiology. Willis provided detailed descriptions of brain structures, introduced the term "neurology," and laid the foundation for the functional localization of brain regions. Additionally, the Circle of Willis—a critical arterial structure that supplies blood to the brain—was named in his honor. Nicolaus Steno's *Lecture on the Anatomy of the Brain* (1669) expanded comprehension of the brain's structure and its relationship to mental processes, setting the stage for the scientific study of neuroanatomy (Finger 1994).

The Meanings of the Word Soul

The word soul has encompassed many meanings throughout the history of Western civilization, embodying concepts such as spirit, breath, vital force, life, creature, reason, intellect, thought, will, moral or emotional nature, and the entirety of personal being (Taylor 1989). Exploring its significance leads us back to foundational terms in early Western culture, rooted in Hebrew: basar, nefesh, ruah, and leb—terms later translated into Greek as sarx, psyche, pneuma, and kardia, and into Latin as caro, anima, spiritus, cor, or mens (Barr, 1961). These translations reflect the soul's layered meanings concerning body, spirit, and personal identity, and we will revisit these terms further in subsequent discussions.

A pre-Aristotelian tradition located the soul in various bodily organs, from the diaphragm—a muscle believed to mediate between the thorax and abdomen—to the liver, heart, and later, the brain, depending on cultural and philosophical context (Longrigg 1993). In Homeric thought and the pre-Aristotelian materialist tradition, thinkers likened the soul to a breath, a force, or a ghostly vapor—a subtle material essence they considered essential to life (Snell 1960). As Karl Popper noted, materialist thinkers in this tradition upheld a form of dualism, regarding the soul or mind as the true essence of the body. Though viewed in physical terms, such as air, fire, or atoms, the soul's essence had a dual nature: it was part of the physical body but also possessed an incorporeal quality that transcended physical existence (Popper and Eccles 1977). Pythagoras and the Pythagoreans later advanced this view by proposing that the soul's essence was an abstract reality—expressed as a number, harmony, or balance—foreshadowing later philosophical and metaphysical conceptions of the soul as an organizing principle beyond mere material existence (Kirk, Raven, and Schofield 1983). These transitions profoundly shaped Western intellectual traditions. The shift from naturalistic to idealistic conceptions of the soul influenced metaphysical philosophies and theological discourses, with terms like psyche and spiritus preparing the ground for developments in later Christian thought.

Plato

Plato conceived of the soul within his theory of Forms, where the soul exists in a transcendent, immaterial realm beyond the physical world. His conception of the mind includes three distinct aspects: reason, appetite, and energy, each governing different dimensions of human behavior (Plato, *Republic* IV). Plato viewed the body as inferior to the soul, often describing it as a prison from which the soul seeks liberation. This counterposition between body and soul underpins his anthropology and cosmology, characterizing his dualistic view of human nature as antagonism or tension (Plato, *Phaedo*). Plato's metaphor of the soul as the pilot of the body suggests a guiding force. However, it mentions but does not elaborate on the biological connection as later Aristotle does in his hylomorphic view, where the soul is the form and essence of the body. In Plato's view, death brings about the soul's liberation from the body, enabling the survival of human beings' ideal, incorporeal aspect. This view illustrates Plato's dualist-interactionist conception of personhood, where the body and soul are distinct yet interdependent entities.

Plato's recognition of a transphysical domain inherent to human existence forms a fundamental contrast to modern physicalism. Central to his metaphysics is an eternal realm of Forms or Ideas, a transcendent reality that surpasses the physical world. For Plato, these Forms are immutable, perfect, and abstract entities that embody true reality, contrasting the flawed, ever-changing physical world we perceive through the senses (Plato, *Timaeus*). This notion sets him apart from early Greek materialists like Democritus, who posited that reality consisted solely of atoms and void. Plato's departure from materialism, expressed in dialogues like the *Phaedo* and the *Republic*, emphasizes a dualistic approach. The eternal soul, distinct from the perishable body, pre-exists birth and endures after death. This view challenges reductionist perspectives that equate mental and psychological phenomena with physical processes, advocating instead for human existence's enduring, transcendent aspects (Burnyeat 1982).

Plato's framework has contributed significantly to ongoing philosophical debates surrounding consciousness, selfhood, and the limitations of materialism. While modern physicalism generally reduces mental phenomena to physical entities and processes, Plato's transphysical perspective suggests there are dimensions of reality beyond material explanations. For example, his influence is seen in the work of contemporary thinker Roger Penrose, a British mathematician, physicist, and philosopher of science. Penrose, known for his contributions to mathematical physics, posits that abstract mathematical forms and concepts exist independently of the physical world and human cognition. He argues that the mathematical structures underlying the universe are not inventions but discoveries of pre-existing truths, reflecting a Platonic belief in the existence of a realm of abstract Forms (Penrose 1994). In a later chapter on Theism, we will explore Penrose's belief in three distinct worlds—Physical, Mental, and Platonic Mathematical—and the implications of his stance on physicalist perspectives in philosophy.

The enduring relevance of Plato's ideas demonstrates how Neo-Platonism continues to inform contemporary debates in the philosophy of science, providing a basis for arguments that reject purely materialist explanations of consciousness and cognition. By acknowledging Plato's insights, we gain a richer philosophical foundation to explore the nature of consciousness, the self, and the limits of physicalism in comprehending the profound complexities of human existence (Rist 1967). Plato's legacy points to human experience involving dimensions beyond the reach of empirical science, inviting philosophical inquiry into the transcendent and eternal aspects of the human person.

Aristotle

In contrast to Plato's dualistic view, Aristotle developed a unitary concept of the soul within a framework he termed *hylomorphism*, emphasizing the inseparable unity of form (morphe) and matter (hyle). While Plato viewed the soul as transcendent and separate from the body, Aristotle considered it the organizing and unifying principle of the organism, intrinsically linked to its material structure. According to Aristotle, earlier philosophers had characterized the soul by properties like movement, sensation, and incorporeality, noting movement as a distinguishing quality between living and inanimate things (Aristotle, *De Anima*). However, Aristotle expanded this notion of movement to include internal processes such as growth and change in shape, making the soul not only the origin of external motion but also the source of inherent physiological transformations. In this view, the soul embodies the form that gives life and structure to the organism, a stark contrast to the dualism prevalent in his time.

Aristotle's view of the soul as the *unifying principle* of life has had a lasting influence on philosophy and biology. He was among the first to approach biology systematically, observing organisms as unified entities whose form and matter are interdependent (Aristotle, *Parts of Animals*). Hylomorphism encapsulates this approach by positing that matter cannot be separated from form but constitutes an integrated reality. In hylomorphism, the philosopher can theoretically separate form and matter, yet in biological reality, they are one. This notion of unity has profoundly influenced history, particularly in the Middle Ages, where Aristotelian ideas shaped Scholastic thought. It remains relevant today, challenging modern reductionist views (Nussbaum & Rorty 1995). Aristotle's integration of biology and philosophy introduced a holistic view that still resonates in contemporary scientific discussions, especially in disciplines examining the relationship between structure and function.

In modern biology, Aristotle's hylomorphic doctrine finds renewed relevance as scientists recognize the interconnectedness of form and function within living systems. Aristotle challenges reductionist approaches by emphasizing that one can only understand an organism's parts and processes within the context of the whole. This perspective is evident in recent biological models that explore top-down causation, where larger organizational structures in organisms influence lower-level processes. Such models align with Aristotle's assertion that the unity of form and function is essential in understanding the complexity of life (Lennox 2001). The interplay between genetic information, cellular architecture, and physiological function highlights the integrated nature of living organisms, underscoring the continued influence of Aristotle's thought on current biological perspectives.

Additionally, there is a resurgence of interest in Aristotle's concept of *telos*—the inherent purpose or end that drives living organisms. This notion suggests that each organism possesses an intrinsic goal or purpose, which resonates with ecological and holistic biology's current emphasis on the interconnectedness of biological components. Rather than viewing organisms as mere aggregates of parts, *telos* encourages us to see each species and individual as directed toward specific ends within broader ecosystems. This holistic perspective on purpose and function speaks to a growing trend in modern biology that seeks to move beyond reductionism, emphasizing the organization, sustainability, and purposefulness of living systems (Ariew, Cummins, & Perlman 2002). Aristotle's insights into purpose inspire biologists and philosophers alike, offering a framework that honors the complexity and intentionality present in nature.

Aristotle's Concept of Substantial Form

Aristotle's concept of *substantial form* is the vital principle that organizes matter into a cohesive, functional organism. This organization reflects a relational disposition among the different parts, forming a unity and totality fundamentally irreducible to the sum of its components (Aristotle, *Metaphysics*). For Aristotle, life expresses itself in a tripartite nature: plants have a vegetative soul that enables them to obtain nutrients, grow, and reproduce; animals have a sensitive soul that allows sensation and movement; and human beings uniquely possess an intellectual soul that enables reasoning, self-awareness, and the knowledge of mortality. This soul hierarchy reflects Aristotle's view that human beings are a synthesis of these aspects—vegetative, sensitive, and intellectual—which collectively form an integrated whole (Aristotle, *De Anima*).

Aristotle also laid the groundwork for psychology, psychophysics, and psychosomatic studies. His writings describe the interplay between body and soul: "The soul and the body react one on the other by sympathy. A change in the psychic state produces a change in the form of the body, and inversely, a change in the form of the body produces a change in the state of the soul" (Aristotle, *De Anima*). This psychosomatic relationship hints at a mutual influence between mental and physical states, an idea that has enduring relevance in modern psychology and medicine. Philosopher Karl Popper later remarked on Aristotle's concept of substantial form, noting that it anticipates the contemporary theory of genetic information. Just as Aristotle saw the soul as guiding the organism toward its telos or purpose, scientists view genes as encoded information that directs the organism's development and functions (Popper 1972). Popper's observation emphasizes the intricate, encoded complexity of every cell, protein, and biological component, echoing Aristotle's view of life as directed toward natural perfection. In essence, Aristotle's philosophy presents the human being as a unity, where the soul is not only the conscious aspect but the vital, integrating principle of existence.

Descartes

René Descartes is a central figure in the philosophy of dualism, having developed a mechanistic concept of the body and an immaterial view of the mind. Gilbert Ryle famously coined the term "the ghost in the machine" to describe Descartes' dualism, which views the human body as a mechanistic automaton capable of voluntary movement yet fundamentally distinct from the mind (Ryle 1949). For Descartes, extension—or spatial occupation—is the essence of materiality, a concept similar to the Aristotelian idea of *prime matter* and Plato's notions of space (*Meditations on First Philosophy*). Following ancient tradition, Descartes posited that the mind is incorporeal, devoid of spatial extension, existing as an inextensive "thinking substance" independent of place and matter. This assumption led him to postulate that mind-body interaction could only occur through a mechanism analogous to physical causation. As an interactionist dualist, Descartes devised a mechanistic theory of mind-body interaction to account for this interrelation (Descartes 1985).

In Descartes' view, the inextensive soul interacts with the body through an extensive structure: the pineal gland, which he proposed as the point of contact between mind and body. In modern neo-Cartesian interpretations, renowned figures have proposed various brain regions to mediate this interaction, including Wilder Penfield and Sir John Eccles, offering unique perspectives (Penfield 1975; Eccles 1989). Penfield, a prominent Canadian neurosurgeon, proposed that the "centroencephalon" was critical for mind-body interrelation and suggested that disanimation, or the separation of soul and body, would require the destruction of most of the brain. Eccles, an Australian neurophysiologist awarded the Nobel Prize for his work on synaptic transmission, theorized that the primary language zone in the dominant hemisphere was essential for this interaction. Each theory presented a different threshold for disanimation: Penfield's model required widespread brain damage, while Eccles' centered on a more localized area.

Descartes' mechanistic worldview and problematic dualism present the mind-body relationship as an extrinsic juxtaposition, allowing for a potential dissociation between the immaterial "soul" and the physical "machine." This view raises a speculative issue beyond medicine: determining when the "immaterial" soul abandons the physical body. Cartesian dualism thus introduces a separation between the biological basis of death and its philosophical and theological implications, implying that the soul's presence or absence is physiologically irrelevant. This dualism laid the foundation for subsequent positions that could separate personhood and death from biological functioning alone. Unfortunately, this perspective is the most associated with alternatives to physicalism in the ongoing tension between physicalist and dualist perspectives in contemporary debates about consciousness and the nature of personhood.

The Definition of Person

The question of what constitutes personhood has long been at the heart of philosophical, ethical, and legal debates, influencing views on identity, human rights, and moral worth. Kant linked personhood to the capacity—or potential, in the case of those not yet mature—for moral reasoning and action. Other perspectives have emphasized attributes like rationality and self-consciousness. However, bioethics continues to grapple with the challenge of establishing clear and non-circular criteria that align with our intuitions about personhood (Farah and Heberlein 2007). While performance-based views define personhood through external capacities such as self-awareness and rationality, substance-based perspectives argue for an intrinsic essence that grounds personhood independently of functional abilities.

The naturalist view of personhood, rooted in the tradition of thinkers like Descartes, Locke, and Hume, posits that personhood centrally depends on possessing continuous consciousness and its associated faculties (Descartes 1985; Locke 1975). The "performance criterion" perspective defines personhood by distinguishing persons from non-persons by functional attributes, such as self-awareness, memory, and rationality. However, this view departs from a longer tradition that considers a unifying metaphysical foundation, emphasizing an intrinsic nature over functional abilities. Taylor critiques the naturalist view of reducing personhood to observable behaviors, which risks overlooking human existence's moral and metaphysical dimensions. He warns that such criteria could blur distinctions between persons and non-person entities, such as advanced AI or animals, undermining the unique dignity attributed to human beings (Taylor 1989).

Other contemporary philosophers like J. P. Moreland defend a substance-based view of personhood, often grounded in Christian thought. Moreland argues that a person possesses an intrinsic essence that confers unity across various dispositions, capacities, and properties, allowing identity continuity over time. He notes, "It is because an entity has an essence and falls within a natural kind that it can possess a unity of dispositions, capacities, parts, and properties at a given time and can maintain identity through change" (Moreland 1998). Similarly, Francis J. Beckwith asserts that personhood is inherent in the being itself, not contingent upon function: "What is crucial morally is the being of a person, not his or her functioning. A human person does not come into existence when human function arises. Still, a human possesses the natural inherent capacity to give rise to human functions, whether or not those functions are ever attained" (Beckwith 2007).

In an era where technology, and transhumanist and physicalist philosophies challenge traditional notions of personhood, substance-based views offer a robust framework that preserves human dignity by grounding identity in a transphysical foundation rather than transient capacities. The distinction between performance and substance-based views has profound implications for bioethics, influencing debates on abortion, end-of-life decisions, and the treatment of individuals with severe cognitive impairments. Substance-based views safeguard the moral consideration of individuals regardless of their functional state, emphasizing the intrinsic value of human life. As advances in science and technology challenge traditional definitions of personhood, the philosophical foundations laid by thinkers like Moreland and Beckwith, aligned with a long tradition in Western civilization, remain vital. Their perspectives counterbalance reductionist trends, ensuring that debates about personhood remain anchored in a framework that values the human person's unity, essence, and dignity. By grounding personhood in intrinsic essence rather than external performance, these perspectives emphasize the enduring value of human life, offering a robust foundation for philosophical and ethical inquiry.

The Uniqueness of Christianity

Christianity offers a distinctive perspective on personhood and human unity. It integrates spiritual, intellectual, and physical dimensions to emphasize humanity's divine image. This holistic view challenges dualistic philosophies, presenting a coherent understanding of human nature rooted in the doctrine of the Imago Dei and the transformative hope of the resurrection. Tertullian, an influential early Christian theologian from the second to third centuries AD, was the first to introduce the term persona in a technical theological context, particularly in his work Against Praxeas (Tertullian 1977). He employed this term to differentiate the three persons of the Trinity, making Christian thought the first tradition to use "person" in a sense approaching its contemporary meaning. Christian theology deeply roots the unity of the human person in the doctrine of Imago Dei, the belief that humanity is created in the image of God. This doctrine emphasizes that every person reflects the divine image in a harmonious integration of spiritual, intellectual, emotional, and physical dimensions, suggesting an intrinsic interconnectedness within the person. As articulated through Semitic categories, Biblical anthropology presents a holistic view of human beings, portraying the individual as an indivisible unity rather than a collection of separable parts (Wenham 1987).

Unlike Platonic thought, which often viewed the body as a transient vessel, Christianity upholds the intrinsic value of the body as an essential part of human personhood. This perspective challenges dualistic tendencies by affirming the sanctity and eternal destiny of the physical body, positioning the Incarnation as a central event that embodies divine affirmation of human nature. Central to Christian thought on human unity is the doctrine of the Incarnation. In the person of Jesus Christ, divinity and humanity are united, with God taking on human flesh. This radical union affirms the intrinsic value of the body, rejecting any view that diminishes it as mere material. Through the Incarnation, Christianity underscores the sanctity of embodiment, emphasizing its profound significance in the divine-human relationship.

The crucifixion of Jesus Christ represents the depth of God's engagement with human suffering and mortality. Christ's voluntary sacrifice encompasses human experience, transforming suffering, sin, and death. As the Apostle Paul articulates, believers partake spiritually and physically in this victory, underscoring a comprehensive redemption that affirms the unity of body and soul (Romans 8:11; 1 Corinthians 15:42-44). The Resurrection of Jesus Christ further solidifies this holistic view, demonstrating the restoration of the entire person—spirit, soul, and body. Christ triumphed over death through His resurrection, offering believers a prototype of the future resurrection and affirming that the body itself is destined for transformation, not dissolution (Wright 2003). This eschatological promise affirms the intrinsic value of embodiment and envisions human personhood as fully restored and glorified in eternal communion with God.

This holistic vision of personhood addresses modern debates on identity and embodiment, including bioethics, artificial intelligence, and end-of-life care. The distinction between biological and philosophical approaches to the unity of the human person invites deeper contemplation of the sanctity of life and the relationship between body and soul. Christianity's unique synthesis of spiritual and physical realities provides a profound response to these questions, offering a robust framework for navigating personhood's ethical and technological complexities. In an era of unprecedented challenges, it affirms the unity and sacredness of human life as an enduring gift of divine creation and redemption.

Two Central Theological Aspects

Theology must address two central problems with coherence and depth: the profound identity and the meaning and truth of human existence. In Christian thought, the human being is the summit of God's Creation, integrating material and spiritual aspects in a unique synthesis and unity. Ratzinger (1991) captures this idea, describing the human person as "the horizon of Creation, where the heavens and the earth are united; as a link between time and eternity; as a synthesis of the Creation." Aquinas views humanity as a "synthesis of the material universe," emphasizing that the human situation in the cosmos is not accidental but essential to human identity. The book of Genesis reveals this foundational unity: "Then the Lord God formed a man from the dust of the ground and breathed into his nostrils the breath of life, and the man became a living being" (Genesis 2:7, NIV). This passage expresses the human constitution through two co-principles—matter and spirit—while affirming humanity's unity with the rest of Creation.

Christ's life and teachings offer profound insights into contemporary anthropological questions by addressing the complexities of human identity, dignity, and purpose. In a world grappling with issues like alienation, inequality, and the loss of meaning, Christ's life reveals the inherent value of every person, rooted in their relational nature and capacity for love. His life, death, resurrection, and teachings challenge reductionist views of humanity, emphasizing forgiveness, service, and the interconnectedness of all people, thereby providing a transformative framework for understanding human flourishing in personal and communal contexts.

Biologically, humans do not possess a super-specialized functional constitution. Instead, their unique faculties emerge from what might be termed a "transphysically enabled" brain, capable of creating a distinct cultural environment. Humanity exerts a transformative influence on the world through technological and intellectual pursuits, both constructively and destructively. This capacity for cultural creation reflects humanity's potential to transcend the physical world and reach toward a transcendent and infinite personal reality (Boff 1980). The doctrine of *Imago Dei* speaks directly to modern challenges, particularly in the realms of ethics and technology. In an age of artificial intelligence, robotics, genetic engineering, transhumanist ideas, and global trends that threaten to unravel civilizational foundations, the belief that humans are created in the image of God calls for a renewed focus on identity, responsibility, stewardship, and relationality. It underscores the unique capacity and the risks of humanity transcending biological limits, urging individuals and societies to uphold dignity, respect for human life, and responsible stewardship as central to their engagement with the world (Cortez 2010). As finite beings, humans are loved by God wholly and individually, endowed with freedom and spontaneity that enable continual synthesis and creativity. This unity of the human person—spanning spiritual, intellectual, and physical dimensions—allows individuals to find fulfillment within the created world, their relationship with the Creator, and their interactions with others.

The essential foundation of this unity is theological and ultimately Christological. In Christian thought, humanity's identity, freedom, dignity, harmony, and supernatural destiny are fully understood in relation to Christ, as only Christ reveals man to man himself (Second Vatican Council, 1965, §22). Through Christ's example and teachings, humanity recognizes its unique anthropological status, potential for communion with God, and eternal destiny. This Christological perspective is the anchor for understanding the unity of the human person, emphasizing that human existence is not merely a biological or social reality but an expression of divine purpose and relationship. Recognizing humanity's supernatural destiny in Christ brings theological clarity to human dignity, freedom, and the search for ultimate meaning, grounding these aspects in the relational framework between God and humanity.

The Semitic Anthropology

In the Judeo-Christian tradition, an understanding of human beings emerges that captures the complexity of human existence while providing a holistic vision of personhood. This perspective, rooted in Semitic anthropology, resonates with contemporary discussions on identity, the mind-body relationship, and the integration of spiritual and physical dimensions in human life. Scripture offers a portrayal of humanity that highlights the unity of the person, affirming the capacity for communion with God—encapsulated in the Latin *capax Dei*—as central to human identity. A review of anthropological terms in the Bible highlights personhood's diverse yet interconnected aspects. These terms—*basar, nefesh, ruah,* and *leb*—reflect different facets of human existence and are translated into Greek as *sarx, psyche, pneuma,* and *kardia,* and into Latin as *caro, anima, spiritus,* and *cor* (Eichrodt 1967; Wolff 1974). Together, these terms provide a comprehensive framework for understanding human nature, reflecting a unified vision that integrates physical, emotional, intellectual, and spiritual dimensions. By exploring *basar, nefesh, leb,* and *ruah,* we see how these aspects collectively contribute to the depth and richness of personhood, emphasizing the inseparable unity of human experience.

Basar, or flesh-man (Greek: *sarx*), emphasizes humanity's connection to the material world and mortality. This aspect reflects humans' biological existence, with embodiment as a core quality. In this dimension, humans are grounded in earthly experiences, vulnerable to suffering, disease, temptation, and sin. The term also connotes a self-centered, isolated existence, focusing solely on empirical life without openness to others or God. This concept is sometimes aligned with *soma* (Greek for body), highlighting the relational aspect of embodiment and personhood, resembling what we might now call personality (Westermann 1984). *Nefesh*: In biblical language, the soul-man (Greek: *psyche*) refers to conscious identity—the "I," or the entire human being. It captures the essence of life, not merely as something humans possess but as something they are. Here, the human soul encompasses consciousness, psychosocial dimensions, and involvement in historical and relational contexts, reflecting individual responsibility and personal identity (Moreland & Rae 2000). It also hints at an immaterial dimension beyond bodily existence, serving as the ground of conscious experience and moral responsibility.

Leb, often translated as "heart," encompasses emotion, reason, and synthesizing the physical and spiritual dimensions. It integrates these facets, showing how each contributes to the unity of the person. *Leb* is thus the center of thought, emotion, and will, forming the moral and spiritual core of the individual (McKenzie 1965). This term highlights the inseparable intertwining of the human person's rational, emotional, and physical aspects, resonating with the contemporary understanding of psychosomatic unity. *Ruah:* The spirit-man (Greek: *pneuma*) denotes humanity's affinity with God, pointing to the capacity for communion with the divine and engagement with the most profound aspects of reality. This dimension captures the transcendence of the human being, enabling a relationship with God and embodying a higher, spiritual calling. This aspect of human nature aligns with the view of humanity as both body and soul, yet capable of life in a new dimension beyond mere biological existence (Kittel 1985).

This holistic view of personhood offers valuable insights for addressing contemporary issues, such as the ethical implications of artificial intelligence, genetic engineering, and the mind-body problem in neuroscience. Semitic anthropology challenges reductionist frameworks by emphasizing relationality, responsibility, and transcendence, offering a robust alternative grounded in theological and metaphysical depth. In the Hebrew anthropological view, the human person is a uni-totality, a cohesive whole capable of centering life in spirit, soul, or body. This unity does not imply uniformity; instead, it acknowledges that humans can live in two fundamental existential modes. As a "flesh-man," one lives a self-enclosed life without openness to love and freedom for God or others. Conversely, as a "spirit-man," one is open to the divine and others, finding fulfillment, relational depth, and immortality (von Rad 1962). By bridging ancient wisdom and modern challenges, Semitic anthropology invites a renewed appreciation of human dignity and purpose in the context of Creation and divine relationship. It provides a vision of human existence that unites the body, soul, and spirit. It offers a profound understanding of personhood that continues to resonate across theological, philosophical, and scientific landscapes. This framework emphasizes human existence's depth and relational essence, uniting the physical and spiritual in a coherent vision of personhood.

The Unity of the Human Person does not Imply Uniformity.

The Apostle Paul, in his letters, presents a nuanced view of human nature, distinguishing among three types of individuals based on their responses to apostolic teaching: the spiritual person (pneumatikos), the soulish person (psychikós), and the carnal or fleshy person (sarkikós) (1 Corinthians 2:12–14; 3:1-3, NASB/ESV). A particular aspect of their being predominantly guides each type of person—spirit, soul, or flesh. This categorization highlights Paul's view of human complexity, wherein different faculties—spirit, soul, and flesh—play roles in shaping the individual's moral and spiritual orientation. Notably, the spirit is seen here as a distinct yet integral faculty of the soul, allowing the terms "soul" and "spirit" to be used interchangeably in Scripture while maintaining a subtle distinction when they appear together. For instance, Paul writes, "May God himself, the God of peace, sanctify you through and through. May your whole spirit, soul and body be kept blameless at the coming of our Lord Jesus Christ" (1 Thessalonians 5:23, NIV). Similarly, Hebrews 4:12 distinguishes these terms: "For the word of God is alive and active. Sharper than any double-edged sword, it penetrates even to dividing soul and spirit, joints and marrow; it judges the thoughts and attitudes of the heart" (Hebrews 4:12, NIV).

Judeo-Christian Scriptures emphasize the concept of a unified, multifaceted human being. For instance, the Hebrew term lēb, often translated as "heart," represents a person's center, connecting will, intellect, emotions, and spirit in a unified whole (Sarna 1993). In Psalms, the heart is described as the seat of wisdom and truth: "Behold, You desire truth in the innermost being, and in the hidden part You will make me know wisdom" (Psalm 51:6, NASB). The Book of Proverbs underscores the intellect's rootedness in the heart and its association with identity: "As he thinks in his heart, so is he" (Proverbs 23:7, KJV/NKJV). Similarly, Daniel exemplifies the link between heart and will: "Daniel purposed in his heart..." (Daniel 1:8, NASB). Jesus also aligns the heart with emotional resilience, saying, "Let not your heart be troubled" (John 14:1, KJV/NKJV). Here, the Gospel writer uses the Greek term kardia, and in the book of Hebrews, the heart is the source of intention: "For the word of God is living and active, piercing to the division of soul and spirit… discerning the thoughts and intentions of the heart" (Hebrews 4:12, NIV). Finally, Proverbs 4:23 underscores the heart's foundational role: "Above all else, guard your heart, for everything you do flows from it" (Proverbs 4:23, NIV), while 1 Peter 3:4 speaks of "the hidden person of the heart" (1 Peter 3:4, NASB).

These passages reveal a richly layered concept of human unity in Judeo-Christian thought. The heart is the inner center of an individual's faculties, integrating intellect, will, emotions, and spirit. This understanding does not reduce the person to a single aspect. Still, it affirms that human nature involves a harmonious interplay of diverse dimensions, all centered in the heart as a unifying element. This Scriptural vision of human unity provides valuable insights for modern debates on personhood, particularly those engaging philosophical, theological, and scientific perspectives. In a time when questions about the mind-body relationship, consciousness, and identity dominate discussions, the Judeo-Christian concept of the human person as a cohesive yet multifaceted being offers a robust alternative to reductionist frameworks. This understanding affirms personhood's inherent dignity and relational essence by emphasizing the integration of spiritual, emotional, intellectual, and physical dimensions, challenging fragmented or overly mechanistic views. It invites a deeper exploration of what it means to be human, rooted in a vision that embraces individuality, interconnectedness, and the transcendent. The compatibility of this Scriptural view with the holistic understanding of the brain functions central to this book will remain with the reader.

The Resurrection, Unity of Body and Soul, and the Fulfillment of Personhood in Christian Anthropology

Christ's death and resurrection, undeniably the central events in the Christian faith, are framed through the lens of Semitic anthropology. Rather than centering on the soul's immortality, as in Platonic philosophy—which often neglects the dignity of the body—the Christian message emphasizes redemption and the resurrection of the whole person, body included. In this view, the apostles recognized the body as "the temple of the Spirit" (1 Corinthians 3:16, NIV), indicating a clear distinction from the Platonic separation of soul and body. As Cullmann (1958) highlights, the Christian emphasis on resurrection fundamentally departs from the Greek dualistic view of the soul's immortality, affirming the redemption of the whole person. The classic definition of death, influenced by Platonic thought, describes it as the separation of the soul from the body. While this view acknowledges humanity's spiritual, transcendent dimension, it fails to capture the full anthropological richness of the resurrection. It implies that death affects only the body, while the soul remains unaffected, which misses the essential anthropological concept of corporeality.

Christ's resurrection signifies that death affects the entire person, not just the body. Embodiment is intrinsic to human nature; a non-animated body is a corpse. Thus, a living human body is an essential constituent of the soul. The heart of Christian teaching encompasses the whole person—not only one aspect. Redemption does not imply abandoning the body but rather its complete assumption and liberation in God and for God. The apostle Paul speaks of a "spiritual body," suggesting that spirit and "flesh" may conflict, but spirit and body do not (Boff 1980). In the resurrection, the natural body is transformed into a glorified or "pneumatic" body (1 Corinthians 15, NIV). Wright (2003) elaborates on the resurrection as the pivotal event affirming the unity of body and soul, offering a holistic vision of salvation. Individuals receive the fullness of salvation by being filled with divine reality, overcoming all forms of alienation. This entry into the kingdom of God represents the complete realization of the human person. The concept of the soul also hints at this dynamic drive in humans to seek the infinite, a transcendent journey that culminates after death. Humans then transcend the physical realm, attaining a "transfigured body."

Furthermore, the Christian message expands to encompass the redemption not only of humanity but of all Creation, culminating in a transformed "cosmos" where "God may be all in all" (1 Corinthians 15:28, NIV). Humanity's unity with God and the universe awaits consummation (Romans 8:21, NIV). Building on these theological foundations, we turn to Thomas Aquinas's conception of the unity of the human person, demonstrating how these principles align with the brain metafunction framework, a perspective first proposed over two decades ago and now bolstered by contemporary biomedical and neuroscientific knowledge.

The Unity of the Human Person in Thomist Thought

Thomas Aquinas revisited and redefined Aristotelian categories, especially the concept of hylomorphism, which explores the relationship between matter (*hyle*) and form (*morphe*). In doing so, Aquinas developed a comprehensive view of human nature that upholds the soul and body's distinctness and unity within the person. This conception, deeply aligned with the Biblical model, has become integral to the Christian intellectual inheritance. Aquinas's anthropology centers on two fundamental principles: first, the unity of soul and body, which provides insight into the intelligibility of the human being and his history, and second, the doctrine of the person, which holds ethical significance, guiding our understanding of humanity's place within

Creation and concerning Christian salvation. Both aspects carry profound axiological, or value-oriented, significance. As Kretzmann (1993) notes, Aquinas's natural theology provides a robust metaphysical framework for understanding the unity of soul and body, reflecting a theistic vision that integrates human nature with divine purpose.

Building on Boethius's definition of the person as *"individua substantia rationalis naturae"*—an individual substance of a rational nature—Aquinas developed a nuanced view of personhood (Boethius, 1918). This perspective encompasses several essential elements. Substantiality refers to the person as a complete ontological entity greater than the sum of any particular components. Aquinas argued that a person's identity is not reducible to individual expressions or acts but remains an enduring whole. Individuality points to the uniqueness of each person, presenting each human as a singular, unrepeatable living being with both a biological and corporeal nature. This individuality signifies that each person holds a distinct place within Creation. MacDonald (1990) emphasizes the centrality of goodness in Thomist metaphysics, illustrating how Aquinas's anthropology aligns human dignity with the intrinsic value of Creation. According to Aquinas, rationality is an inherent attribute of human beings, transcending the capacity for expression at any given moment or circumstance. For Aquinas, rationality is more than a faculty—it embodies the metaphysical status of humans, indicating their participation in a higher order of being. Lastly, Aquinas described personhood as a quality deeply embedded in each person's constitution. This personhood does not arise from specific acts but is an essential, enduring quality that affirms each person's unique value and dignity.

The Soul is the "Substantial Form of the Body."

The Aristotelian tradition understands humanity as a synthesis of three aspects: vegetative life, the sensitive soul, and the intellectual soul (*De Anima*, Aristotle). Thomas Aquinas refines this model, advancing a unified theory that integrates these three conceptually distinct levels within a single, indivisible soul (*Summa Theologica*, Aquinas, I, q. 76, a. 1). According to Aquinas, the unity and totality of the body transcend the sum of its components. The ontological dimension of the human person thus converges in these aspects, not as a sum of separate essences but as a unified whole where soul and body, or spirit and matter, are co-principles that constitute the person (*Summa Theologica*, Aquinas, I, q. 76, a. 2). The soul, then, is the "substantial form of the body," with the body serving as the realization and expression of the soul.

Modern terms such as information and configuration resonate with the idea of the soul as the body's substantial form, though they do not entirely capture its depth. Alan Shewmon noted an affinity between modern biological concepts and the Aristotelian-Thomist understanding of the soul as the "substantial form" of the body (Shewmon 1985). Living organisms, including all their tissues and organs, undergo a constant and dynamic exchange of atoms and molecules, even in seemingly static structures like bones and nervous tissue (Shewmon 1985). Over time, the organism replaces its original atoms and molecules, yet it maintains its identity by preserving relational structures at atomic, molecular, and biological levels. This persistence of relationships—or, in modern terms, codified information—sustains the form of the body over time, reflecting an essential or "substantial" aspect of living beings (Shewmon 1985).

Shewmon (1985) builds on the Aristotelian-Thomist tradition, equating brain death with the loss of the conditions that support the substantial form of the body. By arguing that the cessation of brain function signifies the loss of the integrative principle essential for the body's unity, Shewmon bridged classical metaphysical principles with contemporary medical criteria. This earlier work demonstrates the enduring relevance of Thomistic thought in addressing complex bioethical questions, even as Shewmon's views diverged from this understanding in later years. Shewmon highlights the parallel between Aristotelian-Thomist concepts and specific modern analyses of vital processes, specifically referencing Varela's "autopoiesis" or "autopoietic mechanism," which describes the essence of living organisms (Varela, Maturana, and Uribe 1974). According to Varela, "An autopoietic machine continuously generates and specifies its own organization throughout its operation as a system of production of its own components, and it does this in an endless exchange of components under conditions of continuous perturbations and compensations of the perturbations. Therefore, an autopoietic machine is a homeostatic system (or rather a relation-static system) that has its own organization (defining networks of relations) as the fundamental invariant" (Varela, Maturana, and Uribe 1974). From a biological standpoint, this description and the organizational principle it implies align closely with the Aristotelian-Thomist concept of the soul as the "substantial form of the body." Increasingly, biologists are interested in understanding organisms through a holistic and teleological lens, reflecting a renewed engagement with ideas central to the Aristotelian-Thomist tradition (Jonas 1966; Rosen 1991).

The Head is the Vital Center of Life and Unity in the Human Person

Thomas Aquinas regarded the head as essential for sustaining life and enabling the unity of body and soul, though he understood these functions within the framework of medieval anatomy. For Aquinas, the intellect (*intellectus*) and the will (*voluntas*) are immaterial faculties of the rational soul (*Summa Theologica*, Part I, Q. 75, Art. 2). While these faculties do not directly depend on a bodily organ for their operations, they rely on sensory input provided through the body to perform their earthly functions (Pasnau 2011, 120). In Aquinas's view, the intellect requires "phantasms," or sensory-based mental images processed by the imaginative faculty (*phantasia*), to abstract universal knowledge from particular experiences (*Summa Theologica*, Part I, Q. 84, Art. 7) (Stump 2003). These phantasms are formed and stored through the body's sensory powers, which medieval thinkers associated with the brain (*cerebrum*) and its imaginative and mnemonic functions (Des Chene 2000, 150). Although Aquinas did not identify the brain as the seat of the intellect or will, he recognized its indirect yet indispensable role in enabling intellectual abstraction and volitional acts (Finger 1994, 52).

Aquinas emphasized the importance of respiration and bodily vitality to sustain the body-soul union. He argued that the cessation of respiration disrupts the conditions necessary for this union, not because respiration itself unites the soul and body. Still, because it maintains the body's disposition to receive the soul (*Summa Theologica*, Part I, Q. 76, Art. 5). This perspective reflects Aquinas's broader understanding of the human person as an integrated unity of body and soul, where the body is an instrument of the soul's spiritual faculties (Moreland and Rae 2000, 120). While Aquinas lacked detailed anatomical knowledge, his insights align with the medieval view that the brain, housed in the head, plays a vital role in supporting sensory processing and imagination, which are prerequisites for intellectual operations (Friston 2010). Thus, the head is not merely a physical structure but a critical component of human life and unity in Thomistic thought.

A Division at the Level of Unity Results in the Complete Absence of the Living Substance

The Thomistic view on human unity presents a profound understanding of the integration of body and soul, acknowledging both material and transcendent dimensions. Aquinas asserts that the rational soul is not an independent substance; rather, humans constitute a single, unified substance (*Summa Theologica*, I, q. 75, a. 4). The rational soul alone does not define the human person; Aquinas identifies the person as the composite of matter and form (*Summa Contra Gentiles*, II, ch. 57). Although this form transcends materiality, without the body, it is incomplete—a state exemplified between death and resurrection (Pieper 1964, 46). Aquinas further emphasizes that the more a form transcends materiality, the greater the unity it creates with its matter (*Summa Contra Gentiles*, II, ch. 57).

Given the human form's spiritual nature, surpassing all other natural forms, human unity is unparalleled. As totalities, humans represent the highest level of natural substances (Stump 2003). This complexity suggests that humans are less substantially divisible than other beings. Unlike non-living substances, any division at the level of human unity results in the complete loss of the living substance (Kretzmann 1993). The human soul's intrinsic qualities—substantiality, individuality, rationality, simplicity, indivisibility, and incorruptibility—define this unity beyond physical wholeness. The soul's inherent selfhood, always present to itself, establishes a completeness fundamental to human nature (Pieper 1964, 50). Rational self-awareness and self-mastery, exclusive to humans, signify the apex of substantial forms embodying true selfhood. This transcendence allows the soul to communicate its being into the body, organizing human matter and manifesting unity through self-awareness and intellectual reflection (*Summa Theologica*, I, q. 76, a. 1).

Aquinas describes the body as "the organ of the soul," emphasizing that human unity is natural, with every power and structure oriented towards rational activity (*Summa Theologica*, I, q. 76, a. 1). Spiritually informed human acts are thus expressions of the person, demonstrating the soul-body unity (Stump 2003, 202). In Thomistic and Neo-Thomistic thought, the concepts of totalities or wholes provide critical insights into human unity, particularly concerning the brain's role (Pieper 1964, 47). Here, the whole is ontologically before its parts, meaning that the individual's complete unity—soul and body—takes precedence over any single component. Each part exists dependently within the organism's unity. For humans, this implies that organs and functions find meaning within the unified whole, ultimately operating as "instruments" that contribute to expressing the complete person (*Summa Contra Gentiles*, II, ch. 57; Stump 2003, 204).

Unity of Substance

The Thomistic view defines the unity of the human person by integrating a specific form with a particular matter, creating a completeness or wholeness unique to human beings (Summa Theologica, I, q. 76, a. 1). By these standards, the human person exemplifies a metaphysical unity surpassing all other natural substances (Pieper 1964, 55). The human person, encompassing both soul and body, exists as a unified whole through a singular act of spiritual being originating from the soul and expressed in the body. This spiritual principle, the source of unity, fully actualizes selfhood and completeness, realized through self-awareness and self-governance (Summa Contra Gentiles, II, ch. 57).

The human form uniquely unites with a particular material body while also participating in a transcendent unity that exceeds that of lesser beings (Stump 2003, 199). In this framework, the soul does not simply reside in the body; instead, as the substantial form, it actualizes and organizes the body as a unified whole (Summa Theologica, I, q. 75, a. 4). Thomistic philosophy holds that the soul permeates the entire body, presupposing the unity of the body as a whole (Kretzmann 1993, 201). Modern neuroscientific knowledge has furthered this understanding by identifying the brain as central to the unity of the human person. The brain, acting as a critical hub, facilitates integrating immaterial aspects of selfhood and the material body, harmonizing the individual as one cohesive substance (Finger 1994, 52). Modern neuroscience highlights the brain's pivotal role as a locus of sensory, cognitive, and bodily function integration, enabling the expression of a unified human essence. From a biological perspective, the brain has been approached as an optimizer and the Free-Energy Principle suggests the possible unification of several global brain theories within a free-energy framework (Friston 2010). The integration of Thomistic philosophy with contemporary scientific insights reflects an enduring commitment to understanding the unity and wholeness of the human person.

Biological Unity

In biological terms, the brain is the central organ that integrates and regulates the entire organism, enabling the life and unity of the self. As a singular, functionally unifying synthesis of the body, the brain ensures homeostasis and coordinates interactions across physiological systems, thus sustaining the biological unity of the person. It is an irreplaceable hub that allows for the unique interaction between the spiritual soul and the material body, aligning with the human person's complex unity. By orchestrating various bodily functions and facilitating the expression of the soul's powers through embodiment, the brain fundamentally contributes to the coherence and unity of the entire organism.

Thomas Aquinas regarded the head as vital, acknowledging it as central to consciousness, intellect, and the soul's faculties. While Aquinas lacked modern neuroscience, his philosophical anthropology prefigures contemporary insights into the brain's integrative role. Aquinas affirmed that the soul is the substantial form of the body, actualizing it and unifying its powers into a cohesive whole (*Summa Theologica*, I, q. 76, a. 1). This emphasis on the head's role resonates with modern neuroscience, which recognizes the brain's centrality in the unity of the human person. From the soul's metaphysical unity to its powers' experiential coherence, the brain is pivotal in maintaining human coherence and functioning (Pieper 1964, 55). Aquinas observed an "overflow" (*redundantia*) of one power into another, highlighting how the lower human powers, though distinct, remain unified and rational in their human expression (*Summa Contra Gentiles*, II, ch. 57). In this sense, even immaterial spiritual faculties rely on the brain's functioning for their expression in the embodied person (Kretzmann 1993, 201). Thus, Aquinas's perspective aligns with contemporary neuroscience, reinforcing the brain's indispensable role in supporting the unity and life of the human person across multiple levels of existence and experience.

Psychological Unity

The Thomistic standpoint expresses the unity of the human person by integrating sensory perception, rational thought, emotion, and the physical and metaphysical. Today, we understand the brain as a crucial organ in coordinating these functions. Aquinas's perspective acknowledges that human faculties—sensation, cognition, and affect—work harmoniously within the unity of body and soul. While he acknowledged the role of sensory organs and the brain in facilitating imagination and memory, which support the intellect and will, Aquinas did not explicitly conceptualize the brain as a central organ unifying human faculties as modern neuroscience does (*Summa Theologica*, I, q. 76, a. 1). In contemporary cognitive science, theories such as "4E cognition" (embodied, embedded, enacted, and extended cognition) suggest that psychological unity extends beyond the brain, involving interactions with the environment and bodily states. While this framework differs from Thomistic philosophy, it resonates with Aquinas's recognition of emotions as integral to human consciousness and intelligence. Aquinas understood emotions, or passions, not merely as physiological responses but as essential to rational life. In *Summa Theologiae* I-II, q. 22-48, he examines how passions operate within the soul and influence reason, understanding them as morally relevant and integral to moral virtues. Aquinas held that passions could be morally good when aligned with reason and the will, indicating their significant role in pursuing virtue. Far from being obstacles, rightly ordered passions contribute to moral excellence and the exercise of intelligence.

In *Summa Theologiae* I, q. 81, a. 3, Aquinas identifies the sensitive appetite (associated with passions) as subordinate to reason but capable of participating in rational life through the will's governance. He emphasized that when adequately oriented, passions contribute to the exercise of intelligence and the perfection of virtue. Unlike the later Cartesian view, which regarded emotions as mechanical or physiological responses, Aquinas considered passions deeply connected to the intellectual and moral life. As Eleonore Stump (2003) observes, Aquinas's approach to emotions avoids the extremes of devaluing emotions as irrational or allowing them to dominate reason. Diana Fritz Cates (1999) emphasizes that Aquinas viewed emotions as responses informed by cognition, integral to human experience and interaction with the world. This nuanced view places emotions within a rational and moral framework, recognizing their role in personal and ethical development.

Unity of Selfhood

Thomistic and Neo-Thomistic perspectives emphasize the brain's importance in contributing to personal unity, providing a coherent sense of self while grounding the essential foundation of selfhood in the soul. Aquinas refers to the body as "the organ of the soul," emphasizing that human unity is natural, with all powers and physical structures directed toward rational activity (*Summa Theologiae*, I, q. 76, a. 1). For Aquinas, though spiritually informed, human actions are not merely acts of a spirit but instead of a unified person, reflecting the integration of soul and body. The persistence of self-identity and self-awareness over time illustrates the complex unity of sensory and rational faculties in human experience. The unity of selfhood, while rooted in the soul, is closely related to psychological unity. The brain plays a dynamic role in constructing and sustaining a cohesive sense of identity through executive functions like self-awareness and self-governance, which support the individuality and selfhood of the person. Thomistic scholars have noted that Aquinas described the brain as the "primary seat of sensation" (*Summa Theologiae,* I, q. 76, a. 5), emphasizing its role in sensory integration. While Aquinas did not directly associate the brain with selfhood or intellectual acts, its function in mediating sensory data is indispensable to the soul's higher faculties. Neo-Thomist interpretations, such as those by Jacques Maritain, emphasize the brain's role in facilitating sensory and rational functions. However, they uphold that the soul is the ultimate principle unifying these functions into a cohesive self.

Aquinas's description of the body as the "organ of the soul" (*Summa Theologiae,* I, q. 76, a. 1) reflects his holistic view, where body and soul are not separate entities but a composite unity, with bodily actions serving as expressions of the whole person. This view supports the notion that, while spiritually informed, human actions are a unified person's acts. The persistence of identity and self-awareness, as discussed in Aquinas's framework, derives from the rational soul, which is the seat of intellectual and volitional acts (*Summa Theologiae,* I, q. 75, a. 6). Contemporary Neo-Thomistic scholars have extended this to consider the brain's role in enabling continuous self-awareness through executive functions. However, Aquinas did not conceptualize selfhood with these neurological functions in mind. Aquinas's view that the soul unifies the body aligns with the idea that selfhood is rooted in the spiritual soul, allowing the integrated experience of a singular self. This perspective aligns with interpretations by scholars such as Robert Pasnau (2002, 105), who explains that, for Aquinas, the soul integrates the powers of sensation and intellect, unifying them into a cohesive self.

Unity of Action

Unity of action is essential to understanding how humans function within natural and social contexts, including moral and ethical ones. According to Thomas Aquinas, moral actions and decisions emerge from an integrated harmony of cognitive and emotional processes within the human person, guided by the rational soul. In *Summa Theologiae* I-II, q. 24, Aquinas discusses how moral decisions involve both reason and passions, with reason guiding passions toward good actions. While Aquinas did not attribute a central role to the brain as modern neuroscience does, he emphasized that moral unity involves cognitive and affective dimensions aligned with reason and will. Contemporary neuroscience demonstrates that the brain supports coherent decision-making through brain-wide neural networks and executive functions. These physical processes align with Aquinas's recognition of the body's role in enabling the rational soul to govern ethical actions. Moral decisions, then, are not isolated events but products of the brain's integrated functions, deeply connected to the transphysical dimension of the person—an aspect Aquinas would ascribe to the soul's influence.

Drawing from Thomistic principles, Jacques Maritain suggests that the brain functions as an instrumental cause through which the rational soul integrates sensory and cognitive processes, ensuring moral unity. While the brain mediates perceptual and intellectual activity, the rational soul remains the ultimate source of moral direction and personal integrity. In this view, the unity of action reflects the broader unity of the human person, encompassing metaphysical, biological, psychological, and bodily dimensions. Aquinas posits that the rational soul, as the form of the body, influences moral and cognitive unity, though he does not assign this function to the brain specifically. As Robert Pasnau (2002) observes, Aquinas regarded ethical action as a holistic expression of the soul's rational and volitional powers, underscoring that moral unity is not solely a product of physical processes but of the rational soul's governance.

Unity of Simplicity (Spiritual Unity)

Aquinas posits that the soul is "always actually present to itself" by virtue of its simple existence (*Summa Theologiae*, I, q. 76, a. 1). While the brain is a complex organ and the pinnacle of biological synthesis, it is an essential substrate through which the spiritual soul expresses itself. In Thomistic philosophy, the unity inherent in the human soul—characterized by simplicity, indivisibility, and incorruptibility—transcends any form of physical wholeness, emphasizing the profound integration of spiritual and material aspects within the human person. This simplicity and indivisibility distinguish the soul from the body, underscoring a metaphysical unity that transcends physicality. Scholars such as Eleonore Stump (2003) discuss how Aquinas's view of the soul's simplicity supports his broader view of human unity. Although the soul itself is simple and incorruptible, its faculties find expression through the organized complexity of the body, particularly the brain, which Aquinas identified as instrumental for sensory processing (*Summa Theologiae*, I, q. 76, a. 5).

The brain's coordinated functions and connectivity facilitate the expression of the soul's unity and capacities, reflecting the human person's metaphysical unity in its embodied state. Neo-Thomistic interpretations suggest that the complete and irreversible loss of brain function signifies the breakdown of the person's embodied unity, rendering the soul's animation of the body impossible. While Aquinas himself did not frame the brain in these terms, his principles imply that bodily integrity is necessary for the soul to actualize its powers. Aquinas does not assign the brain a central role as the substrate of the soul's unity. As Robert Pasnau (2002) notes, Aquinas situated the source of personal unity in the rational soul, which harmonizes sensory and intellectual faculties into a cohesive whole. This view underscores that while the brain mediates sensory and cognitive functions, the soul remains the ultimate principle of unity, simplicity, and integration within the human person.

The Neural Basis of the Unity of the Human Person

While theological and philosophical perspectives offer foundational insights into the unity of the human person, neuroscience provides additional clarity on the brain's essential role in coordinating this unity. The brain—the central organ of the nervous system—governs consciousness and integrates bodily functions at the highest hierarchical level, aligning traditional perspectives with contemporary neuroscientific understanding. Neuroscientific research affirms the brain's integral connection to consciousness and the regulation of bodily functions. It is the epicenter of cognitive processes, housing neural networks that underlie subjective experience, thought, and self-awareness. Various brain regions contribute to facets of consciousness, forming a complex interplay that provides the biological foundation for a unified sense of self essential to the human person.

Beyond consciousness, the brain integrates and regulates diverse bodily functions. It coordinates motor control, sensory perception, and communication across physiological systems through large-scale neural networks, maintaining homeostasis and enabling adaptive responses to internal and external demands. Hippocrates, the father of medicine, recognized the brain as the seat of consciousness and the center of intelligence, perception, and volition. His insights foreshadow modern neuroscience's emphasis on the brain's role in shaping identity, as memories, emotions, a sense of embodiment, and personality traits relate to its complex circuitry. These neural patterns contribute to the coherence and uniqueness of each individual, underscoring the brain's role in the unity of the human person. This unity extends beyond consciousness to include bodily integration, with consciousness and physiological processes sharing a unified neural substrate. As the hub of consciousness and the regulator of bodily functions, the brain mediates the holistic nature of human existence.

Toward a Brain Metafunction Framework

Recent advancements in neuroscience and technology have revealed limitations in traditional frameworks for understanding death and personhood, sparking bioethical debates. These debates often divide into biologicist views, emphasizing the permanent loss of bodily integration, and mentalist or neocorticalist views, prioritizing the loss of consciousness. The brain metafunction framework (García, 1995a, 1995b, 1998, 1999, 2001) offers a unifying solution by recognizing the inseparability of the neural substrates supporting consciousness and bodily integration. This framework proposes:

- *Differentiated Hierarchical Levels*: The human brain operates across complex hierarchical levels, integrating functions that maintain the unity of the human organism.
- *Functional Integration*: Advances in neuroscience highlight the brain's role in unifying sensory, cognitive, homeostatic / allostatic, and motor functions, forming the foundation for human experience and action.
- *Distributed Consciousness Substrate*: Consciousness not supported by isolated regions but from the collaboration of brain-wide networks, including unconscious processes that inform conscious content.
- *Integration with Bodily Systems*: Neural processes supporting cognitive processes and bodily integration are closely linked to communication systems such as the endocrine, immune, circulatory, and autonomic nervous systems, forming a unified physiological network.

The brain metafunction framework enables a multidisciplinary articulation of the unity of the human person. Neuroscience illuminates the brain's integrative functions, theology and philosophy provides a metaphysical understanding of personhood, and bioethics bridges these insights to address practical issues such as the criteria for death, disorders of consciousness, and the ethical implications of emerging technologies. Together, these disciplines form a cohesive understanding of the human person as a unified being, where consciousness and bodily integration are inseparably intertwined. Let us look at these proposals in more detail.

Differentiated Hierarchical Levels in the Human Organism

A *univocal* treatment of hierarchical levels in the human organism risks overlooking substantial neuroscientific evidence highlighting the brain's pivotal role in orchestrating and unifying the functions of the human body and person. The brain is not just one organ among many; it is the essential structure synthesizing biological, psychosocial, and metaphysical dimensions of human existence. Neglecting this role fosters reductionist views that fail to capture the profound integrative processes underlying human experience. As Gerald Edelman (1989) argued in his theory of neural Darwinism, the brain continuously adapts, integrating sensory and cognitive experiences to generate a unified perception of self. This aspect aligns with Antonio Damasio's (1999) view that brain structures, particularly the brainstem and higher cortical areas, maintain the core unity of the person by linking bodily regulation and emotions with self-awareness.

Recognizing the brain's hierarchical primacy is also essential for interdisciplinary discourse, particularly at the intersection of neuroscience, ethics, theology, and philosophy. Not acknowledging this privileged role limits the depth of discussions on bioethical issues and human personhood. For instance, Alan Shewmon (1985) upheld the conventional view of brain death, portraying the brain as irreplaceable in coordinating somatic integration—a position he later revised. Similarly, Francisco Varela, Evan Thompson, and Eleanor Rosch (1991) emphasize the enactive approach to cognition, showing how the brain dynamically integrates bodily and environmental inputs to support action and subjective experience. Many neuroscientists and philosophers increasingly regard the brain as the central organ of human unity, a perspective that is crucial for fostering a comprehensive understanding of human nature (Friston 2010; Pasnau 2002; Tononi and Koch 2015; Damasio 1999; Bernat 2018, S19-S21; Condic 2016, 257-278; Moschella 2019, 437-453; McGee et al. 2023, 434-452; Edelman 1989; Varela et al. 1991).

Acknowledging the brain's integrative role provides a foundation for addressing pressing bioethical concerns with greater depth and nuance. Moving beyond a uniform treatment of bodily hierarchies and emphasizing the brain's coordinating function enriches discourse on human personhood. For example, Stanislas Dehaene's (2014) research on consciousness reveals neural mechanisms related to how the brain binds diverse cognitive processes into a unified experience, reinforcing the idea of a centralized governing system. Additionally, frameworks such as the *brain metafunction model* reconcile the autonomy of individual organs with the brain's higher-order integration of person-level functions. This approach underscores the interplay between specialized brain regions and whole-brain activity, linking localized neural tasks with the holistic experiences that define personhood. As neuroscience advances our understanding of brain structure and function, this perspective offers a robust model for addressing ethical and philosophical questions concerning the unity of the human person.

What Function or Set of Functions Determines the Criticality of the Brain?

The lack of a precise definition of which functions determine the brain's criticality poses challenges across philosophical, theological, bioethical, and medical domains. These challenges are particularly evident in debates surrounding neocortical (reductionist, mentalist) perspectives, whole-body integration models, and cardiorespiratory (biologicist) rationales. *Criticality* refers to the brain's essential role in sustaining life, enabling consciousness, and supporting the capacities defining personhood. However, failing to account for the interdependence of these functions has led to conceptual gaps and ethical dilemmas, particularly in discussions of brain death and neurological impairment. The brain metafunction framework provides a unified perspective integrating cognitive, physiological, and regulatory functions to address these challenges. This framework, supported by advances in neuroscience and systems biology, challenges reductionist models by emphasizing the brain's holistic role in coordinating interconnected processes—including memory, sensory perception, motor coordination, autonomic regulation, and the integration of bodily functions. Rather than isolating specific neural functions, it demonstrates how disruptions in neural connectivity can lead to cascading effects on consciousness, self-awareness, and bodily integration.

Defining which brain functions are critical is not merely a theoretical question; it has profound ethical implications. Ambiguities in defining critical brain functions impact end-of-life decisions, including the determination of death and personhood—particularly in cases such as a persistent vegetative state, advanced dementia, and minimally conscious states. Without a consistent framework, inconsistencies arise in determining brain death and assessing moral and legal personhood. The brain metafunction framework helps resolve these ambiguities by offering a scientifically and ethically robust model. Research on brain networks and functional connectivity supports a more comprehensive understanding of the brain's role in sustaining life and consciousness, bridging mentalist and biologicist perspectives. By emphasizing the integrative functions of the brain, this approach provides a more precise conceptual foundation for defining death, personhood, and related ethical concerns. Unlike incomplete or compartmentalized definitions, the *metafunction model* captures the interconnected nature of neural activity, fostering greater clarity in interdisciplinary debates.

The "Content of Consciousness": Beyond the Cerebral Neocortex

Recent decades of neuroscience research underscore the extensive role of brain structures beyond the cerebral neocortex in supporting consciousness and integrating bodily functions. Foundational observations have long indicated that subcortical regions significantly contribute to sensory experiences such as pain, thirst, and hunger, with the cortex acting in a modulatory rather than exclusive capacity (Bromm & Desmedt, 1995; Crick, 1984). For instance, the thalamus is critical in managing attentional focus and sensory filtering before cortical processing, demonstrating the importance of subcortical structures in shaping conscious experience (Crick, 1984; Llinás & Paré, 1991). More recent research reinforces this distributed model, showing that brainstem structures—including the reticular formation and midbrain—play a crucial role in core consciousness and wakefulness (Parvizi & Damasio, 2001; Merker, 2007).

Differentiation within motor systems, such as between the pyramidal and extrapyramidal pathways, further illustrates the coordination of cortical and subcortical systems in both conscious and unconscious motor functions (Freeman, 1995). This distributed perspective aligns with the Integrated Information Theory (IIT) (Tononi & Edelman, 1998) and Global Neuronal Workspace Theory (GNWT) (Dehaene & Changeux, 2011), both of which argue that consciousness requires the dynamic interaction of multiple brain regions, not just cortical processing. The interoceptive nervous system (INS) exemplifies this integrative model. Spanning both peripheral and central pathways, the INS includes structures such as the vagus nerve and lamina I, which connect autonomic regulation with cortical regions like the insula and cingulate cortex. The INS continuously monitors the body's internal state, integrating physiological regulation, emotional states, and cognitive functions (Craig, 2002). This more profound understanding of autonomic regulation supports Damasio's (1999) argument that emotion (related to homeostatic regulation) is foundational to subjective awareness. Additionally, predictive models of interoception suggest that bodily awareness is a continuous process of inference, where the brain actively constructs perceptions of internal states based on physiological feedback (Seth, 2013).

Relying exclusively on the cerebral neocortex as the sole basis of consciousness reflects a reductionist view that oversimplifies neural complexity. While the neocortex specialization relates to advanced reasoning, sensory perception, and linguistic abilities, consciousness requires the coordinated activity of subcortical structures such as the thalamus, limbic system, and brainstem. These subcortical regions are integral to emotional processing, memory formation, and sensory integration. For instance, the limbic system—including the amygdala and hippocampus—plays a central role in emotional and memory processing, reinforcing the importance of subcortical regions in shaping conscious experience.

Studies of global brain networks, such as the default mode network (DMN) and salience network, further demonstrate that nearly the entire brain participates in conscious experience (Raichle, 2015; Seeley et al., 2007; Buckner et al., 2008). These networks integrate cortical and subcortical regions, reinforcing the view that consciousness requires large-scale neural interactions rather than isolated cortical activity. Foundational insights into distributed brain processes by Mesulam (1990) and Freeman (1995) provide historical context for understanding the integrative nature of these networks. Recognizing this interplay enhances our understanding of consciousness as a dynamic and unified phenomenon, integrating bodily, cognitive, and emotional dimensions to enrich human experience.

The Subcortical Brain Structures: Beyond Arousal

The neurobiology of arousal encompasses complex mechanisms involving the brainstem's reticular activating system (RAS), neurotransmitter pathways, and cortical modulation. Understanding these interconnected processes is essential for a comprehensive view of how neural mechanisms sustain consciousness. Arousal is not a simple, isolated function; instead, it is a dynamic state, integrating wakefulness, sensory processing, and bodily regulation into the broader framework of conscious experience. The development of the human brain, particularly the expansion of the cerebral cortex, is marked by deep anatomical and functional interconnections between cortical and subcortical structures. Recent findings highlight the functional interplay between the brainstem and cortex, demonstrating how subcortical structures contribute not only to arousal but also to conscious awareness itself (Hansen et al., 2024). This research reinforces the growing recognition that subcortical regions, particularly the thalamus and brainstem, are integral to the persistence and quality of conscious states.

The brain metafunctional framework redefines arousal as a multi-dimensional component of consciousness rather than merely a trigger for wakefulness. Arousal extends beyond basic neural activation, engaging widespread networks that integrate neurotransmitter pathways, sensory processing, attention, homeostasis, and alertness. These mechanisms shape perceptual awareness, cognitive processing, and behavioral responses, significantly enriching the depth and content of conscious experience. Overlooking arousal's integrative role in consciousness leads to an incomplete understanding of how neurobiological processes shape human experience. A comprehensive approach to arousal, particularly through the lens of cortical and subcortical network integration, is crucial for both philosophical inquiries into consciousness and biomedical approaches to arousal-related disorders. For example, Hansen et al. (2024) underscore the interdependence of the brainstem and cortical networks, illustrating the pivotal role of subcortical structures in sustaining conscious states. This integrative model refines existing interpretations of consciousness and informs medical advancements in treating disorders of consciousness such as coma, minimally conscious states, and arousal dysfunctions.

Beyond neurobiology, unconventional cases such as anencephaly and hydrocephaly challenge traditional cortical-centered models of consciousness. Some individuals with severely reduced or absent neocortical structures still exhibit behavioral responses or signs of awareness, suggesting that subcortical structures play a more significant role than previously assumed (Merker, 2007). These cases raise important theoretical and interdisciplinary questions: If consciousness can manifest without extensive cortical involvement, what does this imply about the neural basis and nature of conscious experience? Such questions invite broader discussions, including interdisciplinary perspectives on consciousness that extend beyond strictly neurobiological models. The chapter on Consciousness explores contemporary contributions by Damasio (1999), Parvizi, and Damasio (2001), and others, pointing to neurobiological underpinnings of consciousness involving broader neural substrates. These perspectives open avenues for dialogue between neuroscience, philosophy, and consciousness studies, allowing new interdisciplinary models to emerge without necessarily contradicting established neuroscientific findings.

Brain-Related Bodily Integration Processes Are Not Brainstem-Centered

Contemporary neuroscience underscores the complexity of autonomic, endocrine, immune, and neural integration, revealing intricate connections between higher limbic and cortical structures (Benarroch, 1993; Critchley & Harrison, 2013; Hansen et al., 2024). This new understanding highlights the need for a unified sense of brain function. While the brainstem plays a foundational role in regulating essential autonomic processes such as heartbeat and respiration, viewing autonomic integration as solely brainstem-centered overlooks the hierarchical and distributed nature of bodily communication and autonomic control systems (Parvizi & Damasio, 2001; Porges, 2007). Neuroscientific models emphasize that the brainstem, essential for basic autonomic and arousal functions, does not operate in isolation (Parvizi & Damasio, 2001). Instead, its interaction with limbic and cortical structures contributes to higher-order autonomic regulation and cognitive-affective integration (Thayer & Lane, 2009). Hierarchically higher limbic structures, such as the amygdala and hypothalamus, along with cortical regions—particularly the prefrontal cortex—play integral roles in modulating autonomic responses to emotional, perceptual, and cognitive stimuli (Critchley & Harrison, 2013; Hansen et al., 2024). The limbic system is a crucial intermediary, linking emotional processing to bodily regulation (Damasio, 1999). Emotional experiences and behaviors shape autonomic responses through limbic-cortical interactions, while the prefrontal cortex exerts top-down control, influencing autonomic function through executive processes such as decision-making, attention, and stress regulation (Damasio et al., 1991; Thayer & Lane, 2009). Restricting autonomic control to the brainstem fails to capture the extensive influence of cognitive processes on autonomic regulation, resulting in an incomplete understanding of the interaction between mental and bodily processes (Benarroch, 1993; Berntson et al., 1991).

Research on large-scale brain networks, such as the Central Autonomic Network (CAN), further highlights the integration of limbic and cortical structures with the brainstem in modulating autonomic responses (Benarroch, 1993). This distributed network supports a holistic view of autonomic function, emphasizing that the brain operates as a cohesive system rather than a set of isolated modules. Parvizi and Damasio (2001) argue that while the brainstem is indispensable for consciousness and homeostatic regulation, its role is inseparable from broader cortico-limbic networks that provide the contextual and cognitive modulation necessary for complex autonomic functions. Historically, compartmentalizing autonomic functions to the brainstem alone has oversimplified the deeply interconnected nature of brain structures involved in autonomic regulation (Hansen et al., 2024; Thayer & Lane, 2009). Hansen et al. (2024) demonstrated that multiple organizational features of cortical activity can be traced back to the brainstem. A more comprehensive approach reveals the brain as an integrated system, where dynamic interactions among cortical, limbic, and brainstem regions collectively shape the regulation of internal states (Damasio, 1994, 1999; Critchley & Harrison, 2013).

This integrated model extends beyond basic homeostatic reflexes to include allostatic responses, which connect abstract cognitive processes with internal bodily states (Damasio, 1999; Porges, 2007). Autonomic regulation, influenced by cognitive domains, often manifests in conscious thoughts and feelings that predictively adjust bodily responses to anticipated challenges (Damasio, 1994; Thayer & Lane, 2009). The *somatic marker hypothesis* (Damasio et al., 1991) provides evidence that bodily states, including autonomic responses, shape decision-making processes through interactions between the prefrontal cortex and limbic structures. Within a *brain metafunctional framework*, the dynamic interplay between cognition and internal states illuminates the complex, bidirectional coupling essential to human psychophysiology. The perspective offered by Parvizi & Damasio (2001) reinforces this view by demonstrating that consciousness—and, by extension, sophisticated autonomic integration—depends not only on the integrity of the brainstem but also on the continuous interaction between subcortical and cortical structures.

Brain-Related Bodily Integration Processes and Consciousness Are Coextensive

Contemporary neuroscience underscores the complexity of autonomic, endocrine, immune, and neural integration, highlighting the interconnected roles of brainstem, limbic, and cortical structures in bodily regulation. While the brainstem plays a foundational role in autonomic processes such as heartbeat and respiration, viewing autonomic integration as solely brainstem-centered oversimplifies autonomic control's hierarchical and distributed nature (Benarroch, 1993). Higher-order limbic structures, such as the amygdala and hypothalamus, and cortical regions—particularly the prefrontal cortex—are integral to modulating autonomic responses to emotional, perceptual, and cognitive stimuli (Hansen et al., 2024). The limbic system is a crucial intermediary, linking emotional processing to bodily regulation (Critchley & Garfinkel, 2017). Emotional experiences influence autonomic responses, while the prefrontal cortex exerts top-down control, shaping autonomic function through executive processes such as decision-making, attention, and stress regulation (Thayer & Lane, 2009). Reducing autonomic control of the brainstem alone overlooks the extensive influence of cognitive processes on bodily regulation, leading to an incomplete understanding of mind-body interactions.

Research on large-scale brain networks, such as the Central Autonomic Network (CAN), highlights the integration of limbic and cortical structures with the brainstem in modulating autonomic responses (Benarroch, 1993). The CAN comprises key regions—including the anterior cingulate cortex, insula, hypothalamus, amygdala, and brainstem nuclei—that work together to regulate autonomic activity in response to emotional, cognitive, and environmental stimuli (Kleckner et al., 2017). Kleckner et al. (2017) further emphasize how large-scale brain systems support allostasis and interoception, integrating bodily and cognitive processes to maintain internal balance. This distributed network supports a holistic view of autonomic function, reinforcing that the brain operates as an interconnected system rather than a series of isolated modules.

Historically, compartmentalizing autonomic functions to the brainstem has oversimplified the deeply interconnected nature of brain structures involved in autonomic regulation. A more comprehensive approach reveals the brain's role as an integrated system in which dynamic interactions among cortical, limbic, and brainstem regions collectively shape the regulation of internal states (Hansen et al., 2024). This integrated model extends beyond basic homeostatic reflexes, encompassing allostatic responses that link cognitive processes and mental representations with internal bodily states (Sterling & Eyer, 1988). Autonomic regulation, influenced by cognitive domains, often manifests as conscious thoughts and feelings that predictively adjust bodily responses to anticipated challenges (Critchley & Garfinkel, 2017). Within a brain metafunctional framework, the dynamic interplay between cognition and internal states illuminates the complex bidirectional coupling essential to human psychophysiology, offering new insights into how the brain achieves holistic bodily integration (Thayer & Lane, 2009).

Unified Brain-Body Regulation

Heart-brain regulatory interactions exemplify the intricate interconnectedness of brain functions. Recent research identifies a specialized network, the Complex Central Autonomic Network (CCAN), which provides sophisticated autonomic control over the heart. The CCAN spans the frontal, temporal, parietal, and occipital lobes, incorporating key regions such as the cingulate gyrus, orbitofrontal cortex, hippocampus, and precuneus. Notably, at rest, the CCAN excludes traditional sympathovagal control regions, such as the insular cortex, amygdala, and thalamus, suggesting that it plays a distinct role in facilitating complex vagal mechanisms linked to emotional processing (Hansen et al., 2024). The Central Autonomic Network (CAN) further elucidates the reciprocal interactions between cortical activity and cardiovascular function, emphasizing the foundational role of the brainstem. Within the brainstem, the nucleus of the solitary tract (NTS) integrates vagal input and relays signals to higher-order brain regions, including the insular cortex, anterior cingulate cortex, and amygdala. These forebrain structures, in turn, modulate CAN activity, contributing to homeostatic and interoceptive functions essential for autonomic nervous system (ANS) regulation (Benarroch, 1993; Kleckner et al., 2017).

Cortical-subcortical interactions, particularly those involving the orbitofrontal cortex, medial prefrontal cortex, and amygdala, further influence brainstem mechanisms that regulate vital bodily functions. Autonomic regulation engages multiple subcortical regions, including the dorsal raphe nuclei, periaqueductal gray, and hypothalamus. Additionally, data about lateralization in ANS control indicates that the right insular cortex predominantly influences sympathetic functions, whereas left-hemisphere regions regulate parasympathetic responses (Thayer & Lane, 2009). The bidirectional pathways between the brainstem and higher-order regions underscore the intricate communication required for homeostasis and adaptive responses to environmental challenges.

Antonio and Hanna Damasio (2024) argue that conscious experiences emerge from the continuous monitoring and regulation of internal bodily states. They propose that homeostatic feelings form the foundation of consciousness, arising from the integration of cortical and subcortical structures within networks such as the CAN and CCAN. This perspective challenges outdated dichotomies that separate consciousness from autonomic regulation, presenting them as interdependent processes that incorporate homeostatic and emotional dimensions (Critchley & Garfinkel, 2017). Understanding how consciousness integrates with bodily regulation has profound clinical implications. Treating neurological and psychiatric conditions may benefit from interventions informed by the brain metafunction model, which addresses both cognitive and autonomic well-being. For instance, targeted therapies for anxiety, depression, and autonomic dysfunction could leverage this unified framework to restore brain-body integration. As evidence grows, the historical separation between consciousness and autonomic regulation becomes increasingly untenable, giving way to a more cohesive view of the brain's response to internal and external demands.

Disintegration at the Level of Unity is Crucial

Disintegration at the level of unity is fundamental to defining the loss of personhood and death, as these phenomena are inherently interconnected. A key objection to brain death arises from arguments suggesting that remaining bodily functions may still retain meaningful activity. However, analyses by scholars such as Bernat, Condic, Moschella, and McGee et al. (Bernat 2018, S19-S21; Condic 2016, 257-278; Moschella 2019, 437-453; McGee et al. 2023, 434-452) offer compelling rationales regarding the hierarchical organization of the human organism. These frameworks differentiate between "the whole organism" and the "organism as a whole," emphasizing the brain's central role in systemic unity. Condic identifies a hierarchy within the body, ranging from molecular and cellular to systemic levels. While lower levels can exhibit autonomous activity, the brain orchestrates the highest level of systemic integration. Bernat's mereological framework complements this view by distinguishing between the brain's role in maintaining organismal unity and the localized activity of its individual parts. These perspectives underscore the brain's hierarchical primacy in integrating consciousness and systemic functions, thus providing a neuroscientific foundation for personhood.

The brain metafunction framework builds on this understanding, defining the brain as the critical locus of organismal unity. The irreversible loss of brain metafunction—such as consciousness and systemic integration, understood as inseparable and irreducible—marks the disintegration of the organism as a unified whole and signifies death. In philosophical terms, this loss represents a substantial change: the complete cessation of the human person as a single living substance. From a neuroscientific perspective, this rationale highlights the inseparability of personhood and human death, reinforcing the brain's role as the nexus of metaphysical, psychosocial, and biological unity. For example, Moschella (2019) emphasizes that even seemingly autonomous bodily functions, like heartbeat or digestion, rely on the brain's integrative mechanisms for coherence within the living organism. This perspective has profound bioethical, philosophical, and theological implications, providing a robust platform for addressing debates on brain death and personhood.

The brain metafunction framework addresses traditional conceptual challenges by unifying these perspectives. It posits that the permanent cessation of brain metafunction corresponds to the loss of both personhood and life, resolving long-standing controversies about the definition of death. This framework also underscores the inseparability of consciousness and bodily regulation, positioning the brain as the biological cornerstone of human existence. By examining the biological foundations of personhood through this lens, we can refine the conceptual and practical approaches to ethical issues in medicine, theology, and philosophy.

What Defines the Loss of Personhood?

The loss of personhood is the irreversible [1] loss of the brain metafunction.

Conceptual Criterion

The brain metafunction is the dynamic and unified activity of brain systems, sub-regions, regions, and neurons, connected through structural and functional pathways. This activity integrates synchronized dynamics across spatial and temporal scales, supporting the unity of interdependent domains in the human person. It encompasses the following subfunctions [2]:

1. Consciousness and cognitive processes, including perception, attention, memory, and emotions.
2. Neural integration and regulation of the body's communication systems, including the nervous, endocrine, immune, and circulatory systems.
3. Higher-level regulation of the internal environment, encompassing homeostasis and allostasis.
4. Processing afferent input from and coordinating efferent responses to the organism and its environment.

5. Behavioral and adaptive interactions with the external world.
6. Integration of physiological processes and mental states.
7. Unified synthesis of these subfunctions into an irreducible whole.

The brain metafunction is an extraordinary example of intricate brain activity, reflecting its material complexity and systemic integration.[2] Some perspectives, including this author's, suggest it also reveals a transphysical dimension characterized by a unique and inherent causal interdependence. While this transphysical component offers significant explanatory power, its inclusion remains optional, providing a biomedical foundation compatible with diverse philosophical and theological viewpoints.

Note on Irreversible and Subfunctions

[1] The term "irreversible" is preferred in this context due to its conceptual and ontological clarity. While "permanent loss" may be used in clinical settings where all brain death criteria are met, "irreversible" more precisely denotes the non-contingent cessation of the brain's integrative role as defined in the Brain Metafunction Framework.

[2] The described brain metafunction subfunctions represent interconnected, dynamically integrated processes that span the entire system. These functions are not isolated but operate cohesively to maintain the unity and identity of the human person.

The biological substrate for the conceptual criterion

The loss of personhood is the permanent failure of the brain's critical metafunctional system, which integrates diverse neural processes into a unified whole. This metafunctional system is not a collection of isolated components but an intricately interconnected network with seamlessly integrated functional specialization across subsystems. These subsystems sustain consciousness, homeostasis, and the cohesive operation of the human person. This system's complete and permanent failure provides a sufficient criterion for determining the loss of personhood and death. The brain's metafunctional system is composed of the following subsystems:

Subsystem I: The Ascending Reticular Activation System (ARAS)

The ARAS forms the foundation for higher-order brain activity by maintaining wakefulness and alertness, which are essential for the brain's engagement in conscious processes. This subsystem encompasses:

a. The tegmentum of the rostral pons and midbrain,
b. The intralaminar and reticular nuclei of the thalamus, and
c. The posterior hypothalamus.

These interconnected structures ensure the brain remains receptive to sensory stimuli and capable of coordinating neural responses. While the ARAS specializes in arousal, its proper function depends on continuous input from and communication with higher cortical and limbic regions.

Subsystem II: The Limbic System and Associated Brainstem Nuclei

This subsystem integrates emotional, autonomic, and memory functions as a critical mediator between physiological regulation and behavioral adaptation. Its components include:

a. The hypothalamus, basal forebrain, and nucleus basalis of Meynert,
b. The amygdala, hippocampal complex, and cingulate cortex,
c. The orbitofrontal cortex, cingulum, olfactory cortex, and temporal pole,
d. Portions of the basal ganglia, reticular nuclei in the midbrain and rostral pons, and the parabrachial nucleus,
e. Additional nuclei include the pedunculopontine nucleus rostral raphe nuclei, periaqueductal gray, and nucleus locus ceruleus.

While this subsystem specializes in emotional and autonomic regulation, its functionality is deeply integrated with the ARAS for maintaining arousal and the cerebral cortex for processing and acting upon environmental stimuli. For example, the amygdala's role in processing fear and emotional learning requires bidirectional communication with the prefrontal cortex and brainstem autonomic centers to generate appropriate physiological and behavioral responses.

Subsystem III: The Cerebral Cortex, Thalamus, and Basal Ganglia

In collaboration with the thalamus and basal ganglia, the cerebral cortex supports higher-order cognitive processes, including perception, memory, decision-making, and voluntary motor control. These structures do not operate in isolation but rely on reciprocal interactions with the ARAS and the limbic system to execute complex tasks. Key distinguishing elements include:

a. The prefrontal cortex for executive functions and decision-making,
b. The thalamus for relaying and integrating sensory information,
c. The basal ganglia involve motor planning, adequate execution, and reward-based learning.

This subsystem exemplifies the interconnectedness of brain functions. For instance, the cerebral cortex depends on the thalamus for sensory input and the limbic system for emotional context. At the same time, the basal ganglia enable smooth coordination of motor responses and goal-directed behavior.

The Interconnected Nature of Subsystems

The brain's metafunctional system operates as a cohesive whole, with each subsystem contributing specialized capabilities indispensable to the person's unified operation. However, these subsystems are functionally inseparable. For instance:

The ARAS not only maintains wakefulness but also provides the foundational neural tone that supports the limbic system's emotional processing and the cerebral cortex's cognitive activity. Moreover, this subsystem contribution is intrinsic to consciousness as previously discussed. The limbic system, responsible for emotional regulation and memory, continuously interacts with the ARAS to modulate arousal and the cortex to influence decision-making and behavior. Additionally, it contributes to autonomic regulation. While specialized in higher-order cognition, the cerebral cortex relies on input from the ARAS for sustained attention and modulation and consciousness itself, and the limbic system for the emotional and motivational context of its activities.

This mutual dependence underscores that the brain's metafunction is not localized to any single region but emerges from the dynamic interplay among these subsystems. A failure in any one or more subsystems disrupts this interplay, leading to severe dysfunction or even the disintegration of the unified whole that defines personhood. The brain's metafunctional system exemplifies hypercomplex integration, where specialized subsystems merge to form an indivisible unity. The failure of this interconnected system represents the loss of personhood, as it signals the cessation of the brain's ability to sustain the fundamental dynamics of consciousness, homeostasis, and unified interaction with the environment. This framework provides a robust biological foundation for defining death, encompassing the full scope of personhood within its interconnected processes.

Conclusion

This final section has provided a neuroscientific perspective closely aligned with the unity of the human person by emphasizing the brain's unifying role in the body and person. This role is directly relevant to fundamental bioethical issues. The brain metafunction framework offers a compelling rationale, identifying the brain as the essential and highest hierarchical level in human organization. This framework clarifies the brain's irreplaceability and criticality, particularly regarding personhood and the definition of death. It designates the brain as the privileged organ where the human organism's essential "functional synthesis" occurs. The permanent destruction of the system responsible for generating this critical brain metafunction results in the loss of the organism's and person's fundamental unity and identity—a "uni-totality," where personhood and life are inseparable.

Initially proposed by the present author over two decades ago, the brain metafunction concept now finds increasing support in fundamental neuroscientific research, affirming the brain's unique status as the body's critical organ. This framework encapsulates the idea that consciousness, expressed through the brain in the living human, is inseparable from the brain's integrative processes, enabling the organism to function "as a whole." We have aimed to describe this essential, irreplaceable system with sufficient detail to encompass its central role in personhood. The brain systems supporting this brain metafunction are thus critical for maintaining personhood. Temporary or partial disruptions of this capacity, such as those seen in a coma, vegetative state, dementia, or minimally conscious state, do not equate to a permanent loss of personhood. This framework underscores the importance of distinguishing reversible dysfunctions from permanent loss in clinical and ethical decision-making.

We have demonstrated that the brain metafunction's comprehensive "whole-system" integration constitutes a qualitatively higher function—greater than the sum of its parts or subfunctions. This dynamic interplay among neural processes generates a higher-order functional level unique to the living brain. Through interdependence and mutual enrichment, these processes form a distinctive brain metafunction that serves as a crucial hub, uniting transphysical and material components of the human person. The permanent loss of this metafunction corresponds to the loss of personhood and coincides with human death. This conclusion provides a unified, robust foundation for addressing bioethical and philosophical debates about the definition of death and personhood. Building on this foundation, the next chapter will explore the implications of this framework for defining human death in greater detail, offering further clarity and interdisciplinary insights.

Note. Unless otherwise indicated, Scripture quotations are drawn from the King James Version (KJV), English Standard Version (ESV), or New International Version (NIV) of the Bible.

REFERENCES

Aquinas, T. (1920–1922). *Summa Theologica* (Fathers of the English Dominican Province, Trans.). (2nd rev. Ed.). London: Burns Oates & Washbourne. (Original work published 1273).

Aquinas, T. (1947). *Summa Theologica* (Fathers of the English Dominican Province, Trans.). New York: Benziger Bros. (Original work published 1273).

Aquinas, T. (1955). *Summa Contra Gentiles* (A. C. Pegis, Trans.). Vol. 2, Chapter 68, Section 2. Notre Dame, IN: University of Notre Dame Press. (Original work published 1259–1265).

Ariew, A., Cummins, R., & Perlman, M. (2002). *Functions: New Essays in the Philosophy of Psychology and Biology*. Oxford University Press.

Aristotle. (1924). *De Anima* (J. A. Smith, Trans.). Oxford University Press.

Aristotle. (1893). *Nicomachean Ethics* (F. H. Peters, Trans.) (5th ed.). London: Kegan Paul, Trench, & Trübner.

Aristotle. (n.d.). *Metaphysics* (W. D. Ross, Trans.). Oxford University Press.

Aristotle. (n.d.). *Parts of Animals* (W. Ogle, Trans.). Oxford University Press.

Assmann, J. (2002). *The mind of Egypt: History and meaning in the time of the Pharaohs* (A. Author, Trans.). Metropolitan Books/Henry Holt and Company. (Original work published 1996)

Avicenna. (1329). *The Canon of Medicine* [Manuscript]. Library of Congress. Retrieved from https://www.loc.gov/item/2021667339/

Avena-Koenigsberger, A., Misic, B., & Sporns, O. (2017). Communication dynamics in complex brain networks. *Nature Reviews Neuroscience, 19*, 17–33. https://doi.org/10.1038/nrn.2017.149

Barnard, C. N. (1967). The operation: A human cardiac transplant: An interim report of a successful operation performed at Groote Schuur Hospital, Cape Town. *South African Medical Journal, 41*, 1271–1274.

Barr, J. (1961). *The semantics of biblical language.* Oxford University Press. https://archive.org/details/semanticsof bibli0000barr

Barrett, L. F. (2017). *How Emotions Are Made: The Secret Life of the Brain.* Boston: Houghton Mifflin Harcourt.

Barrett, L. F., & Simmons, W. K. (2015). Interoceptive predictions in the brain. *Nature Reviews Neuroscience, 16*, 419–429. https://doi.org/10.1038/nrn3950

Beckwith, F. J. (2007). Defending life: A moral and legal case against abortion choice. Cambridge University Press. https://doi.org/10.1017/CBO9780511804885

Benarroch, E. E. (1993). The central autonomic network: Functional organization, dysfunction, and perspective. *Mayo Clinic Proceedings, 68*(10), 988–1001. https://doi.org/10.1016/S0025-6196(12)62272-1

Bernat, J. L. (2018). A conceptual justification for brain death. *Hastings Center Report, 48*(4), S19–S21. https://doi.org/ 10.1002/hast.946

Berntson, G. G., Cacioppo, J. T., & Quigley, K. S. (1991). Autonomic determinism: The modes of autonomic control, the doctrine of autonomic space, and the laws of autonomic constraint. *Psychological Review, 98*(4), 459–487. https://doi.org/10.1037/0033-295X.98.4.459

Berthelsen, P. G., & Cronqvist, M. (2003). The first intensive care unit in the world: Copenhagen 1953. *Acta Anaesthesiologica Scandinavica, 47*(10), 1190–1195. https://doi.org/10.1046/j.1399-6576.2003.00256.x

Boethius. (1918). *Theological Tractates: The Consolation of Philosophy* (H. F. Stewart & E. K. Rand, Trans.). Harvard University Press. (Loeb Classical Library) https://www.loebclassics.com/view/boethius-theological_tractates_contra_eutychen/1973/pb_LCL074.73.xml.

Boff, L. (1980). *La Resurrección de Cristo: Nuestra Resurrección en la Muerte* [The Resurrection of Christ: Our Resurrection in Death]. Santander: SAL TERRAE.

Bromm, B., & Desmedt, J. E. (1995). *Pain and the Brain: From Nociception to Cognition.* Raven Press.

Brugger, E. C. (2016). Are brain-dead individuals dead? Grounds for reasonable doubt. *Journal of Medicine and Philosophy, 41*(3), 329–350. https://doi.org/10.1093/jmp/jhw 003

Buckner, R. L., Andrews-Hanna, J. R., & Schacter, D. L. (2008). The brain's default network: Anatomy, function, and relevance to disease. *Annals of the New York Academy of Sciences, 1124*(1), 1–38. https://doi.org/10.1196/annals.1440.011

Burnyeat, M. F. (1982). The philosophical sense of Plato's *Phaedo*. *Proceedings of the British Academy*.

Cates, D. F. (1999). *The Emotions in the Philosophy of Aquinas*. Washington, DC: Georgetown University Press.

Chanes, L., & Barrett, L. F. (2016). Redefining the role of limbic areas in cortical processing. *Trends in Cognitive Sciences, 20*(2), 96–106. https://doi.org/10.1016/j.tics.2015.11.005

Chein, J. M., Moore, A. B., & Conway, A. R. A. (2011). Domain-general mechanisms of complex working memory span. *Neuroimage, 54*(1), 550–559. https://doi.org/10.1016/j.neuroimage.2010.07.067

Chiong, W. (2005). Brain death without definitions. *Hastings Center Report, 35*(6), 20–30. https://doi.org/10.1353/hcr.2005 .0105

Cohen, J. R., & D'Esposito, M. (2016). The segregation and integration of distinct brain networks and their relationship to cognition. *Journal of Neuroscience, 36*.

Cole, M. W., et al. (2013). Multi-task connectivity reveals flexible hubs for adaptive task control. *Nature Neuroscience, 16*(9), 1348–1355. https://doi.org/10.1038/nn.3470

Condic, M. L. (2016). Determination of death: A scientific perspective on biological integration. *Journal of Medicine and Philosophy, 41*(3), 257–278. https://doi.org/10.1093/jmp/jhw00 4

Cortez, M. A. (2010). *Theological anthropology: A guide for the perplexed*. T&T Clark International.

Cowan, N., et al. (2011). A neural region of abstract working memory. *Journal of Cognitive Neuroscience, 23*(10), 2852–2863. https://doi.org/10.1162/jocn.2011.21625

Craig, A. D. (2002). How do you feel? Interoception: The sense of the physiological condition of the body. *Nature Reviews Neuroscience, 3*(8), 655–666. https://doi.org/10.1038/nrn894

Crick, F. (1984). Function of the thalamic reticular complex: The searchlight hypothesis. Proceedings of the National Academy of Sciences of the United States of America, 81(14), 4586–4590. https://doi.org/10.1073/pnas.81.14.4586

Critchley, H. D., & Harrison, N. A. (2013). Visceral influences on brain and behavior. *Neuron, 77*(4), 624–638. https://doi.org/10.1016/j.neuron.2013.02.008

Critchley, H. D., & Garfinkel, S. N. (2017). Interoception and emotion. *Current Opinion in Psychology, 17*, 7–14. https://doi.org/10.1016/j.copsyc.2017.04.020

Cullmann, O. (1958). *Immortality or Resurrection? Christianity Today*, July 7.

Damasio, A. R. (1994). *Descartes' error: Emotion, reason, and the human brain.* Putnam.

Damasio, A. R. (1996). The somatic marker hypothesis and the possible functions of the prefrontal cortex. *Philosophical Transactions of the Royal Society of London. Series B, Biological Sciences, 351*(1346), 1413–1420. https://doi.org/10.1098/rstb.1996.0125

Damasio, A. R., Tranel, D., & Damasio, H. (1991). Somatic markers and the guidance of behavior: Theory and preliminary testing. In H. S. Levin, H. M. Eisenberg, & A. L. Benton (Eds.), *Frontal lobe function and dysfunction* (pp. 217–229). Oxford University Press.

Damasio, A. (1999). *The feeling of what happens: Body and emotion in the making of consciousness.* Harcourt Brace & Company.

Damasio, A., & Damasio, H. (2024). *Feeling and knowing: Making minds conscious.* Pantheon Books.

Dehaene, S., & Changeux, J. P. (2011). Experimental and theoretical approaches to conscious processing. *Neuron, 70*(2), 200–227. https://doi.org/10.1016/j.neuron.2011.03.018

Dehaene, S. (2014). *Consciousness and the brain: Deciphering how the brain codes our thoughts*. Viking.

Descartes, R. (1985). *Meditations on first philosophy* (J. Cottingham, Trans.). Cambridge University Press.

Des Chene, D. (2000). *Life's form: Late Aristotelian conceptions of the soul.* Cornell University Press.

Eccles, J. C. (1989). *Evolution of the brain: Creation of the self.* Routledge.

Edelman, G. M. (1989). *The remembered present: A biological theory of consciousness.* Basic Books.

Eichrodt, W. (1967). *Theology of the Old Testament* (Vol. 1, J. A. Baker, Trans.). Westminster John Knox Press.

Farah, M. J., & Heberlein, A. S. (2007). Personhood and neuroscience: Naturalizing or nihilating? *The American Journal of Bioethics, 7*(1), 37–48. https://doi.org/10.1080/1526 5160601064199

Finger, S. (1994). *Origins of neuroscience: A history of explorations into brain function.* Oxford University Press.

Finley, J. (2016). The extraordinary unity of the human being. Part of the 2016–17 St. Vincent de Paul Lecture and Concert Series. Retrieved January 1, 2024, from https://www.thomasaquinas.edu/news/dr-john-finley-extraordinary-unity-human-being

Freeman, W. J. (1995). *Societies of brains: A study in the neuroscience of love and hate.* Lawrence Erlbaum Associates.

Friston, K. (2010). The free-energy principle: A unified brain theory? *Nature Reviews Neuroscience, 11*(2), 127–138. https://doi.org/10.1038/nrn2787

Galen. (1916). *On the natural faculties* (A. J. Brock, Trans.). Harvard University Press.

García, O. D. (1995a). Reformulating death: Overcoming whole-brain inconsistencies. In C. Machado (Ed.), *Brain death* (pp. 79–92). Elsevier.

García, O. D. (1995b). Whole brain? Higher brain? A new formulation of death. In C. Machado (Ed.), *Brain death* (pp. 93–104). Elsevier.

García, O. D. (1998). Una reformulación completa y precisa de la muerte humana [A complete and precise reformulation of human death]. *Revista de Neurología, 26*(154), 1048–1053. https://pubmed.ncbi.nlm.nih.gov/9659470

García, O. D. (1999). A new formulation of death and its relevance to medical law. *Medicine and Law, 17*(4).

García, O. D. (2001). The unity of the human person: A central concept in the interface between medicine and theology. *The Linacre Quarterly, 68*(1), Article 5.

Gu, X., & FitzGerald, T. H. (2014). Interoceptive inference: Homeostasis and decision-making. *Trends in Cognitive Sciences, 18*(6), 269–270. https://doi.org/10.1016/j.tics.2014.02.001

Halevy, A., & Brody, B. (1993). Brain death: Reconciling definitions, criteria, and tests. *Annals of Internal Medicine, 119*(6), 519–525. https://doi.org/10.7326/0003-4819-119-6-199309150-00013

Hansen, J. Y., Cauzzo, S., Singh, K., García-Gomar, M. G., Shine, J. M., Bianciardi, M., & Mišić, B. (2024). Integrating brainstem and cortical functional architectures. *Nature Neuroscience, 27*(12), 2500–2511. https://doi.org/10.1038/s41 593-024-01787-0.

Hardie, W. F. R. (1976). Concepts of consciousness in Aristotle. *Mind, 85*(July), 388–411. https://doi.org/10.1093/mind/LXXXV.339.388

Hippocrates. (1923). *On the sacred disease*. In *Hippocrates: Volume II* (W. H. S. Jones, Trans.). Harvard University Press.

Honorary Secretary of the Conference of Medical Royal Colleges and Their Faculties in the United Kingdom. (1976). Diagnosis of brain death. *British Medical Journal, 2*(6045), 1187–1188. https://doi.org/10.1136/bmj.2.6045.1187

Honorary Secretary of the Conference of Medical Royal Colleges and Their Faculties in the United Kingdom. (1979). Diagnosis of death. *The Lancet, 313*(8110), 261–262. https://doi.org/10.1016/S0140-6736(79)90781-5

Jennett, B., & Plum, F. (1972). Persistent vegetative state after brain damage: A syndrome in search of a name. *The Lancet, 299*(7753), 734–737. https://doi.org/10.1016/S0140 -6736(72)90229-5

Johnson, L. S. M. (2017). Death by neurological criteria: Expert definitions and lay misgivings. *QJM: An International Journal of Medicine, 110*(5), 267–270. https://doi.org/10.1093/qjmed/hcw181

Jonas, H. (1966). *The phenomenon of life: Toward a philosophical biology.* Harper & Row.

Katsumi, Y., et al. (2023). Correspondence of functional connectivity gradients across human isocortex, cerebellum, and hippocampus. *Communications Biology, 6*(1), 401. https://doi.org/10.1038/s42003-023-04796-0

Kauffman, S. A. (1993). *The origins of order: Self-organization and selection in evolution.* Oxford University Press.

Kelly, W. D., et al. (1967). Allotransplantation of the pancreas and duodenum along with the kidney in diabetic nephropathy. *Surgery, 61*(6), 827–837. https://doi.org/10.1097/00007890-19 6801000-00034

Kelso, J. S. A. (2012). Multistability and metastability: Understanding dynamic coordination in the brain. *Philosophical Transactions of the Royal Society B, 367*(1591), 906–918. https://doi.org/10.1098/rstb.2011.0351

Kirk, G. S., Raven, J. E., & Schofield, M. (1983). *The Presocratic philosophers: A critical history with a selection of texts.* Cambridge University Press.

Kittel, G. (1965). *Theological dictionary of the New Testament.* Eerdmans.

Kleckner, I. R., et al. (2017). Evidence for a large-scale brain system supporting allostasis and interoception in humans. *Nature Human Behaviour, 1*, 0069. https://www.nature.com/articles/s41562-017-0069

Koelsch, S., et al. (2009). Functional architecture of verbal and tonal working memory: An fMRI study. *Human Brain Mapping, 30*(3), 859–873. https://doi.org/10.1002/hbm.205 50

Kretzmann, N. (1993). *The metaphysics of theism: Aquinas's natural theology in Summa Contra Gentiles.* Oxford University Press.

Lennox, J. G. (2001). *Aristotle's philosophy of biology: Studies in the origins of life science.* Cambridge University Press.

Lizza, J. P. (2004). The conceptual basis for brain death revisited. In C. Machado & D. A. Shewmon (Eds.), Brain death and disorders of consciousness (pp. 51–59). Advances in Experimental Medicine and Biology (Vol. 550). Springer. https://doi.org/10.1007/978-0-306-48526-8_5

Llinás, R. R., & Paré, D. (1991). Of dreaming and wakefulness. *Neuroscience, 44*(3), 521–535. https://doi.org/10.1016/0306-4522(91) 900 75-Y

Lloyd, G. E. R. (1975). *Magic, reason and experience: Studies in the origin and development of Greek science.* Cambridge University Press.

Locke, J. (1975). *An essay concerning human understanding.* Oxford University Press.

Longrigg, J. (1993). *Greek rational medicine: Philosophy and medicine from Alcmaeon to the Alexandrians.* Routledge.

MacDonald, S. (1990). *Being and goodness: The concept of the good in metaphysics and philosophical theology.* Cornell University Press.

Machado, C. (1999). Consciousness as a definition of death: Its appeal and complexity. *Clinical Electroencephalography, 30*(4), 156–164. https://doi.org/10.1177/155005949903000408

Majerus, S., et al. (2010). The commonality of neural networks for verbal and visual short-term memory. *Journal of Cognitive Neuroscience, 22*(11), 2570–2593. https://doi.org/10. 1162/jocn.2009.21378

Maritain, J. (1947). *The person and the common good* (J. J. Fitzgerald, Trans.). University of Notre Dame Press.

Maritain, J. (1995). *The degrees of knowledge* (G. B. Phelan, Trans.). University of Notre Dame Press. (Original work published 1932).

McGee, A., et al. (2023). A new defense of brain death as the death of the human organism. *Journal of Medicine and Philosophy, 48*(5), 434–452. https://doi.org/10.1093/jmp/jhac040

McKenzie, J. L. (1965). *Dictionary of the Bible.* Bruce Publishing Company.

McMahan, J. (1995). The metaphysics of brain death. *Bioethics, 9*(2), 91–126. https://doi.org/10.1111/j.1467-8519.1995.tb00305.x

McMahan, J. (2002). *The ethics of killing: Problems at the margins of life.* Oxford University Press.

Merker, B. (2007). Consciousness without a cerebral cortex: A challenge for neuroscience and medicine. *Behavioral and Brain Sciences, 30*(1), 63–81. https://doi.org/10.1017/S0140525X07000891

Merrill, J. P., et al. (1956). Successful homotransplantation of the human kidney between identical twins. *JAMA, 160*(4), 277–282. https://doi.org/10.1001/jama.1956.029603 90027008

Mesulam, M.-M. (1990). Large-scale neurocognitive networks and distributed processing for attention, language, and memory. *Annals of Neurology, 28*(5), 597–613.

Mollaret, P., & Goulon, M. (1959). The depassed coma (preliminary memoir). *Revue Neurologique, 101*, 3–15.

Moreland, J. P. (1998). *Body and soul: Human nature and the crisis in ethics.* InterVarsity Press.

Moreland, J. P. (2018). In defense of a Thomistic-like dualism. In J. Loose, A. Menuge, & J. Moreland (Eds.), *The Blackwell companion to substance dualism* (pp. 102–122). Wiley-Blackwell.

Moreland, J. P., & Rae, S. B. (2000). *Body & soul: Human nature & the crisis in ethics.* InterVarsity Press.

Moschella, M. (2019). The human organism is not a conductorless orchestra: A defense of brain death as true biological death. *Theoretical Medicine and Bioethics, 40*(5), 437–453. https://doi.org/10.1007/s11017-019-09501-z

Murphy, N. B., et al. (2023). Rationale for revisions to the definition of death and criteria for its determination in Canada. *Canadian Journal of Anaesthesia, 70*(4), 558–569. https://doi.org/10.1007/s12630-023-02407-4

Nutton, V. (2004). *Ancient medicine.* Routledge.

Nussbaum, M. C., & Rorty, A. O. (Eds.). (1995). *Essays on Aristotle's De Anima.* Clarendon Press.

O'Collins, G. (1995). *Christology: A biblical, historical, and systematic study of Jesus.* Oxford University Press.

Panksepp, J., & Biven, L. (2012). *The archaeology of mind: Neuroevolutionary origins of human emotions.* W. W. Norton & Company.

Papez, J. W. (1937). A proposed mechanism of emotion. *Archives of Neurology and Psychiatry, 38*(4), 725–743. https://doi.org/10.1001/archneurpsyc.1937.02260220069003

Parvizi, J., & Damasio, A. (2001). Consciousness and the brainstem. *Cognition, 79*(1–2), 135–159. https://doi.org/10.1016/S0010-0277(00)00127-X

Pasnau, R. (2011). *Metaphysical themes: 1274–1671.* Oxford University Press.

Pasnau, R. (2002). *Thomas Aquinas on human nature.* Cambridge University Press.

Pellegrino, E. D., & Thomasma, D. C. (1993). *The Christian virtues in medical practice.* Georgetown University Press.

Penfield, W. (1975). *The mystery of the mind: A critical study of consciousness and the human brain.* Princeton University Press.

Penrose, R. (1994). *Shadows of the mind: A search for the missing science of consciousness.* Oxford University Press.

Pernick, M. S. (1999). Brain death in a cultural context: The reconstruction of death, 1967–1981. In S. Youngner, R. Arnold, & R. Schapiro (Eds.), *The definition of death: Contemporary controversies* (pp. 3–33). Johns Hopkins University Press.

Pezzulo, G., Rigoli, F., & Friston, K. (2015). Active inference, homeostatic regulation, and adaptive behavioural control. *Progress in Neurobiology, 134*, 17–35. https://doi.org/10.1016/j.pneurobio.2015.09. 001

Pieper, J. (1964). *The Christian idea of man*. St. Augustine's Press.

Plato. (n.d.). *Phaedo* (G. M. A. Grube, Trans.). Hackett Publishing.

Plato. (n.d.). *Republic* (C. D. C. Reeve, Trans.). Hackett Publishing.

Plato. (n.d.). *Timaeus* (D. J. Zeyl, Trans.). Hackett Publishing.

Plum, F. (1991). Coma and related global disturbances of the human conscious state. In A. Peters & E. G. Jones (Eds.), *Cerebral cortex: Normal and altered states of function* (pp. 307–317). Springer.

Popper, K. (1972). *Objective knowledge: An evolutionary approach*. Oxford University Press.

Popper, K., & Eccles, J. (1977). *The self and its brain: An argument for interactionism*. Springer.

President's Commission for the Study of Ethical Problems in Medicine and Biomedical and Behavioral Research. (1981). *Defining death: Medical, legal and ethical issues in the determination of death*. U.S. Government Printing Office. Retrieved from https://repository.library.georgetown.edu/handle/10822/559345

Raichle, M. E. (2015). The brain's default mode network. *Annual Review of Neuroscience, 38*, 433–447. https://doi.org/10.1146/annurev-neuro-071013-014030

Ratzinger, J. (1991). El misterio de la mente humana [The mystery of the human mind]. In *Actas de la Quinta Conferencia Internacional: La Mente Humana* (Vol. 16, pp. 18–20). Vatican City: Dolentium Hominum.

Rist, J. M. (1967). *Plotinus: The road to reality*. Cambridge University Press.

Rosen, R. (1991). *Life itself: A comprehensive inquiry into the nature, origin, and fabrication of life*. Columbia University Press.

Ryle, G. (1949). *The concept of mind.* Hutchinson.

Sarna, N. M. (1993). *Exploring Exodus: The origins of Biblical Israel.* Schocken Books.

Sarti, A. J., et al. (2023). When is a person dead? The Canadian public's understanding of death and death determination: A nationwide survey. *Canadian Journal of Anaesthesia, 70*(4), 617–627. https://doi.org/10.1007/s12630-023-02409-2

Second Vatican Council. (1965). *Gaudium et spes: Pastoral constitution on the Church in the modern world.* Vatican City: Libreria Editrice Vaticana.

Seeley, W. W., Menon, V., Schatzberg, A. F., Keller, J., Glover, G. H., Kenna, H., Reiss, A. L., & Greicius, M. D. (2007). Dissociable intrinsic connectivity networks for salience processing and executive control. *The Journal of Neuroscience, 27*(9), 2349–2356. https://doi.org/10.1523/JNEUROSCI.5587-06.2007

Seth, A. K. (2013). Interoceptive inference, emotion, and the embodied self. *Trends in Cognitive Sciences, 17*(11), 565–573. https://doi.org/10.1016/j.tics.2013.09.007

Seth, A. K. (2014). A predictive processing theory of sensorimotor contingencies: Explaining the puzzle of perceptual presence and its absence in synesthesia. *Cognitive Neuroscience, 5*(2), 97–118. https://doi.org/10.1080/17588928.2013.877880

Shewmon, D. A. (1985). The metaphysics of brain death, persistent vegetative state, and dementia. *The Thomist, 49*(1), 24–80.

Snell, B. (1960). *The discovery of the mind: The Greek origins of European thought.* Harper & Row.

Siraisi, N. G. (1990). *Medieval and early Renaissance medicine: An introduction to knowledge and practice.* University of Chicago Press.

Sporns, O. (2010). *Networks of the brain.* MIT Press.

Sporns, O. (2013). Network attributes for segregation and integration in the human brain. *Current Opinion in Neurobiology, 23*(2), 162–171. https://doi.org/10.1016/j.conb .2012.11.015

Starzl, T. E., et al. (1963). Homotransplantation of the liver in humans. *Surgery, Gynecology & Obstetrics, 117*, 659–676.

Sterling, P., & Eyer, J. (1988). Allostasis: A new paradigm to explain arousal pathology. In S. Fisher & J. Reason (Eds.), *Handbook of life stress, cognition and health* (pp. 629–649). John Wiley & Sons.

Stump, E. (2003). *Aquinas*. Routledge.

Suárez, F. (1597). *Disputationes metaphysicae*. In J. P. Moreland (Ed.), *The Blackwell companion to substance dualism*. Wiley-Blackwell.

Taylor, C. (1989). *Sources of the self: The making of the modern identity*. Harvard University Press.

Tertullian. (1977). *Against Praxeas*. In A. Cleveland Coxe (Ed.), *The Ante-Nicene Fathers* (Vol. 3). Hendrickson Publishers.

Thayer, J. F., & Lane, R. D. (2009). Claude Bernard and the heart-brain connection: Further elaboration of a model of neurovisceral integration. *Neuroscience & Biobehavioral Reviews, 33*(2), 81–88. https://doi.org/10.1016/j.neubiorev.2008.08.004

Tononi, G. (2004). An information integration theory of consciousness. *BMC Neuroscience, 5*, 42. https://doi.org/10. 1186/1471-2202-5-42

Tononi, G., & Koch, C. (2015). Consciousness: Here, there and everywhere? *Philosophical Transactions of the Royal Society B, 370*(1668), 20140167. https://doi.org/10.1098/rstb.2014.0167

Tononi, G., & Edelman, G. M. (1998). Consciousness and complexity. *Science, 282*(5395), 1846–1851. https://doi.org/10.1126/science.282. 5395.1846

Truog, R. D. (1997). Is it time to abandon brain death? *Hastings Center Report, 27*(1), 29–37.

Truog, R. D., & Miller, F. G. (2012). Brain death: Justifications and critiques. *Clinical Ethics, 7*(3), 128–132. https://doi.org/10.1258/ce.2012. 012m09

van den Heuvel, M. P., & Sporns, O. (2011). Rich-club organization of the human connectome. *Journal of Neuroscience, 31*(44), 15775–15786. https://doi.org/10.1523/JNEUROSCI.3539-11.2011

Varela, F. G., Maturana, H. R., & Uribe, R. (1974). Autopoiesis: The organization of living systems, its characterization and a model. *Current Modern Biology, 5*(4), 187–196. https://doi.org/10.1016/0303-2647(74)90031-8

Varela, F. J., Thompson, E., & Rosch, E. (1991). *The embodied mind: Cognitive science and human experience.* MIT Press.

Veatch, R. M. (1977). The definition of death: Ethical, philosophical, and policy confusion. *Annals of the New York Academy of Sciences, 315*, 307–321. https://doi.org/10.1111/j.1749-6632. 1978.tb50348.x

von Rad, G. (1962). *Old Testament theology* (Vol. 1, D. M. G. Stalker, Trans.). Harper & Row.

Von Staden, H. (1989). *Herophilos: The art of medicine in early Alexandria.* Cambridge University Press.

Wenham, G. J. (1987). *Genesis 1–15.* Word Books.

Westermann, C. (1984). *Genesis 1–11: A continental commentary.* Fortress Press.

Willis, T. (1681). *The anatomy of the brain and nerves* (S. Pordage, Trans.). Dring, Harper, Leigh, & Martyn.

Wolff, H. W. (1974). *Anthropology of the Old Testament.* Fortress Press.

Wright, N. T. (2003). *The resurrection of the Son of God.* Fortress Press.

Youngner, S. J., & Bartlett, E. T. (1983). Human death and high technology: The failure of the whole-brain formulations. *Annals of Internal Medicine, 99*(2), 252–258. https://doi.org/10.7326/0003-4819-99-2-252

Zheng, K. et al. (2022). Public understandings of the definition and determination of death: A scoping review. *Transplantation Direct, 8*(5), e1300. https://doi.org/10.1097/TXD.0000000000001300

14

A Refined Formulation of Human Death

Defining the moment of human death remains one of the most challenging and consequential questions in medicine, ethics, and philosophy. This chapter introduces the Brain Metafunction Framework, a refined formulation grounded in contemporary neuroscience and rooted in classical philosophical anthropology. It seeks to clarify the conceptual foundation for determining death by identifying the irreversible loss of a critical integrative function within the human brain—one that sustains both personhood and organismal unity. In doing so, it offers a comprehensive response to ongoing conceptual and clinical challenges surrounding the definition of human death.

What defines human death? What critical function in humans is irreplaceable by technological advances? Is it even possible to define human death conclusively? How can we improve existing formulations of death? These questions have been pondered for millennia and gained renewed urgency in the second half of the 20th century with advances in science and technology (Mollaret & Goulon, 1959; Berthelsen & Cronqvist, 2003). Dissociating the diagnosis of death from traditional cardiorespiratory criteria, coupled with advancements in organ transplantation, introduced profound bioethical dilemmas (Merrill et al., 1956; Starzl et al., 1963; Barnard, 1967; Kelly et al., 1967). The shift toward brain-centered definitions of death has raised persistent questions about their conceptual and clinical foundations. This chapter responds to those concerns by presenting a rationale grounded in the hierarchical unity of the human organism and person, supported by both philosophical argument and neuroscientific evidence.

While a detailed history of brain death's concept development is beyond this chapter's scope, it is essential to note key milestones. The concept was formally introduced by the Harvard Committee in 1968, proposing irreversible coma as a criterion for death (Report 1968, 337-340). This report was followed by the landmark 1981 report by the President's Commission for the Study of Ethical Problems in Medicine, which endorsed the neurological criteria for death and established the

Uniform Determination of Death Act (UDDA). The UDDA defines death as the irreversible cessation of all functions of the entire brain, also known as the Whole-Brain Death criterion (President's Commission 1981). In the UK, the Brainstem Death criterion was recognized earlier (Conference of Medical Royal Colleges 1976, 1187-1188 and 1979, 261-262), focusing on the brainstem's loss of function as indicative of death, implying the death of the brain "as a whole." Philosophical debates have also influenced the discourse on death. The Higher-Brain definition focuses on the loss of neocortical functions, such as the "content of consciousness," arguing that this loss represents the essence of human death (Veatch 1975; Youngner and Bartlett 1983; McMahan 1995). These formulations emphasize the irreplaceable capacities unique to human beings, including higher-order cognition and self-awareness.

The Origins of the Brain Metafunction Framework and Other Contributions

In 1992, the First International Symposium on Coma and Death in Havana brought together prominent experts, including Cristopher Pallis (UK), Earl Walker, Allan Shewmon (US), and other notable contributors. Around this time, Cuba officially adopted a brain-based definition of death as proposed by a national commission (Machado and García 1995). Machado subsequently advanced a concept emphasizing the loss of consciousness including the cerebral neocortex and subcortical structures, aligning with whole-brain operational criteria (Machado 1995, 1999). In parallel, publications by the present author and symposium co-organizer (García 1995, 1997) emphasized the anatomical and physiological inseparability of consciousness and bodily function integration at the brain's highest hierarchical level of unification. Additionally, research on heart rate variability in coma and brain death (García, Machado, et al. 1995) (Machado, García, et al. 2005) suggested to the present author that the brain's multisystemic role in heart regulation, a function traditionally attributed solely to the lower brainstem, appeared to extend to other physiological processes. These findings and developments in neuroscience led to a new perspective on *global brain function* and *integration within the brain.*

Further academic contributions elaborated on a brain-centered definition of death, conceptualizing integration within the brain as the hierarchically highest level of integration in the human body (García 1998a, b, c, 1999, 2001). This perspective posited that consciousness and bodily integration are anatomically and functionally coextensive within the critical brain metafunction. The framework also critiqued implicit "biological Cartesianism" present in prior approaches. Proponents of the higher-brain concept focused narrowly on the cerebral neocortex and some aspect of consciousness, or consciousness in general, while whole-brain approach, while sound as a criterion, provided a biologicist rationale. The separation of consciousness from biological integration creates a false dichotomy. In contrast, the brain metafunction framework emphasizes the unity of the human person and the critical role of brain function in maintaining systemic integration. It also argues that the irreversible loss of brain metafunction is the most coherent rationale supporting the whole-brain criterion of death. The subsequent two decades have witnessed significant advances in neuroscience, corroborating and expanding the brain metafunction framework. These developments support this integrative conceptualization of death, which we will explore further below.

Recently in Canada

In Canada, the 2023 guidelines published by Murphy et al. in the *Canadian Journal of Anaesthesia* refined criteria for determining death to align with contemporary clinical practices. These revisions address ethical considerations and advancements in medical technology but primarily focus on diagnostic consistency rather than conceptual criteria (Murphy et al. 2023). The historical evolution of death definitions reflects an ongoing effort to balance medical advances with ethical, philosophical, and practical considerations. A neurologically grounded framework rooted in the brain's critical role as the integrator of both human person and organismal unity provides a foundation for addressing these challenges. By defining death as the permanent loss of this critical metafunctional system, we gain a clearer understanding of the intersection between personhood, biological unity, and the end of life.

Methodological Approach

The Brain Metafunction framework includes a multidisciplinary analysis synthesizing insights from neurology, neuroscience, bioethics, philosophy, and theology. This methodology is structured into the following components:

1. *Literature Review:* This examines existing definitions of death, identifying the most robust conceptual criteria for brain death. The review addresses these definitions' medical, bioethical, and

interdisciplinary implications, highlighting areas of agreement and controversy. Emphasis is on identifying gaps in current frameworks and their implications for clinical and legal standards.

2. *Philosophical Analysis:* The framework critically evaluates the concept of "unity" in the human organism and person, drawing from Western philosophical traditions, including Aristotle, Aquinas, and contemporary thinkers. It examines how neuroscientific advances, such as understanding distributed neural networks and integrative brain functions, challenge or reinforce classical notions of unity. This analysis explores the consequences of disintegration at the organismal and personal *unity level,* proposing a refined formulation of human death that bridges metaphysical and scientific perspectives.

3. *Bioethical Considerations:* The Brain Metafunction framework provides a conceptual platform for addressing bioethical issues such as:
   a. The definition and determination of death,
   b. The loss of personhood (See the chapter on The Unity of the Human Person)
   c. Medical and legal standards for diagnosing death,
   d. Ethical dilemmas in organ transplantation and end-of-life care and
   e. Public understanding and communication of death-related criteria.

The framework aims to resolve bioethical challenges and foster greater clarity in clinical and societal practices by integrating philosophical rigor with neuroscientific evidence.

The Unity of the Human Person

Western Tradition

The concept of the unity of the human person has been a central theme in Western philosophy and theology, evolving through Semitic anthropology, Greek philosophy, and Christian thought. Plato's concept of the soul emphasizes its cognitive and intellectual aspects but leaves unanswered how the soul relates to the functions of life. Aristotle, in contrast, provides a more integrated view. In the *Nicomachean Ethics* (1.13, 1102b11–12), he introduces the concept of the 'substantial form,' the essence of a living organism embodied in the organization of matter. This organization creates a relational arrangement among various parts, resulting in a unified whole greater than the sum of its components. Aristotle's reinterpretation of Plato's analogy of the soul as the pilot of the body carries profound biological and metaphysical significance. For Aristotle, the soul is the unifying principle of the organism and the source of consciousness (*De Anima* Book I 412a27, Book II 738b4–739a2). He posits that the soul is both the form and substance of a particular body, making it the essence and source of unity for a living organism (Hardie 1976).

Semitic anthropology complements Aristotle's view by presenting the human person as a unified totality capable of life centered "as a whole" in spirit, soul, or body. This unity, however, does not imply uniformity. Terms such as *basar, nefesh, ruah,* and *leb* (translated into Greek as *sarx, psuche, pneuma,* and *kardia*) represent different dimensions of the person as a unified whole (Kittel & Friedrich, 1965; Boff 1980; Ratzinger 1991). In Christian theology, the doctrine of the *Imago Dei* affirms the integrated nature of humanity—spiritual, intellectual, emotional, and physical. The Incarnation further illustrates this unity by showing how divinity and humanity, spirit, and matter coalesce in the person of Jesus Christ. The Resurrection underscores that redemption encompasses the entire person—spirit, soul, and body—transforming the natural body into a glorified "spiritual body" (1 Cor. 15).

Thomas Aquinas expanded these ideas in his *Summa Theologica* (1266–1273, Pt. 1, Q. 76) and *Summa Contra Gentiles* (1259–1265, SCG.2 C68.2). He argued that the soul and body are united as form and matter, creating a single, unified substance: "The human soul is an intellectual substance united to the body as its form." This *hylomorphic* view emphasizes that humans are not merely souls inhabiting bodies but are a profound unity of body and soul—a single substance, not two. Francisco Suárez (1548–1617) built on this by noting that the rational soul must be the principle of both intellectual and bodily operations to serve as the body's true form (Suárez 1597, quoted in Moreland 2018).

The unity of the human person was later challenged by René Descartes, whose *Meditations on First Philosophy* ([1641] 2008) introduced a dualistic view separating the mind and body. Descartes' dictum, "I think, therefore I am," emphasized the mind as a distinct substance, fragmenting the unity of the human person. This Cartesian dualism, rooted in Platonic thought, has significantly influenced modern Western philosophy and has been critiqued for undermining the integration of the human person. Contemporary Thomists and neo-Thomists argue for a return to an Aristotelian-Thomistic view of the soul as the principle of life, consciousness, and spirit, integrated within a unified human person (Des Chene 2000; Pasnau 2011; Moreland 2018). Finley (2016) succinctly notes: "If a living substance is divided at the level of unity, the substance itself is gone." This perspective affirms that the human person is an irreducible unity, with the soul and body fundamentally interdependent, forming a single, indivisible whole.

The Unity of the Human Organism and Brain Death as Death

The concept of the unity of the organism "as a whole" has been central to the rationale for the "whole brain" criteria of death. Recent proponents have further developed this rationale, emphasizing the brain's irreplaceable role in maintaining systemic unity. James L. Bernat, a long-standing advocate of a brain-centered concept of death, provides a compelling argument in his article *"A Conceptual Justification for Brain Death"* (2018). Bernat highlights the brain's essential role in maintaining the body as an integrated whole rather than a mere collection of parts. Using concepts from mereology—the study of parts and their relationships to the whole—and emergent functions, Bernat argues that the brain orchestrates a new level of biological order necessary for life. He introduces the thought experiment of "physiological decapitation," likening brain death to the physiological equivalent of decapitation. This analogy underscores the distinction between the whole organism and the organism "as a whole," highlighting the centrality of the brain in defining death (Bernat 2018, S19–S21). In her 2019 article *"The Human Organism is Not a Conductorless Orchestra: A Defense of Brain Death as True Biological Death,"* Melissa Moschella expands on this theme by likening the brain to a conductor orchestrating the body's complex interactions. She argues that the loss of brain function leads to the disintegration of the organism as a unified whole. Moschella emphasizes the hierarchical unity of the organism, where the brain serves as the critical regulator of bodily functions. Without the brain's integrative activity, the organism ceases to function as a coordinated system and thus cannot sustain life (Moschella 2019, 437–453).

Maureen L. Condic, in her paper *"Determination of Death: A Scientific Perspective on Biological Integration"* (2016), analyzes the hierarchical levels of integration within the human body. She explains that while lower levels (e.g., cellular or organ functions) can operate autonomously, the brain is essential for coordinating these functions into a unified whole. Condic argues that brain death signifies the irreversible loss of this higher-order integration, marking the end of the organism's biological life. Her analysis reinforces the brain's role as the irreplaceable organ for maintaining systemic unity (Condic 2016, 257–278). In their 2023 article *"A New Defense of Brain Death as the Death of the Human Organism,"* Andrew McGee, Dale Gardiner, and Melanie Jansen argue that brain death represents the end of the human organism due to the irreversible loss of the brain's integrative functions. They emphasize the brain's role in regulating vital processes, including consciousness and autonomic functions like breathing. Their work highlights the importance of the brain's integrative activity in sustaining the organism and defining the boundary between life and death (McGee et al. 2023, 434–452).

The Critique of Brain Death as Death: There is No Hierarchical Level of Unity

D. A. Shewmon, once a proponent of brain death (Shewmon 1997), is now recognized for critically analyzing its definitions and criteria. In his article *"Constructing the Death Elephant: A Synthetic Paradigm Shift for the Definition, Criteria, and Tests for Death"* (2010), Shewmon critiques existing paradigms for defining and determining death, particularly brain death. He argues for a paradigm shift that reflects the complexity of human death, emphasizing that many biological functions can persist even after brain function ceases, suggesting the body remains alive in a meaningful way. Shewmon distinguishes between two death-related moments: "passing away" (the end of the organism's relation to the external world) and "disanimation" (the metaphysical end of the organism as a unified entity). He argues that "civil death," the cessation of external relations, is a more appropriate concept in contexts like organ donation after circulatory death. This view situates any critical level of unity, if considered, at the metaphysical level rather than the biological.

Robert D. Truog, in his 2020 article in *Pediatrics*, critiques the assumption that brain death is equivalent to biological death. He highlights that many bodily functions persist after brain function ceases, suggesting the organism remains alive significantly. Truog also raises ethical concerns about whether organ procurement from brain-dead donors complies with the dead donor rule, which mandates procurement of vital organs only after the donor is biologically dead. Truog concludes by proposing the framing of brain death as a "social construct," requiring societal acceptance of treating brain-dead patients as legally and ethically equivalent to biologically dead individuals (Truog 2020, S75–S80).

Halevy and Brody (1993) explore alternative approaches to defining death. They propose a "non-definition" model that avoids sharp dichotomies between life and death. Instead, they suggest practical criteria for determining when care can be withdrawn, organs harvested, or other end-of-life decisions made, independent of a single definition of death. This approach challenges the necessity of a universal biological or philosophical standard. Other critiques advocate for a return to the cardio-circulatory standard (Truog 1997; Truog and Miller 2012; Brugger 2016), emphasizing the irreversible cessation of circulation and respiration as definitive markers of death. Some views argue that death is fundamentally a social convention developed for practical purposes like organ procurement without a strictly biological or philosophical basis. These perspectives often separate concepts such as "personal," "social," or "philosophical" death from "biological" death (Shewmon 2010; Truog 2020).

The Brain Metafunction Framework

Recent shifts in neuroscience challenge the traditional view of the brain as composed of region-specific functions. Emerging models emphasize the following features:

1. Domain-General Neural Populations
Brain networks demonstrate multi-use, domain-general functions rather than single-task specificity (Avena-Koenigsberger, Misic, and Sporns 2018, 17–33; Cowan et al. 2011, 2852–2863; Chein, Moore, and Conway 2011, 550–559; Majerus et al. 2010, 2570–2593; Koelsch et al. 2009, 859–873).

2. Dense Interconnectivity Across Regions
Cortical and subcortical regions exhibit extensive interconnectivity, forming a cohesive neural network (Sporns 2010; Cole et al. 2013, 1348–1355; van den Heuvel and Sporns 2011, 15775–15786).

3. Universal Connectivity
Connectivity spans all brain sectors, including the brainstem and cerebellum, integrating sensory, motor, and regulatory functions. One "connectivity core" may include regions distributed across all major brain sectors (e.g., all cortical lobes, thalamus, and subcortical regions).

4. Anatomical and Functional Characteristics
The brain functions as a highly interconnected network, with dense neural connections forming a tightly integrated system that enables

efficient communication across different regions. This massive interconnectivity allows for seamless information transfer, supporting complex cognitive processes. Additionally, the brain's network structure ensures high global accessibility, meaning signals can travel across vast neural circuits, facilitating coordination between distant brain areas. At the core of this connectivity is what is known as the "connectivity core" or "rich club"—a hub of highly connected regions that spans multiple cortical lobes, subcortical structures, and the thalamus. This specialized network is crucial in integrating and distributing information, allowing the brain to function as a cohesive and dynamic system.

5. Distributed Core Regions
The connectivity core includes distributed regions across all major brain sectors, enabling communication across diverse neural systems.

6. Hub Connectivity for Predictions: Key hub regions exhibit connectivity patterns, allowing them to generate predictive signals that influence sensory and motor systems.

7. Interaction Between Networks
Sensory and motor networks interact with the default mode and salience networks via hub regions, facilitating inter-system communication (Kleckner et al. 2017).

8. Integration via Default Mode and Salience Networks
These networks serve as high-capacity backbones for integrating interoceptive and allostatic information, supporting both psychological functions and bodily regulation (Tononi, 2004; Seeley et al., 2007). In particular, dynamic connectivity within the Default Mode Network has been shown to track fluctuations in affective experience, reflecting its integrative role in internal state regulation.

9. Total Brain States
Total brain states refer to the large-scale functional configurations of neural activity that shape cognition, perception, and behavior at any moment. These states arise from dynamic interactions among distributed brain networks, allowing the brain to integrate sensory information, internal experiences, and regulatory mechanisms into a cohesive whole. A significant example is the default mode network (DMN), primarily active during introspective activities such as self-referential thinking, memory retrieval, and mind-wandering. The DMN plays a

crucial role in integrating internal representations and maintaining a continuous sense of self. Complementing the DMN, the salience network is responsible for detecting and filtering relevant stimuli, helping to direct attention between the external environment and internal cognitive states.

The interaction between these networks highlights the significance of total brain states in unifying cognitive and regulatory processes. For instance, transitions between different states—from rest (dominated by the DMN) to active problem-solving (involving task-positive networks)—reflect how the brain dynamically reconfigures itself to adapt to situational demands. Disruptions in these global states relate to neurological and psychiatric disorders; for example, altered DMN activity has been associated with depression, while impaired salience network function with schizophrenia and autism. Understanding total brain states provides insight into how the brain orchestrates its vast functions, ensuring the balance between internal cognition, external engagement, and physiological regulation.

The brain metafunction framework anticipated these general characteristics, emphasizing the anatomical and functional co-extensiveness of consciousness and bodily regulation processes. This unified system integrates cognitive, motor, visceromotor, homeostatic, and other functions, underscoring the interconnected nature of brain activity. As we examine the growing body of research, these findings increasingly validate the framework's conceptual predictions and highlight its importance for advancing our understanding of the human brain.

Towards a Holistic Understanding of the Human Brain

The human brain exhibits diverse spatiotemporal dynamics—activity patterns across different brain regions and over time—characterized by integrated and segregated functional states (Dehaene and Changeux 2011). These dynamics involve simultaneous information processing by specialized (domain-specific) systems and global (integrated) networks, enabling adaptive behaviors and complex functions. For instance, the brain employs domain-specific systems to process visual, auditory, motor, and homeostatic information independently while integrating outputs to produce coherent, goal-directed behaviors. This balance between segregation and integration underpins the brain's ability to respond adaptively to environmental demands (Sporns 2013, 162–171).

Research by Dehaene and Changeux highlights how integration facilitates adaptive behaviors, such as decision-making and problem-solving, by synthesizing diverse information streams into unified action plans. Maintaining equilibrium between segregated and integrated states is critical for local and global neural communication, supporting diverse cognitive and bodily functions. Studies by Kelso (2012), Sporns (2013), Cohen and D'Esposito (2016), and Wang et al. (2021) demonstrate that this dynamic balance shapes the brain's functional architecture, enabling perception, attention, memory, and decision-making. The brain's flexibility allows the coexistence of segregated and integrated states. Segregated systems handle tasks like sensory processing with relative independence, while integrated states enable cross-system communication. This flexibility facilitates adaptive responses to changing demands, as noted in studies by Kelso and Wang et al., which emphasize the brain's dynamic integration of information across systems to support cognitive processes.

Allostasis, a concept introduced by Sterling and Eyer (1988), describes how the brain achieves stability through change by predicting and adjusting to varying conditions. Accurate predictions of interoceptive signals—sensory inputs from the body's internal state—are essential for maintaining allostasis (Kleckner et al. 2017). Interoception involves processing autonomic, neuroendocrine, and neuroimmune signals (Friston 2010; Barrett and Simmons 2015). Networks responsible for these processes also govern social affiliation, empathy, reward processing, and decision-making. Kleckner et al. (2017) demonstrate that these systems, extending beyond basic bodily regulation, underpin complex psychological phenomena like emotion, memory, and judgment. The salience and default mode networks, which exhibit domain-general characteristics, play a pivotal role in integrating interoceptive and allostatic information. Papez's (1937) hypothesis regarding the limbic system's role in emotion finds modern validation in these networks, which connect regions like the dorsomedial prefrontal cortex and hippocampus to visceromotor control systems. This interconnectedness underscores the brain's role as a generative system, modeling the world through predictive mechanisms rather than merely reacting to sensory inputs (Seth 2014).

Emerging evidence challenges the traditional modular view of brain function, revealing the domain-general nature of many neural systems. Multisensory processing in regions previously thought unimodal, such as the auditory cortex responding to visual stimuli, illustrates the brain's flexibility (Koelsch et al. 2009; Majerus et al. 2010; Sporns 2010).

These findings align with the brain metafunction framework, which integrates these insights to address medical, bioethical, and philosophical challenges. This framework advances our understanding of the intricate interplay between cognition, emotion, and bodily regulation by conceptualizing the brain as a unified system that integrates internal and external inputs to guide adaptive behaviors.

The Brain Metafunction Framework builds on contemporary cognitive neuroscience studies that demonstrate the fundamentally distributed and interconnected nature of brain function. Research by Avena-Koenigsberger et al. (2018), Cowan et al. (2011), and Chein et al. (2011) shows that even basic cognitive functions like working memory rely on domain-general, dynamically interacting networks rather than isolated regions. Similarly, studies by Sporns (2010) and Cole et al. (2013) reveal the brain's dense interconnectivity, which sustains consciousness by integrating information from sensory, motor, and regulatory systems. The collapse of these integrative processes aligns with medical criteria for brain death, where irreversible loss of global brain connectivity marks the end of the human organism's integrative functioning. Thus, cognitive neuroscience provides strong support for defining death in terms of total brain state failure, consistent with the Brain Metafunction Framework.

What is a Formulation of Death?

1. A formulation of death, as outlined by Veatch (1978), consi ts of several distinct levels:
2. A concept or definition of what it means to die.
3. A conceptual criterion that provides the anatomical and physiological basis supporting the definition.
4. Appropriate operational criteria to determine the fulfillment of the conceptual criterion.
5. Specific medical tests to demonstrate whether the operational criteria are, without doubt, present.

This formulation moves from a philosophical level to a clinical and technological one. Since definitions of death are fundamentally conceptual and philosophical, technology cannot replace them. However, advances in neuroscience and bioethics enable us to refine these formulations, aligning them more closely with the realities of human biology. This section examines a refined formulation of death grounded in the concept of the brain metafunction. The subsequent section will explore its implications.

A Refined Formulation of Death

Building on an intuition developed in the 1990s (García 1995, 1997, 1998a,b,c) and supported then by anatomical and physiological research (Plum 1991; Damasio 1996), advances in neuroscience over the past two decades allow for a refined formulation of death. This understanding emphasizes the brain's role as a unified, irreducible whole, articulated here as the *brain metafunction framework*.

Definition

Human death is the irreversible [1] *loss of the brain metafunction, the dynamic, irreducible, unified activity of the brain's integrated systems that form the functional and anatomical substrate of consciousness, cognitive processes, and higher-order bodily regulation, thereby sustaining the human person and the organism as an integrated whole.*

Conceptual Criterion

The conceptual criterion underlying this definition is the brain metafunction, the dynamic and unified activity that integrates all levels of brain function into a single, cohesive system sustaining the unity of the human person. This metafunction is characterized by the dynamic and unified activity of brain systems, sub-regions, regions, and neurons, linked by structural and functional connectivity. It integrates local-global synchronized dynamics across space and time, unifying interdependent domains of the human person. The brain metafunction includes the following subfunctions [2]:

1. Consciousness and Cognitive Processes: Perception, attention, memory, and emotions.
2. Neural Integration of Communication Systems: Coordi- nation of the nervous, endocrine, immune, and circulatory systems.
3. Regulation of the Internal Environment: Homeostasis and allostasis at the highest hierarchical level.
4. Processing Afferent and Efferent Flows: Integration of inputs and outputs to and from the whole organism and its environment.
5. Behavioral and Adaptive Interactions: Coordinating responses to external stimuli.
6. Integration of Mental and Bodily Processes: Harmonizing physiological functions with mental states with massive unconscious processing.
7. Unified Functionality: A fundamental, irreducible whole that synthesizes all other subfunctions.

[1] *The term "irreversible" underscores that the brain metafunction is not merely the sum of its component operations or structures, but a unified, emergent process. This aligns with systems theory and emergentist neuroscience and supports the author's view that the brain metafunction constitutes the foundational basis for both personhood and biological integration. Its loss therefore marks not only the end of consciousness, cognitive processes, and systemic regulation, but also the collapse of personhood itself. On this view, death and the loss of personhood are coincident and inseparable—not because any one function fails, but because the indivisible whole ceases to exist. Nonetheless, the definition remains compatible with frameworks that define death in terms of the inseparability of consciousness and organismal integration, even if personhood is not explicitly invoked.*

[2] *The specified brain metafunction subfunctions are system-wide, richly interconnected, dynamically integrated functions characterized by rich, hypercomplex functional integration. While biologically supported, they also display attributes pointing to a transphysical domain, which offers the required explanatory power. However, this component is optional in defining death, allowing for a shared biomedical framework across philosophical and theological perspectives.*

Operational Implications

This refined framework offers a precise basis for diagnosing death by identifying the irreversible loss of unified brain functions, specifying integration within the brain, and providing conceptual clarity for applying operational criteria. It bridges philosophical definitions with clinical applications, clarifying end-of-life care, organ donation protocols, and bioethical considerations.

The Biological Substrate for the Conceptual Criterion.

Death is defined as the permanent loss of function of the brain's critical metafunctional system, which is, in principle, irreplaceable. This critical system integrates and sustains the unified activity of the human organism. Conceptually, the destruction of the brain's metafunctional system is sufficient to define death. This system comprises the following three subsystems, each of which contributes uniquely to the brain's integrative functions:

Subsystem I: Ascending Reticular Activation Subsystem
This subsystem consists of nonspecific populations of neurons located in the *tegmentum of the rostral pons and midbrain*, the *intralaminal and reticularis nuclei of the thalamus*, and the *posterior hypothalamus*. These

structures play a critical role in maintaining arousal, wakefulness, intrinsic contributions to consciousness, and the foundational level of consciousness necessary for higher-order brain functions.

Subsystem II: Limbic System Structures and Brainstem Nuclei

This subsystem includes the *hypothalamus, basal forebrain, nucleus basalis of Meynert, amygdala, hippocampal complex, cingulate cortex, orbitofrontal cortex, cingulum, olfactory cortex, temporal pole,* and portions of the *basal ganglia.*

Brainstem nuclei, including the *reticularis nuclei in the midbrain and rostral pons, parabrachial nucleus, pedunculopontine nucleus, rostral raphe nuclei, periaqueductal gray,* and *nucleus locus ceruleus,* contribute to consciousness, emotional regulation, memory processing, and autonomic control. These structures work in concert to mediate the connection between cognitive processes, internal bodily states and external environmental interactions.

Subsystem III: Cerebral Cortex with the Thalamus and Basal Ganglia

The *cerebral cortex, thalamus,* and *basal ganglia* support higher-order cognitive processes, including perception, attention, memory, reasoning, voluntary motor, and modulation of autonomic and other physiological processes. These regions integrate information from Subsystems I and II, enabling coherent, adaptive behaviors and the experience of selfhood.

Interconnectedness of Subsystems

While each subsystem within the brain's critical metafunction has distinct and specialized roles, their functional interdependence, mutual enrichment, and inseparability ensure the unified operation of the brain as an integrated system:

Functional Interdependence:

Subsystem I (Ascending Reticular Activation) ensures wakefulness and arousal, providing the foundation for Subsystems III (Cerebral Cortex and Thalamus) and II to perform higher-order cognitive functions like attention, reasoning memory, emotional integration, and other functions. Without the baseline activity and intrinsic contribution to consciousness provided by Subsystem I, Subsystem III's advanced processing capabilities, and subsystem II functions would not be possible.

Subsystem II (Limbic System and Brainstem Nuclei) acts as a bridge, linking the physiological regulation of Subsystem I with the cognitive and perceptual functions of Subsystem III. This connection allows for the integration of emotional, sensory, and autonomic inputs into coherent responses and behaviors.

Mutual Enrichment:

Subsystem II enriches Subsystem III by providing emotional and contextual information critical for decision-making and adaptive behavior. Similarly, Subsystem III enhances Subsystem II's processing through top-down control, such as modulating stress responses via structures like the prefrontal cortex or refining emotional regulation.

Subsystem I amplifies the functional efficacy of Subsystems II and III by sustaining the alertness and arousal necessary for

their activities. At the same time, Subsystem III ensures that Subsystem I's output is meaningfully directed toward adaptive behavior and survival.

Inseparability:

The subsystems operate in a tightly coupled manner, where the disruption of one subsystem affects the functionality of the others. For example, damage to Subsystem I can lead to loss of consciousness, effectively disabling the cognitive and regulatory capacities of Subsystems II and III. Conversely, dysfunction in Subsystem III can impair top-down modulation, leading to dysregulation of emotional and autonomic processes governed by Subsystem II.

This interdependence underscores the holistic nature of the brain's metafunction, where no subsystem operates in isolation. Together, these subsystems form an inseparable, dynamic network that underpins the unity of the human organism and enables the coherent integration of bodily, emotional, and cognitive domains. This interconnectedness is central to maintaining the "as-a-whole" function of the brain and the human person. The permanent loss of function in any of these subsystems disrupts the brain's capacity to maintain the metafunction necessary for life. Clinically, understanding these subsystems provides a robust framework and conceptual rationale for diagnosing death based on the irreversible loss of the brain's integrative functions.

Operational criteria to determine the fulfillment of the conceptual criterion and ancillary tests

The whole-brain *criterion* for death remains the operational standard, aligning with the conceptual framework and definition outlined above.

In 2020, the World Brain Death Project established the minimum clinical criteria necessary for determining brain death under various conditions, drawing on a comprehensive review of the literature and the consensus of a broad multidisciplinary, international panel (Greer et al., 2020). These criteria emphasize that before assessing a patient for brain death or death by neurologic criteria (BD/DNC), the patient must have a confirmed neurological diagnosis capable of causing the complete and irreversible cessation of all brain functions. Additionally, confounding conditions and diseases that could mimic brain death must be ruled out. BD/DNC determination relies on a clinical examination demonstrating coma, absence of brainstem reflexes, and apnea. Specific indicators include:

1. No arousal or awareness in response to maximal external stimuli, including visual, auditory, and tactile noxious stimuli.
2. Pupils that are fixed, midsize, or dilated and nonreactive to light.
3. Absence of corneal, oculocephalic, and oculovestibular reflexes.
4. Lack of facial movement in response to noxious stimuli.
5. Absence of the gag reflex upon bilateral posterior pharyngeal stimulation.
6. No cough reflex during deep tracheal suctioning.
7. Absence of brain-mediated motor responses to limb stimulation.
8. No spontaneous respirations during apnea testing, as evidenced by a pH of less than 7.30 and a Paco2 level of 60 mm Hg or higher.

When clinicians cannot fully assess these criteria, they may use ancillary tests such as blood flow studies or electrophysiologic testing. Special considerations apply to children, patients undergoing extracorporeal membrane oxygenation, those receiving therapeutic hypothermia, and in cases influenced by cultural, legal, and resource-related factors. While existing guidelines set widely accepted minimum criteria for determining brain death, confounding factors in certain situations may complicate assessments. Advances in ancillary testing methods, including *computed tomography (CT) perfusion, CT angiography, transcranial Doppler,* and *radionuclide brain perfusion studies* using lipophilic radiopharmaceuticals, provide enhanced sensitivity for evaluating brain blood flow. As

Canada's 2023 guidelines recommended, these ancillary tests offer additional tools for confirming brain death in complex cases, thereby minimizing the risk of misdiagnosis (Murphy et al., 2023).

The refined conceptual framework addresses limitations in other approaches by offering a rationale that avoids rigidly requiring proof of absent activity in every cell or cell group. The framework allows reliance on clinical sampling, including profound unconsciousness, absence of brainstem reflexes, and various confirmatory tests, as valid indicators. As emphasized during the 1968 Sydney meeting, no single technological test is sufficient, nor can any replace the physician's comprehensive judgment (Gilder, 1968). Observation periods should be neither arbitrary nor excessively prolonged. Instead, confirmatory tests should be interpreted within a clearly defined conceptual framework, ensuring moral certainty for medical professionals and family members. This framework promotes systematic documentation and analysis of patient history, neurological examination findings, and ancillary test results. Ultimately, it emphasizes the destruction of the brain's critical system as indicative of the absence of relevant brain functions, moving beyond the mere loss of isolated cellular activity to identify the systemic disintegration of the brain's integrative essence. Ultimately, clinical judgment remains indispensable in interpreting all operational criteria and ancillary tests, ensuring that the determination of death rests on a comprehensive understanding of the brain's unified functioning as defined by the Brain Metafunction Framework.

Discussion

Theoretical Rationale and Implications

The Brain Metafunction framework provides a refined formulation of human death that resolves conceptual and practical limitations of existing brain death criteria. Its key implications include:

*1. **Unified Conceptual Foundation:***
The framework reconciles consciousness and bodily-integration-related brain processes, demonstrating their inherent interconnectedness and inseparability.

*2. **Defined Critical Integration:***
It explicitly identifies the brain as the critical level of unity in human beings by defining integration within the brain.

3. Consistency in Defining Death:
The framework offers a consistent and rational formulation of death, addressing ethical dilemmas in organ transplantation, timing of death determination, and related medical practices.

4. Interdisciplinary Applicability:
Its cohesive and unifying model is broadly relevant to bioethics, philosophy, theology, and legal medicine, fostering interdisciplinary dialogue and application.

Practical Implications of Conceptual Refinement

The conceptual refinement offered by the *Brain Metafunction Framework* provides practical solutions to long-standing challenges in the clinical determination of death:

1. Support for Practical Criteria:
The framework supports reliance on clinically validated observations, such as the absence of brainstem reflexes and confirmatory tests, without necessitating proof of inactivity in every cell or brain micro region.

2. Role of Clinical Judgment:
As highlighted during the 1968 Sydney meeting (Gilder 1968), no single technological criterion can fully replace comprehensive clinical judgment. Observation periods should remain contextually informed and avoid arbitrary extensions.

3. Enhanced Moral Certainty:
By grounding confirmatory tests within a well-defined *conceptual framework*, the framework offers moral certitude to physicians, medical personnel, and families, alleviating uncertainty in difficult end-of-life decisions.

4. Explicit definition of Integration Within the Brain:
The permanent loss of the brain metafunctional system logically implies the absence of genuinely relevant brain functions. This integrated holistic perspective shifts focus away from individual "islands" of cellular activity and toward the systemic disintegration of the brain's unified and integrative essence. This refined framework advances the conceptual and practical understanding of death. It offers a robust rationale for clinical application while addressing the ethical and philosophical challenges surrounding the definition and determination of death. Further interdisciplinary dialogue and empirical validation will strengthen its integration into medical practice and ethical discourse.

Disintegration at the Level of Unity is Crucial

A central objection to the concept of brain death is the argument that certain bodily functions can persist in a meaningful and relevant sense. However, recent contributions by authors such as Bernat, Condic, Moschella, and McGee et al. (Bernat 2018 S19-S21; Condic 2016 257-278; Moschella 2019 437-453; McGee et al. 2023 434-452) provide a robust defense of brain death as a loss of the organism's critical hierarchical level of integration. These analyses differentiate between "the whole organism" and the "organism as a whole," offering clarity in understanding the brain's unique role in systemic integration. Condic identifies distinct hierarchical levels in the human body—molecular, cellular, tissue, organ, and systemic—and emphasizes that while lower levels (e.g., cellular or organ functions) may exhibit autonomous activity, the brain orchestrates the highest level of systemic integration. Bernat's mereological framework complements this by applying concepts of the "body as a whole" to the "brain as a whole." Together, these rationales substantiate that the brain's integrative functions are irreplaceable for maintaining organismal unity.

This formulation integrates the loss of consciousness with the regulation of bodily functions, presenting them as inseparable at the highest level of integration. It underscores the brain's hierarchical position as the critical unifying organ in the human person and organism. The brain metafunction framework articulates this principle, demonstrating that the permanent loss of this unified and unifying function marks the disintegration of the human organism and person. In classical philosophical terms, such disintegration constitutes a "substantial change," signifying the complete absence of the organism or person as one living substance. From a neuroscientific perspective, this rationale aligns the loss of personhood with human death, offering profound implications for bioethical, philosophical, and theological discussions. At this level of unity, the brain acts as the capacity for converging and expressing various unities—metaphysical, psychosocial, and biological—cementing its role as the defining criterion of life and death. By establishing the brain as the critical hierarchical level of integration, this framework reinforces the scientific and philosophical coherence of brain death. It provides a conceptual foundation for addressing contemporary ethical and clinical medical challenges.

A Unifying Rationale

As scholars like Pernick (1999) and Lizza (2004) have observed, the Harvard Committee alternated between emphasizing the loss of consciousness and the loss of bodily integration as the conceptual basis for the new brain-centered death criterion they ultimately proposed. Over the decades, diverse rationales have emerged to support brain-centered death formulations, including the "body as a whole" (Condic 2016; Bernat 2018; Lizza 2018; Moschella 2019; McGee, Gardiner, and Jansen 2023), "higher brain" functions (Youngner and Bartlett 1983), and broader definitions of consciousness (Machado 1999). However, these perspectives have often been treated as mutually exclusive, resting on a false dichotomy between consciousness and bodily integration.

The Brain Metafunction framework resolves this dichotomy by recognizing that these two rationales are inseparable once the level of systemic unity is explicitly acknowledged. Consciousness and bodily integration are not isolated processes but interdependent dimensions of a hierarchical and holistic unity in the human organism. This framework unifies these perspectives by highlighting the interconnectedness of brain functions and their role in maintaining the integrated unity of the person. This approach provides an elegant and cohesive rationale, incorporating the rich interplay of consciousness, bodily integration, and systemic regulation as fundamental aspects of human life. Addressing the interconnected dimensions of human beings offers a unifying model that transcends the limitations of prior fragmented formulations, paving the way for a more precise and comprehensive understanding of human death.

Integration Within the Brain and Residual Brain Activity

A key objection to brain death centers on the argument that isolated residual brain functions are meaningful and relevant. Canada's *2023 Clinical Practice Guidelines* defines death as "the permanent cessation of brain function (i.e., brain function is lost, will not resume spontaneously, and will not be restored through intervention)" and characterized by the absence of consciousness and brainstem reflexes, including independent breathing (Murphy et al. 2023). However, unlike the refined formulation proposed here, these guidelines lack a robust conceptual basis for distinguishing between essential brain functions critical to sustaining life and residual functions that do not reflect systemic integration.

Veatch (1992) highlighted this issue decades ago and criticized the whole-brain criterion for failing to differentiate between neurological integration occurring outside versus within the brain. The refined

Brain Metafunction framework addresses this limitation by offering a clear rationale for distinguishing essential brain functions from residual activities irrelevant to determining death. Examples of residual functions deemed immaterial in this framework include:

1. Pituitary Hormone Production:
   The temporary production of hormones like antidiuretic hormone, evident by the absence of diabetes insipidus, reflects localized activity rather than systemic brain integration.
2. Cervical-Spinal Junction Activity:
   Evoked potentials recorded via deep electrodes at the cervical-spinal junction occasionally persist but lack the integration required for brain metafunction.
3. Minimal EEG Activity:
   Electroencephalographic activity recorded at minimal levels does not indicate functional systemic connectivity.
4. Hemodynamic Responses:
   Blood pressure and heart rate variations during organ harvesting that originate from residual bulbar activity and do not indicate higher-level integration.
5. Isolated Reflex Activity:
   Reflexive actions originating in the inferior protuberance do not contribute to global brain integration.

The refined Brain Metafunction framework identifies the brain's massive "subtotal but supra-critical" global subsystems as the focal point for determining death. By doing so, it clarifies the distinction between functions necessary for sustaining life and those that are merely residual or reflexive. This approach ensures *conceptual coherence and operational fit* and explains the exclusion of certain functions from death determination criteria. Moreover, this framework facilitates consistency in clinical practice. It highlights the interconnected and dynamically integrated functions that define brain metafunction, enabling practitioners to discern between ongoing relevant brain activity and residual reflexive functions. By articulating these distinctions, the framework provides clarity and moral certitude for medical professionals, patient's families, and ethical oversight bodies, addressing potential controversies arising from the detection of non-essential functions. The refined formulation of human death aligns with advances in neuroscience and bioethics, providing a robust and practical framework that transcends the limitations of current conceptual criteria while promoting clarity and consistency in clinical and ethical decision-making.

The Loss of All Relevant Functions

A recent analysis of Canada's 2023 clinical practice guidelines highlights the need for precise articulation of what is encompassed by the cessation of brain function (Murphy et al., 2023; Shemie, Wilson, and Hornby, 2023). Unlike the often misinterpreted phrase "the absence of all brain functions" (President's Commission 1981), brain death refers to the cessation of the *critical system*—a loss encompassing all *relevant functions* necessary for the brain's integrative role. While total brain infarction is rare, clinical evidence indicates that massive subtotal but supra-critical infarction results in the disintegration of this system. This understanding is derived from a comprehensive evaluation of the clinical context, not merely from detecting isolated residual activity. The refined formulation offered here explicitly articulates the meaning of *cessation of brain function*, enhancing conceptual clarity and guiding practical implementation.

This approach emphasizes that isolated brain activities, such as reflexive or non-integrative functions, do not indicate continued brain metafunction. For instance, minimal electroencephalographic activity or transient hormonal outputs do not undermine the diagnosis of brain death when the brain's integrative systems have irreversibly ceased functioning. By clearly delineating these parameters, the refined formulation improves the reliability and consistency of death determination practices, reducing the potential for misinterpretation or error. Moreover, this conceptual refinement mitigates the risk of both false positives and false negatives in death determination. By focusing on the cessation of critical system functions, the formulation minimizes reliance on exhaustive technological investigations to confirm the absence of all residual activity, offering a balanced and efficient approach. This specificity ensures that individuals are neither erroneously identified as deceased nor mistakenly excluded from a valid determination of death. Consequently, the refined formulation enhances death determination's theoretical and practical dimensions, providing a robust and ethically sound framework that aligns with contemporary medical, legal, and bioethical standards.

Consciousness and the Rest of the Brain's Metafunction Subfunctions are Inseparable

In their article *"Human death and high technology: the failure of the whole-brain formulation"* (1983), Youngner and Bartlett reasoned that death entails the permanent cessation of functions that are inherently irreplaceable and vital for sustaining life. These functions must arise spontaneously from within the organism, reflecting its intrinsic capabilities rather than external technological interventions. At present, as Shemie and Gardiner (2018) observe, "It is now possible to sustain and/or replace organ function for all vital organs and systems except the brain." This understanding highlights the brain as the singular organ whose functions—integrated within the brain metafunction framework—are essential and irreplaceable for the human organism's unity and survival. The brain metafunction framework uniquely fulfills these criteria by conceptualizing brain activity as a unified system of interdependent subfunctions, including consciousness, cognitive processes, autonomic regulation, homeostasis, and allostasis. Contrary to the "higher-brain" definition of death, which isolates consciousness from other subfunctions, this refined formulation emphasizes that consciousness is inseparable from the rest of the brain's metafunction subfunctions. These subfunctions operate as a cohesive, irreducible whole, integrating the organism's cognitive, physiological, and regulatory processes.

The higher-brain definition falters because it artificially segregates consciousness, treating it as an isolated criterion for determining death. However, consciousness is deeply intertwined with other brain functions, such as the regulation of homeostasis and the integration of sensory and motor inputs. For example, consciousness relies on the same neural networks that support autonomic functions, demonstrating the inseparability of these processes. By disregarding this interdependence, the higher-brain definition fails to account for the integrated nature of brain activity. The brain metafunction framework, in contrast, offers a holistic perspective that aligns with the organism's intrinsic capabilities and technological limitations. It provides a cohesive rationale for defining death based on the permanent cessation of the brain's critical unified function, encompassing all system-wide, integrated subfunctions necessary for sustaining life and personhood. This framework resolves conceptual and practical inconsistencies by emphasizing the inseparability of consciousness and other metafunction subfunctions, offering a robust foundation for understanding and determining human death.

Death as a Transition Between the Process of Dying and the Process of Disintegration

The debates about the concept of death have led some to argue that death is a "fuzzy" concept, inherently difficult or even impossible to define (Chiong 2005). However, the Brain-Metafunction-centered formulation recognizes death as a fundamental change: a singular event marking the transition between the process of dying and the process of disintegration. This perspective builds on Kass's (1971) assertion that death should not be reduced to a gradual process, as doing so risks eroding the conceptual clarity necessary to distinguish the living from the dead. This formulation addresses the conceptual tension between continuity and discontinuity by framing death as a distinct event within a dynamic process. Dying involves the process of loss of vital functions, while disintegration refers to the irreversible breakdown of the organism's structure and unity. Death, situated between these stages, serves as a pivotal moment when both the organism and the person's unified function ceases definitively. This understanding avoids the conceptual pitfalls of viewing death as purely processual or static.

Recognizing death as a transition event has critical implications. It reinforces the importance of pinpointing the moment of transition from life to death with precision. This recognition is particularly significant in organ donation, end-of-life care, and legal determinations of death. In cases of brain death, this framework facilitates the distinction between residual biological activity and the loss of unified brain metafunction, ensuring ethical and medically sound decisions. Moreover, this nuanced understanding of death challenges reductive perspectives that treat it solely as a biological endpoint or a social construct. It underscores the interconnectedness of biological, metaphysical, and ethical dimensions in defining human death. By preserving the conceptual boundary between life and death, this framework enables a more profound engagement with the existential, moral, and clinical realities surrounding human mortality. The Brain-Metafunction-centered formulation offers a refined understanding of death as a singular event within a broader process. By emphasizing death's transitional nature, this perspective maintains clarity and coherence in defining the boundary between life and death, with far-reaching implications for medical, ethical, and philosophical discourse.

Should Death be Viewed as Several Separate Events?

Shewmon critiques the concept of brain death by arguing that significant biological functions can persist even after brain activity ceases, suggesting that the body retains a meaningful form of life. He proposes a broader definition of death that integrates biological, psychological, and social dimensions, viewing death not as a single event but as two distinct moments: "passing away" (civil death) and "deanimation" (the metaphysical end) (Shewmon 2010). Similarly, Truog (2020) advocates for reframing brain death as a "social construct," which, while legally and ethically treating brain-dead individuals as deceased, separates this designation from the biological concept of death. In contrast to these views, the refined formulation of death here reconciles such perspectives by emphasizing the inseparability of biological, psychosocial, and metaphysical dimensions through the brain metafunction framework. This approach highlights two crucial aspects:

1. *Unity of the Human Organism and Person are inseparable*:
   The brain metafunction serves as the critical level of unity, integrating functions across biological, psychological, and metaphysical levels. This unified framework underscores that death signifies the irreversible loss of personhood and the organism's systemic coherence.

2. *Death as a Definitive Event*:
   The refined formulation identifies the cessation of brain activity as the ultimate disruption at the brain metafunction level, marking the definitive end of human life. This approach conceptualizes death as a singular, unified event rather than distinct stages.

By acknowledging the unity of the human organism and person, this formulation addresses the concerns of Shewmon, Truog, and others regarding biological, psychological, and social dimensions while grounding the definition of death in contemporary neuroscience. Over the past two decades, advances have reinforced the brain's role as the pinnacle of bodily integration. The cessation of brain metafunction indicates the loss of all relevant integrative capacities, culminating in the transition from life to death. This unified perspective ensures conceptual clarity while offering practical medical, ethical, and legal guidance. By resolving debates surrounding the fragmentation of death into separate events, the refined formulation promotes consistency and coherence in defining the boundary between life and death.

Public Perception

One of the key criticisms of brain-centered concepts of death, as noted by L. Syd M. Johnson, is "the significant public skepticism about the concept" (Johnson 2017). Public understanding and acceptance of neurologic criteria for determining death remain contentious in some contexts. However, a recent study of Canadian perspectives found that most Canadians accept both neurologic and circulatory criteria for death determination, indicating openness to brain-centered concepts (Sarti et al. 2023). Importantly, this study revealed strong public interest in learning about death determination, highlighting opportunities for targeted public engagement and education. However, empirical research on public perception of death determination, particularly concerning cardiorespiratory standards, is lacking. In a recent review, only two Canadian and American studies examined public understanding of death based on circulatory determination (Zheng et al. 2022). This review underscores the need for further research to comprehensively assess and address public attitudes and misconceptions. Clear and consistent communication from healthcare professionals is critical to fostering public trust. Conceptual clarity among medical experts provides the foundation for standardized messaging and supports public confidence in the determination of death. Developing communication protocols that explain brain death's rationale transparently and ethically is essential. Moreover, ensuring moral certitude in determining death requires that medical professionals possess deep, independent knowledge of death's conceptual and physiological aspects, separate from its implications for organ procurement.

The refined formulation of brain death, grounded in the brain metafunction framework, offers a coherent and unifying rationale for declaring individuals brain-dead as deceased. By integrating biological, psychosocial, and spiritual dimensions, this framework presents death as the permanent loss of the brain's capacity to sustain systemic unity and personhood. This nuanced understanding reconciles divergent perspectives in medicine, bioethics, and philosophy, enabling more precise explanations to patients, families, and the public. Education rooted in this coherent rationale is essential for fostering informed decision-making and ethical practices within the medical profession. Public confidence in the healthcare system depends on perceiving medical decisions as grounded in sound reasoning and guided by ethical principles. By demonstrating that brain death is both scientifically validated and ethically justified, healthcare professionals can reinforce trust in medical practices surrounding death determination. Furthermore, such education extends beyond professionals to public awareness campaigns and accessible information for families navigating these sensitive issues. Transparent, evidence-based, and conceptually sound education can promote ethical conduct in medicine and strengthen public trust in healthcare practices, particularly in areas as complex and emotionally charged as determining human death.

Conclusion

The Brain Metafunction framework offers a refined and robust formulation of human death, grounded in the neurological basis of the level of unity in the human organism and person. By refining the concept of death, this framework addresses practical and ethical challenges in medicine and bioethics that previous formulations have left unresolved. Crucially, the framework conceptualizes death as the dissolution of unity at the brain metafunction level, integrating consciousness and systemic bodily regulation into a cohesive understanding of personhood. This perspective aligns with Western philosophical and theological traditions, where death is "a single event, consisting in the total disintegration of that unitary and integrated whole that is the personal self. It results from the separation of the life-principle (or soul) from the corporal reality of the person. The death of the person, understood in this primary sense, is an event which no scientific technique or empirical method can identify directly" (John Paul II 2000). The Brain Metafunction framework provides a compelling rationale for understanding death by reconciling this metaphysical view with modern neuroscientific insights.

Future research should explore the framework's application in bioethics, including its implications for organ transplantation, end-of-life care, and public education on death determination. Additionally, interdisciplinary studies could investigate how this formulation informs legal, medical, and theological discussions about human dignity, personhood, and the ethical boundaries of life-sustaining interventions. By bridging conceptual clarity with practical applications, the Brain Metafunction framework has the potential to unify diverse perspectives on death and enhance trust in medical practices, ensuring ethically and scientifically grounded approaches to one of humanity's most profound questions.

REFERENCES

Ad Hoc Committee of the Harvard Medical School to Examine the Definition of Brain Death. (1968). A definition of irreversible coma. *JAMA, 205*(6), 337–340.

Aquinas, T. (1265–1274). *Summa Theologica* (Vol. 4, Treatise on Man, Questions LXXV–CII; Fathers of the English Dominican Province, Trans.). Christian Classics.

Aquinas, T. (1259–1265). *Summa Contra Gentiles* (Book 2, Chapter 68, Section 2; A. C. Pegis, Trans.). University of Notre Dame Press.

Aquinas, T. (c. 1265). *Quaestiones Disputatae de Anima*, q. 11, resp. (cited in Finley, J. (2016). *The extraordinary unity of the human being.* St. Vincent de Paul Lecture and Concert Series).

Aristotle. (1893). *Nicomachean ethics* (F. H. Peters, Trans., 5th ed.). London: Kegan Paul, Trench, & Trübner.

Aristotle. (1986). *De Anima* (H. Lawson-Tancred, Trans.). Penguin Books.

Avena-Koenigsberger, A., Mišić, B., & Sporns, O. (2018). Communication dynamics in complex brain networks. *Nature Reviews Neuroscience, 19*(1), 17–33.

Barnard, C. N. (1967). The operation: A human cardiac transplant: An interim report of a successful operation performed at Groote Schuur Hospital, Cape Town. *South African Medical Journal, 41*(48), 1271–1274.

Barrett, L. F. (2017). *How emotions are made: The secret life of the brain.* Houghton Mifflin Harcourt.

Barrett, L. F., & Simmons, W. K. (2015). Interoceptive predictions in the brain. *Nature Reviews Neuroscience, 16*, 419–429.

Bernat, J. L. (2018). A conceptual justification for brain death. *Hastings Center Report, 48*(Suppl. 4), S19–S21.

Berthelsen, P. G., & Cronqvist, M. (2003). The first intensive care unit in the world: Copenhagen 1953. *Acta Anaesthesiologica Scandinavica, 47*(10), 1190–1195.

Boff, L. (1980). *La resurrección de Cristo: Nuestra resurrección en la muerte* [The resurrection of Christ: Our resurrection in death]. Santander: SAL TERRAE.

Brugger, E. C. (2016). Are brain-dead individuals dead? Grounds for reasonable doubt. *Journal of Medicine and Philosophy, 41*(3), 329–350.

Chanes, L., & Barrett, L. F. (2016). Redefining the role of limbic areas in cortical processing. *Trends in Cognitive Sciences, 20*, 96–106.

Chein, J. M., Moore, A. B., & Conway, A. R. A. (2011). Domain-general mechanisms of complex working memory span. *NeuroImage, 54*, 550–559.

Chiong, W. (2005). Brain death without definitions. *Hastings Center Report, 35*(6), 20–30.

Cohen, J. R., & D'Esposito, M. (2016). The segregation and integration of distinct brain networks and their relationship to cognition. *Journal of Neuroscience, 36*, 12083–12094.

Cole, M. W., Reynolds, J. R., Power, J. D., Repovs, G., Anticevic, A., & Braver, T. S. (2013). Multi-task connectivity reveals flexible hubs for adaptive task control. *Nature Neuroscience, 16*(9), 1348–1355.

Condic, M. L. (2016). Determination of death: A scientific perspective on biological integration. *Journal of Medicine and Philosophy, 41*(3), 257–278.

Cowan, N., Li, D., Moffit, A., Becker, T. M., Martin, E. A., Saults, J. S., & Christ, S. E. (2011). A neural region of abstract working memory. *Journal of Cognitive Neuroscience, 23*, 2852–2863.

Damasio, A. R. (1996). The somatic marker hypothesis and the possible functions of the prefrontal cortex. *Philosophical Transactions of the Royal Society of London. Series B, Biological Sciences, 351*(1346), 1413–1420.

Dehaene, S., & Changeux, J. P. (2011). Experimental and theoretical approaches to conscious processing. *Neuron, 70*, 200–227.

Des Chene, D. (2000). *Life's form: Late Aristotelian conceptions of the soul.* Ithaca, NY: Cornell University Press.

Finley, J. (2016, October). The extraordinary unity of the human being [Lecture transcript]. St. Vincent de Paul Lecture and Concert Series. https://www.thomasaquinas.edu/news/dr-john-finley-extraordinary-unity-human-being#_ftnref2.

Friston, K. (2010). The free-energy principle: A unified brain theory? *Nature Reviews Neuroscience, 11*, 127–138.

García, O. D. (1995). Whole brain? Higher brain? A new formulation of death. In C. Machado (Ed.), *Brain death* (pp. 191–200). Amsterdam: Elsevier.

García, O. D. (1997). Una reformulación completa y precisa de la muerte humana [A complete and accurate reformulation of human death]. *Cuadernos de Bioética, 1997*(2).

García, O. D. (1998a). La unidad de la persona y su fundamentación neurológica [The unity of the person and its neurological foundation]. *Cuadernos de Bioética*, 1998(2).

García, O. D. (1998b). A new formulation of death and its relevance to medical law. *Medicine and Law, 17*(4), 493–501.

García, O. D. (1998c). Una reformulación completa y precisa de la muerte humana [A complete and accurate reformulation of human death]. *Revista de Neurología, 26*(154), 1048–1053.

García, O. D. (1999). Commentary on the article on the death of Dr. Alejandro Serani: "Death" of the brain and the practical determination of death: Another dissenting opinion. *Cuadernos de Bioética, 37*(1).

García, O. D. (2001). The unity of the human person: A central concept in the interface between medicine and theology. *The Linacre Quarterly, 68*(1), 71–89. https://doi.org/10.1080/20508549.2001.11877600.

García, O. D., Machado, C., Román, J. M., Cabrera, A., Díaz-Comas, L., Rivera, B., & Grave de Peralta, R. (1995). Heart rate variability in coma and brain death. In C. Machado (Ed.), *Brain death (Proceedings of the Second International Symposium on Brain Death), Developments in Neurology* (Vol. 9, pp. 191–200). Amsterdam: Elsevier.

Gilder, S. S. B. (1968). Twenty-second World Medical Assembly. *British Medical Journal, 3*, 493–494.

Greer, D. M., Shemie, S. D., Lewis, A., et al. (2020). Determination of brain death/death by neurologic criteria: The World Brain Death Project. *JAMA*, 324(11), 1078–1097. https://doi.org/10.1001/jama. 2020.11586

Gu, X., & FitzGerald, T. H. (2014). Interoceptive inference: Homeostasis and decision-making. *Trends in Cognitive Sciences, 18*, 269 –270.

Halevy, A., & Brody, B. (1993). Brain death: Reconciling definitions, criteria, and tests. *Annals of Internal Medicine, 119*, 519–525.

Hardie, W. F. R. (1976). Concepts of consciousness in Aristotle. *Mind, 85*(July), 388–411.

Honorary Secretary of the Conference of Medical Royal Colleges and their Faculties in the United Kingdom. (1976). Diagnosis of brain death. *British Medical Journal, 2*(6045), 1187–1188.

Honorary Secretary of the Conference of Medical Royal Colleges and their Faculties in the United Kingdom. (1979). Diagnosis of death. *Lancet, 313*, 261–262.

John Paul II, Pope. (1996). La doctrina tomista sobre el alma en relación con los valores de nuestro tiempo [The Thomistic doctrine on the soul in relation to the values of our time]. *L'Osservatore Romano, January 4*, 23-VI-1886, p. 23.

John Paul II, Pope. (2000, August 29). Address to the 18th International Congress of the Transplantation Society. Vatican City. Retrieved from https://www.vatican.va/content/john-paul-ii/en/speeches/2000/jul-sep/documents/hf_jp-ii_spe_20000829_transplants.html.

Johnson, L. S. M. (2017). Death by neurological criteria: Expert definitions and lay misgivings. *QJM: An International Journal of Medicine, 110*(5), 267–270.

Katsumi, Y., Zhang, J., Chen, D., et al. (2023). Correspondence of functional connectivity gradients across human isocortex, cerebellum, and hippocampus. *Communications Biology, 6*, 401.

Kelly, W. D., Lillehei, R. C., Merkel, F. K., Idezuki, Y., & Goetz, F. C. (1967). Allotransplantation of the pancreas and duodenum along with the kidney in diabetic nephropathy. *Surgery, 61*(6), 827–837.

Kelso, J. S. (2012). Multistability and metastability: Understanding dynamic coordination in the brain. *Philosophical Transactions of the Royal Society B: Biological Sciences, 367*, 906–918.

Kittel, G., & Friedrich, G. (Eds.). (1965). *Theological dictionary of the New Testament* (Trans. G. W. Bromiley). Eerdmans.

Kleckner, I. R., Zhang, J., Touroutoglou, A., Chanes, L., Xia, C., Simmons, W. K., Quigley, K. S., Dickerson, B. C., & Barrett, L. F. (2017). Evidence for a large-scale brain system supporting allostasis and interoception in humans. *Nature Human Behaviour, 1*, 0069.

Koelsch, S., Schulze, K., Sammler, D., Fritz, T., Müller, K., & Gruber, O. (2009). Functional architecture of verbal and tonal working memory: An fMRI study. *Human Brain Mapping, 30*, 859–873.

Lizza, J. P. (2004). The conceptual basis for brain death revisited. In C. Machado & D. A. Shewmon (Eds.), *Brain death and disorders of consciousness* (Vol. 550, pp. 45–58). *Advances in Experimental Medicine and Biology*. Boston, MA: Springer.

Machado, C. (1995). A new definition of death based on the basic mechanisms of consciousness generation. , In C. Machado (Ed.), *Brain death* (pp. 57-66). Amsterdam: Elsevier.

Machado, C. (1999). Consciousness as a definition of death: Its appeal and complexity. *Clinical Electroencephalography, 30*(4), 156–164.

Machado, C., & García, O. D. (1995). Guidelines for the determination of death. In C. Machado (Ed.), *Brain death* (pp. 75-80). Amsterdam: Elsevier.

Machado, C., García, O. D., Gutiérrez, J., Portela, L., & García, M. C. (2005). Heart rate variability in comatose and brain-dead patients. *Clinical Neurophysiology, 116*, 2859–2860.

Majerus, S., D'Argembeau, A., Martinez Perez, T., Belayachi, S., Van der Linden, M., Collette, F., Salmon, E., Seurinck, R., Fias, W., & Maquet, P. (2010). The commonality of neural networks for verbal and visual short-term memory. *Journal of Cognitive Neuroscience, 22*, 2570–2593.

McGee, A., Gardiner, D., & Jansen, M. (2023). A new defense of brain death as the death of the human organism. *Journal of Medicine and Philosophy, 48*(5), 434–452.

McMahan, J. (1995). The metaphysics of brain death. *Bioethics, 9*(2), 91–126.

Merrill, J. P., Murray, J. E., Harrison, J. H., & Guild, W. R. (1956). Successful homotransplantation of the human kidney between identical twins. *JAMA, 160*(4), 277–282.

Mollaret, P., & Goulon, M. (1959). [The depassed coma (preliminary memoir)]. *Revue Neurologique, 101*, 3–15.

Moreland, J. P. (2018). In defense of a Thomistic-like dualism. In J. J. Loose, A. J. L. Menuge, & J. P. Moreland (Eds.), *The Blackwell companion to substance dualism* (pp. 102–122). Wiley Blackwell.

Moschella, M. (2019). The human organism is not a conductorless orchestra: A defense of brain death as true biological death. *Theoretical Medicine and Bioethics, 40*(5), 437–453.

Murphy, N. B., Hartwick, M., Wilson, L. C., Simpson, C., Shemie, S. D., Torrance, S., & Chandler, J. A. (2023). Rationale for revisions to the definition of death and criteria for its determination in Canada. *Canadian Journal of Anaesthesia, 70*(4), 558 –569.

Papez, J. W. (1937). A proposed mechanism of emotion. *Archives of Neurology and Psychiatry, 38*(4), 725–743.

Pasnau, R. (2011). *Metaphysical themes: 1274–1671.* Oxford: Clarend on.

Pernick, M. S. (1999). Brain death in a cultural context: The reconstruction of death, 1967–1981. In S. J. Youngner, R. M. Arnold, & R. Schapiro (Eds.), *The definition of death: Contemporary controversies* (pp. 3–33). Baltimore, MD: Johns Hopkins University Press.

Pezzulo, G., Rigoli, F., & Friston, K. (2015). Active inference, homeostatic regulation, and adaptive behavioural control. *Progress in Neurobiology, 134*, 17–35.

Plum, F. (1991). Coma and related global disturbances of the human conscious state. In A. Peters & E. G. Jones (Eds.), *Normal and altered states of function* (Vol. 9, pp. 307–317). *Cerebral Cortex.* Boston, MA: Springer.

President's Commission for the Study of Ethical Problems in Medicine and Biomedical and Behavioral Research. (1981). *Defining death: Medical, legal and ethical issues in the determination of death.* Washington, DC: Government Printing Office. Available at https://repository.library.georgetown.edu/bitstream/handle/ 10822/559345/defining_death.pdf?sequence=1.

Ratzinger, J. (1991). El misterio de la mente humana ["The mystery of the human mind"]. In *Actas de la Quinta Conferencia Internacional: La Mente Humana* (Vol. 16, pp. 18–20). *Dolentium Hominum.*

Sarti, A. J., Honarmand, K., Sutherland, S., Hornby, L., Wilson, L. C., Priestap, F., Carignan, R., Joanne, B., Brewster, H., Elliott-Pohl, K., Weiss, M., & Shemie, S. D. (2023). When is a person dead? The Canadian public's understanding of death and death determination: A nationwide survey. *Canadian Journal of Anaesthesia, 70*(4), 617–627.

Seth, A. K. (2013). Interoceptive inference, emotion, and the embodied self. *Trends in Cognitive Sciences, 17*, 565–573.

Seth, A. K. (2014). A predictive processing theory of sensorimotor contingencies: Explaining the puzzle of perceptual presence and its absence in synesthesia. *Cognitive Neuroscience, 5*(2), 97–118. https://doi.org/10.1080/17588928.2013.877880

Shemie, S. D., & Gardiner, D. (2018). Circulatory arrest, brain arrest and death determination. *Frontiers in Cardiovascular Medicine, 5*, 15.

Shemie, S. D., Wilson, L. C., Hornby, L., Baker, A. J., et al. (2023). A brain-based definition of death and criteria for its determination after arrest of circulation or neurologic function in Canada: A 2023 clinical practice guideline. *Canadian Journal of Anaesthesia, 70*(4), 483–557.

Shewmon, D. A. (1997). Recovery from 'brain death': A neurologist's apologia. *The Linacre Quarterly, 64*(1), 30–96.

Shewmon, D. A. (2001). The brain and somatic integration: Insights into the standard biological rationale for equating brain death with death. *Journal of Medicine and Philosophy, 26*(5), 457–478.

Shewmon, D. A. (2004). The "critical organ" for the organism as a whole: Lessons from the lowly spinal cord. *Advances in Experimental Medicine and Biology, 550*, 23–42.

Shewmon, D. A. (2009). Brain death: Can it be resuscitated? *Hastings Center Report, 39*(2), 18–24.

Shewmon, D. A. (2010). Constructing the death elephant: A synthetic paradigm shift for the definition, criteria, and tests for death. *The Journal of Medicine and Philosophy: A Forum for Bioethics and Philosophy of Medicine, 35*(3), 256–298.

Shewmon, D. A. (2018). The case of Jahi McMath: A neurologist's view. In *Defining death: Organ transplantation and the fifty-year legacy of the Harvard report on brain death*, special report, *Hastings Center Report, 48*(6), S74–S76.

Sporns, O. (2010). *Networks of the brain.* MIT Press.

Sporns, O. (2013). Network attributes for segregation and integration in the human brain. *Current Opinion in Neurobiology, 23*, 162–171.

Starzl, T. E., Marchioro, T. L., Vonkaulla, K. N., Hermann, G., Brittain, R. S., & Waddell, W. R. (1963). Homotransplantation of the liver in humans. *Surgery, Gynecology & Obstetrics, 117*, 659–676.

Sterling, P., & Eyer, J. (1988). Allostasis: A new paradigm to explain arousal pathology. In S. Fisher & J. Reason (Eds.), *Handbook of life stress, cognition and health* (pp. 629–649). John Wiley & Sons.

Suárez, F. (1597). *Disputationes metaphysicae.* As cited in Moreland, J. P. (2018). In J. J. Loose, A. J. L. Menuge, & J. P. Moreland (Eds.), *The Blackwell companion to substance dualism* (pp. 102–122). Wiley Blackwell.

Tognoli, E., & Kelso, J. S. (2014). The metastable brain. *Neuron, 81*, 35–48.

Tononi, G. (2004). An information integration theory of consciousness. *BMC Neuroscience, 5*, 42.

Truog, R. D. (2020). Defining death: Lessons from the case of Jahi McMath. *Pediatrics, 146*(Supplement_1), S75–S80.

Truog, R. D. (1997). Is it time to abandon brain death? *Hastings Center Report, 27*(1), January–February.

Truog, R. D., & Miller, F. (2012). Brain death: Justifications and critiques. *Clinical Ethics, 7*(3), 128–132.

van den Heuvel, M. P., & Sporns, O. (2011). Rich-club organization of the human connectome. *Journal of Neuroscience, 31*, 15775–15786.

Veatch, R. M. (1978). The definition of death: Ethical, philosophical, and policy confusion. In J. Korein (Ed.), *Brain death: Interrelated medical and social issues* (Vol. 315, pp. 307–321). New York, NY: New York Academy of Sciences.

Veatch, R. M. (1992). Brain death and slippery slopes. *Journal of Clinical Ethics, 3*(3), 181–187.

Veatch, R. (1975). The whole-brain-oriented concept of death: An outmoded philosophical formulation. *Journal of Thanatology, 3*, 13–23.

Wang, R., Liu, M., Cheng, X., Wu, Y., Hildebrandt, A., & Zhou, C. (2021). Segregation, integration and balance of large-scale resting brain networks configure different cognitive abilities. *arXiv preprint arXiv:2103.00475.*

Youngner, S. J., & Bartlett, E. T. (1983). Human death and high technology: The failure of the whole-brain formulations. *Annals of Internal Medicine, 99*, 252–258.

Zheng, K., Sutherland, S., Hornby, L., Shemie, S. D., Wilson, L., & Sarti, A. J. (2022). Public understandings of the definition and determination of death: A scoping review. *Transplant Direct, 8*, e1300.

15

Metafunctional Free Will

One of the most profound and enduring philosophical inquiries is whether we possess free will. Defining free will is challenging yet significant for individuals and society. Free will prompts exploration into the nature of human agency, the complexities of decision-making, and the interplay between determinism, indeterminism, and personal choice (O'Connor 2019, 2000). In general terms, free will refers to a person's capacity to make decisions not solely determined by external forces, pre-existing conditions, or random events (Kane 2011). It suggests that humans possess a degree of autonomy and agency over their actions, often associated with conscious deliberation, intentionality, and the ability to choose alternatives (Kane 2016). Our understanding of free will is intimately tied to our worldview—the fundamental beliefs, assumptions, and perspectives about the nature of reality, existence, morality, and the universe. This chapter provides a concise historical perspective on free will, from antiquity and Christianity to modern science and, more recently, neuroscience. Additionally, it addresses a significant philosophical issue: a "sui generis" form of inherent causality that does not fit the conventional category of physical causality. Lastly, we explore free will within the brain metafunction framework and its relation to the unity of the person by introducing the concept of *metafunctional free will*, a multidimensional model of human agency that integrates biological, psychological, and transphysical elements. Unlike reductive accounts that locate decision-making solely within neural mechanisms or abstract metaphysical assertions, this concept proposes that free will emerges from the coordinated interaction of conscious deliberation, unconscious processes, and a transcendent causal domain that undergirds personhood. Metafunctional free will emphasizes that genuine autonomy arises not from escaping causality altogether, but from participating in layered, hierarchically integrated causes across functional domains.

The concept of metafunctional free will advances a novel and integrative account of human agency, uniting biological, psychological, and transphysical dimensions within a single explanatory framework.

Rather than locating free will solely in neural determinism, cognitive intentionality, or metaphysical dualism, this model introduces metafunctional causation: a higher-order causal structure in which mental and agent causation interact dynamically and coherently. This integrated approach provides a structured account of how conscious deliberation, brain processes, and moral responsibility are mutually implicated in human volition. Additionally, by presenting free will as a properly basic belief, a foundational cognitive intuition not invalidated by current empirical findings, this model affirms the epistemic and existential necessity of agency. In doing so, it resists the reductive logic of materialist eliminativism and the fatalism of strict determinism, offering instead a scientifically grounded, philosophically coherent, and theologically resonant understanding of human autonomy.

Free Will in Antiquity

In antiquity, particularly in Western civilization, the term free will does not explicitly appear. Still, it is implicit in philosophical reflections on choice and control over actions, as seen in the works of Plato, Aristotle (Irwin, 1992), and other Greek thinkers. In *The Republic* (Book IV), Plato identifies three elements of the human soul: reason, spiritedness, and appetite. He envisions an ideal state of justice within the individual, where reason guides, spiritedness supports reason's judgments, and the passions are subdued under reason's authority. The absence of this inner justice leads to enslavement by base desires. For Plato, freedom is a form of self-mastery achieved through cultivating virtues like wisdom, courage, and temperance. This process liberates individuals from the tyranny of desires, enabling them to comprehend and pursue the good. Plato's concept suggests freedom for excellence instead of the modern notion of freedom of indifference (Pinckaers 1995) (Hecht 2014).

Aristotle believed that, while upbringing influences human actions, individuals still possess the capacity for rational deliberation. In his *Nicomachean Ethics* (Book III), Aristotle argued that humans are responsible for their voluntary actions and have free will within the boundaries of their inherent nature. Human freedom manifests in two ways: the ability to choose among alternatives presented by one's dispositions and the capacity to shape and cultivate those dispositions through conscious effort. These two aspects suggest that introspection and rational contemplation imply a transphysical element that transcends causal determinism (Irwin, 1999). Though open to interpretation, Aristotle's view aligns with a form of libertarianism regarding free will (Broadie, 2002).

While Plato and Aristotle dominate the classical Greek discourse on free will, other ancient perspectives contribute to its depth. The Stoics, such as Epictetus and Marcus Aurelius, offered a deterministic worldview, emphasizing alignment with the *logos* (universal reason). For the Stoics, freedom was internal, achieved by mastering one's judgments and desires to live in harmony with nature, a concept that complements and contrasts with Plato and Aristotle's more libertarian inclinations (Long & Sedley, 1987). In contrast, Epicurus introduced a materialistic framework where atoms' random swerves (*clinamen*) provided a basis for human freedom, challenging strict determinism. This early naturalistic approach emphasized spontaneity and the rejection of divine or external control over human agency (O'Keefe, 2005).

These Greek philosophies collectively laid the foundation for later discussions on free will, especially their emphasis on moral development, rationality, and self-mastery. The contrast between ancient and modern notions of freedom is particularly notable. Plato's *freedom for excellence*—rooted in virtue and self-mastery—differs markedly from the contemporary *freedom of indifference*, which prioritizes the ability to choose freely among alternatives without moral or rational constraints (Hecht, 2014; Taylor, 2012). For Plato, true freedom entails acting under reason and the good, while the modern perspective often emphasizes autonomy as an end in itself. Similarly, Aristotle's concept of freedom as rational choice conditioned by virtue underscores ethical responsibility, whereas modern interpretations may focus on autonomy divorced from external moral or rational guidance (Broadie, 2002). Christian thought, beginning with early Church Fathers like Augustine, reinterpreted free will in light of theological concerns, such as divine omniscience, grace, and human sinfulness. This interpretation introduced significant questions about the compatibility of free will with God's foreknowledge and the role of grace in human freedom, laying the foundation for centuries of theological inquiry about the relationship between free will, moral responsibility, and divine providence (Matthews 2001; Stump 2001).

Free Will in Christianity

"Liberum arbitrium" (free will) originated in Latin Christian and philosophical traditions. One of the early Latin Christian writers, Tertullian (c. 155–220 CE), discussed concepts related to free will, though he did not fully systematize them. In works like *Adversus Marcionem* (Against Marcion), he defended human freedom in moral choices (Tertullian, 1972). Similarly, Lactantius (c. 250–325 CE), another early Christian thinker, elaborated on the idea of human freedom and moral responsibility in his work *Divinae Institutiones* (Divine Institutes), where he emphasized that humans, created by God, possess liberum arbitrium to pursue virtue or vice (Lactantius, 2002). Ambrose of Milan (c. 340–397 CE), one of Augustine's major influences, also used the term when discussing human will and divine justice (Ambrose of Milan, 1961).

Although Augustine (354–430 CE) did not introduce the term, he significantly refined and theologically developed it. His work *De libero arbitrio* (On Free Choice of the Will, c. 395 CE) systematically explores the relationship between human will, divine grace, and moral responsibility (Augustine, 1998). Augustine later introduced a key distinction between liberum arbitrium (the mere capacity to choose) and *libertas* (true freedom, which he associated with choosing rightly through God's grace). This distinction became central to later Christian thought. His theological contributions laid the groundwork for the medieval and modern Christian understanding of free will, particularly concerning grace, sin, and divine foreknowledge (Teske 2006). His influence shaped later figures such as Anselm, Aquinas, and the Reformers, making him the most pivotal thinker in the development of the concept. The concept of free will in Christian theology initially denoted the absence of necessity in human choice, implying that the will was not constrained to a specific course of action. Both incompatibilists and compatibilists accepted this essential aspect, though the Enlightenment later introduced alternative interpretations. Augustine asserts that "man's ability to live rightly" is grounded in the freedom of the will. He emphasized that our capacity to act righteously hinges on the freedom of choice inherent in the will. For Augustine, free will was essential for moral responsibility, as it underpins the capacity to choose the virtuous path.

Furthermore, Augustine argued that evil in the world stems from the misuse of free will, as seen in the deliberate disobedience of Adam and Eve (Augustine, 1998). This emphasis on the connection between moral responsibility and free will would profoundly influence later Christian theologians, including Aquinas and Calvin. Augustine closely tied his perspective on free will to his theodicy and the problem of evil. He argued that evil is not an independent entity but a privation of good—a distortion of the natural order. "Evil has no positive nature; but the loss of good has received the name 'evil,'" he wrote in *The City of God*. Thus, evil arises from human misuse of free will, leading to moral wrongdoing and suffering (Matthews, 2001). Augustine's theological framework addressed the origin of evil and reinforced the need for divine grace. He emphasized that although free will enables moral responsibility, the human will is weakened by sin and requires God's grace for moral restoration. This Augustinian framework significantly influenced later Christian thought, particularly during the Reformation through John Calvin. Calvin built upon Augustine's view that human nature was corrupted by sin, arguing that divine grace is essential for ethical action. Without God's grace, Calvin contended, the human will alone is insufficient to choose the good. This interplay between grace and free will became central to Protestant theology, especially in debates over predestination and moral responsibility (Calvin 1960).

Aquinas synthesized these earlier ideas while addressing objections drawn from biblical passages, such as Romans 7:19 ("For I do not do the good I want, but the evil I do not want is what I keep on doing"). In his *Summa Theologiae,* written between 1265 and 1274, Aquinas defended free will by arguing that human actions, moral judgments, and counsel would be meaningless without it. He distinguished human free will from instinctive behavior, noting that while animals act according to instinct, humans act from reason. This rational deliberation allows humans to choose freely, making free will necessary for moral responsibility (Summa Theologiae, I, Q.83, Art.1) (Kretzmann, 1993). Aquinas's work reflects Augustine's influence, especially in the role of divine grace in enabling free will. However, Aquinas emphasized the harmony between grace and human nature, asserting that grace perfects, rather than overrides, free will.

The role of divine grace remains a central theme in Christian theology. In Eastern Orthodox thought, synergism emphasizes the cooperation between human free will and divine grace in salvation. This view, rooted in the writings of John Chrysostom and Gregory Palamas, maintains that salvation involves both human response and God's enabling power, with grace taking the initiative (Lossky, 1957). In contrast, monergism, particularly in Calvinist theology, emphasizes that salvation is initiated solely by divine grace, with human free will playing no independent role in starting the process. However, Calvinism does acknowledge that believers actively respond to God's grace — a response that is itself enabled by grace — and participate in the ongoing process of sanctification and Christian living. These theological distinctions highlight the broader diversity within Christian traditions regarding the relationship between free will, moral responsibility, and divine providence.

In modern times, Alvin C. Plantinga adapted Augustine's ideas in his free will defense, addressing the logical problem of evil. Unlike Augustine, Plantinga did not seek to prove God's existence but aimed to show that the presence of evil is not logically inconsistent with the reality of an omnipotent, benevolent God (Plantinga, 1974). According to Plantinga, free will is a necessary condition for moral responsibility, and the existence of evil results from human choices that deviate from the good. In this framework, human freedom is crucial for understanding the persistence of evil without compromising divine goodness. Plantinga's argument includes the concept of "Transworld depravity," which posits that even an omnipotent God cannot ensure that free creatures always choose good without negating their freedom. His ideas support Augustine's theodicy, connecting ancient theological concerns with contemporary philosophy (Goetz, 2011).

Another vital contribution to understanding free will is Mortimer Adler's framework, which identifies three types of freedom. The first, circumstantial freedom, refers to freedom from external coercion or restraint, allowing individuals to exercise their will. Second, natural freedom is the inherent ability to make decisions, plans, and choices. Adler argues that biblical texts, such as Deuteronomy 30:19 ("Choose life, so that you and your descendants may live") and Joshua 24:15 ("Choose for yourselves this day whom you will serve"), assume this type of freedom. The third type, acquired freedom, involves the ability to live as one ought to live. In Christian thought, this transformation occurs through divine grace, which perfects human freedom by aligning the will with righteousness (Adler, 1991; McInerny, 1982).

Recent scholarship, such as Michael Frede's work, has questioned the presence of free will in early Christian texts. Frede argued that "freedom and free will cannot be found in either the Septuagint or the New Testament" (Frede, 2011). Similarly, Alister McGrath noted that neither the language nor the concept of free will appear in these texts (McGrath, 2004). Both scholars contend that the idea of free will was introduced into Christian thought by Stoic philosophy and later developed by theologians like Tertullian. Despite these claims, many theologians and philosophers find implicit evidence of free will within biblical narratives. For example, the story of Adam and Eve reflects deliberate choices to disobey God, embodying the exercise of free will. This brief review illustrates that discussing free will in Christianity involves intricate theological, philosophical, and practical considerations. The historical perspectives outlined here—from Augustine, Aquinas, and

Calvin to modern thinkers like Plantinga and Adler—lay a foundation for understanding how Christian thought introduced the term and has treated free will. This exploration of free will sets the stage for further examination of its role in contemporary thought and neuroscience.

Free Will and Modern Science

Newtonian Determinism and the Rise of Mechanistic Thought

Sir Isaac Newton, the brilliant English mathematician, physicist, astronomer, theologian, and author, was pivotal in the Scientific Revolution and in shaping the subsequent Enlightenment period. His work in physics became a cornerstone of deterministic thought in Western philosophy, particularly in developing Newtonian mechanics. Based on his laws of motion and universal gravitation, Newtonian mechanics describes the material universe as functioning according to a fixed set of physical laws. This framework introduces a deterministic view of the cosmos, suggesting that the universe's initial conditions determine all future events with predictable certainty. In this model, if one had perfect knowledge of all the positions and velocities of particles at any given moment, one could theoretically predict the outcome of every future event, much like the movement of billiard balls on a table (Newton 1687). This deterministic view influenced Enlightenment thinkers such as Pierre-Simon Laplace, who famously articulated the notion of a hypothetical "Laplace's Demon"—an entity that, if it knew the precise state of all particles in the universe at a given time, could predict the entire future with complete certainty (Laplace 1814). This classical determinism led many to question the possibility of free will, as the universe appeared to operate mechanistically and pre-determinedly, leaving no room for spontaneous or undetermined human actions.

Quantum Mechanics and Indeterminism

In the late 19th and early 20th centuries, cracks began to appear in this deterministic view of the universe. Physicists encountered challenges within Newton's theory that eventually gave rise to two revolutionary new theories—relativity and quantum mechanics. While Einstein's theory of relativity modified our understanding of space, time, and gravity, quantum mechanics posed the most significant challenge to classical determinism. Quantum mechanics introduced probabilistic laws at the subatomic level, where particles such as electrons do not follow predictable paths but behave probabilistically, governed by Heisenberg's uncertainty principle (Heisenberg 1927). This principle states that the more precisely one measures the position of a particle, the less precisely one can know its momentum, and vice versa, leading to inherent unpredictability in the behavior of particles. The implications of quantum mechanics for free will are profound. Unlike Newtonian mechanics' deterministic universe, quantum mechanics allows for indeterminism at the foundational level of physical reality. Some philosophers argue that quantum indeterminacy allows human free will to exist, as not all events are strictly determined by prior causes (Kane 1999). However, critics contend that randomness does not equate to meaningful autonomy. Instead, the challenge lies in connecting subatomic indeterminacy to the complex processes underlying human decision-making.

Determinism and Philosophical Implications

Philosophically, determinism is the view that every event or action, including human decision-making, is causally determined by an unbroken chain of prior causes. This worldview suggests that the state of the universe at any given time, combined with the laws of nature, dictates only one possible future (Honderich 2002). Determinism has led to divergent positions in the free will debate:

1. *Compatibilism:* Compatibilism seeks to reconcile determinism with free will. Compatibilists argue that free will does not require the absence of causation but rather the ability to act according to one's desires and values without external coercion (Hume 1748). They emphasize internal factors, such as rational deliberation and personal motivations, asserting that they can constitute meaningful freedom even if the universe operates deterministically (Frankfurt 1969). Critics, however, challenge this perspective by pointing out that determinism negates the possibility of alternative choices, which many consider essential for genuine free will (van Inwagen 1983).

2. *Incompatibilism and Libertarianism:* Incompatibilists reject compatibilism's premise, asserting that free will and determinism are fundamentally incompatible. Among incompatibilists, libertarians argue that genuine free will exists and depends on the ability to make choices that are not entirely determined by prior events or laws of nature (Clarke 2003). Libertarianism often invokes the concept of agent causation, wherein individuals, as agents, can initiate actions independently of deterministic causal chains

(O'Connor 2000). Robert Kane's idea of self-forming actions (SFAs) is central to this view, suggesting that key decision-making moments shape a person's character and future actions in a non-determined, autonomous way (Kane 1996).

3. *Hard Determinism:* Hard determinists, on the other hand, deny the existence of free will entirely, arguing that all actions are the inevitable result of preceding causes. Philosophers like Galen Strawson contend that because individuals do not control the circumstances of their birth or upbringing, they cannot be held morally responsible for their actions in the traditional sense (Strawson 1994).

Synthesis: Determinism, Indeterminism, and Free Will

The scientific shift from classical determinism to quantum mechanics challenges deterministic views and raises questions about whether indeterminism can support meaningful free will. Quantum mechanics introduces randomness at the subatomic level, but randomness alone does not equate to human autonomy. Similarly, philosophical debates about compatibilism and libertarianism highlight differing conceptions of freedom, with libertarianism emphasizing genuine alternatives and compatibilism focusing on internal motivations. The implications of these debates extend beyond academic philosophy to ethics, law, and social responsibility. A deterministic perspective might challenge traditional notions of moral accountability, while an indeterministic view rooted in agent causation might support a more vigorous defense of human autonomy. Ultimately, the interplay between scientific insights and philosophical theories deepens our understanding of free will, even as it complicates the search for definitive answers.

The Neuroscience of Free Will

The Readiness Potential

Before returning to the problem of determinism, indeterminism, and human freedom, let us review some aspects regarding the neural correlations of free will and some theoretical implications. Here, we will consider the compatibility of scientific understanding with the notion of free will, whether intellectual or moral and later examine it through the brain metafunction framework. In recent decades, findings from neuroscience have sparked debates about the existence of human free will, with some asserting that these findings challenge the notion of autonomous human agency. One of the most significant discoveries in this context was made in 1965 when Hans Kornhuber and Lüder Deecke reported their pioneering results on the readiness potential (*Bereitschaftspotential*) (Kornhuber and Deecke 1965). They observed a gradual buildup of electrical activity in the brain's motor regions before subjects initiated a voluntary movement, such as pressing a button. This neural activity began approximately one second before the movement occurred and was interpreted as the brain's preparatory process for voluntary actions.

What made this discovery particularly groundbreaking was the absence of any external stimulus or sensory cue determining when the subjects should act. Instead, the timing of the movement was determined solely by the subjects' volition. Kornhuber and Deecke described this buildup of electrical potential as "the electrophysiological manifestation of planning, preparation, and instigation of volitional actions" (Kornhuber and Deecke 1990). Researchers understood this readiness potential as reflecting the brain's preparation for self-initiated movements, providing critical insights into the neural processes underlying voluntary action. Some have interpreted the temporal gap between the onset of the readiness potential and the execution of a movement as potentially problematic for traditional notions of free will. The initial discovery suggested that brain activity related to movement initiation begins before individuals become consciously aware of their decision to move. This finding leads some to question whether volitional actions are genuinely free (Libet 1985). Was this a case of premature conclusions?

The Libet Experiment

The research paper titled "Time of Conscious Intention to Act in Relation to Onset of Cerebral Activity (Readiness Potential): The Unconscious Initiation of a Freely Voluntary Act," authored by Benjamin Libet and his colleagues, published in *Brain* in 1983, is a pivotal contribution to the debate on the nature of free will and conscious agency (Libet et al. 1983). In this study, Libet and his team explored the temporal relationship between the conscious intention to act and the onset of related neural activity, specifically the readiness potential, discovered by Kornhuber and Deecke in 1965. The controversial findings suggested that the readiness potential emerged approximately 350 milliseconds before participants became conscious of their intention to move. This preceding brain activity would imply that the brain initiated the movement unconsciously before the conscious decision to act. Conscious intention appeared to lag brain activity, which sparked significant debate regarding the implications of free will. Libet and colleagues proposed that while the brain initiates actions unconsciously, individuals retain the ability to veto or inhibit these actions before they are carried out—a concept he referred to as *free won't*. According to Libet, this ability to inhibit actions without a corresponding unconscious neural precursor represents a crucial aspect of free will (Libet 2004). He argued that the role of conscious will might not lie in initiating actions but in preventing them.

Challenges and Philosophical Counterarguments

While some have interpreted Libet's findings as evidence against free will, others, such as Alfred Mele, have questioned the extent to which the readiness potential predicts actions. Mele argues that Libet's results do not definitively demonstrate that unconscious neural activity determines voluntary decisions, suggesting that readiness potential reflects a general state of preparation rather than a specific intention to act (Mele 2009). Philosopher Daniel Dennett defends a compatibilist perspective, arguing that free will can still exist meaningfully in a deterministic framework. Dennett emphasizes that conscious will involves complex deliberations and feedback loops, which neuroscientific experiments cannot fully capture (Dennett 2003). This view aligns with critiques that caution against overinterpreting the implications of Libet's experiments for free will and moral responsibility. The findings surrounding the readiness potential and Libet's experiments appeared to challenge traditional views of conscious agency. However, philosophical counterarguments and alternative neuroscientific perspectives highlight the complexity of this issue, suggesting that the debate is far from settled. These discussions underscore the need for a nuanced understanding of the interplay between neural processes, conscious intention, and human autonomy.

Subsequent Research on Neural Determinants of Decision -Making

Adding another piece to the puzzle, research conducted in the 2000s and 2010s suggested that unconscious neural processes play a significant role in shaping decisions and actions before conscious awareness. This line of inquiry would support the view that the conscious sensation of having willed an action and being the instigator of that action is appended retrospectively. Such findings seemed to challenge traditional notions of free will by undermining the relationship between consciousness, decision-making, and the initiation of action. Mainstream media popularized this perspective, such as in the article "There's No Such Thing as Free Will," published by *The Atlantic* in 2016 (Berns). In 2008, Chun Siong Soon and colleagues published a study in *Nature Neuroscience* exploring the relationship between unconscious neural processes and conscious decision-making using fMRI. They revealed that brain activity in the prefrontal and parietal cortex could predict decisions up to seven seconds before participants were consciously aware of their choice (Soon et al. 2008). This finding suggests that unconscious neural precursors significantly shape decisions, challenging the idea that free will is driven solely by conscious intention. In 2009, Michel Desmurget and Angela Sirigu's work at CNRS in France furthered this understanding. Their experiments involved direct electrical stimulation of brain regions in patients undergoing surgery. By stimulating the posterior parietal cortex, patients experienced an intention to move without actual movement. Conversely, stimulating the premotor cortex caused actual movements, even when patients denied performing them (Desmurget and Sirigu 2009). These findings reveal distinct neural regions governing the sense of intention and the execution of movements, suggesting that free will may arise from complex neural interactions rather than solely from conscious deliberation.

Alternative Interpretations of Readiness Potential

Aaron Schurger and colleagues (2012) offered a paradigm-shifting reinterpretation of Libet's readiness potential. Their stochastic accumulator model suggests that the readiness potential represents the gradual accumulation of neural noise reaching a threshold, triggering spontaneous movement. This model challenges the traditional view that the readiness potential directly indicates motor preparation (Schurger, Sitt, and Dehaene 2012). Their findings highlight the complex neural dynamics underlying self-initiated actions, emphasizing that the readiness potential may not be a definitive marker of free will. In the same year, Filevich, Kühn, and Haggard (2013) published research showing that unconscious brain activity precedes both actions and decisions to inhibit actions. This finding challenges Libet's concept of "free won't," suggesting that even inhibitory decisions have unconscious precursors. More recently, Maoz et al. (2019) investigated the readiness potential (RP) in relation to decision-making, challenging its role as evidence against conscious control. While the RP was present in arbitrary decisions—those made without purpose or consequence—it was notably absent in deliberate decisions involving meaningful choices, such as allocating $1000 donations to non-profit organizations. Their findings suggest that the RP reflects random neural fluctuations that influence arbitrary decisions rather than deliberate, goal-directed ones. This distinction highlights that studies arguing against conscious control in decision-making may not apply to real-life, intentional decisions.

Broader Implications and Counterarguments

Although many studies suggest that unconscious neural activity precedes conscious decisions, some researchers argue that these findings do not necessarily undermine free will. For instance, Alfred Mele contends that the readiness potential reflects general preparatory states rather than specific intentions to act (Mele 2009). Similarly, Iain McGilchrist argues that unconscious processes are as integral to personal identity as conscious deliberation, suggesting that free will may involve both conscious and unconscious elements (McGilchrist 2009). Aaron Schurger emphasizes that the complexity of brain dynamics renders deterministic interpretations speculative at best. He notes, "no hard neuroscientific data... seriously undermines the belief in free will" (Schurger, 2021). Research on neural determinants of decision-making reveals intricate interactions between unconscious and conscious processes. While these findings challenge simplistic views of free will as a solely conscious agency in a Cartesian ghost in the machine scenario, they do not eliminate the possibility of meaningful volition. Instead, alternative interpretations highlight the complexity of these dynamics, supporting models such as the metafunctional free will framework. These insights underscore the need for a nuanced understanding of free will that integrates empirical findings with broader philosophical perspectives. Rather than disproving free will, neuroscientific evidence invites a broader framework—one in which human agency arises through the complex interaction of neural, psychological, and transphysical factors. Having considered the empirical findings surrounding neural precursors of action, we now turn to the philosophical implications of these findings for determinism and indeterminism.

The Problem Surrounding Free Will, Determinism, and Indeterminism

The complex debate surrounding free will presents a paradoxical challenge. On the one hand, if our actions are entirely determined by brain processes, determinism appears to undermine the possibility of genuine free will. On the other hand, if actions arise from random processes, the absence of intentional control likewise questions our capacity for agency. Despite this apparent dilemma, the persistent and widespread belief in free will suggests that neither pure determinism nor randomness fully explains human volition. The brain metafunction framework offers a more comprehensive perspective, recognizing agency as rooted not merely in neural processes but also in a transphysical dimension inherent to the human person. This dimension transcends the material constraints of brain function while remaining integrated with conscious and unconscious cognitive and physiological processes. From this perspective, agency reflects the unity of diverse dimensions—biological, psychological, and transphysical—enabling conscious intention, self-reflection, and purposeful action. By emphasizing this integrative view, the brain metafunction framework reconciles determinism and randomness with the human experience of volition, offering a coherent account of how free will operates within the broader context of personhood.

Determinism and Its Challenges

Determinism posits that every event, including human actions and decisions, is causally determined by prior events or conditions. Within this deterministic framework, the universe's trajectory is governed by a chain of cause-and-effect relationships, leaving no room for actions genuinely chosen or initiated by conscious agents (van Inwagen, 1983). A crucial problem within determinism revolves around the idea that if factors beyond our control predetermine all actions, the concept of free will seems incompatible. Determinism challenges notions of autonomy and moral responsibility, raising questions about whether individuals can be held accountable for choices merely resulting from pre-established causal sequences (Pereboom, 2001). This perspective undermines the traditional notion of free will, which involves conscious deliberation and genuine agency. However, compatibilists, such as Daniel Dennett, argue that freedom can exist within deterministic systems if free will is understood as the capacity to act in accordance with one's desires and rational deliberations without external coercion (Dennett, 2003). This view redefines free will, shifting the focus from metaphysical independence to practical autonomy.

Indeterminism and Its Implications

Indeterminism asserts that some events are random and not determined by prior causes. At first glance, indeterminism appears to offer more room for free will by breaking the strict causal chain found in deterministic models. However, this view introduces its own challenges: reconciling randomness with meaningful choice remains difficult. If decision-making were entirely governed by random processes, the notion of rational, intentional action would be undermined (Kane, 1996). Moreover, if our brains fundamentally operate under quantum mechanical laws, where outcomes are probabilistic rather than strictly determined, it raises questions about the degree to which we exercise control over our decisions (Libet, 1985). While quantum indeterminacy introduces unpredictability at the micro-level, translating this randomness into meaningful agency at the macro-level of conscious decision-making remains complex and unresolved. Libertarian perspectives, such as Kane's concept of self-forming actions (SFAs), attempt to bridge this gap. Kane argues that free will emerges in moments of deep deliberation, where individuals encounter conflicting motivations and must make difficult choices that shape their character and future intentions. Although these decisions may involve elements of randomness, they are not reducible to chance alone. Instead, Kane emphasizes that such choices are guided by the agent's values, goals, and conscious effort, preserving a coherent sense of agency and moral responsibility (Kane, 1996).

Materialist Eliminativism and the Challenge to Free Will

The materialist perspective, particularly eliminative materialism, often denies the existence of free will by asserting that consciousness and agency are illusory. According to this view, what we experience as intentional thought, decision-making, and voluntary action is reducible to underlying neurobiological processes. Proponents such as Paul Churchland and Patricia Churchland argue that traditional concepts like "beliefs," "desires," and "intentions" are outdated folk psychological constructs that will ultimately be replaced by neuroscientific explanations (Churchland, 1981; Churchland, 1986). From this standpoint, consciousness is not a metaphysical entity but an emergent property of complex neural interactions — a byproduct rather than a causal agent. Consequently, free will is considered an illusion, arising from the brain's predictive and interpretative mechanisms rather than from genuine autonomy. Eliminativists contend that human actions are determined by deterministic causal chains shaped by neural activity, genetic predispositions, environmental stimuli, and prior experiences.

This perspective fundamentally challenges traditional notions of agency and moral responsibility. If volition is merely an epiphenomenon — a subjective experience with no causal power — the foundations of personal accountability and ethical frameworks are called into question.

Critics argue, however, that while eliminative materialism highlights the role of unconscious neural processes in shaping behavior, it overlooks the complexity of conscious deliberation, intentionality, and the integrated nature of cognitive control. Emerging research in neuroscience continues to explore these complexities, revealing that conscious awareness, influenced by unconscious mechanisms, appears to play a meaningful role in guiding behavior and decision-making (Schurger et al., 2012; Maoz et al, 2019).

Toward a Third Perspective

Exploring higher-order theories offers a promising approach to resolving the tension between determinism and indeterminism. Agent causation, for example, posits that free will emerges from agents themselves as causal initiators, distinct from both deterministic and random processes. While emergentist theories suggest that free will arises as a higher-order phenomenon shaped by the complex interactions of neural, psychological, and social systems, this view alone may not fully account for the depth of human agency. Building on cumulative evidence from diverse disciplines, a more comprehensive framework that integrates a transphysical dimension offers a robust explanation for genuine free will. This framework acknowledges that human agency involves not only complex brain dynamics but also transcendent aspects of personhood that enable conscious intention, rational deliberation, and moral responsibility. Such an integrated view transcends the dichotomy of determinism and indeterminism by uniting insights from neuroscience, psychology, theology, and philosophy while affirming the unique depth of human experience.

Reconciling Determinism, Indeterminism, and the Role of Agency in Free Will

The problem of free will presents a persistent paradox. On one hand, if our actions are solely the result of brain processes, determinism implies that free will cannot exist. On the other hand, if actions depend entirely on randomness, the absence of intention similarly undermines the concept of free will. Yet, the prevailing belief in free will remains deeply ingrained in human experience. Addressing this challenge requires a higher-order perspective that reconciles these conflicting aspects and provides a coherent understanding of the relationship between the brain, determinism, randomness, and our sense of autonomy. Philosopher Colin McGinn, a proponent of new mysterianism, highlights the limits of human understanding in grappling with the problem of free will. He argues that the human mind may be inherently incapable of fully comprehending phenomena such as consciousness and free will (McGinn, 1993). While committed to a materialist framework, McGinn maintains that consciousness must arise from natural properties of the brain, yet he acknowledges the inherent dilemma: if free will is entirely a product of brain processes, all actions are determined; yet if free will results from randomness, it lacks meaningful intentionality. Despite this paradox, McGinn emphasizes our enduring perception of autonomy and proposes that understanding causation may require recognizing mental causation—the idea that conscious processes can causally influence human actions (McGinn, 1999). This notion offers a potential bridge between deterministic processes and the subjective sense of control.

Cognitive scientist John Searle similarly defends the reality of mental causation while rejecting both strict determinism and Cartesian dualism. In his theory of biological naturalism, Searle asserts that consciousness, including intentionality and rational decision-making, is an emergent property of the brain's physical processes. While rejecting the idea of a non-material soul, he argues that mental states—such as beliefs, desires, and intentions—possess causal efficacy. According to Searle (2001), rational action depends on an agent's capacity to deliberate, weigh reasons, and make choices, underscoring the essential role of mental causation. He critiques deterministic models that reduce decision-making to mechanistic brain processes, emphasizing that rational agency and moral responsibility require that mental states genuinely influence behavior. His stance thus navigates between reductionist physicalism and dualism, advocating for a non-reductive materialist view in which the mind exerts downward causation on neural processes.

In *Mind and Cosmos* (2012), Thomas Nagel challenges the adequacy of materialist explanations for consciousness, reason, and morality. He contends that the materialist neo-Darwinian conception of nature is insufficient to explain these fundamental aspects of human experience. Nagel emphasizes that first-person experiences (qualia) and subjective consciousness cannot be reduced to purely physical processes. He further argues that evolutionary processes alone cannot reliably account for rational cognition. According to Nagel, if human reasoning abilities emerged strictly through evolutionary pressures, there would be no reason to trust that those faculties accurately track objective truth rather than merely supporting survival. To address these gaps, Nagel proposes that teleological principles—some inherent directionality or purpose in nature—may be necessary to explain consciousness, reason, and moral realism. He maintains that moral truths are real and binding, challenging the notion that they can be reduced to social constructs or evolutionary byproducts.

The Case for Free Will: Convergence Between Nagel's Critique of Materialism and Plantinga's Theistic Argument

The debate over free will is often framed as a choice between deterministic materialism and theistic libertarianism. Both Thomas Nagel, a secular philosopher, and Alvin Plantinga, a Christian theist, critique materialist naturalism for its inability to explain key aspects of human experience—particularly rationality and moral responsibility, which presuppose free will. While Nagel does not endorse theism, his rejection of strict materialist determinism aligns with Plantinga's theistic argument for human freedom as a necessary condition for rational thought and moral accountability. This unlikely convergence highlights a shared concern: the inadequacy of materialism to account for the conscious, intentional nature of human cognition. Materialist determinism, which views human thoughts and decisions as the inevitable byproducts of physical processes, leaves little room for genuine agency. If all mental activity is reducible to neurochemical reactions governed by physical laws, then our sense of free will becomes an illusion—mere epiphenomenal byproducts of a causally closed system. Nagel (2012) critiques this view, arguing that materialism cannot fully explain consciousness, rationality, or moral truths, all of which presuppose the reality of free choice. Although Nagel remains committed to a naturalistic worldview, he proposes that an expanded framework—one that includes teleological principles or some unexplored feature of reality—may be necessary to account for these aspects of human existence.

Similarly, Plantinga (2011) argues that materialist naturalism is self-defeating because if our cognitive faculties are shaped purely by natural selection, they are optimized for survival, not truth. In such a scenario, rational thought would be unreliable, as our beliefs would reflect adaptive advantages rather than objective reasoning. According to Plantinga's Evolutionary Argument Against Naturalism (EAAN), the reliability of human cognition—and by extension, genuine free will—can only be sustained if human minds are created with the capacity for rational reflection and moral awareness, features best explained by a theistic framework. Although Nagel and Plantinga arrive at different metaphysical conclusions, their critiques of materialist determinism underscore the challenges materialism faces in accounting for rationality, agency, and moral responsibility. Their overlapping insights suggest that the existence of genuine free will may require a framework that transcends purely physical explanations, one that acknowledges the irreducibility of conscious experience and intentional thought.

Free Will and Rationality

Both Nagel and Plantinga emphasize that rationality itself depends on the existence of free will. Nagel argues that the process of reasoning requires the ability to deliberately reflect and choose between competing ideas — a capacity that strict determinism struggles to explain. If our reasoning were entirely the result of physical causes — such as neural impulses or biochemical reactions — distinguishing between rational persuasion and mere causal influence would become impossible. In this scenario, beliefs would not result from thoughtful evaluation but from automatic, mechanistic processes beyond our control. To illustrate, imagine two individuals reaching the same conclusion — one after carefully weighing evidence and forming a reasoned judgment, and the other simply because a chain of neural events forced that belief. Without free will, both outcomes would be equally predetermined, reducing what appears to be rational thought to an automatic response — much like a domino falling because it was pushed. In this view, the mind would not be engaging in reasoning but rather being caused to believe by forces over which it has no control.

Plantinga (2011) reinforces this argument with his Evolutionary Argument Against Naturalism (EAAN). He asserts that if materialism were true — meaning human cognition is purely shaped by evolutionary survival pressures — we would have little reason to trust our own reasoning abilities. Since natural selection prioritizes adaptive behavior over truth-tracking processes, our cognitive faculties might develop beliefs that promote survival but fail to reliably produce true conclusions. Both thinkers argue that genuine rational deliberation requires more than passive causation; it presupposes the ability to consciously evaluate evidence, assess competing ideas, and freely choose conclusions based on reason. This capacity they maintain, is only possible if free will is real.

Free Will and Moral Responsibility

Beyond rationality, moral responsibility also hinges on the existence of free will. If our actions were entirely dictated by physical causes, moral accountability would lose its foundation — much like blaming a river for flooding a village. Without the capacity to make genuine choices, concepts such as praise, blame, guilt, or moral duty would be rendered meaningless. Nagel (2012) argues that moral realism — the view that certain actions are objectively right or wrong — is incompatible with a deterministic materialist framework. In a purely deterministic system, moral behavior would be the inevitable result of preceding events, leaving no room for deliberate moral choice. For morality to have genuine significance, humans must possess the freedom to choose between right and wrong. Plantinga expands this argument within a theistic framework, contending that moral responsibility is best explained by God-given human freedom. He argues that meaningful moral accountability requires not just the capacity to act, but the autonomy to choose in accordance with moral principles. In this view, Christian theism offers a coherent explanation for both objective moral values and the free agency necessary to uphold them — an explanation that materialist determinism struggles to provide.

The Theistic Explanation: Free Will as a Gift from God

While Nagel avoids drawing a theological conclusion, his critique of materialist determinism raises a profound question: What kind of reality best accounts for true rationality, moral accountability, and free will? Alvin Plantinga offers a compelling answer: Christian theism uniquely explains these foundational aspects of human experience. According to Christian doctrine, humans are created in God's image (Genesis 1:27), endowed with rationality, self-awareness, and the genuine capacity to choose between good and evil. This view situates free will as an essential aspect of human nature, not as a random byproduct of evolutionary processes or an illusion created by deterministic forces. In this framework, free will reflects the nature of a free and rational Creator, giving human choices both moral significance and existential meaning. Nagel's and Plantinga's arguments collectively highlight the limitations of deterministic materialism. While Nagel stops short of endorsing theism, his rejection of strict determinism, alongside his defense of reason and moral realism, opens the door to a theistic understanding of free will. Plantinga builds upon this foundation, demonstrating that Christian theism not only affirms human freedom but also provides a coherent basis for its existence. Without God, free will remains inexplicable; with God, it becomes both essential and meaningful — a gift that underpins our capacity for reason, moral judgment, and personal responsibility.

Free Will is Crucial to Science

If scientists assert that their scientific theories necessitate a deterministic worldview, they face a profound paradox. The rational acceptance of scientific theories inherently presupposes the existence of free will. Scientific reasoning depends on the assumption that individuals freely evaluate evidence, assess competing hypotheses, and arrive at conclusions based on rational deliberation — not merely as the result of predetermined physical processes (Searle, 2001). If beliefs were solely the byproduct of neural activity or atomic interactions governed by fixed causal laws, there would be no guarantee that those beliefs correspond to truth or evidence. In such a framework, rational inquiry itself becomes unreliable, since conclusions would be dictated by mechanistic forces rather than reasoned judgment. This paradox reveals that scientific inquiry inherently requires freedom of thought — the capacity to weigh evidence, reflect critically, and choose between competing explanations. Without this freedom, scientific reasoning collapses into the very deterministic framework it attempts to describe. Philosopher Karl Popper argued that a deterministic worldview, which reduces all thoughts to physical causation, contradicts the rational basis of science itself (Popper, 1972). He maintained that the scientific enterprise depends on an open-ended process of hypothesis testing and critical thinking, which requires intellectual freedom.

C. S. Lewis similarly critiqued deterministic views in Miracles, arguing that naturalism undermines the validity of reason if taken to its logical conclusion. He observed that if human thoughts were merely accidental byproducts of atomic motion, there would be no reason to trust those thoughts as truth-tracking:

"If the solar system was brought about by an accidental collision, then the appearance of organic life on this planet was also an accident, and the whole evolution of Man was an accident too. If so, then all our thought processes are mere accidents—the accidental byproduct of the movement of atoms. And this holds for the thoughts of the materialists and astronomers as well as for anyone else's. But if their thoughts — i.e., of materialism and astronomy — are merely accidental byproducts, why should we believe them to be true?" (Lewis, 1984).

Lewis's critique emphasizes that without free will; rational inquiry loses its foundation. If thoughts are fully determined by non-rational physical causes, the reliability of any conclusion — including the claims of materialism itself — becomes suspect. In this sense, free will is not merely a philosophical concern but a fundamental prerequisite for science itself. By affirming free will, scientists acknowledge the rational agency necessary for objective inquiry, critical evaluation, and the pursuit of knowledge. Any deterministic framework that denies this freedom risks undermining the very integrity of scientific reasoning.

Mental Causation and Neuroscientific Research

Mental causation refers to the idea that cognitive elements of our mental life—such as thoughts, beliefs, desires, intentions, and memories—play a causal role in shaping our actions and behaviors. In everyday life, this relationship is commonly reflected in statements like "After thinking about it, I decided to take my holidays next week" or "I believe studying diligently leads to good grades, so I've decided to spend more time studying." Such examples illustrate how mental states influence decisions, behaviors, and outcomes. For instance, a desire to improve one's health, coupled with the belief that exercise promotes well-being, can motivate consistent physical activity and healthier choices. These intuitive connections reflect our lived experience, where conscious deliberation appears to have real causal power. Consequently, those who argue that mental causation is illusory bear the burden of proof to demonstrate that our experiences of rational deliberation and intentional action are deceptive.

Neuroscientific research on neuroplasticity offers compelling empirical support for mental causation by showing that conscious thought can actively shape brain function. For example, Jeffrey M. Schwartz (Schwartz and Begley 2002) demonstrated that patients with obsessive-compulsive disorder (OCD) could alter their neural activity through mindfulness and intentional cognitive strategies. By consciously redirecting their attention away from compulsive urges, patients induced measurable changes in brain function, revealing that deliberate mental effort can physically alter neural pathways. Norman Doidge (2007, 2015) expanded on this, illustrating how psychotherapy, focused attention, and conscious effort facilitate brain recovery and restructuring. His work emphasized the brain's remarkable capacity for self-directed change, reinforcing the role of intentional thought in shaping neural processes.

Alvaro Pascual-Leone (2005) further demonstrated that mental imagery and visualization exercises induce observable changes in brain structure. In one study, participants who mentally rehearsed a piano exercise showed nearly identical motor cortex changes as those who physically practiced the task. This finding underscores the mind's capacity to influence the brain's physical state without direct sensory or motor input. Similarly, Richard J. Davidson (2012) found that meditation and emotional regulation can reshape neural circuitry, supporting the idea that conscious intention can positively influence mental well-being and resilience. Mario Beauregard (2007) explored how belief, spiritual experiences, and the placebo effect generate measurable changes in brain activity. His research concluded that subjective mental states, including faith, hope, and expectation, directly impact neural functioning, further affirming the mind's causal influence on the brain. These findings and others challenge the strict materialist notion that brain states are entirely determined by physical processes alone. Instead, they support the concept of top-down mental causation, where conscious intention actively directs neural changes. This growing body of evidence reinforces the view that the mind possesses genuine causal power, strengthening the case for human agency and free will as scientifically preferable concepts. To prevent conceptual overlap, the following delineation clarifies how mental causation, agent causation, and metafunctional causation function within the framework of human agency:

Clarifying Three Dimensions of Human Causation

1. Mental Causation

- *Description:* The influence of mental states—such as beliefs, intentions, and desires—on physical behavior.
- *Scope:* Psychological and neurobiological.
- *Distinction:* Highlights top-down processes within the mind-brain system. Emphasizes that mental states can have causal power without invoking metaphysical assumptions.

2. Agent Causation

- *Description:* The view that a person, as a unified subject, can initiate actions that are not entirely determined by prior events.
- *Scope:* Philosophical and metaphysical.
- *Distinction:* Positions the agent as the originator of causal chains, distinct from mechanistic or random processes. Closely associated with libertarian theories of free will.

3. Metafunctional Causation

- *Description:* An integrative model of causation in which human agency arises from the dynamic interaction of biological, psychological, and transphysical dimensions.
- *Scope:* Multidisciplinary—bridging neuroscience, philosophy, theology, and anthropology.
- *Distinction*: Synthesizes mental and agent causation while moving beyond both. Rejects reductionism and Cartesian dualism in favor of a unified, multidimensional conception of personhood and volition.

The Necessity of Mental Causation

Mental causation is not only intuitive but also philosophically necessary for preserving key aspects of human experience such as rationality, agency, and moral responsibility. It posits that mental states—such as beliefs, desires, and intentions—can causally influence physical actions in ways that are not reducible to purely physical processes. Philosopher Colin McGinn (1999) acknowledges that while the precise mechanisms of mental causation remain elusive, this form of causation is distinct from standard physical causation. As a proponent of new mysterianism, McGinn suggests that certain aspects of consciousness and mental influence may remain beyond human comprehension due to the inherent limits of our cognitive faculties. However, rather than dismissing mental causation as illusory, mysterian philosophers maintain that its reality is undeniable—even if the underlying mechanisms are not yet fully understood.

The brain metafunction framework offers an alternative perspective that integrates mental and physical causation through a transphysical domain. This framework suggests that conscious intentions and mental states participate in a broader, multidimensional causal system—one that transcends the limitations of deterministic physical processes and random chance. Unlike reductionist materialism, which often denies mental causation by attributing all behavior to neural mechanisms, this model affirms the mind's active role in guiding behavior. This integrative perspective aligns with theistic perspectives, which maintain that consciousness and agency reflect a dimension of reality that cannot be reduced to material processes alone. In this view, mental causation is central to understanding the unity of mind and body, reinforcing the notion that humans possess genuine autonomy and moral accountability.

The debate over mental causation extends far beyond academic discourse; it speaks to fundamental questions about what it means to be human. If mental causation is real, it validates the conscious experience of deliberation, choice, and responsibility that underpins personal identity, social ethics, legal frameworks, and the scientific enterprise. Conversely, if mental causation is an illusion—as Eliminativists argue—our understanding of selfhood and moral agency faces profound challenges. Resolving this issue demands an expanded understanding of causation, as McGinn suggests, or the adoption of integrative models like the brain metafunction framework, which acknowledges the unified interplay between mental states, brain processes, and a transphysical dimension. By embracing these possibilities, we move closer to a comprehensive account of human agency that honors both scientific insights and the lived reality of conscious experience.

Metafunctional Causation: A Broader Perspective

Criteria for Free Will in Neuroscience, Beyond Libet's Experiment

Patrick Haggard from University College London proposed key criteria for identifying free will in neuroscientific research (Haggard, 2008). These include:

1. The brain activity involves motor expression, producing observable consequences in the external world.
2. There is no immediate external trigger for the brain activity.
3. The ensuing action is connected to a reason and is not random or purposeless.
4. The action is goal-directed.
5. The action is not automatic or habitual.
6. The activity involves a conscious decision.

These criteria emphasize conscious initiation, deliberate control, and reasoned deliberation, highlighting the multifaceted nature of free will. Haggard's criteria align with the brain metafunctional framework, which incorporates cognitive, intentional, and motivational factors into volitional action. This perspective underscores that free will arises through highly integrated processes, including reasoning, choice, and behavioral expression (Schurger et al., 2015). Emerging research supports the brain metafunctional framework's emphasis on distributed processes. For instance, de Jong (2011) argues that free will is not localized to a single brain region but emerges from dynamic interactions between cortical and subcortical networks. His findings suggest that voluntary action is not a strictly linear sequence of conscious decisions followed by motor execution; rather, it involves complex feedback mechanisms between cognitive control systems and motor pathways. This expanded view challenges simplistic notions of free will as a purely conscious process, reinforcing the idea that volition arises from distributed neural activity rather than a singular executive command (de Jong, 2011). Importantly, this multi-stage process illustrates that early brain signals, such as the readiness potential (RP), do not negate free will but rather contribute to its complexity.

Andrea Lavazza highlights the concept of "capacity" as a crucial factor in defining free will. He describes capacity as the mechanisms enabling individuals to adjust actions based on internal goals and environmental stimuli (Lavazza & Inglese, 2015). Lavazza (2016) argues that free will should not be dismissed as an illusion but understood as an emergent property that arises from the interaction between neural correlates and higher-order cognitive functions. He highlights several neural mechanisms that contribute to volitional action:

1. Sensorimotor processes regulate perception, movement, and motor planning, influencing how decisions are executed.
2. Subcortical structures (e.g., the basal ganglia and thalamus) play a role in action selection and initiation, shaping voluntary behavior.
3. Functional connectivity across brain networks, particularly involving the prefrontal cortex, parietal lobes, and the default mode network, integrates conscious intentions with environmental stimuli.
4. Neural oscillations and synchronization, such as gamma and beta waves, coordinate activity across brain regions, supporting goal-directed action.

Moreover, Lavazza emphasizes the interaction between bottom-up automatic responses and top-down executive control, demonstrating that free will is not merely a product of deterministic neural activity but a complex, adaptive process. By emphasizing flexibility and adaptability, Lavazza challenges reductionist interpretations of neuroscience and affirms the scientific viability of human agency (Lavazza, 2016).

In parallel, Schurger and Uithol propose that voluntary actions arise from a dynamic causal web involving internal neural activity, external stimuli, personal preferences, and accumulated information (Schurger et al., 2015; Bode et al., 2014). Their work challenges deterministic interpretations by reframing the readiness potential (RP)—once considered evidence of unconscious decision-making—as a reflection of evidence accumulation rather than predetermined intent. Expanding on this, Bode et al. (2014) demonstrate that prior mental states—such as learned associations, unconscious preferences, and goal-directed tendencies—can bias decision outcomes even before conscious deliberation occurs. Their findings suggest that early neural activity is shaped by previously acquired information rather than being purely random or predetermined. This reinforces the idea that neural processes do not rigidly dictate behavior but emerge through the integration of past experiences and present cognitive states. Collectively, the perspectives of Schurger, Uithol, and Bode reinforce that voluntary actions are neither mechanistically determined nor random. Instead, free will emerges from a flexible, context-sensitive process—one that integrates internal cognition, environmental demands, and accumulated knowledge. This aligns with compatibilist interpretations of free will, which emphasize that genuine agency can exist within a causally structured framework.

The Brain Metafunction Framework and Metafunctional Causation

The concept of *"metafunctional causation"* extends these insights by offering a broader perspective that transcends both Cartesian dualism and reductionist physicalism. While Cartesian dualism separates mind and body, and physicalism reduces consciousness to mechanistic brain activity, metafunctional causation emphasizes the interplay of biological, psychosocial, and transphysical dimensions. By integrating neuroscientific insights with broader frameworks of human experience, the brain metafunction framework affirms that human agency is neither an illusion nor reducible to unconscious neural processes. Instead, it emerges from a complex interplay of mental states, social influences, and adaptive cognitive mechanisms. This view supports the notion that free will is both scientifically viable and philosophically coherent—a conclusion that reinforces the reality of genuine human agency and moral responsibility.

The concept of metafunctional free will highlights the dynamic interplay between biological, psychosocial, and transphysical domains, presenting a multidimensional understanding of human agency and volition. While studies on neural and cognitive processes provide valuable insights into the mechanisms underlying decision-making, the brain metafunction framework extends these ideas by situating them within a broader systemic context. This integrative perspective acknowledges that free will emerges not solely from isolated neural events but from the coordinated influence of conscious intentions, environmental interactions, and underlying transphysical dimensions. While mental causation, agent causation, and metafunctional causation all address how human beings initiate actions, each concept operates at a different explanatory level. Mental causation refers to the influence of mental states—such as beliefs, desires, or intentions—on physical behavior, often framed within debates about whether the mind can exert causal force over the body. Agent causation, by contrast, shifts the focus from mental events to the agent as a whole, positing that persons themselves—rather than prior events or psychological states—can be the originators of actions. Metafunctional causation expands this further by framing agency as grounded in integrated biological, psychological, and transphysical processes. It synthesizes internal causation across multiple dimensions of personhood, suggesting that free will arises from a unified but multilayered system of causation that is neither reducible to neural mechanics nor to mental states alone. This layered clarification helps avoid conflation while allowing the three terms to be viewed as nested or complementary rather than redundant. By emphasizing this holistic integration, the term brain metafunctional free will reflects a more comprehensive model that accounts for the richness and complexity of human autonomy.

Free Will as a Properly Basic Belief

Alvin Plantinga's concept of properly basic beliefs offers a compelling framework for understanding free will as an epistemic foundation in daily life. According to Plantinga, a properly basic belief is one that can be rationally accepted without inferential justification unless significant defeaters arise. Free will, experienced directly as part of conscious agency, aligns with this concept. Individuals naturally assume their capacity for free choice unless strong evidence suggests otherwise. Alvin Plantinga's concept of *properly basic belief* offers a compelling epistemological foundation for affirming free will. A properly basic belief is one that is rationally held without inferential justification—unless strong defeaters emerge. The immediate experience of agency and volition, central to daily decision-making and moral responsibility, exemplifies such a belief. Critics might argue that empirical findings in neuroscience undermine this belief, suggesting that decisions are predetermined by unconscious neural processes. However, properly basic beliefs are not immune to challenge; rather, they are held rationally until significant, undefeated evidence dislodges them. As such, empirical critiques—while informative—do not automatically override the lived, first-person experience of agency. Instead, they invite careful interpretation within a broader philosophical framework. In this context, free will remains epistemically justified unless empirical findings provide conclusive defeaters—a standard that, as discussed in this chapter, has not been met. At best, the empirical evidence remains contested and open to interpretation. The argument for a transphysical dimension in human cognition further supports the default position that free will is not an illusion. While the precise mechanisms of mental causation remain under debate, the existence of intentional agency offers a coherent basis for understanding human decision-making. Neuroscience can identify the neural correlates of volitional processes, but interpreting such data ultimately depends on one's underlying philosophical and anthropological views of human nature (Lavazza & Inglese, 2015; Lavazza, 2016).

Plantinga's epistemology, notably in *Warrant: The Current Debate* (1993) and *Warranted Christian Belief* (2000), asserts that certain beliefs—such as belief in God or the reliability of our cognitive faculties—are rationally maintained without inferential proof because they emerge from immediate experience or fundamental cognitive structures. Similarly, the experience of volition—the felt sense of making intentional choices—qualifies as a properly basic belief. This assumption underlies key aspects of human life, including moral responsibility, rational deliberation, and practical decision-making. While ongoing debates about the nature and origins of free will persist, the default epistemic position holds that free will is real unless compelling evidence decisively refutes it. This aligns with Plantinga's broader claim that properly basic beliefs can remain rational in the absence of clear defeaters, preserving the coherence of human agency as a foundational truth.

Conclusion

In a naturalist-materialist societal framework—where human beings are primarily seen as complex biological organisms governed by physical laws, the implications of neuroscience for free will and moral responsibility become particularly significant. This perspective often rejects *a priori* metaphysical notions such as the soul or a transcendent self, focusing instead on observable neural activity as the primary basis for human behavior. Within this framework, deterministic neural processes are viewed as the driving force behind behavior, potentially undermining the concept of freely made choices. Consequently, this reductionist view may diminish the notion of moral responsibility, as actions are perceived as inevitable outcomes of prior causes rather than intentional decisions. This shift in understanding could have far-reaching ethical and societal implications. Individuals may increasingly attribute their actions to brain activity rather than moral deliberation, potentially altering core societal institutions. For example, in law and education, the emphasis could shift from personal accountability to managing behavior through neurological interventions or environmental adjustments (Greene & Cohen, 2004). Such developments risk weakening the ethical foundations of justice, personal responsibility, and moral agency, potentially eroding the moral framework that underpins social cohesion.

The concept of *metafunctional free will* represents a novel contribution to the ongoing discourse on human agency by advancing a multidimensional model that integrates biological, psychological, and transphysical elements within a unified framework. Unlike traditional views that isolate free will within either neuroscientific determinism, psychological intentionality, or metaphysical dualism, this approach posits *metafunctional causation*—a higher-order causal structure through which mental and agent causation are coordinated rather than compartmentalized. This structured integration allows for a coherent account of how conscious intentions, neural processes, and moral agency interrelate. Moreover, by framing free will as a *properly basic belief*, this model responds to empirical challenges while affirming the epistemological necessity of free will for rational inquiry, ethical responsibility, and personal identity. In doing so, it avoids both the reductive tendencies of materialist eliminativism and the logical inconsistencies of strict determinism, offering instead a robust alternative that is scientifically informed, philosophically rigorous, and theologically resonant.

Building on this foundation, a deeper account of human agency emerges—one that recognizes the limitations of purely physical explanations and affirms the need for a richer perspective. In contrast, recognizing that humans possess dimensions beyond mere physical processes opens the door to a richer understanding of human agency. The concept of metafunctional free will expands this view by emphasizing volition as a multidimensional capacity, integrating biological, psychosocial, and transphysical domains. This broader perspective transcends the deterministic or random constraints of materialist views, emphasizing the richness of human autonomy and intentionality. Both neuroscientific and philosophical insights increasingly suggest that human freedom reflects something distinctive about our nature—a capacity that resonates with the divine image within us. This recognition reinforces our ability to engage in rational deliberation, make moral choices, and take responsibility for our actions (Plantinga, 2011). By acknowledging the complexity and multidimensionality of free will, we safeguard essential aspects of human dignity and ethical responsibility. By developing the concept of metafunctional free will, this chapter provides a coherent alternative to reductionist materialism and defends a multidimensional account of human agency—one that remains consistent with scientific evidence while preserving the epistemic and ethical foundations of autonomy. As we look ahead to the chapter on the soul, this exploration will deepen our understanding of the interplay between human nature's transphysical and physical dimensions, offering further insights into what it means to be fully human.

REFERENCES

Adler, M. J. (1991). *The Four Dimensions of Philosophy: Metaphysical, Moral, Objective, Categorical.* Macmillan.

Ambrose of Milan. (1961). *On the Duties of the Clergy* (H. de Romestin, Trans.). In P. Schaff & H. Wace (Eds.), *Nicene and Post-Nicene Fathers, Second Series, Vol. 10.* Christian Literature Publishing Co. (Original work published c. 391 CE).

Aquinas, T. (1947). *Summa Theologica* (Fathers of the English Dominican Province, Trans.). New York: Benziger Bros. (Original work published 1273).

Augustine. (1998). *On Free Choice of the Will* (T. Williams, Trans.). Hackett Publishing. (Original work published c. 395 CE).

Bode, S., He, A. H., Soon, C. S., Trampel, R., Turner, R., & Haynes, J. D. (2011). Tracking the unconscious generation of free decisions using ultra-high field fMRI. *PLOS ONE, 6*(6), e21612. https://doi.org/10.1371/journal.pone.0021612

Berns, G. S. (2016, June). There's no such thing as free will. *The Atlantic.* https://www.theatlantic.com/magazine/archive/2016/06/theres-no-such-thing-as-free-will/480750/

Broadie, S. (2002). *Ethics with Aristotle.* Oxford University Press.

Calvin, J. (1960). *Institutes of the Christian Religion* (J. T. McNeill, Ed.; F. L. Battles, Trans.; 2 vols.). Westminster John Knox Press. (Original work published 1559).

Clarke, R. (2003). *Libertarian accounts of free will.* Oxford University Press.
Churchland, P. M. (1981). Eliminative materialism and the propositional attitudes. Journal of Philosophy, 78(2), 67–90. https://doi.org/10.2307/2025900

Churchland, P. S. (1986). Neurophilosophy: Toward a unified science of the mind/brain. MIT Press.

Dennett, D. C. (2003). *Freedom Evolves.* Viking Penguin.

Desmurget, M., & Sirigu, A. (2009). A parietal-premotor network for movement intention and motor awareness. *Trends in Cognitive Sciences, 13*(10), 411–419. https://doi.org/10.1016/j.tics.2009.08.001

Filevich, E., Kühn, S., & Haggard, P. (2013). There is no free won't: Antecedent brain activity predicts decisions to inhibit. *PLoS ONE, 8*(2), e53053. https://doi.org/10.1371/journal.pone.0053053

Frankfurt, H. G. (1969). Alternate possibilities and moral responsibility. *The Journal of Philosophy, 66*(23), 829–839.

Frede, M. (2011). *A Free Will: Origins of the Notion in Ancient Thought* (A. A. Long, Ed.). University of California Press.

Goetz, S. (2011). Freedom, teleology, and evil. *European Journal for Philosophy of Religion, 3*(2), 460–465.

Greene, J., & Cohen, J. (2004). "For the Law, Neuroscience Changes Nothing and Everything." *Philosophical Transactions of the Royal Society B: Biological Sciences*, 359(1451), 1775–1785.

Haggard, P. (2008). "Human Volition: Towards a Neuroscience of Will." *Nature Reviews Neuroscience*, 9(12), 934-946.

Hecht, J. M. (2014). *Stay: A History of Suicide and the Philosophies Against It.* Yale University Press.

Heisenberg, W. (1927). "Über den anschaulichen Inhalt der quantentheoretischen Kinematik und Mechanik." *Zeitschrift für Physik*, 43(3-4), 172-198. [Translated as: *The Physical Principles of the Quantum Theory*.]

Heisenberg, W. (1927). Über den anschaulichen Inhalt der quantentheoretischen Kinematik und Mechanik. Zeitschrift für Physik, 43(3–4), 172–198. https://doi.org/10.1007/BF01 397280 (Original work published 1927) [Translated as The Physical Principles of the Quantum Theory].

Honderich, T. (2002). How Free Are You? The Determinism Problem. Oxford University Press.

Hume, D. (1748). *An enquiry concerning human understanding* (Modern edition). Oxford University Press. (Original work published 1748).

Irwin, T. H. (1992). *Plato's Ethics*. Oxford University Press.

Irwin, T. H. (1999). *Aristotle: Nicomachean Ethics*. Hackett Publishing.

Kane, R. (1996). *The Significance of Free Will*. Oxford University Press.

Kane, R. (1999). "Responsibility, Luck, and Chance: Reflections on Free Will and Indeterminism." *The Journal of Philosophy*, 96(5), 217-240.

Kane, R. (2011). *The Oxford Handbook of Free Will* (2nd ed.). Oxford University Press.

Kane, R. (2016). *A Contemporary Introduction to Free Will*. Oxford University Press.

Kornhuber, H. H., & Deecke, L. (1965). "Changes in the Brain Potential in Voluntary Movements and Passive Movements in Humans: Readiness Potential and Reafferent Potentials." *Pflügers Archiv*, 284(1), 1-17.

Kornhuber, H. H., & Deecke, L. (1990). "Readiness for Movement: The Bereitschaftspotential Story." *Current Content Clinical Medicine*, 10(6), 48-53.

Kretzmann, N. (1993). *The Metaphysics of Creation: Aquinas's Natural Theology in Summa Contra Gentiles II*. Clarendon Press.

Lactantius. (2002). *Divine Institutes* (A. Bowen & P. Garnsey, Trans.). Liverpool University Press. (Original work published c. 304–311 CE).

Laplace, P.-S. (1814). *Essai philosophique sur les probabilités* [A philosophical essay on probabilities] (F. W. Truscott & F. L. Emory, Trans.). Springer. (Original work published 1814).

Lavazza, A., & Inglese, S. (2015). *Operationalizing and measuring (a kind of) free will (and responsibility): Towards a new framework for psychology, ethics, and law*. Rivista Internazionale di Filosofia e Psicologia, 6(1), 37–55. https://doi.org/10.4453/rifp.2015.0004

Lavazza, A. (2016). "Free Will and Neuroscience: From Explaining Freedom Away to New Ways of Operationalizing and Measuring It." *Frontiers in Human Neuroscience*, 10, Article 262.

Lewis, C. S. (1947). *Miracles: A Preliminary Study.* Geoffrey Bles.

Lewis, C. S. (1984). *The Business of Heaven: Daily Readings from C. S. Lewis.* Edited by Walter Hooper. Harcourt Brace Jovanovich.

Libet, B. (1985). "Unconscious Cerebral Initiative and the Role of Conscious Will in Voluntary Action." *Behavioral and Brain Sciences*, 8(4), 529-566.

Libet, B. (2004). *Mind Time: The Temporal Factor in Consciousness.* Harvard University Press.

Libet, B., Gleason, C. A., Wright, E. W., & Pearl, D. K. (1983). "Time of Conscious Intention to Act in Relation to Onset of Cerebral Activity (Readiness Potential): The Unconscious Initiation of a Freely Voluntary Act." *Brain*, 106(3), 623-642.

Long, A. A., & Sedley, D. N. (1987). *The Hellenistic Philosophers.* Cambridge University Press.

Lossky, V. (1957). *The Mystical Theology of the Eastern Church.* James Clarke & Co.

Maoz, U., Yaffe, G., Koch, C., & Mudrik, L. (2019). Neural precursors of decisions that matter: An ERP study of deliberate and arbitrary choice. *eLife, 8*, e39787. https://doi.org/10.7554/eLife.39787

McGilchrist, Iain. 2009. *The Master and His Emissary: The Divided Brain and the Making of the Western World.* New Haven, CT: Yale University Press.

Matthews, G. B. (2001). *Augustine.* Blackwell Publishers.

McGinn, C. (1993). *Problems in Philosophy: The Limits of Inquiry.* Blackwell Publishers.

McGinn, C. (1999). *The Mysterious Flame: Conscious Minds in a Material World.* Basic Books.

McGrath, A. E. (2004). *The Intellectual Origins of the European Reformation.* Blackwell Publishing.McInerny, R. (1982).

Ethica Thomistica: The Moral Philosophy of Thomas Aquinas. The Catholic University of America Press.

Mele, A. R. (2009). *Effective Intentions: The Power of Conscious Will.* Oxford University Press.

Moreland, J. P. (2009). *The Recalcitrant Imago Dei: Human Persons and the Failure of Naturalism.* SCM Press.

Newton, I. (1687). *Philosophiæ Naturalis Principia Mathematica* [Mathematical principles of natural philosophy] (I. Bernard Cohen & A. Whitman, Trans.). University of California Press. (Original work published 1687).

O'Connor, T. (2000). *Persons and Causes: The Metaphysics of Free Will.* Oxford University Press.

O'Connor, T. (2019). *The Routledge Companion to Free Will.* Routledge.

O'Keefe, Tim. 2005. *Epicurus on Freedom.* Cambridge: Cambridge University Press.

Pereboom, D. (2001). *Living without free will.* Cambridge University Press.

Plantinga, A. (1974). *God, freedom, and evil.* Harper & Row.

Plantinga, A. (1993). *Warrant: The current debate.* Oxford University Press.

Plantinga, A. (2000). *Warranted Christian belief.* Oxford University Press.

Plantinga, A. (2011). *Where the Conflict Really Lies: Science, Religion, and Naturalism.* Oxford University Press.

Popper, K. (1972). *Objective Knowledge: An Evolutionary Approach.* Oxford University Press.

Servais Theodore Pinckaers. (1995). *The Sources of Christian Ethics.* The Catholic University of America Press.

Schurger, A., et al. (2015). "The Dynamics of Volition: A Dynamical Systems Approach to the Perception of Free Will." *Neuroscience*, 7, 234-245.

Schurger, A., Sitt, J. D., & Dehaene, S. (2012). An accumulator model for spontaneous neural activity prior to self-initiated movement. *Proceedings of the National Academy of Sciences, 109*(42), E2904–E2913. https://doi.org/10.1073/pnas.121 0467109

Schurger, A., Mylopoulos, M., & Rosenthal, D. (2016). Neural antecedents of spontaneous voluntary movement: A new perspective. Trends in Cognitive Sciences, 20(2), 77–79. https://doi.org/10.1016/j.tics.2015.11.003

Schwartz, J. M., & Begley, S. (2002). *The mind and the brain: Neuroplasticity and the power of mental force.* Regan Books/HarperCollins Publishers.

Searle, J. R. (2001). *Rationality in Action.* MIT Press.

Schurger, A. (2021, Oct. 31). What does neuroscience have to say about free will? *Society of Catholic Scientists.* https://catholicscientists.org/articles/what-does-neuroscience-have-to-say-about-free-will/

Soon, C. S., Brass, M., Heinze, H.-J., & Haynes, J.-D. (2008). Unconscious determinants of free decisions in the human brain. Nature Neuroscience, 11(5), 543–545. https://doi.org/10.1038/nn.2112

Stump, Eleonore. "Augustine on Free Will." *The Cambridge Companion to Augustine*, edited by Eleonore Stump and Norman Kretzmann, Cambridge University Press, 2001, pp. 124–147.

Strawson, G. (1994). "The Impossibility of Moral Responsibility." *Philosophical Studies*, 75(1-2), 5-24.

Taylor, C. (2012). *Sources of the Self: The Making of the Modern Identity.* Harvard University Press.

Tertullian. (1972). *Against Marcion* (E. Evans, Trans.). Oxford University Press. (Original work published c. 207 CE).

Teske, R. J. (2006). *To Know God and the Soul: Essays on the Thought of Saint Augustine.* The Catholic University of America Press.

van Inwagen, P. (1983). *An Essay on Free Will.* Oxford University Press.

Wegner, D. M. (2003). *The Illusion of Conscious Will.* MIT Press.

16

Brain Metafunctional Ports: A Novel Framework for Neural Interfaces

The human brain is an extraordinary network of interconnected systems, seamlessly integrating local and global dynamics to support the cognitive, emotional, and physiological processes that define human life. Recent advances in neurotechnology have begun to harness this intricate connectivity, enabling direct interactions between the brain and external devices. Central to these developments is the concept of metafunctional ports—specialized neural structures that act as conduits for modulating and accessing the brain's interconnected systems. The foundation of this concept lies in the brain metafunction framework, which emphasizes the synchronized activity across neural systems, integrating structural and functional dynamics across various scales of space and time. This framework highlights both the intrinsic integration within the brain and the potential for targeted technological engagement through metafunctional ports.

Brain metafunction is defined as: *"The dynamic and unified activity of brain systems, sub-regions, regions, and neurons, linked together by structural or functional connectivity, that integrates the local-global synchronized dynamics across different scales of space and time, unifying interdependent domains in the living human person."*

Building on this definition, the brain metafunction framework characterizes a *metafunctional port* as a specialized brain structure that enables seamless information exchange between external devices and the brain's metafunctional architecture. These ports function as privileged conduits for interaction with technologies such as brain-computer interfaces (BCIs), enabling precise modulation of neural processes for therapeutic or assistive purposes (Nicolelis, 2001; Lebedev & Nicolelis, 2006; Krucoff et al., 2016). Recent developments in AI-driven BCI adaptation and multimodal sensing further underscore the importance of identifying structurally and functionally accessible metafunctional nodes in the brain's network architecture (Edelman et al., 2024).

A metafunctional port is a structurally and functionally privileged neural node that enables dynamic, bidirectional information flow between the brain's local and global systems and external technologies. These ports:

- *Integrate input and output functions* (e.g., sensory inflow and motor or modulatory outflow)
- Operate across *hierarchical levels* (e.g., brainstem → limbic → cortical systems)
- Exhibit *cross-network accessibility*, influencing multiple domains (affective, cognitive, autonomic)
- Are *accessible by natural (e.g., cranial nerves) or artificial (e.g., BCI electrodes) means*

To enhance conceptual clarity, it is useful to contrast *metafunctional ports* with simpler *functional nodes*. A functional node refers to a brain region that participates in a network, contributing to its localized or distributed processing. These nodes are critical for internal neural operations but typically lack the structural or functional capacity to support external interfacing or multimodal integration. In contrast, a metafunctional port goes beyond mere participation in a neural network. It facilitates interfacing between the brain and external systems—such as neurotechnological devices—and supports dynamic reconfiguration and integration across multiple domains, including cognitive, affective, sensory, autonomic, and behavioral functions. This distinction underscores the transformative role of metafunctional ports in bridging internal neural activity with broader technological and physiological systems.

Natural metafunctional ports include sensory pathways such as the visual and auditory systems, which process perceptual information through large-scale interconnected neural networks. In contrast, artificial metafunctional ports—accessed via BCIs or neuromodulation technologies—enable targeted interventions in the central nervous system. A prominent example is vagus nerve stimulation (VNS), in which an implanted device delivers electrical or biochemical signals to the vagus nerve. This modulation influences diverse structures within the brainstem, limbic system, and cortex, producing systemic effects such as improved mood, reduced inflammation, enhanced cognition, and seizure control (Groves & Brown, 2005). These examples illustrate how metafunctional ports enable advanced therapeutic interventions by leveraging the brain's inherent connectivity and dynamic architecture.

Comparative Table (Natural vs. Artificial Ports)

Property	Natural Port (e.g., Vagus Nerve, Amygdala)	Artificial Port (e.g., BCI Electrode, DBS Lead)
Access	Intrinsic to anatomy; physiologically embedded	Surgically implanted or externally applied
Bidirectionality	Yes (afferent/efferent, interoceptive/exteroceptive feedback)	Yes (recording + stimulation; feedback loop possible)
Hierarchy Level	Spans subcortical to cortical structures	Often cortical or focal subcortical, expanding via modeling
Plasticity	High; modifiable through experience, stimulation, and training	Increasing via AI, neurofeedback, and adaptive algorithms
Systemic Reach	Broad (affective, autonomic, cognitive integration)	Variable; depends on depth of integration and multimodal input
Energy Source	Endogenous neural activity	Exogenous (electrical, magnetic, or optical stimulation)
Clinical Use	Approved for epilepsy, depression, inflammation, etc.	Approved for paralysis, depression, motor prosthetics, etc.
Risk Profile	Low (noninvasive) to moderate (implants like VNS)	Moderate to high (invasive implants; risk of infection, drift)

Artificial metafunctional ports aim to emulate the integrative roles of natural ports by enabling access to complex, system-level brain functions through engineered interfaces. While natural ports like the vagus nerve are intrinsically embedded within regulatory hierarchies, artificial interfaces such as BCIs or DBS leads increasingly replicate these functions through closed-loop adaptation, multimodal sensing, and network-aware targeting. The convergence of biological and synthetic access points under the brain metafunction framework highlights the evolution of neurotechnology from localized stimulation toward systemic engagement.

Despite significant progress in BCI development, current technologies often lack the spatial precision and adaptive scalability required to integrate seamlessly with the brain's complex, dynamic networks. The concept of metafunctional ports addresses these limitations by offering a cohesive framework for enhancing neural connectivity and systemic control. By aligning with the principles of brain metafunction, these ports provide a conceptual foundation for future neurotechnological innovation. This chapter explores the theoretical foundations of brain metafunctional ports, their implications for refining brain-computer interfaces, and their transformative potential across disciplines including neurology, psychiatry, rehabilitation, and cognitive enhancement. As research advances, this framework offers promising pathways for harnessing the brain's extraordinary capabilities.

Core Technological Context — Brain-Computer Interfaces

A brain-computer interface (BCI) is a technological system that captures brain signals, processes them, and converts them into instructions that control external devices, enabling the execution of specific tasks. By harnessing the brain's synchronized neural dynamics, as described in the brain metafunction framework, BCIs have enabled unprecedented restoration of function and autonomy for individuals with motor impairments. BCIs empower paralyzed individuals to control computers, navigate wheelchairs, or manipulate prosthetic limbs—all through neural intent alone (Birbaumer et al., 1999; Chaudhary, Birbaumer, & Ramos-Murguialday, 2016). Beyond motor restoration, BCIs demonstrate significant promise in sensory, cognitive, and communicative domains. For example, BCI systems have been developed to restore visual input in individuals with blindness (Brunner et al., 2011), and research into auditory BCIs offers potential breakthroughs for hearing loss (Mak & Wolpaw, 2009). BCIs also enable direct communication for individuals with severe speech or language impairments, establishing a real-time interface between thought and expression (Nijboer et al., 2008). For older adults and individuals with neurodegenerative conditions, BCIs offer tools for preserving autonomy by facilitating complex tasks via mental control alone (Kübler & Birbaumer, 2008).

Recent developments in adaptive BCI systems—which integrate real-time feedback, neural state modeling, and artificial intelligence—further enhance the precision and personalization of BCI interventions (Herbozo Contreras et al., 2024). These next-generation systems utilize multimodal inputs (e.g., EEG, fMRI, biosignals) and deep learning algorithms to dynamically adjust control parameters, improving the scalability and user-specific adaptability of BCI performance (Liu et al., 2025). This evolution reflects a growing convergence between BCIs and the metafunctional architecture of the brain, reinforcing their shared emphasis on synchronized network integration and self-organizing complexity. As BCI research advances, its potential to transform healthcare, education, and human-computer interaction continues to grow, redefining the boundaries of cognition, embodiment, and communication.

The Vagus Nerve as a Metafunctional Port: A Natural Conduit for Systemic Neural Integration

The vagus nerve is not merely as a peripheral autonomic pathway but a metafunctional port—integrating visceral-sensory, emotional, and regulatory dynamics across brainstem, limbic, and cortical circuits. It functions as a bidirectional conduit between visceral physiological states and distributed neural circuits, integrating local autonomic signals with network-wide affective and cognitive regulation. This integration enables targeted engagement with neural systems through vagal nerve stimulation (VNS) and its transcutaneous variants (tVNS). Originating in the medulla oblongata, the tenth cranial nerve (CN X) extends through the neck, thorax, abdomen, playing a central role in parasympathetic regulation of cardiac rhythm, respiration, digestion, and immune function (Kandel, Schwartz, & Jessell, 2000; Butt et al., 1994). It transmits sensory information from organs such as the heart, lungs, and gastrointestinal tract to the brain, thereby supporting homeostatic balance and emotional stability (Craig, 2002; Damasio, 1999; Berthoud & Neuhuber, 2000). Within the brain metafunction framework, the vagus nerve exemplifies a natural metafunctional port, linking visceral inputs with higher-order cognitive and affective processes. Its afferent and efferent projections to the brainstem, limbic system, and cortex allow dynamic modulation of circuits involved in mood, cognition, and immune function (Thayer & Sternberg, 2010). This bidirectional interface between body and brain underscores the vagus nerve's distinctive role in supporting self-regulation and embodied personhood.

Historical Milestones in Vagal Nerve Stimulation (VNS)

Vagal nerve stimulation (VNS) has evolved from a pioneering but speculative technique into an established neuromodulatory intervention with widespread clinical applications. In the 1880s, Dr. John Leonard Corning conducted early experiments using electrical stimulation of the vagus nerve in animals and humans to manage seizures (Lanska, 2002). While these preliminary efforts lacked mechanistic clarity, they foreshadowed the modern clinical utility of VNS. Significant advancement occurred in the 1980s when Professor Jacob Zabara (1992) demonstrated that VNS could terminate chemically induced seizures in canine models within seconds. This research catalyzed further exploration into VNS as a treatment for epilepsy. In 1997, the U.S. Food and Drug Administration (FDA) approved VNS for the treatment of medically refractory epilepsy in adults (Ben-Menachem, 2001). As clinical usage expanded, researchers observed secondary benefits including mood enhancement, improved cognitive performance, and enhanced quality of life. These unanticipated effects sparked interest in the broader psychophysiological potential of vagus nerve modulation (Bunck et al., 2004; Groves & Brown, 2005).

Current Advances in Vagal Nerve Stimulation (tVNS)

Building on the foundational success of implanted VNS, recent research has led to the development of transcutaneous auricular VNS (tVNS), a noninvasive method that delivers stimulation to accessible branches of the vagus nerve, particularly in the auricular concha and tragus (Bauer et al., 2016; Ellrich, 2019). This approach eliminates the need for surgical implantation, increasing safety, comfort, and accessibility for patients. tVNS is now being investigated for a broad range of indications, including epilepsy, migraine, tinnitus, depression, and stroke rehabilitation (Cecchini et al., 2009; Lehtimäki et al., 2013; Ma et al., 2019). Beyond neurological disorders, tVNS may enhance cognitive performance by modulating neurotransmitter systems such as norepinephrine and GABA (Sellaro et al., 2015). Notably, recent studies have highlighted closed-loop tVNS systems that monitor physiological biomarkers and adapt stimulation in real time—a development aligned with broader trends in bioadaptive neurotechnology.

The vagus nerve's role as a metafunctional port is underscored by its capacity to influence multiple brain systems simultaneously, including the brainstem, limbic structures, and neocortex. These systemic effects resonate with the principles of the brain metafunction framework, which emphasizes hierarchical and network-wide integration across spatial and temporal domains.

Cranial Nerves as Metafunctional Ports: Expanding Neuro- modulation Pathways

Cranial nerve stimulation engages natural metafunctional ports—biological conduits like the vagus nerve that link peripheral input with deep regulatory systems in the brainstem, limbic network, and cortex. While the vagus nerve has attracted significant attention as a metafunctional port, emerging research suggests that other cranial nerves also offer viable neuromodulatory pathways. These nerves provide minimally invasive access to brainstem circuits involved in homeostatic regulation, consciousness, and affective processing, thereby expanding the landscape of neuromodulation for neurological and psychiatric disorders (Simon & Blake, 2020; Fridman et al., 2014). A recent meta-analysis examined the effects of vagus nerve stimulation on cognitive function in epilepsy, revealing mixed results that underscore the need for further research (Kong et al., 2024). Cranial nerves originate in the brainstem and project to various sensory and motor structures in the head and neck. Because they traverse both peripheral and central nervous systems, they offer unique entry points to modulate distributed brain networks—particularly those associated with autonomic regulation, sensory integration, and limbic-affective processing.

Recent studies indicate that cranial nerve stimulation can indirectly influence higher-order brain regions by modulating subcortical and thalamocortical relay pathways. For instance, cranial nerve inputs to the limbic system alter autonomic responses and emotional states, a mechanism with therapeutic relevance for mood and anxiety disorders (Adair et al., 2020). Similarly, by affecting thalamic relay circuits, cranial nerve stimulation may influence sensorimotor integration and higher-order cognitive functions—a pathway of interest in treating Parkinson's disease and sensory processing disorders (Iodice et al., 2019). A review by Adair and colleagues (2021) highlighted the capacity of cranial nerve stimulation to selectively target neural modules from early sensory cortices to executive regions of the frontal lobe. Stimulating nerves with access to these regions may enhance information processing in the early stages of perception while also providing a therapeutic entry point for disorders marked by sensory dysregulation and impaired top-down control.

Emerging stimulation protocols explore both bottom-up and top-down pathways, leveraging the bidirectional influence of cranial nerve activity on cortical function. Adair et al. (2020) emphasize that cranial nerve stimulation offers a promising dual mechanism: modulating afferent sensory input and simultaneously regulating higher-order cortical networks through distributed circuitry. This positions cranial nerves as flexible and accessible metafunctional ports, capable of orchestrating integrative functions across emotional, sensory, and cognitive domains. As clinical research advances, cranial nerve stimulation is being refined into a precise therapeutic tool—one that may revolutionize mental health interventions, cognitive enhancement protocols, and neurorehabilitation strategies. By accessing these neural gateways, cranial nerve stimulation facilitates whole-brain modulation, reinforcing its role as a biomimetic metafunctional port suited for system-wide therapeutic intervention. Viewed through the brain metafunctional lens, cranial nerves offer minimally invasive ports that enable distributed modulation of affective-autonomic circuits—aligning bottom-up sensory input with top-down cortical regulation.

Deep Brain Ports: Deep Brain Stimulation (DBS)

DBS for Parkinson's disease

Deep Brain Stimulation (DBS) operates as a deep metafunctional port—providing direct access to subcortical hubs that regulate distributed neural circuits. By stimulating key integration nodes such as the subthalamic nucleus and anterior cingulate cortex, DBS enables systemic recalibration across motor, emotional, and cognitive domains. DBS has demonstrated efficacy in the treatment of Parkinson's disease and essential tremor, significantly improving motor symptoms and quality of life. A recent systematic review with meta-analysis and trial sequential analysis highlights the potential benefits of deep brain stimulation for Parkinson's disease but emphasizes that the certainty of evidence is very low and that the risk of serious adverse events may outweigh the benefits due to methodological limitations (Petersen et al., 2024). Robot-assisted deep brain stimulation offers a promising approach to improving surgical precision and safety for patients with Parkinson's disease, as highlighted in a recent meta-analysis (Huang et al., 2024). As the field matures, DBS is being explored as a platform for broader therapeutic applications across neurological, psychiatric, and even cognitive domains. The technique involves the surgical implantation of electrodes into specific subcortical regions, where controlled electrical impulses modulate dysfunctional neural activity. In Parkinson's disease, DBS commonly targets:

- Subthalamic nucleus (STN), which regulates basal ganglia output and suppresses abnormal oscillations to restore voluntary movement (Benabid et al., 2009).
- Globus pallidus internus (GPi), which restores balance in motor circuits through deep inhibitory modulation (Okun, 2012).
- Ventrolateral thalamus (VL), which influences thalamocortical activity to reduce tremor.
- Pedunculopontine nucleus (PPN), which is under investigation for treating gait and balance deficits unresponsive to standard DBS (Thevathasan et al., 2011).

These targets highlight DBS's capacity to modulate interconnected circuits rather than isolated brain regions. The network-wide effects of DBS reflect its function as a deep brain metafunctional port—an interface through which systemic neural modulation becomes possible (Ashkan et al., 2017). Encouraging new developments in this field include patient-specific connectivity maps and biomarkers—like functional MRI—that can support the programming of deep brain stimulation (DBS). Additionally, optimization algorithms, such as particle swarm approaches, show potential for automating DBS settings. Other cutting-edge methods involve adaptive DBS driven by local field potential (LFP) recordings, as well as coordinated reset DBS (Kremer, 2021).

DBS for Chronic Pain Management

DBS is also being applied to chronic pain—an area where conventional treatments often fall short due to the multifactorial and network-based nature of pain. Key DBS targets include:

- Periaqueductal gray (PAG) and periventricular gray (PVG): Nodes in descending pain inhibitory pathways that modulate pain perception and endogenous opioid release (Bittar et al., 2005).
- Thalamic nuclei (VPM and VPL): Modulate the sensory -discriminative dimensions of pain via thalamocortical projections (Levy et al., 2010).
- Anterior cingulate cortex (ACC): The ACC functions as a bidirectional hub interfacing affective evaluation with cognitive control networks. It modulates the affective and cognitive aspects of pain, including emotional distress and attention to pain signals (Boccard et al., 2013).

The multifocal and affective-somatic nature of chronic pain makes DBS uniquely suitable as a multifaceted neuromodulatory strategy. By synchronizing disparate neural nodes through metafunctional ports, DBS facilitates a systemic recalibration of dysfunctional circuits underlying pain persistence and amplification. Recent developments in closed-loop DBS systems, which dynamically adjust stimulation based on neural biomarkers, promise to increase efficacy and reduce adverse effects in pain and mood disorders (Herbozo Contreras et al., 2024). These systems reflect a broader shift toward personalized, adaptive neuromodulation grounded in real-time neural feedback.

Expanding DBS Applications

DBS is FDA-approved for dystonia, essential tremor, medication-resistant epilepsy, Parkinson's disease, and obsessive-compulsive disorder (OCD). However, emerging research points to its potential utility across an expanded spectrum of conditions, including:

- Addiction, through modulation of reward-related networks (Salling & Martinez, 2016)
- Major depressive disorder (MDD), via stimulation of the subgenual cingulate cortex (Mayberg et al., 2005). The subgenual cingulate serves as a metafunctional port regulating affective circuitry, and its modulation via DBS provides a gateway for systemic emotional recalibration.
- Alzheimer's disease, targeting memory-related circuits (Laxton et al., 2010; Cheyuo et al., 2022).
- Anxiety, eating disorders, schizophrenia, Tourette syndrome, and beyond (McIntyre, Miocinovic & Butson, 2007; Kringelbach et al., 2011).

This expansion is made possible by an increasingly nuanced understanding of large-scale brain networks and the identification of hub nodes—regions that exert outsized influence over distributed circuit function. DBS, as a metafunctional intervention, directly engages these hubs to modulate network-wide dynamics.

Despite its promise, DBS remains an invasive and resource-intensive procedure. Expanding its clinical application will require rigorous long-term studies to assess safety, efficacy, and neurocognitive outcomes. Ethical considerations—including informed consent, potential for personality changes, and neuroethical boundaries of cognitive modulation—must be thoroughly addressed. Emerging research into connectomic-guided DBS and machine learning–enhanced stimulation offers promising directions for future personalization. These approaches aim to tailor interventions to individual neural signatures and disease trajectories—enhancing the precision and resilience of neuromodulatory outcomes. Rather than a mere focal intervention, DBS functions as a conduit for hierarchical control within and between networks—solidifying its role as a deep metafunctional interface bridging brainstem, limbic, and cortical systems.

The DLPFC as a Metafunctional Port: Modulation via TMS

Within the metafunctional framework, the dorsolateral prefrontal cortex (DLPFC) operates as a port that serves as a metafunctional conduit linking executive regulation with emotional salience networks, and attention. Although Transcranial Magnetic Stimulation (TMS) is a localized neuromodulatory technique, it exerts systemic effects by targeting the DLPFC. TMS provides a noninvasive means of modulating this port, engaging large-scale networks through targeted cortical excitation. Its capacity to engage large-scale systems makes it an ideal conduit for interfacing with the brain's affective-cognitive axis. TMS is a noninvasive neuromodulation technique that employs fluctuating magnetic fields to induce localized electric currents in the cerebral cortex. A magnetic coil placed on the scalp generates a dynamic electromagnetic field that penetrates the skull, producing cortical excitation or inhibition depending on the parameters of stimulation (Hallett, 2007). This allows precise and region-specific modulation of neural activity. In the US, TMS achieved regulatory validation with its 2008 FDA approval for major depressive disorder (MDD), followed by broader international endorsement (Perera et al., 2016). These milestones catalyzed extensive research into TMS's therapeutic potential across psychiatric, neurological, and cognitive domains. Its noninvasive profile and flexible targeting capabilities position TMS as a key translational tool in modern neuromodulation.

TMS Applications

Diagnostic Applications

TMS is used in clinical neurology to assess corticomotor function, interhemispheric connectivity, and central motor conduction time (CMCT). It aids in diagnosing demyelinating diseases and motor pathway dysfunctions and provides pharmacodynamic insights in disorders like ALS (Vucic et al., 2013). TMS-based biomarkers are increasingly incorporated in precision medicine approaches to stratify patients and evaluate therapeutic response.

Therapeutic Applications

TMS demonstrates robust efficacy across a spectrum of conditions. FDA-approved for migraines with aura and depression, it has also shown benefit in:

- Chronic pain, by modulating the primary motor cortex and descending pain pathways (Lefaucheur et al., 2014; 2020a)
- Parkinson's disease, with studies reporting modest improvements in motor function (Chou et al., 2015)
- Stroke rehabilitation, where it facilitates cortical plasticity and motor recovery (Di Lazzaro et al., 2014; Kielar et al., 2022)
- Epilepsy, with ongoing research exploring seizure suppression during interictal periods (Lefaucheur et al., 2014)
- Cognitive and speech rehabilitation, particularly in Alzheimer's and primary progressive aphasia (Freitas et al., 2011)
- Obsessive-compulsive disorder (OCD) and post-traumatic stress disorder (PTSD), via stimulation of symptom-specific networks (Carmi et al., 2019)
- Emerging research is investigating applications in autism spectrum disorder, tinnitus, and functional neurological disorders, with a focus on individualized protocols.

TMS in Stroke Rehabilitation

Stroke frequently results in hemispheric imbalance: hypoactivity in the lesioned (ipsilesional) hemisphere and compensatory hyperactivity in the contralesional hemisphere. TMS protocols aim to restore interhemispheric equilibrium. Low-frequency repetitive TMS (rTMS) applied to the contralesional side reduces inhibitory influence, while high-frequency rTMS to the ipsilesional side enhances excitability (Hao et al., 2013; Lefaucheur et al., 2020b). A recent systematic review suggests that rTMS improves motor function recovery in stroke patients (Xie et al., 2025). Meta-analyses confirm that TMS improves upper limb function when paired with physical therapy, particularly in the subacute phase (Zhang et al., 2017). Recent evidence indicates that rTMS can significantly improve language outcomes in patients with post-stroke aphasia (Kielar et al., 2022). Advances in theta burst stimulation (TBS)—a patterned form of rTMS—show promise in producing longer-lasting neuroplastic effects (Suppa et al., 2016; Liu et al., 2024). TBS protocols are being adapted based on neuroimaging and neurophysiological markers to enhance personalization (Di Pino et al., 2014).

Expanding Horizons in TMS Research

Contemporary TMS research emphasizes (Siebner et al., 2022):

- Multimodal integration with EEG, fMRI, and behavioral assessments to refine targeting

- Individualized stimulation guided by connectomics and cortical biomarkers
- Combining TMS with BCIs and pharmacological agents for synergistic gains

These trends reflect a shift from region-specific stimulation toward network-based neuromodulation, aligning with the brain metafunction framework. TMS is not merely altering local excitability but engaging dynamic systems-level reorganization.

TMS and Metafunctional Ports

As a noninvasive technology capable of modulating multiple interconnected regions, TMS exemplifies a neuromodulatory approach that aligns with metafunctional dynamics. By influencing focal hubs that connect to distributed circuits—such as the dorsolateral prefrontal cortex (DLPFC) for mood or the primary motor cortex for movement—TMS activates cascades of local and global effects. TMS engages cortical conduits to recalibrate excitability across distributed executive-emotional networks. Moreover, research confirms that TMS engages: (1) Local cortical microcircuits; (2) Long-range corticocortical pathways; and (3) Subcortical relays; and (4) Peripheral nervous system feedback loops. This breadth of action underscores TMS's potential as a metafunctional port: a tool that leverages the brain's intrinsic capacity for synchronized local-global activity, neural plasticity, and integrative function (Siebner et al., 2022). In this light, TMS is best understood not as an isolated cortical modulator, but as a dynamic metafunctional port capable of synchronizing cognitive-affective networks across the brain.

Transcranial Electrical Stimulation (tES) and CES

Historical Context

Transcranial Electrical Stimulation (tES) exemplifies a noninvasive metafunctional port—subtly tuning large-scale oscillatory activity through surface-level access to functionally rich hubs. Its distributed effects across cortical and subcortical systems position it as a broad-spectrum tool for modulating network synchrony and neuroplasticity. The use of electricity for therapeutic purposes traces back to ancient practices involving electric fish for pain relief. Though lacking mechanistic insight, these methods anticipated modern electrical neuromodulation. The 20th century saw renewed interest in electrical sedation and relaxation, culminating in the development of cranial electrical stimulation (CES) (Brunyé et al., 2021). Devices like the Neurotone 101 reflected a paradigm shift from rudimentary experimentation to clinically oriented technologies (Parent, 2004).

Modern Transcranial Electrical Stimulation (tES)

tES encompasses several noninvasive modalities:

- Transcranial Direct Current Stimulation (tDCS): Applies constant current to modulate cortical excitability (Nitsche & Paulus, 2000)
- Transcranial Alternating Current Stimulation (tACS): Uses sinusoidal currents to entrain brain oscillations (Antal & Herrmann, 2016)
- Transcranial Random Noise Stimulation (tRNS): Delivers stochastic stimulation to enhance neuroplasticity

These techniques influence neural membrane potential and synaptic plasticity without inducing action potentials. Clinical and experimental protocols are refining dosage, electrode montages, and session timing to optimize outcomes.

Mechanisms and Effects of tES

tES primarily affects spike timing and oscillatory dynamics rather than bulk firing rates (Bikson et al., 2004). Entrainment of endogenous rhythms, particularly via tACS, is a key mechanism in enhancing cognitive performance, learning, and memory (Helfrich et al., 2014). These effects are task- and network-specific, often modulated by individual differences in cortical excitability and neurochemical profile. A meta-analysis by Gao et al. (2020) revealed that tES can modulate motor performance and brain network efficiency. However, outcomes vary due to stimulation parameters, cortical targets, and inter-individual variability. The future of tES lies in personalized protocols informed by neuroimaging, genetics, and baseline electrophysiological states. When strategically deployed, tES can synchronize or desynchronize circuits that underlie affect, attention, and executive control—thereby serving as a low-intensity but systemically impactful metafunctional interface.

Cranial Electrical Stimulation: Promise and Challenges

CES has gained interest for treating depression, anxiety, insomnia, and cognitive deficits (Razza et al., 2020; Fregni et al., 2007, 2021). Despite low-intensity current delivery, studies report modulation of mood-regulatory networks such as the default mode network (DMN) and limbic regions (Bouchard et al., 2023). However, evidence remains mixed. Computational modeling (Bikson et al., 2009) and neuroimaging studies suggest subtle but variable changes. Proposed mechanisms include modulation of neurotransmitters like serotonin, dopamine, and GABA, but robust biochemical markers are lacking (Brunyé et al., 2021). Challenges to CES include Inconsistent reproducibility across studies; low focality and inter-individual variability; and need for integration with behavioral, electrophysiological, and neurochemical data. As methodologies develop, CES may emerge as a low-risk adjunctive tool for affective and cognitive regulation.

Prefrontal Cortex Interfaces and Stress Modulation

Noninvasive Prefrontal Cortex Interfaces

The prefrontal cortex (PFC) is a central metafunctional port that orchestrates high-level cognition, emotion regulation, and adaptive behavior. It is a major target for TMS and tES in psychiatric and cognitive interventions (Brunoni et al., 2017). The PFC's connectivity with the limbic system and sensorimotor networks makes it a key hub in mediating top-down modulation of internal states. Most tES studies focus on behavioral outputs, but the lack of focality in tES necessitates integration with neuroimaging to identify true neural effects (Polanía, Nitsche, & Ruff, 2018). Combining tES with EEG or fMRI helps clarify how neuromodulation influences specific circuits involved in emotional and cognitive regulation (Gilam et al., 2018).

Multimodal Investigations

Multimodal studies have shown that anodal tDCS to mPFC increases emotion regulation and reduces negative affect (Abend et al., 2019). Temporoparietal Junction (TPJ) stimulation affects empathy-related EEG activity, although effects may bypass traditional physiological markers (Coll et al., 2017; Donaldson et al., 2019). These findings affirm that tES can selectively modulate distinct social-emotional circuits, though neurophysiological corroboration is essential.

Stress Modulation via Prefrontal Cortex Interfaces

Stress exerts deleterious effects on executive function and emotional regulation, particularly in high-pressure settings (Arora et al., 2010; Regehr & LeBlanc, 2017). Chronic stress is linked to affective disorders and cognitive impairment (McLaughlin et al., 2010; Swartz et al., 2015). The ventromedial PFC (vmPFC) inhibits amygdala reactivity and HPA-axis activation (Etkin, Egner, & Kalisch, 2011), while DLPFC and VLPFC regulate reappraisal, inhibition, and attentional control (Ochsner & Gross, 2012). Chronic stress can impair these functions, contributing to emotional dysregulation (Arnsten, 2015). tES and TMS targeting the PFC have shown promise in enhancing stress resilience, supporting the metafunctional concept. These tools affect interconnected circuits involving: (1) Amygdala-PFC loops; (2) Hippocampus-PFC pathways for memory-stress integration; (3) Autonomic centers regulating arousal. Multimodal approaches combining neuromodulation with behavioral therapy, imaging, and biomarkers are advancing interventions for stress-related conditions (Bishop, 2007). These strategies exemplify the integration of metafunctional architecture with applied therapeutic innovation.

The Amygdala as a Metafunctional Port: Integrating Emotion, Memory, and Autonomic Regulation

The amygdala functions not merely as an emotion-processing center but as a metafunctional port—integrating affective, memory, and arousal dynamics across temporal and prefrontal networks. This integration enables targeted engagement with neural systems through deep brain stimulation, behavioral modulation, and emotion-focused interventions. As a central hub in the coordination of threat detection, emotional salience, and autonomic responses, the amygdala plays a vital role in the neural architecture of personhood. It processes emotionally salient stimuli—particularly those related to fear, anger, and uncertainty—and contributes to the formation and consolidation of emotional memories in close interaction with the hippocampus (LeDoux, 2000).

Interactions with Key Brain Regions

The amygdala's regulatory functions rely on dynamic interactions with several cortical and subcortical regions:

- *Ventromedial Prefrontal Cortex* (vmPFC): The vmPFC exerts top-down inhibitory control over the amygdala, dampening excessive reactivity and facilitating emotional regulation. Increased vmPFC activity is associated with improved fear extinction and emotional resilience (Etkin, Egner, & Kalisch, 2011; Diorio et al., 1993).
- *Dorsolateral and Ventrolateral Prefrontal Cortex* (DLPFC/VLPFC): These regions support executive strategies for emotion regulation, such as reappraisal, cognitive control, and attentional

redirection (Ochsner & Gross, 2012; Buhle et al., 2014). Chronic stress and affective disorders often involve reduced functional connectivity between these cortical regions and the amygdala, compromising regulatory control (Arnsten, 2015).

- *Hippocampus*: The hippocampus enables contextual encoding and recall of emotional experiences, allowing for past events to inform current affective responses (LeDoux, 2000). This integration facilitates adaptive behavior in complex environments.
- *Brainstem Structures* (PAG, LC, DRN): The amygdala's connections with the periaqueductal gray, locus coeruleus, and dorsal raphe nucleus coordinate physiological correlates of stress—such as arousal, respiration, and hormone release (Bandler et al., 2000; Sara & Bouret, 2012).

These multidirectional pathways highlight the amygdala's role in synchronizing emotion, memory, cognition, and somatic regulation, aligning with the brain metafunctional framework's emphasis on integrated local-global neural dynamics.

The Amygdala as a Brain Metafunctional Port

By virtue of its extensive connectivity and modulatory influence, the amygdala functions as a metafunctional port—processing emotionally salient input, coordinating systemic physiological responses, and interfacing with higher-order cognitive systems. It bridges subcortical affective activity with prefrontal regulatory mechanisms, supporting vertical integration across emotional, cognitive, and behavioral domains. Positioned at the nexus of the emotional-cognitive axis, the amygdala has become a key target for neuromodulation strategies aimed at emotion regulation, trauma recovery, and stress adaptation. Neuroimaging studies have shown that enhanced connectivity between the ventromedial prefrontal cortex (vmPFC) and amygdala is associated with emotional resilience, whereas dysregulated patterns are characteristic of conditions such as PTSD, generalized anxiety disorder, and major depression (Gold et al., 2015). These findings underscore the role of functional synchrony in emotional well-being and support targeted modulation of amygdala circuits.

Clinical Implications of the Amygdala as a Metafunctional Port

Recognizing the amygdala as a metafunctional port provides critical insights for developing interventions targeting maladaptive emotional reactivity. Clinical applications increasingly focus on restoring prefrontal-amygdala balance through cognitive-behavioral strategies, neuromodulation, or pharmacological modulation:

- Cognitive Behavioral Therapy (CBT): CBT strengthens top-down regulation of affect through techniques like cognitive restructuring and exposure therapy, enhancing vmPFC-mediated control (Ressler & Mayberg, 2007).
- Transcranial Magnetic Stimulation (TMS): TMS targeting the DLPFC or vmPFC has shown efficacy in downregulating amygdala hyperactivity in patients with treatment-resistant depression or anxiety (Baeken et al., 2019).
- Pharmacological Interventions: SSRIs and NRIs may indirectly reduce amygdala hyperreactivity through long-term modulation of serotonergic and noradrenergic pathways (Harmer et al., 2006).

Together, these findings point toward a transdisciplinary approach to emotional and stress regulation that encompasses cognitive, affective, neural, and systemic dimensions.

Hippocampal Stimulation Ports and Deep Brain Stimulation

Memory Function and Psychiatric Potential

The hippocampus functions as a cross-scale integrator, linking localized encoding processes with distributed memory and contextual representation systems across temporal and spatial domains. As a core structure for episodic memory, spatial navigation, and contextual learning, it also acts as a mnemonic interface—integrating spatial, emotional, and episodic inputs into long-term memory architecture. Increasingly recognized as a promising site for therapeutic intervention, the hippocampus has been targeted in deep brain stimulation (DBS) studies addressing memory impairment and mood disorders. Within the brain metafunction framework, it exemplifies a candidate metafunctional port—dynamically interfacing with neocortical and limbic structures to encode, consolidate, and retrieve memory representations (Eichenbaum, 2017). DBS targeting hippocampal circuits has been investigated in conditions such as Alzheimer's disease, major depressive disorder, and post-traumatic stress disorder (Lozano & Lipsman, 2013), with the aim of restoring temporal-spatial integrity in memory processing and stabilizing networks involved in emotional regulation.

Emerging Technologies: The Hippocampal Prosthesis

A novel advancement in this area is the hippocampal prosthesis, which seeks to replicate the hippocampus's role in encoding memory via patterned electrical stimulation. Building on the nonlinear modeling of hippocampal input-output transformations, these devices decode activity in the CA3 and entorhinal cortex and deliver tailored signals to support memory formation in CA1 neurons (Berger et al., 2011; Hampson et al., 2018). Recent studies have demonstrated successful short-term memory augmentation using hippocampal neural prosthetics in humans, informed by real-time encoding states and adaptive stimulation (Roeder et al., 2024).

Challenges in Hippocampal Stimulation

Despite promising findings, the translation of hippocampal prostheses faces several technical and conceptual hurdles: (1) Decoding high-dimensional hippocampal signals with sufficient temporal fidelity; (2) Avoiding interference with endogenous memory consolidation processes; (3) Preventing unintended modulation of affective circuits via entorhinal-hippocampal interactions; (4) Respecting cognitive integrity and personhood, especially in prosthetic applications. These complexities necessitate robust interdisciplinary collaboration, involving systems neuroscience, machine learning, clinical neuroethics, and the philosophy of mind. Additionally, as hippocampal stimulation technologies develop, ethical concerns intensify. Restoration of memory may blur the line between authentic recollection and technologically mediated reconstructions, with implications for personal identity, legal testimony, and therapeutic consent. Moreover, the potential for enhancement (rather than just restoration) raises distributive justice concerns (Clark, 2013). To responsibly integrate these technologies, researchers and clinicians must adopt a framework that balances innovation with moral discernment—one that acknowledges the hippocampus not merely as a memory hub but as a port into the experiential narrative of the person.

Thalamic Metafunctional Ports or Interfaces

The Thalamus and Epilepsy Treatment

The thalamus is a structurally and functionally centralized hub that mediates corticocortical and subcortical communication (Redinbaugh 2020). Its bidirectional connections with nearly all brain regions allow it to dynamically gate sensory input, modulate consciousness, and facilitate large-scale synchronization. These attributes position the thalamus as a prime metafunctional port, especially in disorders characterized by dysregulated network synchrony such as epilepsy. As conceptual models of epilepsy shift from focal excitation-inhibition paradigms to network-based interpretations, the thalamus has emerged as a key modulator in seizure propagation, particularly in generalized and bilateral cortical epilepsies (Englot et al., 2016).

Thalamic Neuromodulation in Epilepsy

DBS and responsive neurostimulation (RNS) targeting thalamic nuclei—particularly the centromedian (CM) and anterior (AN) nuclei—have demonstrated efficacy in reducing seizure frequency and improving quality of life for patients with drug-resistant epilepsy (Fisher et al., 2010; Gummadavelli et al., 2015). A recent study (Yang et al., 2022) highlights the efficacy of adaptive thalamic stimulation using closed-loop algorithms that respond to ictal patterns in real time. The integration of wearable EEG with AI-driven seizure prediction algorithms further enhances this personalized modulation approach (Xu et al., 2024).

Looking ahead, the convergence of BCI, neuroimaging, and bioadaptive electrodes is paving the way for noninvasive or minimally invasive interfaces targeting thalamic networks. Advanced modeling of seizure networks using fMRI-ECoG fusion has improved localization of pre-ictal states, enabling preemptive intervention. These systems offer promise not only in epilepsy but also in disorders of consciousness, where the thalamus plays a central integrative role.

The Default Mode Network (DMN) and Mind-Wandering

The discovery of the default mode network (DMN) by Raichle and colleagues (2001) catalyzed a paradigm shift in neuroscience. Unlike task-positive networks activated during goal-directed behavior, the DMN becomes more active during rest, self-referential thinking, and stimulus-independent thought (Smallwood & Schooler, 2006, 2015). Commonly associated with "mind-wandering," DMN activity encompasses spontaneous, introspective cognition, including future planning, episodic memory recall, and moral reflection. Mind-wandering (MW) is not inherently maladaptive. In fact, it plays a pivotal role in creative insight, emotional processing, and conceptual integration. However, dysregulated MW—marked by excessive or ruminative activity—can promote perseverative negative thought patterns and has been linked to disorders such as depression and anxiety (Marchetti et al., 2014). The brain metafunctional framework offers a theoretical lens through which to interpret large-scale neural networks like the DMN, salience network (SN), and central executive network (CEN).

It proposes that cognitive-emotional well-being depends not on isolated node activity, but on the integrative synchronization of networks that modulate attention, memory, emotion, and consciousness. Dysfunction within or between these networks—especially the DMN's dominance over the CEN—can lead to the internalization of affective distress, persistent rumination, and cognitive inflexibility. This aligns with findings in Alzheimer's disease, depression, and PTSD, where DMN hyperactivity correlates with cognitive and emotional decline (Whitfield-Gabrieli et al., 2020; Buckner et al., 2008). Emerging studies demonstrate that neuromodulation can recalibrate these networks. For example, transcranial magnetic stimulation (TMS) or direct current stimulation (tDCS) applied to dorsolateral prefrontal cortex regions has been shown to reduce DMN hyperconnectivity while enhancing CEN-SN coordination (Liston et al., 2014).

Promising Targets for Neuromodulation: Subgenual Cingulate Cortex and ALIC

Two major hubs of therapeutic interest are the subgenual cingulate cortex (sgACC) and the anterior limb of the internal capsule (ALIC). The sgACC, densely interconnected with limbic and autonomic centers, plays a core role in emotional regulation and homeostatic adaptation. Dysfunction in this region is associated with treatment-resistant depression and affective dysregulation. The ALIC, by contrast, acts as a conduit between the prefrontal cortex and deeper limbic areas. It is a target in DBS for OCD, major depressive disorder, and other mood-cognitive disorders, where maladaptive loops between cognition and emotion require targeted interruption. Both structures serve as metafunctional ports—nodal points where modulation can influence wide-ranging brain networks in distributed but coordinated ways (Siebner et al., 2022).

TMS as a Metafunctional Tool in Network Modulation

Though often framed as a localized cortical intervention, TMS demonstrates non-local effects. Electrical stimulation applied to prefrontal regions propagates along axonal pathways, influencing: (1) Local inhibitory/excitatory circuits; (2) Distant cortical hubs; (3) Subcortical nodes, including the thalamus, basal ganglia, and limbic areas. Recent studies show that TMS targeting the DLPFC not only modulates mood circuits but also enhances executive functioning, emotion regulation, and network coherence (Siebner et al., 2022; Liu et al., 2024). These effects validate the inclusion of TMS within the brain metafunctional framework, positioning it as a method for synchronizing or disrupting dysfunctional dynamics across interconnected cognitive-affective systems.

Clinical Implications

Understanding these network dynamics has direct clinical relevance for neuropsychiatric and neurodegenerative conditions. Therapeutic strategies that engage metafunctional ports (e.g., sgACC, ALIC) offer: (1) Restoration of cognitive control in Alzheimer's and MCI; (2) Modulation of emotion circuits in mood disorders; (3) Stabilization of salience detection in schizophrenia and autism spectrum disorder. As clinical protocols advance, the strategic selection of network access points—guided by network topology and individualized functional mapping—may offer a new frontier in precision neuromodulation.

Opportunities and Challenges in Brain Metafunctional Ports and Neural Interfaces

As neural technologies rapidly advance, they present both unprecedented opportunities and complex challenges for engaging with the brain's dynamic architecture. At the heart of this development is the concept of brain metafunctional port, strategic access points to integrated neural systems that support therapeutic, assistive, and augmentative interventions. Let us outline current advances, implementation challenges, and the ethical imperatives that shape the future of these technologies. Innovations in neural interface materials are dramatically improving the longevity, compatibility, and precision of brain-machine communication. Technologies such as flexible electrodes, conductive hydrogels enriched with neurotrophic factors, and organic electrochemical devices that mimic biological synapses have emerged as key enablers of stable, long-term integration with neural tissue (Guo & Ma, 2018). These biocompatible systems reduce inflammation, enhance durability, and promote functional connectivity with cortical and subcortical regions—facilitating sustained interaction with metafunctional ports.

In parallel with hardware innovations, advances in signal processing and computational modeling are reshaping the capabilities of brain-computer interface (BCI) systems. Neuromorphic computing architectures replicate synaptic plasticity, allowing real-time learning and adaptive control (Indiveri & Liu, 2015). The integration of multimodal inputs—such as EEG, fMRI, biosignals, and behavioral markers—has improved neural decoding accuracy and personalization. These developments support multi-site modulation, enabling coordinated access to distributed neural hubs and shifting BCI paradigms from single-target control to systemic engagement with brain-wide networks. With greater neurotechnological power comes greater ethical responsibility. As neural interfaces extend beyond clinical rehabilitation into the realm of cognitive enhancement, pressing concerns arise around cognitive liberty, neuroprivacy, and the authenticity of experience. Interventions that alter memory, emotion, or personality challenge fundamental notions of identity and consent. Equitable access also remains a critical issue, as these technologies risk amplifying existing disparities. As Ienca and Andorno (2017) emphasize, the development of neurotechnology must be guided by values that uphold human dignity, agency, and justice.

The Integrative Role of the Brain Metafunction Framework

To navigate this terrain, the brain metafunction framework provides a critical theoretical foundation. Emphasizing vertical integration (across neural, cognitive, and emotional domains) and horizontal integration (across social, ecological, and existential dimensions), the framework situates neural interfaces within a holistic view of the human person. Rather than treating neural systems in isolation, it conceptualizes metafunctional ports as gateways that mediate complex interactions between internal states and external environments. This model encourages designers and clinicians to think beyond raw functionality—fostering innovations that are technically robust, experientially coherent, and ethically informed.

REFERENCES

Abend, R., Sar-El, R., Gonen, T., Jalon, I., Vaisvaser, S., Bar-Haim, Y., & Hendler, T. (2019). Modulating emotional experience using electrical stimulation of the medial-prefrontal cortex: A preliminary tDCS-fMRI study. *Neuromodulation*, 22(8), 884–893. https://doi.org/10.1111/ner.12787.

Ackerley, S. J., Stinear, C. M., Barber, P. A., & Byblow, W. D. (2010). Combining theta burst stimulation with training after subcortical stroke. *Stroke, 41*(7), 1568–1572. https://doi.org/10.1161/STROKEAHA.110.583278

Adair, D., Xie, Y., Havercamp, S., et al. (2021). Cranial nerve stimulation: Bridging lower-level sensory systems and higher-order cognitive functions. Neuroscience Reviews, 42(6), 123–135.

Adair, D., Truong, D., Esmaeilpour, Z., Gebodh, N., Borges, H., Ho, L., Bremner, J. D., Badran, B. W., Napadow, V., Clark, V. P., & Bikson, M. (2020). Electrical stimulation of cranial nerves in cognition and disease. *Brain Stimulation*, 13(3), 717–750. https://doi.org/10.1016/j.brs.2020.02.019

Alexander, G. E., DeLong, M. R., & Strick, P. L. (1986). Parallel organization of functionally segregated circuits linking basal ganglia and cortex. Annual Review of Neuroscience, 9(1), 357–381.

American Headache Society. (2022). *Vagus Nerve Stimulation for Migraine and Cluster Headache.* Retrieved from https://americanheadachesociety.org/research/library/vagus-nerve-stimulation-for-migraine-and-cluster-headache.

Antal, A., & Herrmann, C. S. (2016). Transcranial alternating current and random noise stimulation: Possible mechanisms. Neural Plasticity, 2016, Article ID 3616807.

Arnsten, A. F. T. (2015). Stress weakens prefrontal networks: Molecular insights and new treatments. Nature Neuroscience, 18(10), 1376–1385.

Arora, S., Sevdalis, N., Nestel, D., Woloshynowych, M., Darzi, A., & Kneebone, R. (2010). The impact of stress on surgical performance: A systematic review of the literature. *Surgery, 147*(3), 318–330. https://doi.org/10.1016/j.surg. 2009.10.007

Ashkan, K., Rogers, P., Bergman, H., & Ughratdar, I. (2017). Insights into the mechanisms of deep brain stimulation. Nature Reviews Neurology, 13(9), 548–554.

Awuah, W. A., Ahluwalia, A., Darko, K., Sanker, V., Tan, J. K., Tenkorang, P. O., Ben-Jaafar, A., Ranganathan, S., Aderinto, N., Mehta, A., Shah, M. H., Chun, K. L. B., Abdul-Rahman, T., & Atallah, O. (2024). Bridging minds and machines: The recent advances of brain-computer interfaces in neurological and neurosurgical applications. *World Neurosurgery, 189*, 138–153.

Bandler, R., Keay, K. A., Floyd, N., & Price, J. (2000). Central circuits mediating patterned autonomic activity during active vs. passive emotional coping. Brain Research Bulletin, 53(1), 95–104.

Baud, M. O., Perneger, E., Racz, A., et al. (2018). Intrinsic excitability mechanisms determine the role of individual neurons in networks. Journal of Neuroscience, 38(8), 1800–1813.

Bauer, S., Baier, H., Baumgartner, C., et al. (2016). Transcutaneous vagus nerve stimulation (tVNS) for treatment of epilepsy: a clinical study. Seizure, 39, 90–93.

Ben-Menachem, E. (2001). Vagus-nerve stimulation for the treatment of epilepsy. The Lancet Neurology, 357(9253), 2100–2105.

Benabid, A. L. (2007). Deep brain stimulation for Parkinson's disease. Current Opinion in Neurobiology, 13(6), 696–706.

Benabid, A. L., Pollak, P., Louveau, A., Henry, S., & de Rougemont, J. (2009). Combined (thalamotomy and stimulation) stereotactic surgery of the VIM thalamic nucleus for bilateral Parkinson disease. Applied Neurophysiology, 50(1–6), 344–346.

Berger, T. W., Hampson, R. E., Song, D., et al. (2011). A hippocampal cognitive prosthesis: Multi-input, multi-output nonlinear modeling and VLSI implementation. IEEE Transactions on Neural Systems and Rehabilitation Engineering, 20(2), 198–211.

Bergey, G. K., Morrell, M. J., Mizrahi, E. M., et al. (2015). Long-term treatment with responsive brain stimulation in adults with refractory partial seizures. Neurology, 84(8), 810–817.

Berthoud, H. R., & Neuhuber, W. L. (2000). Functional and chemical anatomy of the afferent vagal system. Autonomic Neuroscience, 85(1–3), 1–17.

Bikson, M., Datta, A., & Elwassif, M. (2009). Establishing safety limits for transcranial direct current stimulation. *Clinical Neurophysiology, 120*(6), 1033–1044. doi: 10.1016/j.clinph.2009.03.018.

Bikson, M., Inoue, M., Akiyama, H., et al. (2004). Effects of uniform extracellular DC electric fields on excitability in rat hippocampal slices in vitro. The Journal of Physiology, 557(1), 175–190.

Birbaumer, N., & Cohen, L. G. (2007). Brain-computer interfaces: Communication and restoration of movement in paralysis. The Journal of Physiology, 579(3), 621–636.

Birbaumer, N., Ghanayim, N., Hinterberger, T., et al. (1999). A spelling device for the paralysed. Nature, 398(6725), 297–298.

Bishop, S. J. (2007). Neurocognitive mechanisms of anxiety: An integrative account. Trends in Cognitive Sciences, 11(7), 307–316.

Bittar, R. G., Kar-Purkayastha, I., Owen, S. L., et al. (2005). Deep brain stimulation for pain relief: A meta-analysis. Journal of Clinical Neuroscience, 12(5), 515–519.

Blumenfeld, H. (2014). What is a seizure network? Long-range network consequences of focal seizures. Advances in Experimental Medicine and Biology, 813, 63–70.

Boccard, S. G. J., Pereira, E. A., & Aziz, T. Z. (2013). Deep brain stimulation for chronic pain. Journal of Clinical Neuroscience, 20(7), 935–943.

Boddeti, U., McAfee, D., Khan, A., Bachani, M., & Ksendzovsky, A. (2022). Responsive neurostimulation for seizure control: Current status and future directions. *Biomedicines, 10*(11), Article 2677. https://doi.org/10.3390/biomedicines10112677

Borton, D. A., Yin, M., Aceros, J., & Nurmikko, A. (2013). An implantable wireless neural interface for recording cortical circuit dynamics in moving primates. Journal of Neural Engineering, 10(2), 026010.

Bouchard, A. E., Renauld, E., & Fecteau, S. (2023). Changes in resting-state functional MRI connectivity during and after transcranial direct current stimulation in healthy adults. *Frontiers in Human Neuroscience, 17*, 1229618. https://doi.org/10.3389/fnhum.2023.1229618

Brunner, P., Ritaccio, A. L., Emrich, J. F., et al. (2011). Brain-computer interfaces: the future of neuroprosthetics. Epilepsy & Behavior, 22(Supplement 1), S79–S83.

Brunoni, A. R., Moffa, A. H., Fregni, F., et al. (2017). Transcranial direct current stimulation for major depressive disorder: A meta-analysis of randomized controlled trials. JAMA Psychiatry, 73(6), 511–519.

Brunyé, T. T., Patterson, J. E., Wooten, T., & Hussey, E. K. (2021). A critical review of cranial electrotherapy stimulation for neuromodulation in clinical and non-clinical samples. *Frontiers in Human Neuroscience, 15*, 625321. https://doi.org/10.3389/fnhum.2021.625321

Buhle, J. T., Silvers, J. A., Wager, T. D., et al. (2014). Cognitive reappraisal of emotion: A meta-analysis of human neuroimaging studies. Cerebral Cortex, 24(11), 2981–2990.

Burdette, D., Patra, S., & Johnson, L. (2024). Corticothalamic responsive neurostimulation for focal epilepsy: A single-center experience. *Journal of Clinical Neurophysiology, 41*(7), 630–639. https://doi.org/10.1097/WNP.0000000000001060

Butt, H. R., Thomas, J. E., & Ahlquist, R. P. (1994). Cranial nerves: Functional anatomy. American Journal of Medicine, 96(4), 439–444.

Carmi, L., Tendler, A., Bystritsky, A., et al. (2019). Repetitive transcranial magnetic stimulation for obsessive-compulsive disorder: A double-blind, placebo-controlled study. American Journal of Psychiatry, 176(11), 931–938.

Cecchini, A. P., LoGiudice, G., Sances, G., et al. (2009). Noninvasive vagus nerve stimulation for migraine prophylaxis. Neurology, 72(14), 1370–1373.

Chaudhary, U., Birbaumer, N., & Ramos-Murguialday, A. (2016). Brain-computer interfaces for communication and rehabilitation. Nature Reviews Neurology, 12(9), 513–525.

Cheyuo, C., Germann, J., Yamamoto, K., Vetkas, A., Loh, A., Sarica, C., ... & Lozano, A. M. (2022). Connectomic neuromodulation for Alzheimer's disease: A systematic review and meta-analysis of invasive and non-invasive techniques. *Translational Psychiatry, 12*, 490. https://doi.org/10.1038/s41 398-022-02246-9

Chou, Y.-H., Hickey, P. T., Sundman, M., Song, A. W., & Chen, N.-K. (2015). Effects of repetitive transcranial magnetic stimulation on motor symptoms in Parkinson disease: A systematic review and meta-analysis. *JAMA Neurology, 72*(4), 432–440. https://doi.org/10.1001/jamaneurol.2014. 4380

Christoff, K., Gordon, A. M., Smallwood, J., et al. (2016). Mind-wandering as spontaneous thought: A dynamic framework. Nature Reviews Neuroscience, 17(11), 718–731.

Chung, S. W., Hoy, K. E., & Fitzgerald, P. B. (2016). The use of repetitive transcranial magnetic stimulation in neurorehabilitation. Journal of NeuroEngineering and Rehabilitation, 13(1), 1–12.

Clark, A. (2013). Natural-born cyborgs: Minds, technologies, and the future of human intelligence. Oxford University Press.

Coll, M. P., Hobson, H., Bird, G., et al. (2017). Transcranial direct current stimulation over the temporoparietal junction reduces perceptions of others' pain. Journal of Neuroscience, 37(36), 8895–8905.

Craig, A. D. (2002). How do you feel? Interoception: the sense of the physiological condition of the body. Nature Reviews Neuroscience, 3(8), 655–666.

Damasio, A. R. (1999). The feeling of what happens: Body and emotion in the making of consciousness. Harcourt Brace.

Di Lazzaro, V., Oliviero, A., Profice, P., et al. (2014). Transcranial magnetic stimulation and stroke rehabilitation. Brain Stimulation, 7(4), 585–594.

Diekhof, E. K., Geier, K., Falkai, P., et al. (2011). Fear extinction and emotion regulation in schizophrenia: The role of prefrontal-amygdala interactions. NeuroImage, 54(1), 299–310.

Diorio, D., Viau, V., & Meaney, M. J. (1993). The role of the medial prefrontal cortex (mPFC) in the regulation of hypothalamic–pituitary–adrenal responses to stress. The Journal of Neuroscience, 13(9), 3839–3847.

Di Pino, G., Pellegrino, G., Assenza, G., Capone, F., Ferreri, F., Formica, D., Ranieri, F., Tombini, M., Ziemann, U., Rothwell, J. C., & Di Lazzaro, V. (2014). Modulation of brain plasticity in stroke: A novel model for neurorehabilitation. *Nature Reviews Neurology, 10*, 597–608. https://doi.org/10.1038/nrneurol.2014.162.

Donaldson, P. H., Rinehart, N. J., & Enticott, P. G. (2019). Social cognition and transcranial direct current stimulation: A meta-analytic review. Neuroscience & Biobehavioral Reviews, 105, 46–62.

Doux, J. E. (2000). Emotion circuits in the brain. Annual Review of Neuroscience, 23(1), 155–184.

Edelman, B. J., Zhang, S., Schalk, G., Brunner, P., Müller-Putz, G., Guan, C., & He, B. (2025). Non-invasive brain-computer interfaces: State of the art and trends. *IEEE Reviews in Biomedical Engineering, 18.*

Eichenbaum, H. (2017). The role of the hippocampus in navigation is memory. The Journal of Neuroscience, 37(6), 1227–1237.

Englot, D. J., Modi, B., Mishra, A. M., et al. (2016). Cortical deactivation induced by subcortical network dysfunction in limbic seizures. Brain, 139(1), 98–112.

Ellrich, J. (2019). Transcutaneous auricular vagus nerve stimulation. *Journal of Clinical Neurophysiology*, 36(6), 437–442. https://doi.org/10.1097/WNP.0000000000000576

Fang, J., Rong, P., Hong, Y., et al. (2021). Transcutaneous auricular vagus nerve stimulation for stroke rehabilitation: a randomized controlled trial. Neurorehabilitation and Neural Repair, 35(4), 349–357.

Fisher, R., Salanova, V., Witt, T., et al. (2010). Electrical stimulation of the anterior nucleus of thalamus for treatment of refractory epilepsy. Epilepsia, 51(5), 899–908.

Fregni, F., & Pascual-Leone, A. (2007). Technology insight: Noninvasive brain stimulation in neurology—perspectives on the therapeutic potential of rTMS and tDCS. Nature Clinical Practice Neurology, 3(7), 383–393.

Fregni, F., El-Hagrassy, M. M., Pacheco-Barrios, K., Carvalho, S., Leite, J., Simis, M., Brunelin, J., Nakamura-Palacios, E. M., Marangolo, P., Venkatasubramanian, G., San-Juan, D., Caumo, W., Bikson, M., & Brunoni, A. R. (2021). Evidence-based guidelines and secondary meta-analysis for the use of transcranial direct current stimulation in neurological and psychiatric disorders. *International Journal of Neuropsychopharmacology, 24*(4), 256–313. https://doi.org/10.1093/ijnp/pyaa051

Freitas, C., Mondragon-Llorca, H., & Pascual-Leone, A. (2011). Noninvasive brain stimulation in Alzheimer's disease: Systematic review and perspectives for the future. Alzheimer's Research & Therapy, 3(1), 2.

Fridman, E. A., & Schiff, N. D. (2014). Neuromodulation of the conscious state following severe brain injuries. *Current Opinion in Neurobiology, 29*, 172–177. https://doi.org/10.1016/j.conb.2014.09.008

Gao, Y., Cavuoto, L., Schwaitzberg, S., Norfleet, J. E., Intes, X., & De, S. (2020). The effects of transcranial electrical stimulation on human motor functions: A comprehensive review of functional neuroimaging studies. *Frontiers in Neuroscience, 14*, 744. https://doi.org/10.3389/fnins.2020 .00744

Giacino, J. T., Fins, J. J., Laureys, S., & Schiff, N. D. (2014). Disorders of consciousness after acquired brain injury: The state of the science. *Nature Reviews Neurology, 10*(2), 99–114. https://doi.org/10.1038/nrneurol.2013.279

Gazzaniga, M. S. (2009). The cognitive neurosciences IV. MIT Press.

Gilam, G., Abend, R., Shani, H., et al. (2018). Neuroimaging-aided transcranial direct current stimulation in social and affective neuroscience. NeuroImage, 181, 690–702.

Gold, A. L., Morey, R. A., & McCarthy, G. (2015). Amygdala-prefrontal cortex functional connectivity during threat-induced anxiety and goal distraction. *Biological Psychiatry, 77*(4), 394–403. https://doi.org/10.1016/j.biopsych.2014.03.030

Green, R. A., Lovell, N. H., Wallace, G. G., & Poole-Warren, L. A. (2008). Conducting polymers for neural interfaces: Challenges in developing an effective long-term implant. Biomaterials, 29(24-25), 3393–3399.

Groves, D. A., & Brown, V. J. (2005). Vagal nerve stimulation: A review of its applications and potential mechanisms that mediate its clinical effects. Neuroscience & Biobehavioral Reviews, 29(3), 493–500.

Groves, D. A., & Brown, V. J. (2005). Vagal nerve stimulation: A review of its applications and potential mechanisms that mediate its clinical effects. Neuroscience & Biobehavioral Reviews, 29(3), 493–500. https://doi.org/10. 1016/j.neubiorev.2005.01.004

Gummadavelli, A., Krook-Magnuson, E., & Lüders, H. O. (2015). Responsive neurostimulation targeting the centromedian nucleus of the thalamus for seizure control. Epilepsy Research, 110, 207–213.

Guo, B., & Ma, P. X. (2018). Conductive polymers for tissue engineering. Biomacromolecules, 19(6), 1764–1782.

Hallett, M. (2007). Transcranial magnetic stimulation: A primer. Neuron, 55(2), 187–199.

Hallett, M. (2007). Transcranial magnetic stimulation: A primer. Neuron, 55(2), 187–199.

Hampson, R. E., Song, D., Robinson, B. S., Fetterhoff, D., Dakos, A. S., Roeder, B. M., She, X., Wicks, R. T., Witcher, M. R., Couture, D. E., Laxton, A. W., Munger-Clary, H., Popli, G., Sollman, M. J., Whitlow, C. T., Marmarelis, V. Z., Berger, T. W., & Deadwyler, S. A. (2018). Developing a hippocampal neural prosthetic to facilitate human memory encoding and recall. Journal of Neural Engineering, 15(3), 036014. https://doi.org/10.1088/1741-2552/aaaed7

Hao, Z., Wang, D., Zeng, Y., & Liu, M. (2013). Repetitive transcranial magnetic stimulation for improving function after stroke. *Cochrane Database of Systematic Reviews, 2013*(5), CD008862. https://doi.org/10.1002/14651858. CD008862.pu b2

Helfrich, R. F., Schneider, T. R., Rach, S., et al. (2014). Entrainment of brain oscillations by transcranial alternating current stimulation. Current Biology, 24(3), 333–339.

Herbozo Contreras, L. F., Truong, N. D., Eshraghian, J. K., Xu, Z., Huang, Z., Bersani-Veroni, T. V., Aguilar, I., Leung, W. H., Nikpour, A., & Kavehei, O. (2024). Neuromorphic neuromodulation: Towards the next generation of closed-loop neurostimulation. *PNAS Nexus, 3*(11), Article pgae488.

Huang, Y., Li, S., Chen, H., & Chen, J. (2024). Efficacy and safety of robot-assisted deep brain stimulation for Parkinson's disease: A meta-analysis. *Frontiers in Aging Neuroscience, 16*, 1419152. https://doi.org/10.3389/fnagi.2024.1419152

Ienca, M., & Andorno, R. (2017). Towards new human rights in the age of neuroscience and neurotechnology. Life Sciences, Society and Policy, 13(1), 1–27.

Indiveri, G., & Liu, S. C. (2015). Memory and information processing in neuromorphic systems. Proceedings of the IEEE, 103(8), 1379–1397.

Iodice, V., Masi, G., & Farneti, M. (2019). Thalamic modulation through cranial nerve stimulation in sensory processing disorders. Cognitive Neuroscience Reports, 17(3), 75–82.

Jobst, B. C., Kapur, R., Barkley, G. L., Bazil, C. W., Berg, M. J., Bergey, G. K., Boggs, J. G., Cash, S. S., Cole, A. J., Duchowny, M. S., Duckrow, R. B., Edwards, J. C., Eisenschenk, S., Fessler, A. J., Fountain, N. B., Geller, E. B., Goldman, A. M., Goodman, R. R., Gross, R. E., ... Morrell, M. J. (2017). Brain-responsive neurostimulation in patients with medically intractable seizures arising from eloquent and other neocortical areas. *Epilepsia, 58*(6), 1005–1014. https://doi.org/10.1111/epi.13739

Kandel, E. R., Schwartz, J. H., & Jessell, T. M. (2000). Principles of Neural Science. McGraw-Hill.

Kielar, A., Deschamps, T., & Meltzer, J. A. (2022). Efficacy of repetitive transcranial magnetic stimulation in treating stroke aphasia: A systematic review and meta-analysis. *Clinical Neurophysiology, 141*(8), 2023–2034. https://doi.org/10.1016/j.clinph.2022.04.017

Kokkinos, V., Urban, A., Sisterson, N. D., & Others. (2020). Responsive neurostimulation of the thalamus improves seizure control in idiopathic generalized epilepsy: A case report. *Neurosurgery, 87*, E578–E583.

Kong, Y., Zhao, K., Zeng, D., Lu, F., Li, X., Wu, Y., Jiang, Z., & Wen, W. (2024). Effects of vagus nerve stimulation on cognitive function in patients with epilepsy: A systematic review and meta-analysis. *Frontiers in Neurology, 15*. https://doi.org/10.3389/fneur.2024.1332882

Kramer, M. A., & Cash, S. S. (2012). Epilepsy as a disorder of cortical network organization. The Neuroscientist, 18(4), 360–372.

Kremer, N. I., Pauwels, R. W. J., Pozzi, N. G., Lange, F., Roothans, J., Volkmann, J., & Reich, M. M. (2021). Deep Brain Stimulation for Tremor: Update on Long-Term Outcomes, Target Considerations and Future Directions. *Journal of Clinical Medicine, 10*(16), 3468. https://doi.org/10.3390/jcm10163468

Kringelbach, M. L., Green, A. L., & Aziz, T. Z. (2011). Balancing the brain: Resting state networks and deep brain stimulation. *Frontiers in Integrative Neuroscience, 5*, 8. https://doi.org/10.3389/fnint.2011.00008

Krucoff, M. O., Rahimpour, S., Slutzky, M. W., Edgerton, V. R., & Turner, D. A. (2016). Enhancing nervous system recovery through neurobiologics, neural interface training, and neurorehabilitation. Frontiers in Neuroscience, 10, 584. https://doi.org/10.3389/fnins.2016.00584

Kwan, P., Arzimanoglou, A., Berg, A. T., Brodie, M. J., Hauser, W. A., Mathern, G., ... & French, J. (2010). Definition of drug-resistant epilepsy: Consensus proposal by the ad hoc Task Force of the ILAE Commission on Therapeutic Strategies. Epilepsia, 51(6), 1069–1077. https://doi.org/10.1111/j.1528-1167.2009.02397.x

Kübler, A., & Birbaumer, N. (2008). Brain–computer interfaces and communication in paralysis: extinction of goal-directed thinking in completely paralysed patients? Clinical Neurophysiology, 119(11), 2658–2666.

Lanska, D. J. (2002). J. L. Corning and vagal nerve stimulation for seizures in the 1880s. *Neurology, 58*(3), 452–459. https://doi.org/10.1212/wnl.58.3.452

Laxton, A. W., Tang-Wai, D. F., McAndrews, M. P., Zumsteg, D., Wennberg, R., Keren, R., Wherrett, J., Naglie, G., Hamani, C., Smith, G. S., & Lozano, A. M. (2010). A phase I trial of deep brain stimulation of memory circuits in Alzheimer's disease. *Annals of Neurology, 68*(4), 521–534. https://doi.org/10.1002/ana.22089

Lebedev, M. A., & Nicolelis, M. A. (2006). Brain–machine interfaces: past, present and future. Trends in Neurosciences, 29(9), 536–546.

Lebedev, M. A., & Nicolelis, M. A. L. (2006). Brain-machine interfaces: Past, present, and future. Trends in Neurosciences, 29(9), 536–546. https://doi.org/10.1016/j.tins.2006.07.004

Lefaucheur, J. P., Aleman, A., Baeken, C., Benninger, D. H., Brunelin, J., Di Lazzaro, V.,... & Ziemann, U. (2014). Evidence-based guidelines on the therapeutic use of repetitive transcranial magnetic stimulation (rTMS). *Clinical Neurophysiology, 125***(11), 2150–2206.** https://doi.org/10.1016/j.clinph.2014.05.021

Lefaucheur, J. P., Aleman, A., Baeken, C., et al. (2014). Evidence-based guidelines on the therapeutic use of repetitive transcranial magnetic stimulation (rTMS). Clinical Neurophysiology, 125(11), 2150–2206.

Lefaucheur, J. P., Drouot, X., & Nguyen, J. P. (2020a). Transcranial magnetic stimulation in pain management. Pain, 157(3), 1597–1610.

Lefaucheur, J.-P., Aleman, A., Baeken, C., Benninger, D. H., Brunelin, J., Di Lazzaro, V., Filipović, S. R., Grefkes, C., Hasan, A., Hummel, F. C., Jääskeläinen, S. K., Langguth, B., Leocani, L., Londero, A., Nardone, R., Nguyen, J.-P., Nyffeler, T., Oliveira-Maia, A. J., Oliviero, A., ... Ziemann, U. (2020b). Evidence-based guidelines on the therapeutic use of repetitive transcranial magnetic stimulation (rTMS): An update (2014–2018). *Clinical Neurophysiology, 131*(2), 474-528. https://doi.org/10.1016/j.clinph.2019.11.002

Lehtimäki, J., Hyvärinen, P., Ylikoski, M., Bergholm, M., Mäkelä, J. P., Aarnisalo, A., Pirvola, U., Mäkitie, A., & Ylikoski, J. (2013). Transcutaneous vagus nerve stimulation in tinnitus: A pilot study. Acta Oto-Laryngologica, 133(4), 378–382. https://doi.org/10.3109/00016489.2012.750736

Lempka, S. F., & McIntyre, C. C. (2013). Theoretical analysis of the local field potential in deep brain stimulation applications. PLoS One, 8(3), e59839. https://doi.org/10.1371/journal.pone.0059839

Levy, R. M., Deer, T. R., & Henderson, J. (2010). Intracranial neurostimulation for pain control: A review. Neurosurgery, 67(3), 439–455.

Li, K. P., Wu, J. J., Zhou, Z. L., Xu, D. S., Zheng, M. X., Hua, X. Y., & Xu, J. G. (2023). Noninvasive brain stimulation for neurorehabilitation in post-stroke patients. *Brain Sciences, 13*(3), 451. https://doi.org/10.3390/brainsci13030451

Lipton, R. B., Dodick, D. W., Silberstein, S. D., et al. (2010). Single-pulse transcranial magnetic stimulation for acute treatment of migraine with aura: A randomized controlled trial. The Lancet Neurology, 9(4), 373–380.

Liu, M., Jin, S., Liu, M., Yang, B., Wang, Q., Fan, C., Li, Z., & Wu, L. (2024). Global research hotspots and trends of theta burst stimulation from 2004 to 2023: A bibliometric analysis. *Frontiers in Neurology, 15*, 1469877. https://doi.org/10.3389/fneur.2024.1469877

Liu, X. Y., Wang, W. L., Liu, M., & et al. (2025). Recent applications of EEG-based brain-computer-interface in the medical field. *Military Medical Research, 12*, 14. https://doi.org/10.1186/s40779-025-00598-z

Lowry, C. A., Johnson, P. L., Hay-Schmidt, A., et al. (2008). Modulation of anxiety circuits by serotonergic systems. Stress, 11(5), 389–400.
Lozano, A. M., & Lipsman, N. (2013). Probing and regulating dysfunctional circuits using deep brain stimulation. Neuron, 77(3), 406–424.

Lu, Y., Xia, Y., Wu, Y., et al. (2022). Repetitive transcranial magnetic stimulation for upper limb motor function and activities of daily living in patients with stroke: A protocol of a systematic review and Bayesian network meta-analysis. *BMJ Open, 12*, e051630. https://doi.org/10.1136/bmjopen-2021-051630

Luigjes, J., van den Brink, W., Feenstra, M., van den Munckhof, P., Schuurman, P. R., Schippers, R., Mazaheri, A., de Vries, T. J., & Denys, D. (2012). Deep brain stimulation in addiction: A review of potential brain targets. Molecular Psychiatry, 17(6), 572–583. https://doi.org/10.1038/mp.2011 .114

Mak, J. N., & Wolpaw, J. R. (2009). Clinical applications of brain-computer interfaces: current state and future prospects. IEEE Reviews in Biomedical Engineering, 2, 187–199.

Marchetti, I., Van de Putte, E., & Koster, E. H. (2014). Self-generated thoughts and depression: From daydreaming to depressive symptoms. *Frontiers in Human Neuroscience, 8*, 1–10. https://doi.org/10.3389/fnhum.2014.00331

Mayberg, H. S., Lozano, A. M., Voon, V., et al. (2005). Deep brain stimulation for treatment-resistant depression. Neuron, 45(5), 651–660.

McEwen, B. S., & Morrison, J. H. (2013). The brain on stress: Vulnerability and plasticity of the prefrontal cortex over the life course. Neuron, 79(1), 16–29.

McIntyre, C. C., Miocinovic, S., & Butson, C. R. (2007). Computational analysis of deep brain stimulation. *Expert Review of Medical Devices, 4*(5), 615–622. https://doi.org/10. 1586/17434440.4.5.615

McLaughlin, K. A., Conron, K. J., Koenen, K. C., & Gilman, S. E. (2010). Childhood adversity, adult stressful life events, and risk of past-year psychiatric disorder: A test of the stress sensitization hypothesis in a population-based sample of adults. *Psychological Medicine, 40*(10), 1647–1658. https://doi.org/10.1017/S0033291709992121.

Nicolelis, M. A. (2001). Actions from thoughts. Nature, 409(6818), 403–407. https://doi.org/10.1038/35053191.

Nijboer, F., Sellers, E. W., Mellinger, J., et al. (2008). A P300-based brain-computer interface for people with amyotrophic lateral sclerosis. Clinical Neurophysiology, 119(8), 1909–1916.

Nitsche, M. A., & Paulus, W. (2000). Excitability changes induced in the human motor cortex by weak transcranial direct current stimulation. The Journal of Physiology, 527(3), 633–639.

O'Reardon, J. P., Solvason, H. B., Janicak, P. G., et al. (2007). Efficacy and safety of transcranial magnetic stimulation in the acute treatment of major depression: A multi-site randomized controlled trial. Biological Psychiatry, 62(11), 1208–1216.

Ochsner, K. N., & Gross, J. J. (2012). The cognitive control of emotion. Trends in Cognitive Sciences, 9(5), 242–249.

Ochsner, K. N., Silvers, J. A., & Buhle, J. T. (2012). Functional imaging studies of emotion regulation: A synthetic review and evolving model of the cognitive control of emotion. Annals of the New York Academy of Sciences, 1251(1), E1–E24.

Okun, M. S. (2012). Deep-brain stimulation for Parkinson's disease. New England Journal of Medicine, 367(16), 1529 – 1538.

Ottaviani, C., Shapiro, D., Davydov, D. M., et al. (2016). Cognitive, behavioral, and autonomic correlates of mind-wandering and perseverative cognition in major depression. Frontiers in Psychology, 7, 433.

Owen, S. L. F., Green, A. L., Davies, P., et al. (2007). Deep brain stimulation for neuropathic pain. Acta Neurochirurgica, 97(1), 111–120.

Parent, A. (2004). Giovanni Aldini: From animal electricity to human brain stimulation. *The Canadian Journal of Neurological Sciences, 31*(4), 576–584. https://doi.org/10.1017/S0317167100003851

Perera, T., George, M. S., Grammer, G., et al. (2016). The clinical TMS society consensus review and treatment recommendations for TMS therapy for major depressive disorder. Brain Stimulation, 9(5), 336–346.

Petersen, J. J., Kamp, C. B., Faltermeier, P., Juul, S., Løkkegaard, A., Gluud, C., & Jakobsen, J. C. (2024). Deep brain stimulation for Parkinson's disease: Systematic review with meta-analysis and trial sequential analysis. *BMJ Medicine*, 3, e000705. https://doi.org/10.1136/bmjmed-2023-000705

Polanía, R., Nitsche, M. A., & Ruff, C. C. (2018). Studying and modifying brain function with non-invasive brain stimulation. *Nature Neuroscience*, 21(2), 174–187. https://doi.org/10.1038/s41593-017-0054-4

Radley, J. J., Morilak, D. A., Viau, V., & Campeau, S. (2015). Chronic stress and brain plasticity: Mechanisms underlying adaptive and maladaptive changes and implications for stress-related CNS disorders. Neuroscience & Biobehavioral Reviews, 58, 79–91.

Radley, J. J., Morilak, D., Viau, V., & Campeau, S. (2015). Chronic stress and brain plasticity: Mechanisms underlying adaptive and maladaptive changes and implications for stress-related CNS disorders. *Neuroscience & Biobehavioral Reviews, 58*, 79–91. https://doi.org/10.1016/j.neubiorev.2015.06.018

Raichle, M. E., MacLeod, A. M., Snyder, A. Z., et al. (2001). A default mode of brain function. Proceedings of the National Academy of Sciences, 98(2), 676–682.

Razza, L. B., Palumbo, P, Moffa, A. H., Carvalho, A. F., & Fregni, F. (2020). A systematic review and meta-analysis on the effects of transcranial direct current stimulation in depressive episodes. *Depression and Anxiety, 37*(7), 594–608. https://doi.org/10.1002/da.23004

Redinbaugh, M. J., Phillips, J. M., Kambi, N. A., Mohanta, S., Andryk, S., Dooley, G. L., Afrasiabi, M., Raz, A., & Saalmann, Y. B. (2020). Thalamus modulates consciousness via layer-specific control of cortex. *Neuron, 106*(1), 66–75.e12. https://doi.org/10.1016/j.neuron.2020.01.005.

Regehr, C., & LeBlanc, V. R. (2017). PTSD, acute stress, performance and decision-making in emergency service workers. *Journal of the American Academy of Psychiatry and the Law, 45*(2), 184–192. https://jaapl.org/content/45/2/184

Ressler, K. J., & Mayberg, H. S. (2007). Targeting abnormal neural circuits in mood and anxiety disorders: From the laboratory to the clinic. *Nature Neuroscience, 10*(9), 1116–1124. https://doi.org/10.1038/nn1944

Roeder, B. M., She, X., Dakos, A. S., Moore, B., Wicks, R. T., Witcher, M. R., Couture, D. E., Laxton, A. W., Clary, H. M., Popli, G., Liu, C., Lee, B., Heck, C., Nune, G., Gong, H., Shaw, S., Marmarelis, V. Z., Berger, T. W., Deadwyler, S. A., Song, D., & Hampson, R. E. (2024). Developing a hippocampal neural prosthetic to facilitate human memory encoding and recall of stimulus features and categories. *Frontiers in Computational Neuroscience, 18*, Article 1263311. https://doi.org/10.3389/fncom.2024.1263311

Salling, M. C., & Martinez, D. (2016). Brain stimulation in addiction. *Neuropsychopharmacology, 41*(12), 2798–2809. https ://doi.org/10.1038/npp.2016.80

Sara, S. J., & Bouret, S. (2012). Orienting and reorienting: The locus coeruleus mediates cognition through arousal. Neuron, 76(1), 130–141.

Sellaro, R., van Leusden, J. W., Tona, K. D., Verkuil, B., Nieuwenhuis, S., & Colzato, L. S. (2015). Transcutaneous vagus nerve stimulation enhances post-error slowing. *Journal of Cognitive Neuroscience, 27*(11), 2126–2132. https://doi.org/10.1162/jocn_a_00851

Schulz, H., Blake, R. W., & Simon, C. (2021). Neuromodulation techniques in cranial nerve research: Emerging insights for therapeutic development. Neurotherapeutics, 18(4), 780–798.

Shields, G. S., Sazma, M. A., & Yonelinas, A. P. (2016). The effects of acute stress on decision-making. Psychological Bulletin, 142(2), 132–142.

Schiff, N. D. (2010). Recovery of consciousness after brain injury: A mesocircuit hypothesis. *Trends in Neurosciences, 33*(1), 1–9. https://doi.org/10.1016/j.tins.2009.10.002

Siebner, H. R., Bergmann, T. O., Bestmann, S., Massimini, M., Johansen-Berg, H., Mochizuki, H., ... & Ziemann, U. (2022). Transcranial magnetic stimulation of the brain: What is stimulated?—A consensus and critical position paper. *Clinical Neurophysiology, 140*, 59–97. https://doi.org/10.1016/j.clinph.2022.04.022

Sisterson, N. D., Kokkinos, V., Urban, A., & Others. (2022). Responsive neurostimulation of the thalamus improves seizure control in idiopathic generalized epilepsy: Initial case series. *Journal of Neurology, Neurosurgery & Psychiatry, 93*, 491–498.

Simon, C., & Blake, R. W. (2020). Cranial nerve neuromodulation: Principles, practice, and future directions. Neurology Advances, 23(7), 567–580.

Smallwood, J., & Schooler, J. W. (2006). The restless mind. *Psychological Bulletin, 132*(6), 946–958. https://doi.org/ 10.1037/0033-2909.132.6.946

Smallwood, J., & Schooler, J. W. (2015). The science of mind wandering: Empirically navigating the stream of consciousness. *Annual Review of Psychology, 66*, 487–518. https://doi.org/10.1146/annurev-psych-010814-015331

Spencer, S. S. (2002). Neural networks in human epilepsy: Evidence of and implications for treatment. Epilepsia, 43(3), 219–227.

Stinear, C. M., Lang, C. E., Zeiler, S., & Byblow, W. D. (2020). *Advances and challenges in stroke rehabilitation.* The Lancet Neurology, 19(4), 348-360. https://doi.org/10.1016/S1474-4422(19)30415-6

Suppa, A., Huang, Y.-Z., Funke, K., Ridding, M. C., Cheeran, B., Di Lazzaro, V., Ziemann, U., & Rothwell, J. C. (2016). Ten years of theta burst stimulation in humans: Established knowledge, unknowns, and prospects. *Brain Stimulation, 9*(3), 323–335. https://doi.org/10.1016/j.brs.2016.01.006

Swartz, J. R., Knodt, A. R., & Hariri, A. R. (2015). A neural biomarker of psychological vulnerability to future life stress. *Neuron, 85*(1), 144–152. https://doi.org/10.1016/j.neuron.2014.12.055.

Thayer, J. F., & Sternberg, E. M. (2010). Neural aspects of immunomodulation: Focus on the vagus nerve. *Brain, Behavior, and Immunity, 24*(8), 1223–1228. https://doi.org/10.1016/j.bbi.2010.07.247

Thevathasan, W., Coyne, T. J., Hyam, J. A., et al. (2011). Pedunculopontine nucleus stimulation improves gait freezing in Parkinson's disease. Brain, 134(1), 59–72.

Thut, G., Scharnowski, F., Gross, J., et al. (2017). Guiding transcranial brain stimulation by EEG/MEG to interact with ongoing brain activity and associated functions: A position paper. Clinical Neurophysiology, 128(5), 843–857.

Vonck, K., Thadani, V., Gilbert, K., Dedeurwaerdere, S., De Groote, L., De Herdt, V., Goossens, L., Gossiaux, F., Achten, E., Thiery, E., Vingerhoets, G., Van Roost, D., Caemaert, J., De Reuck, J., Roberts, D., Williamson, P., & Boon, P. (2004). Vagus nerve stimulation for refractory epilepsy: A transatlantic experience. *Journal of Clinical Neurophysiology, 21*(4), 283–289. https://doi.org/10.1097/01.wnp.0000139654.32974.4e

Vucic, S., Ziemann, U., Eisen, A., Hallett, M., & Kiernan, M. C. (2013). Transcranial magnetic stimulation and amyotrophic lateral sclerosis: Pathophysiological insights. *Journal of Neurology, Neurosurgery & Psychiatry*, 84(10), 1161–1170. https://doi.org/10.1136/jnnp-2012-304019
Xie, G., Wang, T., Deng, L., Zhou, L., Zheng, X., Zhao, C., Li, L., Sun, H., Liao, J., & Yuan, K. (2025). Repetitive transcranial magnetic stimulation for motor function in stroke: A systematic review and meta-analysis of randomized controlled studies. *Systematic Reviews, 14*, Article 47. https://doi.org/10.1186/s13643-025-02794-3

Xu, S., Liu, Y., Lee, H., & Li, W. (2024). Neural interfaces: Bridging the brain to the world beyond healthcare. *Exploration*, *4*(5), Article 20230146. https://doi.org/10.1002/EXP.20230146

Yang, J. C., Bullinger, K. L., Isbaine, F., Alwaki, A., Opri, E., Willie, J. T., & Gross, R. E. (2022). Centromedian thalamic deep brain stimulation for drug-resistant epilepsy: Single-center experience. *Journal of Neurosurgery, 137*(6), 1591–1600. https://doi.org/10.3171/2022.2.JNS212237

Zabara, J. (1992). Inhibition of experimental seizures in canines by repetitive vagal stimulation. *Epilepsia*, 33, 1005–1012. https://doi.org/10.1111/j.1528-1157.1992.tb01751.x.

Zhang, L., Xing, G., Fan, Y., Guo, Z., Chen, H., & Mu, Q. (2017). Short- and long-term effects of repetitive transcranial magnetic stimulation on upper limb motor function after stroke: A systematic review and meta-analysis. *Clinical Rehabilitation, 31*(9), 1137–1153. https://doi.org/10.1177/0269215517692386

17

Bioethics

Bioethics is an academic discipline and professional practice concerned with the ethical dimensions of life and healthcare. Originating from the Greek words bios (life) and ēthikós (relating to one's character), bioethics initially focused on human medical ethics but has since expanded to encompass issues such as animal protection, ecological concerns, and the ethical implications of biotechnology (Beauchamp & Childress, 1997, 2013). This expansion reflects the novel challenges of modern society, including scientific innovation, environmental degradation, and health crises. Bioethics examines moral assessments within society, exploring what qualifies as "good" or "bad" decisions and the ethical frameworks that justify those choices. While closely linked to medical policies and practices, bioethics influences individual well-being, environmental stewardship, and public health (Engelhardt, 1996).

Medical ethics, the oldest subdiscipline of bioethics, traces back to the Hippocratic Oath, a collection of texts outlining physician conduct. Central to the Oath is the physician's duty to prioritize patient well-being, expressed through the principle of beneficence. The Oath also emphasizes nonmaleficence ("do no harm"), patient confidentiality, and justice in medical practice (Gillon, 1985; Pellegrino, 1985, 1990). The original Oath prohibits actions such as euthanasia and abortion, and these principles continue to inform modern debates in medical ethics. The Oath's enduring influence reflects a broader ethical principle: physicians are called to uphold moral standards that transcend individual judgment or cultural trends. Its commitment to patient welfare, privacy, and fairness remains a cornerstone of contemporary healthcare ethics.

While rooted in classical medical traditions, bioethics has also been profoundly shaped by religious and philosophical frameworks. A fundamental contribution comes from Judeo-Christian ethics, emphasizing the sanctity of life, care for the vulnerable, and the pursuit of justice, which has significantly influenced Western bioethical thought (Pellegrino & Thomasma, 1993). Non-Western traditions have contributed meaningful insights. Confucianism emphasizes familial obligations and collective well-being, while Buddhism emphasizes compassion and non-harm. These perspectives contribute to bioethical discussions, particularly when navigating cross-cultural healthcare challenges.

Modern Bioethics

Modern bioethics has expanded to address new ethical dilemmas emerging from scientific and technological advances, secularism, and ideological and geopolitical perspectives. Topics such as genetic engineering, artificial intelligence in healthcare, and robotics present complex ethical challenges. For example, CRISPR technology raises questions about human genetic modification. At the same time, discussions surrounding the recent pandemic origin and responses highlight the tension between individual rights, public health priorities, and the ever-present risk of government overreach in censorship, and disproportionate impositions. This growing complexity reinforces bioethics' potential role in guiding medical innovation and promoting social responsibility. Bioethics' interdisciplinary nature connects it with law, philosophy, sociology, and technology. This broad applicability enables bioethics to address ethical questions that extend beyond individual patient care, shaping policies on international health policies, environmental stewardship, and medical research. By exploring these intersections, bioethics must adapt to new realities, guiding decision-making in ways that uphold human dignity and moral responsibility. Bioethics should play a vital role in ensuring that scientific progress aligns with respect for human dignity, social justice, and ethical integrity. As technology and medical advancements accelerate, bioethics remains essential in balancing innovation with moral reflection. By engaging with the complexity of healthcare ethics, bioethics equips society to confront emerging challenges while safeguarding core principles that promote human life and well-being.

Love and Compassion as Central to the Judeo-Christian Worldview

At the foundation of Western civilization lies the transformative ideal of love. Saint Paul, in his letter to Timothy, expresses this ideal: "For God has not given us a spirit of fear, but of power and of love and of a sound mind" (2 Timothy 1:7). This verse holds deep religious significance, but it also reflects three enduring values—power, love, and moderation—that have shaped Western ethics. While power and moderation played key roles in Roman and Greek traditions, the ideal of love, central to the Judeo-Christian worldview, has had a profound influence, particularly in shaping Western medical ethics and approaches to care. The foundations of Christian ethics emerged during the early Christian era, a pivotal period that began with the ministry of Jesus (27–30 AD) and culminated with the First Council of Nicaea in 325 AD. Christian ethics draw from a shared heritage with Judaism while incorporating Greek and Hellenistic philosophical insights, resulting in a distinctive moral philosophy that has profoundly influenced Western civilization (Niebuhr, 1951). Over the two millennia since Christianity's expansion, Judeo-Christian values have substantially shaped the principles of medical ethics in Western civilization, guiding how we treat patients and conceptualize care. However, it is essential to recognize that many of these principles, such as compassion and justice, resonate across cultures and religious traditions, reflecting universal ethical ideals. The Christianization of the Greco-Latin world introduced a transformative understanding of love, expressed in multiple dimensions: Eros (sexual love), Agape (selfless, charitable love), Philia (friendship), and Storge (parental love) (Lewis, 1960). Among these, Agape stands out in Christian ethics, emphasizing "willing the good of the other" (Aquinas, Summa Theologica II-II, Q. 26, Art. 3). This selfless love has influenced Western institutions such as hospitals, universities, and charitable organizations, demonstrating love's profound impact on medical care and social responsibility.

Love is not merely an emotional sentiment but a profound force that undergirds ethics and morality. The Christian call to love extends beyond family and friends to include even one's enemies (Matthew 5:44). This love compels us to resist evil, pursue knowledge, and share that knowledge with others. It drives the core institutions of Western civilization—marriage, family, government, universities, church, and charities—each embodying the values of love, forgiveness, mercy, compassion, and creativity (Pelikan, 1988). Without the richness of the Judeo-Christian worldview, expressed in sacred Scripture, architecture, literature, and art, many of the essential treasures that define Western civilization would be profoundly diminished. However, it is equally important to acknowledge that Enlightenment thought and secular humanism have also contributed to the ethical frameworks of Western society, at times building upon—and at other times challenging—its religious foundations.

In Western civilization, the ideal of love serves as a moral compass, differentiating right from wrong. This ideal transcends love as a mere feeling; it is an act of will and points to the divine origin of human existence. As Paul of Tarsus beautifully expresses in 1 Corinthians 13:4-7: "Love is patient, love is kind. It does not envy, it does not boast, it is not proud...

It always protects, always trusts, always hopes, always perseveres." This vision of love has inspired the moral philosophy of Western civilization for centuries and contributed significantly to its ethical frameworks (Meilaender, 2005). Historically, the Christian spirit of mutual love and communal charity has been fundamental to ethical development. Early Christians, often at significant personal risk, cared for the poor, the sick, and the dying, establishing some of the first hospitals. Unlike earlier medical institutions that often served elite members of society, Christian hospitals aimed to care for all—regardless of social status, ethnicity, or religious background. By 580 AD, hospitals such as the one in Merida, Spain, served Christians, Jews, and enslaved people alike (Brockliss & Jones, 1997). This inclusive model spread across Europe, with over 300 hospitals established by 1000 AD. Christianity also transformed society's approach to death, encouraging acceptance of mortality with a meaningful purpose, a shift that profoundly influenced medical ethics (Porter, 1999).

The parable of the Good Samaritan (Luke 10:25–37) exemplifies this ethic of compassion and love. Jesus' endorsement of medical aids such as oil and wine illustrates the importance of practical care, while His command to "go, and do likewise" underscores the moral responsibility to care for the vulnerable. This ethic continues to inform modern medical ethics, inspiring approaches to humanitarian aid, global health equity, and compassionate caregiving. Prominent theologians like Augustine have provided profound interpretations of this parable, suggesting that the Samaritan represents Christ, bearing the wounds of fallen humanity and offering redemption through the Church (Augustine, Sermon on the Good Samaritan). The Samaritan's example offers both theological reflection and practical guidance in addressing social justice issues and ethical medical practices today.

The Judeo-Christian worldview has had a profound influence on Western ethics, and parallel themes of compassion and care resonate in other traditions. For instance, the Buddhist concept of karuṇā (compassion) and the Islamic emphasis on rahma (mercy) exemplify the role of conscience and a universal moral imperatives to serve others. Recognizing these universal aspiration and values enriches our understanding of love and compassion as essential ethical foundations that transcend cultural and religious boundaries. The Judeo-Christian emphasis on love and compassion has profoundly shaped Western civilization's ethical framework, particularly in the realm of healthcare. As emerging medical technologies and bioethical debates develop, these enduring principles continue to offer guidance in navigating complex moral challenges. By embracing love as a central ethical value, medical practitioners can uphold the dignity, care, and well-being of all individuals in an increasingly diverse and complex world.

Ethics as a Component of a Coherent Worldview

In the Middle Ages, Christian scholars and theologians played a pivotal role in shaping ethical guidelines for medical practice. Among the most influential figures in this development was Thomas Aquinas, whose synthesis of Aristotelian philosophy and Christian theology profoundly shaped Western medical ethics. Aquinas emphasized moral virtues—such as prudence and justice—that were foundational in shaping medical ethics codes. He believed ethical principles are rooted in natural law, an objective and universal moral order accessible through reason (Finnis, 1980). For Aquinas, natural law is rooted in God's eternal law and inscribed in human nature, guiding moral reasoning through innate principles accessible to reason. This understanding aligns with the Apostle Paul's teaching in Romans: "For when Gentiles, who do not have the law, do by nature what the law requires, they are a law to themselves... They show that the work of the law is written on their hearts, their consciences also bearing witness..." (Romans 2:14-15, ESV). In this view, natural law serves as a universal guide for ethical reasoning, bridging the gap between faith and reason.

Aquinas's teleological ethics emphasized that the ultimate purpose (telos) of human life is to achieve beatitude — a state of perfect happiness and union with God (MacIntyre, 1998). This ultimate end serves as the moral compass guiding human actions. In a pluralistic modern context, this raises questions about how shared ethical principles can emerge from differing conceptions of telos. Central to Aquinas's ethics is the cultivation of virtuous character traits. Virtues such as courage, justice, prudence, and temperance are essential for a morally upright life. Aquinas viewed these virtues as habitual dispositions that enable individuals to make ethically sound decisions consistently. His affirmation of the inherent dignity of every human being reflects the Judeo-Christian belief that humans are created in the image of God (Clarke, 2001). This theological conviction underpins the ethical imperative to treat others with respect and justice.

One of Aquinas's most influential contributions to ethical theory is *the principle of double effect,* a guideline often invoked in complex ethical dilemmas. The principle of double effect asserts that an action producing both good and harmful outcomes may be morally permissible if the agent's intention is good, the harmful consequence is unintended, and the good effect outweighs the bad (Boyle, 1980). This principle remains highly relevant in contemporary medical ethics, such as palliative care, where treatments may alleviate suffering but inadvertently hasten death. Such applications demonstrate the enduring relevance of Aquinas's ethical framework.

Aquinas's ethical system also integrates reason and divine revelation, emphasizing a balance between the two. Practical reason guides actions under natural law, while divine commandments provide additional moral guidance (Porter, 2004). This interplay between reason and faith illustrates how Aquinas viewed morality as integral to a coherent worldview. Despite critiques of its theological assumptions, Aquinas's system offers a foundational framework for exploring universal ethics. By integrating natural law, virtue ethics, and practical reasoning, Aquinas's framework has left a lasting mark on Christian moral theology and continues to shape contemporary ethical discussions, particularly in medical ethics. His approach demonstrates how ethics is a cornerstone of a worldview that seeks coherence between human action and the divine order.

The Reformation Contributions

Building on earlier Christian traditions, key figures such as Martin Luther and John Calvin reshaped ethical thinking in Western civilization during the Protestant Reformation. The Reformation emphasized several transformative principles that redefined ethical thought: the centrality of divine grace, the sanctity of everyday actions, the universal priesthood of believers, and the separation of Church and state. These developments influenced religious and moral thought, laid the groundwork for enduring principles of religious freedom, and gave continuity to virtue ethics in Western ethical discourse (McGrath, 2001). The Reformation era significantly affected how individuals perceived morality and their relationship with God. In On the Freedom of a Christian (1520/2008), Martin Luther argued that ethical behavior is a response to divine grace rather than a means of attaining righteousness. According to Luther, moral actions flow naturally from faith, as believers inspired by divine love become inclined toward virtuous deeds. This emphasis on grace fundamentally transformed how Christians understood morality and virtue, shifting the focus from human efforts to divine initiative (Bainton, 1950).

Building upon Luther's ideas, John Calvin further systematized this ethical framework by emphasizing God's sovereignty and the sacredness of everyday work. Calvin introduced the concept of vocation, arguing that all legitimate occupations hold spiritual significance when aligned with God's will. This perspective contributed to what later became known as the Protestant work ethic (Weber, 2001). Calvin also introduced the notion of the separation of Church and state, arguing that God governs spiritual matters through the Church and worldly affairs through civil government. This distinction laid the groundwork for modern ideas about religious and secular institutions and their respective roles in society (Witte, 2008). In addition to emphasizing grace and vocation, the Reformers incorporated natural law principles and Old Testament covenant themes into their ethical framework. This emphasis on individual responsibility and community engagement positioned Protestant Christians among the earliest advocates for religious liberty, human rights, and freedom of conscience (Berman, 1983). Despite theological shifts, virtue ethics remained central to Protestant moral thought. Aristotle's Nicomachean Ethics continued to influence Protestant universities until the late seventeenth century. Numerous Protestant commentaries on Aristotle's work emerged during this period, underscoring the Reformation's nuanced engagement with classical virtue ethics while introducing new theological interpretations. This continuity highlights how the Reformation preserved virtue ethics while reinterpreting it through the lens of divine grace and individual conscience.

Christianity and Bioethics: Sacred and Secular Principles

Christianity offers a distinctive moral framework that continues to shape contemporary bioethics through both theological insights and universally recognized ethical principles. Its influence on medical ethics includes both sacred principles—rooted in theological teachings—and secular principles, which often share those same foundations. Together, these principles complement each other, offering a holistic ethical framework. Christianity asserts that medicine's proper role is to prevent and alleviate suffering by ethical means while recognizing that theological dimensions of suffering are beyond medicine's scope. Central to modern bioethical theories are four guiding principles: beneficence, nonmaleficence, autonomy, and justice. While some of these principles trace their roots to the Hippocratic Oath, Christianity has enriched their meaning by situating them within a broader moral and theological framework (Pellegrino, 1990).

Imago Dei

Central to Christian ethics is the concept of Imago Dei, or the belief that human beings are made in the image of God: "So God created mankind in his own image, in the image of God he created them; male and female he created them" (Genesis 1:27). This doctrine asserts that every person, regardless of abilities or disabilities, carries the divine image—affirming the intrinsic dignity that shapes Christian approaches to healthcare and bioethics. Disabilities or incapacities do not diminish this profound truth. Consequently, every human being possesses inherent dignity and worth that transcends legal systems or societal opinions. This belief has profoundly influenced medical ethics by emphasizing the preservation of human life and the importance of caring for the vulnerable and marginalized (Meilaender, 2005).

The Sanctity of Life

The doctrine of the Sanctity of Life stands as a cornerstone of Christian bioethics, affirming that human life, created in God's image, holds inherent dignity and worth. This doctrine asserts the inviolability of human life, advocating for its protection from intentional harm or destruction. It opposes practices such as abortion, euthanasia, and the taking of innocent life (John Paul II, 1995). While humans are granted dominion over creation (Genesis 1:28), this does not give them the authority to end human life intentionally. The Fifth Commandment, "You shall not murder" (Exodus 20:13), underscores the inviolability of life and calls for the protection of the innocent.

The Sanctity of Life doctrine promotes a consistent ethical stance that values all stages of life—from conception to natural death—and advocates for compassionate care for those facing end-of-life issues. As a moral response to suffering, palliative care aligns with Christian teachings on human dignity and the value of life. In contrast, secular bioethicists like Peter Singer, a prominent proponent of Utilitarianism, advocate for practices such as infanticide when a child is perceived to have diminished capacities, such as in cases of Down syndrome. Singer's framework prioritizes cognitive ability as the basis for moral worth, sharply diverging from the Christian belief that all individuals possess inherent dignity as bearers of God's image. Consequently, Christian ethics maintain that preborn infants, comatose patients, and individuals with disabilities are entitled to full human rights (Singer, 1993; Cahill, 1995).

Free Will

Christianity's emphasis on free will has profoundly shaped medical ethics, particularly through the principle of autonomy, which recognizes an individual's moral agency and right to informed decision-making. In this ethical context, the doctrines of Imago Dei and the Sanctity of Life provide a vital foundation for understanding autonomy, asserting that free will is essential for fostering a loving relationship with God and neighbor (Ramsey, 2002). In medical practice, the principle of autonomy closely aligns with informed consent, respecting a patient's right to make decisions about their treatment. This practice reflects Christian teachings on personal responsibility and moral accountability. However, free will is not to be understood as mere independence. In the Christian framework, it is a relational gift—one that involves choosing according to God's will. John Paul II eloquently emphasized that true freedom is relational, grounded in self-giving and responsibility toward others: "Freedom... includes an essential relational dimension. It is a great gift from God, the Creator, put in the service of a person and his/her accomplishment through self-giving and accepting others" (Evangelium Vitae, No. 19).

While Christian ethics views autonomy as intertwined with responsibility and moral purpose, modern culture often emphasizes individual autonomy as the ability to make decisions without constraints. This understanding, particularly prevalent in North American bioethical discussions, sees autonomy as "first among equals" in Beauchamp and Childress's four core bioethical principles: autonomy, beneficence, nonmaleficence, and justice (Beauchamp & Childress, 2013). However, this modern view of freedom is incomplete. It often emphasizes personal choice without recognizing the role of virtue, goodness, and moral responsibility. In this framework, individuals may pursue actions that harm rather than liberate, reducing freedom to a self-centered pursuit rather than a path toward true fulfillment.

In contrast, a Christian understanding of autonomy finds its fullest expression in Christ. True freedom is not merely the absence of constraints but the ability to pursue what is good, true, and life-giving. As Jesus said: "If you abide in my word, you are truly my disciples, and you will know the truth, and the truth will set you free" (John 8:31-32). Aquinas echoed this teaching, emphasizing that in His divinity, Jesus is the Truth and the Life, and in His humanity, He is the Way. Thus, freedom is not merely about self-determination but about living in accordance with God's truth. In bioethics, this distinction becomes crucial: true freedom is not merely about the absence of constraints but about the pursuit of moral excellence. To better understand this, we can distinguish between "freedom from" (also known as freedom of indifference) and "freedom for excellence" (Pinckaers, 1995). This understanding underscores that true autonomy flourishes not in isolation but in relationship—with God, with others, and in alignment with moral truth.

Freedom for Excellence

Building upon the Christian understanding of free will, freedom for excellence emphasizes that true freedom is inseparable from truth, virtue, and a properly ordered life—a freedom personified in Christ. Great thinkers throughout Western history have recognized this concept, emphasizing the role of discipline, knowledge, and virtue as pathways to true freedom. Rather than freedom being an end in itself, freedom for excellence is a freedom directed toward the good, enabled by the grace of God and the life of the Spirit received through Christ.

This vision of freedom finds practical expression in everyday activities such as learning a language, creating art, or starting a business—each requiring discipline and perseverance to develop excellence. For instance, if someone wishes to learn another language, they must invest time and practice. Once achieved, the freedom to speak the language fluently is a reward for their effort and discipline. Servais Pinckaers, a Belgian moral theologian, illustrates this idea with the metaphor of playing a musical instrument: true musical freedom—the ability to perform beautifully in concert—arises only after disciplined practice has transformed skill into second nature (Pinckaers, 1995). This idea of freedom—achieved through discipline and growth—applies to all areas of life, from artistic endeavors to developing personal virtues.

Theologians and philosophers have linked freedom for excellence to concepts such as discipline, purpose, virtue, and the actualization of human potential. As Bishop Robert Barron encapsulates, "Freedom for excellence is the shaping of desire so as to make the achievement of the good first possible and then effortless" (Barron, 2007). This concept suggests that true freedom allows individuals to realize their highest aspirations through the gifts given by the Creator. The path to this freedom is through self-discipline and moral excellence. In contrast, the popular notion of "I can do what I want" reflects a freedom of indifference—focused solely on choice without regard for virtue or truth. True freedom, by contrast, is the ability to choose what is good and fulfilling— "I can do what I ought"—requiring serious engagement with life's higher purposes.

Quality of Life

When considering the doctrines of Imago Dei, the Sanctity of Life, and the concept of free will, it is essential to adopt a balanced approach that acknowledges the importance of "quality of life." While some within faith communities may express reservations about emphasizing quality of life, it is a significant reality in human existence. In the context of these doctrines, it is crucial to recognize that when the burdens associated with prolonging life become overwhelming—so much so that they hinder an individual's ability to fulfill God's purpose—there is no moral obligation to employ disproportionate means to extend life artificially (Ramsey, 2002). This perspective respects the sanctity of life while acknowledging that life should not be prolonged needlessly when it no longer aligns with God's plan or leads to undue suffering.

The Catholic tradition has long held that while human life is sacred, there is no moral imperative to use extraordinary or disproportionate medical interventions when they offer no reasonable hope of benefit and only add to the patient's suffering (John Paul II, 1995). This balanced understanding highlights the importance of discerning when medical interventions become disproportionate and potentially counterproductive, particularly in preserving the quality of life and aligning with God's will. In this context, the Church strongly affirms the value of palliative care as a compassionate and morally sound response to suffering. As emphasized in *Samaritanus Bonus* (2020), palliative care is considered an act of charity that neither hastens nor unduly prolongs death. By providing pain relief, emotional support, and spiritual care, palliative care reflects the Church's commitment to preserving dignity while alleviating unnecessary suffering. This approach offers a moral alternative to both euthanasia and overly aggressive treatment. By focusing on both the sanctity of life and the quality of life, this perspective allows individuals to consider the broader purpose of life and God's divine plan. It emphasizes that the ultimate goal of life is not mere survival but the fulfillment of one's spiritual vocation. Consequently, medical ethics in this context call for careful consideration of when life-prolonging interventions no longer serve the person's well-being or spiritual flourishing (Meilaender, 2005).

The Fall

The Fall is a foundational concept in Christian theology, profoundly shaping its understanding of suffering, disease, and death. In the biblical narrative, Adam and Eve's disobedience in the Garden of Eden (Genesis 3) disrupted humanity's original harmony with God. This rupture introduced sin into the world, resulting in suffering, disease, and death—consequences that have marked human experience ever since (Romans 5:12). This theological understanding profoundly shapes Christian perspectives on healthcare, underscoring the moral duty to alleviate suffering and uphold human dignity. In Christian ethics, the response to suffering calls for acts of compassion, empathy, and care for those afflicted. Suffering encompasses physical, emotional, and spiritual pain, and the Christian ethical tradition emphasizes the importance of alleviating suffering through medical care and acts of charity. From this perspective, healthcare and pursuing medical advancements are vital expressions of compassion and love for one's neighbor (Meilaender 2005).

Christian ethics encourages efforts to mitigate suffering and views medical treatment as an integral aspect of the call to care for the sick and the vulnerable (Matthew 25:36). At the same time, Christian ethics affirms the moral duty to preserve and protect life while acknowledging that death is an inevitable part of the human condition. Yet, Christian theology offers a unique hope beyond earthly existence—the resurrection and eternal life (1 Corinthians 15:20-22). This belief provides comfort in the face of death, emphasizing that human mortality is not the end but a passage toward ultimate renewal and redemption (Romans 8:18-21). In light of this hope, Christians are called to trust in God's providence while actively engaging in compassionate efforts to alleviate suffering and uphold the sanctity of life. At the same time, recognizing the limits of medical intervention reflects faith in God's ultimate plan for humanity (John Paul II, 1995). This balanced understanding affirms the value of preserving life while acknowledging that life-prolonging interventions may not always align with God's will or the well-being of the patient.

Justice

In the Judeo-Christian tradition, justice transcends mere adherence to rules or the notion of fairness. Within the Christian ethical framework, justice is a profound moral imperative that reflects God's perfect righteousness, calling individuals to pursue fairness, compassion, and mercy in their interactions with others. Christian justice is not merely about ensuring that people receive what they deserve; it is a commitment to align human systems with God's flawless standard of justice (Wolterstorff, 2008). To fully grasp this concept of justice, it is important to explore three key dimensions: accountability, mercy, and grace.

Accountability in Christian ethics underscores that individuals are responsible for their actions and must accept appropriate consequences. As Scripture teaches, "A man reaps what he sows" (Galatians 6:7). This principle reflects God's justice, which affirms moral responsibility and upholds the dignity of human choice. Accountability ensures that justice is rooted in truth, recognizing that individuals must face the outcomes of their actions while preserving their inherent dignity. While accountability affirms moral responsibility, Christian ethics extends justice by emphasizing mercy—the call to forgive and extend compassion even when others fall short. As Jesus taught, "Blessed are the merciful, for they shall receive mercy" (Matthew 5:7). This emphasis on mercy tempers justice with compassion, promoting healing and restoration. Mercy does not negate accountability but acknowledges the human condition—marked by weakness, suffering, and the need for forgiveness. By extending mercy, Christians reflect God's patient love, which desires repentance and restoration rather than mere punishment.

Christian justice is also inseparable from grace—God's unmerited favor given freely despite human failings. As Paul writes, "All are justified freely by his grace through the redemption that came by Christ Jesus" (Romans 3:24). Grace transforms justice, encouraging Christians to practice generosity, forgiveness, and humility. By extending grace to others, individuals recognize that they themselves are recipients of divine mercy and forgiveness. Grace not only transcends human merit but calls believers to cultivate a spirit of compassion and charity in their interactions with others (Meilaender, 2005). These three dimensions—accountability, mercy, and grace—demonstrate that Christian justice is not only about ensuring fair outcomes but also about nurturing a compassionate and redemptive moral order that mirrors God's perfect love and righteousness. By balancing accountability with mercy and grace, Christian ethics offers a justice model that emphasizes both responsibility and restoration—one that seeks the good of the individual while promoting the common good.

Hope and Faith

Hope and faith hold profound significance in the Christian ethical framework, particularly in the context of eternal life. For Christians, the challenges and trials of earthly existence—including suffering, illness, and mortality—are not seen as the final chapter. Instead, believers find comfort and strength in the hope of eternal life in God's presence. This hope transcends present hardships, offering a sense of meaning and purpose that provides solace during suffering and loss (Meilaender, 2005). Christian faith actively nurtures this hope, encouraging believers to draw upon their trust in God to sustain resilience and courage when facing disease or adversity. The hope of eternity provides a broader perspective, reminding believers that earthly trials are temporary, and that a greater, unending reality awaits them. In this context, hope is not mere optimism but is deeply anchored in faith—an unwavering trust in God's promises of healing, restoration, and eternal life (Romans 8:18). Central to this understanding is the role of prayer, which reflects both hope and faith. Prayer may involve seeking physical healing, spiritual strength, or the grace to endure suffering with patience and courage. For Christians, prayer is not only a means of requesting outcomes but also an act of trusting in God's presence and wisdom. This interplay between hope and faith reinforces the belief that God's plan is ultimately one of goodness, redemption, and restoration (Hebrews 11:1).

At the same time, Christian ethics emphasizes the importance of accepting God's will, even when it diverges from personal desires. Faith invites believers to trust in God's wisdom and sovereignty, recognizing that while healing and recovery are deeply desired, they may not always align with divine providence. In these moments, hope directs believers toward the ultimate promise of eternal life—a reality that transcends earthly suffering and assures believers of God's unwavering presence and love (Benedict XVI, 2007). This dual understanding—hope for earthly healing and the hope of eternity—guides the Christian response to suffering. While believers maintain hope for positive outcomes in this life, they also find comfort in the deeper assurance that God's eternal plan offers lasting peace and fulfillment. This balanced perspective strengthens resilience in adversity, fostering trust that, whether in healing or in death, God's love ultimately prevails.

Stewardship

Within the Christian ethical framework, stewardship is a profound concept that emphasizes humanity's responsibility as caretakers of the gifts and resources entrusted to us by God. This responsibility extends to various aspects of life, including decisions related to contemporary medicine and the timing of death. Stewardship is grounded in the belief that God has granted humanity both freedom and the ability to choose; yet with these gifts comes the serious responsibility to act wisely, in alignment with God's divine plan (Cunningham, 2008). This moral responsibility is particularly significant in healthcare, where medical advancements have given individuals greater control over treatment decisions, including those concerning end-of-life care.

Stewardship calls individuals, healthcare providers, and families to thoughtfully evaluate medical decisions, carefully considering their moral and ethical implications. This responsibility is especially crucial when assessing interventions such as cardiopulmonary resuscitation, ventilator support, or chemotherapy. Stewardship requires a conscientious balance—one that seeks to preserve life while respecting human dignity and alleviating suffering. Informed by ethical principles like beneficence and nonmaleficence, stewardship encourages decisions that neither hasten death nor prolong it through disproportionate or futile interventions. In this context, stewardship urges careful discernment to ensure that medical treatments align with the patient's well-being, values, and spiritual convictions. It acknowledges that there are limits to medical interventions and that aggressive treatments may become counterproductive if they merely extend suffering without meaningful benefit (Meilaender, 2005). Good stewardship involves recognizing when to embrace medical care to preserve life and when to allow natural processes to take their course with dignity and peace.

Both Catholic and Protestant traditions emphasize that stewardship does not imply preserving life at all costs. Instead, it affirms life as a sacred gift to be respected and nurtured according to God's will (John Paul II, 1995). The Catholic Church, in particular, advocates for palliative care as an act of charity that neither accelerates nor unduly prolongs death, offering compassionate care as a moral alternative to euthanasia or excessive medical intervention (Samaritanus Bonus, 2020). In practical terms, stewardship calls for thoughtful collaboration between healthcare providers, patients, and families. It encourages open, compassionate communication to ensure medical decisions honor the patient's wishes while upholding Christian ethical principles. Stewardship ultimately reflects moral accountability—guiding individuals to make healthcare decisions that align with faith, respect human dignity, and promote holistic well-being (Hauerwas, 1988). As a guiding principle, stewardship calls for a balanced approach that cherishes the sanctity of life, acknowledges the natural limits of medicine, and promotes compassion, responsibility, and trust in God's providence.

Hope, Faith, and Miracles in Practice

Hope and faith provide a profound lens for confronting suffering, illness, and mortality. While eschatological hope points to eternal life, Christian healthcare institutions cultivate hope in the present through pastoral support, community prayer, and holistic treatment approaches. For example, chaplaincy services help patients and families draw upon their faith during challenging times, reinforcing the Christian emphasis on resilience and trust in God's providence. The reality of miracles is a source of hope. A practical definition of miracles is a special divine action—a unique intervention in the natural course of events. While believers often recognize the universe's creation and ongoing existence as miracles of immeasurable magnitude, other extraordinary phenomena also reveal divine action. Examples include the fine-tuning of the cosmos, the origin of life, the complexity of DNA as a digital code, the intelligibility of the universe, and the mystery of consciousness—each standing as foundational miracles that shape both cosmic and human existence (Meyer, 2009). These grand-scale miracles reflect God's sustaining presence in creation, inspiring awe and reverence.

New Testament scholar Craig Keener has extensively documented miracles in his two-volume work, Miracles: The Credibility of the New Testament Accounts (Keener, 2011). Keener's research highlights the reliability of miraculous accounts in the New Testament, underscoring their purpose within the biblical narrative: to draw attention to God's message and revelation. While miracles may not always follow predictable patterns, their extraordinariness often leaves a profound and unmistakable impression. Such events, marked by their divine signature, serve to reveal God's presence and transformative power. In addition to exploring biblical miracles, Keener's research bridges the gap between the ancient and modern worlds by documenting well-attested contemporary miracles. His extensive records demonstrate continuity between the miraculous events described in Scripture and those witnessed in the present day. By including modern testimonies—many corroborated by medical professionals and other credible witnesses—Keener reinforces the claim that God continues to intervene in extraordinary ways. These accounts lend credibility to the Christian conviction that God remains actively engaged with humanity throughout history (Keener, 2011).

The ethical implications of contemporary miracles extend to both believers and non-believers. For believers, witnessing or hearing about miracles can deepen faith, strengthen resilience, and inspire a greater sense of responsibility toward others. Such testimonies remind Christians to approach life's challenges with hope, trusting that God's power can bring unexpected restoration and healing. They also encourage compassion, empathy, and a renewed commitment to serving others while relying on divine guidance. For non-believers, Keener's detailed documentation offers an opportunity for reflection. Contemporary accounts of healing, restoration, or remarkable outcomes following prayer in the name of Jesus provide compelling evidence of God's active presence in the world. These testimonies may prompt reflection on the life, death, and resurrection of Jesus Christ, opening pathways for spiritual inquiry and belief. For Christians facing seemingly impossible situations, awareness of contemporary miracles can rekindle hope and reinforce trust in God's power. These accounts testify to the living, dynamic nature of the Christian faith, offering encouragement in times of uncertainty. Believers are invited to approach life's challenges with expectant trust, confident that God can intervene in extraordinary ways (Acts 2:22).

Bioethics as a Contemporary Discipline

In their seminal work *Principles of Biomedical Ethics*, Beauchamp and Childress outline Principlist bioethics, a widely adopted framework for ethical decision-making in healthcare. This model identifies four foundational principles: respect for *autonomy*, *beneficence*, *nonmaleficence*, and *justice*. These principles guide healthcare professionals in analyzing and resolving moral dilemmas in clinical practice and medical research. Respect for autonomy emphasizes the right of individuals to make informed decisions about their healthcare. Beneficence obligates healthcare providers to act in the patient's best interest, while nonmaleficence requires avoiding harm. Justice pertains to the fair distribution of healthcare resources and equitable access to medical services (Beauchamp & Childress, 2013).

While *Principlist bioethics* offers a structured and pragmatic model, *Personalist bioethics* introduces a more comprehensive philosophical framework that places the dignity and value of the human person at its core. Personalist bioethics integrates the four principles of Principlism within a broader vision that emphasizes the unity of the human person, encompassing physical, psychological, and spiritual dimensions. By addressing not only medical practices but also the ontological and metaphysical aspects of human existence, Personalist bioethics provides a holistic approach to complex moral dilemmas, particularly in matters of life and death (Sgreccia, 2012). In this model, respect for autonomy is viewed not merely as individual freedom but as an expression of the person's inherent dignity, rooted in a broader moral order. Beneficence and nonmaleficence are likewise reframed—not merely as utilitarian obligations but as expressions of love and care for each individual. Justice, in this context, transcends the equitable distribution of resources; it incorporates solidarity, mutual responsibility, and the recognition of every human being's intrinsic worth (Melé, 2009). For instance, this perspective may influence decisions regarding scarce medical resources during a global pandemic by emphasizing not only fairness but also the importance of fostering community solidarity and protecting the most vulnerable.

Three core principles rooted in ontology distinguish Personalist bioethics. The principle of *freedom and responsibility* emphasizes the balance between respecting individual autonomy and recognizing the moral obligations that accompany freedom. *Proportionate therapy* advocates medical interventions that carefully weigh benefits against burdens, ensuring that treatments align with the patient's overall well-being rather than merely prolonging life at all costs. Lastly, the principle of *sociality and subsidiarity* underscores the communal nature of human existence, encouraging healthcare decisions to be made at the most appropriate level of authority, such as within families, communities, or institutions (Sgreccia, 2012).

While Principlist bioethics offers a widely accepted and practical guide for ethical decision-making in healthcare, Personalist bioethics enriches this approach by grounding ethical choices in a deeper understanding of the interconnectedness of human life. By emphasizing the dignity of the person and addressing spiritual and psychological dimensions alongside physical well-being, Personalist bioethics ensures that healthcare decisions reflect a truly human-centered approach. Together, these frameworks provide complementary insights, offering healthcare professionals valuable tools to navigate the ethical complexities of modern medical practice.

The Principles of Bioethics and the Bioethics of Principles

The Hippocratic Oath marks one of the earliest recognitions of ethical principles in medical practice, establishing foundational commitments such as promoting good (beneficence), avoiding harm (nonmaleficence), and preserving patient confidentiality. Over the centuries, these principles became closely tied to Christianity, shaping the moral framework of healthcare institutions. Faith-based hospitals, hospices, and care missions reflected this union of medicine, ethics, and compassion. However, significant cultural changes and rapid scientific advancements over the past two centuries have reshaped healthcare, presenting new and complex ethical challenges (Jonsen, 1998). As medical innovation expanded treatment possibilities, concerns emerged about the potential "dehumanization" of medicine. In response, bioethics formally emerged around 1970, emphasizing the need to safeguard human dignity amid technological progress. During this period, the biopsychosocial model also gained prominence, promoting a holistic approach to healthcare by integrating biological, psychological, and social factors (Engelhardt, 1996). As medicine developed, faith-based institutions adapted in diverse ways—some maintaining their religious missions, while others aligned more closely with secular healthcare models. In 1979, Beauchamp and Childress published Principles of Biomedical Ethics, introducing four foundational principles: respect for autonomy, beneficence, nonmaleficence, and justice. Known as Principlism, this framework offered a structured and adaptable approach to resolving moral dilemmas in healthcare, gaining prominence for its practicality in diverse cultural and clinical settings.

Despite its influence, Principlism faced notable critiques in the 1990s. In their essay "A Philosophical Critique of Bioethics," Clouser and Koppelman argued that the four principles, while useful, risked oversimplifying the complexity of real-world ethical dilemmas. They noted that Principlism lacked a clear mechanism for resolving conflicts between competing principles, often leaving healthcare providers to rely on intuition or subjective judgment (Clouser & Gert, 1990). These critiques prompted discussions about alternative frameworks such as virtue ethics, care ethics, and narrative ethics. For example, in culturally diverse settings, prioritizing autonomy may conflict with community-centered values, requiring a more nuanced approach. Similarly, health challenges, such as resource allocation in underserved regions, often demand justice frameworks that emphasize solidarity and equity beyond traditional Western interpretations. While these critiques revealed Principlism's limitations, the framework remains widely used in healthcare ethics due to its flexibility and practical utility (Veatch, 2011). As bioethics continues to address evolving challenges, integrating broader ethical frameworks — rooted in human dignity and cultural sensitivity — can enhance its ability to navigate the complexities of modern medicine.

A Critique of Principlism

In their 1995 publication, Principlism and the Ethical Appraisal of Clinical Trials, Meslin, Sutherland, Lavery, and Till introduced the concept of context sensitivity in bioethics. They highlighted the difficulty of applying universal principles — autonomy, beneficence, nonmaleficence, and justice — without considering the complex realities of clinical practice. Factors such as patient vulnerability, risk-benefit analysis, and societal implications often require a more adaptable approach. Clouser and Gert further critiqued Principlism for functioning more as a checklist than an actionable ethical guide. They argued that because the four principles often conflict, resolving dilemmas frequently depends on intuition or retrospective justification, potentially leading to ethically inconsistent outcomes. For example, a scenario that pits autonomy against beneficence lacks a clear mechanism to determine which principle should take priority. Critics contend that alternative frameworks — such as Kantian ethics or Utilitarianism — offer clearer guidance for moral decision-making.

In a 1997 lecture, Gonzalo Herranz argued that Principlism's flexibility, while advantageous, also makes it vulnerable to manipulation. He warned that its adaptability risks justifying virtually any action, making it "sophisticated and sophistic," unable to provide reliable guidance in ethically complex situations. More recently, Rosamond Rhodes and K. Danner Clouser with Bernard Gert have echoed these concerns. Rhodes critiqued Principlism for oversimplifying complex medical dilemmas, particularly by rigidly prioritizing its four principles without accounting for contextual factors (Rhodes, 2015). Similarly, Clouser and Gert argued that Principlism lacks a coherent theoretical foundation, leading to inconsistencies in ethical decision-making (Clouser & Gert, 1990). They contend that the four principles are often applied without a unifying moral theory, resulting in conflicts that are difficult to resolve. While Principlism remains a foundational framework in bioethics, these critiques highlight the need for a more nuanced approach. Integrating elements of virtue ethics, casuistry, and patient-centered care offers a more comprehensive strategy for navigating the complexities of contemporary medicine.

Value and Limits of the Bioethics of Principles

In his essay "A Defense of Universal Principles in Biomedical Ethics," Tom L. Beauchamp defended Principlism as a universal ethical foundation that transcends cultural, religious, and philosophical boundaries. He argued that the four core principles provide a common language for navigating ethical dilemmas in diverse healthcare settings, offering practical guidance for clinicians and policymakers alike (Beauchamp & Childress, 2013). While Principlism's flexibility enables it to adapt to modern healthcare challenges, critics have warned that it lacks a clear method for resolving conflicts between competing principles. This limitation has driven calls for interdisciplinary approaches that incorporate insights from fields such as law, theology, anthropology, economics, and psychology. For example, legal frameworks shape decisions on patient rights, while theological perspectives address ethical concerns in religiously sensitive cases.

Moreover, postmodern critiques of bioethics challenge Western assumptions about autonomy and individualism. Scholars such as Madan Sarup and Paul Michael Lützeler have explored how postmodern thought and its challenges, especially following the 1960s social revolution, redefined long-held values regarding family, love, and social institutions, with significant societal implications and for bioethics (Sarup, 1988). Subrata Chattopadhyay has criticized bioethics for maintaining a Western-centric bias that often excludes non-Western moral frameworks. He argues that integrating perspectives such as Confucian ethics, emphasizing familial responsibility, or Islamic ethics, prioritizing communal care, can enhance bioethical discourse and counterbalance Western individualism (Chattopadhyay & De Vries, 2008). However, to counterbalance Western individualism the West need not look any other way than to its roots and find the richness of those traditional values. To meet the demands of healthcare, bioethics must embrace an interdisciplinary approach. Incorporating civilizational cultural values, alongside frameworks like virtue ethics and case-based reasoning, ensures that bioethics remains relevant and responsive to the complexities of modern medicine. By recognizing the interplay between universal principles and contextual realities, bioethics can better address the ethical challenges facing a rapidly evolving world.

Beyond Principles: Integrating Christian Ethics into Bioethics

Christianity offers a wealth of moral and spiritual resources that extend beyond the four core bioethics principles: autonomy, beneficence, nonmaleficence, and justice. Rooted in Scripture, tradition, and community, Christian ethics emphasizes love, mercy, justice, and the sanctity of human life. At the heart of this framework are the two greatest commandments Jesus gives—love for God and love for neighbor (Matthew 22:37-39)—which form the cornerstone of Christian moral teaching. These guiding principles encourage believers to act compassionately, seek divine wisdom through prayer, and rely on the Holy Spirit's guidance when confronting complex bioethical dilemmas (Meilaender, 2005).

Christianity's transcendent worldview also shapes its unique approach to end-of-life issues. From a secular, materialist perspective, death is often seen as the definitive end of consciousness, with ethical decisions focusing on minimizing suffering and maximizing quality of life. In contrast, the Christian tradition views death as a transition to eternal life, offering hope and comfort by affirming that death is not the final chapter but a gateway to reunion with God (John Paul II 1995). This belief fundamentally influences bioethical decision-making by upholding the sanctity of life while acknowledging the limitations of medical interventions. For instance, Christian ethics may support palliative care that relieves suffering without hastening death, emphasizing compassion, human dignity, and the natural process of dying. This approach contrasts with euthanasia or overly aggressive treatments that risk undermining the spiritual significance of life's final stages.

While Principlism offers a structured framework for ethical analysis, its limitations become apparent when addressing healthcare's relational, existential, and spiritual dimensions. Personalist Bioethics, by contrast, offers a more comprehensive approach grounded in the dignity and unity of the human person. Integrating insights from Christian ethics, Personalist Bioethics emphasizes love, mercy, and justice as guiding principles. This approach allows bioethical decision-making to transcend secular frameworks, incorporating interdisciplinary insights from law, philosophy, and theology to address moral complexities more holistically. Christianity's contributions to bioethics are particularly relevant as emerging fields like Neuroethics present new moral challenges. Advances in neuroscience and brain-related technologies raise complex questions about neuroenhancement, brain-computer interfaces, and cognitive privacy. As this chapter concludes with a broad examination of bioethics, the next chapter will explore how Christian moral principles can illuminate pathways forward in this rapidly developing discipline.

REFERENCES

Aquinas, T. (n.d.). *Summa Theologica, II-II, Q. 26, Art. 3.* Retrieved from https://www.newadvent.org/summa/

Augustine. (n.d.). *Sermon on the Good Samaritan.* Retrieved from https://www.earlychurchtexts.com

Bainton, R. H. (1950). *Here I stand: A life of Martin Luther.* Nashville, TN: Abingdon-Cokesbury Press.

Barron, R. (2007). *The priority of Christ: Toward a postliberal Catholicism.* Grand Rapids, MI: Brazos Press.

Beauchamp, T. L., & Childress, J. F. (1979). *Principles of biomedical ethics.* New York, NY: Oxford University Press.

Beauchamp, T. L., & Childress, J. F. (2013). *Principles of biomedical ethics* (7th ed.). New York, NY: Oxford University Press.

Benedict XVI. (2007). *Spe salvi (Saved in hope).* Vatican Press. Retrieved from https://www.vatican.va/content/benedict-xvi/en/encyclicals/documents/hf_ben-xvi_enc_2007 1130_spe-salvi.html

Berman, H. J. (1983). *Law and revolution: The formation of the Western legal tradition.* Cambridge, MA: Harvard University Press.

Boyle, J. (1980). Toward understanding the principle of double effect. *Ethics, 90*(4), 527–538. https://doi.org/10.1086/292207

Brockliss, L., & Jones, C. (1997). *The medical world of early modern France.* New York, NY: Oxford University Press.

Cahill, L. S. (1995). *Love your enemies: Discipleship, pacifism, and just war theory.* Minneapolis, MN: Fortress Press.

Chattopadhyay, S., & De Vries, R. (2008). Bioethical concerns are global, bioethics is Western. *Eubios Journal of Asian and International Bioethics, 18*(4), 106–109.

Clarke, W. N. (2001). *The one and the many: A contemporary Thomistic metaphysics.* Notre Dame, IN: University of Notre Dame Press.

Clouser, K. D., & Gert, B. (1990). A critique of Principlism. *The Journal of Medicine and Philosophy, 15*(2), 219–236.

Cunningham, L. S. (2008). *Christian ethics: The end of the law.* Continuum.

Edelstein, L. (1943). *The Hippocratic Oath: Text, translation and interpretation.* Baltimore, MD: Johns Hopkins Press.

Engelhardt, H. T. (1996). *The foundations of bioethics* (2nd ed.). New York, NY: Oxford University Press.

Finnis, J. (1980). *Natural law and natural rights.* Oxford, UK: Clarendon Press.

Gillon, R. (1985). Medical ethics: Four principles plus attention to scope. *BMJ (Clinical Research Edition), 290*(6483), 191–193. https://doi.org/10.1136/bmj.290.6483.191

Hauerwas, S. (1988). *Suffering presence: Theological reflections on medicine, the mentally handicapped, and the church.* Burns & Oates.

Herranz, G. (1997). *The malleability of principles: A lecture on the limits of Principlism.* University of Navarra.

Herranz, G. (1997). *The exhaustion of principled bioethics: In search of new solutions* [Lecture]. Escuela de Medicina de Málaga, Department of Bioethics, University of Navarra.

John Paul II. (1995). *Evangelium Vitae* (The gospel of life). Vatican.

Jones, W. H. S. (1924). *The doctor's Oath: An essay in the history of medicine.* Cambridge, UK: Cambridge University Press.

Jonsen, A. R. (1998). *The birth of bioethics.* New York, NY: Oxford University Press.

Jonsen, A. R., Siegler, M., & Winslade, W. J. (1982). *Clinical ethics: A practical approach to ethical decisions in clinical medicine.* New York, NY: McGraw-Hill.

Keener, C. S. (2011). *Miracles: The credibility of the New Testament accounts* (Vols. 1–2). Grand Rapids, MI: Baker Academic.

Lewis, C. S. (1960). *The four loves*. New York, NY: Harcourt.

Luther, M. (2008). *On the freedom of a Christian* (T. L. Frazier, Trans.). Missionary Society of St. Paul. (Original work published 1520)

MacIntyre, A. (1998). *A short history of ethics*. New York, NY: Routledge.

McGrath, A. E. (2001). *Christian theology: An introduction* (3rd ed.). Malden, MA: Wiley-Blackwell.

Meilaender, G. (2005). *Bioethics: A primer for Christians* (2nd ed.). Grand Rapids, MI: Eerdmans Publishing.

Meslin, E. M., Sutherland, H. J., Lavery, J. V., & Till, J. E. (1995). Principlism and the ethical appraisal of clinical trials. *Bioethics, 9*(5), 399–418. https://doi.org/10.1111/j.1467-8519.1995.tb00 314.x

Melé, D. (2009). Integrating personalism into virtue-based business ethics: The personalist and the common good principles. *Journal of Business Ethics, 88*(1), 227–244. https://doi.org/10.1007/s10551-009-01 08-y

Meyer, S. C. (2009). *Signature in the cell: DNA and the evidence for intelligent design*. New York, NY: HarperOne.

Niebuhr, H. R. (1951). *Christ and culture*. New York, NY: Harper & Row.

Pelikan, J. (1988). *The melody of theology: A philosophical dictionary*. Cambridge, MA: Harvard University Press.

Pellegrino, E. D. (1990). The relationship of autonomy and integrity in medical ethics. *Bulletin of the Pan American Health Organization, 24*(4), 361–371.

Pellegrino, E. D. (1985). The virtuous physician, and the ethics of medicine. *Journal of Medical Ethics, 16*(4), 205–211.

Pellegrino, E. D., & Thomasma, D. C. (1993). *The virtues in medical practice*. New York, NY: Oxford University Press.

Pinckaers, S. (1995). *The sources of Christian ethics*. Washington, DC: Catholic University of America Press.

Porter, J. (2004). *Nature as reason: A Thomistic theory of the natural law.* Grand Rapids, MI: Eerdmans.

Porter, R. (1999). *The greatest benefit to mankind: A medical history of humanity.* New York, NY: W. W. Norton & Company.

Ramsey, P. (2002). *The patient as person: Explorations in medical ethics.* New Haven, CT: Yale University Press.

Rhodes, R. (2015). Good and not so good medical ethics. *Journal of Medical Ethics*, 41(1), 71–74. https://doi.org/10.1136/medethics-2014-102312

Sarup, M. (1988). *An introductory guide to post-structuralism and postmodernism.* Athens, GA: University of Georgia Press.

Singer, P. (1993). *Practical ethics.* New York, NY: Cambridge University Press.

Sgreccia, E. (2012). *Personalist bioethics: Foundations and applications.* Philadelphia, PA: The National Catholic Bioethics Center.

Veatch, R. M. (2012). *Hippocratic, religious, and secular medical ethics: The points of conflict.* Washington, DC: Georgetown University Press.

Veatch, R. M. (2011). *The basics of bioethics* (3rd ed.). New York, NY: Routledge.

Weber, M. (2001). *The Protestant ethic and the spirit of capitalism.* New York, NY: Routledge.

Witte, J. Jr. (2008). *The Reformation of rights: Law, religion, and human rights in early modern Calvinism.* New York, NY: Cambridge University Press. https://doi.org/10.1017/CBO9780511819377

Wolterstorff, N. (2008). *Justice: Rights and wrongs.* Princeton, NJ: Princeton University Press.

CHAPTER 18

Neuroethics

Neuroethics, a subfield of bioethics, explores the ethical, legal, and social implications of advancements in neuroscience. This emerging discipline addresses technologies that manipulate or enhance the human brain, raising profound questions about identity, autonomy, and the boundaries of human capability. Neuroethics extends traditional bioethical principles — beneficence, nonmaleficence, justice, and autonomy — into the complex and often uncharted terrain of neuroscientific innovation (Illes & Racine, 2005). Among the most significant advancements examined within neuroethics are Brain-Computer Interfaces (BCIs) and Brain-to-Brain Interfaces (BTBIs). BCIs enable direct communication between the brain and external technologies, allowing individuals to control devices or communicate using neural signals alone. These systems offer extraordinary potential, from medical therapies for individuals with severe disabilities to immersive entertainment systems and military applications (Matamala-Gomez & Malighetti, 2022; Niketeghad & Pouratian, 2019). BTBIs expand on this concept by enabling direct neural communication between individuals — or even between humans and animals — unlocking novel possibilities for collaboration, learning, and enhanced sensory perception (Rao et al., 2014). However, these innovations introduce pressing ethical concerns. How can meaningful informed consent be ensured when dealing with experimental or poorly understood technologies? What are the privacy risks of devices capable of accessing and transmitting neural data? Should cognitive enhancement be regulated, and to what extent? These questions demand careful ethical scrutiny and the development of robust regulatory frameworks to ensure the responsible deployment of neurotechnologies.

The Emergence of Neuroethics

Neuroethics emerged as a distinct discipline in the early twenty-first century. In 2002, the Dana Foundation convened a groundbreaking meeting titled "Neuroethics: Mapping the Field," where neuroscientists, ethicists, and policymakers discussed the ethical implications of neuroscientific advancements (Marcus, 2002). This event marked a pivotal moment, solidifying neuroethics as a recognized field of study.

Renowned columnist William Safire, a participant in the meeting, famously defined neuroethics as "the examination of what is right and wrong, good and bad, about the treatment of, perfection of, or unwelcome invasion of and worrisome manipulation of the human brain" (Roskies, 2024). Scholars such as Judy Illes and Eric Racine have significantly contributed to the field, addressing ethical dilemmas posed by rapid developments in neuroscience (Illes & Racine, 2005). Today, neuroethics provides a vital framework for evaluating technologies that push the boundaries of cognition, privacy, and human agency. BCIs are one of the most transformative technologies in neuroscience, offering unprecedented opportunities to enhance human capabilities. In medicine, BCIs enable individuals with paralysis to control prosthetic limbs, operate wheelchairs, and regain lost communication abilities (Matamala-Gomez & Malighetti, 2022; Barbara et al., 2019). For older adults, BCIs offer enhanced independence by facilitating the performance of physically demanding tasks through mental control. Beyond therapeutic applications, BCIs are expanding into entertainment, military, and household domains. For example, gamers can now interact with virtual worlds using neural signals, creating immersive experiences that transcend traditional gaming systems (Vasiljevic & De Miranda, 2020). In military contexts, BCIs are being developed to help soldiers pilot drones with mental commands, enhancing operational precision and efficiency (Friedrich et al., 2021). Meanwhile, mind-controlled appliances are transforming home automation, simplifying daily routines. While the potential of BCIs is vast, their adoption raises significant ethical questions. How can neural data be protected from unauthorized access or exploitation? The ability to extract thought-related information through BCIs heightens concerns about surveillance and privacy violations. Ensuring that users fully understand the risks associated with BCIs is critical, particularly as these technologies become more complex (Friedrich et al., 2021). Additionally, questions about equitable access must be addressed to prevent BCIs from deepening socioeconomic inequalities.

The Ethical Implications of BTBIs

Building on the success of BCIs, BTBIs represent a step toward direct neural connectivity between individuals. This technology enables thought-driven communication and collaboration, as demonstrated in experiments where one person's thoughts triggered physical actions in another (Rao et al., 2014). BTBIs have also been extended to human-animal interactions, allowing humans to influence animal movements or leverage enhanced sensory capabilities, such as a dog's acute sense of smell (Leconstant & Spitz, 2022). However, these advancements raise significant ethical dilemmas. Obtaining meaningful consent becomes particularly complex in neural connectivity scenarios, especially in experimental contexts. As neural boundaries blur, concerns arise about individual identity, free will, and control over one's thoughts and actions. The potential for BTBIs to be misused for manipulation or coercion underscores the urgent need for robust ethical guidelines and regulatory oversight. The rapid evolution of neurotechnologies like BCIs and BTBIs holds extraordinary promise but requires careful ethical consideration. Neuroethics plays a critical role in addressing the profound questions these technologies raise about human identity, privacy, and equity. By fostering interdisciplinary collaboration among neuroscientists, ethicists, policymakers, and the broader public, society must harness these innovations responsibly while safeguarding fundamental rights and values. As the boundaries of neuroscience continue to expand, neuroethics will remain indispensable in navigating the complex interplay between technological innovation and ethical responsibility.

One Illustrative Example of BMI Medical Use

In 2022, Pat Bennett demonstrated the remarkable potential of brain-machine interfaces (BMIs) for individuals with motor impairments. Pat, who has been living with Amyotrophic Lateral Sclerosis (ALS), faced the immense challenge of losing her ability to communicate effectively (Henderson, 2023). To address this, she underwent a groundbreaking procedure that involved implanting four tiny sensors — each the size of a baby aspirin — into her brain. These sensors formed part of an advanced intracortical brain-computer interface (iBCI) system, designed to transmit signals from the language regions of Pat's brain to sophisticated software capable of decoding her thoughts into text. Pat's journey involved 25 intensive training sessions, each lasting approximately four hours. During these sessions, she attempted to repeat randomly selected sentences drawn from a dataset of recorded speech samples. The iBCI system relied on an autocorrect algorithm that meticulously decoded Pat's brain activity into phoneme streams, which were then converted into words and displayed on a computer screen. Through this iterative process — involving 260 to 480 sentences per session — the system gradually adapted to Pat's unique neural patterns, achieving greater precision and effectiveness over time. This innovative iBCI system offered a significant advancement over less invasive yet less precise BCIs, underscoring the potential of intracortical technologies to restore communication in individuals with ALS and other motor impairments.

The Ethical Implications of BCIs

While brain-computer interfaces (BCIs) hold extraordinary promise for restoring lost functions — such as enabling movement in paralyzed individuals — they also introduce complex ethical concerns about autonomy and agency. As BCIs increasingly blur the boundary between human thought and machine control, questions arise about how these technologies might alter an individual's sense of self-governance — their ability to initiate and control their actions. A major concern is the potential for "agency erosion" — a scenario where the BCI, rather than the individual, becomes the primary driver of decision-making or behavior. Because BCIs intervene directly in neural processes, they may influence thoughts or actions without the user's full awareness or consent. For instance, BCIs designed to enhance cognitive or physical performance could inadvertently alter an individual's authentic desires, raising questions about personal responsibility and whether the individual remains in meaningful control. This ambiguity creates significant ethical dilemmas regarding the authenticity of decisions made through BCI systems (Clausen, 2011).

The risks of agency erosion are magnified in materialist-naturalist worldviews, which often reduce human consciousness, will, and identity to mere neural processes. Within such frameworks, BCIs are viewed as tools for optimizing biological function, potentially diminishing the deeper dimensions of human autonomy. This reductionist perspective may prioritize technological efficiency over human dignity, leading to scenarios in which individuals are encouraged — or even coerced — into adopting BCIs to enhance productivity or conform to social expectations. For example, workplaces may pressure employees to adopt BCIs to improve performance or maintain competitiveness, creating ethical concerns about freedom and consent. In response to these concerns, some neuroscientists and ethicists are developing strategies to mitigate the risks of agency erosion. User-centered design approaches prioritize ensuring that individuals retain meaningful control over their neural interfaces. Developers are also advancing transparent algorithms that make the BCI's decision-making processes clearer and more predictable for users. Additionally, scholars have proposed ethical frameworks that emphasize the intrinsic value of human freedom and personal agency. These frameworks seek to ensure that BCIs empower individuals rather than mechanize their decisions. Without such ethical safeguards, BCIs risk becoming tools for manipulation, exploitation, or social conformity — diminishing individual freedom rather than enhancing it (Schermer, 2009). Recognizing the inherent dignity and moral worth of every individual is essential to ensuring that BCIs fulfill their intended role: to restore and enhance human capabilities without compromising personal autonomy. By grounding BCI development in robust ethical principles, society can harness this transformative technology responsibly, preserving the fundamental rights and freedoms that define human personhood.

Artificial Intelligence

Artificial Intelligence (AI), a subfield of computer science, emerged in the mid-20th century when researchers sought to create machines capable of simulating human intelligence. The term "AI" was first introduced at the Dartmouth Conference in 1956, marking a pivotal milestone in the field's development (McCarthy et al., 2006). Early AI systems relied heavily on symbolic reasoning and rule-based algorithms, but progress remained limited for several decades. However, significant breakthroughs emerged in the 21st century with the rise of machine learning and neural networks. These advancements revolutionized AI, enabling systems to excel in tasks like natural language processing and computer vision. Today, AI systems have achieved remarkable feats, such as defeating human champions in complex games like chess and Go and powering autonomous vehicles. AI's transformative potential spans diverse sectors, including healthcare, finance, education, and transportation. In healthcare, AI aids in diagnostics, drug discovery, and personalized treatment plans. In education, AI offers tailored learning experiences, adapting to individual students' needs. Meanwhile, in finance, AI optimizes investment strategies and enhances fraud detection. By improving decision-making processes and automating complex tasks, AI has the potential to enhance human well-being and reshape industries worldwide.

Ethical Concerns and Potential Risks

Despite its transformative potential, AI poses significant ethical concerns and risks that demand careful consideration. One major concern is the ethical use of AI, particularly regarding algorithmic bias, surveillance, and privacy infringements. AI systems, if trained on biased data, may inadvertently reinforce discrimination in areas like hiring, law enforcement, or lending practices. Furthermore, concerns about data security and the misuse of AI for intrusive surveillance pose serious threats to individual privacy and human rights. The potential emergence of superintelligent AI—machines that could surpass human intelligence and operate autonomously—raises existential concerns. If such systems are developed without robust safeguards, they could act in unpredictable ways, potentially bypassing human control and compromising humanity's best interests. In addition to existential risks, AI also presents economic challenges, particularly regarding job displacement caused by automation. As industries adopt AI-driven solutions, certain sectors may experience significant labor shifts, potentially deepening economic inequalities if these changes are poorly managed.

Human-AI Collaboration and Responsible Innovation

Amid these concerns, AI also offers profound opportunities to augment rather than replace human capabilities. When used responsibly, AI can serve as a powerful tool that complements human skills, enhancing our ability to solve complex problems, analyze vast datasets, and improve decision-making. This synergy — often called human-AI collaboration — holds enormous potential for fostering innovation and addressing societal challenges more effectively. Ensuring the responsible development and deployment of AI requires well-defined ethical frameworks, transparent governance, and robust accountability mechanisms. Developers and policymakers must prioritize fairness, transparency, and accountability to prevent harmful outcomes. Balancing innovation with moral responsibility is essential to ensure AI serves the common good while minimizing unintended consequences.

Navigating the Future of AI: Key Challenges

The future of AI presents two particularly urgent challenges: censorship and control and autonomy with unintended consequences. The first concern arises from the potential misuse of AI by undemocratic governments, which may exploit these technologies for state-sponsored censorship, information manipulation, and mass surveillance. Such misuse could suppress dissent, restrict access to information, and infringe upon civil liberties, posing significant threats to personal freedom and democratic values. The second challenge involves the increasing autonomy of AI systems and the risks associated with uncontrolled AI behavior. As AI technologies become more independent, concerns grow about their potential to make harmful decisions without human oversight. Without robust safeguards, autonomous systems could inadvertently cause catastrophic errors or even act against humanity's best interests. The prospect of AI controlling critical infrastructure, influencing healthcare decisions, or managing military operations highlights the urgent need for stringent oversight and ethical governance frameworks to mitigate these risks. Balancing innovation with accountability will be essential to ensuring AI serves society's best interests. AI's future hinges on society's ability to strike a delicate balance between innovation and ethical responsibility. While AI offers immense potential for advancing human well-being, failure to implement thoughtful safeguards risks introducing unprecedented dangers. Addressing these concerns requires global cooperation, interdisciplinary collaboration, and a commitment to ensuring that AI development aligns with humanity's moral and ethical values. By prioritizing responsible innovation, society can harness AI's transformative power for the betterment of humankind.

The Asilomar AI Principles

The Asilomar AI Principles, established by the Future of Life Institute in 2017, provide a foundational framework for the ethical development of artificial intelligence. These principles aim to address key challenges posed by AI under the category of Ethics and Values, which include Safety, Failure Transparency, Judicial Transparency, Responsibility, Value Alignment, Human Values, Personal Privacy, Liberty and Privacy, Shared Benefit, Shared Prosperity, Human Control, Non-subversion, and AI Arms Race. While these principles offer a vital ethical framework, significant challenges remain regarding their effective implementation in a rapidly developing technological landscape.

One major concern involves the autonomy of AI systems and the widening gap between AI advancements and regulatory oversight. For example, self-driving cars have progressed faster than the development of legal frameworks governing liability in accidents, illustrating the difficulty in ensuring safety, transparency, and accountability in autonomous systems. The value alignment principle, which emphasizes aligning AI systems with human values, presents additional complexity. Defining and codifying universal human values is a formidable challenge due to cultural and societal differences. Privacy expectations, for instance, vary widely across cultures, complicating AI's deployment in global markets. Achieving a balance between customization and standardization in AI value alignment remains an ongoing dilemma.

The principle of shared benefit and prosperity further underscores concerns about AI-driven inequality. Despite AI's potential to enhance global prosperity, technological advancements are often concentrated in wealthier nations, contributing to a widening digital divide that risks marginalizing developing countries. Ensuring equitable access to AI's economic benefits requires strategic collaboration and proactive policymaking. Another pressing concern is the non-subversion principle, which emphasizes that AI should enhance rather than undermine social processes. Yet, in practice, AI technologies have been exploited for malicious purposes, such as propaganda campaigns and cyberattacks, posing a direct threat to democratic institutions. Managing the potential for AI subversion, particularly in context where internal or external bad actors seek to exploit vulnerabilities—remains a significant ethical challenge.

While the Asilomar AI Principles offer an essential ethical framework, concerns persist about their practical enforcement. Achieving these ideals demands robust regulatory mechanisms, international cooperation, and ongoing ethical dialogue. As AI continues to advance at an unprecedented pace, ensuring that ethical considerations remain central to development will require vigilance, adaptability, and a proactive approach. The potential misuse of AI for population control systems or geopolitical manipulation further underscores the urgency of implementing effective safeguards. These challenges reflect the broader need to guide AI's growth in a manner that upholds human dignity, freedom, and societal well-being.

Four Ethical Priorities in Neurotechnologies and AI

In their 2017 Nature article, Rafael Yuste and colleagues outlined four ethical priorities essential for guiding the responsible development and use of neurotechnologies and artificial intelligence (AI). These priorities address the growing ethical challenges posed by advancements in brain-computer interfaces (BCIs), neural implants, and AI-driven cognitive interventions. The first priority emphasizes privacy protection and the necessity of obtaining informed consent when using neurotechnologies. As these technologies advance, robust ethical and legal frameworks—such as those modeled after the General Data Protection Regulation (GDPR)—are crucial to safeguarding personal data. Ensuring individuals maintain control over their neural information is vital for protecting autonomy and preventing exploitation. Ethical dilemmas in this area include the unauthorized use of neural data by corporations or the risk of data breaches exposing sensitive mental information.

The second priority highlights the need for ethical responsibility among researchers, developers, and users of neurotechnologies and AI. Establishing clear accountability mechanisms, such as ethics boards and institutional review processes, is crucial to mitigating potential misuse, harm, or unintended consequences. Given the complexity of attributing responsibility for AI-driven outcomes—especially in autonomous systems—proactive ethical guidelines are necessary to ensure those developing these technologies are held accountable for their impacts. The third priority concerns the ethical distinction between enhancement and therapy. While neurotechnologies offer valuable therapeutic potential, they also raise concerns when applied to enhance cognitive or physical capabilities beyond normal function. For example, neural implants designed to improve memory in healthy individuals differ fundamentally from those intended to treat Alzheimer's disease. Drawing clear boundaries between therapeutic interventions and enhancement is vital to prevent misuse, unjust disparities, or the emergence of a privileged "neuro-elite." Ensuring equitable access to beneficial neurotechnological advancements is key to addressing these concerns. The fourth priority focuses on the governance of neurodata, stressing the importance of comprehensive policies regulating the collection, storage, and sharing of neural information. As neurotechnologies generate vast amounts of personal data, clear guidelines are essential to protect individuals from exploitation and privacy violations. Without proper governance, there is a risk that corporations or governments could misuse neural data for surveillance, manipulation, or targeted advertising.

While these four priorities provide a crucial ethical foundation, Yuste and his colleagues emphasize that they alone are insufficient to address the multifaceted challenges posed by neurotechnologies and AI. Achieving meaningful oversight requires robust regulatory mechanisms, international cooperation, and ongoing ethical dialogue to balance innovation with accountability. Initiatives like the EU's GDPR exemplify how policies can uphold ethical principles while encouraging technological advancement. Furthermore, engaging the public in these discussions and addressing global disparities in neurotechnological access are essential steps toward ensuring these powerful technologies serve humanity's best interests. As neurotechnologies continue to advance, fostering a proactive and adaptive ethical framework will be critical in navigating their far-reaching implications.

Transhumanism

Transhumanism is a philosophical movement that advocates using advanced science and technology—particularly neuroscience, biotechnology, artificial intelligence (AI), and nanotechnology—to enhance and augment human characteristics and abilities (Bostrom, 2005). Proponents envision a future where humans transcend their physical and cognitive limitations, achieving enhanced longevity, intelligence, and physical capabilities. By integrating technology with the human body and mind, transhumanism seeks to redefine what it means to be human (Kurzweil, 2005). Yet this vision of radical progress raises profound concerns that challenge the very essence of what it means to be human. Central to the transhumanist narrative is the notion that consciousness, free will, and identity can be reduced to mechanistic processes—manipulable and subject to technological optimization. This reductionist view distorts the complexity of personhood, treating humanity as a collection of biological components to be upgraded rather than beings endowed with intrinsic dignity and spiritual depth.

Moreover, transhumanism's fixation on enhancement risks undermining social cohesion. Technologies such as brain implants and AI-driven therapies may disproportionately benefit wealthier populations, creating a privileged "neuro-elite" while marginalized communities are left behind (Harari, 2018). This growing technological divide threatens deepening inequality, fostering new forms of discrimination and exclusion. Efforts to radically extend human life or alter cognitive functions further blur the boundaries of identity and agency, raising troubling questions about what it means to be human. By portraying suffering, aging, and mortality as technical problems to be "solved," transhumanism risks promoting a vision of humanity devoid of moral and existential depth. The movement's ideological pursuit of a "post-human" state—a technological leap toward self-perfection—introduces both utopian ambitions and dystopian dangers. While transhumanists present this as an inevitable evolutionary step, such aspirations often ignore the risks of technological overreach. Technologies like AI systems can amplify biases, manipulate public opinion, or be exploited for surveillance and control (Miskelley, 2024). The increasing autonomy of AI also raises serious moral concerns about accountability, particularly in contexts such as autonomous weapons systems or corporate data exploitation (The Royal Society, 2019).

Paradoxically, transhumanism's materialist underpinnings assign technology a quasi-spiritual role, promising "perfection" through mechanization. In seeking to transcend human limitations, transhumanism risks devaluing the inherent dignity of human beings, replacing compassion, empathy, and vulnerability with a fixation on efficiency and control. As Elon Musk warned, unchecked technological advancement could threaten the preservation of human consciousness itself, reducing individuals to programmable entities shaped by corporate or political agendas (Isaacson, 2023). Critics stress that society must balance scientific progress with ethical responsibility. While technologies like brain implants may offer therapeutic benefits, their misuse could undermine personal autonomy and jeopardize mental privacy. Yuval Noah Harari bluntly states, "Humans are now hackable animals" (Harari, 2018). Are advances in bioengineering and neural data collection stripping away freedom of thought and reducing individuals to subjects of manipulation? In a report titled "iHuman: Blurring the lines between mind and machine," the Royal Society notes that neural interfaces bring exciting possibilities and risks. Corporations or governments could misuse these technologies by accessing neural and shared data. For instance, corporations could exploit neural data to monitor employee moods or thoughts, raising significant privacy and human rights concerns (The Royal Society, 2019). This misuse risks widening social inequalities further, creating a future where only the privileged few benefit from transhumanist advancements. Without firm ethical boundaries, transhumanism's promise of human enhancement risks descending into exploitation and control. As society stands at the threshold of unprecedented technological change, fostering interdisciplinary dialogue and prioritizing equity will be essential to ensure that advancements in neuroscience, AI, and biotechnology serve humanity's well-being rather than undermine its fundamental values. By integrating scientific progress with respect for human dignity, society can foster innovation while safeguarding the moral and existential dimensions that define humanity.

In *The Abolition of Man*, C. S. Lewis offers a prophetic critique of emerging trends in scientific and philosophical thought that closely align with contemporary transhumanist ideals. Lewis warns that by pursuing the manipulation of human nature through scientific and technological advancements, humanity risks undermining its own moral foundations. He argues that when values are reduced to subjective constructs and human nature becomes a target for redesign, the result is not the enhancement of human dignity but its erosion. Lewis foresees a future where a select few, armed with unprecedented power, impose their will upon future generations, effectively reducing them to mere products of conditioning. His concern lies not only in the loss of individual freedom but in the abandonment of objective moral order, which he deemed essential for preserving what it means to be truly human. In this way, *The Abolition of Man* offers a profound warning about the potential dehumanizing consequences of unchecked transhumanist ambitions.

From a Christian perspective, transhumanism's vision of self-perfection through technological enhancement stands in stark contrast to the Christian understanding of human dignity, which is rooted in the belief that each person is created in the image of God (Genesis 1:27). Christianity affirms that human value is not contingent on physical strength, cognitive ability, or technological augmentation but is intrinsic to every individual. While transhumanism seeks to overcome suffering, aging, and mortality through science, the Christian worldview seeking to mitigate suffering also views these realities as part of the human condition—challenges that invite compassion, spiritual growth, and dependence on God. Rather than viewing death as a technical failure to be "solved," Christianity sees it as a passage to eternal life, where true fulfillment and perfection are found not through human engineering but through union with God (John 11:25-26). In this light, the pursuit of limitless enhancement risks distorting the meaning of life itself, reducing human identity to a project of self-optimization while ignoring the deeper spiritual truths that define authentic personhood. True human flourishing, from a Christian standpoint, lies not in transcending our limitations through technology but in embracing our dependence on God, cultivating virtue, and recognizing that love, self-giving, and spiritual transformation are the highest forms of human fulfillment.

Cognitive Enhancement

Cognitive enhancement is one of the most pressing topics in neuroethics, raising significant ethical concerns about the use of neurotechnologies and pharmaceuticals to augment cognitive abilities in healthy individuals. Proponents argue that cognitive enhancement could yield societal benefits, such as improved productivity, better decision-making, and enhanced learning capacities, potentially advancing medicine, education, and technology. Researchers are actively exploring nootropics and brain-computer interfaces to help individuals overcome cognitive limitations, with some suggesting these tools could level the playing field in competitive environments (Bostrom & Roache, 2008). However, cognitive enhancement also presents substantial risks. Wealthy individuals and organizations may have privileged access to such technologies, exacerbating existing social disparities. Additionally, concerns about coercion are growing, as societal pressures may compel individuals to undergo cognitive enhancement to remain competitive in academic or professional spheres. This raises critical questions about personal autonomy, the authenticity of achievements, and the fairness of enhancement in a society increasingly driven by performance metrics (Hughes, 2004).

The distinction between enhancement and therapy presents further ethical challenges. While medical therapies aim to restore brain function in conditions such as Parkinson's disease, depression, or traumatic brain injuries, cognitive enhancement seeks to augment abilities beyond typical human functioning. As this boundary blurs, there is a risk that enhancements may shift from being optional advantages to perceived necessities. This could create social pressures where those who refuse enhancement are disadvantaged in education, employment, or social standing, fostering inequity and discrimination. In such a climate, individuals may feel coerced into pursuing enhancements to meet evolving standards of productivity and intelligence (Sandel, 2007).

A deeper ethical concern arises within the materialist-naturalist view of the human person, which sees humans primarily as biological organisms governed by neural processes. This reductionist framework risks reducing consciousness, morality, and identity to mere cognitive mechanics. In this context, cognitive enhancement may be treated as a simple upgrade to a biological machine, ignoring the profound implications such interventions can have on human nature. Without recognizing the metaphysical or spiritual dimensions of personhood, there is a heightened risk of commodifying human abilities and experiences, undermining core values such as authenticity, relationality, and human dignity (Hughes, 2004). In contrast, perspectives like the brain metafunction framework, which acknowledges a transphysical dimension in human beings, emphasize the intrinsic worth of individuals beyond their cognitive capacities. This view helps safeguard against reducing personhood to neurological efficiency alone. Without such guiding principles, cognitive enhancement risks driving society toward a transhumanist vision, where the distinction between humans and machine becomes increasingly unclear. This raises deeper questions about identity, moral agency, and the societal impact of creating a privileged cognitive elite (Bostrom & Roache, 2008). Balancing scientific progress with respect for human dignity, diversity, and ethical responsibility will be essential to ensuring cognitive enhancement advances society without compromising what it means to be authentically human.

Privacy and Neurosecurity

Advancements in neuroscience and neurotechnology have significantly improved our ability to decode neural activity, yet the capacity to fully interpret thoughts, emotions, and intentions remains limited. Brain-computer interfaces (BCIs), such as those developed by Neuralink, Synchron, and Blackrock Neurotech, have demonstrated success in translating neural signals into digital outputs, enabling paralyzed individuals to control computers and robotic devices (Ajiboye et al., 2017; Elon Musk's Neuralink, 2023). However, these systems primarily decode motor intentions—such as the desire to move a limb—rather than complex thoughts or abstract reasoning. For instance, Neuralink's ongoing human trials focus on restoring motor function rather than comprehensive mind reading (Neuralink, 2024).

Similarly, advanced neuroimaging tools like functional MRI (fMRI) have shown promise in reconstructing perceived speech or visualized images when combined with artificial intelligence (Huth et al., 2016; Takagi & Nishimoto, 2023). While AI models trained on fMRI data have translated some brain activity into recognizable stimuli, these methods require extensive individual-specific training and are confined to narrow, controlled tasks. Emerging technologies such as magnetoencephalography (MEG), electrocorticography (ECoG), and optogenetics offer improved spatial and temporal resolution but remain insufficient for decoding complex thoughts or subjective emotional states with precision (Chang et al., 2010; Schalk, 2020). Despite these technological limitations, concerns about unauthorized access to neural data are growing. Governments, corporations, or malicious actors could potentially misuse neural information, threatening personal autonomy and privacy (Ienca & Andorno, 2017). Such misuse may include manipulation, coercion, or exploitation—posing significant risks to individual freedom. As neurotechnology continues to advance, robust legal and ethical safeguards are essential to protect neural data from breaches and ensure mental privacy remains a fundamental human right.

The ethical risks become particularly concerning when viewed through a materialist-naturalist lens, which often reduces human consciousness to a set of biological processes. This reductionist view risks commodifying thoughts and emotions, overlooking the deeper dimensions of human personhood, moral agency, and dignity. Without a framework that acknowledges the intrinsic value of the individual, mental surveillance could become normalized, reducing human freedom to data points and enabling control mechanisms that threaten both privacy and autonomy (Lavazza & Massimini, 2018). In contrast, Christian theology asserts that mental privacy is not merely a matter of autonomy but a sacred right that protects the mind as a space for moral reflection, self-determination, and divine inspiration. Rooted in the belief that each person is created in the image of God (Imago Dei), this view upholds the sanctity of human thought as essential to preserving dignity and freedom. Violating this sanctity risks dehumanizing individuals by reducing them to objects of control (Tham, Kwan, & Garcia, 2017; Tham, Durante, & García Gómez, 2018). John Paul II underscored this principle, emphasizing that ethical reflection must prioritize the safeguarding of human dignity, particularly when emerging technologies risk undermining core aspects of identity, freedom, and moral agency (John Paul II, 1995). As neurotechnologies evolve, they must be governed by frameworks that respect the spiritual and moral dimensions of the human person, ensuring mental privacy remains inviolable and protected.

Informed Consent in Neuroscience Research

Informed consent is a cornerstone of ethical research, yet it presents unique complexities in neuroscience, particularly when involving participants with cognitive impairments or altered states of consciousness. Neuroscience research frequently targets conditions such as Alzheimer's disease, schizophrenia, or traumatic brain injuries—disorders that can impair memory, reasoning, and judgment. Since these very faculties are essential for informed decision-making, obtaining valid consent becomes challenging. Research involving interventions like deep brain stimulation for Parkinson's disease or experimental drugs for Alzheimer's further illustrates these difficulties.

In addition, neurotechnologies such as brain stimulation and neuroimaging are often highly technical, making them difficult for participants—even those without cognitive impairments—to fully understand. This complexity raises concerns about whether participants can genuinely comprehend the risks, benefits, and objectives of the research. Consequently, there is a heightened risk that participants may provide consent without fully grasping the implications. To mitigate these risks, strategies such as clear communication, simplified language, visual aids, and ongoing assessments of participants' understanding are essential. When necessary, involving guardians or proxies can also help protect individuals with diminished cognitive capacity (Zuk & Lázaro-Muñoz, 2019).

Ethical concerns are further amplified in materialist-naturalist frameworks, which often view human beings primarily as biological entities without intrinsic metaphysical dignity. Within this reductionist framework, participants with cognitive impairments may be regarded as less than fully autonomous agents, increasing the risk that their involvement in high-risk research could be justified on utilitarian grounds. This mindset may encourage superficial consent practices, where researchers inadequately assess participants' understanding or minimize their capacity to make meaningful decisions (Fins, 2010). Such an approach risks reducing vulnerable individuals to mere research instruments rather than recognizing them as persons with inherent moral worth. To address these risks, neuroscience research must adopt robust safeguards that uphold the dignity and autonomy of all participants, particularly those with cognitive impairment. This requires interdisciplinary oversight, transparent communication strategies that simplify complex concepts, and a commitment to ongoing consent processes that account for participants' changing cognitive states. By combining scientific progress with ethical responsibility, researchers can ensure that even the most vulnerable participants are treated with the respect, care, and dignity they deserve.

Neuroimaging and Prediction

Neuroimaging technologies, such as functional MRI (fMRI) and electroencephalography (EEG), have provided valuable insights into brain activity associated with mental illnesses and behavioral traits. Research using fMRI has identified neural patterns linked to conditions like schizophrenia and depression, particularly altered connectivity in the default mode network and prefrontal-limbic dysfunction (Anticevic et al., 2011; Schmaal et al., 2020). Similarly, EEG studies have explored differences in frontal lobe activity associated with impulsivity and aggression, traits that may correlate with antisocial tendencies (Gao et al., 2009; Raine, 2018). While these findings suggest that neuroimaging could potentially identify biomarkers for mental health risks, the technology is not yet reliable enough for individual diagnosis or prediction. Mental illnesses and behavioral tendencies arise from a complex interplay including genetic, environmental, and social factors, making it difficult to establish clear causal links based on brain scans alone (Insel, 2022). Although machine learning applied to neuroimaging data has shown promise in distinguishing at-risk individuals, accuracy remains limited due to high variability between individuals and significant rates of false positives (Woo et al., 2017). These challenges highlight the need for further research before neuroimaging can be confidently applied to individual risk assessment or early intervention in psychiatric or forensic contexts.

In addition to technical limitations, predictive neuroimaging raises serious ethical concerns about privacy, autonomy, and potential misuse. Without robust legal frameworks grounded in liberty and due process principles, predictive neuroimaging could enable a dangerous shift toward preemptive justice—where individuals are judged and restricted based on perceived neural risks rather than their actual behavior. Such practices would undermine the presumption of innocence, replacing accountability with neuro-determinism and reducing personal agency to brain-based probabilities (Farah, 2012). The misuse of neural data in legal or social contexts could also exacerbate social inequalities. Marginalized groups may face heightened stigmatization and surveillance if predictive neuroimaging disproportionately targets communities already vulnerable to systemic bias. This risk underscores the importance of ethical safeguards that protect individual rights while ensuring the responsible use of neuroimaging technologies. To mitigate these risks, comprehensive guidelines must prioritize autonomy, privacy, and fairness. Legal protections should explicitly prohibit using neural data to justify discrimination or undermine personal freedoms. Transparency in how predictive neuroimaging data is interpreted and applied will be essential to maintaining public trust. By striking a balance between scientific advancement and the protection of fundamental rights, society can responsibly explore neuroimaging's potential without compromising individual dignity and justice.

Identity and Neuroplasticity

Neuroplasticity, the brain's remarkable ability to reorganize itself by forming new neural connections, has revolutionized our understanding of cognitive flexibility and neurological recovery. Therapies that harness neuroplasticity, such as deep brain stimulation (DBS) and cognitive-behavioral therapy (CBT), have shown considerable success in treating conditions like Parkinson's disease, depression, and obsessive-compulsive disorder (OCD) (Baylis, 2013). While these interventions offer substantial therapeutic benefits, they also raise complex ethical concerns, particularly regarding personal identity and autonomy. Since neuroplasticity involves modifying neural pathways, interventions like DBS can influence not only symptoms but also personality traits, emotional responses, and decision-making patterns. Some patients undergoing DBS have reported shifts in temperament, altered motivation, and changes in self-perception—raising concerns about the continuity of self and whether individuals remain the same "person" post-intervention (Glannon, 2016, 2022). These concerns echo long-standing debates in the philosophy of mind and neuroethics. Scholars such as Derek Parfit (1984) have argued that identity is not a fixed entity but rather a continuum of psychological experiences. However, if a neuroplastic intervention fundamentally alters an individual's emotional landscape, moral reasoning, or decision-making, does this undermine their continuity of self? This dilemma becomes especially troubling in cases where interventions are involuntary or externally imposed, challenging traditional notions of agency and moral responsibility (Baylis, 2013).

A naturalist-materialist perspective, which views humans as biological systems governed by neural processes, further complicates these ethical questions. In societies that prioritize optimization and behavioral conformity, neuroplastic interventions risk being misused as tools for social engineering rather than individual well-being. Without adequate legal and ethical safeguards, authorities could exploit neuroplastic treatments for coercive purposes, whether to enforce rehabilitation programs, modify behavior in prisoners, or enhance productivity in vulnerable populations. Historical precedents highlight this danger: in the Soviet Union, psychiatric treatments were misused to suppress political dissidents under the guise of "rehabilitation" (van Voren, 2010). Future applications of DBS, neurostimulation, or brain-computer interfaces may similarly risk becoming instruments of control, especially if decision-makers reduce individuals to modifiable neural circuits optimized for societal demands (Parens, 2015). This utilitarian mindset could diminish respect for personal identity and agency, turning neuroplasticity from a tool for empowerment into a means of manipulation (Glannon, 2016).

To prevent such ethical pitfalls, policymakers must implement robust legal frameworks that protect individual autonomy and safeguard the right to preserve one's sense of self. Critical measures include clear informed consent protocols, independent ethics review boards, and strict limitations on involuntary interventions. Additionally, healthcare professionals must adopt patient-centered practices that prioritize well-being over societal pressure to conform or perform. Interdisciplinary collaboration between neuroscientists, ethicists, legal experts, and policymakers will be essential to developing ethical guidelines that balance scientific innovation with human rights protections. Harnessing neuroplasticity's potential to restore function, improve cognition, and enhance well-being must be guided by ethical foresight rather than unchecked technological ambition. Respecting the intrinsic dignity of human identity, society must ensure that neuroplastic interventions remain a tool for healing and empowerment rather than control.

Memory Manipulation

Memory manipulation technologies, designed to modify, weaken, or erase memories, offer significant therapeutic potential for individuals suffering from post-traumatic stress disorder (PTSD), severe anxiety, and traumatic experiences. Techniques such as optogenetics and pharmacological interventions are being explored to selectively alter neural circuits associated with specific memories. For example, propranolol—a beta-blocker—has shown promise in reducing the emotional intensity of traumatic memories, potentially offering relief to PTSD patients (Parsons & Ressler, 2013). While these advancements may alleviate suffering, they also raise profound ethical concerns regarding identity, justice, and autonomy. Memory is integral to personal identity, shaping one's sense of self, values, and life narrative. As John Locke (1690/1975) argued, personal identity is closely tied to the continuity of memories, suggesting that altering or erasing key recollections may disrupt the essence of selfhood. Similarly, Marya Schechtman (2011) emphasizes that identity is constructed through a coherent life narrative, meaning memory alteration could fragment one's sense of self. Beyond individual identity, memories are deeply embedded in social relationships and collective history. Modifying traumatic memories could inadvertently alter interpersonal bonds, particularly those formed through shared hardship or emotional experiences. This raises the ethical question of whether a person who undergoes memory alteration remains the same "self" as before (Glannon, 2011).

The ethical risks of memory manipulation extend beyond personal identity to broader societal and political concerns. In a naturalist-materialist framework that views humans primarily as biological systems governed by neural processes, memory may be reduced to a manipulable data set, increasing the risk of misuse. Without clear legal safeguards and ethical oversight, authorities or organizations may exploit memory technologies for purposes beyond therapy, such as social control or ideological conformity. Historical precedents warn of this danger: during the Cold War, the CIA's MKUltra program experimented with pharmacological and psychological techniques to erase or modify memories in an attempt to manipulate behavior (Marks, 1979). Similar risks emerge when considering the potential for governments or corporations to exploit memory manipulation technologies to suppress dissent, erase inconvenient recollections, or engineer compliance within a population. Such practices would undermine personal autonomy and authenticity, replacing individual agency with manufactured narratives.

The implications for justice are equally concerning. The legal system relies heavily on memory as evidence in criminal investigations, witness testimony, and accountability. Research by Elizabeth Loftus (2005) demonstrates that even natural memory distortions can lead to wrongful convictions. If memory manipulation technologies enable individuals to erase or alter recollections of crimes, criminals could potentially evade responsibility, while victims might be coerced into forgetting traumatic experiences. Such interventions could compromise justice for survivors and undermine the integrity of legal proceedings (Farahany, 2023). Moreover, in the absence of clear regulatory frameworks, powerful entities may exploit these technologies to rewrite historical narratives or suppress inconvenient truths, creating a society where technology manipulates reality rather than preserving objective facts.

To mitigate these risks, robust ethical and legal frameworks must be established to protect memory integrity and individual autonomy. Critical measures should include clear informed consent protocols, independent oversight bodies, and strict prohibitions on non-consensual interventions. Additionally, interdisciplinary collaboration between neuroscientists, ethicists, policymakers, and legal experts is essential to ensure that memory manipulation technologies are developed and applied ethically. By prioritizing ethical foresight over unchecked technological acceleration, society can responsibly harness the potential of these interventions—ensuring they are used to heal, not to harm.

Access to Neurotechnology

The rapid development of neurotechnologies, including brain-computer interfaces (BCIs), neural implants, and cognitive enhancement therapies, offers exciting possibilities for improving cognitive abilities, mental health, and overall quality of life. These innovations hold significant promise for individuals with neurological disorders, potentially restoring lost functions and enhancing mental well-being. However, a major ethical concern arises regarding equitable access to these technologies. Without proactive measures, access may become limited to those with financial means, deepening existing social and economic inequalities. Wealthier individuals and communities are likely to gain early access to neural enhancements, improving their educational, professional, and social opportunities while marginalized groups risk exclusion. This imbalance could create a widening "neuro-divide," where cognitive and mental health benefits become privileges of the affluent, reinforcing systemic disparities (Jotterand & Ienca, 2023).

While concerns about healthcare access are immediate, the implications of unequal neurotechnology distribution extend far beyond medical treatment. As neural enhancements integrate into sectors like education, business, and governance, those with access may gain substantial societal advantages. This growing neuro-divide aligns with broader concerns in biomedical ethics and distributive justice, where emerging technologies risk reinforcing existing hierarchies if access is unequally distributed (Buchanan, 2011). Scholars such as Schermer (2008) caution that cognitive enhancements could shift societal notions of meritocracy, making achievement less about effort or skill and more about access to expensive neurotechnologies. For example, students using memory-enhancing implants or professionals leveraging BCIs for multitasking may outperform peers without such modifications. As a result, cognitive performance may no longer reflect talent, effort, or dedication but rather financial privilege.

This pattern mirrors historical trends in education, healthcare, and digital access, where early adopters gain disproportionate benefits while marginalized groups face systemic barriers to entry. Cognitive meritocracy risks evolving into a system where professional success depends heavily on access to neural enhancements. This shift raises fundamental questions of fairness and justice: Should enhanced individuals receive preferential treatment in hiring, politics, or education simply because they could afford cognitive upgrades? Without safeguards, neurotechnologies risk entrenching intellectual hierarchies in which enhanced individuals hold disproportionate influence over society.

International disparities in access to neurotechnology further compound these concerns. Just as wealthier nations have gained an advantage through access to advanced computational resources and AI, neurotechnologies could become disproportionately available in the industrialized countries while low-income regions face continued barriers to basic healthcare and infrastructure (Eubanks, 2018). This imbalance risks widening global inequalities, leaving under-resourced populations further behind in cognitive health, education, and workforce competitiveness. To prevent neurotechnologies from becoming tools of exclusion and privilege, robust regulatory frameworks and ethical oversight are essential. National policies must promote affordability and accessibility to ensure that cognitive enhancements and therapeutic neurotechnologies benefit diverse populations. International collaboration is equally crucial, ensuring that advancements in neurotechnology reduce inequalities rather than deepen them. Policymakers must ensure that the neurotechnology revolution becomes a force for greater social inclusion and opportunity rather than a driver of division and disparity.

The Ethical Aspects of Treating Mental Health Disorders

Neurotechnological interventions are transforming the treatment of mental health disorders, offering new hope for individuals with conditions such as treatment-resistant depression, obsessive-compulsive disorder (OCD), and bipolar disorder. These interventions range from invasive techniques like deep brain stimulation (DBS), which involves surgically implanting electrodes to modulate neural activity, to noninvasive methods such as transcranial magnetic stimulation (TMS) and neurofeedback, which use external devices to stimulate or suppress specific brain regions. While these therapies have shown remarkable potential in alleviating symptoms and improving quality of life, they raise significant ethical concerns regarding identity, autonomy, coercion, and commercialization. Altering brain function through neurotechnological interventions can profoundly impact cognition, personality, and behavior. For example, individuals undergoing DBS for depression have reported shifts in temperament, altered decision-making patterns, and changes in emotional responses. Such effects challenge the notion of a stable identity and raise questions about whether individuals remain the same "person" they were before treatment (Kraemer, 2013). If personal identity is understood as a continuity of psychological experiences (Parfit, 1984), interventions that disrupt emotional and cognitive consistency could fragment one's sense of self. While some scholars view these changes as part of a therapeutic transformation, others warn they may blur the boundaries of personal agency and moral responsibility (Baylis, 2013). The issue of informed consent becomes particularly complex in mental health contexts, where individuals with severe psychiatric conditions may have impaired decision-making capacities. Ensuring that patients fully understand the risks, benefits, and long-term implications of interventions is crucial—especially when treatments have the potential to reshape their identity in unpredictable ways. Meaningful consent requires clear communication strategies, ongoing assessments of patient understanding, and safeguards to prevent patients from being pressured into treatments they may not fully comprehend.

Coercion and the Risk of Involuntary Treatment

Beyond individual identity concerns, the broader societal implications of neurotechnological interventions warrant scrutiny. In a materialist framework that views mental disorders purely as neural malfunctions, the risk of involuntary or coercive treatment becomes particularly troubling. A reductionist approach to mental health may emphasize "correcting" brain dysfunction without considering the broader ethical implications for personal autonomy. Such thinking risks normalizing coercive treatments in the name of societal welfare or productivity. Historically, psychiatric interventions have been misused for social control, from the forced institutionalization of marginalized individuals to the widespread administration of electroconvulsive therapy (ECT) without consent (van Voren, 2010). If neurotechnologies are adopted to modify behavior, reduce addiction, or address psychiatric conditions, they could become tools in criminal justice systems or psychiatric institutions without adequate patient protection. This risk echoes concerns in biomedical ethics, where medical interventions risk shifting from means of healing to instruments of social compliance (Ienca & Andorno, 2017).

Commercialization and Neurocapitalism

The commercialization of noninvasive neurotechnologies, such as TMS, neurofeedback, and consumer-grade brain-stimulation devices, further complicates these ethical challenges. In a commodified mental health landscape, companies may market these interventions not only as therapeutic tools but also as performance enhancers or lifestyle products. For example, TMS may be promoted as a means to boost productivity, or neurofeedback may be advertised as a stress-management tool, blurring the line between medical treatment and consumer enhancement. This trend raises concerns about neurocapitalism, where market forces, rather than medical necessity, shape mental health interventions (Ienca & Andorno, 2017). In such a landscape, individuals might feel pressured to undergo neurotechnological treatments to meet professional or societal expectations—even when these interventions are not medically necessary. The potential for employers or institutions to incentivize—or even require—cognitive or emotional enhancements raises serious ethical concerns. Without robust oversight, neurotechnologies could be misused to alter personality, behavior, or decision-making for economic or social gain, further compromising personal autonomy and moral agency (Farah, 2013).

Artificial Intelligence and Neuroscientific Research

The integration of artificial intelligence (AI) with neuroscientific research presents transformative opportunities to model brain activity, decode neural patterns, and develop advanced neurotechnologies such as neuroimaging tools and brain-computer interfaces (BCIs). These innovations hold immense promise for improving our understanding of cognition and developing new treatments for neurological disorders. However, they also raise significant ethical concerns, particularly regarding their impact on human autonomy, cognitive liberty, and moral agency. As AI increasingly interprets and modifies neural data, the boundary between human decision-making and machine intervention becomes blurred. This raises critical questions about personal responsibility, informed consent, and free will. For example, AI systems capable of predicting behaviors or mental health outcomes may be used to justify preemptive interventions that undermine individual autonomy. If neural data suggests an individual is "at risk" for criminal behavior or psychological instability, institutions might implement restrictive measures, potentially limiting personal freedom and self-determination (Ienca & Andorno, 2017; Russell, Dewey, & Tegmark, 2015). These concerns are especially pressing in contexts such as mental health care, legal proceedings, and criminal justice, where individuals often have limited agency. When AI-driven risk assessments shape decisions about sentencing, mandatory treatments, or institutionalization, there is a danger that individuals will be judged and managed based on algorithmic predictions rather than their conscious choices, posing serious challenges to moral responsibility and personal agency (Danaher, 2019).

AI, Neuroscience, and the Risk of Cognitive Control

These ethical concerns become even more pronounced in societies that prioritize technological control over individual freedom. A reductionist framework that views cognition purely as a neural system to be optimized risks treating individuals as biological entities to be controlled rather than as moral agents. In such a model, institutions, corporations, or governments could misuse AI to manipulate cognitive processes, regulate thought patterns, or suppress undesirable behaviors—not to promote individual well-being but to enforce social conformity or political agendas (Bostrom & Yudkowsky, 2014). One alarming possibility is the use of AI to influence thought conformity or behavioral compliance through neurotechnological interventions. For example, AI-enhanced brain stimulation could be leveraged to suppress emotional responses deemed counterproductive to workplace efficiency or reinforce cognitive patterns aligned with ideological goals. Without robust legal safeguards that recognize the intrinsic dignity of the human person, AI-driven neuroscience could shift from empowering individuals to becoming a tool for social engineering, threatening the fundamental principles of moral autonomy and self-governance (Floridi & Cowls, 2019).

The commercialization of AI-driven neurotechnologies introduces additional ethical risks. Corporations may market AI-enhanced neurostimulation devices as tools for productivity, cognitive enhancement, or emotional regulation, effectively commodifying cognitive functions. As seen in past technological revolutions, individuals may feel pressured to undergo AI-driven neuroenhancements to remain competitive in professional, academic, or social environments. This trend raises concerns about neurocapitalism, where brain optimization becomes a market commodity rather than a medical intervention (Zuk & Lázaro-Muñoz, 2019; Zuk et al., 2020). Left unchecked, such developments risk deepening socioeconomic inequalities, creating a cognitive divide where only those with access to enhancement technologies benefit from heightened mental capabilities.

Theological Reflections on AI and Human Dignity

From a Christian theological perspective, these ethical concerns take on deeper significance. Christianity affirms that humans are created in the image of God (Imago Dei), endowed with free will, moral responsibility, and an intrinsic spiritual dimension. Unlike materialist perspectives that reduce cognition to mere neural processes, Christian anthropology emphasizes that human identity transcends biology and is rooted in reason, moral conscience, and divine purpose (Kilner, 2015). Pope John Paul II (1995) warned against idolizing technological progress, cautioning that an uncritical embrace of enhancement technologies risks distorting humanity's moral and spiritual integrity. When AI modifies cognition, influences decision-making, or predicts behavioral tendencies, it risks reducing individuals to mechanistic entities, undermining the moral agency that defines human personhood. Christian ethicists emphasize that AI's role in neuroscience must be evaluated through the lens of free will, dignity, and divine purpose, ensuring that technological advancements enhance human flourishing rather than compromise it (Tham, Kwan, & Garcia, 2017; Tham, Durante, & García Gómez, 2018). A central concern in Christian ethics is that moral choice is fundamental to human nature—an essential quality that distinguishes humans from animals and machines. If AI overrides, shapes, or dictates cognitive processes, it risks compromising moral accountability and undermining personal freedom (Coeckelbergh, 2020). From this perspective, it is essential to ensure that AI functions as an assistive rather than a controlling force, preserving human sovereignty over cognition and respecting the inherent dignity that defines what it means to be truly human.

Ethical and Policy Recommendations: A Christian Perspective

Addressing the ethical challenges posed by AI-driven neuroscience requires establishing robust frameworks that safeguard autonomy, dignity, and transparency. These frameworks must align with Christian moral principles, which emphasize the sanctity of the human person as created in the image of God (Imago Dei) (Genesis 1:27). In Christian ethics, technological progress must serve the integral good of the human person rather than reducing individuals to neural data or cognitive functions optimized for efficiency. When applied to neuroscience, AI should be a tool for healing and empowerment—not control and manipulation. A primary safeguard must involve implementing transparent AI decision-making processes to prevent algorithmic bias, discrimination, and opaque neural interventions that compromise autonomy. AI systems in neuroscience should function with accountability and fairness, ensuring that individuals are not subject to automated decisions that override their moral agency. Truth and justice—fundamental theological values (Proverbs 12:22; Proverbs 21:3)—demand that AI models operate with clarity and integrity. Developing explainable AI models is essential so that individuals understand how AI-generated decisions influence their neurological treatment and cognitive assessment.

Policymakers must also enforce legal protections to prevent the coercive application of AI in mental health care, governance, and criminal justice. Without such safeguards, vulnerable individuals—particularly those facing mental health challenges—could be subjected to non-consensual interventions that violate their God-given dignity and free will. The Catholic Church has consistently affirmed the inviolability of human dignity, warning that any attempt to impose neurological modifications or classify individuals as "high-risk" based on AI predictions undermines both human worth and moral responsibility (Catechism of the Catholic Church, §1738). Governments and institutions should never weaponize AI to justify involuntary treatment, predictive policing, or cognitive regulation—practices that resemble dystopian systems of control rather than respecting the sacredness of human choice. To ensure that AI-driven neurotechnologies remain tools of empowerment rather than control, interdisciplinary dialogue is critical. Collaboration between neuroscientists, ethicists, theologians, policymakers, and human rights advocates must shape the development of these technologies. Christianity teaches that wisdom is found in counsel (Proverbs 11:14), and collective discernment is essential to ensuring that technological advancements are ethically sound. Integrating Christian moral theology into AI policy discussions can provide a strong ethical foundation that prioritizes the soul, moral conscience, and individual freedom over mere cognitive efficiency.

Public engagement is also crucial for ensuring societal awareness and democratic oversight in AI ethics discussions. Christian social teaching emphasizes the principle of subsidiarity, which holds that decision-making should be as localized as possible and that individuals should actively participate in shaping the moral direction of society (Pope Pius XI, *Quadragesimo Anno*, 1931). Ensuring that AI-driven neurotechnologies are governed not solely by corporations, governments, or scientific elites but also by communities of faith and civil society will help protect against technological dominance that could suppress intellectual diversity or impose uniform neurological standards for productivity and social order. Ultimately, aligning technological innovation with human dignity is paramount to ensuring that AI in neuroscience remains a force for good rather than a mechanism of control. From a Christian perspective, the true goal of technological progress is not the relentless pursuit of intelligence augmentation or neural efficiency, but the holistic flourishing of individuals in body, mind, and spirit. Jesus emphasized that human fulfillment is found not in material power or cognitive superiority but in love, virtue, and spiritual wisdom (Matthew 22:37-40). Thus, the ethical application of AI in neuroscience must uphold this sacred balance, ensuring that technological advancements serve humanity without compromising the spiritual integrity and moral agency of the person. By fostering moral discourse, prioritizing human dignity, and implementing ethical safeguards, society can harness AI to reflect God's vision for human life—one rooted in justice, freedom, and authentic human flourishing.

Neuroeducation: Promise and Ethical Challenges

Neuroeducation, the interdisciplinary field that integrates cognitive neuroscience, psychology, and educational science, seeks to apply neuroscientific insights to optimize learning strategies, inform pedagogy, and shape education policy (Ansari, Coch, & De Smedt, 2011). By leveraging brain-based learning strategies, cognitive training, and neurostimulation devices, researchers aim to enhance attention, memory, and problem-solving abilities. While these innovations hold promise, they also introduce significant ethical concerns related to reductionism, inequality, and the commercialization of neuroenhancement tools. Addressing these challenges requires an ethically grounded approach that upholds human dignity, educational equity, and moral responsibility.

Ethical Concerns: Reductionism and Educational Inequality

One significant risk in neuroeducation is an over-reliance on neuroscientific models, which can lead to overly mechanistic approaches that overlook the emotional, social, cultural, and moral dimensions of learning. Neuroscientific reductionism, which frames intelligence and learning as purely neurobiological processes, risks diminishing the role of creativity, ethical reasoning, and character development (Bruer, 1997). This reductionist approach aligns with technocratic models of education, where success is measured solely by cognitive performance rather than holistic human development. The commercialization of neuroeducation tools, such as brain-training software and neurostimulation devices, raises further concerns. Despite limited empirical evidence supporting their long-term efficacy, these tools are often marketed aggressively to schools, leading to misallocated resources and unfulfilled promises (Howard-Jones, 2014). Wealthier institutions often gain earlier access to these emerging technologies, exacerbating the "cognitive divide" between privileged and disadvantaged students. Without intervention, neuroenhancement tools could reinforce educational inequalities, granting privileged students disproportionate academic and career advantages.

A Christian Perspective on Human Dignity in Education

From a Judeo-Christian perspective, the risks posed by materialist approaches to neuroeducation are particularly concerning. Secular frameworks that reduce cognition to neural processes risk neglecting the spiritual and moral dimensions of learning. The Christian doctrine of *Imago Dei* (Genesis 1:27) affirms that each person is created in God's image, possessing intrinsic dignity, moral agency, and free will. Consequently, an educational model that prioritizes cognitive efficiency over moral and spiritual formation risks producing students who excel intellectually but lack the ethical maturity to navigate complex moral decisions—a concern echoed by Christian thinkers such as Augustine, John Henry Newman, and C.S. Lewis. As Lewis warns, "Education without values, as useful as it is, seems rather to make man a more clever devil" (Lewis, 1943, p. 25). Catholic education has long emphasized the harmonious integration of faith, reason, and virtue (Pieper, 1952). This holistic approach stands in contrast to materialist paradigms that prioritize academic performance over character formation. Christian ethics calls for neuroeducation strategies that respect human dignity and ensure that technological advancements serve the flourishing of the whole person rather than reducing individuals to instruments of societal utility.

Ethical and Policy Recommendations

To navigate the ethical challenges associated with neuroeducation, educators and policymakers must adopt transparent, evidence-based strategies that prioritize fairness and student well-being. First, transparency is essential, schools must clearly communicate the efficacy and limitations of neuroenhancement tools before implementing them. Scientific claims must be rigorously evidence-based to prevent the premature adoption of interventions lacking empirical support. Without transparency, educational institutions risk misallocating resources and potentially harming students. Equitable access to neuroeducation technologies is equally crucial. If only privileged students and institutions can afford cutting-edge cognitive tools, educational inequalities will deepen. Ensuring that these resources are widely available and affordable will help prevent the emergence of a "cognitive elite," where neuroenhanced individuals enjoy disproportionate social and economic advantages.

Additionally, integrating cognitive training with social, moral, and emotional learning is vital. Education that prioritizes academic performance while neglecting character formation risks producing individuals who excel cognitively but lack empathy, integrity, and social responsibility. By embedding virtue, ethical reasoning, and emotional intelligence into the curriculum, educators can ensure that students are not only knowledgeable but also morally equipped to contribute positively to society. Finally, fostering interdisciplinary dialogue among neuroscientists, educators, ethicists, and theologians will guide neuroeducation policies toward ethical and effective outcomes. Such collaboration ensures that students are recognized not merely as cognitive entities but as beings created for intellectual growth, moral virtue, and spiritual wisdom. By embedding ethical principles into educational frameworks, neuroeducation can fulfill its transformative potential without compromising human dignity or individuality.

Neuroethics and the Larger Context

Neuroethics examines the ethical, legal, and societal implications of advances in neuroscience, addressing both the ethics of neuroscience (responsible research and application) and the neuroscience of ethics (how brain science informs moral reasoning and decision-making) (Roskies, 2024). As neuroscience increasingly intersects with artificial intelligence (AI), governance, and global cultural shifts, Neuroethics faces increasingly complex challenges that extend beyond science to political, economic, and ideological concerns. Supranational institutions such as the United Nations and the World Economic Forum have exerted a growing influence over bioethical discourse. While international cooperation is necessary to establish ethical standards for emerging technologies, top-down decision-making risks detaching bioethics from local cultural values and religious traditions (Moreno, 2010). Ethical frameworks imposed without public engagement may undermine public trust and lead to ethical norms that favor elite interests rather than promoting justice and fairness. Historically, bioethics emerged within a Judeo-Christian moral framework that emphasizes human dignity, autonomy, and justice (Engelhardt, 1996). However, as societies grow more secular, the consensus on shared moral principles has weakened. In a climate where utilitarian ethics increasingly dominates bioethical discourse, protecting human dignity and ensuring moral accountability in technological advancement becomes even more urgent.

Technological Acceleration and the Crisis of Ethical Oversight

The rapid evolution of AI, neurotechnology, and data-driven governance has outpaced traditional bioethical frameworks. For instance, advancements in brain-computer interfaces (BCIs) and neural implants raise serious concerns about cognitive liberty, mental privacy, and social inequality (Yuste et al., 2017). Meanwhile, AI-driven decision-making in criminal justice, mental health, and employment risks embedding algorithmic biases, potentially violating individual rights and deepening social divisions (Floridi & Cowls, 2019). Unchecked technological acceleration risks commodifying cognitive performance, encouraging coercive neuroenhancement practices, and subordinating human dignity to economic and political agendas (Ienca & Andorno, 2017). Without proactive ethical oversight, emerging technologies may exploit vulnerable populations, blur the boundaries between human identity and machine control, and reshape social norms to prioritize efficiency over moral responsibility.

A Christian Bioethical Perspective

From a Christian bioethical perspective, addressing the ethical challenges of neurotechnology requires a dynamic and morally grounded approach that upholds the sanctity of human life, the dignity of the person, and the responsible stewardship of scientific progress. Ethical frameworks must prioritize transparent governance, ensuring that corporations and institutions remain accountable to moral law and the common good (Racine, 2010). Christian bioethics emphasizes that technological advancements must respect the inherent dignity of the human person, safeguarding free will and moral agency rather than reducing individuals to biological mechanisms for optimization (John Paul II, 1995). As stewards of God's creation, bioethicists must ensure that scientific advancements promote the holistic flourishing of individuals and communities according to God's design rather than fueling transhumanist ambitions (Ienca & Andorno, 2017).

Conclusion: The Crisis of Ethical Oversight

The accelerating pace of neurotechnological innovation demands ethical frameworks that can adapt to rapid change while safeguarding human dignity. One of the most significant challenges ahead is the *crisis of ethical oversight.* As advancements in brain-computer interfaces (BCIs), artificial intelligence, and cognitive enhancement tools outpace traditional regulatory mechanisms, there is an urgent need to establish proactive safeguards. Without robust oversight, corporate interests, ideological agendas, or political forces may exploit these innovations, compromising individual autonomy, privacy, and moral agency. Society must combine scientific expertise, philosophical reflection, and theological insights to address this crisis and ensure that neurotechnological advancements remain rooted in justice, human dignity, and moral responsibility. Ethical frameworks must resist utilitarian approaches that prioritize efficiency over human worth, instead emphasizing respect for the person's intrinsic value. Fostering interdisciplinary collaboration, promoting public engagement, and reinforcing transparent governance, society must ensure that neuroscience and AI serve the common good rather than becoming instruments of control or exploitation. Only through sustained ethical vigilance and a commitment to safeguarding human dignity can Neuroethics fulfill its crucial role in protecting the integrity of the human person amid unprecedented technological change.

REFERENCES

Ajiboye, A. B., Willett, F. R., Young, D. R., Memberg, W. D., Murphy, B. A., Miller, J. P., ... & Kirsch, R. F. (2017). *Restoration of reaching and grasping movements through brain-controlled muscle stimulation in a person with tetraplegia: A proof-of-concept demonstration.* The Lancet, 389(10081), 1821-1830. https://doi.org/10.1016/S0140-6736(17)30601-3

Ansari, D., Coch, D., & De Smedt, B. (2011). Connecting education and cognitive neuroscience: Where will the journey take us? *Educational Philosophy and Theory, 43*(1), 37–42. https://doi.org/10.1111/j.1469-5812.2010.00705.x

Anticevic, A., Repovs, G., & Barch, D. M. (2011). Working memory encoding and maintenance deficits in schizophrenia: Neural evidence for activation and deactivation abnormalities. *Schizophrenia Bulletin, 39*(1), 168–178.

Barbara, M., Covelli, E., Filippi, C., & Margani, V. (2019). Transitions in auditory rehabilitation with bone conduction implants (BCI). *Acta Oto-Laryngologica.*

Baylis, F. (2013). "'I Am Who I Am': On the Perceived Threats to Personal Identity from Deep Brain Stimulation." *Neuroethics, 6*(3), 513–526.

Bostrom, N. (2005). Transhumanist values. *Journal of Philosophical Re- search, 30* (Supplement), 3–14. https://doi.org/10 5840/jpr_2005 _26

Bostrom, N., & Roache, R. (2008). Ethical issues in human enhancement. In J. Ryberg, T. Petersen, & C. Wolf (Eds.), *New Waves in Applied Ethics,* 120–152. Palgrave Macmillan. Available at https://elearning.ntvt.nl/uploads/ntvt/artikel/bostrom_human_enhancement.pdf.

Bostrom, N., & Yudkowsky, E. (2014). The ethics of artificial intelligence. In K. Frankish & W. M. Ramsey (Eds.), *The Cambridge handbook of artificial intelligence* (pp. 316–334). Cambridge University Press.

Buchanan, A. (2011). *Beyond Humanity? The Ethics of Biomedical Enhancement.* Oxford University Press.

Bruer, J. T. (1997). Education and the Brain: A Bridge Too Far. *Educational Researcher, 26*(8), 4–16. https://doi.org/10.23 07/1176301

Catechism of the Catholic Church, §1738: *"Freedom is exercised in relationships between human beings. Every human person, created in the image of God, has the natural right to be recognized as a free and responsible being. All owe to each other this duty of respect. The right to the exercise of freedom, especially in moral and religious matters, is an inalienable requirement of the dignity of the human person. This right must be recognized and protected by civil authority within the limits of the common good and public order."*

Chang, E. F., Rieger, J. W., Johnson, K., Berger, M. S., Barbaro, N. M., & Knight, R. T. (2010). *Categorical speech representation in human superior temporal gyrus.* Nature Neuroscience, 13(11), 1428-1432. https://doi.org/10.1038/nn.2641

Clausen, J. (2011). *Conceptual and ethical issues with brain-hardware interfaces. Current Opinion in Psychiatry*, 24(6), 495–501. https://doi.org/10.1097/YCO.0b013e32834bb8ca.

Coeckelbergh, M. (2020). *AI Ethics*. MIT Press.

Danaher, J. (2019). *Automation and Utopia: Human Flourishing in a World Without Work*. Harvard University Press.

Edelman, B. J., Zhang, S., Schalk, G., Brunner, P., Müller-Putz, G., Guan, C., & He, B. (2025). Non-invasive brain-computer interfaces: State of the art and trends. *IEEE Reviews in Biomedical Engineering, 18*, 26–49. https://doi.org/ 10.1109/RBME.2024.3449790.

Elon Musk's Neuralink. (2023). *Neuralink's update on brain implant technology and human trials.* Retrieved from https://neuralink.com

Engelhardt, H. T. (1996). *The foundations of bioethics* (2nd ed.). Oxford University Press.

Eubanks, V. (2018). *Automating Inequality: How High-Tech Tools Profile, Police, and Punish the Poor.* St. Martin's Press.

Farah, M. J. (2012). Neuroethics: The ethical, legal, and societal impact of neuroscience. *Annual Review of Psychology, 63*, 571–591. https://doi.org/10.1146/annurev.psych.093008.100438

Farah, M. J. (2013). Neuroscience and neuroethics in the 21st Century. In J. Illes & B. J. Sahakian (Eds.), *The Oxford Handbook of Neuroethics* (pp. 761–782). Oxford: Oxford University Press.

Farahany, N. (2023). *The Battle for Your Brain: Defending the Right to Think Freely in the Age of Neurotechnology*. St. Martin's Press.

Fins, J. J. (2010). Deep brain stimulation, free markets, and the scientific commons: Is it time to revisit the Bayh-Dole Act of 1980? *Neuromodulation: Technology at the Neural Interface.*

Floridi, L., & Cowls, J. (2019). A unified framework of five principles for AI in society. *Harvard Data Science Review, 1*(1). https://doi.org/10.1162/99608f92.8cd550d1

Foucault, M. (1977). *Discipline and punish: The birth of the prison.* Pantheon Books.

Friedrich, J., Wolkenstein, A., Bublitz, C., Jox, R. J., & Racine, E. (2021). Clinical Neurotechnology Meets Artificial Intelligence. *Springer.*

Fuchs, T. (2021). *In defence of the human being: Foundational questions of an embodied anthropology*. Springer.

Fukuyama, F. (2002). *Our Posthuman Future: Consequences of the Biotechnology Revolution.* Farrar, Straus, and Giroux.

Future of Life Institute. (2017). The Asilomar AI Principles. Retrieved from https://futureoflife.org/ai-principles/

Gao, Y., Glenn, A. L., Schug, R. A., Yang, Y., & Raine, A. (2009). The neurobiology of psychopathy: A neurodevelopmental perspective. *Canadian Journal of Psychiatry, 54*(12), 813–823. https://doi.org/10.1177/070674370905401204

Glannon, W. (2011). *Brain, body, and mind: Neuroethics with a human face.* Oxford University Press.

Glannon, W. (2017). Brain implants to erase memories. *Frontiers in Neuroscience, 11*, 584. https://doi.org/10.3389/fnin s.2017.00584

Glannon, W. (2016). Ethical issues in neuroprosthetics. *Journal of Neural Engineering, 13*(2), 021002. https://doi.org/ 10.1088/1741-2560/13/2/021002

Glannon, W. (2022). *The Ethics of Consciousness*. Cambridge University Press.

Harari, Y. N. (2018). *21 Lessons for the 21st Century*. Spiegel & Grau.

Howard-Jones, P. A. (2010). *Introducing neuroeducational research: Neuroscience, education, and the brain from contexts to practice*. Routledge.

Harris, J. (2007). *Enhancing Evolution: The Ethical Case for Making Better People*. Princeton University Press.

Henderson, J. M. (2023). Brain implant gives ALS patient ability to speak again. *Stanford Medicine News*. Retrieved from https://med.stanford.edu/neurosurgery/news/news-archive/2023/henderson-brain-implant-speech-als.html

Howard-Jones, P. (2014). Neuroscience and education: Myths and messages. *Nature Reviews Neuroscience, 15*(12), 817–824. https://doi.org/10.1038/nrn3817

Hughes, J. (2004). *Citizen Cyborg: Why Democratic Societies Must Respond to the Redesigned Human of the Future*. Westview Press.

Huth, A. G., de Heer, W. A., Griffiths, T. L., Theunissen, F. E., & Gallant, J. L. (2016). *Natural speech reveals the semantic maps that tile human cerebral cortex*. Nature, 532(7600), 453-458. https://doi.org/10.1038/nature17637

Ienca, M., & Andorno, R. (2017). Towards a new definition of mental privacy: Ethical challenges and legal implications of neurotechnology. *Life Sciences, Society and Policy, 13*(1), 5. https://doi.org/10.1186/s40504-017-0050-1

Ienca, M., & Andorno, R. (2017a). Towards new human rights in the age of neuroscience and neurotechnology. *Life Sciences, Society and Policy, 13*(1), 5. https://doi.org/10.1186/s40504-017-0050-1

Illes, J. (2006). Neuroethics: Defining the issue in theory, practice, and policy. Oxford: Oxford University Press.

Illes, J., & Racine, E. (2005). Imaging or imagining? A neuroethics challenge informed by genetics. *The American Journal of Bioethics, 5*(2), 5–18. doi: 10.1080/15265160590923358.

Insel, T. R. (2022). *Healing: Our path from mental illness to mental health.* Penguin Random House.

Isaacson, W. (2023). Inside Elon Musk's Struggle for the Future of AI. *TIME.* Retrieved from https://time.com/6310076/elon-musk-ai-walter-isaacson-biography/

Jotterand, F., & Ienca, M. (2023). *The Routledge handbook of the ethics of human enhancement.* Routledge.

John Paul II. (1995). *Evangelium Vitae: The Gospel of Life.* Libreria Editrice Vaticana. https://www.vatican.va/content/john-paul-ii/en/encyclicals/documents/hf_jp-ii_enc_25031995_evangelium-vitae.html

Koutsouleris, N., Meisenzahl, E. M., Borgwardt, S., Riecher-Rössler, A., Frodl, T., Kambeitz, J., Köhler, Y., Falkai, P., Möller, H. J., Reiser, M., & Davatzikos, C. (2015). Individualized differential diagnosis of schizophrenia and mood disorders using neuroanatomical biomarkers. *Brain, 138*(Pt 7), 2059–2073. https://doi.org/10.1093/brain/awv111

Kilner, J. (2015). *Dignity and Destiny: Humanity in the Image of God.* Eerdmans.

Kraemer, F. (2013). Authenticity or autonomy? When deep brain stimulation causes a dilemma. *Journal of Medical Ethics,* 39(12), 757–760. https://doi.org/10.1136/medethics-2011-100427

Kragel, P. A., & LaBar, K. S. (2016). *Decoding the nature of emotion in the brain.* Trends in Cognitive Sciences, 20(6), 444-455. https://doi.org/10.1016/j.tics.2016.03.011

Kurzweil, R. (2005). *The Singularity is Near: When Humans Transcend Biology.* Viking Penguin.

Lavazza, A., & Massimini, M. (2018). Cerebral organoids and consciousness: How far are we willing to go? *Journal of Medical Ethics,* 44(9), 613–621. https://doi.org/10.1136/medethics-2018-104976.

Newman, J. H. (1852). *The Idea of a University.*

Leconstant, C., & Spitz, E. (2022). Integrative Model of Human-Animal Interactions: A One Health–One Welfare Systemic Approach to Studying HAI. *Frontiers in Veterinary Science.* https://doi.org/10.3389/fvets.2022.656833

Lewis, C. S. (1943). *The abolition of man.* Oxford University Press.

Locke, J. (1690/1975). *An Essay Concerning Human Understanding* (P. H. Nidditch, Ed.). Oxford University Press.

Loftus, E. F. (2005). "Planting misinformation in the human mind: A 30-year investigation of the malleability of memory." *Learning & Memory, 12*(4), 361–366.

MacIntyre, A. (1984). *After virtue: A study in moral theory* (2nd ed.). University of Notre Dame Press.

Marcus, S. (2002). Neuroethics: Mapping the field. In D. Kennedy (Ed.), *Neuroethics: Defining the Issues in Theory, Practice, and Policy* (pp. 313-328).

Marks, J. (1979). *The Search for the Manchurian Candidate: The CIA and Mind Control.* W. W. Norton & Company.

Matamala-Gomez, M., Malighetti, C., Mancuso, V., Bernini, S., & Bottiroli, S. (2022). Non-invasive technologies in neurorehabilitation: Novel neurorehabilitative treatments for motor and cognitive disorders. In *Analyzing multidisciplinary uses and impact of innovative technologies* (p. 36). IGI Global. https://doi.org/10.4018/978-1-6684-6015-3.ch005.

McCarthy, J., Minsky, M. L., Rochester, N., & Shannon, C. E. (2006). A proposal for the Dartmouth Summer Research Project on Artificial Intelligence, August 31, 1955. *AI Magazine, 27*(4), 12–14.

Mikołajewska, E., Mikołajewski, D., Komendziński, T., Dreszer-Drogorób, J., Lewandowska, M., & Wolak, T. (2014). Chances for and Limitations of Brain-Computer Interface Use in Elderly People. In *Emerging Theory and Practice in Neuroprosthetics* (pp. 11–22). IGI Global. https://doi.org/10.4018/978-1-4666-6094-6.ch007

Miskelley, A. (2024). Innovation and international influence: China's artificial intelligence pursuit in a geopolitical context. *American University Repository.* Retrieved from https://aura.american.edu/ndownloader/files/46308160

Moreno, J. D. (2010). *The Body Politic: The Battle Over Science in America.* Bellevue Literary Press.

Niketeghad, S., & Pouratian, N. (2019). Brain machine interfaces for vision restoration: The current state of cortical visual prosthetics. *Neurotherapeutics, 16*(1), 134–149. https://doi.org/10.1016/j.nurt.2019.05.004

Neuralink. (2024). *Neuralink's first-in-human clinical trial progress.* Retrieved from https://neuralink.com

Parfit, D. (1984). *Reasons and persons.* Oxford University Press.

Parsons, R., & Ressler, K. (2013). Implications of memory modulation for post-traumatic stress and fear disorders. *Nature Neuroscience, 16*(2), 146–153. https://doi.org/10.1038/nn.3296

Parens, E. (2015). *Shaping our selves: On technology, flourishing, and a habit of thinking.* Oxford University Press.

Pieper, J. (1952). *Leisure, the basis of culture* (A. Dru & J. Pieper, Eds.). Liberty Fund.

Racine, E. (2010). *Pragmatic neuroethics: Improving treatment and understanding of the mind-brain.* MIT Press. https://doi.org 10.7551/mitpress/8027.001.0001

Raine, A. (2013). *The anatomy of violence: The biological roots of crime.* Pantheon Books.

Rao, R. P. N., Stocco, A., Bryan, M., Sarma, D., Youngquist, T. M., Wu, J., & Prat, C. S. (2014). A Direct Brain-to-Brain Interface in Humans. *PLOS ONE*, 9(11), e111332. https://doi.org/10.1371/journal.pone.0111332

Roskies, A. (2024). Neuroethics. In E. N. Zalta & U. Nodelman (Eds.), The Stanford encyclopedia of philosophy (Winter 2024 ed.). StanfordUniversity. https://plato.stanford.edu/archives/win2024/entries/neuroethics/).

Russell, S., Dewey, D., & Tegmark, M. (2015). Research priorities for robust and beneficial artificial intelligence. *AI Magazine, 36*(4), 105–114.

Sandel, M. J. (2007). *The case against perfection: Ethics in the age of genetic engineering.* Cambridge, MA: Harvard University Press.

Schechtman, M. (2011). *The Constitution of Selves.* Cornell University Press.

Schermer, M. (2009). The mind and the machine: On the conceptual and moral implications of brain-machine interaction. *Nanoethics, 3*(3), 217–230. https://doi.org/10.1007/s11569-009-0076-9

Schermer, M. (2008). On the argument that enhancement is "cheating." *Journal of Medical Ethics, 34*(2), 85–88. https://doi.org/10.1136/jme.2006.019646

Schmaal, L., Pozzi, E., Ho, T. C., van Velzen, L. S., Veer, I. M., Opel, N., Van Someren, E. J. W., Han, L. K. M., Aftanas, L., Aleman, A., Baune, B. T., Berger, K., Blanken, T. F., Capitão, L., Couvy-Duchesne, B., Cullen, K. R., Dannlowski, U., Davey, C., Erwin-Grabner, T., ... Veltman, D. J. (2020). ENIGMA MDD: Seven years of global neuroimaging studies of major depression through worldwide data sharing. *Translational Psychiatry, 10*(1), 172. https://doi.org/10.1038/s41398-020-0842-6

Takagi, Y., & Nishimoto, S. (2023). *High-resolution image reconstruction with latent diffusion models from human brain activity.* Nature Communications, 14(1), 1-12.

Tham, J., Durante, C., & García Gómez, A. (Eds.). (2018). *Interreligious perspectives on mind, genes and the self: Emerging technologies and human identity* (1st ed.). Routledge. https://doi.org/10.4324/9780429456145.

Tham, J., Kwan, K. M., & Garcia, A. (Eds.). (2017). *Religious perspectives on bioethics and human rights.* Springer.

Royal Society. (2019). *iHuman: Blurring lines between mind and machine.* Royal Society. https://royalsociety.org/-/media/policy/projects/ihuman/report-neural-interfaces.pdf

Yang, J. (2024). Ethical Issues in Memory Modification Technology: A Scoping Review. *Journal of Bioethical Inquiry.* https://doi.org/10.1007/s11673-024-10371-x

Yuste, R., Goering, S., Bi, G., Carmena, J. M., Carter, A., Fins, J. J., ... & Rommelfanger, K. S. (2017). Four ethical priorities for neurotechnologies and AI. *Nature, 551*(7679), 159–163. https://doi.org/10.1038/551159a

van Voren, R. (2010). Political abuse of psychiatry—An historical overview. *Schizophrenia Bulletin, 36*(1), 33–35. https://doi.org/10.1093/schbul/sbp119

Vasiljevic, G. A. M., & De Miranda, L. C. (2020). Brain–computer interface games based on consumer-grade EEG devices: A systematic literature review. *International Journal of Human-Computer Interaction.* https://doi.org/10.1080/10447318.2019.1612213

Woo, C. W., Chang, L. J., Lindquist, M. A., & Wager, T. D. (2017). Building better biomarkers: Brain models in translational neuroimaging. *Nature Neuroscience, 20*(3), 365–377. https://doi.org/10.1038/nn.4478

Zuk, P., & Lázaro-Muñoz, G. (2019). "Ethical Analysis of 'Mind Reading' or 'Neurotechnological Thought Apprehension': Keeping Potential Limitations in Mind." *AJOB Neuroscience, 10*(1), 32–34.

Zuk, P., Sanchez, C. E., Kostick, K., Torgerson, L., Muñoz, K. A., Hsu, R., Kalwani, L., Sierra-Mercado, D., Robinson, J. O., Outram, S., Koenig, B. A., Pereira, S., McGuire, A. L., & Lázaro-Muñoz, G. (2020). Researcher perspectives on data sharing in deep brain stimulation. *Frontiers in Human Neuroscience*, 14, 578687. https://doi.org/10.3389/fnhum.2020.578687

19

Twelve Brain Metafunction Framework Predictions

Constrained by reductionist frameworks and controlled environments, traditional neuroscience struggles to capture the brain's richness and interconnected nature. Advances in measurement techniques, computational modeling, and analytical approaches now allow researchers to study brain processes in real-world, dynamic settings. Unlike traditional methods that isolate variables but overlook systemic interactions, these advancements reveal how brain functions operate as integrated, multidimensional systems. Emerging technologies, such as wearable devices, functional neuroimaging, and advanced sensors, provide ecologically valid data, while AI, machine learning, and neural simulations uncover patterns and relationships previously undetectable. This chapter outlines predictions from the brain metafunction framework, emphasizing integration, systemic analysis, and real-world complexity.

Predictions

1. Integration Across Brain Regions

The brain metafunction framework highlights brain function's dynamic, interdependent nature, moving beyond traditional models of isolated specialization. Increasingly, complex cognitive functions—including perception, memory, attention, and emotion—are understood as emerging from cross-regional and multifunctional integration rather than localized activity. This shift underscores how sensory modalities, cognitive systems, and predictive processes work together to construct perception, regulate behavior, and maintain homeostasis and allostasis. As neuroscience advances, this holistic and predictive framework will offer an increasingly refined and comprehensive understanding of how cross-regional neural networks dynamically coordinate perception, cognition, and physiological regulation.

2. Advancing the Understanding of Consciousness Neural Correlates

The brain metafunction framework will advance the study of consciousness neural correlates as a distributed, systemic phenomenon supported by large-scale neural networks. This approach challenges purely reductionist and physicalist models by emphasizing that we cannot reduce conscious experience to isolated neural events or physical processes. The framework will also illuminate social cognition—interactions and understanding of others—as inherently integrated with cognitive and emotional processes, with implications for conditions like autism spectrum disorder (ASD) and the development of socially intelligent AI.

3. Enhanced Insights into Binding and Memory as Networked Phenomena

The framework will expand our understanding of neural binding and metabinding problems and memory by viewing them as distributed, networked processes rather than localized functions. Memory storage, retrieval, and reconsolidation will be recognized as dynamically connected to real-time perceptual and emotional states, providing new approaches for understanding and treating memory-related disorders.

4. Unified Emotional-Cognitive Processing

The framework will emphasize the interdependence of emotional and cognitive processes in a larger brain metafunctional context, revealing how emotional states shape decision-making, memory, and problem-solving in real-world contexts. This holistic perspective will improve mental health treatments, targeting emotional and cognitive systems together for more effective therapies for mood and cognitive disorders.

5. Predictive Coding and Brain-Environment Interaction

The framework will refine predictive coding models, highlighting the brain's role in adapting to and anticipating environmental changes. These predictions will advance neuroadaptive technologies and human-computer interactions by integrating sensory, motor, and cognitive domains, enabling machines to respond intuitively to user needs.

6. Brain-Body Integration and Interoception as a Core Brain Meta-function

Interoception—the brain's ability to sense and regulate internal bodily states—will emerge as a core brain metafunction. Recognizing its foundational role in self-awareness, emotional regulation, and health will transform treatments for chronic illness, stress resilience, and emotional well-being.

7. Neuroplasticity as a Systemic Function

Neuroplasticity will be redefined as a systemic property enabling large-scale reorganization across brain regions. This insight will revolutionize rehabilitation for brain injuries, neurodegenerative diseases, and cognitive decline by promoting therapies targeting network-level plasticity rather than isolated regions.

8. Brain Metafunctional Applications in Medicine and Society

Advances in the brain metafunction framework will drive innovations in Brain-Computer Interfaces (BCIs) and neuroprosthetics, enabling seamless brain-device communication for individuals with motor impairments. Ethical and regulatory frameworks must evolve to address potential cognitive modification and neuroenhancement challenges, ensuring human dignity and autonomy.

9. A New Model for Psychosomatic Health Dynamics

The framework will establish a new model for psychosomatic health, demonstrating feedback loops between mental states and physical health. Insights into how brain metafunction influence immune responses, inflammation, and healing will transform approaches to chronic pain, stress management, and overall well-being.

10. Network-Based Diagnostics in Neurology and Psychiatry

Diagnostics will shift from symptom-based assessments to network-based evaluations, identifying dysregulated brain processes across interconnected systems. This precision will enable more personalized treatments, focusing on restoring network balance rather than merely addressing symptoms.

11. Bioethical Implications of Brain Metafunction

The framework's nuanced understanding of brain integration will inform ethical debates on human death and personhood. This perspective will continue integrating insights from medicine, law, theology, and philosophy by highlighting the brain's role in unifying bodily and cognitive functions.

12. A Paradigm Shift in Neuroscience

The brain metafunction framework heralds a paradigm shift in neuroscience. It legitimizes exploring human experience phenomena that may require consideration of a transphysical domain and fosters interdisciplinary collaboration. This holistic perspective will drive advances in multiple areas of research that challenge the materialistic, scientistic model. This new perspective will impact diagnostics, therapy, and machine learning while broadening our understanding of the brain as an integrated, irreducible system.

Together, these predictions reflect a transformative shift in neuroscience that emphasizes the brain's systemic complexity, practical applications, and profound implications for medicine, ethics, and society. By embracing this integrative framework, unconstrained by *a priori* materialist assumptions, researchers can better understand the intricate dynamics of brain function while promoting innovations that enhance human well-being.

Predictions

Prediction 1: Toward a Unified Understanding of Cross-Regional and Multifunctional Brain Integration

Multisensory and Cognitive Integration: A Metafunctional Perspective

The brain metafunction framework emphasizes that perception and cognition depend on the dynamic interplay between neural regions and functional systems rather than isolated, modular processes. This large-scale integration is particularly evident in multisensory perception, where sensory modalities and higher-order cognitive functions cooperate to refine perception and behavior. Traditional neuroscience has often conceptualized brain regions as functionally specialized, yet accumulating evidence suggests that perceptual, attentional, mnemonic, and conceptual processes dynamically interact, forming a predictive network that continuously refines our experience of reality (Friston, Stephan, Montague, & Dolan, 2014; Friston 2023). This view aligns with the "perception as controlled hallucination" theory, which proposes that the brain actively generates predictions about the world, compares them against incoming sensory data, and adjusts accordingly (Seth, 2021).

Cooperation Between Auditory and Visual Perception

The interplay between auditory and visual systems enhances perception beyond single-sense processing. Auditory input can amplify early visual processing, improving accuracy in visual motion detection tasks (Vroomen & De Gelder, 2004; Alais & Burr, 2004). Naturalistic sounds facilitate perceptual alignment, increasing response speed and accuracy in visual target detection (Williams, Markov, Tiurina, & Störmer, 2022). These findings highlight how cross-regional neural communication enables predictive processing, allowing the brain to anticipate and optimize responses to environmental stimuli.

Conversely, visual cues significantly shape auditory perception, particularly in ambiguous contexts. The McGurk effect (McGurk & MacDonald, 1976) illustrates how conflicting visual and auditory inputs create a fused perceptual experience, demonstrating the brain's reliance on visual information when resolving discrepancies. Additionally, visual cues improve sound localization and rhythm perception, enhancing spatial awareness (Bolognini et al., 2007). The ventriloquist effect (Alais & Burr, 2004) exemplifies how visual input can "capture" auditory spatial perception, overriding sound location cues. Moreover, crossmodal dynamic capture occurs when visual motion influences auditory motion perception, highlighting the adaptive nature of sensory integration (Soto-Faraco et al., 2002; Beauchamp et al. 2008). These examples demonstrate that sensory perception is not merely a passive response to stimuli but an active, predictive process shaped by multimodal integration. This predictive framework reflects the brain's sophisticated ability to anticipate and refine perception in response to environmental demands.

Other Sensory Modalities in Metafunctional Cooperation

Vision and touch collaborate in object recognition and spatial awareness. In low-visibility environments, haptic exploration (touch-based perception) improves visual object recognition (Ernst & Banks, 2002). Additionally, in blind individuals, the visual cortex adapts to process tactile information, demonstrating the brain's remarkable crossmodal plasticity (Sadato et al., 1996). The olfactory and gustatory systems are deeply intertwined, as 80% of flavor perception is driven by olfactory input (Small & Prescott, 2005). The orbitofrontal cortex integrates smell and taste signals, producing a unified perception of flavor (Spence, 2015). This integration explains why food loses its richness in flavor when olfactory input is blocked, such as during nasal congestion. In another example, the vestibular system, responsible for balance and spatial orientation, integrates with visual and somatosensory inputs to maintain postural stability and navigation (Angelaki & Cullen, 2008). Vestibular dysfunctions disrupt this coordination, leading to motion sickness and spatial disorientation. In extreme cases, such as astronauts adapting to microgravity, the brain recalibrates sensory integration to maintain balance (Mergner, Schweigart, Fennell, & Botti, 2009). These diverse examples underscore the brain's remarkable flexibility in merging sensory data, demonstrating that multisensory integration is essential to perception and adaptive behavior.

Integration of Sensory Perception with Attention, Memory, and Thought

Attention as a Modulator of Predictive Perception

Attentional selection mechanisms actively shape perception and cognition by prioritizing relevant sensory inputs and suppressing distractions rather than passively responding to the environment. Predictive models propose that the brain continuously generates hypotheses about incoming sensory data and compares them against real-time inputs, adjusting attention accordingly (Summerfield & Egner, 2009). In multisensory environments, attentional shifts influence crossmodal integration, such as when auditory cues direct visual attention, enhancing object recognition and spatial awareness (Talsma et al., 2010). This dynamic interplay between attention and prediction reinforces that perception is not merely reactive but actively constructed through goal-directed cognitive processes.

Memory as the Foundation of Predictive Processing

Memory plays a crucial role in shaping perception by providing the historical context necessary for inference (Bar, 2009). The brain maintains hierarchical perceptual priors based on long-term memory, allowing it to anticipate patterns and meaning in sensory inputs

(Clark, 2013). This predictive mechanism ensures that perception is not merely a response to immediate stimuli, but a context-dependent process influenced by past experiences. Additionally, prior auditory or tactile experiences alter subsequent visual perception through cross-modal memory effects, continuously shaping perception based on past sensory events (Ernst & Bülthoff, 2004).

Thought and Conceptual Knowledge in Perception

Perception is not solely a bottom-up sensory process; it is deeply intertwined with abstract thought, language, and conceptual knowledge. Linguistic influences on perception demonstrate that speakers of languages with distinct color terms perceive finer color gradations, illustrating how language structures sensory interpretation (Winawer et al., 2007). Similarly, conceptual priming effects show that ambiguous figures, such as the duck-rabbit illusion, are more likely to be perceived as consistent with recently activated concepts, underscoring the brain's capacity to integrate prior knowledge into perception (Lupyan & Ward, 2013). Collectively, these findings reinforce the view that perception is a dynamic, cognitively informed process rather than a passive reception of external stimuli. The brain's predictive mechanisms, influenced by memory, language, and conceptual knowledge, actively shape how individuals perceive and respond to their surroundings.

2. Advancing the Understanding of Consciousness

Real-World Neuroscience and Consciousness

Real-world neuroscience explores brain function in dynamic, naturalistic settings, revealing how sensory, cognitive, and emotional processes interact to guide real-life behavior (Hari & Kujala, 2009). Unlike controlled laboratory conditions, this approach considers the full spectrum of brain metafunction subfunctions, including homeostatic and allostatic regulation. Consciousness is not an isolated neural phenomenon but an adaptive, context-sensitive process that continuously integrates external stimuli and internal states to generate coherent experiences and actions (Fuster, 2005). This perspective aligns with the Brain Metafunction Framework, which emphasizes causal interactions and functional integration across brain regions. Friston's dynamic causal modeling (2010, 2013) supports this view by illustrating how neural networks dynamically update predictions in response to changing sensory and cognitive inputs. This predictive framework, which suggests that the brain actively constructs reality by anticipating sensory input, indicates that consciousness is not limited to passively reacting to sensory data but instead involves actively generating internal models of the world. Future neuroscience will explore how predictive models extend beyond neural computations, considering bodily, social, and transphysical influences on conscious experience.

The Embodied Mind and the 4E Cognition Framework

The embodied mind theory challenges traditional views of cognition as purely neural, arguing that cognition is deeply rooted in bodily and environmental interactions (Thompson & Varela, 2001; Clark, 2008). Conscious experience is not confined to brain activity alone but is influenced by proprioception, interoception, and physiological states, reinforcing the unity between mind and body (Damasio, 1999; Park & Tallon-Baudry, 2014). The 4E Cognition Framework, which describes cognition as embodied, embedded, extended, and enactive, proposes that conscious processes are distributed across the brain, body, and environment (Rowlands, 2010). This perspective reframes consciousness as supported by an adaptive system that dynamically interweaves perception, action, and cognition with bodily states and external contexts. For instance, recent research into the gut-brain axis further supports this integrated view, demonstrating how microbiome activity influences mood, cognition, and even self-awareness (Mayer, 2011; Dinan & Cryan, 2017). These findings suggest that consciousness is not an isolated cortical function but a systemic property of the entire organism, dependent on biological, sensory, and environmental interactions. While embodied cognition emphasizes physical interactions with the environment, emerging theories suggest that consciousness may extend beyond material processes alone. Transphysical cognition explores whether consciousness involves non-material dimensions, potentially interfacing with informational or nonlocal processes. While speculative, these theories increasingly point to post-materialist neuroscience (Hameroff & Penrose, 2014; Fisher, 2015; Kastrup, 2019).

Social Cognition as a Core Aspect of Brain Metafunction

Social cognition—critical for empathy, moral reasoning and decision-making—is a fundamental component of consciousness, reflecting the brain's ability to integrate sensory, emotional, and cognitive processes dynamically and in a context-dependent manner. The neural correlates of this process are not localized to a single brain region but distributed across functionally interconnected neural networks that dynamically adjust based on social and environmental contexts. Key brain regions, including the medial prefrontal cortex (mPFC) and temporoparietal junction (TPJ), serve as hubs for social cognition, processing social and emotional information to enable perspective-taking, theory of mind, and self-other distinctions (Schilbach et al., 2008; Saxe & Kanwisher, 2003). These regions interact with the anterior cingulate cortex (ACC) and amygdala, which modulate emotional salience and conflict resolution in social decision-making, as well as the insula, which integrates interoceptive awareness into social perception, grounding social cognition in bodily states.

From a Brain Metafunction Framework perspective, social cognition exemplifies the brain's capacity for predictive processing and cross-modal integration. It continuously compares incoming social cues against internal models of the self and others. Consciousness, in this sense, does not relate solely to individual cognitive processes but to the dynamic interplay between internal states and external social realities. This aspect aligns with the concept of the predictive brain, where social interactions serve as feedback loops that refine and update neural configurations and mental representations of social reality. Furthermore, the ability to model the mental states of others is an extension of broader neural mechanisms for simulation and inference, allowing the brain to anticipate outcomes, regulate emotional responses, and optimize adaptive behavior. These processes are evident in moral reasoning and ethical decision-making, where neural circuits involved in reward processing (ventromedial prefrontal cortex, VMPFC) and cognitive control (dorsolateral prefrontal cortex, DLPFC) interact with social prediction models to guide context-sensitive responses to social and moral dilemmas. Thus, the Brain Metafunction Framework highlights social cognition as a property of integrated neural systems, illustrating how consciousness is individually constructed and socially embedded.

The interdependence of social cognition, predictive modeling, and affective regulation underscores the brain's metafunctional organization, where perception, emotion, and cognition continuously shape and reshape conscious awareness through interpersonal and societal interactions. Furthermore, this socially embedded model of consciousness has significant implications for neurological and psychiatric disorders. Conditions such as autism spectrum disorder (ASD) and schizophrenia involve disruptions in social and predictive processing, reinforcing the idea that conscious experience is inherently relational (Pelphrey et al., 2011; Uddin et al., 2007). Interventions targeting these social-cognitive networks could enhance self-awareness and adaptive functioning, further supporting the idea that intrinsic neural processes and external social contexts dynamically shape consciousness. Additionally, the study of social cognition informs the development of AI systems capable of interpreting human emotions, social cues, and decision-making contexts (Złotowski, Yogeeswaran, & Bartneck, 2017). As AI models increasingly incorporate predictive processing and social reasoning mechanisms, they offer new insights into the computational foundations of consciousness and its role in social adaptation.

Beyond Materialist Constraints: The Future of Consciousness Research

The Brain Metafunction Framework suggests that consciousness is not confined to isolated neural activity but arises from the dynamic integration of sensory, cognitive, emotional, and social processes. Insights from real-world neuroscience, embodied cognition, and social cognitive neuroscience emphasize that consciousness is:

1. Context-dependent, dynamically integrating external stimuli with internal states.
2. Predictive, shaped by hierarchical inference and continuous model updating.
3. Embodied and extended, emerging from brain-body-environment interactions.
4. Socially embedded, structured by interpersonal networks and shared meaning-making processes.
5. Transphysical, extending beyond purely material processes, implying that consciousness encompasses dimensions beyond traditional physicalist explanations.

By embracing a post-materialist framework, researchers may unlock deeper insights into consciousness—one that recognizes subjective experience as an essential aspect of reality itself, bridging scientific rigor with philosophical depth. This expansion would advance theoretical models of consciousness and reshape fields such as artificial intelligence, cognitive science, and mental health research, offering new insights into the nature of self-awareness, perception, and reality itself.

3. Enhanced Insights into Binding and Memory as Networked Phenomena

The Metabinding Problem: An Exponential Megaproblem

The metabinding problem expands on traditional binding by addressing the brain's ability to unify diverse processes—such as perception, memory, attention, and emotional regulation—into a cohesive conscious experience (Damasio, 1999; Pessoa, 2013). While traditional binding models focus on synchronizing sensory features within specific modalities (e.g., visual or auditory processing), metabinding requires integrating distributed processes across neural networks operating at varying spatial and temporal scales. This integration enables real-time, context-sensitive cognition. For instance, real-world decision-making demands that the brain dynamically synthesize perception, memory retrieval, emotional evaluation, and motor planning in milliseconds to generate rapid, goal-directed responses (Miller & Cohen, 2001). Unlike classical feature binding, which relies on synchrony and local connectivity, metabinding requires large-scale coordination across the entire brain, linking sensory, cognitive, and autonomic functions. This complexity suggests the need for a comprehensive framework that may extend beyond strictly physicalist explanations. From a predictive processing perspective, metabinding aligns with hierarchical Bayesian inference, where distributed neural circuits continuously update predictions about the internal and external world (Friston, 2010, 2013). In this model, memory, perception, and decision-making are not isolated modules but parts of an anticipatory system that integrates multimodal information in real time.

Toward Integrative Models for Metabinding

Addressing the metabinding problem requires theoretical models capable of capturing both local and global brain dynamics. Local dynamics involve specialized circuits handling distinct cognitive functions, such as hippocampal pattern encoding or prefrontal executive processing. Global dynamics, on the other hand, refer to the large-scale coordination of these distributed functions into an integrated neural network that underlies conscious experience. Several computational frameworks attempt to model this large-scale integration. Dynamic Causal Modeling (DCM) examines how directed neural interactions contribute to cognition (Friston, 2011). Graph theoretical approaches reveal network-level organization, showing how different brain regions self-organize into functional subnetworks that support perception, memory, and attention (Sporns, 2010). Integrated Information Theory (IIT) proposes that consciousness emerges from highly integrated, differentiated information processing (Tononi & Edelman, 1998). While these models describe connectivity patterns, they do not fully explain how diverse processes such as memory, attention, and emotion merge into a singular, unified subjective experience. This limitation suggests the need for an expanded theoretical framework—one that considers both the physical mechanisms of binding and its emergent, potentially transphysical properties.

The Case for a Transphysical Perspective

The complexity of metabinding raises questions about whether a strict physicalist paradigm can fully explain conscious integration. Physicalist models successfully describe neural correlates of specific cognitive and emotional functions. However, they struggle to account for the unity of subjective experience—the seamless coalescence of distributed processes into a coherent whole. Emerging perspectives suggest that consciousness may extend beyond purely material interactions. Non-reductive materialism (Chalmers, 1996) and panpsychism propose that consciousness may be a fundamental aspect of reality rather than a byproduct of neural computation alone. Complex systems theory reinforces this view, highlighting how emergent properties often display behaviors that individual components cannot predict or explain (Anderson, 1972).

Moreover, the seamless unity of conscious experience implies the presence of a continuous and coherent self—one that transcends fragmented neural processes. This view aligns with arguments for an immaterial soul, which provides an intrinsic source of unity that computational binding models struggle to achieve. An immaterial soul also accounts for the continuity of personal identity despite neural changes, brain trauma, or structural remodeling driven by plasticity. Additionally, it addresses the "hard problem" of consciousness—the unresolved question of how subjective awareness and qualia arise from physical processes. This transphysical perspective aligns with empirical anomalies, such as near-death experiences, cases of self-awareness despite extreme brain damage, and the inability of artificial intelligence to replicate conscious experience. If metabinding involves both emergent and transphysical properties, neuroscience may require frameworks that integrate material and non-material dimensions.

Future Directions: Bridging Neuroscience and AI

The convergence of neuroscience and AI presents new opportunities to study binding and memory integration at scale. Advanced AI models that mimic neural architectures can simulate how memory, perception, and emotion dynamically integrate across hierarchical networks. Deep learning models provide insights into schema-building and memory retrieval, shedding light on how the brain reconstructs past experiences (Tulving, 1983). Meanwhile, neuromorphic computing—which mimics brain-like information processing—offers experimental platforms for studying how distributed networks achieve unified cognitive states (Hassabis et al., 2017). AI-driven graph analysis further models how functional brain networks self-organize, allowing researchers to explore how engram-related processes contribute to binding and conscious integration. This interdisciplinary approach holds promise for expanding our understanding of memory and consciousness while challenging conventional neuroscientific paradigms. Future research may reveal that metabinding reflects both emergent neural properties and transphysical dimensions, opening new avenues for studying the nature of subjective experience beyond reductionist frameworks.

4. Unified Emotional-Cognitive Processing

Emotion and Cognition as an Integrated Brain Metafunction

Recent research demonstrates that emotional and cognitive processes are deeply interconnected, challenging traditional dual-process models that treat these domains as separate. The Brain Metafunction Framework posits that emotions and cognition operate as a unified, adaptive system, particularly in real-world contexts requiring rapid decision-making, problem-solving, and social interaction. This perspective aligns with predictive processing models, which propose that the brain continuously generates expectations about incoming sensory data, compares them to actual sensory input, and adjusts cognition and behavior accordingly (Pessoa, 2013). Rather than being a reactive system, emotional-cognitive processing functions as a dynamic, context-sensitive regulator of behavior. Emotional states influence cognitive flexibility, decision-making strategies, and learning, enabling organisms to navigate uncertain environments.

Impact on Memory and Problem-Solving

Emotion is fundamental to memory consolidation and problem-solving by modulating neural activity across interconnected brain networks. Emotionally salient events are often more vividly remembered due to the amygdala's modulation of hippocampal activity, enhancing long-term encoding and retrieval (McGaugh, 2004; Phelps, 2006). This adaptive process allows individuals to prioritize emotionally relevant information for future decision-making. Moreover, different emotional states influence cognitive strategies in problem-solving. Positive emotions promote cognitive flexibility, creativity, and exploratory thinking, facilitating divergent problem-solving (Fredrickson, 2001, 2004). Negative emotions, particularly anxiety or sadness, enhance focus and analytical precision, improving detail-oriented reasoning (Schwarz & Clore, 1983). This dynamic interaction suggests that emotion is not a secondary influence on cognition but a core driver of adaptive thinking and decision-making.

Applications to Mental Health Treatment

The integrated nature of emotion and cognition has critical implications for mental health treatment. Disorders such as depression, anxiety, and PTSD involve disruptions in both emotional regulation and cognitive processing, resulting in maladaptive thought patterns, memory distortions, and attentional biases. Modern interventions increasingly leverage emotional-cognitive integration for more effective treatments. Cognitive Behavioral Therapy (CBT) restructures maladaptive cognitive patterns by incorporating emotional regulation techniques (Beck, 2011). Mindfulness-based Cognitive Therapy (MBCT) enhances meta-awareness, helping patients disentangle emotions from rigid cognitive schemas (Teasdale et al., 2000; Arch & Craske, 2006). PTSD treatments—such as trauma-focused cognitive therapies—focus on recontextualizing traumatic memories and fostering healthier cognitive-emotional associations to improve outcomes (Foa et al., 2007). This evidence reinforces the Brain Metafunction Framework's emphasis on unified processing by demonstrating that emotion and cognition are interdependent factors in psychological health.

Emotional-Cognitive Integration in Social Cognition

Social cognition—the ability to understand and interact with others—emerges as a fundamental brain metafunction within this framework. Social interactions require continuous coordination between emotional states, cognitive reasoning, and adaptive behavioral responses. Social cognition integrates emotional, cognitive, homeostatic, and allostatic systems. Empathy requires emotional resonance and cognitive perspective-taking, which are essential for ethical decision-making and cooperative behavior (Decety & Jackson, 2004). Moral reasoning includes emotional intuitions (e.g., disgust, empathy) that interact with deliberate cognitive judgments in ethical dilemmas (Haidt, 2001). Group dynamics studies emotional contagion and collective cognition that shape decision-making in social groups (Amodio & Frith, 2006). Together, these processes illustrate how the brain integrates emotional and cognitive functions to navigate complex social environments. Neuroimaging studies show that regions such as the medial prefrontal cortex (mPFC) and temporoparietal junction (TPJ) are activated during social reasoning tasks, reinforcing the brain's reliance on integrated emotional-cognitive processing to navigate complex interpersonal interactions.

Future Directions and Broader Implications

The recognition of unified emotional-cognitive processing has profound implications across disciplines, influencing artificial intelligence, education, and organizational behavior. Integrating emotional and cognitive models in AI could produce socially responsive machines capable of understanding and responding to human cues with greater nuance. Advancements in emotionally intelligent AI (e.g., sentiment-aware algorithms and human-AI empathy modeling) could enhance human-computer interactions (Ziemke, 2016). In education, fostering emotional intelligence alongside cognitive skills improves learning outcomes, resilience, and social engagement, enhancing adaptive decision-making in complex educational settings (Immordino-Yang & Damasio, 2007). Recognizing emotional-cognitive integration in leadership and decision-making improves team dynamics, strategic planning, and workplace well-being. Emotional intelligence relates to higher leadership performance and greater crisis management adaptability (Goleman, 1995). As the Brain Metafunction Framework advances, it promises to reshape our understanding of consciousness, human behavior, and intelligent systems. By integrating emotional and cognitive processes into research, clinical practice, and AI development, this framework offers a holistic approach that enriches mental health treatments, social dynamics, and technological advancements—ultimately promoting human flourishing.

5. A Comprehensive Model of Predictive Coding and Brain-Environment Interaction

Extending Predictive Coding: The Brain as an Adaptive, Metafunctional System

The Brain Metafunction Framework extends predictive coding by positioning the brain as a fully adaptive, anticipatory system that continuously aligns with environmental, cognitive, and affective changes. Predictive coding theory suggests that the brain generates hypotheses about incoming stimuli, compares them with sensory input, and refines internal models to minimize prediction error (Friston, 2010; Clark, 2013). This process does not merely apply to perception and governs motor planning, cognitive decision-making, and emotional regulation within a unified, hierarchical system.

Traditional predictive coding models primarily focus on sensory processing, yet real-world behavior demands a metafunctional integration of multiple domains. The Brain Metafunction Framework expands predictive coding to encompass sensory and motor responses, cognitive reasoning, affective states, and goal-directed behavior. This expansion aligns with active inference theory, where the brain does not passively interpret sensory data but actively shapes interactions with the environment through top-down predictions (Friston, 2010, 2013). This interdisciplinary approach positions prediction as the foundation of real-world cognition, helping to bridge the gap between biological intelligence, computational modeling, and adaptive learning systems.

Implications for Neuroadaptive Technologies and Machine Learning

Advances in predictive coding research have profound implications for neuroadaptive technologies, artificial intelligence, and human-computer interaction. Neuroadaptive systems, inspired by the brain's error-minimization strategies, could predict user intent, infer mental states, and dynamically adjust to real-time feedback. These systems would enhance brain-computer interfaces (BCIs), neurorehabilitation, and intelligent assistive technologies (Zander et al., 2016; Millidge et al., 2021; Salvatori et al., 2023; Hassabis et al., 2017). Real-time predictive coding in BCIs could improve motor prosthetics by anticipating movement intentions before execution, offering seamless neural control over assistive devices (Lebedev & Nicolelis, 2006).

In cognitive neurofeedback systems, closed-loop AI interactions could optimize performance and recovery in neurological disorders. Emotion-sensitive AI, where affective computing models track physiological and neural signals, could adjust interface responses based on user emotional states. Predictive AI could also revolutionize augmented reality (AR) and virtual reality (VR) by anticipating gaze direction, motor actions, and user intent to create ultra-immersive, lag-free experiences (Pezzulo, 2017; Ciria et al., 2021). Similarly, adaptive learning systems could predict student engagement levels, personalizing educational content in real time to optimize cognitive retention and motivation (Carey, 2014). By incorporating brain-inspired prediction strategies, neuroadaptive technologies could transform human-machine interaction, making artificial systems more intuitive, context-sensitive, and capable of real-time adaptation.

Enhancing Machine Learning and Real-Time Adaptivity

Traditional machine learning models operate reactively, processing inputs only after receiving them, which limits adaptability in dynamic environments. However, predictive coding offers a computational framework for anticipatory processing, enabling AI to predict and adapt to new data streams rather than simply responding to past inputs (Hohwy, 2016; Kriegeskorte & Douglas, 2018). Predictive AI could allow robots to anticipate human actions and adjust movement trajectories preemptively in robotics and autonomous systems, improving collaboration. In personalized AI applications, predictive modeling could refine recommendation algorithms, making virtual assistants more proactive in anticipating user needs. Predictive AI could forecast and mitigate cyber threats before they fully materialize, improving digital security infrastructures. By mirroring the brain's predictive architecture, AI systems could enhance efficiency in decision-making, uncertainty management, and real-time response.

Toward a Unified Framework for Prediction and Adaptation

The Brain Metafunction Framework positions prediction as a universal mechanism that integrates sensory, motor, cognitive, and emotional processing. This perspective bridges neuroscience, psychology, AI, and adaptive technology, creating a cohesive model of brain-environment interaction. By treating prediction as a domain-general function, this framework deepens our understanding of human cognition and inspires technological advancements that intuitively respond to human needs. This move marks a pivotal step in theoretical exploration and practical applications, pushing beyond traditional reactive models toward anticipatory, self-regulating biological and artificial intelligence systems. Integrating predictive processing into AI, robotics, and adaptive technology could reshape how machines understand and interact with humans, creating a future where technology is responsive and proactively aligned with human cognition, emotion, and intent.

6. Brain-Body Integration and Interoception as a Core Aspect of Brain Metafunction

Interoception and Brain-Body Integration in the Brain Metafunction Framework

The Brain Metafunction Framework highlights interoception as a fundamental process integrating brain-body interactions, shaping cognition, emotional regulation, and overall well-being. Interoception refers to the brain's capacity to sense, interpret, and regulate internal bodily states, including heart rate, respiration, and metabolic balance (Craig, 2002; Critchley & Garfinkel, 2017). While traditionally viewed as a mechanism for homeostatic regulation, emerging research suggests that interoception plays a crucial role in higher-order cognition, self-awareness, and emotional processing (Seth, 2013).

Unlike exteroceptive processes that interpret external sensory input, interoception operates as an internal prediction system, continuously updating the brain's body representation to maintain homeostasis and allostasis. This predictive regulation of bodily states supports adaptive behavior, influencing stress resilience, emotional stability, and self-perception (Tsakiris & De Preester, 2018). Interoception bridges physical and mental health by grounding cognition in bodily signals, forming the foundation for self-regulation, affective experiences, and subjective awareness.

Interoception as a Pillar of Emotional Regulation and Self-Awareness

Interoception is deeply intertwined with emotional regulation and self-awareness, as bodily states directly shape affective experiences. Emotional states are often grounded in physiological changes—for example, anxiety correlates with increased heart rate and hyperarousal, while calmness aligns with balanced autonomic activity. Recognizing and interpreting these bodily cues allows individuals to regulate emotions more effectively, enhancing stress resilience and adaptive decision-making (Barrett & Simmons, 2015; Seth, 2013). This perspective aligns with embodied cognition theories, which argue that bodily sensations are not secondary to emotions but integral to their formation (Damasio, 1999). Interoception thus forms a continuous feedback loop where bodily awareness informs emotional states, shaping how individuals perceive and respond to environmental challenges. Disruptions in interoception are linked to various psychiatric and neurological disorders, including anxiety, depression, and PTSD, reinforcing its role as a core component of mental health (Khalsa et al., 2018).

Advancing Research in Brain-Body Integration and Interoceptive Technologies

Recognizing interoception as a core function of the brain has profound implications for neuroadaptive technologies, biofeedback, and AI-driven health monitoring systems. Modern biofeedback techniques leverage interoceptive awareness to improve emotional regulation, reduce anxiety, and optimize cognitive performance (Lehrer & Gevirtz, 2014). Wearable devices and brain-computer interfaces (BCIs) now track physiological signals such as heart rate variability (HRV), galvanic skin response (GSR), and respiration patterns, providing users with real-time feedback for self-regulation and stress management.

Technological advancements in interoception-based AI further align with the Brain Metafunction Framework, allowing intelligent systems to integrate physiological data into adaptive health interventions. For example, neuroadaptive computing platforms can personalize meditation, sleep, and focus-enhancement programs by dynamically adjusting based on the user's physiological state. AI-driven mental health applications could leverage predictive models of interoceptive regulation, identifying patterns that signal stress or emotional dysregulation before conscious awareness. Integrating interoception-based biofeedback, neuroadaptive AI, and personalized health systems marks a new frontier in neuroscience and technology. By harnessing the power of interoceptive awareness, these innovations will empower individuals to achieve better emotional control, resilience, and overall well-being.

7. Neuroplasticity as a Systemic Function

Redefining Neuroplasticity as a Whole-Brain, Network-Level Process

The Brain Metafunction Framework redefines neuroplasticity as a network-level, systemic function, emphasizing large-scale brain reorganization rather than isolated synaptic changes. Traditionally, neuroplasticity refers to the brain's ability to rewire in response to learning, experience, or injury, often conceptualized as localized changes in neural connectivity. However, this framework expands neuroplasticity beyond individual neuronal connections, viewing it as a coordinated, whole-brain response that integrates sensory, motor, cognitive, and emotional processes to optimize immediate and long-term adaptability (Pascual-Leone et al., 2005; Sporns, 2010).

The systemic nature of neuroplasticity is evident in experience-dependent reorganization, such as musical training, bilingualism, and cognitive skill acquisition. Studies on musicians and multilingual individuals reveal widespread cortical adaptations, where motor, auditory, and executive networks undergo structural and functional modifications to support complex, multi-system tasks (Zatorre et al., 2012; Bamidis et al., 2014). These findings demonstrate that neuroplasticity is not confined to individual brain regions but is a dynamic, integrative mechanism for whole-brain coordination. The Brain Metafunction Framework aligns neuroplasticity with lifespan adaptability, shifting focus from synaptic-level mechanisms to large-scale neural networks that sustain cognitive and functional resilience. While early theories emphasized neuroplasticity during critical developmental periods, modern research confirms that plasticity persists throughout life, enabling the brain to compensate for functional declines, trauma, or environmental challenges.

Across the lifespan, neuroplasticity enables adaptive compensation through alternative neural recruitment strategies. In aging populations, studies show that bilateral activation and distributed processing strategies help maintain cognitive performance despite regional declines (Cabeza et al., 2002; Park & Reuter-Lorenz, 2009). The compensation-related utilization of neural circuits hypothesis (CRUNCH) suggests that the aging brain actively reorganizes functional networks to optimize performance under increased cognitive load. This systemic adaptability highlights neuroplasticity as a primary driver of cognitive resilience, sustaining mental agility and functional independence in later life. The predictive processing perspective further supports this systemic view, proposing that neuroplasticity is reactive and proactive. The brain continuously updates its internal models based on sensory input and experience, allowing for anticipatory adaptation to new challenges (Friston, 2010, 2013). This interaction between neuroplasticity and active inference ensures that the brain remains dynamically responsive to environmental changes, reinforcing its role as an adaptive, self-organizing system.

Implications for Rehabilitation, Cognitive Resilience, and Neurodegenerative Disorders

The concept of cognitive reserve further supports the systemic nature of neuroplasticity. Research suggests that lifelong learning, multilingualism, and mentally stimulating activities build a "cognitive buffer" that allows individuals to maintain functional independence despite neurodegenerative pathology (Stern, 2002). Studies on cross-domain training programs combining motor and cognitive tasks reveal that simultaneously engaging multiple brain systems enhances executive function, attentional control, and working memory (Valenzuela & Sachdev, 2006). These findings reinforce the need for holistic, network-based interventions that harness systemic plasticity to promote long-term resilience.

Driving Technological Advancements in Neuroadaptive Systems

The systemic nature of neuroplasticity is also shaping the development of neuroadaptive technologies, brain-computer interfaces (BCIs), and AI-driven learning models. By providing real-time neurofeedback, BCIs can enhance motor and cognitive training by actively engaging widespread neural networks to support functional recovery (Lebedev & Nicolelis, 2006). Advances in machine learning and neuro-inspired AI architectures mirror neuroplastic principles, fostering systems capable of continuous learning and environmental adaptation (Hassabis et al., 2017). Adaptive systems can emulate the brain's ability to restructure and optimize its networks by incorporating neuroplasticity-inspired AI. AI-driven rehabilitation platforms could personalize cognitive training based on real-time neurophysiological feedback, while context-aware learning models could dynamically reconfigure computational architectures, improving autonomous AI adaptability. These technological advancements reflect the convergence of neuroscience, artificial intelligence, and neuro-engineering, with profound implications for human cognition and machine intelligence.

Toward a Holistic Understanding of Neuroplasticity

Viewing neuroplasticity as a systemic function shifts our perspective toward a holistic understanding of the brain's adaptability. The Brain Metafunction Framework positions the brain as an integrated, self-organizing network, where plasticity is not a localized property but a whole-brain mechanism for maintaining cognitive, emotional, and physical resilience. This systemic approach will guide advancements in rehabilitation, neuroadaptive technology, and cognitive training, offering a comprehensive framework for human adaptation, growth, and well-being.

8. Brain Metafunctional Ports in Medical and Societal Applications

Brain Metafunctional Ports: A New Frontier in Brain-Technology Integration

As our understanding of brain metafunction deepens, brain-computer interfaces (BCIs) and brain metafunctional ports emerge as transformative tools in medicine, rehabilitation, entertainment, and beyond. The Brain Metafunction Framework informs the design of more efficient, seamless, and adaptive brain-device interactions, extending beyond conventional BCI models by integrating cognitive, motor, and emotional functions into a unified neural interface. These advancements hold tremendous promise for enabling individuals with motor impairments to regain control, enhancing human-computer interaction, and even expanding how we define intelligence, selfhood, and human augmentation. However, the rapid evolution of BCIs and neural interfaces raises ethical concerns surrounding autonomy, mental integrity, and cognitive enhancement (Farah, 2012; Goering et al., 2017). The ability to read, interpret, and potentially modify brain activity presents profound philosophical and legal challenges, requiring careful consideration of privacy, access, and human dignity in an increasingly neurotechnological world.

Medical and Rehabilitation Applications: The Brain as an Adaptive System

BCIs have already demonstrated transformative potential in medical treatment and neurorehabilitation. They allow individuals with disabilities to control assistive devices, restore lost functions, and enhance cognitive performance. BCIs empower users to interact with prosthetics, robotic limbs, and communication devices by decoding brain signals into actionable commands, improving mobility and autonomy (Lebedev & Nicolelis,

2006; Hochberg et al., 2012). Beyond assistive technologies, BCIs are revolutionizing stroke recovery and neurorehabilitation by retraining neural circuits and fostering neuroplasticity. Adaptive BCIs engage affected and intact brain regions, promoting compensatory plasticity and restoring motor and cognitive functions after brain injuries (Mane, Ang, & Guan, 2023). These neural interfaces also enhance diagnostic precision in neurological disorders such as Parkinson's, epilepsy, and ALS, offering real-time insights into brain activity and disease progression (Birbaumer et al., 2008). In non-verbal conditions such as locked-in syndrome, BCIs serve as communication tools, bridging the gap between thought and action to enable social engagement and educational participation (Brumberg et al., 2010). This expanding role underscores how brain metafunctional ports could reshape the future of disability, accessibility, and cognitive augmentation, positioning the human brain as an actively adapting system capable of direct interaction with external technology.

Beyond Medicine: BCIs in Entertainment, Learning, and Human Enhancement

BCIs are not limited to medical rehabilitation but are increasingly shaping entertainment, education, and human augmentation. In the gaming and entertainment industries, thought-controlled interfaces enable users to navigate virtual environments, manipulate digital objects, and engage with immersive simulations using only their neural activity (Piszcz et al., 2024; Nijboer et al., 2015; Lotte et al., 2015). These advancements blur the line between physical and virtual reality, introducing novel forms of interactive engagement and human-computer collaboration. BCIs could personalize learning experiences in education by detecting real-time cognitive load, attention levels, and mental fatigue. Adaptive learning platforms could adjust instructional difficulty, recommend study breaks, and optimize learning strategies based on neural feedback. Similarly, neuroenhancement applications could enhance memory consolidation, creativity, and problem-solving skills, raising critical ethical questions about fairness, accessibility, and cognitive modification.

The increasing integration of brain interfaces into daily life presents serious privacy and security concerns. Unlike traditional biometric information, Neural data is deeply personal, involuntary, and uniquely revealing. Researchers and policymakers must consider who should access brain data, how to store it securely, and whether corporations, employers, or governments might misuse it (Yuste et al., 2017). Regulatory standards must ensure that neural privacy, cognitive autonomy, and informed consent remain at the core of BCI development. Beyond privacy, cognitive enhancement through BCIs introduces societal challenges. If BCI-enabled enhancements—such as faster learning, heightened concentration, or AI-assisted cognition—become available, should they be accessible to everyone or restricted based on economic or regulatory constraints? The emergence of "neuro-inequality," where financial status limits access to cognitive augmentation, could exacerbate societal divisions, necessitating frameworks prioritizing ethical and equitable innovation (Ienca & Andorno, 2017).

Future Directions: The Ethical and Existential Challenges of Brain Metafunctional Ports

As brain-computer interfaces and brain metafunctional ports become increasingly sophisticated, their integration into medicine, education, and daily life will redefine the relationship between humans and technology. These interfaces will likely become integral to personalized medical treatments, improving motor and cognitive rehabilitation while offering neuroscientists deeper insights into neuroplasticity, learning, and memory formation. However, as BCIs and neural interfaces become more prevalent, society must grapple with their implications for identity, agency, and autonomy. The ability to alter cognitive processes, enhance mental performance, and merge biological intelligence with artificial systems challenges traditional concepts of individuality and selfhood. If thought-controlled interfaces enable seamless interaction with AI, will human cognition become indistinguishable from machine-assisted reasoning? If technology augments memory and perception, how do we redefine the limits of natural intelligence? These existential questions highlight the urgency of establishing ethical frameworks that ensure brain metafunctional ports remain tools for human flourishing rather than instruments of coercion or inequality. Striking a balance between innovation and ethical responsibility will be essential in shaping a future where neurotechnology enhances, rather than compromises, human dignity and autonomy.

9. A New Model for Psychosomatic Health Dynamics

Redefining Health: The Brain Metafunction Framework and Psychosomatic Integration

The Brain Metafunction Framework introduces a transformative model of psychosomatic health, emphasizing the continuous interplay between cognitive, emotional, and physiological processes.

This framework positions health not as a static state but as an adaptive, self-regulating system, where psychological states actively shape immune function, inflammation, and overall physical resilience. By reframing health as a networked, bidirectional process, this model reveals that mental and physical well-being are fundamentally interconnected (McEwen, 2007; Dhabhar, 2014). Unlike traditional psychosomatic models, which often treat mind-body interactions as secondary health influences, the Brain Metafunction Framework suggests that mental states and physiological processes co-regulate each other through hierarchical prediction and self-organization. This perspective aligns with predictive processing theories, where the brain continuously updates its internal model of bodily states, shaping perceptions of pain, fatigue, and healing (Seth, 2013). This model lays the foundation for a holistic approach to human health and adaptability by integrating neuroscience, immunology, and cognitive-affective regulation.

Mental States, Immune Function, and Health Resilience

A core insight of the Brain Metafunction Framework is the role of mental states—such as stress, anxiety, and emotional resilience—in immune function and inflammation regulation. Stress activates the hypothalamic-pituitary-adrenal (HPA) axis, leading to cortisol release, immune suppression, and increased vulnerability to infection and chronic disease (Glaser & Kiecolt-Glaser, 2005). In contrast, positive psychological states, including gratitude, social connection, and emotional regulation, enhance immune efficiency, with research linking these emotions to increased natural killer (NK) cell activity and improved immune resilience (Kemeny, 2003). The vagus nerve is a key mediator between mental states and immune function, regulating parasympathetic activity, inflammatory signaling, and autonomic balance. Studies suggest high vagal tone is associated with lower systemic inflammation, better cardiovascular health, and improved stress resilience (Thayer & Sternberg, 2006). Interventions that increase vagal tone, such as deep breathing, meditation, and biofeedback, enhance immune system regulation and emotional stability, reinforcing the self-regulatory nature of psychosomatic health.

The Psychosomatic Basis of Chronic Inflammation and Disease

The framework also highlights the role of chronic psychological stress in driving inflammatory conditions, linking mental health and systemic disease progression. Chronic stress triggers dysregulation of cytokine activity, sustaining low-grade inflammation, which contributes to cardiovascular disease, diabetes, and autoimmune disorders (Miller et al., 2009). Persistent emotional distress disrupts neuroimmune communication, producing heightened pro-inflammatory signaling and accelerated biological aging (Segerstrom & Miller, 2004). Conversely, psychosomatic interventions can modulate inflammatory responses, reducing disease risk and improving clinical outcomes. Studies show that cognitive-behavioral therapy (CBT), social support, and mindfulness-based stress reduction (MBSR) lower inflammatory biomarkers, reinforcing the link between mental health and immune regulation (Black & Slavich, 2016). These findings underscore the necessity of integrating mental health strategies into chronic disease management, treating inflammation as a physiological and psychological phenomenon.

Enhancing Healing and Recovery Through Psychosomatic Regulation

The Brain Metafunction Framework also illuminates the role of psychological states in influencing healing and recovery. Research demonstrates that positive emotions, optimism, and stress reduction accelerate post-injury healing and surgical recovery, while stress and anxiety impair immune responses and delay tissue repair (Broadbent et al., 2003). High stress levels disrupt wound healing by impairing immune cell mobilization, whereas psychological interventions—such as guided imagery, hypnosis, and meditation—enhance recovery rates by modulating immune activity and reducing pain perception (Schoultz et al., 2016). By integrating neuroplasticity, autonomic regulation, and psychosomatic feedback, this model supports a holistic framework for healing, emphasizing the importance of mental state optimization in physical recovery.

Applications in Pain Management, Stress Reduction, and Immune Health

Psychosomatic health dynamics have far-reaching applications in chronic pain management, stress resilience, and immune optimization. Chronic pain, for example, is not solely a nociceptive process but is shaped by cognitive, emotional, and learned responses. The Brain Metafunction Framework highlights how pain perception emerges from predictive models, where past experiences, emotional states, and expectation biases shape the intensity of pain signals (Melzack, 2001). CBT, neurofeedback, and somatic therapy have been shown to retrain pain-processing circuits, reducing pain intensity and enhancing quality of life (Apkarian et al., 2009). Similarly, stress-reduction techniques—including mindfulness, biofeedback, and breathing exercises—modulate neuroimmune pathways, lowering cortisol levels, reducing inflammation, and strengthening immune resilience (Goyal et al., 2014). These approaches recognize stress as a systemic health determinant, positioning psychosomatic interventions as critical tools for disease prevention and holistic well-being.

The Brain Metafunction Framework promotes a paradigm shift in healthcare, moving beyond symptom-focused treatments to a network-based, self-regulating model of health. This model integrates psychological, neural, and physiological dynamics to emphasize the inherent connection between mental and physical health. Viewing health as an emergent, dynamic system provides a roadmap for personalized, proactive interventions, guiding the development of next-generation mind-body therapies, bioadaptive technologies, and integrative medicine practices. By embracing the interplay of brain, body, and environment, this approach redefines how we prevent, manage, and optimize human health, fostering resilience, adaptability, and long-term well-being.

10. Expanded Role of Network-Based Diagnostics in Neurology and Psychiatry

Redefining Mental and Neurological Disorders: A Shift from Symptoms to Networks

The Brain Metafunction Framework introduces a transformative shift in neurology and psychiatry, moving beyond traditional symptom-based classifications to network-based diagnostics. Rather than viewing mental and neurological disorders as discrete conditions, this approach conceptualizes them as dysregulated brain network states, offering greater diagnostic precision and personalized treatment pathways. Traditional diagnostic systems such as the DSM and ICD categorize disorders based on symptom clusters without addressing the underlying neural mechanisms. However, advances in connectomics, functional neuroimaging, and computational modeling reveal that conditions such as schizophrenia, depression, and Alzheimer's disease arise from disruptions in functional and structural connectivity rather than isolated regional deficits (Buckholtz & Meyer-Lindenberg, 2012). By shifting the diagnostic focus from symptom checklists to dynamic network dysfunctions, clinicians can assess individualized connectivity profiles and design targeted interventions that restore network balance.

Network-Based Diagnostics: A New Paradigm for Under -standing Brain Disorders

The network-based paradigm reconceptualizes mental and neurological disorders as dysregulated brain states, where symptoms emerge from network-wide imbalances rather than localized deficits. For example, depression involves disruptions in the default mode network (DMN), salience network (SN), and central executive network (CEN), affecting self-referential thought, emotional regulation, and cognitive control (Hamilton et al., 2015). Similarly, ADHD is linked to atypical connectivity in the frontoparietal network and DMN, explaining difficulties in sustained attention and impulse control (Castellanos & Proal, 2012). In schizophrenia, disrupted connectivity between sensory integration, executive control, and salience networks contributes to hallucinations, cognitive fragmentation, and executive dysfunction (Friston et al., 2016). This systemic approach clarifies why psychiatric and neurological conditions often share overlapping symptoms and provides individualized treatment pathways based on specific network dysfunctions. Rather than applying broad diagnostic labels, clinicians can tailor interventions to a patient's unique neural connectivity profile, improving treatment precision and outcomes.

Precision Diagnostics and Personalized Treatment: Targeting Neural Networks

Network-based diagnostics align with precision medicine by allowing clinicians to identify individual connectivity imbalances and select targeted interventions. Functional neuroimaging and EEG biomarkers detect patterns of dysconnectivity, guiding neuromodulation techniques such as transcranial magnetic stimulation (TMS) and transcranial direct current stimulation (tDCS), which have demonstrated efficacy across diverse psychiatric and neurological disorders (Fox et al., 2014). Clinicians can personalize cognitive-behavioral therapy (CBT) to strengthen disrupted brain networks and optimize cognitive and emotional regulation circuits (Mayberg, 2009). Additionally, network-based pharmacology offers an alternative to conventional medication trials by tailoring psychiatric treatments to individual neurobiological profiles, improving efficacy while minimizing side effects. This model enhances clinical precision, reduces trial-and-error prescribing, and accelerates recovery rates by shifting from broad-spectrum interventions to individualized, network-based treatments.

Advancing Diagnostics with Machine Learning and Computational Models

Machine learning and computational models have revolutionized network-based diagnostics by analyzing complex, large-scale neural data. AI-powered algorithms trained on neuroimaging datasets can classify mental health disorders based on brain connectivity profiles, distinguishing conditions such as bipolar disorder and major depression (Drysdale et al., 2017). Graph-theoretic models track dynamic changes in functional networks, providing predictive insights into neurodegenerative disease progression (Seeley et al., 2009). Computational models enable early detection of neurodegenerative diseases, such as Alzheimer's and Parkinson's, allowing for interventions before cognitive decline becomes irreversible. They also facilitate real-time tracking of psychiatric symptoms, predicting mood episodes in conditions such as bipolar disorder and major depression. Additionally, AI can personalize neuromodulation strategies by identifying optimal brain stimulation targets based on an individual's unique connectivity fingerprint. By integrating AI, neuroinformatics, and predictive modeling, network-based diagnostics will provide unprecedented mental health and neurological care precision.

Moving Toward Integrated, Interdisciplinary Care

A network-based approach fosters integrated care models by merging neurology, psychiatry, and psychology to offer a holistic understanding of mental and neurological health. For example, treating anxiety disorders may involve psychiatric therapies targeting limbic dysregulation alongside behavioral interventions that strengthen cognitive control networks (Menon, 2011). This framework encourages interdisciplinary collaboration, bridging the gap between psychiatry and neurology to enhance diagnostic accuracy, treatment effectiveness, and long-term recovery (Insel, 2014). As our understanding of brain networks deepens, network-based diagnostics will redefine clinical practice in neurology and psychiatry, shifting from symptom-centered diagnoses to a systems-level, precision-based approach.

Challenges and Future Directions: Ethical and Clinical Considerations

Despite its potential, network-based diagnostics present ethical, practical, and clinical challenges. The complexity of interpreting neural connectivity patterns raises concerns about over-complicating diagnostic procedures and patient care. Additionally, as machine learning algorithms analyze large-scale neuroimaging data, ensuring patient confidentiality and preventing data misuse become critical concerns. High-cost neuroimaging and computational diagnostics may widen healthcare disparities unless policymakers take measures to integrate them into standard clinical settings. Addressing these challenges requires ongoing collaboration between neuroscientists, clinicians, AI researchers, and policymakers to ensure that network-based approaches enhance, rather than replace, patient-centered care. As neuroimaging, AI, and network neuroscience develop, network-based diagnostics will redefine how we understand, diagnose, and treat mental health and neurological disorders, shaping a future where precision, integration, and innovation drive clinical excellence.

11. Bioethical Implications of the Brain Metafunction Framework

Rethinking Death and Personhood: A Unified Perspective

The Brain Metafunction Framework provides a unifying model for defining death, resolving the longstanding divide between definitions based on consciousness alone (Youngner and Bartlett 1983; Machado 1995; Veatch 2005) and those centered on whole-body integration (President's Commission 1981; Bernat 2018). Historically, these perspectives have resulted in philosophical, medical, and legal inconsistencies, particularly in cases involving persistent vegetative states, severe brain injuries, and residual isolated functions. The Brain Metafunction Framework introduces a crucial clarification: death and the loss of personhood are coincident and inseparable. In this view, death is not merely the cessation of biological function but the definitive loss of the brain's ability to maintain the unity of the organism and the person: the permanent loss of the brain metafunction.

This framework makes a key distinction between the process of dying, the event of death, and the process of disintegration. While dying is a progressive decline in systemic function, and disintegration refers to the postmortem breakdown of bodily and neural structures, death itself is an event—the moment when the brain irreversibly loses its metafunction. This conceptual refinement ensures that death is neither reducible to the irreversible loss of consciousness nor to the gradual breakdown of the body but instead marks the precise boundary between life and non-life, where the unity of the person and the organism ceases to exist.

Defining Death as the Loss of the Brain Metafunction: Resolving Conceptual Divides

Defining death is a clinical and ethical imperative, requiring precise, scientifically valid, and philosophically coherent criteria. The Brain Metafunction Framework situates the continuity of life within the brain's ability to orchestrate bodily, cognitive, and emotional integration. By doing so, it resolves the false dichotomy between models that focus on neurological function alone and those that define death based on whole-body physiological breakdown. By identifying the loss of brain metafunction as the definitive boundary between life and death, this model also strengthens public trust in death declarations, particularly in end-of-life decision-making and organ donation ethics. Clear neurological markers of irreversible brain dysfunction reinforce medical integrity while respecting human dignity and autonomy.

Beyond clinical practice, the Brain Metafunction Framework informs legal and philosophical discussions on personhood and death. Legally, it provides science-based criteria for end-of-life decisions, advance directives, and medical futility determinations, ensuring that laws governing life support, organ transplantation, and assisted dying remain ethically and scientifically sound (President's Council on Bioethics, 2008). Philosophically, the loss of brain integration aligns with traditional views of personhood, and theologically, this concept resonates with the soul's departure as articulated in Aquinas's metaphysics of life (1947) and Pope John Paul II's reflections on medical ethics (2000). This model bridges disciplinary divides by integrating scientific insights with legal, ethical, and theological considerations, offering a coherent and unified approach to death determination.

By defining death as the event that separates the process of dying from the process of disintegration, the Brain Metafunction Framework strengthens ethical standards in medical practice. This approach ensures that clinical determinations respect biological and existential aspects of personhood, fostering dignity, transparency, and public confidence in neurological death criteria. The framework advances a holistic understanding of death and personhood by integrating neuroscience, ethics, philosophy, and theology perspectives. This interdisciplinary approach aligns with theology's emphasis on life's sacred unity, philosophy's exploration of existence and identity, and bioethics' commitment to protecting human dignity in medical decision-making. By offering a scientifically coherent and ethically responsible framework, this model refines death determinations, legal definitions, and end-of-life care protocols, ensuring a consistent and equitable approach to death determination in modern medicine. The framework also intersects with philosophical and theological perspectives on selfhood and the afterlife while maintaining medical accuracy, which will require ongoing interdisciplinary dialogue.

A Unified and Ethically Robust Model of Death

The Brain Metafunction Framework provides a scientifically rigorous, ethically consistent, and philosophically informed approach to defining death. By conceptualizing death as the permanent loss of the brain metafunction, this model advances neurological, legal, and theological discussions on personhood and end-of-life care. Critically, this model clarifies that death is not a gradual decline but a singular event. Unlike previous models that treat biological death and the loss of personhood as separate issues, the Brain Metafunction Framework asserts that they are inseparable. This perspective resolves longstanding conceptual disputes, establishing a clear, unified criterion for death that aligns with scientific rigor and ethical responsibility. By recognizing that death and the loss of personhood coincide, this framework ensures greater clarity in death determination, improved clinical protocols, and enhanced respect for human dignity at life's end. As medical technology develops, a robust and ethically sound definition of death becomes increasingly necessary. Through continued interdisciplinary collaboration, the Brain Metafunction Framework has the potential to reshape bioethics, inform public policy, and refine the future of medical science and end-of-life care.

12. A Paradigm Shift in Neuroscience: Embracing the Brain Metafunction Framework's Transphysical Domain

Revolutionizing Neuroscience: Beyond Materialism

The Brain Metafunction Framework anticipates a transformative shift in neuroscience, integrating the transphysical domain as a legitimate area of inquiry (Beauregard 2007). This paradigm shift extends beyond traditional materialist assumptions, encouraging interdisciplinary collaboration across neuroscience, philosophy, theology, bioethics, medicine, and psychology. By embracing a holistic model of brain function, consciousness, and personhood, this framework expands beyond the limitations of reductionist approaches while advancing network-based diagnostics (Friston, 2011; Tononi & Edelman, 1998) and personalized treatments. For decades, materialism has constrained neuroscience by promoting the belief that physical neural activity alone gives rise to all cognitive and conscious processes. While materialism is compatible with discoveries in brain structure, connectivity, and function, it fails to explain the subjective experience of consciousness—a challenge known as the "hard problem" of consciousness (Chalmers, 1995). The Brain Metafunction Framework acknowledges that consciousness and specific aspects of cognition—self-awareness, perception, and intentionality—resist reduction to purely material processes. This recognition expands the scope of neuroscience, encouraging exploration into transphysical dimensions of cognition and consciousness.

The Transphysical Domain as a Legitimate Inquiry

Thomas Kuhn's The Structure of Scientific Revolutions (1962) describes paradigm shifts as transformative changes in scientific perspective that redefine foundational assumptions. The Brain Metafunction Framework represents such a shift by integrating consciousness, perception, memory, emotion, and bodily regulation into a unified system. Despite remarkable progress in neuroimaging and computational modeling, materialist neuroscience still falls short of explaining how neural activity gives rise to subjective experience. Phenomena such as veridical near-death experiences (NDEs), savant abilities, and altered states of consciousness continue to elude conventional models (Greyson, 2010; Treffert, 2009). NDEs, for example, involve vivid perception and memory during periods of clinical unconsciousness—raising the possibility that aspects of consciousness may operate independently of typical brain function (van Lommel, 2010). Likewise, reports of spiritual insight, veridical perception, and extrasensory awareness challenge strictly reductionist paradigms, suggesting that consciousness may extend beyond conventional neural mechanisms (Radin, 1997). The transphysical domain thus offers a conceptual framework for investigating these outlier phenomena—previously marginalized by mainstream neuroscience—without abandoning scientific rigor. It opens the door to considering consciousness as more than a computational byproduct, with far-reaching implications for diagnostics, therapeutic innovation, and ethical AI development. Philosophical models such as non-reductive physicalism, panpsychism, and dual-aspect monism offer additional ontological grounding for this broader epistemology (Chalmers, 1996; Spinoza, 1677/1985).

Materialist neuroscience has successfully mapped the architecture of cognition, but it remains unable to explain the origins of qualia or the first-person perspective. Broader questions—such as the origin of the universe, fine-tuning, emergence of life, and the informational complexity of biological systems—continue to challenge purely physicalist accounts (Penrose, 2004; Meyer, 2021). Similarly, empirical anomalies such as NDEs and savant syndromes remain unresolved within reductionist frameworks. Veridical NDEs, where individuals recall detailed experiences during periods of clinical brain inactivity, defy current explanatory models (van Lommel, 2010). Accounts of spiritual experiences and non-ordinary perception suggest cognitive access to dimensions not easily accounted for by conventional neurobiology (Radin, 1997). Rather than dismissing such cases, the transphysical approach calls for their systematic inclusion in neuroscience. This emerging domain does not reject empirical rigor but proposes a more inclusive ontology—one that recognizes consciousness as potentially fundamental to reality and worthy of inquiry.

Implications for Neuroscience, Ethics, and Medical Practice

Adopting the Brain Metafunction Framework has far-reaching implications for neuroscience, neuroethics, and medical practice. By recognizing brain networks as integrated systems rather than isolated regions, this model seeks to enhance precision diagnostics and personalized treatments. For example, network-based analyses of depression, schizophrenia, and neurodegenerative diseases will allow clinicians to target dysregulated brain networks rather than focusing on isolated symptoms (Friston, 2011; Hamilton et al., 2015). Ethically, this framework supports responsible medical practices (Goering et al., 2017) that honor the unity of the human person. It provides a foundation for defining death as losing integrative brain functions, ensuring clarity in end-of-life decisions, cognitive enhancement ethics, and mental autonomy. Additionally, it aligns with philosophical and theological views of personhood, bridging science and ethics to address questions of identity, consciousness, and selfhood (Aquinas, 1947; John Paul II, 2000).

Shaping the Future of Neuroscience and AI

The Brain Metafunction Framework transforms neuroscience and influences artificial intelligence, education, and human cognition research. By modeling brain networks as dynamic, adaptive, holistic systems, this framework inspires AI architectures that emulate human cognition, bridging biological and artificial intelligence (Hinton, 2007; Kriegeskorte, 2015). Advances in connectomics and computational neuroscience will enable researchers to map neural dynamics at an unprecedented level, fostering breakthroughs in neurotechnology, brain-computer interfaces (BCIs), and cognitive augmentation (Sporns, 2010). As neuroscience embraces the transphysical domain, it will reshape public perceptions of consciousness, mental health, and human potential. Acknowledging the brain's integrative role across biological and transphysical dimensions fosters new ethical paradigms and cautionary restraint in applying cognitive enhancement and transhumanist philosophical concepts. By integrating insights from neuroscience, AI, bioethics, and philosophy, the Brain Metafunction Framework offers a powerful model for advancing science while preserving the dignity and complexity of the human person. This shift positions the Brain Metafunction Framework as a central model for understanding brain function, human cognition, and personhood in the 21st century.

REFERENCES

Alais, D., & Burr, D. (2004). The ventriloquist effect results from near-optimal bimodal integration. *Current Biology, 14*(3), 257–262. https://doi.org/10.1016/j.cub.2004.01.029

Anderson, P. W. (1972). More is different: Broken symmetry and the nature of the hierarchical structure of science. *Science, 177*(4047), 393–396. https://doi.org/10.1126/science. 177.4047.393

Angelaki, D. E., & Cullen, K. E. (2008). Vestibular system: The many facets of a multimodal sense. *Annual Review of Neuroscience, 31*(1), 125–150. https://doi.org/10.1146/annurev.neuro.31.060407.125555

Apkarian, A. V., Baliki, M. N., & Geha, P. Y. (2009). Towards a theory of chronic pain. *Progress in Neurobiology, 87*(2), 81–97. https://doi.org/10.1016/j.pneurobio.2008.09.018

Aquinas, T. (1947). *Summa Theologica* (Fathers of the English Dominican Province, Trans.). Benziger Bros.

Arch, J. J., & Craske, M. G. (2006). Mechanisms of mindfulness: Emotion regulation following a focused breathing induction. *Behaviour Research and Therapy, 44*(12), 1849–1858. https://doi.org/10.1016/j.brat. 2005.12.007

Bamidis, P. D., Vivas, A. B., Styliadis, C., Frantzidis, C. A., Klados, M. A., Schlee, W., & Paraskevopoulos, E. (2014). A review of physical and cognitive interventions for older adults. *Frontiers in Aging Neuroscience, 6*, 1–22. https://doi.org /10.3389/fnagi.2014.00110

Bar, M. (2009). Predictions: A universal principle in the operation of the human brain. *Philosophical Transactions of the Royal Society B: Biological Sciences, 364*(1521), 1181–1182. https://doi.org/10.1098/rstb.2008.0321

Barrett, L. F., & Simmons, W. K. (2015). Interoceptive predictions in the brain. *Nature Reviews Neuroscience, 16*(7), 419–429. https://doi.org/10.1038/nrn3950

Beauchamp, M. S., Yasar, N. E., Frye, R. E., & Ro, T. (2008). Touch, sound, and vision in the human superior temporal sulcus. *NeuroImage, 41*(3), 1011–1020. https://doi.org/10.1016/j.neuroimage.2008.03.015

Beauregard, M., & O'Leary, D. (2007). *The spiritual brain: A neuroscientist's case for the existence of the soul.* HarperCollins.

Beck, A. T. (2011). *Cognitive therapy of depression.* Guilford Press.

Bernat, J. L. (2018). A conceptual justification for brain death. *Hastings Center Report, 48*(Suppl. 4), S19–S21. https://doi.org/10.1002/hast.962

Birbaumer, N., Ghanayim, N., Hinterberger, T., Iversen, I., Kotchoubey, B., Kübler, A., ... & Flor, H. (2008). Brain-computer interfaces for communication in paralysis. *Nature Reviews Neuroscience, 9*(7), 511–522. https://doi.org/10.1038/nrn2412

Black, D. S., & Slavich, G. M. (2016). Mindfulness meditation and the immune system: A systematic review. *Annals of the New York Academy of Sciences, 1373*(1), 13–24. https://doi.org/10.1111/nyas.12998

Bolognini, N., Frassinetti, F., Serino, A., & Làdavas, E. (2007). Acoustical vision of below-threshold stimuli: Interaction among spatially converging audiovisual inputs. *Experimental Brain Research, 179,* 595–605. https://doi.org/10.1007/s00221-006-0813-2

Bonnet, L., Lotte, F., & Lécuyer, A. (2013). Two brains, one game: Design and evaluation of a multiuser BCI video game based on motor imagery. *IEEE Transactions on Computational Intelligence and AI in Games, 5*(2), 185–198. https://doi.org/ 10.1109/TCIAIG.2012.2237173

Broadbent, E., Petrie, K. G., Alley, J. M., & Booth, J. R. (2003). Psychological stress impairs wound healing in humans. *Psychosomatic Medicine, 65*(4), 698–701. https://doi.org/10.1097/01.PSY.0000078374. 74448.49

Buckholtz, J. W., & Meyer-Lindenberg, A. (2012). Psychopathology and the human connectome. *Neuron, 74*(6), 990–1004. https://doi.org/ 10.1016/j.neuron.2012.06.002

Cabeza, R., Anderson, N. D., Locantore, J. K., & McIntosh, A. R. (2002). Aging gracefully: Compensatory brain activity in high-performing older adults. *NeuroImage, 17*(3), 1394–1402. https://doi.org/ 10.1006/nimg.2002.1280

Carey, B. (2014). *How we learn: The surprising truth about when, where, and why it happens.* Random House.

Castellanos, F. X., & Proal, E. (2012). Large-scale brain systems in ADHD: Beyond the prefrontal-striatal model. *Trends in Cognitive Sciences, 16*(1), 17–26. https://doi.org/10.1016/j.tics.2011.11.007

Chalmers, D. J. (1995). Facing up to the problem of consciousness. *Journal of Consciousness Studies, 2*(3), 200–219.

Chalmers, D. J. (1996). *The conscious mind: In search of a fundamental theory.* Oxford University Press.

Ciria, A., Schillaci, G., Pezzulo, G., Hafner, V. V., & Lara, B. (2021). Predictive processing in cognitive robotics: A review. *arXiv preprint* arXiv:2101.06611. https://doi.org/10.48550/arXiv.2101.06611

Clark, A. (2008). *Supersizing the mind: Embodiment, action, and cognitive extension.* Oxford University Press.

Clark, A. (2013). Whatever next? Predictive brains, situated agents, and the future of cognitive science. *Behavioral and Brain Sciences, 36*(3), 181–204. https://doi.org/10.1017/S0140525X 12000477

Craig, A. D. (2002). How do you feel? Interoception: The sense of the physiological condition of the body. *Nature Reviews Neuroscience, 3*(8), 655–666. https://doi.org/10.1038/nrn894

Critchley, H. D., & Garfinkel, S. N. (2017). Interoception and emotion. *Current Opinion in Psychology, 17,* 7–14. https://doi.org/10.1016/j.copsyc. 2017.06.020

Damasio, A. R. (1999). *The feeling of what happens: Body and emotion in the making of consciousness.* Harcourt Brace.

Decety, J., & Jackson, P. L. (2004). The functional architecture of human empathy. *Behavioral and Cognitive Neuroscience Reviews, 3*(2), 71–100. https://doi.org/10.1177/1534582304267187

Dhabhar, F. S. (2014). Effects of stress on immune function: The good, the bad, and the beautiful. *Immunologic Research, 58*(2–3), 193–210. https://doi.org/10.1007/s12026 -014-8517-0

Dinan, T. G., & Cryan, J. F. (2017). Gut instincts: Microbiota as a key regulator of brain development, aging, and neurodegeneration. *Journal of Physiology, 595*(2), 489–503. https://doi.org/10.1113/JP273106

Drysdale, A. T., Grosenick, L., Downar, J., Dunlop, K., Mansouri, F., Meng, Y., ... & Liston, C. (2017). Resting-state connectivity biomarkers define neurophysiological subtypes of depression. *Nature Medicine, 23*(1), 28–38. https://doi.org/10.1038/nm.4246

Ernst, M. O., & Banks, M. S. (2002). Humans integrate visual and haptic information in a statistically optimal fashion. *Nature, 415*(6870), 429–433. https://doi.org/10.1038/415429a

Farah, M. J. (2012). Neuroethics: The ethical, legal, and societal impact of neuroscience. *Annual Review of Psychology, 63*(1), 571–591. https://doi.org/10.1146/annurev.psych.093008.100438

Fisher, M. P. A. (2015). Quantum cognition: The possibility of processing with nuclear spins in the brain. *Annals of Physics, 362*, 593–602. https://doi.org/10.1016/j.aop.2015.08.020

Foa, E. B., Hembree, E. A., & Rothbaum, B. O. (2007). *Prolonged exposure therapy for PTSD: Emotional processing of traumatic experiences: Therapist guide.* Oxford University Press. https://doi.org/10.1093/med:psych/9780195308501.001.0001

Fox, M. D., Pascual-Leone, A., & Snyder, L. G. (2012). Network-based approaches to brain stimulation. *Nature Reviews Neuroscience, 13*(6), 426–439. https://doi.org/10.1038/nrn3206

Fredrickson, B. L. (2001). The role of positive emotions in positive psychology: The broaden-and-build theory of positive emotions. *American Psychologist, 56*(3), 218–226. https://doi.org/10.1037/0003-066 X.56.3.218

Fredrickson, B. L. (2004). The broaden-and-build theory of positive emotions. *Philosophical Transactions of the Royal Society B: Biological Sciences, 359*(1449), 1367–1378. https://doi.org/10.1098/rstb.2004.1512

Friston, K. (2010). The free-energy principle: A unified brain theory? *Nature Reviews Neuroscience, 10*(11), 752–763. https://doi.org/10.1038/nrn2787

Friston, K. (2013). Active inference and free energy. *Behavioral and Brain Sciences, 36*(3), 212–213. https://doi: 10.1017/S01405 25X12002142.

Friston, K. (2011). Functional and effective connectivity: A review. *Brain Connectivity, 1*(1), 13–36. https://doi.org/10. 1089/brain.2011.0008

Friston, K. J., Stephan, K. E., Montague, R., & Dolan, R. J. (2014). Computational psychiatry: The brain as a phantastic organ. *The Lancet Psychiatry, 1*(2), 148–158. https://doi.org/10.1016/S2215-0366(14)70275 -9

Friston, K. J., Brown, H. R., Siemerkus, J., & Stephan, K. E. (2016). The dysconnection hypothesis (2016). *Schizophrenia Research, 176*(2–3), 83–94. https://doi.org/10.1016/j.schres.20 16.07.014

Friston, K. (2023). Computational psychiatry: From synapses to sentience. *Molecular Psychiatry, 28*, 256–268. https://doi.org/10.1038/s41380 -022-01743-z

Fuster, J. M. (2005). *Cortex and mind: Unifying cognition.* Oxford University Press.

Ghazanfar, A. A., & Schroeder, C. E. (2006). Is neocortex essentially multisensory? *Trends in Cognitive Sciences, 10*(6), 278–285. https://doi.org/10.1016/j.tics.2006.04.008

Glaser, R., & Kiecolt-Glaser, J. K. (2005). Stress-induced immune dysfunction: Implications for health. *Nature Reviews Immunology, 5*(3), 243–251. https://doi.org/10.1038/nri1571

Goering, S., Klein, E., Dougherty, D. D., & Widge, A. S. (2017). Staying in the loop: Relational agency and identity in next-generation DBS for psychiatry. *AJOB Neuroscience, 8*(2), 59–70. https://doi.org/10.1080/21507740.2017.1293065

Goff, P. (2019). *Galileo's error: Foundations for a new science of consciousness.* Pantheon.

Goleman, D. P. (1995). *Emotional intelligence: Why it can matter more than IQ for character, health, and lifelong achievement.* Bantam Books.

Goyal, M., Singh, S., Sibinga, E. M., Gould, N. F., Rowland-Seymour, A., Sharma, R., ... & Haythornthwaite, J. A. (2014). Meditation programs for psychological stress and well-being: A systematic review and meta-analysis. *JAMA Internal Medicine, 174*(3), 357–368. https://doi.org/10.1001/jamainternmed.2013.13018

Greyson, B. (2010). Seeing dead people not known to have died: "Peak in Darien" experiences. *Anthropology and Humanism, 35*(2), 159–171. https://doi.org/10.1111/j.1548-1409.2010.01079.x

Haidt, J. (2001). The emotional dog and its rational tail: A social intuitionist approach to moral judgment. *Psychological Review, 108*(4), 814–834. https://doi.org/10.1037 /0033-295X.108.4.814

Hameroff, S., & Penrose, R. (2014). Consciousness in the universe: A review of the 'Orch OR' theory. *Physics of Life Reviews, 11*(1), 39–78. https: //doi.org/10.1016/j.plrev.2013.08.002

Hamilton, J. P., Farmer, M., Fogelman, P., & Gotlib, I. H. (2015). Depressive rumination, the default-mode network, and the dark matter of clinical neuroscience. *Biological Psychiatry, 78*(4), 224–230. https://doi. org/10.1016/j.biopsych .2015.02.020

Hari, R., & Kujala, M. V. (2009). Brain basis of human social interaction: From concepts to brain imaging. *Physiological Reviews, 89*(2), 453–479. https://doi.org/10.1152/physrev.00 041.2007

Hassabis, D., Kumaran, D., Summerfield, C., & Botvinick, M. (2017). Neuroscience-inspired artificial intelligence. *Neuron, 95*(2), 245–258. https: //doi.org/10.1016/j.neuron.2017.06.011

Hinton, G. (2007). Learning multiple layers of representation. *Trends in Cognitive Sciences, 11*(10), 428–434. https://doi.org/10.1016/j.tics.2007 .09.004

Hochberg, L. R., Bacher, D., Jarosiewicz, B., Masse, N. Y., Simeral, J. D., Vogel, J., ... & Donoghue, J. P. (2012). Neuronal ensemble control of prosthetic devices. *Nature, 485*(7398), 372–375. https://doi.org/10.1038/nature11076

Hohwy, J. (2013). *The predictive mind.* Oxford University Press.

Ienca, M., & Andorno, R. (2017). Towards new human rights in the age of neuroscience and neurotechnology. *Life Sciences, Society, and Policy, 13*(1), 1–27. https://doi.org/10.1186/s40504-017-0050-1

Immordino-Yang, M. H., & Damasio, A. (2007). We feel, therefore we learn: The relevance of affective and social neuroscience to education. *Mind, Brain, and Education, 1*(1), 3–10. https://doi.org/10.1111/j.1751-22 8X.2007.00004.x

Insel, T. R. (2014). The NIMH Research Domain Criteria (RDoC) project: Precision medicine for psychiatry. *American Journal of Psychiatry, 171*(4), 395–397. https://doi.org/10.1176/appi.ajp.2014.14020138

John Paul II, Pope. (2000, August 29). *Address to the 18th International Congress of the Transplantation Society*. Vatican City. Retrieved from https://www.vatican.va/content/john-paul-ii/en/speeches/2000/jul-sep/documents/hf_jp-ii_spe_20000829_transplants.html

Kastrup, B. (2019). *The idea of the world: A multi-disciplinary argument for the mental nature of reality*. Iff Books.

Kemeny, M. E. (2003). The psychobiology of stress. *Current Directions in Psychological Science, 12*(4), 124–129. https://doi.org/10.1111/1467-8721 .01246

Khalsa, S. S., Adolphs, R., Cameron, O. G., Critchley, H. D., Davenport, P. W., Feinstein, J. S., ... & Zucker, N. (2018). Interoception and mental health: A roadmap. *Biological Psychiatry: Cognitive Neuroscience and Neuroimaging, 3*(6), 501–513. https://doi.org/10.1016/j.bpsc.2017.12.004

Kriegeskorte, N., & Douglas, P. K. (2018). Cognitive computational neuroscience. *Nature Neuroscience, 21*(9), 1148–1160. https://doi.org/10.1038/s41593-018-0210-5

Kriegeskorte, N. (2015). Deep neural networks: A new framework for modeling biological vision and brain information processing. *Annual Review of Vision Science, 1*, 417–446. https://doi.org/10.1146/annurev-vision-082114-035447

Kuhn, T. S. (1962). *The structure of scientific revolutions*. University of Chicago Press.

Laureys, S. (2005). Death, unconsciousness, and the brain. *Nature Reviews Neuroscience, 6*(11), 899–909. https://doi.org/10.1038/nrn1789

Lebedev, M. A., & Nicolelis, M. A. (2006). Brain-machine interfaces: Past, present and future. *Trends in Neurosciences, 29*(9), 536–546. https://doi.org/10.1016/j.tins.2006.07.004

Lehrer, P. M., & Gevirtz, R. (2014). Heart rate variability biofeedback. *Applied Psychophysiology and Biofeedback, 39*(3–4), 203–211. https://doi.org/10.1007/s10484-014-9266-8

Macaluso, E., & Driver, J. (2005). Multisensory spatial interactions: A window onto functional integration in the human brain. *Trends in Neurosciences, 28*(5), 264–271. https://doi.org/10.1016/j.tins.2005.03.008

Machado, C. (1995). A new definition of death based on the basic mechanisms of consciousness generation. In C. Machado (Ed.), *Brain death*. Elsevier.

Mane, R., Ang, K. K., & Guan, C. (2023). Brain-computer interface for stroke rehabilitation. In N. V. Thakor (Ed.), Handbook of neuroengineering (pp. [insert page numbers if available]). Springer. https://doi.org/10.1007/978-981-16-5540-1_33

McEwen, B. S. (2007). Physiology and neurobiology of stress and adaptation. *Physiological Reviews, 87*(3), 873–904. https://doi.org/10.1152/physrev.00041.2006

McGaugh, J. L. (2004). The amygdala modulates the consolidation of memories of emotionally arousing experiences. *Annual Review of Neuroscience, 27*, 1–28. https://doi.org/10.1146/annurev.neuro.27.07020 3.144157

McGurk, H., & MacDonald, J. (1976). Hearing lips and seeing voices. *Nature, 264*(5588), 746–748. https://doi.org/ 10.1038/264746a0

Melzack, R. (2001). Pain and the neuromatrix in the brain. *Journal of Dental Education, 65*(12), 1378–1382. https://doi.org/10.1002/j.0022-03372001.65 .12.tb03497.x

Mergner, T., Schweigart, G., Fennell, L., & Botti, F. (2009). Posture control in vestibular-loss patients. *Annals of the New York Academy of Sciences, 1164*(1), 206–215. https://doi.org/10.1111/j.1749-6632.2008.03736.x

Meyer, S. C. (2021). *Return of the God hypothesis: Three scientific discoveries that reveal the mind behind the universe*. HarperOne.

Miller, G. E., Chen, E., & Cole, S. (2009). Health psychology: Developing biologically plausible models linking the social world and physical health. *Annual Review of Psychology, 60*, 501–524.https://doi.org/10.1146/annurev.psych.60110707.16355 1

Millidge, B., Seth, A. K., & Buckley, C. L. (2021). Predictive coding: A theoretical and experimental review. *arXiv preprint* arXiv:2107.12979. https://arxiv.org/abs/2107.12979

Mossbridge, J., Tressoldi, P., & Utts, J. (2012). Predictive physiological anticipation preceding seemingly unpredictable stimuli: A meta-analysis. *Frontiers in Psychology, 3*, 390. https://doi.org/10.3389/fpsyg.2012.00390

Nagel, T. (2012). *Mind and cosmos: Why the materialist neo-Darwinian conception of nature is almost certainly false.* Oxford University Press.

Nijboer, F., Clausen, J., Allison, B. Z., & Haselager, P. (2013). The Asilomar survey: Stakeholders' opinions on ethical issues related to brain–computer interfacing. *Neuroethics, 6*(3), 541–578. https://doi.org/10.1007/s12152-011-9132-6

Park, H. D., & Tallon-Baudry, C. (2014). The neural subjective frame: From bodily signals to perceptual consciousness. *Philosophical Transactions of the Royal Society B: Biological Sciences, 369*(1641), 20130208. https://doi.org/10.1098/rstb. 2013.0208

Pascual-Leone, A., Amedi, A., Fregni, F., & Merabet, L. B. (2005). The plastic human brain cortex. Annual Review of Neuroscience, 28, 377–401. https://doi.org/10.1146/annurev.neuro.27.070203.144216

Penrose, R. (2004). *The road to reality: A complete guide to the laws of the universe.* Vintage.

Pelphrey, K. A., Shultz, S., Hudac, C. M., & Vander Wyk, B. C. (2011). Research review: Constraining heterogeneity: The social brain and its development in autism spectrum disorder. *Journal of Child Psychology and Psychiatry, 52*(6), 631–644. https://doi.org/10.1111/j.1469-7610.2010.02349.x

Pessoa, L. (2013). *The cognitive-emotional brain: From interactions to integration.* MIT Press.

Pezzulo, G. (2017). Tracing the roots of cognition in predictive processing. In T. Metzinger & W. Wiese (Eds.), *Philosophy and predictive processing: 20.* MIND Group. https://doi.org/10.15502/9783958573215

Piszcz, A., Rojek, I., & Mikołajewski, D. (2024). *Impact of virtual reality on brain–computer interface performance in IoT control—Review of current state of knowledge. Applied Sciences, 14*(22), 10541. https://doi.org/10.3390/app142210541

President's Commission for the Study of Ethical Problems in Medicine and Biomedical and Behavioral Research. (1981). *Defining death: Medical, legal, and ethical issues in the determination of death.* Government Printing Office. Retrieved from https://repository.library.georgetown.edu/bitstream/handle/10822/559345/defining_death.pdf?sequence=1

President's Council on Bioethics. (2008). *Controversies in the determination of death: A white paper by the President's Council on Bioethics*. Retrieved from https://bioethicsarchive.georgetown.edu/pcbe/reports/death/

Radin, D. I. (1997). *The conscious universe: The scientific truth of psychic phenomena*. HarperEdge.

Rowlands, M. (2010). *The new science of the mind: From extended mind to embodied phenomenology*. MIT Press. https://doi.org/10.7551/mitpress/9780262014557.001.0001

Sadato, N., Pascual-Leone, A., Grafman, J., Ibañez, V., Deiber, M. P., Dold, G., & Hallett, M. (1996). Activation of the primary visual cortex by Braille reading in blind subjects. *Nature, 380*(6574), 526-528. https://doi.org/10.1038/ 380526a0

Salvatori, T., Mali, A., Buckley, C. L., Lukasiewicz, T., Rao, R. P. N., Friston, K., & Ororbia, A. (2023). A survey on brain-inspired deep learning via predictive coding. *arXiv preprint* arXiv:2308.07870. https://arxiv.org/abs/2308.07870

Saxe, R., & Kanwisher, N. (2003). People thinking about thinking people: The role of the temporoparietal junction in "theory of mind." *NeuroImage, 19*(4), 1835–1842. https://doi.org/10.1016/S1053-8119(03)00230-1

Schilbach, L., Eickhoff, S. B., Rotarska-Jagiela, A., Fink, G. R., & Vogeley, K. (2008). Minds at rest? Social cognition as the default mode of cognizing and its putative relationship to the "default system" of the brain. *Consciousness and Cognition, 17*(2), 457–467. https://doi.org/10.1016/j.concog. 2008.03.013

Searle, J. R. (1992). *The rediscovery of the mind*. MIT Press.

Seeley, W. W., Crawford, R. K., Zhou, J., Miller, B. L., & Greicius, M. D. (2009). Neurodegenerative diseases target large-scale human brain networks. *Neuron, 62*(1), 42–52. https://doi.org/10.1016/j.neuron.2009.03.024

Segerstrom, S. C., & Miller, G. E. (2004). Psychological stress and the human immune system. *Psychological Bulletin, 130*(4), 601–630. https://doi.org/10.1037/0033-2909.130.4.601

Seth, A. (2013). Interoceptive inference, emotion, and the embodied self. *Trends in Cognitive Sciences, 17*(11), 565–573. https://doi.org/10.1016/j.tics.2013.09.007

Seth, A. (2021). *Being you: A new science of consciousness.* Faber & Faber.

Seth, A. K., & Tsakiris, M. (2018). Being a beast machine: The somatic basis of selfhood. *Trends in Cognitive Sciences, 22*(11), 969–981. https://doi.org/10.1016/j.tics.2018.08.008

Soto-Faraco, S., Lyons, J., Gazzaniga, M., Spence, C., & Kingstone, A. (2002). The ventriloquist in motion: Illusory capture of dynamic information across sensory modalities. *Cognitive Brain Research, 14*(1), 139-146. https://doi.org/ 10.1016/S0926-6410(02)00068-X

Spinoza, B. (1677/1985). *Ethics* (E. Curley, Trans.). Princeton University Press. (Original work published 1677)

Sporns, O. (2010). Networks of the brain. The MIT Press. https://doi.org/10.7551/mitpress/8476.001.0001

Thayer, J. F., & Sternberg, E. M. (2006). Beyond heart rate variability: Vagal regulation of allostatic systems. *Annals of the New York Academy of Sciences, 1088*, 361–372. https://doi.org/10.1196/annals.1366.014

Thompson, E., & Varela, F. J. (2001). Radical embodiment: Neural dynamics and consciousness. *Trends in Cognitive Sciences, 5*(10), 418–425. https://doi.org/10.1016/S1364-6613 (00)01750-2

Tononi, G., & Edelman, G. M. (1998). Consciousness and complexity. *Science, 282*(5395), 1846–1851. https://doi.org/10.1126/science.282.5395.1846

Tononi, G. (2008). Consciousness as integrated information: A provisional manifesto. *Biological Bulletin, 215*(3), 216–242. https://doi.org/10.2307/25470707

Treffert, D. A. (2009). The savant syndrome: An extraordinary condition. A synopsis: Past, present, future. *Philosophical Transactions of the Royal Society B: Biological Sciences, 364*(1522), 1351–1357. https://doi.org/10.1098/rstb.2008.0326

Tsakiris, M., & De Preester, J. (Eds.). (2018). *The interoceptive mind: From homeostasis to awareness.* Oxford University Press.

Tulving, E. (1983). *Elements of episodic memory.* Clarendon Press.

Uddin, L.Q., Iacoboni, M., Lange, C., & Keenan, J.P. (2007). The self and social cognition: the role of cortical midline structures and mirror neurons. Trends in Cognitive Sciences, 11, 153-157.

van Lommel, P. (2010). *Consciousness beyond life: The science of the near-death experience.* HarperCollins.

Veatch, R. M. (2005). The death of whole-brain death: The plague of the disaggregators, somaticists, and mentalists. *Journal of Medicine and Philosophy, 30*(4), 353–378. https://doi.org/10.1080/03605310591008504

Vroomen, J., & de Gelder, B. (2004). Perceptual effects of cross-modal stimulation: The cases of ventriloquism and the freezing phenomenon. In G. Calvert, C. Spence, & B. E. Stein (Eds.), *Handbook of multisensory processes* (pp. 141–150). MIT Press.

Yuste, R., Goering, S., Arcas, B. A. Y., Bi, G., Carmena, J. M., Fins, J. J., ... & Wolpaw, J. (2017). Four ethical priorities for neurotechnologies and AI. *Nature, 551*(7679), 159–163. https://doi.org/10.1038/551159a

Youngner, S. J., & Bartlett, E. T. (1983). Human death and high technology: The failure of the whole-brain formulations. *Annals of Internal Medicine, 99*(2), 252–258. https://doi.org/10.7326/0003-4819-99-2-252

Zander, T. O., Krol, L. R., Birbaumer, N. P., & Gramann, K. (2016). Neuroadaptive technology enables implicit cursor control based on medial prefrontal cortex activity. *Proceedings of the National Academy of Sciences, 113*(52), 14898–14903. https://doi.org/10.1073/pnas.1605155114

Ziemke, T. (2016). The body of knowledge: On the role of the living body in grounding embodied cognition. *Biosystems, 148*, 4–11. https://doi.org/10.1016/j.biosystems.2016.08.005

Złotowski, J., Yogeeswaran, K., & Bartneck, C. (2017). Can we control it? Autonomous robots threaten human identity, uniqueness, safety, and resources. *International Journal of Human-Computer Studies, 100*, 48–54. https://doi.org/10. 1016/j.ijhcs.2016.12.008

20

The Human Soul and the Brain Metafunction

What is the nature of the soul? How does it relate to consciousness, the body, human behavior, and decision-making? Does it survive death? These questions have captivated human thought for millennia, inviting profound reflection on existence and the essence of being. From the philosophical musings of Plato and Aristotle to the theological inquiries of St. Thomas Aquinas, the soul has remained central to debates about identity, morality, and the divine. While traditional perspectives often emphasize the soul's immortality and its relationship to the body, materialist thinkers of the last two centuries have dismissed the soul as an outdated concept, favoring explanations rooted in physical matter—particularly the brain. Despite this skepticism, the soul challenges our understanding of human nature, offering insights into the intangible yet manifest essence that shapes our identity and humanity.

Throughout this book, we have examined the brain metafunction framework, revealing a transphysical dimension in attributes of the mind—such as consciousness, qualia, intentionality, memory, and free will—that defy reduction to purely physical processes. These facets of human experience hint at a reality that transcends the brain's material structure, pointing toward the profound depth of the human soul. In this chapter, we will delve deeper into this relationship, situating the soul within the philosophical and theological context of Christian Theism. By framing the soul as central to the transphysical qualities of the mind, we aim to present a unified view of human experience—one that acknowledges our transcendent nature and illuminates the soul's vital role in shaping consciousness, identity, and the unity of the person.

Christian Anthropology

The Unity of the Human Person Does Not Imply Uniformity.

The chapter on "The Unity of the Human Person" explored the nature of unity, which presupposes the existence of various dimensions within a single being. In theological anthropology, some traditions propose a tripartite view of human nature, comprising spirit, soul, and body (Origen 1979; cf. *De Principiis*). Others, including Aquinas, argue for a bipartite view, distinguishing between spirit and matter—configured as a body-soul unity—while acknowledging the spirit as a faculty or aspect of the soul (*Summa Theologiae*, I, Q75-76). The immaterial soul, without parts, is understood to possess capacities or powers rather than distinct components. This allows it to manifest the person's essence across different dimensions. Without the soul, as the *Catechism of the Catholic Church* (1994) notes, the body becomes a corpse, a mere material form.

The Apostle Paul's writings reflect this unity through the lens of Semitic anthropology, which views the human person as an indivisible whole while recognizing distinct dimensions. Paul identifies three existential categories based on how a person responds to divine teaching: the spiritual person (*pneumatikos*), the soulish person (*psychikós*), driven by intellect, and the carnal or fleshy person (*sarkivós*) (1 Corinthians 2:12–3:4). Each category is governed by the dominant aspect within the individual—spirit, soul, or flesh. In this view, the spirit functions as a crucial faculty of the soul, explaining the Bible's often interchangeable use of "soul" and "spirit," while preserving a nuanced distinction when both terms appear together. As Paul writes, "May God himself, the God of peace, sanctify you through and through. May your whole spirit, soul, and body be kept blameless at the coming of our Lord Jesus Christ" (1 Thessalonians 5:23). The writer of Hebrews further delineates the interconnectedness of soul and spirit: "For the word of God is alive and active...it penetrates even to dividing soul and spirit, joints and marrow; it judges the thoughts and attitudes of the heart" (Hebrews 4:12).

Christian Anthropology: Scripture, the Significance of the Heart, and Neuroscience

Scripture frequently highlights humanity's unified dimensions, mainly through the Hebrew term *lēb* (heart), seen as the center that encompasses the will, intellect, emotions, and spirit (Barr 1961). The psalmist captures this unity, proclaiming, "Behold, You desire truth in the innermost being, And in the hidden part You will make me know wisdom" (Psalm 51:6). Proverbs underscores the intellect's connection to identity with, "As he thinks in his heart, so is he" (Proverbs 23:7). Likewise, Daniel's resolve, "Daniel purposed in his heart" (Daniel 1:8), reflects the link between the will and the heart. In the New Testament, Jesus' statement, "Let not your heart be troubled" (John 14:1), associates *kardia* with emotions, while Hebrews further specifies "the thoughts and intents of the heart" (Hebrews 4:12). This connection resonates with Augustine's reflections in *Confessions* (2002), where he identifies the heart as the locus of spiritual and emotional unity. He observes, "You have made us for yourself, and our heart is restless until it rests in you," encapsulating the soul's desire for divine harmony and wholeness. Augustine portrays restlessness as a spiritual condition—human longing misdirected until reoriented toward God. Aquinas expands this understanding in *Summa Theologiae* by framing the soul as the substantial form of the body, with faculties such as intellect and will uniting spiritual and physical aspects in a cohesive whole (I, Q75, Art. 6).

The human person is a unified whole, composed of both material and spiritual principles, as suggested in several Bible passages indicating that death involves the departure of the animating principle of life, highlighting the profound connection between the body and spirit as essential aspects. Scripture presents the human person as a single, unified substance brought into being through the union of matter and spirit. Death, then, is not the division of two distinct things but the disintegration of this unity, as the life-giving spirit departs while the body returns to the earth. Genesis 2:7 (English Standard Version) illustrates this relationship, describing how God formed man from dust and breathed into him the breath of life, making him a living being. This passage suggests that the human person is not merely "physical matter" but a unified reality composed of corporeal and spiritual principles. Ecclesiastes 12:7 reinforces this idea, stating that at death, "the dust returns to the earth as it was, and the spirit returns to God who gave it." Here, death is understood as the loss of the soul's vital presence, leading to the body's dissolution.

The New Testament further emphasizes this understanding. In Luke 23:46, Jesus, in his final moments, proclaims, "Father, into your hands I commit my spirit!" Similarly, Acts 7:59 records Stephen's dying words, "Lord Jesus, receive my spirit." These passages reveal that life is not sustained by the body alone but by the presence of the spirit, which, once withdrawn, results in death. James 2:26 makes this explicit, stating, "For as the body apart from the spirit is dead, so also faith apart from works is dead." This analogy underscores the idea that the human person exists as a single, embodied reality, dependent on both matter and spirit for life. Jesus affirms this unity in Matthew 10:28, warning, "Do not fear those who kill the body but cannot kill the soul." This verse acknowledges that human life is more than physical existence and does not describe two separate beings. There is a spiritual principle that sustains and defines it. These passages present a view of the human person as a single, integrated being, created from matter and spirit as co-principles of life. In this framework, death is not the separation of two substances but the dissolution of the harmonious unity that makes a person fully alive. It highlights that death does not destroy the person but dissolves the bodily-spiritual unity, awaiting its ultimate restoration.

The book of Proverbs echoes the importance of the heart as a unifying aspect: "Above all else, guard your heart, for everything you do flows from it" (Proverbs 4:23). Peter also emphasizes this unity, referring to "the hidden person of the heart" (1 Peter 3:3–4). This profound interconnection of human faculties aligns with contemporary neuroscience, which suggests an integrated model of human identity and mind. Neuroscientists like Antonio Damasio (1999) argue that the brain's processing of emotions and consciousness reflects a deeply intertwined relationship between physical and non-physical elements of human experience. This insight aligns with theological anthropology, where the brain metafunction framework illustrates how the unity of the human person—encompassing spirit, soul, and body—resonates with both theological and scientific perspectives. Through this integration of theological, philosophical, and contemporary scientific approaches, the unity of the human person is revealed as a complex but cohesive reality, bridging material and immaterial dimensions. This model underscores the profound depth of Christian anthropology and highlights the enduring relevance of theological insights in modern discourse.

Thomistic Principles

Thomas Aquinas regarded substances as unified entities of the highest order within his metaphysical framework. In his ontology, substances are cohesive wholes with intrinsic unity and identity, distinguishing them from mere aggregates or collections of parts (Aquinas 1947; Wippel 2000). Substances are irreducible to lower-level entities, embodying a high degree of integration within the metaphysical hierarchy. This intrinsic integrity allows substances to maintain their essential nature despite undergoing accidental changes, reflecting their dynamic unity over time (Klubertanz 1953; Stump 2003). Central to Thomistic philosophy is the human person, understood as a unified whole composed of corporeal (material) and incorporeal (immaterial) aspects. The soul, as the substantial form of the body, unifies and animates the entire person (Aquinas 1947; McInerny 1999). The soul provides the organizing principle and faculties that endow humans with identity and intellectual capacities, aligning with Aristotelian metaphysics, where form gives matter structure and purpose (Aristotle 1987). Within this framework, the soul's immaterial nature is essential to the body, its configuration, and its function while providing a metaphysical foundation, as explored in contemporary discussions (Moreland 2018). This unity of body and soul underscores the human person's indivisibility as "unum simpliciter"—a single, unified being (Gilson 1952).

Aquinas describes the soul as incorporeal, subsistent, and incorruptible. As a spiritual substance, the soul exists independently of the body yet remains naturally oriented toward it (Aquinas 1947). While subsistent, the soul is incomplete without the body, as it achieves its full expression through sensory perception, physical actions, and intellectual faculties. The soul's incorruptibility further distinguishes it from material entities, affirming its exemption from decay and capacity to persist beyond physical death (Stump 2003). These qualities highlight the soul's unique ontological status, intimately connected to the body while retaining independence. Aquinas's nuanced understanding of the human person offers insights into contemporary Neo-Thomistic perspectives (Des Chene 2000; Pasnau 2001, 2011; Moreland 2018) and interdisciplinary discussions, including the brain metafunction framework. This synthesis of metaphysical principles with modern scientific insights enriches the understanding of human identity, emphasizing the integral unity of body and soul as created in the image of God (Kenny 2002).

The Brain Metafunction Framework

Concept of the Human Soul:

The human soul is an immaterial, transphysical principle that actualizes the body, serving as the metaphysical foundation for the brain's metafunctional capacities. As the substantial form of the body, it provides unity, identity, developmental guidance, and continuity to the organism as a parts-preceding whole, integrating the human person into a cohesive living being.

Brain Metafunction Transphysicalism is an original concept introduced by the present author to describe a framework that integrates Thomistic and neo-Thomistic metaphysical principles with insights from contemporary neuroscience. This framework posits that the soul — as the substantial form of the body — operates through the brain's metafunction and the entire body, enabling consciousness, perception, and both physiological and transphysical phenomena. By emphasizing the unity of soul, mind, and body, *Brain Metafunction Transphysicalism* offers a comprehensive model for understanding human experience that transcends materialist explanations.

Integrated Embodiment and Cognitive Capacities

Emotions, intimately tied to bodily experiences, are expressed through gestures, physiological responses, and facial expressions. This integrated sense of embodiment aligns the body and mind, regulating emotional states and enabling coherent, goal-directed action (Craig 2003). Cognitive functions such as memory, attention, and executive processes further shape embodied experiences, supporting decision-making, problem-solving, and self-awareness (Fuchs 2018). Beyond embodiment, abstract conceptualization highlights the soul's higher capacities, allowing humans to synthesize complex ideas and construct frameworks for understanding reality (McGilchrist 2009; Spitzer 2015). These intellectual capacities, grounded in self-consciousness, enable reflection on thoughts and actions, fostering moral deliberation. Guided by conscience, the soul embodies God's moral presence within the self, imparting a sense of ethical responsibility and directing actions toward love and goodwill (Newman 1874/1999).

Transphysical Dimensions of the Brain Metafunction

The soul's immaterial nature ensures its metaphysical independence while actively unifying and animating the body as its substantial form. Under normal conditions, the soul operates through the brain's metafunctional mediation. However, in exceptional circumstances such as near-death experiences, the soul demonstrates its capacity to operate independently of physical mediation (Beauregard & O'Leary 2007; Greyson 2003; Holden 2009; Kelly et al. 2000, 2007; Van Lommel 2010). These phenomena illustrate the soul's enduring influence as the unifying reality behind consciousness and bodily processes. The brain's metafunction bridges biological processes with transphysical realities by integrating material and immaterial dimensions, grounding consciousness in the human person as a unified whole (Aquinas 1947; Gilson 1952). For instance, perception is a privileged domain within this framework, blending neural mechanisms with transphysical qualities. This process reveals the soul's active role in mediating between sensory input and the subjective phenomenology of experience. Brain Metafunction Transphysicalism provides a comprehensive conceptual framework for contextualizing consciousness, perception, and spiritual experiences. It enriches our understanding of the human person, merging theological insights with advances in neuroscience. By affirming the unity of the human person as a single substance composed of body and soul, this framework highlights the brain's critical role in expressing the transphysical domain. Ultimately, this framework offers a parsimonious yet profound approach that bridges metaphysical insights and empirical research, advancing our understanding of human nature across disciplines.

Human Beings' Constitution is One Substance

A central tenet of Christian anthropology and Thomistic philosophy (discussed in *The Unity of the Human Person*) is that human beings are one unified substance, not two. This unity accommodates a differentiation of dimensions and faculties without dividing them into separate entities. Within this framework, the spirit is understood as a faculty of the soul. Thus, the term "one substance" signifies that humans are not composed of distinct, independent entities such as soul and body but are instead a cohesive whole in which soul and body are intimately connected. Thomas Aquinas emphasized this unity, asserting that humanity is not a composite of separate parts but a singular, complete substance. He described the soul as the substantial form of the body, animating and directing it without existing separately (Aquinas, Summa Theologica, I, q. 76, a. 1; Gilson 1952). Aquinas clarified this principle, explaining: "Body and soul are not two actually existing substances, but from these two is made one actually existing substance." He further noted: "The body is not actually the same in the soul's presence and absence, but the soul makes it exist actually" (Summa Theologica, I, q. 76, a. 1).

This unified constitution implies that humans are not merely collections of parts but an integrated being with interrelated faculties. This perspective has profound implications for understanding personhood and the integrated nature of human experience (Stump, 2003). Although Aquinas did not equate the brain with consciousness or self-awareness as modern neuroscience does, he acknowledged its critical role in sensory coordination, which supports intellectual activity. He recognized the brain's function in processing sensory information as essential for cognition and facilitating the soul's expression of intellect and will (Pasnau, 2001). While no part of the body relates to bodily integrity in the same way as the head, Aquinas maintained that it is the soul, as the form of the body, that unifies the human person, with the brain complementing this unity. Regarding respiration, Aquinas clarified that while necessary for life, respiration is not the means of union between body and soul. Instead, respiration conditions the body for its union with the soul (Des Chene, 2000). This distinction highlights the soul's role as the animating principle, with the brain as a vital organ that enables the soul to interact with the physical world.

While Aquinas maintained that the soul unifies the human person as its substantial form, he recognized the brain's vital role in coordinating sensory information and supporting imaginative processes essential for cognition. The brain serves as an instrument through which the soul gathers the material data required for intellectual activity, while the faculties of intellect and will remain immaterial powers rooted in the soul itself (Summa Theologica, I, q. 84, a. 7). Aquinas emphasized that the intellect depends on the imagination, a sensory power tied to the brain and bodily processes. Thus, proper bodily functioning, including a "certain disposition of the brain," is necessary for the intellect to operate effectively in this life (Summa Theologica, I, q. 84, a. 7; Summa Theologica, I, q. 75, a. 2). This understanding underscores that while the brain mediates sensory input and imaginative processes, the intellect and will transcend purely material functions, belonging to the soul as its immaterial faculties.

Aquinas further elaborated that the human soul is not merely joined with the body in fact; rather, it is the kind of soul which, though capable of separate existence, nevertheless by its nature presupposes union with a body — a body with organs — to exercise all its powers, even reason (Des Chene, 2000). This illustrates that human intellectual activities, though ultimately immaterial, rely on sensory and bodily processes when the soul is united with the body. These insights underscore that a soul without bodily expression is incomplete; a human being is fundamentally a unified substance. This unity is not merely an amalgamation of parts but an intrinsic essence of the human condition, encompassing a dynamic, transphysical complexity. The soul's immaterial essence is fully present throughout the body, guiding and unifying the individual at every level of being. This intrinsic unity of body and soul defines the human person, aligning with Thomistic principles and contemporary frameworks like the Brain Metafunction Framework.

The Brain Metafunction Results from and Expresses a Primitive, Transphysical Domain: The Human Soul.

The Brain Metafunction Framework posits that the brain's metafunction results from and expresses a primitive, transphysical domain: the human soul. Drawing from Neo-Thomist principles, this framework emphasizes the soul's intrinsic nature and organizing role in the body's development (Moreland 2014, 2018). The soul embodies a unified arrangement of capacities and dispositions that shape the body's formation and growth. This configuration permeates the body holomerically—fully present in every part without being spatially divided—acting as the primary principle of activity. Unlike physical substances that occupy space, the soul has no measurable dimensions; instead, it functions as the intrinsic principle that directs the body's vital processes. It guides the sequential changes required for growth and development, orchestrating bodily processes according to its intrinsic design. Organs, including the brain, function as instruments sustaining bodily operations that are rooted in the soul. Thus, the soul is not merely a passive presence but an active principle that shapes and sustains the body's integration and coherence. Acting as both design and blueprint, the soul contains the essential information that regulates bodily structure and function, ensuring purposeful organization (Gilson 1952; Des Chene 2000).

From a teleological perspective, the soul directs the body's developmental trajectory toward a specific end. This purpose-driven orientation governs the orderly progression of growth and maturation, reflecting the soul's inherent essence. Although the soul can exist independently of the body, the body's vitality and integration rely on its constant presence. This dual aspect highlights the unity of the physical and transphysical dimensions of human nature, affirming the body as both material and immaterial (Oderberg 2005; Moreland 2018; Aquinas, Summa Theologica, I, q. 76, a. 1). The soul's pervasive presence throughout the body underscores its role as the animating principle, harmonizing physical and transphysical aspects into a cohesive whole. By informing and animating every facet of the body, the soul distinguishes the human organism as a purposeful and living entity. This intrinsic unity reflects the profound relationship between soul and body, wherein the soul imparts its unique nature and guides the individual toward fulfillment.

The Brain Metafunction Level of Unity and the Expression of the Human Soul's Capacities and Dispositions.

The soul embodies a fundamental unity expressed through a complex structural arrangement of capacities and dispositions. These capacities guide the body's formation, with the brain serving as the highest biological expression of the soul's influence. The Brain Metafunction Framework offers insight into how the human soul's capacities are expressed through the brain's complex integrative processes. The brain metafunction, defined in earlier chapters, refers to the dynamic, unified activity of brain systems, sub-regions, and neurons, interlinked through structural and functional connectivity. It integrates local-global dynamics across different spatial and temporal scales. Through a unique, inherent causal connection to a transphysical domain, the brain metafunction unifies the following key subfunctions:

1. Consciousness and cognition, including perception, attention, memory, and emotions;
2. Neural integration and regulation of the organism's communication systems (nervous, endocrine, immune, and circulatory);
3. Homeostatic and allostatic regulation of the internal environment;
4. Coordination of afferent input and efferent responses across the organism and in interaction with the environment;
5. Behavioral adaptation and environmental interaction;
6. Synthesis of physiological processes with mental states;
7. Unified integration of the previous subfunctions.

The unified coordination of these subfunctions creates an irreducible and cohesive whole, representing the body's highest level of integration. In this context, separating the person from the human organism is impossible. The brain metafunction reflects the profound unity and structural complexity inherent in the metaphysical nature of the soul. As the primary organ of the human body, the brain's metafunctional activities create a higher-level operation irreducible to the sum of its parts. Permanent and irreversible loss of the brain's metafunction results in the disintegration of personhood, marking death as the definitive severance from the transphysical domain (García, 1995a, 1995b, 1998a, 1998b, 2001).

Recent neuroscientific developments increasingly validate the Brain Metafunction Framework, demonstrating that consciousness is inseparable from the brain's broader integrative processes. Conditions involving partial loss of function—such as coma, vegetative states, and dementia—illustrate severe but incomplete damage to the brain's metafunction. Complete and irreversible destruction of this function results in a definitive severance from the transphysical domain, marking the disintegration of personhood. The Brain Metafunction Framework addresses complex philosophical and theological issues, such as personal identity, consciousness, and the mind-body relationship. By illuminating the interaction between the physical brain and the transphysical aspects of human existence, it deepens our understanding of the soul's integral role in human life. This framework bridges the physical and spiritual dimensions of human existence, offering valuable insights into neuroscience's broader theological implications.

Perception is a Privileged Window into the Transphysical Quality of Consciousness

Perception offers a distinctive window into the transphysical quality of consciousness, as explored in the chapter on Perception. This process encompasses three primary components:

- *Afferent input*: The brain's reception of sensory data from the body and environment, modulated by top-down processes.
- *Metafunctional processing*: The brain's integration, modulation, and predictive mechanisms that shape perception.
- *Efferent output*: The subjective, phenomenological experience of perception, reflecting consciousness's transphysical dimension.

The efferent loop of visual perception, especially its outward-directed nature, demonstrates how consciousness actively integrates internal and external dimensions. Unlike purely biological functions, this loop reflects a psychological projection, or "extended mind," existing both within the mind and phenomenologically in the external world (Velmans 2000, 2008; Sheldrake 2013). Vision, the most studied sensory modality, illustrates how perception bridges afferent input, brain processing, and conscious experience. Afferent input provides the first link between the physical and transphysical domains, integrating sensory data from the body and the environment. Brain metafunctional processing then filters this data, combining it with prior knowledge and expectations. While some theories localize perceptual "representations" exclusively in the brain, this framework asserts that the efferent loop of visual perception includes a transphysical component, and representations reside within the mind itself beyond the brain, underscoring the mind's irreducibility to brain processes alone.

The efferent output, the subjective experience of perception, offers the most apparent evidence of its transphysical dimension. William James highlighted this paradox, noting that perception appears to exist in "two places at once," both in external reality and within the mind (James 1890). This dual aspect, often overlooked in materialist theories, emphasizes the interplay between sensory input, brain processing, and conscious phenomenology. Through perception, we engage with the external world while simultaneously integrating that experience into our subjective awareness. Contemporary thinkers have expanded on this outward-directed view of perception. Max Velmans describes perception as "psychological projection," situating perceptual experience where it appears rather than confining it to the brain. Alva Noë emphasizes perception as an active, embodied interaction between the individual and the environment, while Rupert Sheldrake extends this notion, proposing that consciousness interacts with external reality through dynamic fields (Velmans 2000; Noë 2009; Sheldrake 2005, 2013).

The Brain Metafunction Transphysicalism framework conceptualizes perception as a unified act of the soul engaging the whole person with the world. By resolving James's paradox, this view underscores perception as a transphysical process that connects internal consciousness with external reality. In this model, perception is not merely a neural representation but a relational and embodied act of awareness—one that actively reveals the soul's engagement with the world. Through its integration of sensory input, brain processing, and phenomenological experience, perception provides profound insights into the nature of consciousness itself.

In Principle, a Naturalist, Physicalist Framework Only Partially Explains Brain Function.

The physicalist framework, grounded in the premise that all mental phenomena are ultimately reducible to physical processes, offers only a limited explanation of brain function. While physicalism has demonstrated significant adequacy in identifying neural correlates and mechanisms, it faces serious challenges in accounting for key phenomena such as consciousness, qualia, intentionality, abstract thinking, memory, the binding problem, savant abilities, and near-death experiences. These persistent gaps collectively weaken the philosophical foundation of physicalism. While previous chapters have explored some of these issues in depth, this section revisits key aspects to underscore the framework's limitations and strengthen the case for the soul.

Consciousness, Qualia, and Intentionality

Consciousness remains one of the most enduring challenges for physicalism, particularly its subjective, first-person nature. While physicalism offers insights into correlations between brain activity and conscious experiences, it cannot adequately explain the origin or nature of subjective qualia—the personal qualities of sensory experiences, such as the redness of red or the sweetness of sugar. David Chalmers (1995) famously described this as the "hard problem" of consciousness, emphasizing that understanding the qualitative aspects of experience cannot be reduced to an objective analysis of brain processes. The concept of qualia underscores the inherent explanatory gap in physicalist accounts (Levine 1983). Intentionality—the mind's capacity to be "about" or directed toward something—presents another major difficulty for physicalism (Brentano 1874/1995). Intentionality allows thoughts to reference objects, events, or concepts outside the mind. However, physical processes within the brain lack intrinsic directedness, relying instead on causal interactions that do not inherently relate to external referents. Memory further complicates this challenge by introducing what philosophers such as Husserl (1931/1964), Searle (1983) and Tallis (2004, 2010, 2011) have described as "double intentionality"—the ability to preserve past experiences while reinterpreting them meaningfully in the present. This dual representation reveals the inadequacy of purely reductionist explanations, as memory's contextual and reconstructive nature exceeds the limits of physicalist models (Schacter 1996; Perler 2022).

While physicalist frameworks can describe neural correlates of consciousness and sensory mechanisms, they fail to account for the qualitative, intentional, and subjective dimensions of mental experience. For example, the binding problem—how the brain integrates information from different sensory modalities into a unified experience—remains unresolved within physicalist paradigms. Similarly, exceptional phenomena such as savant abilities defy simple physicalist explanations, suggesting cognitive capacities that surpass conventional neural processing models. These limitations point to the need for frameworks that transcend the reductionist assumptions that dominate neuroscience and philosophy of mind.

Language: Creativity, Abstract Representation, and the Limits of Physicalism

Human language presents a profound challenge to physicalist explanations of cognition. Its remarkable productivity—the ability to generate and comprehend an infinite range of novel and meaningful expressions—cannot be adequately reduced to physical processes such as memorization or imitation (Chomsky, 1965; Spitzer, 2015). This generative creativity implies underlying cognitive mechanisms that transcend mechanistic brain functions. Language also enables abstract representation, allowing individuals to conceptualize ideas beyond direct sensory input. Humans regularly reason about hypothetical scenarios, metaphysical concepts, moral dilemmas, and mathematical structures. These capabilities, central to human thought, suggest mental faculties that are difficult to reconcile with a purely materialist framework (Chomsky, 2000). Language acquisition further reinforces this view. Children acquire language rapidly, across varied environments and without explicit instruction. The apparent universality of grammatical structures and the child's ability to produce grammatically correct sentences despite limited input support Chomsky's theory of Universal Grammar (UG), a biologically endowed language faculty. The poverty of the stimulus argument underscores that environmental exposure alone cannot account for the depth of linguistic competence, pointing to innate cognitive structures beyond what empirical models can explain (Chomsky, 1965).

Moreover, language's intentionality—the "aboutness" of words and symbols—remains a core philosophical problem for physicalism. Electrochemical brain states lack intrinsic meaning, yet language is inherently meaningful (Brentano, 1874/1995; Searle, 1983; Tallis, 2024). The subjective experience of meaning, metaphor, and irony further complicates reductionist models, as such phenomena remain resistant to quantification or neural localization (Nagel, 1974; Velmans, 2000). These complexities highlight the need for alternative or expanded explanatory frameworks. While cognitive and functionalist theories seek to enhance physicalist accounts by incorporating symbolic reasoning and intentional states, they still fall short of explaining language's full richness. Even Chomsky, while working within a naturalist paradigm, emphasized that language acquisition and creativity defy conventional physicalist explanation (Chomsky, 2000). Recognizing these limitations, transphysical perspectives may offer a more comprehensive account of language. By acknowledging non-material dimensions of cognition—including innate intentionality and subjective awareness, such frameworks better capture the integrative, creative, and conscious nature of human linguistic ability. Ultimately, understanding language may require a synthesis of physical, cognitive, and transphysical insights into the full scope of the human mind and meaning.

Abstract Thinking: Beyond the Physical

Abstract thinking encompasses reasoning, conceptualizing, and engaging in symbolic thought processes that transcend immediate sensory experience. While physicalist frameworks account for some neural correlates of cognitive processes, they face significant challenges in explaining how purely physical interactions within neurons and synapses give rise to the capacity for abstract reasoning and the understanding of non-physical concepts. As previously discussed, abstract thought manifests in language but also extends to non-symbolized forms, adding depth and complexity to human cognition. This capacity enables humans to grapple with concepts such as infinity, justice, and existence—ideas that lack direct physical counterparts. Such abilities suggest cognitive dimensions that exceed the explanatory power of naturalism, which asserts that all phenomena arise solely from physical causes (Nagel 1986).

The human capacity for abstraction is evident in activities such as formulating complex mathematical theories, engaging in philosophical inquiry, appreciating profound artistic expressions, and contemplating spiritual or existential insights. These activities point to a realm of cognition that physicalism struggles to explain comprehensively (Chalmers 1995; Searle 1983). For example, the ability to conceive and apply mathematical principles, such as infinity or multidimensional spaces, occurs in a purely abstract realm. Similarly, philosophical reasoning about metaphysical questions or moral judgments engages cognitive faculties that resist reduction to material processes alone. Robert Spitzer emphasizes that abstract thinking, self-reflection, and moral reasoning point to a transphysical aspect of the mind that transcends mere brain activity. He argues that awareness of ultimate meaning and purpose cannot be easily reconciled with a strictly materialist perspective, instead suggesting a dimension beyond physical processes (Spitzer 2015).

This transphysical component may underlie the human capacity for abstraction, enabling symbolic reasoning and profound introspection. The richness of abstract thinking reflects a cognitive realm not entirely explicable within a physicalist paradigm, further supporting the existence of a non-material dimension of thought. Recognizing the transphysical aspects of cognition provides a broader framework for understanding abstract thinking. This perspective acknowledges the interplay of physical and non-physical dimensions in human cognitive processes, enriching our understanding of human reasoning and creativity. By embracing this integrated view, we can better appreciate abstract thinking as a defining feature of human cognition (Moreland 2018).

Memory: A Challenge to Physicalism

Memory involves encoding, storing, and retrieving information patterns within the brain, yet its mechanisms go far beyond simple physical storage. While physicalist frameworks can address certain aspects of memory—such as the formation of memory-associated patterns or engrams—they struggle to explain the full depth and complexity of memory function. The precise localization of specific memories in the brain remains elusive and is often based on philosophical assumptions rather than conclusive empirical evidence. Memory's dynamic, adaptive, and flexible characteristics challenge reductive models, pointing instead to a transphysical dimension necessary to account for its multifaceted operation. Unlike static storage systems, memory is selective, reconstructive, and context dependent. Recollection often requires specific cues, intentional effort, or associative triggers, suggesting a non-linear, integrative process. Memory retains vast amounts of information—spanning knowledge, experiences, emotions, and personal identity—across the lifespan (Doty, 1998). This capacity implies a brain metafunctional system that synthesizes distributed information through temporally and spatially organized networks (Tulving, 2002). The brain does not merely record data but actively reconstructs it, often blending sensory, affective, and cognitive elements. Encoding and consolidation transform sensory input into patterns that become more stable over time, drawing on multiple systems to support recall of richly layered experiences (Squire & Kandel, 1999).

Memory's durability further defies simple physical explanations. Long-term memories can endure for decades, and individuals often retrieve them with emotional clarity and narrative coherence. Yet memory is also strikingly malleable. Research has shown that memories can be distorted or even implanted through suggestive techniques, undermining the idea of memory as a fixed physical record (Loftus & Pickrell, 1995). This interpretive flexibility supports the idea that memory is less about storage and more about reconstructive synthesis. The phenomenon of hyperthymesia (Highly Superior Autobiographical Memory, or HSAM) adds further weight to this view. Individuals with HSAM recall vast amounts of personal details but often report the burden of intrusive and persistent memories, illustrating that forgetting is not a flaw but an essential aspect of healthy cognitive function (Parker, Cahill, & McGaugh, 2006). This adaptability in memory management underscores the sophistication of its underlying architecture, suggesting processes that extend beyond electrochemical encoding.

Recalling a specific memory from the vast ocean of stored experiences involves far more than retrieving a static data point is an active, reconstructive process shaped by attention, emotional salience, and contextual cues. The brain does not store memories in isolated compartments but in distributed, overlapping networks that encode associations across time, place, emotion, and meaning. The hippocampus plays a central role in coordinating these networks, acting as a dynamic index that reactivates relevant cortical patterns during recall. Yet what determines which memory surfaces at a given moment—especially amid billions of potential traces—remains a mystery. Moreover, memory is not solely cognitive; it is deeply intertwined with emotional significance. Emotionally charged memories are often more vivid, enduring, and influential in shaping behavior and decision-making. This interplay between cognitive and emotional dimensions further complicates reductionist physicalist accounts. Antonio Damasio (1994) emphasizes that emotional memories involve complex interactions between brain activity and subjective experience. Emotional memories often evoke meaning, identity, and purpose—elements that resist materialist explanations.

Some theories propose a cue-dependent resonance mechanism, where internal or external stimuli selectively resonate with matching neural signatures, amplifying certain traces above the background of latent memories. This selective amplification process, shaped by intention, affective context, and neural plasticity, reflects the complexity of memory not as static storage but as a fluid, integrative act of recalling the past in light of the present. Memory's cross-modal integration, temporal continuity, resilience, constructive nature, and recall capacity indicate that purely physicalist models are insufficient. A comprehensive understanding of memory may require expanding beyond neural mechanics into transphysical or metafunctional frameworks that acknowledge memory's role in identity, consciousness, and meaning.

The Neural Binding and Metabinding Problems

The neural binding problem addresses a central challenge in cognitive neuroscience: how distinct features of a sensory experience—such as color, shape, and motion—are processed in separate regions of the brain yet perceived as a unified whole. Despite decades of investigation, no definitive neural convergence point has been identified where these distributed processes coalesce into coherent perceptual experiences (Crick & Koch, 1990; Treisman, 1996; Hardcastle, 1998; Revonsuo, 1999). Physicalist models struggle to explain this phenomenon, lacking a clear mechanism by which disparate neural signals are integrated into a singular percept. Expanding upon this, the metabinding problem introduces a deeper and more complex layer: how not only perceptual features, but also multiple sensory modalities, emotions, thoughts, and cognitive operations are synthesized into a unified, continuous stream of consciousness. While the binding problem addresses sensory integration, the metabinding problem engages with the full scope of subjective awareness—how the mind weaves together the totality of lived experience into a seamless sense of self. This integration spans higher-order processes, involving both vertical (from subcortical to cortical) and horizontal (across domains like emotion, memory, and attention) coordination. Current physicalist frameworks have not offered a satisfactory solution to either problem—particularly the metabinding challenge, which underscores the limitations of purely mechanistic explanations of consciousness. The absence of an explanatory account of how the brain achieves this integrative feat leaves both the binding and metabinding problems as open questions in neuroscience and the philosophy of mind, inviting further exploration into non-reductive and potentially transphysical models of brain function and consciousness.

Savant Abilities and the Limits of Physicalist Models

Savant syndrome refers to individuals who exhibit extraordinary abilities in narrowly defined areas such as mathematics, music, or memory, often in the context of broader cognitive or developmental impairments. Among the most remarkable cases is that of Kim Peek, a so-called "mega-savant" whose cognitive profile challenges conventional models of brain function. Peek could reportedly read an entire book page in 8–10 seconds, one page per eye, and retain over 12,000 books with near-perfect accuracy. His ability to recall intricate details—dates, names, locations, historical facts—was unmatched, and yet he exhibited significant developmental and motor difficulties (Treffert, 2009). Peek's abilities present a direct challenge to physicalist frameworks, which generally assume that structural brain anomalies—such as those found in his case—would impair rather than enhance cognitive function. Physicalist models predict that brain damage or malformation leads primarily to deficits, yet in Peek and similar cases, we observe the emergence of extraordinary, domain-specific cognitive abilities. This paradox has led researchers to propose that savant talents may emerge not merely from compensatory neural mechanisms, but from latent cognitive resources that are typically inaccessible in neurotypical individuals (Treffert, 2010).

One hypothesis suggests that in some savants, especially those with brain malformations or trauma, normal inhibitory processes may be disrupted, effectively "unmasking" latent abilities by lifting neural suppression. However, such models still operate within a purely materialist framework and struggle to account for the qualitative uniqueness, precision, and spontaneity of the abilities observed. The depth and coherence of savant skills, particularly those acquired spontaneously, hint at the possibility of non-material dimensions of cognition—a view aligned with transphysical models of mind. These models posit that human consciousness, and intelligence may involve elements not fully reducible to neural computation. As Treffert (2009, 2010) has argued, savant syndrome, whether congenital or acquired, suggests a need for broader explanatory models. These models must go beyond compensation and consider the activation of innate but dormant capacities, possibly linked to cognitive architectures not yet fully understood. Recognizing this possibility opens the door to new paradigms in neuroscience, psychology, and philosophy of mind—ones that take seriously the mystery and richness of human potential.

Acquired Savant Syndrome

Acquired savant syndrome refers to the sudden emergence of extraordinary abilities—often in music, mathematics, art, or memory—following brain injury, trauma, or neurological disruption. Unlike congenital savant syndrome, which is present from birth and frequently co-occurs with developmental disorders, acquired cases appear in previously neurotypical individuals, often without any formal training or prior aptitude in the newly developed skill. These cases pose a direct challenge to conventional neurological and compensatory models, which attribute unusual talents to the brain reallocating resources from impaired areas. Instead, the abrupt onset of advanced abilities suggests the activation of previously latent cognitive capacities—neural or otherwise—that were inaccessible prior to the triggering event (Treffert, 2010).

One of the most well-documented examples is the case of Orlando Serrell, a boy who developed exceptional calendar calculation abilities after being struck on the left side of his head by a baseball at age 10. Though he experienced lingering headaches, Serrell appeared otherwise unharmed. Over time, he began demonstrating the ability to determine the day of the week for any given date—past or future—without any formal instruction or prior interest in such tasks. Additionally, he developed a remarkable autobiographical memory, retaining detailed recollections of personal experiences and weather patterns for each day since the incident. His overall cognitive functioning remained normal, with no evidence of trade-offs or deficits, undermining the explanatory power of compensatory theories (Treffert, 2010). Such cases raise profound questions about the nature of human cognition. How can complex abilities suddenly surface in the absence of developmental training? What mechanisms—neurological or otherwise—could account for such access to specialized knowledge? These phenomena defy reductionist interpretations and point to the existence of dormant cognitive structures or non-localized informational fields potentially activated under specific conditions (Treffert, 2010).

Acquired savant syndrome thus exemplifies the limitations of materialist frameworks in explaining the full range of human cognitive potential. While neuroscience provides critical insights into brain function and plasticity, it does not yet account for the sudden and selective emergence of such advanced capacities. This invites exploration into transphysical models of cognition, which posit that aspects of mind and knowledge may not be fully reducible to neural activity alone. Recognizing these latent abilities opens new philosophical and scientific frontiers for understanding the brain not only as a biological organ but as a gateway to deeper strata of consciousness and intelligence.

Near-Death Experiences and the Surprise Principle

Near-death experiences (NDEs) are vivid, often transformative events reported by individuals who have been declared clinically dead or were near death. These experiences frequently include sensory and transcendental elements, such as out-of-body perceptions, a sense of peace, encounters with a bright light, or profoundly distressing visions. Remarkably, individuals born blind have reported seeing for the first time during NDEs, describing precise, veridical details about their surroundings—details later confirmed by researchers. Because these individuals lack prior visual memory or neural processing of visual data, researchers cannot attribute their experiences to hallucinations or memory distortions. Such accounts challenge physicalist explanations, revealing detailed perceptions of circumstances and events that would otherwise be unknowable. NDEs present unsurmountable challenges to physicalist explanations due to their occurrence in individuals with severely impaired or non-functioning brains, often during states like cardiac arrest. These experiences reveal vivid, lucid perceptions, demonstrate strong memory formation, lead to profound life-changing effects, and include veridical reports of events and environments that the individuals could not have witnessed under such conditions.

Physicalist Explanations and Their Limitations

Several physicalist theories advanced to explain NDEs fail to account for the full range of phenomena:

Neurochemical Theories: The release of opioids and other neurotransmitters during extreme stress (Martial et al., 2025) may help explain certain aspects of near-death experiences (NDEs), such as feelings of peace or euphoria. However, such explanations fall short in accounting for the coherence, clarity, and occasionally veridical content reported in NDEs. Likewise, theories based on oxygen deprivation, particularly in the visual system, cannot adequately explain the presence of detailed visual experiences in individuals who are congenitally blind.

Neurological/neuro-functional explanation: While Birk Engmann (2014) attributes near-death experiences (NDEs) to altered brain states involving the temporo-parietal junction during critical conditions, this neuro-functional model falls short of explaining key features of NDEs. Specifically, it cannot account for their consistent structure, heightened clarity, and in some cases, veridical perceptions—such as accurate observations of events or details from an out-of-body vantage point, even in the absence of measurable brain activity. Labeling NDEs as hallucination-like phenomena overlooks their profound cognitive coherence and the fact that such experiences often occur when brain function is severely impaired or absent, challenging purely neurological explanations. Theories involving brief electrical spikes in the brain during hypoxia fall short because these spikes are brief, disorganized, and inconsistent with the highly structured, extended, and lucid nature of NDEs (Greyson 2003; Chawla et al. 2009).

Psychological Theories: Psychological theories—such as expectation-based imagery or REM intrusion—struggle to account for the timing and content of near-death experiences (NDEs). These models presuppose an active, functioning brain capable of generating imagery, yet NDEs often occur during states like cardiac arrest or deep anesthesia, when consciousness is profoundly impaired or entirely absent. Moreover, many NDEs occur in individuals who had no prior expectation or belief in such experiences, undermining the idea that they are merely products of anticipation or dream-like phenomena. Thus, these theories fail to explain both the temporal context and the vivid, structured nature of NDEs.

These shortcomings of physicalist models highlight their inability to account for the full range of near-death experience (NDE) phenomena. The presence of lucid, highly structured experiences—and in some cases veridical perceptions—during periods of severe brain dysfunction or apparent non-function poses a serious challenge to the sufficiency of purely naturalistic explanations. Such models struggle not only with the content of NDEs but also with their timing, coherence, and veridical perceptions consistently reported across populations and cultures, suggesting that a more comprehensive framework may be required.

The Surprise Principle and Brain Metafunction Transphysicalism

The surprise principle provides a powerful tool for evaluating competing explanations. It suggests that observations that are highly unexpected under one theory, yet natural or anticipated under another, lend stronger support to the latter framework. Naturalism predicts that all mental phenomena, including NDEs, should be explainable solely through physical brain processes. Yet phenomena like consciousness, qualia, intentionality, abstract thinking, memory, savant abilities, and NDEs frequently defy strict physicalist explanations, making them highly unexpected under this framework. In contrast, Brain Metafunction Transphysicalism posits that the brain cooperates with a transphysical dimension, enabling higher-order functions beyond those achievable by physical processes alone. Under this hypothesis, phenomena like NDEs and other extraordinary cognitive abilities become less surprising, as the transphysical dimension provides a coherent framework for explaining their occurrence. Robin Collins highlights this principle also known as prime principle of confirmation, noting, "we have reason to prefer an explanation that makes surprising facts less surprising" (Collins 2009a). This comparison highlights the need for broader models that incorporate transphysical dimensions, offering a more comprehensive understanding of consciousness, cognition, and extraordinary human experiences.

Mind-Brain Relationship: The Need for and Inherent Form of Causality.

Colin McGinn argues that the mind-body problem reveals a fundamental gap in our understanding of causality, suggesting that traditional physical explanations may be inadequate for explaining consciousness. He proposes that mental states and brain processes may require a sui generis form of causality — a unique causal relationship distinct from standard physical mechanisms. According to McGinn, our existing conceptual tools fail to bridge the explanatory gap because consciousness resists explanation through known scientific frameworks. He suggests that an undiscovered causal connection might explain how mental properties arise in relation to physical brain states. However, this remains beyond the reach of current scientific methods. Such a sui generis causality acknowledges consciousness's "mysterious" character while hinting at principles that transcend standard physical causation (McGinn 1999).

Within a framework incorporating a transphysical dimension, consciousness expresses the entire person's interaction with the world. These interactions are shaped by and reciprocally influence an immaterial aspect of human existence. In this perspective, the brain mediates consciousness, serving as a vital hub for integrating bodily systems and enabling subjective experiences and awareness. However, the brain is not the sole origin of consciousness but is an active participant through which the soul-animated body expresses its experiences. This view recognizes the role of a transphysical domain that interacts with and shapes conscious states, suggesting that consciousness is not merely a byproduct of brain activity but a manifestation of the interplay between physical and immaterial human elements (Aquinas 1947; Craig and Moreland 2003; Spitzer 2015). The transphysical domain functions as a foundational framework that unifies sensory perception, cognition, and subjective experience, integrating the physical and non-material dimensions of human consciousness. Simultaneously, brain activity informs and interacts with the transphysical domain, creating a reciprocal causal relationship in which material and immaterial aspects of human existence dynamically shape one another. This dynamic interplay enriches conscious experiences, highlighting the human person's unity, complexity, and depth (Moreland 2018; Collins 2009b; Damasio 1999).

Mind-Brain Causality and the Soul: A More Plausible Foundation for Free Will and Personal Identity

Jeffrey M. Schwartz's research in neuroplasticity provides compelling evidence for top-down mental influence on brain function, challenging the notion that the mind is merely a byproduct of neural activity. In The Mind and the Brain (Schwartz & Begley, 2002), Schwartz demonstrates how cognitive and volitional processes, particularly mindfulness and directed attention, can reshape neural pathways. His work with obsessive-compulsive disorder (OCD) patients shows that intentional thought can alter brain function, supporting the idea that the mind actively interacts with, rather than being reducible to, the brain. Schwartz's findings align with emerging psychiatry models that emphasize consciousness's causal role in shaping neural activity, reinforcing the case for mental causation. Norman Doidge (2007, 2015) expands on Schwartz's work, showing how thoughts, emotions, and experiences shape brain structure and function. His research on psychotherapy-induced neuroplasticity and the role of conscious thought in brain recovery from trauma and neurological disorders provides additional support for mental causation.

Similarly, Alvaro Pascual-Leone (2005) has demonstrated that mental training, imagination, and directed attention can induce changes in brain structure and function, particularly through his work with transcranial magnetic stimulation (TMS) and mental rehearsal in motor learning. These studies provide strong evidence that the mind exerts control over neural pathways rather than being passively determined by them. Further reinforcing this, Richard J. Davidson (2012) has explored how meditation and emotional regulation reshape neural circuitry, demonstrating that intention-driven mental states can rewire the brain over time. His work supports the idea that directed mental effort leads to measurable changes in neural structures, challenging the materialist notion that brain states are purely the result of bottom-up physical processes. Additionally, Mario Beauregard (2007) has examined the influence of placebo effects, spiritual experiences, and self-directed neuroplasticity, concluding that mental states such as belief and intention directly affect neural activity. His research further bolsters the claim that consciousness plays an active role in shaping brain function, aligning with Schwartz's and others' findings. These studies provide strong empirical support for top-down causation, refuting reductionist materialism and reinforcing the case for a non-reductive understanding of the mind-brain relationship. If mental states can actively modify brain structure and function, then the mind must be more than just an emergent brain property—it must possess real causal power over it. This growing body of evidence supports the view that consciousness is fundamental to human agency rather than merely a passive effect of neural activity, making a strong case for a non-materialist perspective on the mind.

Philosophical Implications

If free will, rationality, and moral responsibility are real, they require an adequate ontological foundation. Materialist naturalism, which reduces the mind to brain processes, fails to provide such a foundation, rendering human decisions nothing more than deterministic physical events. This reductionist view undermines both agency and continuity of personal identity over time. By contrast, mind-brain causality, grounded in the existence of the soul, offers a more coherent explanation. The unlikely convergence between Thomas Nagel, a secular philosopher, and Alvin Plantinga, a Christian theist, in their critique of materialist determinism underscores this point: the mind is irreducible to the brain, and free will necessitates a reality beyond physical causation. Further supporting this, David Chalmers, a non-theist philosopher of mind, and Richard Swinburne, a Christian theist, strengthen the case against materialist reductionism and favor a non-physical aspect of human personhood.

Materialism struggles to explain these three key aspects of human experience:

Free Will: Strict determinism renders genuine choice impossible. Nagel critiques materialism's failure to explain reasoned deliberation, while Plantinga argues that evolutionary processes alone cannot guarantee rationality's reliability.

Personal Identity: Materialism struggles to explain how identity persists despite constant neural and bodily change. Swinburne argues that only a non-physical soul can account for stable personal identity.

Mental Causation: Chalmers argues that subjective experience must be fundamental to reality since consciousness shapes physical states. Swinburne reinforces this by asserting that mental causation cannot emerge solely from material processes.

Since materialism cannot account for free will, rationality, or personal identity, an alternative explanation is needed — one that preserves agency and mental causation. The best candidate is the soul, a non-physical, unified center of identity and will. The soul provides a stable foundation for personal identity, ensuring continuity of self beyond physical changes. Unlike materialist models, which reduce personhood to neural processes, the soul affirms that we remain the same individual over time. Additionally, the soul enables genuine free will. If the mind were purely physical, it would be subject to deterministic or probabilistic laws, making choices mere effects of prior causes. However, if the mind is fundamentally non-physical, it is capable of self-determined action, making free will not just possible but necessary. The combined critiques of Nagel, Plantinga, Chalmers, and Swinburne demonstrate that physical terms are insufficient to explain the mind. By acknowledging the soul as the foundation of agency, rationality, and selfhood, we gain a more complete understanding of human nature and a worldview that preserves moral responsibility and human dignity. The most coherent explanation is that human beings are ensouled agents — unified persons whose consciousness, rationality, and moral responsibility transcend material processes alone.

The Limits of Physicalism and the Case for Transphysical Cognition.

The cumulative evidence discussed thus far presents a compelling case for dimensions of human cognition that exceed purely biological explanation. While physicalist neuroscience has been adequate in offering remarkable insights into neural processes, it continues to face critical limitations in explaining the foundational characteristics of consciousness and higher-order cognition. Brain Metafunction Transphysicalism emerges not merely as a critique of physicalism but as a constructive framework that integrates empirical and philosophical perspectives to account for these phenomena. A central challenge to physicalism lies in its inability to account for qualia—the vivid, subjective texture of experience, such as the redness of red or the feeling of sorrow. These qualitative features resist reduction to electrochemical activity, even when associated with specific neural patterns (Nagel, 1974). Closely related is intentionality, the inherent "aboutness" of thought—the mind's capacity to direct itself toward meanings, objects, and abstract concepts. As Searle (1983) argued, no purely physical state in the brain can account for this directedness, pointing to a deeper level of mental representation.

Further complicating the physicalist picture are the cognitive capacities for abstraction, symbolism, and creativity, which enable humans to engage with philosophical ideas, ethical dilemmas, and aesthetic experiences. These faculties frequently operate independently of immediate sensory input or survival advantage, suggesting a broader cognitive architecture. The savant syndrome illustrates this vividly: individuals with significant neurological impairment have demonstrated exceptional skills in mathematics, music, or memory, often in the absence of typical developmental exposure or training. Even more striking are cases of acquired savant syndrome, where such abilities emerge spontaneously after brain injury—phenomena that resist standard compensatory explanations and hint at latent cognitive potentials not fully accounted for by neural mechanisms alone (Treffert, 2010). These observations reinforce the proposition that the brain's metafunctional activity transcends the biological level. Rather than viewing subjectivity as an anomaly or byproduct of neural complexity, Brain Metafunction Transphysicalism interprets it as a primary datum—one that signals a transphysical dimension essential for understanding consciousness, identity, and meaning. This framework invites a rethinking of cognition as both embodied and transmaterial, grounded in the brain yet extending beyond it. In doing so, it lays the groundwork for an integrative science of the mind that honors the depth and mystery of human awareness.

Toward a Transphysical Framework for the Brain and Self

Brain Metafunction Transphysicalism proposes that consciousness and its highest cognitive expressions cooperate with—but are not reducible to—the brain's material functions. Rather than treating subjectivity as a mere byproduct or anomaly, this view reframes it as a primary datum of human existence—central to understanding how the brain generates meaning, awareness, and agency. Consciousness, in this framework, is both embodied and transmaterial. The brain serves not merely as a computational engine but as a dynamic interface between the physical and transphysical dimensions of human experience. One of the most compelling lines of support for this framework arises from near-death experiences (NDEs). Individuals who have undergone NDEs often report vivid perceptions, lucid awareness, and verifiable observations while in states of minimal or absent brain activity (Greyson, 2007). In many documented cases, individuals have accurately recounted events that occurred during clinical death—while lacking measurable neural function. These accounts not only challenge the assumption that consciousness is wholly dependent on brain activity but also suggest a level of coherence, structure, and integration that defies physicalist explanation. The profound, life-altering nature of these experiences further underscores their phenomenological significance and invites serious philosophical and scientific inquiry into the transphysical dimension of cognition.

To account for these phenomena, Brain Metafunction Transphysicalism situates the brain within a broader metaphysical structure—one that encompasses both material and non-material dimensions. In this model, the transphysical domain operates as both efficient and final cause: guiding the development of the brain and body (efficient cause) and directing their purposes and ends (final cause). The coherence of human cognition and identity is not merely the result of neural organization but reflects a deeper ontological unity between biological structures and the soul or metaphysical self (Spitzer, 2015; Moreland, 2018). The brain's metafunctional activity—its ability to integrate sensory, cognitive, emotional, and volitional domains—is understood here as an expression of the soul's animating influence. Rather than emerging from neural complexity alone, consciousness is sustained through an ongoing interaction between the material brain and a non-material dimension. This challenges reductionist models that attempt to explain the human person exclusively in terms of physical constituents or emergent complexity. Instead, it opens space for an integrative ontology that includes both empirical realities and metaphysical foundations.

By reconciling scientific insights with metaphysical principles, Brain Metafunction Transphysicalism significantly broadens the scope of inquiry in neuroscience and the philosophy of mind. Neural mechanisms remain essential for embodiment and experience—but they are not the whole story. This framework encourages a more expansive interpretation of human ontology—one that honors the biological while acknowledging the irreducible nature of consciousness, intention, and identity. In doing so, it offers a robust paradigm for understanding the person as a unity of matter and meaning, grounded in a reality that extends beyond the merely physical.

The Soul's Richness and Depth: Pathways to Truth, Beauty, and the Sacred

Theological reflections on the soul reveal a profound array of capacities that transcend the physical and engage with the transcendental (Balthasar, 1982; Craig & Moreland, 2003; Moreland, 2014; Ratzinger, 2004; Spitzer, 2017). Among these is the unique capacity for self-consciousness — the ability to reflect inwardly and cultivate an inner world of experience and self-awareness. This introspective capacity provides the foundation for conceptual thinking, facilitating abstract reasoning, syntactical control, and the use of complex language. Beyond intellectual abilities, the soul exhibits an intrinsic desire for perfect truth — an insatiable drive to recognize gaps in knowledge and pursue deeper understanding. This yearning reflects the soul's orientation toward a reality greater than itself, inspiring the search for ultimate meaning.

The soul's spiritual dimension becomes evident in its openness to transcendence, whether experienced as the divine or as a profound sense of sacredness. This capacity fosters awe, worship, and a relational connection with what lies beyond the material world. The soul's longing for a perfect home — a state of ultimate fulfillment — emerges from recognizing the limitations of worldly experience and motivates the search for lasting purpose. Alongside this spiritual dimension, the soul's capacity for empathy and love reflects a deep moral awareness. This capacity transcends self-interest, inspiring self-giving care and fostering a yearning for authentic, perfect love. The moral conscience further reveals the soul's depth, guiding individuals in ethical reflection through an innate sense of justice, goodness, and responsibility — rooted in an awareness of moral truth.

Finally, the soul's capacity for appreciating beauty transcends material aesthetics, embracing the sublime in nature, art, music, and ideas. This pursuit of beauty reflects a desire for perfect splendor, pointing to a transcendent source of harmony and order. Tying these dimensions together is the gift of free will, which enables self-consciousness to direct itself toward others, pursue truth, and orient itself toward the divine. In this capacity, the soul reveals its ultimate synthesis — a profound integration of love, goodness, and transcendence. Together, these capacities highlight the soul's remarkable depth, affirming its intrinsic connection to both the temporal and the eternal.

Brain Metafunction Transphysicalism is Parsimonious

Brain Metafunction Transphysicalism (BMT) offers an economical yet integrative framework for understanding the soul, consciousness, perception, near-death experiences, and other profound aspects of human nature. By incorporating a transphysical domain alongside the brain's metafunctional role, BMT provides a unified explanation for phenomena that have long challenged physicalist models. Unlike physicalism, which confines consciousness to neural activity alone, BMT posits that consciousness emerges from the dynamic interplay between physical and immaterial dimensions. This multi-dimensional approach addresses the conceptual and empirical limitations inherent in strictly physicalist explanations. Adhering to Occam's razor, BMT avoids unnecessary complexity by offering a coherent framework that integrates empirical data with metaphysical insights. This approach effectively accounts for subjective experiences, intentionality, and near-death experiences without resorting to speculative or ad hoc explanations. While physicalist models struggle to reconcile the richness of conscious experience with known neural mechanisms, BMT bridges this gap by recognizing the brain's metafunction as a mediator of transphysical influences. In doing so, it remains parsimonious while encompassing the full scope of human experience (Collins 2009b; Moreland 2018; Nagel 1974; Searle 1983).

Generative Possibilities

By integrating a transphysical domain with the brain's metafunction, Brain Metafunction Transphysicalism (BMT) invites interdisciplinary exploration across neuroscience, psychology, philosophy, and theology. In neuroscience, BMT provides a framework for investigating how the brain's metafunction integrates physical processes with transphysical influences, shedding light on the richness and complexity of conscious experience (Damasio 1999; Chalmers 1995). This framework offers a foundation for studying phenomena such as near-death experiences and intentionality, which resist explanation within traditional physicalist models. In psychology, BMT enhances our understanding of subjective experiences, emotions, and self-awareness by emphasizing their interplay with non-material dimensions. Philosophically, BMT presents a robust account of consciousness, perception, and the mind-body relationship. By uniting physical and immaterial dimensions within a cohesive framework, it transcends reductionist models in the philosophy of mind. Theologically, BMT aligns with longstanding

Christian thought, enriching discussions about the unity of the human person, the relationship between matter and spirit, and the transcendent nature of human existence.

By synthesizing insights across disciplines, BMT encourages collaboration that advances understanding in diverse fields. This integrative model reveals consciousness and human identity as rooted in both biological and transphysical dimensions, positioning BMT as a comprehensive yet parsimonious approach to explaining complex aspects of human nature (García, 1995a, 1995b, 1998a, 1998b, 2001). Ultimately, BMT transcends the limitations of physicalism by providing a unified explanatory model that respects both scientific evidence and philosophical reflection. Its emphasis on parsimony ensures that BMT does not multiply assumptions unnecessarily but instead offers a coherent and evidence-based account of consciousness and human existence. By advancing both scientific and metaphysical inquiry, BMT stands as a valuable resource for exploring the complexities of the human person.

The "Energy Conservation Objection" to the Soul

Physicalists argue that all mental phenomena, including consciousness and subjective experience, can ultimately be explained by physical processes in the brain. According to this view, the brain functions as a complex physical system in which mental states emerge from interactions between neurons, synapses, and electrical and chemical signals. Consequently, physicalists reject non-physical explanations of the mind, deeming them unnecessary and incompatible with fundamental physical principles. One prominent critique against non-physicalist views is the energy conservation objection, also referred to as the causal pairing problem. This objection asserts that if mental or conscious processes are irreducible to physical processes yet exert causal influence, they would necessarily violate the principle of energy conservation, which states that energy cannot be created or destroyed, only transferred or converted. The argument contends that if non-physical processes influence the brain, they will require an additional source of energy, thus contradicting established physical laws (Papineau 2001; Kim 2005).

Brain Metafunction Transphysicalism (BMT) responds to this objection by rethinking the mind-body relationship through a framework informed by Thomistic hylomorphism. BMT posits that the soul and body exist in an intrinsic, non-causal unity, eliminating the need for energy transfer between distinct entities. Unlike classical dualism, which conceptualizes the soul and body as separate substances requiring interaction, BMT asserts that the soul is holenmerically present throughout the body (Moreland 2018). This type of presence means the soul is intrinsic to the entire organism, fully present in every part without occupying space like a material object. Consequently, the soul's influence on consciousness and subjective experience does not require energy-mediated causation, avoiding the conflicts raised by the energy conservation objection. This perspective aligns with contemporary concepts in biology that emphasize the role of information-based processes in guiding complex systems. Just as biological information regulates cellular processes without violating physical laws, the soul's influence on consciousness operates as an intrinsic form of organization rather than as a distinct causal agent requiring energy transfer (Craig and Moreland 2003; Feser 2010). BMT's focus on informational causation rather than mechanistic energy transfer offers a more coherent account of consciousness, subjective experience, and mental causation within the framework of established physics.

Other non-physicalist models, such as classical substance dualism, attempt to address this critique by positing a sui generis form of causation—an interaction distinct from mechanistic energy transfer. Proponents like McGinn (1999) argue that this unique causal connection operates beyond standard physical principles. However, BMT and hylomorphic models offer a more comprehensive solution by situating consciousness within the intrinsic unity of the soul-body relationship rather than relying on discrete causative interactions that demand energy exchange. By redefining the relationship between mind and body as a metaphysical unity rather than a dualistic interaction, BMT avoids the pitfalls of the energy conservation critique while providing a more comprehensive account of consciousness and subjective experience. This framework not only addresses key physicalist objections but also advances a more integrated understanding of the mind-body relationship.

A Neo-Thomist Context for Non-Physical Causation

The concept of non-physical causation is essential to the thesis on the human soul, as presented here. In this neo-Thomist view, the human person is a single substance composed of two co-principles: spirit and matter. These co-principles exist in a relationship of primitive unity rather than as juxtaposed entities. This framework provides a coherent basis for understanding the mind-body relationship in a way that accommodates inherent, non-physical causation. In Thomistic metaphysics, the soul is not a separate substance from the body but rather the substantial form of the body, endowing it with unity, organization, and distinct properties. The soul imparts capacities for rational thought, self-awareness, and spiritual life, thereby grounding the unique aspects of human existence (Aquinas 1947; Feser 2010). This model presents the soul as more than an organizing principle; and emphasizes the soul's active role in integrating the physical and immaterial dimensions of human experience.

Within this framework, the mind, as a capacity of the soul, is not an external agent acting upon the body but an intrinsic aspect of the unified human person. This view supports the concept of non-physical causation, where the mind influences bodily processes as an inherent aspect of the soul-body unity. Since this influence arises from within the unified substance rather than through distinct causal mechanisms, it avoids the need for energy transfer across separate entities. Thus, the mind's influence on the body occurs through an inherent informational dynamic, guiding behavior and cognition without violating the principle of energy conservation. This enriched Thomistic model addresses the principle of energy conservation by positing that mental causation operates as an informational influence that directs the body's natural energy resources rather than introducing new energy. Just as genetic information organizes biological processes without adding physical energy, the mind's non-physical influence shapes the brain's activity without contradicting established physical laws (Collins 2009b; Moreland and Craig 2003). By grounding mental causation in a unified substance that includes both spirit and matter, this neo-Thomist model integrates physical and immaterial dimensions of human experience in a way that transcends the limitations of physicalist and dualist models alike. It presents a coherent philosophical foundation for understanding how consciousness, free will, and rational thought can emerge within a unified yet metaphysically rich framework. This approach offers a reconciliatory model that accounts for mental and physical influences without compromising coherence with established scientific principles (Stump 2003; Kretzmann and Stump 1993).

Causal Interaction without Energy Transfer? Clarifying the Role of Informational Causation\

As a unified entity, the soul and organism represent a holistic perspective in which the soul's influence transcends individual bodily components. This framework emphasizes the intrinsic unity between the soul and body, wherein the soul serves as the substantial form that informs and regulates the body's physical structure and functions. Within this model, the soul's influence does not rely on distinct causal interactions that require energy transfer. Instead, its inherent causal power is intrinsic to the soul-body unity itself. This intrinsic causation is *sui generis* — a fundamental and irreducible form of causation that operates through the soul's role as the animating principle of the body. Consequently, this model avoids the common physicalist critique that non-physical causation would violate the principle of energy conservation. Since the soul's causal influence is inherent to its nature as the substantial form of the body, no distinct "transfer" of energy or information is necessary to explain mental causation.

In his paper *A Scientific Case for the Soul,* philosopher Robin Collins expands on the conceptual landscape by exploring alternative models of causation that incorporate information-based dynamics. While his approach is distinct from the inherent causation proposed in the neo-Thomist framework, it introduces a complementary perspective with broader interdisciplinary applications. In Modern Physics and the Energy Conservation Objection to Mind-Body Dualism, Collins addresses the longstanding energy conservation objection, which claims that non-physical causation would violate established physical laws. Collins challenges this critique by proposing that some causal processes — particularly those involving mental influence — may operate not through energy transfer but via the exchange of information. He argues that information, as a non-energetic entity, can exert causal influence by shaping physical processes without requiring additional energy input (Collins, 2008).

To capture this concept, the present author introduces the term Infomism. Infomism refers to a distinct form of causation characterized by the productive exchange of information without requiring energy transfer. While distinct from the intrinsic causation articulated in the neo-Thomist framework, Infomism offers a useful explanatory model for certain causal processes that involve law-like communication within complex systems. Infomism conceptualizes mental phenomena — such as thoughts, beliefs, and perceptions — as instances of high-order information exchange, particularly in contexts where informational dynamics influence physical processes in non-linear or emergent ways. Etymologically, Infomism combines "information" and "mechanism," underscoring its emphasis on productive causation through information while avoiding reductionist implications often associated with mechanistic views. While Infomism does not replace the inherent causation already established in the neo-Thomist model, it introduces a helpful framework for exploring information-based causality in broader contexts. Infomism may provide insights into interdisciplinary fields such as systems biology, cognitive science, and artificial intelligence, where informational dynamics play a crucial role in complex adaptive processes (Bateson 1972; Gleick 2011).

Definition: *An infomism is a form of productive causation characterized by the exchange of information without necessitating energy transfer.*

While this concept has broad applicability across scientific and philosophical disciplines, it remains distinct from the intrinsic causation articulated in the neo-Thomist framework, which operates independently of information exchange or energy transfer.

Applications

The concept of Infomism holds significant potential for advancing understanding across diverse scientific domains. In information theory, Infomism could inspire models for information transmission and processing that minimize or bypass energy expenditure, offering insights into developing more efficient communication systems. By distinguishing information-based causation from traditional energy-driven interactions, Infomism provides a theoretical framework for exploring non-energetic forms of influence in complex systems. In neuroscience, Infomism may help conceptualize how mental phenomena, such as thoughts, beliefs, or conscious intentions, influence neural activity through informational causation rather than energy transfer. This perspective could inform research on neuroplasticity, cognitive processing, and psychosomatic interactions, where the influence of mental states on physical processes requires alternative explanatory models.

In quantum mechanics, Infomism relates to phenomena like quantum entanglement, where changes in one particle instantaneously affect another, seemingly without an energy exchange (Bell, 1987; Wheeler & Ford, 1998). While quantum entanglement remains an area of ongoing interpretation, Infomism offers a conceptual lens through which such non-local, information-driven interactions can be explored. More broadly, Infomism offers a unifying framework for investigating complex systems across disciplines. By addressing the causal role of information without requiring physical energy transfer, it presents a novel conceptual tool for examining processes in biology, artificial intelligence, and consciousness studies. This integrative approach encourages a holistic perspective, combining physical and informational dimensions to deepen our understanding of causation across scientific and philosophical contexts.

Potential Objections to Infomism and Counterarguments

The concept of Infomism, which proposes productive causation through information exchange without energy transfer, introduces a novel framework that challenges conventional physicalist paradigms. While promising, Infomism faces several objections that warrant careful consideration. A primary objection asserts that Infomism violates the principle of energy conservation, a cornerstone of established physical laws. Critics argue that causation inherently requires energy transfer; without this, Infomism risks contradicting well-established scientific principles. However, Infomism does not violate energy conservation because it posits a distinct form of causation—one based on information, not energy. This aligns with emerging insights into quantum mechanics, information theory, and systems biology, where information appears to exert causal influence without observable energy transfer. For example, quantum entanglement demonstrates that particle states can instantaneously influence one another without detectable energy exchange (Bell, 1987; Wheeler & Ford, 1998). Similarly, in genetic regulation, complex molecular signaling pathways direct gene expression with minimal energetic cost. Moreover, computational models reveal that information flow can shape system outcomes without additional energy input as traditionally conceived (Landauer, 1961). These examples illustrate that Infomism introduces a plausible extension of known causal models rather than contradicting them.

Another major objection claims that Infomism lacks a concrete, mechanistic account of how information could exert causal influence without energy transfer. Since most causal models in science rely on identifiable mechanisms (e.g., chemical reactions or electrical currents), critics may argue that Infomism introduces an explanatory gap. However, Infomism deliberately departs from mechanistic causation in the traditional sense. Instead, it proposes information as a primitive, sui generis form of causation that does not require an intermediary energy exchange. While this may appear unconventional, similar models exist in established frameworks. For instance, quantum mechanics posits information patterns that dictate particle behavior, while software systems demonstrate how encoded instructions can produce complex outcomes independently of physical mechanisms beyond basic signal transmission. Additionally, in epigenetics, gene expression is regulated by informational control mechanisms that function without relying on distinct energy transfers. These examples support the idea that informational causation is both conceptually sound and practically evidenced.

Critics may also argue that Infomism is redundant, offering little beyond established concepts in Shannon's Information Theory or cybernetics. While Infomism builds on these ideas, it contributes something distinct: the assertion that information itself can act as a causal force independent of physical energy. Traditional information theory focuses on data transmission through physical media, whereas Infomism emphasizes that information can influence outcomes even in the absence of energetic interaction. For example, Shannon's model describes signal transmission but does not explain how information might shape material outcomes through non-energetic mechanisms. By contrast, Infomism addresses precisely this gap, offering a theoretical framework for understanding how non-energetic information exchange may shape complex systems.

Another objection concerns the perceived lack of empirical support for Infomism. Critics may claim that without experimental confirmation, Infomism risks being purely speculative. While direct experimental validation may be challenging, several domains provide indirect support. Quantum entanglement demonstrates that information can influence physical states without energy transfer, while research on neuroplasticity suggests that mental effort — conceptualized as an informational rather than energetic influence — can reshape brain function (Schwartz & Begley, 2002; Doidge, 2007). Additionally, placebo effects illustrate instances where beliefs and expectations, arguably informational states, can produce measurable physiological outcomes without an identifiable energy-based mechanism. These observations suggest that Infomism aligns with observed causal effects that resist reduction to traditional energy models.

Lastly, some critics may argue that Infomism resembles a "God-of-the-gaps" argument—an appeal to non-physical causation to fill unexplained gaps in scientific understanding. This criticism implies that Infomism merely substitutes speculation where physicalist models currently fall short. However, Infomism does not rely on ignorance or incomplete data; instead, it emerges from positive evidence for information's distinctive causal role. Rather than invoking immaterial causation to bypass unanswered questions, Infomism builds on observed patterns where information demonstrably influences outcomes. Moreover, Infomism encourages further investigation by proposing testable implications—such as exploring informational patterns in neural activity, examining cognitive influence on bodily states, or assessing non-energetic dynamics in quantum processes. By advancing potential empirical applications, Infomism invites continued inquiry rather than discouraging scientific exploration. While Infomism introduces unconventional ideas that face reasonable objections, it remains a defensible and promising framework. By emphasizing information as a distinct causal factor, Infomism offers a coherent alternative to physicalist models that struggle to explain phenomena like consciousness, mental causation, and non-local interactions. By grounding itself in established scientific insights while extending their implications, Infomism bridges gaps between physics, neuroscience, and philosophy, opening new avenues for understanding complex causal relationships.

Infomism and Foundational Theological Concepts

The concept of Infomism—productive causation through information without energy transfer—resonates with foundational theological principles, particularly those emphasizing the centrality of logos (word, reason, or order) in creation and existence. While Infomism does not equate to divine causation itself, it provides a conceptual model that reflects aspects of how immaterial causation may operate within creation, aligning with theological insights regarding divine order and intentionality. In Christian theology, the logos is presented as the divine reason or ordering principle that sustains creation. The prologue to the Gospel of John declares, "In the beginning was the Word (Logos), and the Word was with God, and the Word was God" (John 1:1). This passage establishes the logos as the ultimate creative power through which all things were made: "Through him all things were made; without him nothing was made that has been made" (John 1:3). While divine causation transcends the created order, Infomism proposes a distinct yet complementary framework for understanding how non-physical causation operates within creation. As a conceptual tool, Infomism reflects certain features of divine causation by emphasizing the formative and organizational role of information. This perspective aligns with the theological view that God's creative word imparts coherence and structure to the universe. However, unlike the logos, which functions as the ultimate metaphysical foundation, Infomism describes information-based causation as a secondary, creaturely process that operates within the natural order. By distinguishing Infomism from divine causation, this framework avoids conflating God's sustaining power with natural processes. Instead, it offers a conceptual model for understanding how non-energetic informational causation may function within the created world, harmonizing with the metaphysical principle that all contingent realities ultimately depend on God's creative will.

The Role of Information as a Foundational Principle

Infomism emphasizes that information is not merely an emergent property of material systems but instead reflects a deeper ontological reality. This view resonates with the classical Thomistic distinction between form and matter, wherein form provides the essential blueprint that organizes material structures. In this sense, Infomism proposes that information functions analogously to form, guiding physical processes without requiring energy transfer. This understanding complements the idea that divine wisdom governs creation through order, meaning, and purpose. By conceiving information as a causal principle distinct from material energy, Infomism aligns with theological insights that emphasize the spiritual nature of divine action. In this context, divine influence is understood not as an energetic force but as a formative principle that guides creation toward its intended purpose. This emphasis on information's formative role offers a conceptual bridge between theological insights and scientific inquiry.

In Christian anthropology, the soul is regarded as the substantial form of the body, imparting organization, unity, and purpose to the human person. Within this framework, the soul's influence is not mediated through energy transfer but through an intrinsic, non-physical causation. Infomism provides a helpful conceptual model for understanding this process. It posits that the soul, as the organizing principle of the body, imparts information that directs physical processes without violating the laws of energy conservation. For example, intentional mental states such as conscious deliberation, moral reflection, and acts of will can influence neural processes without introducing additional energy. Drawing from contemporary neuroscience, this framework aligns with evidence showing how directed mental effort can reshape neural pathways through neuroplasticity (Schwartz & Begley, 2002; Doidge, 2007). In this sense, Infomism provides a model for understanding how the soul's rational and volitional capacities can influence the body through information-driven processes, complementing the Thomistic view of the soul as an intrinsic principle of life and cognition.

Implications for Divine Action, Spiritual Experience, the Nature of Creation

Infomism also offers a helpful conceptual framework for understanding divine action and spiritual experiences. Classical theology teaches that God's influence transcends material causation. As Aquinas explains, God's interaction with creation is immediate and direct, guiding reality not through energy transfer but through His sustaining will. This view aligns with the idea that divine action operates informationally rather than energetically, shaping the material world in a non-physical yet causally effective way (Aquinas, Summa Theologica, I, q. 105, a. 5). By emphasizing the causal role of information, Infomism offers insights into how spiritual experiences—such as moments of profound insight, inspiration, or transformative encounters—may involve informational causation rather than purely neural mechanisms. This framework allows for a richer understanding of grace, prayer, and divine communication, reinforcing the idea that God's influence need not rely on physical energy to manifest in the human person. In Genesis, God's creative word brings forth order and structure: "And God said, 'Let there be light,' and there was light" (Genesis 1:3). This account emphasizes the formative role of divine speech, wherein God's command introduces coherence and purpose to creation. Infomism echoes this concept by proposing that information is an irreducible causal factor in shaping reality. Just as divine wisdom structures creation, Infomism envisions information as the guiding principle that directs complex biological, cognitive, and metaphysical processes.

Furthermore, Infomism invites reflection on the relationship between divine providence and the unfolding of natural processes. By recognizing information as a fundamental causal factor, Infomism opens avenues for understanding how God's guidance can operate within the natural world without disrupting established physical laws. Infomism provides a powerful conceptual tool that resonates with core theological insights about divine order, the soul's influence, and the role of non-physical causation. By presenting information as a causal force independent of energy transfer, Infomism complements the Christian understanding of God's creative logos as the ultimate source of meaning, purpose, and structure. Crucially, Infomism does not conflate divine causation with creaturely processes but instead offers a model for understanding how informational causation may operate within the created order. By integrating philosophical reflection, scientific inquiry, and theological insights, Infomism enriches our understanding of the profound interplay between the physical and the transcendent.

The Soul is not Directly Accessible to Scientific Study. How Do We Explain it?

The question of explaining an entity like the soul, which eludes direct scientific inquiry, often assumes that science is the only valid source of truth—a position known as scientism. Scientism is "an exaggerated trust in the efficacy of the methods of natural science applied to all areas of investigation (as in philosophy, the social sciences, and the humanities)" (Merriam-Webster, n.d.). However, there are realities, like gravity, that we understand through their observable effects despite their elusiveness. Similarly, while we cannot directly observe the soul, we can infer its existence and influence through its observable effects on human life—such as rationality, creativity, and moral consciousness. Just as gravity's influence on celestial bodies reveals its presence, the soul's influence manifests in uniquely human capacities, pointing to a reality that transcends the purely physical (Collins, 2009b; Craig & Moreland, 2003).

While the analogy between the soul and gravity offers some conceptual clarity, a stronger parallel may be drawn from non-material realities such as information or logical principles. Just as mathematical truths shape reality without existing in physical space or possessing material properties, the soul's immaterial nature exerts influence through rationality, moral awareness, and consciousness. These capacities demonstrate effects that are neither reducible to nor explicable by physical causes alone, reinforcing the idea that immaterial realities can produce observable outcomes. Inferring the existence of the soul from its effects is consistent with established scientific reasoning. Scientific practice routinely relies on inference to the best explanation, in which theoretical entities—such as dark matter or gravitational waves—are postulated to account for observable phenomena before direct detection becomes possible. In the same way, the soul's explanatory power lies in its ability to account for consciousness, free will, and moral awareness—features that remain elusive within purely materialist models. Such reasoning builds a cumulative case against materialism and in favor of a transphysical dimension, which aligns more coherently with the breadth of human experience (Collins, 2009b; Craig & Moreland, 2003; Moreland, 2009).

Given the explanatory gaps in materialist models, especially regarding consciousness, intentionality, and the unity of the self, the burden of proof lies with the materialist to demonstrate that purely material causes are sufficient to account for these phenomena. The Thomistic understanding of the soul addresses this challenge by proposing that the soul is sui generis—a unique form of causation that transcends physical mechanisms while remaining intrinsic to the human person (Feser, 2010; Stump, 2003). Rather than interacting with the body as a separate substance requiring energy-based transfer, the soul is the substantial form of the body, unifying and animating it. The soul is not an independent entity that operates in parallel with the body, as in Cartesian dualism; instead, it is the unifying principle that makes the body a living, rational organism. Without the soul, the body ceases to be a human body, it is merely a corpse. Conversely, without the body, the soul is incomplete. This intrinsic unity ensures that mental and physical processes are integrated rather than mechanically linked.

The Thomistic framework also aligns with the theological doctrine of the resurrection. As the apostle Paul writes in 1 Corinthians 15, the resurrection body is both continuous with the present body and transformed into a glorified state. This belief reinforces the Thomistic view that the soul, as the substantial form of the body, requires bodily expression for its fullest realization. While there is conscious existence in the intermediate state (Philippians 1:23), the ultimate completion of human nature occurs in the resurrection, where body and soul are reunited in perfected unity. Thus, while the soul cannot be directly observed through empirical methods, its role in configuring the body's unity and enabling distinctly human capacities can be reasonably inferred through legitimate scientific reasoning. Just as gravity's unseen presence is revealed through observable effects, so too does the soul's influence manifest in the body's configuration, rational thought, moral reflection, and self-awareness. This Thomistic view, enriched by theological insights into the resurrection, offers a coherent and integrative understanding of the human person as a unified whole—body and soul.

The Transphysical Soul: A Coherent and Explanatory Model of Personhood

The Brain Metafunction Framework, which integrates perception, cognition, and embodiment as an active, dynamic process, also provides a compelling case for the existence of the soul as a transphysical reality. In previous discussions of perception, we explored how the efferent loop of perception extends beyond passive reception, engaging actively with reality in a manner that challenges purely materialist explanations. Building upon that insight, this chapter argues that physicalist neuroscience alone cannot adequately explain consciousness, selfhood, and personhood. Instead, the soul emerges as a compelling explanatory model, bridging the material and transphysical dimensions of human experience. In contrast, materialist neuroscience seeks to explain consciousness and selfhood solely in terms of neural activity, emergent properties, and brain computations (Tononi & Edelman, 1998; Friston, 2010). While significant progress has been made in mapping brain networks involved in cognition and perception (the neural correlates), no physicalist theory has successfully explained how subjective experience arises from neural processes—a challenge known as the hard problem of consciousness (Chalmers, 1996). Neural correlates of consciousness (NCCs) indicate brain regions involved in awareness (Koch et al., 2016), but these do not explain why first-person experience exists. The Brain Metafunction Framework acknowledges that while the brain is necessary for conscious processes, it alone cannot fully account for the continuity of selfhood and identity over time (Nagel, 2012).

Materialist theories also struggle to account for the persistence of selfhood despite neuroplasticity. The brain undergoes constant structural and functional changes, yet the sense of self often remains continuous. If identity were solely reducible to neural configurations, one would expect radical changes in selfhood following severe neurodegeneration; yet personal identity often persists despite substantial neural loss (Damasio, 1999). Additionally, cases of terminal lucidity—where individuals with advanced dementia or severe brain degeneration briefly recover remarkable clarity of mind before death—further challenge materialist explanations. Such phenomena suggest an underlying principle that is not solely dependent on brain structures but instead unifies and transcends them—a role classically attributed to the soul.

The transphysical concept, as outlined in the Brain Metafunction Framework, serves as a bridge between the physical and the metaphysical. Unlike metaphysical dualism, which posits a strict separation between soul and body, the transphysical view recognizes a unified yet non-reductive relationship between mind and matter. The soul is not merely an abstract, immaterial essence but an integrative, organizing principle that grounds personhood, self-awareness, and agency while engaging with the physical body. Evidence supporting the transphysical nature of the soul comes from several domains. First, perception and the efferent loop demonstrate that conscious experience actively projects meaning onto the world rather than merely receiving sensory input. This outward-directed nature of perception suggests that cognition and awareness are not confined to neural machinery but extend into reality itself (Noë, 2004). The brain functions as an interface between the self and the world but does not fully account for how we experience being in the world (Merleau-Ponty, 1962). Second, consciousness operates as a seamless, unified, and transparent field of experience despite its multifunctional nature (Tononi, 2008). If cognition were purely a computational process distributed across neural networks, one would expect fragmented or parallel streams of awareness. Instead, experience is singular and coherent, suggesting that a transphysical unifying principle integrates these processes into a continuous self.

Near-death and veridical experiences further challenge materialist assumptions. Cases of veridical perception during clinical death—where individuals report accurate descriptions of events while lacking measurable brain activity—suggest that consciousness may operate independently of brain function (van Lommel, 2010). While some materialist critics attribute these experiences to residual brain activity or hallucinations, the structured, vivid, and accurate nature of many such accounts defies simple neurological explanations. Similarly, spiritual experiences and altered states of consciousness across cultures resist materialist explanations, pointing to dimensions of cognition that extend beyond physical processes (Radin, 1997). By embracing these phenomena, the transphysical domain offers a framework for investigating experiences previously sidelined in mainstream neuroscience.

Free will and intentionality also indicate the insufficiency of materialist explanations. If decision-making were entirely reducible to deterministic neural computations (Libet, 1985), then free will would be an illusion (Gazzaniga, 2011). However, the experience of deliberation, self-reflection, and moral responsibility suggests that the human person operates beyond mechanistic determinism—a role naturally attributed to the soul as the true agent of free will (Searle, 1992). While some materialist theories propose that free will arises as an illusion generated by unconscious brain processes, this view struggles to account for the deep coherence between conscious intention and behavior in everyday life. The Brain Metafunction Framework proposes that personhood requires an integrative center that unifies cognitive and physical processes, transcending purely material explanations. In this framework, the soul is best understood as the integrator of cognitive and physical functions, harmonizing perception, memory, thought, and action into a unified subjective experience. It also serves as the seat of personal identity, ensuring continuity despite material changes in the body. Additionally, the soul provides the foundation for rationality and free will, enabling genuine moral decision-making beyond deterministic processes. Finally, the soul functions as the life-giving principle that ensures the coherence of the organism as a whole, maintaining biological and existential unity.

The Brain Metafunction Framework highlights the limits of materialist explanations and the necessity of a transphysical dimension to consciousness and selfhood. While some critics may claim that transphysical explanations invoke metaphysical assumptions, this framework builds on empirical evidence, philosophical reasoning, and the observed limitations of materialist models. Far from an outdated metaphysical construct, the soul emerges as a compelling and coherent model for unified consciousness, personal identity, free will, and intentionality. By incorporating the transphysical domain into neuroscience, this framework advances a holistic, fitting exploration of human cognition. As research progresses, this paradigm will inspire interdisciplinary collaboration, uniting neuroscience, AI, bioethics, and philosophy to unlock the full nature of personhood.

The Transparency of Consciousness and Its Implications for the Soul

The transparency of consciousness refers to the phenomenon where we experience the world as directly given, without awareness of the complex neural and cognitive processes shaping perception. This idea, central to phenomenology and cognitive science, suggests that while underlying mechanisms structure consciousness, we do not perceive these mechanisms themselves—only their effects. Thomas Metzinger (2003) argues that we experience a "self-model" rather than direct access to the brain's computations, making consciousness appear seamless. Michael Tye (1995) and Fred Dretske (1995) further develop this concept in analytic philosophy, asserting that perception is representational but transparent—we see the world, not the neural processes constructing it. In predictive processing models proposed by Jakob Hohwy (2013) and Andy Clark (2008), consciousness arises from top-down predictions in which the brain constructs experience by minimizing uncertainty. Yet despite its inferential structure, this constructive process remains hidden from conscious awareness. Maurice Merleau-Ponty (1945/2012) and Edmund Husserl (1913/1998) similarly emphasize that perception is embodied and intentional, yet we remain unaware of the structures that give rise to experience.

The transparency of consciousness has profound implications for the nature of the soul. While neuroscientific models effectively map the brain's role in correlating with perception and cognition (Tononi & Edelman, 1998), they cannot explain why subjective experience emerges as a unified field rather than fragmented streams of sensory data. This gap underscores the need for a unifying principle that integrates mental processes into a coherent sense of self. If we only ever experience the world transparently—without access to the underlying mechanisms constructing experience—this suggests that consciousness is anchored in something beyond neural representation alone. This reality aligns with philosophical arguments for the soul, positing that personhood, rationality, and self-awareness require a transphysical organizing principle.

Moreover, if perception actively engages with reality through an efferent loop as argued in this book— in line with enactive perception model (Noë 2004) —this further challenges purely physicalist explanations. Noë's view emphasizes that consciousness is not confined to brain activity alone but extends through bodily engagement with the world. This outward-directed nature of perception suggests that the soul—understood as the integrative center of consciousness—operates not as an immaterial "ghost in the machine," but as the unifying principle that harmonizes bodily, cognitive, and intentional experiences. The Brain Metafunction Framework reflects this understanding by recognizing the soul as the principle that grounds selfhood, ensuring continuity of identity, rational agency, and moral responsibility.

In a word, the transparency of consciousness reveals a striking paradox: while neural processes mediate cognition, our conscious awareness remains blind to these mechanisms. This paradox reinforces the notion that subjective experience requires an explanatory foundation beyond mechanistic processes. The soul, as a transphysical unifying principle, provides the most coherent explanation for the persistence of selfhood, the singularity of experience, and the intentional engagement with reality. Rather than contradicting neuroscientific insights, the concept of the soul complements them, offering a deeper account of the unity, continuity, and richness of human consciousness. By incorporating the transphysical dimension into the study of consciousness, the Brain Metafunction Framework highlights the soul's essential role as the integrator of experience, cognition, and identity. As research continues to expand our understanding of perception and consciousness, recognizing the soul's transphysical nature will deepen our comprehension of what it means to be human.

Conclusion

This chapter has explored Brain Metafunction Transphysicalism as a parsimonious, interdisciplinary framework for understanding the transphysical dimension of the human soul and its profound connection to mind-brain causality, consciousness, perception, and human nature. In developing this framework, the chapter defined the human soul and provided multiple lines of evidence supporting its reality. Drawing from neo-Thomism and the Christian tradition, we examined the soul not as an isolated substance but as the substantial form of the body. Within this unified entity, material and immaterial principles of human existence coexist harmoniously. In this Christian theist perspective, the soul functions as the intrinsic organizational principle and animating force of the human person, imparting rationality, self-awareness, and spiritual depth. The brain, in turn, serves as the critical mediator of this soul-body unity, facilitating higher-level physiological integration, consciousness, and complex mental processes that transcend purely physicalist explanations.

A key insight of this chapter is its emphasis on the soul's relationship with the body as a fundamental unity rather than a juxtaposition of separate entities. Brain Metafunction Transphysicalism reconceptualizes the mind-body problem not as a dualistic opposition between matter and spirit but as an integrated dynamic, where the brain's metafunction expresses the soul's transphysical influence. This perspective highlights the interplay between physical and non-physical dimensions in shaping consciousness and cognitive capacities. By advancing a model that transcends reductionist physicalism while differing from

Cartesian dualism, this framework provides a comprehensive explanation for uniquely human phenomena, such as rationality, moral agency, and creativity (Feser, 2010; Craig & Moreland, 2003).

The analogy to gravity has proven particularly useful as a conceptual tool in this discussion. Just as gravity's nature remains elusive but is inferred from observable effects—like the motion of celestial bodies, the soul's existence can be inferred from distinctly human qualities that elude purely physicalist explanations. Rationality, creativity, moral agency, and spiritual depth each point to a unifying principle that transcends neural processes alone. This reasoning aligns with the method of inference to the best explanation, where the cumulative evidence supports the existence of a transphysical dimension. By adopting this approach, Brain Metafunction Transphysicalism challenges the sufficiency of materialist interpretations, which struggle to account for the full richness of human experience.

In addressing physicalist critiques, this chapter also engaged with the energy conservation objection (sometimes referred to as the causal pairing problem)—a significant challenge to non-physical causation. Brain Metafunction Transphysicalism offers a coherent response by conceptualizing the soul's influence as inherent to the soul-body unity rather than an extrinsic force requiring energy transfer. Additionally, the concept of Infomism provides a coherent mechanism for mind-body interaction that avoids violating physical laws. This inherent form of causation is sui generis, akin to information-based causation rather than relying on energy exchange. Robin Collins's insights further support this view, emphasizing that non-physical causation can occur through informational processes, demonstrating that meaningful causation need not involve energy transfer (Collins, 2008). Compatible with this approach, Brain Metafunction Transphysicalism effectively addresses the causal pairing problem, presenting a scientifically coherent model for transphysical influences.

Ultimately, Brain Metafunction Transphysicalism presents a compelling framework that affirms the brain's indispensable role while recognizing the necessity of a transphysical foundation for human nature. This framework reinforces the understanding of the human person as a fundamentally unified entity, where the soul—as the substantial form of the body—integrates material and immaterial dimensions into a cohesive whole. The brain's metafunction emerges as an expression of this unity, harmonizing physical processes with the soul's immaterial influence. This holistic understanding advances a theistic worldview, offering a synthesis that harmonizes scientific principles with theological and philosophical insights. By grounding the human person in both physical and transphysical dimensions, Brain Metafunction Transphysicalism offers a robust, integrative framework—one that illuminates profound questions of consciousness, personhood, and the essence of human existence.

REFERENCES

1 Corinthians 2:12–3:4: Describes three existential categories of persons—spiritual (pneumatikos), soulish (psychikós), and carnal (sarkivós)—based on how individuals respond to divine teaching.

1 Peter 3:3–4: Refers to "the hidden person of the heart," emphasizing the inner spiritual beauty over outward appearance.

1 Thessalonians 5:23: Paul's prayer for the sanctification of the whole person—spirit, soul, and body—emphasizes humans' holistic nature.

Aquinas, T. (1947). Summa Theologica. Translated by the Fathers of the English Dominican Province. Benziger Brothers.

Aristotle. *De Anima (On the Soul).* Translated by Hugh Lawson-Tancred, Penguin Classics, 1987.

Augustine. (2002). Confessions (Henry Chadwick, Trans.). Oxford University Press.

Barr, James. *The Semantics of Biblical Language.* Oxford University Press, 1961.

Bateson, G. (1972). Steps to an Ecology of Mind. University of Chicago Press.

Beauregard, M., & O'Leary, D. (2007). *The spiritual brain: A neuroscientist's case for the existence of the soul.* Harper Collins.

Bell, J. S. (1987). Speakable and Unspeakable in Quantum Mechanics. Cambridge University Press.

Brentano, F. (1874/1995). *Psychology from an empirical standpoint.* Routledge.

Catechism of the Catholic Church. (1994). Second Edition. Libreria Editrice Vaticana.

Chalmers, D. J. (1995). Facing up to the problem of consciousness. Journal of Consciousness Studies, 2(3), 200–219.

Chalmers, D. J. (1996). *The conscious mind: In search of a fundamental theory.* Oxford University Press.

Chalmers, D. J. (2010). *The character of consciousness.* Oxford University Press.

Chawla, L. S., Akst, S., Junker, C., Jacobs, B., & Seneff, M. G. (2009). Surges of electroencephalographic activity at the time of death: A case series. Journal of Palliative Medicine, 12(12), 1095–1100.

Chomsky, N. (1965). Aspects of the Theory of Syntax. MIT Press.

Chomsky, N. (2000). New Horizons in the Study of Language and Mind. Cambridge University Press.

Clark, A. (2008). *Supersizing the mind: Embodiment, action, and cognitive extension.* Oxford University Press.

Clark, A. (2013). Whatever next? Predictive brains, situated agents, and the future of cognitive science. *Behavioral and Brain Sciences, 36*(3), 181–204. https://doi.org/10.1017/S0140525X12000477

Collins, R. (2008). Modern physics and the energy conservation objection to mind-body dualism. In J. Schloss & M. J. Murray (Eds.), The Believing Primate (pp. 143–162). Oxford University Press.

Collins, R. (2009a). The fine-tuning argument: A scientific explanation of the universe's existence. In W. L. Craig & J. P. Moreland (Eds.), The Blackwell Companion to Natural Theology (pp. 202–281). Wiley-Blackwell.

Collins, R. (2009b). A scientific case for the soul. In W. L. Craig & J. P. Moreland (Eds.), The Blackwell Companion to Natural Theology (pp. 202–281). Wiley-Blackwell.

Craig, A. D. (2003). Interoception: The sense of the physiological condition of the body. Nature Reviews Neuroscience, 3, 655–666.

Craig, W. L., & Moreland, J. P. (2003). Philosophical Foundations for a Christian Worldview. IVP Academic.

Crick, F., & Koch, C. (1990). Toward a neurobiological theory of consciousness. Seminars in the Neurosciences, 2(4), 263–275.

Damasio, A. (1994). Descartes' Error: Emotion, Reason, and the Human Brain. G.P. Putnam's Sons.

Damasio, A. (1999). The Feeling of What Happens: Body and Emotion in the Making of Consciousness. Harcourt.

Daniel 1:8 Depicts Daniel's purposing in his heart, connecting the will and the heart.

Des Chene, D. (2000). Life's Form: Late Aristotelian Conceptions of the Soul. Cornell University Press.

Doty, R. W. (1998). The five mysteries of the mind, and their consequences. *Neuropsychologia*, 36(10), 1069–1076. https://doi.org/10.1016/s0028-3932(98)00060-8.

Dretske, F. (1995). *Naturalizing the mind.* MIT Press.

Encyclopædia Britannica. (2023). *Graviton.* In *Encyclopædia Britannica.* https://www.britannica.com/science/graviton

Engmann, B. (2014). *Near-death experiences: Heavenly insight or human illusion?* Springer. https://doi.org/10.1007/978-3-319 -03728-8

Feser, E. (2010). Aquinas: A Beginner's Guide. Oneworld Publications.

Fuchs, T. (2018). Ecology of the Brain: The Phenomenology and Biology of the Embodied Mind. Oxford University Press. https://doi.org/10.1093/med/9780199646883.001.0001

Gallagher, S. (2005). How the Body Shapes the Mind. Oxford University Press.

García, O. D. (1995a). Whole brain? Higher brain? A new formulation of death. In C. Machado (Ed.), *Proceedings of the Second International Symposium on Brain Death* (pp. 31–38). Elsevier.

García, O. D. (1995b). Reformulating death: Overcoming whole-brain inconsistencies. In C. Machado (Ed.), *Proceedings of the Second International Symposium on Brain Death* (pp. 39–46). Elsevier.

García, O. D. (1998a). Una reformulación completa y precisa de la muerte humana [A complete and accurate reformulation of human death]. *Revista de Neurología, 26*(154), 1048–1053. https://doi.org/10.33588/rn.26154.97176

García, O. D. (1998b). A new formulation of death and its relevance to medical law. *Medicine and Law, 17*(4), 493–501. PMID: 10396910.

García, O. D. (2001). The unity of the human person: A central concept in the interface between medicine and theology. *The Linacre Quarterly, 68*(1), 71–89. https://doi.org/10.1080/20508549.2001.11877600

Gilson, E. (1952). Being and Some Philosophers. Pontifical Institute of Mediaeval Studies.

Gleick, J. (2011). The Information: A History, a Theory, a Flood. Pantheon.

Greyson, B. (2003). Near-death experiences in a psychiatric outpatient clinic population. *Psychiatric Services, 54*(12), 1649–1651. doi: 10.1176/appi.ps.54.12.1649.

Greyson, B. (2007). Consistency of near-death experience accounts over two decades: Are reports embellished over time? *Resuscitation, 73*(3), 407–411. https://doi.org/10.1016/j.resuscitation.2006.10.013

Greyson, B., & Bush, N. E. (1992). Distressing near-death experiences. *Psychiatry, 55*(1), 95–110. https://doi.org/10.1080/00332747.1992.11024583

Hardcastle, Valerie Gray (1998). The Binding Problem. In George Graham & William Bechtel (eds.), A Companion to Cognitive Science. Blackwell. pp. 553–565.

Hebrews 4:12 Explains the Word of God's penetrating power, capable of dividing soul and spirit, joints and marrow, and discerning the thoughts and intents of the heart.

Hohwy, J. (2013). *The predictive mind.* Oxford University Press.

Holden, J. M. (2009). Veridical perception in near-death experiences. In J. M. Holden, B. Greyson, & D. James (Eds.), *The handbook of near-death experiences* (pp. 185–212). Praeger.

Husserl, E. (1931/1964). *The phenomenology of internal time-consciousness* (M. Heidegger, Ed.; J. Churchill, Trans.). Indiana University Press.

Husserl, E. (1913/1998). *Ideas pertaining to a pure phenomenology and to a phenomenological philosophy* (F. Kersten, Trans.). Springer.

James, W. (1890). The Principles of Psychology. Henry Holt and Company.

James, W. (1904). Does 'consciousness' exist? The Journal of Philosophy, Psychology and Scientific Methods, 1(18), 477–491.

Jeeves, M. A., & Brown, W. S. (2009). *Neuroscience, psychology, and religion: Illusions, delusions, and realities about human nature.* Templeton Foundation Press.

Kelly, E. W., Greyson, B., & Stevenson, I. (2000). Can experiences near death furnish evidence of life after death? *Omega: Journal of Death and Dying, 40*(4), 513–519. https://doi.org/10.2190/KNTM-6R07-LTVT-MC6K

Kelly, E. F., Kelly, E. W., Crabtree, A., Gauld, A., Grosso, M., & Greyson, B. (2007). *Irreducible mind: Toward a psychology for the 21st century.* Rowman & Littlefield Publishers.

Kenny, A. (2002). Aquinas on Mind. Routledge.

Kim, J. (2005). Physicalism, or Something Near Enough. Princeton University Press.

Klubertanz, George P. *The Philosophy of Human Nature.* Appleton-Century-Crofts, 1953.

Koch, C., Massimini, M., Boly, M., & Tononi, G. (2016). Neural correlates of consciousness: Progress and problems. *Nature Reviews Neuroscience, 17*(5), 307–321.

Koenig, H. G. (2012). The Handbook of Religion and Health. Oxford University Press.

Kreeft, P. (2005). The Philosophy of Tolkien: The Worldview Behind The Lord of the Rings. Ignatius Press.

Kretzmann, N., & Stump, E. (1993). *The Cambridge companion to Aquinas.* Cambridge University Press.

Levine, Joseph. 1983. "Materialism and Qualia: The Explanatory Gap." *Pacific Philosophical Quarterly* 64 (October): 354–361.

Lewis, C. S. (1940). The Problem of Pain. HarperOne.

Lewis, C. S. (1952). Mere Christianity. HarperOne.

Loftus, E. F., & Pickrell, J. E. (1995). The formation of false memories. Psychiatric Annals, 25(12), 720–725.

MacIntyre, A. (1999). Dependent Rational Animals: Why Human Beings Need the Virtues. Open Court Publishing.

Martial, C., Fritz, P., Gosseries, O., & others. (2025). A neuroscientific model of near-death experiences. *Nature Reviews Neurology*. https://doi.org/10.1038/s41582-025-01072-z

McGilchrist, I. (2009). The Master and His Emissary: The Divided Brain and the Making of the Western World. Yale University Press.

McGinn, C. (1999). The Mysterious Flame: Conscious Minds in a Material World. Basic Books.

McInerny, R. (1999). Ethica Thomistica: The Moral Philosophy of Thomas Aquinas. Catholic University of America Press.

Merleau-Ponty, M. (1945/2012). *Phenomenology of perception* (D. Landes, Trans.). Routledge.

Merriam-Webster. (n.d.). *Scientism.* In *Merriam-Webster.com dictionary.* Retrieved February 15, 2025, from https://www.merriam-webster.com/dictionary/scientism

Metzinger, T. (2003). *Being no one: The self-model theory of subjectivity.* MIT Press.

Moreland, J. P. (2009). The Recalcitrant Imago Dei: Human Persons and the Failure of Naturalism. SCM Press.

Moreland, J. P. (2014). The Soul: How We Know It's Real and Why It Matters. Moody Publishers.

Moreland, J. P. (2018). In defense of a Thomistic-like dualism. In J. J. Loose, A. J. L. Menuge, & J. P. Moreland (Eds.), *The Blackwell Companion to Substance Dualism* (pp. 102–122). Wiley-Blackwell.

Nagel, T. (1974). What is it like to be a bat? The Philosophical Review, 83(4), 435–450.

Nagel, T. (2012). *Mind and cosmos: Why the materialist neo-Darwinian conception of nature is almost certainly false.* Oxford University Press.

Nagel, T. (1986). The View from Nowhere. Oxford University Press.

Newman, J. H. (1874/1999). *An essay in aid of a grammar of assent.* Oxford University Press.

Noë, A. (2004). *Action in perception.* MIT Press.

Noë, A. (2009). Out of Our Heads: Why You Are Not Your Brain, and Other Lessons from the Biology of Consciousness. Hill and Wang.

Oderberg, D. S. (2005). Hylemorphic dualism. *Social Philosophy and Policy, 22*(2), 70–99.

Origen. (1979). De Principiis.

Papineau, D. (2001). The rise of physicalism. In C. Gillett & B. Loewer (Eds.), Physicalism and Its Discontents (pp. 3–36). Cambridge University Press.

Parker, E. S., Cahill, L., & McGaugh, J. L. (2006). A case of unusual autobiographical remembering. *Neurocase, 12*(1), 35–49. https://doi.org/10.1080/13554790500473680

Parnia, S., Spearpoint, K., & Fenwick, P. (2014). Near-death experiences, cognitive function, and psychological outcomes of surviving cardiac arrest. Resuscitation, 85(12), 1799–1805.

Pasnau, R. (2001). Thomas Aquinas on Human Nature: A Philosophical Study of Summa Theologiae 1a 75-89. Cambridge University Press. https://doi.org/10.1017/CBO9780511613180

Pasnau, R. (2011). *Metaphysical themes: 1274–1671.* Clarendon Press.

Perler, D. (2022). Ockham on memory and double intentionality. *Topoi, 41*(1), 133–142. https://doi.org/10.1007/s11245-020-09728-9.

Proverbs 23:7: "As he thinks in his heart, so is he," illustrating the heart's role in shaping identity.

Proverbs 4:23 Advises guarding the heart above all else, as it is the wellspring of life.

Psalm 51:6 Highlights the desire for truth in the innermost being and wisdom in the hidden part of the self.

Ratzinger, J. (Pope Benedict XVI). (1968). Introduction to Christianity. Translated by J. R. Foster. Ignatius Press.

Revonsuo, A. (1999). Binding and the phenomenal unity of consciousness. *Consciousness and Cognition: An International Journal, 8*(2), 173–185. https://doi.org/10.1006/ccog.1999.03 84

Rickabaugh, B. (2018). Against emergent dualism. In J. Loose, A. J. L. Menuge, & J. P. Moreland (Eds.), *The Blackwell companion to substance dualism* (pp. 73–86). Wiley-Blackwell.

Schacter, D. L. (1996). *Searching for memory: The brain, the mind, and the past.* Basic Books.

Searle, J. R. (1983). *Intentionality: An essay in the philosophy of mind.* Cambridge University Press.

Searle, J. R. (1992). *The rediscovery of the mind.* MIT Press.

Sheldrake, R. (2005). The extended mind: Recent experimental evidence. Journal of Consciousness Studies, 12(6), 10–31.

Sheldrake, R. (2013). *Science set free: 10 paths to new discovery.* Deepak Chopra Books.

Spitzer, R. J. (2015). The Soul's Upward Yearning: Clues to Our Transcendent Nature from Experience and Reason. Ignatius Press.

Squire, L. R., & Kandel, E. R. (1999). Memory: From Mind to Molecules. Scientific American Library.

Stump, E. (2003). Aquinas. Routledge.

Swinburne, R. (2004). *The existence of God* (2nd ed.). Oxford University Press.

Swinburne, R. (2013). *Mind, brain, and free will.* Oxford University Press.

Tallis, R. C. (2004). *The knowing animal: A philosophical inquiry into knowledge and truth.* Edinburgh University Press.

Tallis, R. (2010). What neuroscience cannot tell us about ourselves. The New Atlantis. https://www.thenewatlantis.com/publications/what-neuroscience-cannot-tell-us-about-ourselves

Tallis, R. (2011). *Aping Mankind: Neuromania, Darwinitis and the Misrepresentation of Humanity.* Durham: Acumen Publishing.

Tallis, R. (2024). *Extending the mind.* Philosophy Now, (164). https://philosophynow.org/issues/164/Extending_The_Mind

Tononi, G. (2008). Consciousness as integrated information: A provisional manifesto. *Biological Bulletin, 215*(3), 216–242.

Tononi, G., & Edelman, G. M. (1998). Consciousness and complexity. *Science, 282*(5395), 1846–1851.

Treffert, D. A. (2009). The savant syndrome: An extraordinary condition. Philosophical Transactions of the Royal Society B: Biological Sciences, 364(1522), 1351–1357.

Treffert, D. A. (2010). *Islands of genius: The bountiful mind of the autistic, acquired, and sudden savant.* Jessica Kingsley Publishers.

Treisman, A. (1996). The binding problem in perception. Current Opinion in Neurobiology, 6(2), 171–178.

Tulving, E. (2002). Episodic memory: From mind to brain. Annual Review of Psychology, 53(1), 1–25.

Tye, M. (1995). *Ten problems of consciousness: A representational theory of the phenomenal mind.* MIT Press.

Van Lommel, P. (2010). *Consciousness beyond life: The science of the near-death experience.* Harper One.

Velmans, M. (1990). Consciousness, brain, and the physical world. *Philosophical Psychology, 3*(1), 77–99.

Velmans, M. (2000). *Understanding Consciousness.* Taylor & Frances/Routledge. https://doi.org/10.4324/97802034650 28

Velmans, M. (2008). Reflexive monism. *Journal of Consciousness Studies, 15*(2), 5–50.

Wheeler, J. A., & Ford, K. (1998). Geons, Black Holes, and Quantum Foam: A Life in Physics. W. W. Norton & Company.

Wippel, J. F. (2000). The Metaphysical Thought of Thomas Aquinas: From Finite Being to Uncreated Being. Catholic University of America Press.

Wright, N. T. (2003). *The resurrection of the Son of God.* Fortress Press.

21

Christian Theism and the Transphysical: A Coherent Framework for Understanding the Brain Metafunction

Materialism, Naturalism, and Reductionism are influential frameworks, yet they fail to fully capture the complexities of human consciousness, purpose, and morality. Materialism's strict physicalist outlook reduces the mind and consciousness to byproducts of neural processes, denying their autonomy and rendering free will, purpose, and moral truths as illusions. Naturalism, though broader, confines all phenomena to empirical observations, leaving no room for the transcendent. Reductionist approaches in both perspectives fall short when confronted with the intricacies of consciousness, intentionality, and the self—what we identify as the transphysical aspect of the brain metafunction. Christian Theism, however, offers a comprehensive and coherent framework that recognizes the physical and transphysical aspects of human experience. It situates the brain metafunction within a view of humans as imago Dei, created in the image of a purposeful, transcendent God. Unlike reductionist philosophies, which deny or diminish these dimensions, Christian Theism asserts that human consciousness, free will, and moral awareness are intrinsic aspects of our nature, grounded in a Creator who imbues life with purpose and dignity. As we build a cumulative case for Christian Theism, this framework uniquely integrates scientific insights with philosophical and metaphysical depth, addressing the full scope of human experience—both material and transphysical—and providing a compelling, coherent explanation for the origin of the universe, biological life, consciousness, morality, and purpose unmatched by materialist or reductionist views.

Philosophical Implications of Materialism, Naturalism, and Reductionism: Consciousness, Morality, and Meaning

The philosophical implications of Materialism and Naturalism extend into debates about determinism, nihilism, moral relativism, and anti-realism. Determinism, particularly in its hard form, holds that human choices are dictated by prior causes—biochemical and environmental—leaving little room for genuine free will (Harris, 2012; Strawson, 1994). This perspective raises serious questions about moral responsibility and personal accountability. Some critics argue that if our actions are entirely determined, then praise, blame, and justice become conceptually unstable. Within a strictly materialist worldview, some have suggested that nihilism may follow—the idea that life lacks inherent meaning or transcendent purpose because human existence is seen as a byproduct of impersonal physical processes (Nietzsche, 1887/2006; Camus, 1942/1991). However, not all naturalists draw this conclusion; thinkers like Thomas Nagel (1971) have acknowledged the "absurdity" of life without supernatural grounding while still affirming the legitimacy of moral and existential inquiry.

Similarly, the absence of a transcendent standard within materialism often leads to moral relativism—the belief that moral values are subjective, culturally contingent, and not universally binding (Street, 2006). Even some naturalists, such as Ruse (1989), acknowledge this challenge; he argues that morality, though it feels objective, is in fact a product of evolutionary processes, shaped for survival rather than truth. According to Ruse, our moral sense is "illusionary" in the sense that it mimics objectivity but lacks any grounding beyond biological utility. This relativism can challenge the philosophical foundations of universal human rights and the ability to assert absolute moral norms, potentially leading to ethical ambiguity and inconsistencies in moral discourse (Rorty, 1989). In contrast, Christian theism affirms moral objectivity by grounding moral values in the character of a perfectly good and just God, offering a coherent and transcendent basis for human dignity, justice, and moral obligation (Craig, 2008; Adams, 1999; Lewis, 1943).

Challenges to Determinism and the Reductionist View

Twentieth-century developments in physics, particularly in quantum mechanics and chaos theory, have profoundly challenged classical determinism and reductionist accounts of nature. Quantum indeterminacy, exemplified by Heisenberg's Uncertainty Principle, demonstrates that particles do not follow fixed, deterministic paths at the most fundamental level. The principle asserts that it is impossible to simultaneously know a particle's position and momentum with absolute precision (Heisenberg, 1930). As a result, quantum systems are governed by probability distributions rather than exact predictability. For example, the precise moment of a radioactive atom's decay cannot be determined, only the likelihood of when it might occur (Penrose, 1989). This probabilistic framework suggests that the physical universe is not wholly causally closed in the classical sense, opening conceptual space for spontaneity and potential freedom (Stapp, 2001). Chaos theory, though grounded in classical physics, has similarly undermined simplistic deterministic expectations at the macroscopic level. Systems such as weather patterns and planetary orbits, while governed by deterministic equations, display extreme sensitivity to initial conditions—a phenomenon known as the "butterfly effect." Small variations in starting parameters can lead to radically different outcomes, rendering long-term prediction practically impossible (Gleick, 1987). As such, these systems are deterministic in principle but unpredictable in practice, due to the exponential amplification of minor fluctuations. Quantum mechanics and chaos theory reveal that nature operates with a far greater degree of complexity, non-linearity, and unpredictability than early modern science envisioned. These findings challenge the rigid determinism and reductionism often associated with materialism. Instead, they suggest that the universe may include genuine elements of indeterminacy and contingency, which may accommodate a metaphysical framework that allows for human freedom and moral responsibility (Kane, 2005).

The Theistic Implications of Quantum Indeterminacy and Chaos Theory

The philosophical implications of these scientific insights are significant. By undermining the notion that all events are inevitable outcomes of prior states, modern physics supports more nuanced accounts of causality—ones that can include intentionality, agency, and contingency. While indeterminacy alone does not guarantee libertarian free will, it undermines the strict determinism that precludes it and provides a metaphysical environment in which human freedom is at least possible (Kane, 2005; Polkinghorne, 2000). Christian theism, particularly in its Thomistic articulation, affirms a metaphysical understanding of the human person that integrates both material and immaterial dimensions. According to Thomas Aquinas, the soul is the substantial form of the body—an immaterial principle that grounds unity, consciousness, and rational agency (Aquinas, Summa Theologiae, I, Q76). This view provides an ontological basis for free will and moral responsibility that reductionist materialism struggles to account for. Contemporary Christian philosophers like Moreland and Rae (2000) argue that Christian anthropology, by affirming the soul and free will, is better positioned to explain human moral experience, intentionality, and self-transcendence. This synthesis between contemporary physics and Christian theism highlights the limitations of materialist reductionism. While materialism posits that all phenomena can be fully explained by physical processes, modern science suggests otherwise. Indeterminacy in quantum mechanics and unpredictability in chaos theory reveal a world that is not rigidly mechanistic, but open-ended and dynamic. Within this conceptual landscape, Christian theism offers a coherent vision of human nature—one that respects both scientific discovery and the reality of moral agency. It provides a framework in which freedom, dignity, and consciousness are not illusory byproducts of matter, but essential features of a divinely ordered creation.

The Irony and Limitations of Scientism

Scientism—the belief that science is the only or most authoritative source of knowledge—has become a prevailing stance within contemporary naturalism. Yet it suffers from serious philosophical, historical, and existential limitations. At its core, scientism asserts that only empirical, scientific inquiry can yield true knowledge. However, as William Lane Craig (2008) and Alvin Plantinga (2011) both argue, this assertion is self-referentially incoherent: the claim that "science is the only source of truth" is itself not a scientific claim and cannot be tested or verified by scientific methods. It is, instead, a metaphysical statement—thus violating its own standard. This irony exposes a deeper philosophical shortcoming: scientism relies on non-empirical foundations that it cannot justify within its own framework. The laws of logic, mathematical truths, and moral axioms are all presupposed by science, but cannot be established scientifically (Feser, 2019). As Plantinga (2011) notes, science also presumes the reliability of human cognitive faculties—another metaphysical assumption. Science, far from operating in isolation, is embedded within a broader epistemic and philosophical framework that includes assumptions it cannot itself prove.

Scientism also overlooks the historical context of science's origins. The scientific revolution did not emerge in a philosophical vacuum but was profoundly shaped by a Christian theistic worldview. Historians such as Peter Harrison (2007) and Stanley Jaki (1978) have shown that early scientists like Galileo, Boyle, and Newton operated with the conviction that the universe was created by a rational God and thus intelligible to human reason. This conviction is captured in John Polkinghorne's (1998) phrase, "the reason within matches the reason without." The very notion that nature operates according to universal, discoverable laws reflects a theological belief in divine order. Modern scientists have echoed these insights. Charles Townes, Nobel Laureate in Physics, remarked that "faith is necessary for the scientist even to get started... because he must have confidence that there is order in the universe and that the human mind... has a good chance of understanding this order" (Townes, 1966, as cited in Begley, 1988). This resonates with Michael Polanyi's (1958) argument that scientific practice rests on tacit beliefs and personal commitments, including trust in the mind's capacity to uncover truth. Science, therefore, is not a purely objective endeavor; it is shaped by the philosophical and sometimes theological convictions of those who practice it.

Further, scientism's philosophical dependency reveals that it is not a rival to philosophy or theology, but a position that silently borrows from both. Geisler and Turek (2004) rightly point out that science depends on philosophical assumptions about causality, induction, truth, and logic. Edward Feser (2019) reinforces this by arguing that when scientists assume a priori that only natural causes are valid explanations, they are making a metaphysical—not scientific—judgment. Such assumptions risk excluding alternative explanations, including the possibility of intelligence, design, or transphysical causes, before the evidence is even considered. Science does not "speak for itself." As Polanyi (1958) and Plantinga (2011) argue, scientists interpret data through philosophical lenses, often shaped by their own metaphysical commitments. When ideology dictates allowable conclusions, science risks mirroring the dogmatism it often critiques in religion. By dismissing non-empirical sources of knowledge, scientism undermines its own philosophical foundation and restricts its explanatory power.

Scientism is especially inadequate when addressing the existential and moral dimensions of human life. It can describe how the brain processes pain but cannot explain why suffering matters. It may track behavior but cannot determine whether a moral choice is right or wrong. As Thomas Nagel (1997) and Roger Scruton (2012) argue, ethical values, aesthetic experiences, and human subjectivity elude reduction to empirical data. These are irreducible aspects of human experience that point to realities science cannot quantify but which are essential for a full account of the human condition. Ultimately, scientism fails the most critical test of any worldview: its ability to account for and respond to the reality of evil. While science can describe suffering mechanistically, it cannot tell us whether suffering is unjust or why it evokes moral outrage. These are philosophical and theological questions. A worldview limited to the empirical cannot address them without stepping beyond its own boundaries. In contrast, a theistic framework—especially that rooted in Christian theology—offers resources for understanding suffering not just as a natural phenomenon but as a moral and existential reality, one that demands meaning and redemption.

The Problem of Evil

The Problem of Evil is frequently presented by naturalists and atheists as a central challenge to theism: how could a benevolent and omnipotent God allow the existence of evil and suffering? This objection, classically posed by J. L. Mackie (1982), argues that the presence of evil is logically incompatible with the existence of an all-good, all-powerful God. However, a deeper analysis reveals a striking irony: while evil is indeed a serious challenge for theism, it presents an even more profound philosophical problem for Naturalism and Materialism. Under Naturalism, the view that reality is fundamentally physical and governed solely by natural laws, the universe is ultimately indifferent to human suffering. Human consciousness, moral instincts, and experiences are seen as products of evolutionary processes and biochemical mechanisms. In this framework, terms like "good" and "evil" refer not to objective realities but to subjective preferences or social conventions. Without a transcendent standard of morality, moral judgments are reduced to emotional responses or survival-enhancing instincts, and suffering lacks intrinsic significance.

As C. S. Lewis (1940) observed, if the universe has no meaning, we should not expect to find meaning in our moral outrage at evil. Yet naturalists often borrow moral categories from theistic frameworks when arguing against God's existence, assuming that suffering and injustice are truly wrong—not just evolutionarily inconvenient. This creates an internal tension: to claim that evil objectively exists is to affirm a moral standard that Naturalism itself cannot account for (Craig, 2008; Plantinga, 2011). As Alvin Plantinga (1974) demonstrated in his Free Will Defense, the existence of evil is not logically incompatible with a good God if human free will is a morally sufficient reason for allowing it. On the other hand, if moral values are mere byproducts of evolution, then the problem of evil becomes not a philosophical objection to God's existence, but a semantic problem about human emotional responses. In contrast, Christian Theism offers a coherent and meaningful account of evil. It affirms the existence of objective moral values grounded in the nature and character of a just and loving God. Within this worldview, evil is real—a deviation from the good—and thus demands an explanation. Christianity acknowledges the seriousness of evil, not by denying it or redefining it, but by placing it within a purpose-driven moral and redemptive framework.

Theistic responses to evil include several complementary perspectives. Some emphasize the role of free will, suggesting that moral agency necessarily involves the possibility of choosing evil (Plantinga, 1974). Others, like John Hick (1966), propose a "soul-making" theodicy, wherein suffering contributes to moral and spiritual growth. Still others point to the fallen state of creation or view evil as part of a larger, mysterious divine plan oriented toward ultimate justice and restoration (Swinburne, 1998). While these explanations do not eliminate the emotional weight of evil, they render it philosophically intelligible and existentially meaningful. At the heart of Christian theology is a radical claim: God does not merely observe human suffering—He enters into it. In the person of Jesus Christ, God experiences the full depth of human pain, betrayal, and death. As Pope John Paul II (1984) writes in *Salvifici Doloris*, "human suffering itself has been redeemed" through Christ's passion. Suffering becomes a pathway to transformation and communion with God. Protestant theologian John Stott (1986) similarly asserts, "the God who allows us to suffer also suffers with us," emphasizing that divine love is not abstract or distant but deeply personal and participatory. This Christocentric answer to the problem of evil offers not merely a theoretical resolution, but a relational and redemptive one. As N. T. Wright (2006) notes, Christianity does not attempt to "solve" the problem of evil as one solves a math problem. Instead, it tells a story of justice, resurrection, and hope—a story in which evil is real but not ultimate, and where suffering, though agonizing, is not meaningless. Christian Theism, therefore, not only acknowledges the reality of evil—it uniquely confronts it with the resources of divine justice, human dignity, and redemptive suffering. In contrast to Naturalism's silence in the face of moral outrage, Christian Theism dares to offer both an explanation and a promise: that evil will not have the final word.

The Cumulative Case for Theism

While Materialism and Naturalism dominate much of contemporary intellectual discourse, their explanatory limitations—especially concerning consciousness, morality, purpose, and the origin of life and the universe—invite a serious reexamination of Theism. This chapter presents a cumulative case for Christian Theism, drawing on multiple lines of evidence and philosophical analysis to argue that Theism offers a more coherent and comprehensive framework for understanding reality than reductionist alternatives. Unlike worldviews that reduce all phenomena to physical processes, Christian Theism affirms the existence of both material and immaterial realities. It provides a framework in which the mind, morality, and metaphysical meaning are not illusory anomalies but essential components of human existence. Theism affirms a holistic view of the person, where the mind is not reducible to brain chemistry but is integrally linked to a non-physical soul (Moreland & Craig, 2003). This perspective helps account for phenomena such as abstract thought, free will, and self-awareness.

Philosopher Alvin Plantinga (2011) has argued that Naturalism is epistemologically self-defeating. If our cognitive faculties are the result of blind evolutionary processes aimed at survival rather than truth, then we have little reason to trust our beliefs, including our belief in Naturalism. In this way, Naturalism undercuts its own foundation. These limitations are increasingly evident: quantum indeterminacy challenges classical determinism, while consciousness, intentionality, moral experience, and various phenomena remain inexplicable within purely physicalist models (Chalmers, 1996; Nagel, 1974; Lewis, 1943). Christian Theism, by contrast, offers a metaphysical account of reality that not only accommodates these phenomena but affirms them as integral to a purposeful and intelligible cosmos grounded in the mind of God. Throughout this book, we have examined the brain metafunction framework, which suggests that human cognition operates on multiple levels—some of which transcend physical explanation. In this chapter, we integrate those findings into a broader philosophical and theological framework. From the origin of consciousness to the foundations of morality, from cosmology to personhood, each line of inquiry contributes to a cumulative argument for Theism. These strands of evidence converge in a coherent and compelling vision of reality—one in which the human mind, the moral order, and the structure of the universe point beyond the merely physical, toward the divine.

The Nature of Consciousness and Qualia: The Argument for God from Consciousness

Consciousness—especially its first-person, subjective dimension known as qualia—resists reduction to purely physical explanations. Qualia refer to the "what-it-is-like" aspect of experience, such as the redness of red or the pain of a headache. These features cannot be captured by third-person physical descriptions alone (Nagel, 1974).

Christian Theism, grounded in the doctrine that humans are created Imago Dei (in the image of God), affirms the existence of a non-physical soul capable of rational reflection and subjective awareness (Moreland, 2014; Swinburne, 2004). Materialist accounts struggle to explain why consciousness exists at all, or how subjective awareness arises from blind physical processes. In contrast, Christian Theism views consciousness as purposeful and personal, grounded in the creative intent of a rational, transcendent mind—God (Craig, 2008; Kreeft & Tacelli, 1994). The Argument from Consciousness maintains that physical brain states alone cannot account for mental phenomena such as thoughts, desires, and sensations (Plantinga, 2011; Swinburne, 2013). J. P. Moreland and Scott Rae (2000) identify five features of mental states that defy reduction to physical processes:

Qualitative Feel (What-It-Is-Like): Mental states are experienced from a first-person perspective. As Thomas Nagel (1974) asked, "What is it like to be a bat?"—a question no physicalist theory can adequately answer. *Intentionality*: Mental states are "about" something—they refer to objects, ideas, or states of affairs. This directedness lacks an equivalent in physical causation (Searle, 1983).

Privacy of Experience: Unlike physical properties, mental states are directly accessible only to the individual experiencing them (Moreland, 2014).

Subjective Ontology: Conscious experience is only accessible from the first-person point of view, resisting objective, third-person reduction (Swinburne, 1997, 2004).

Lack of Spatial Extension: Thoughts and sensations are not located in space; we can find a neuron in the brain, but not a belief or a memory (Plantinga, 2011).

Reductionist philosophers like Dennett (1991) argue that consciousness is an evolutionary illusion—yet this claim presupposes conscious reasoning, rendering the argument self-defeating. Others, like Churchland (1986), propose eliminating mental language altogether, a move few find satisfying. Instead, consciousness appears best explained by the presence of a non-physical mind or soul, consistent with Christian Theism. Swinburne (2004) develops this further in his Argument from Colors and Flavors, showing that the irreducible qualities of experience reinforce the need for a metaphysical explanation rooted in divine intentionality.

Epistemic Reliability and the Evolutionary Challenge to Naturalism

Another critical problem for Naturalism is its inability to justify the reliability of human cognition. If our cognitive faculties are the product of unguided evolutionary processes aimed solely at survival—not truth—then there is no good reason to believe our reasoning is trustworthy. Evolution selects for behavior, not for accurate belief formation. C. S. Lewis (1947) summarized this paradox:

> "Unless I believe in God, I can't believe in thought: so I can never use thought to disbelieve in God."

Philosopher Alvin Plantinga (1993) formalized this insight into the Evolutionary Argument Against Naturalism (EAAN). He argues that if both Naturalism and evolution are true, the probability that our cognitive faculties are reliable is low. Consequently, belief in Naturalism becomes self-defeating. By contrast, Christian Theism provides a coherent alternative: if human beings are created Imago Dei, our rational faculties are designed for truth—a premise that undergirds the very enterprise of science. Additionally, Theism offers a better foundation for abstract, non-physical realities such as the laws of logic and mathematics. These are universal, immaterial, and necessary, properties that defy physicalist explanations. St. Augustine (c. 400/1991) held that such abstract truths exist in the mind of God. William Lane Craig (2016) argues that the existence and intelligibility of such truths are best explained by their grounding in a rational divine mind. In this way, both reason and rational norms are more plausibly the result of design than of chance evolution.

Freedom and Directedness: Evidence for the Soul

Free will and intentionality further challenge the sufficiency of physicalist accounts. If all mental activity is reducible to neural causation, then personal agency becomes an illusion. Physicalism cannot explain how individuals can form beliefs "about" something or choose between alternatives freely. Christian Theism posits a rational soul created by God—capable of intentionality and moral decision-making (Moreland & Rae, 2000). Classical Christian thought, especially in the Thomistic tradition, sees free will not as randomness but as the capacity to act in accordance with reason and moral truth (Aquinas, Summa Theologica, I-II, Q. 83). Kreeft (1990, 2008) explains that this capacity for moral choice reflects our creation in the image of a moral and relational God. Intentionality, the directedness of mental states toward objects or values, is likewise irreducible to physical processes. Peter Geach and Alvin Plantinga (2011) argue that intentionality implies a mind structured for more than mere stimulus-response. Within Theism, this directedness mirrors the relational, communicative nature of the divine mind—a God who acts with purpose and creates creatures capable of intentional action in response.

The Unity of the Self: Memory and the Soul in Theism

The continuity of personal identity across time poses a final challenge to materialism. If memory is merely stored biochemical data, how do we explain the persistent sense of self—the I who remembers childhood and anticipates the future? Christian Theism teaches that memory is not simply information recall, but a function of the soul—an enduring, non-physical self created by God. This view coheres with classical Christian reflections, such as those by St. Augustine in Confessions, where memory is described as a vast interior realm bound to identity, conscience, and spiritual longing. Swinburne (2004, 2013) and Moreland (2014) argue that the enduring self is not reducible to neural continuity. The soul unifies disparate moments of experience into a cohesive narrative—capable of reflection, repentance, and growth. Within Christian Theism, memory becomes not a survival tool, but a faculty essential to spiritual transformation and eternal significance (Craig, 2008; Kreeft & Tacelli, 1994). Personal identity is not an illusion but a reflection of divine design—an anchor of meaning in the story of redemption.

The Conceivability of Disembodiment

Philosopher J. P. Moreland (2014) argues that Near-Death Experiences (NDEs)—especially those involving Out-of-Body Experiences (OBEs)—pose a serious challenge to physicalist claims that the mind is entirely reducible to brain function. In numerous well-documented cases, individuals report observing their own bodies and surroundings from an external vantage point and describe verifiable events with accuracy that defies explanation by brain-bound consciousness. These experiences suggest that consciousness may function independently of the brain's operations, thereby challenging the core assumption of physicalism: that all mental activity is fully dependent on neural processes. Moreland further contends that even apart from the empirical status of such reports, the conceivability of disembodiment itself presents a significant philosophical issue. If it is logically conceivable that a person can exist without their physical body, then it follows—per principles of modal logic—that it is metaphysically possible for the mind to exist independently of the brain (Moreland, 2014; Chalmers, 2009). This undermines the identity thesis central to materialism. The very act of imagining oneself disembodied implies the existence of a transphysical self—a mind or soul that is not reducible to material substance. Such an ontology aligns with Christian Theism, which holds that human beings are both physical and spiritual, made Imago Dei and endowed with a soul capable of functioning beyond bodily constraints.

Near-Death and Transcendental Experiences

When interpreted through the lens of the brain metafunction framework, Near-Death Experiences provide striking evidence of conscious perception under conditions where normal brain activity is severely diminished or absent. Studies report consistent patterns of vivid awareness, peaceful detachment from the body, encounters with light or divine presence, and life reviews during periods of unconsciousness or clinical death (Greyson, 2003, 2010; van Lommel et al., 2001). These events often result in permanent psychological and moral transformation, including heightened compassion, decreased fear of death, and an increased sense of purpose (Ring & Valarino, 1998; Parnia et al., 2001). Especially compelling are cases involving veridical perception—accurate descriptions of external events made during periods of cardiac arrest, general anesthesia, or even flat EEGs. Perhaps most striking are cases reported by individuals who were blind from birth. In these accounts, people without visual memory or capacity describe their environments with stunning details that are later corroborated by witnesses (Ring & Cooper, 1999). Such phenomena are highly improbable under physicalist models, which assume consciousness arises solely from brain function and sensory input.

Critics have offered several naturalistic explanations for NDEs, such as:

1. Neurochemical models (e.g., endorphin release) that account for peaceful feelings.
2. Hypoxic brain states that produce hallucinations during trauma.
3. Psychological mechanisms, such as REM intrusion or expectation-based imagery.

However, these explanations fail to address:

1. The veridical content of NDEs, including cases in the congenitally blind.
2. The structured and lucid quality of these experiences, which contrasts with disorganized brain activity during hypoxia (Chawla et al., 2009).

3. The timing of experiences during periods when conscious brain function is not expected.

While critics like Blackmore (1993) suggest that NDEs are hallucinations triggered by trauma, this fails to explain their verifiability, transformative effects, and cross-cultural consistency (Greyson, 2003). Using the surprise principle (Collins, 2009)—which holds that evidence is more likely under one hypothesis than another—NDEs clearly favor a theistic explanation. These experiences are surprising under materialism but expected under Christian Theism, which teaches that the soul survives bodily death and can encounter transcendent realities. Statistically, about 12% of adults and up to 80% of children who come near death report NDEs (Gallup & Proctor, 1982; Morse & Perry, 1990). While most are positive, a minority involve distressing imagery such as darkness or isolation (Bush, 2012), suggesting that these events cannot be reduced to wish fulfillment or cultural conditioning. Christian theology accommodates both positive and negative NDEs, viewing them as encounters with spiritual realities—glimpses of divine judgment, mercy, or existential transition.

Transcendental Experiences and the Soul's Spiritual Orientation

Beyond Near-Death Experiences, a broader category of transcendental experiences—including mystical union, divine encounters, or overwhelming sensations of unity and love—further supports the existence of a spiritual domain accessible to the soul. These experiences, often occurring outside of religious settings or institutional influence, lead to profound shifts in worldview, ethical sensitivity, and spiritual awareness (James, 1902/2002; Keener, 2011). From a Christian perspective, such experiences are not anomalies but expressions of the soul's innate capacity for spiritual perception and communion with God. In his classic *The Varieties of Religious Experience*, philosopher and psychologist William James documents numerous firsthand accounts of these transformative encounters. One especially compelling example appears in his discussion of "The Sick Soul," where he recounts a man's experience of divine presence during a period of deep existential despair:

"Suddenly, without any warning, I felt that I was in the presence of God. I cannot describe the ecstasy I felt. I felt that I must stand still, that something great was happening to me, and I knew that it was God. The sense of His presence was so real that I could not see anything else." (James, 1902/2002, p. 218)

James notes that the experience was not merely emotional but left a lasting moral and spiritual transformation. The man reported enduring peace, renewed purpose, and a heightened sense of divine nearness. James concluded that such cases must be judged not by their causal origin, which may forever remain mysterious, but by their "fruits"—the enduring moral and existential effects. This principle is consistent with many reports of transcendental and Near-Death Experiences, which are marked not only by vivid perceptual phenomena but by sustained changes in character and values.

These cases lend empirical and philosophical support to the idea that human consciousness includes a spiritual faculty capable of perceiving transcendent realities. Such phenomena are difficult to explain under naturalism, which often reduces them to hallucination, neurochemistry, or psychological delusion. But under Christian Theism, these experiences are expected: they reflect the human soul's orientation toward divine encounter and spiritual truth. The doctrine of Imago Dei affirms that human beings are not merely physically embodied but also spiritually attuned, created to perceive, know, and relate to God. In this view, experiences of divine presence, overwhelming love, or cosmic unity are not misfires of an evolutionary mechanism, but genuine glimpses into spiritual reality. They resonate with biblical accounts of God's self-revelation—through light, peace, awe, and relational intimacy. Such experiences underscore that humans are ensouled persons, created not merely to survive, but to seek truth, beauty, and communion with the divine. In this way, transcendental experiences—like Near-Death Encounters—constitute another powerful line of evidence within the cumulative case for Christian Theism.

The Universal Human Experience of the Sacred and Belief in God as a "Properly Basic Belief"

Across cultures and historical epochs, anthropologists and psychologists have documented the universal human orientation toward the sacred, characterized by belief in a transcendent reality or divine presence (Eliade, 1959; Otto, 1923). Rudolf Otto (1923) described this as the *mysterium tremendum et fascinans*—an experience of overwhelming awe and fascination in the presence of the holy. Mircea Eliade (1959) argued that the human being is best described as *homo religiosus*, structurally oriented toward the sacred as a basic category of human existence. This widespread religious consciousness aligns with the Christian view that human beings are created with an innate capacity and longing for

God. Augustine famously captured this in his Confessions: "You have made us for yourself, O Lord, and our hearts are restless until they rest in you" (Book I). The consistent appearance of sacred experience across cultures suggests a transphysical aspect of human consciousness—one that physicalist explanations fail to fully address.

Christian Theism uniquely accounts for this phenomenon by affirming that humans are spiritual beings, created with immaterial souls designed for communion with God. Philosopher Alvin Plantinga (2000) provides an epistemological model to support this claim. According to his Reformed Epistemology, belief in God can be "properly basic"—justified without inferential argument—if it arises from properly functioning cognitive faculties, operating according to a design plan successfully aimed at truth, in an appropriate epistemic environment. In this view, belief in God is as rationally warranted as our beliefs in the external world, memory, or other minds. It emerges not from indoctrination or inference but from a natural, cognitive disposition—akin to moral intuitions or sense perception. This challenges the modernist assumption that belief in God must be justified only through deductive or empirical proof.

However, Plantinga's model does not reject argumentation. Rather, it supplements it. As Richard Swinburne (2004) states, "This book is written in deep conviction of the possibility of reaching fairly well-justified conclusions by rational argument on this issue, perhaps the most important of all deep issues which stir the human mind" (p. 1). Similarly, William Lane Craig (2010) emphasizes the role of the inner witness of the Holy Spirit, arguing that while rational arguments support belief in God, a direct spiritual experience offers an epistemically foundational encounter with divine reality. Together, these thinkers affirm that belief in God is not an irrational leap but a natural and epistemologically valid response to both internal spiritual awareness and external transcendent encounter. Within Christian Theism, this belief is not only reasonable—it is expected.

The Moral Argument for God

The next pillar in the cumulative case for Christian Theism involves the grounding of objective moral values and duties. Under Naturalism, morality is typically seen as a product of evolutionary development or social convention. On this view, moral norms are adaptive behaviors or cultural preferences—useful for survival, but lacking binding, universal authority. As a result, terms like "right" and "wrong" lose their objective force and become contingent on biology or history. By contrast, Christian Theism affirms a robust moral ontology, grounding moral values in the nature of a perfectly good and just God. William Lane Craig (2008) formulates the Moral Argument as follows:

1. If God does not exist, objective moral values and duties do not exist.
2. Objective moral values and duties do exist.
3. Therefore, God exists.

This syllogism follows a valid logical form (modus tollens). The argument's persuasive power lies in its second premise: the shared human intuition that certain acts—such as murder, torture, or racism—are objectively wrong, regardless of cultural context or personal opinion. As Craig (2008) explains, moral objectivity means that certain truths are "valid and binding whether anyone believes them or not." For example, the Holocaust would remain morally wrong even if the Nazis had won and indoctrinated the world to believe otherwise.

Yet under Naturalism, this kind of objective moral realism is difficult to sustain. Bertrand Russell (1935) acknowledged that morality, on naturalistic terms, arises from social pressures and is therefore conventional. Michael Ruse (1989) goes further, describing morality as a "biological adaptation," an "illusion fobbed off on us by our genes to get us to cooperate." Under this view, moral claims are not grounded in reality but serve only survival functions. While this may explain moral behavior, it does not account for moral truth. By contrast, Christian Theism asserts that moral values are grounded in God's eternal and unchanging nature. God is not merely good—He is the standard of goodness. This view avoids the classic Euthyphro dilemma by asserting that God's nature—not arbitrary will nor independent moral facts—is the standard of goodness. This concept echoes Plato's notion of "the Good" (Plato, 2004/380 BCE), but in Christian theology, it finds its fulfillment in the personal God of Scripture, who is righteous, just, and loving. Moral norms are not arbitrary commands but expressions of God's essential character, giving them both universality and authority.

Furthermore, Theism affirms that humans, made in the image of God (Imago Dei), possess inherent dignity and moral responsibility. Our capacity to recognize moral truth and live according to it reflects our relational design—to know and mirror the moral nature of our Creator. Thus, the Moral Argument not only supports the existence of God—it illuminates the moral structure of reality, which Theism uniquely and coherently explains.

Human Dignity

Flowing directly from the moral argument is the concept of human dignity. Within Naturalism, human beings are viewed as accidental byproducts of impersonal evolutionary processes. On this view, humans are biologically advanced animals, but possess no intrinsic worth beyond instrumental or functional value. Without God or transcendent meaning, human life becomes ultimately insignificant—a conclusion explored by Thomas Nagel (1971), who argued that Naturalism offers no satisfying account of human purpose or value. Christian Theism, in contrast, affirms that humans are created in the image of God (Imago Dei) and are therefore endowed with inherent dignity, moral agency, and purpose. This theistic view undergirds core moral convictions—such as human rights, equality, and moral responsibility—which are difficult to justify on materialist grounds. Without a transcendent source, moral equality becomes a social construct rather than a universal truth—making human rights contingent and potentially revocable.

This issue becomes particularly acute in light of Naturalistic critiques. For example, James Rachels (1990) argues that the concept of speciesism—assigning higher moral worth to humans over animals—is unjustified if humans are merely animals themselves. In Created from Animals, he contends that biological continuity challenges any claim to unique human value. While intended to encourage ethical consistency, this view ultimately undermines the basis for assigning special moral status to human beings. In contrast, Alvin Plantinga (2000) argues that belief in intrinsic human worth fits more coherently within a theistic worldview. If humans are created by a good and purposeful God, then each person—regardless of status, ability, or circumstance—possesses inviolable and objective value. Christian Theism thus grounds justice, compassion, and human rights not in cultural preference or social evolution, but in the eternal character of God. The moral experience of humanity—our belief in right and wrong, our sense of obligation, and our conviction in human worth—finds its most coherent explanation in Theism. These moral intuitions are not illusions or evolutionary byproducts—they reflect the moral structure of reality as created by a righteous and personal God.

Extending the Case for Theism: More Scientific and Philosophical Challenges to Naturalism

Philosophical critiques of Materialism—such as the hard problem of consciousness and the issue of mental causation—expose the limits of a purely physicalist account of the mind (Chalmers, 1996; Searle, 1983). Yet these limitations are not confined to the philosophy of mind. Naturalism, Physicalism, and Materialism face serious challenges when addressing the most fundamental questions of existence: Why is there anything at all? Why does the universe exhibit order, rationality, and beauty? How do we explain consciousness, moral obligation, and human dignity? Christian Theism offers a more coherent and comprehensive explanatory framework. It presents plausible answers to foundational issues such as the beginning and fine-tuning of the universe, the origin of life and biological complexity, the existence of consciousness and free will, and the grounding of objective morality and human worth. It also addresses the problem of evil, the evidential value of near-death and mystical experiences, and the human capacity to perceive objective beauty, rational norms, and universal rights. Furthermore, it accounts for the foundations of logic and mathematics, the prevalence of spiritual experience, fulfilled prophecy, and the historical evidence for the life, death, and resurrection of Jesus Christ. This broad explanatory power across disciplines—from metaphysics to physics—makes Christian Theism a compelling worldview. As we build this cumulative case, we apply the "surprise principle" (Collins, 2009), which states that if a phenomenon is far more likely on one hypothesis than another, it supports the more predictive hypothesis. Many features of the universe and human experience are surprising under Naturalism but anticipated within a Theistic framework—lending cumulative evidential weight to the truth of Christian Theism.

Why Is There Anything at All?

One of the most profound and enduring philosophical questions is: Why is there something rather than nothing? This is the foundation of the Contingency Argument, which maintains that all contingent entities require an external explanation. Since the universe itself is contingent—it could have failed to exist—it must have a cause outside of itself. Christian Theism answers this by positing a necessary, self-existent being—God—whose existence is not contingent but grounded in His own nature. Unlike the universe, God does not depend on anything else to exist. This classical doctrine, affirmed by Augustine, Aquinas, and others, teaches that God created the world not out of necessity, but out of love (Augustine, Confessions, Book XIII). Accordingly, creation is intentional, relational, and meaningful. Humans, made in the image of this self-existent Creator, are capable of reason, love, and relationship—traits that reflect not randomness but divine intentionality. As Plantinga (2000) and Craig (2010) argue, the fact that we exist as conscious, moral, and rational beings points toward a purposive mind behind reality. The universe, on this view, is not an accident, but the result of a wise and personal Creator.

In the Beginning, God

For centuries, materialists assumed an eternal, uncaused universe. However, the discovery of the Big Bang and the expanding universe fundamentally challenged that assumption. The prevailing scientific consensus now affirms that the universe had a definite beginning—a claim long anticipated by Theism. The fact that all space, time, and matter came into existence simultaneously raises a profound metaphysical question: What caused the universe to begin? In A Universe from Nothing, physicist Lawrence Krauss (2012) suggests that quantum fluctuations in a vacuum could account for the universe's origin. However, as Craig (2008) and Swinburne (2004) point out, Krauss's "nothing" is not metaphysical nothing but a quantum vacuum—an entity governed by physical laws and possessing structure. This "nothing" is still very much "something" and therefore does not answer the question of ultimate origin. Christian Theism offers a more robust account: the universe was brought into being by a transcendent, immaterial, uncaused cause—a being outside of space and time. Given the nature of the effect—an orderly, life-permitting cosmos—it is reasonable to infer that the cause is intelligent and personal, hallmarks of the Theistic conception of God.

The Fine-Tuning of the Universe

Among the most astonishing discoveries in modern physics is the fine-tuning of the universe's fundamental constants and quantities. Values such as the gravitational constant, electromagnetic force, and strong and weak nuclear forces are precisely calibrated to permit the existence of life. Even minute deviations would render the universe uninhabitable. For example: The cosmological constant, governing the universe's expansion, is fine-tuned to 1 part in 10^{120}. The initial entropy of the universe is fine-tuned to 1 part in 10^ (10^123), a degree of precision beyond human comprehension (Penrose, 1989). This extraordinary precision forms the basis for the Fine-Tuning Argument for God's existence. Such delicate calibration is highly improbable under Naturalism but expected under Theism, where a rational Creator designed the cosmos with life in mind.

Physicists Barrow and Tipler (1986) observed that the Anthropic Principle reinforces this conclusion: we live in a universe that appears remarkably tailored for observers like us. Some naturalists respond with the multiverse hypothesis, suggesting that countless universes exist with different constants, and we just happen to live in one that supports life. But this is speculative and lacks empirical confirmation. Moreover, it only pushes the explanatory problem back, what explains the existence of the multiverse itself? Physicist and theologian John Polkinghorne (1998) argues that mind, not mechanism, lies behind the order of the universe. He writes: "The theist can believe that there is only one universe, whose anthropic character simply reflects the endowment of potentiality given it by its Creator so that it should have a fruitful history... The anthropic specificity of our world... makes an insightful contribution to a cumulative case for Theism, regarded as the best explanation of the nature of the world that we inhabit" (Polkinghorne, 1998, p. 59). This inference to design is not a "God of the gaps" argument, but an appeal to the best explanation of the highly specific, life-permitting conditions of the universe. The precision and intelligibility of physical laws point to an intelligent, transcendent Designer.

From Cosmology to Biology: The Origin and Complexity of Life

The Origin of Life from Non-Life

Naturalism faces significant challenges in explaining abiogenesis—the origin of life from non-living matter. In contrast, Christian Theism offers a coherent and purpose-driven account: life began through the intentional act of a Creator. Materialism, which asserts that all phenomena, including life and consciousness, can be explained through unguided physical processes, must account for how non-living molecules assembled into functionally integrated biological systems. Life requires not only specific molecular components—such as amino acids and nucleotides—but also their precise arrangement into functional proteins and nucleic acid sequences. These requirements are so stringent that the odds of spontaneous formation approach impossibility.

As synthetic chemist James Tour (2020) states: "We have no idea how the molecules that compose living systems could have been generated, much less assembled."

Abiogenesis models have yet to account for the specificity and interdependence required for even the most basic forms of life. DNA, for example, contains highly specified information, functioning like digital code that stores the instructions for building and sustaining cellular systems. The origin of such functional information—ordered, non-redundant, and purposeful—suggests intelligence rather than chance (Meyer, 2009). Within cells, molecular machines like ATP synthase and ribosomes operate with astonishing efficiency. These components are embedded within interactomes—hierarchical networks of interactions spanning from molecules to organelles to entire cells. This organized complexity is not easily reducible to blind chemistry and points instead toward design (Behe, 2006).

The Development of Complex Biological Life

The emergence of complex multicellular organisms presents an even greater challenge for Naturalism. It asserts that all biological complexity arose through random mutation and natural selection, yet this mechanism struggles to account for the origin of new functional genetic and epigenetic information, especially when entirely new body plans or organ systems emerge. Stephen Meyer (2013) highlights this through the Cambrian explosion, a period during which dozens of distinct animal phyla appeared abruptly in the fossil record. Meyer argues that the rapid emergence of these diverse body plans necessitated an infusion of biological information that undirected processes cannot explain: "The origin of this new information—and the rapid appearance of diverse body plans it enabled—remains a profound mystery within materialistic frameworks."

Michael Behe (2019) adds to this critique with his concept of irreducible complexity—systems like the bacterial flagellum or ATP synthase that require multiple interdependent parts to function. Such systems cannot evolve gradually, as intermediate forms offer no survival advantage: "Random mutations are overwhelmingly more likely to degrade rather than build functional biological structures." Tour reinforces this point from a chemical standpoint: "The statistical odds of obtaining such precise molecular configurations through random events are so astronomically low that naturalistic explanations for the origin of life appear untenable" (Tour, 2016). Together, Meyer, Behe, and Tour show that the specified complexity, interdependence, and molecular precision in living systems point far more plausibly to intelligent design than to chance-based mechanisms. Christian Theism, which affirms a rational Creator, offers a coherent explanation for the informational architecture and purposeful structure of living organisms.

A Paradigm Shift in Biology?

Recent developments in systems biology suggest a growing dissatisfaction with reductionist models. In his 2024 Nature article, Oxford physiologist Denis Noble argues: "It's time to admit that genes are not the blueprint for life" (Noble, 2024). In his review of Philip Ball's How Life Works (2024), Noble explains that life arises from a dynamic interplay of genetic, epigenetic, and environmental factors. Genes function within feedback-regulated networks, not as deterministic programs. He proposes a shift toward organism-centered biology, where systems are viewed holistically, and introduces the idea of biological agency—the capacity of organisms to bring about changes to themselves or their environment in goal-directed ways. Although Noble remains a materialist, his critique underscores how traditional gene-centric models have constrained biological explanation. This openness to agency and goal-directed behavior resonates with Fred Hoyle's earlier intuition: "A commonsense interpretation of the facts suggests that a superintellect has monkeyed with physics, as well as with chemistry and biology" (Hoyle, 1982, p. 12). Thomas Nagel (2012), a prominent atheist philosopher, likewise argues that methodological materialism limits science's capacity to account for consciousness, cognition, and now even life itself. Even Ball (2024), though writing within a broadly naturalistic framework, recognizes the limits of reductionism. He calls for a more expansive biology—one that accounts for emergence, self-organization, and system-level integration. While Ball does not present these findings as evidence for design, they raise questions that are far more intelligible within a theistic metaphysical framework. Christian Theism, with its affirmation of teleology, agency, and intelligibility, provides a conceptual home for such phenomena.

Theistic Implications of Significant Scientific Discoveries

Recent scientific discoveries carry profound implications for Theism, especially regarding the question of purposeful design in the universe and in life. The confirmation that the universe had a definite beginning—as affirmed by Big Bang cosmology—resonates strongly with the Judeo-Christian doctrine of creation, which posits that all space, time, and matter were brought into existence by a transcendent cause. Equally striking is the recognition that the universe's physical laws, constants, and initial conditions are exquisitely fine-tuned to permit the emergence and sustenance of life. This level of calibration is far more consistent with intentional design than with blind chance. As discussed earlier, the probability of such precision arising randomly is astronomically low, and the existence of life-permitting conditions across such a wide range of parameters suggests that the cosmos is not merely habitable—but intentionally designed to be so.

Parallel discoveries in biology further reinforce the case for purpose. The inner workings of the cell reveal digital information systems, complex molecular machines, and biochemical algorithms—features that closely resemble human-engineered systems. The synthesis of functional proteins, for example, requires highly specified sequences of amino acids, encoded in DNA, which are extraordinarily rare among the vast number of possible permutations. This specificity constitutes what intelligent design theorists refer to as functional information—a form of complexity that implies not only order but intentionality (Meyer, 2009; Behe, 2019; Tour, 2016). Notably, the long-held view of "junk DNA" has been overturned. What was once thought to be non-functional residue is now understood to play essential regulatory and structural roles in gene expression and cellular organization. This revelation points to a deeper level of biological sophistication than previously assumed—suggesting not waste or randomness, but optimization.

Moreover, there is growing recognition among biologists and philosophers of science that random mutation and natural selection, though useful in explaining microevolutionary change, appear insufficient as creative mechanisms to generate novel biological systems, body plans, or informational architectures. As the evidence for top-down causation, self-organization, and systems integration increases, so too does openness to alternative metaphysical frameworks—ones that can accommodate agency, intention, and design. While interpretations differ across the scientific and philosophical spectrum, these findings resonate deeply with the theistic worldview, which affirms that the cosmos and life are not the products of undirected processes but reflect the work of a rational, purposeful Creator. Within Christian Theism, these discoveries are not anomalies—they are expected. They reinforce a vision of reality in which design, intelligibility, and purpose are woven into the very fabric of existence.

The Transcendentals

The Meaning and Pull of the Transcendentals

The concept of the transcendentals—truth, love, justice, beauty, and being itself—has long drawn philosophers and theologians toward a higher metaphysical reality. Central to the Western intellectual tradition, these ideals evoke our deepest longings and shape our sense of meaning and purpose. They stir in us a desire for what is perfect, eternal, and beyond the empirical. As C. S. Lewis writes in Mere Christianity: "If I find in myself a desire which no experience in this world can satisfy, the most probable explanation is that I was made for another world... I must keep alive in myself the desire for my true country, which I shall not find till after death" (Lewis, 1952, p. 106). While many philosophies explore human ascent toward the divine, Christianity uniquely asserts that God descends to meet humanity—embodying the transcendentals through the person of Jesus Christ. In Him, truth, love, justice, and beauty are not abstract ideals but personal and incarnate realities. Christianity thus offers not only an explanation of the transcendentals but a relationship in which they are fulfilled.

The Argument from the Transcendentals

The transcendentals—truth, beauty, goodness, love—have served as the metaphysical foundation of Theistic reflection from Plato to contemporary Christian thinkers. Plato conceived of these as objective and eternal Forms—unchanging realities that ground all experience and gesture toward a transcendent source of perfection. For Plato, human knowledge and love are oriented toward these ideals, which suggest a higher realm of being beyond the physical world (Guyer, 1987). Immanuel Kant, building on this tradition, introduced the notion of a transcendental unity of apperception—the a priori cognitive structure by which our experiences are unified into coherent self-consciousness. While Kant did not explicitly affirm a personal God, his framework implied that intelligibility, morality, and freedom have a grounding beyond mere sensory data (Kant, 1998). Many Christian thinkers interpret Kant's philosophy as pointing implicitly toward a rational divine mind, the source of both moral law and epistemic order.

C. S. Lewis further advanced this line of reasoning by tying the transcendentals to the longing of the human soul. He argued that universal human desires—for perfect truth, justice, love, and beauty—cannot be fully met in this life, which suggests that we are oriented toward another world. For Lewis, this longing (*sehnsucht*) is not an evolutionary glitch but an echo of our origin and destiny in God (Lewis, 1952). In this way, the transcendentals form the basis of a powerful argument for Theism: our longing for what is perfect, and eternal implies the existence of a reality in which those longings are satisfied. Christian Theism uniquely identifies this reality not as an abstract principle, but as the personal God who created us in love and invites us to communion with Himself.

The Soul's Transcendental Capacities

The capacities of the human soul offer further evidence that we are designed for transcendence. As philosopher Robert Spitzer (2015) notes, human beings exhibit faculties that surpass material explanations. Chief among these is self-consciousness, which enables inward reflection, moral deliberation, and existential questioning—features not easily reducible to evolutionary fitness or mechanistic causation. Our ability for abstract reasoning, expressed through logic, mathematics, and language, points to an innate longing for perfect truth—an impulse not satisfied by empirical approximation. In Christian Theism, this quest finds its fulfillment in God, who is the source and end of all intelligibility. Likewise, our aesthetic capacity, the ability to perceive and be moved by beauty, suggests a deeper orientation. Whether through art, music, or nature, beauty evokes awe and longing for something more than the material object itself. It functions as what Lewis called a "signpost" to another world.

Our experience of love, especially in sacrificial and unconditional forms, similarly transcends biological survival or utilitarian reciprocity. Christian theology identifies this as rooted in God, who is love (1 John 4:8), and who made us in His image to participate in eternal communion. The same is true for moral experience. Our conviction that justice matters—and that injustice is not just unfortunate but truly wrong—suggests that morality is not a social construct, but a reflection of a moral order grounded in divine goodness. Finally, the human capacity to perceive the numinous—a sense of sacred presence or awe—reinforces the soul's spiritual orientation. Theologian Rudolf Otto (1923) described this as an experience of the mysterium tremendum et fascinans—a paradoxical encounter with the Holy that is both terrifying and compelling. The soul, in this sense, acts like a receiver, tuned to the signal of the divine.

The Transcendentals and Theistic Fulfillment

Taken together, the soul's capacities—its pursuit of truth, longing for beauty, hunger for justice, capacity for love, and awareness of the sacred—form a cohesive argument that human nature is oriented toward the transcendent. These longings are not random or illusory; they are signposts to a deeper metaphysical reality. Christian Theism affirms that these ideals find their fulfillment not in abstractions but in the person of Jesus Christ, in whom truth, beauty, goodness, and love are fully embodied. The Argument from the Transcendentals thus bridges classical philosophy and Christian theology, offering a robust metaphysical framework for understanding the soul's relation to God. From Plato's Forms to Kant's moral reason to Lewis's longing, the trajectory of Western thought suggests that our most profound experiences and aspirations point not inward or downward, but upward—toward God. Christian Theism presents not only a philosophical explanation for this longing, but a personal invitation: to encounter in God the One who is the source and fulfillment of all we most deeply desire.

The Theistic Worldview

A worldview seeks to answer life's most essential questions: Where did we come from? Why are we here? What is right and wrong? Where are we going? While many philosophical systems attempt to address these questions, they differ significantly in coherence, explanatory scope, and existential relevance. Christian Theism affirms a personal, transcendent, and immanent God—a Creator who not only brought the universe into existence but remains actively involved in it. This relational view of God sets Theism apart from other metaphysical systems. Deism, for instance, acknowledges a Creator but denies ongoing divine involvement, portraying God as distant and uninvolved. Pantheism collapses the distinction between God and the world, equating God with the universe itself, while Panentheism views the universe as existing within God, yet fails to uphold a personal and distinct divine being. As physicist and theologian John Polkinghorne (1998) notes, such views depersonalize or distance the divine, thereby undermining the possibility of relational engagement between God and humanity. In contrast to Naturalism and Materialism—which regard human beings as the unintended products of blind evolutionary processes and ultimately without intrinsic purpose—Christian Theism asserts that humans are created in the image of God (Imago Dei), endowed with reason, moral agency, and inherent dignity. As Craig (2010) and Plantinga (2000) argue, these distinctively human attributes are not the result of arbitrary evolutionary developments but intentional reflections of the divine nature. Our capacities for rational reflection, moral judgment, creativity, and spiritual longing point beyond mechanistic explanations and suggest a purposeful design.

Recent philosophical proposals such as Panpsychism and Physicalism attempt to address the inadequacies of reductionist models. Panpsychism proposes that consciousness is a fundamental feature of all matter, yet it cannot explain why rich, unified self-awareness emerges only in complex organisms and not in all material configurations. Physicalism, which claims that all reality—including mental states—is ultimately reducible to physical processes, continues to face the persistent "hard problem of consciousness" identified by Chalmers (1996): how and why subjective experience (qualia) arises from mere matter. Thomas Nagel (1971) famously illustrated this limitation by asking, "What is it like to be a bat?"—a question that reveals the inability of third-person descriptions to capture first-person experience. Christian Theism, by contrast, grounds consciousness in the mind of God, understanding human self-awareness, intentionality, and rationality as reflections of the divine intellect. As Meyer (2021) emphasizes, this theistic model provides a far more coherent and satisfying explanation of human consciousness and reason.

Christian Theism also offers a comprehensive interpretive framework for understanding both scientific discoveries and philosophical insights. The beginning and fine-tuning of the universe point to a transcendent cause; the emergence of consciousness and moral awareness suggests a rational and moral Creator. Unlike Deism or Pantheism, Theism affirms not only the origin of the cosmos but also its ongoing governance by a personal God. It accounts for rationality not as a blind byproduct of evolution but as a faculty granted by a rational source. It explains objective moral truths not as illusions or sociobiological adaptations, but as reflections of God's unchanging moral nature. In doing so, Theism offers not only metaphysical coherence but also existential depth, presenting a unified worldview in which science, reason, morality, and human purpose converge in a personal and relational Creator.

Concluding the Case: The Problem of the One and the Many

Among the oldest and most enduring questions in philosophy is the problem of the one and the many: How can unity and diversity coexist? What underlying principle harmonizes the multiplicity of beings, truths, and experiences within a coherent framework of existence? Mathematician and philosopher Roger Penrose (1994) gave this question a modern form by identifying three distinct but interconnected realms: the physical world, the mental world, and the abstract world. The physical world is the domain of material phenomena governed by physical laws. The mental world includes consciousness, thought, and subjective experience. The abstract world encompasses mathematics, logic, and necessary truths. Penrose highlights the deep mystery of how these three realms relate so precisely—why they appear aligned, mutually intelligible, and profoundly interdependent.

Christian Theism offers a compelling answer. It affirms that the unity among these realms is not coincidental but grounded in God—the uncaused, necessary Being who is not one being among others, but Being itself (esse ipsum). In this view, the physical world is contingent, created by divine will and sustained by divine power. The mental world reflects the divine intellect—human consciousness bears the imprint of God's rational nature (Plantinga, 2011). The abstract world—mathematics, logic, and necessary truths—resides eternally in the mind of God, as taught by Augustine, who saw such truths as divine ideas (Augustine, On the Trinity, XV.10). Thomas Aquinas further developed this metaphysical vision, identifying God as pure act (actus purus) and subsistent being itself—the metaphysical ground of all that exists (Aquinas, Summa Theologiae, I.3.4). When God declares, "I AM WHO I AM" (Exodus 3:14), He is not merely asserting existence but revealing the very fullness of Being—the unity from which all plurality flows. Theism thus unifies the three realms in a single, coherent metaphysical source. The contingent nature of the physical world relies upon the necessary truths of the abstract world, which in turn are grounded in the divine intellect. Human minds, created in God's image, are capable of perceiving both the physical and the abstract because reason itself is a reflection of divine rationality. As Penrose acknowledged, the interconnection among these realms points to "the larger mystery of what everything is all about" (Penrose, 1994). Theism answers this mystery: all things exist through God, are sustained by God, and ultimately return to God.

Nagel and Plantinga: Converging Critiques of Materialist Naturalism

Two of the most influential philosophical critiques of Naturalism, though arising from very different worldviews, come from Thomas Nagel and Alvin Plantinga. Nagel, a secular philosopher, has consistently argued that materialist naturalism fails to account for consciousness, rationality, and moral realism—core elements of human experience that resist reduction to purely physical explanations. In Mind and Cosmos, Nagel (2012) questions whether cognitive faculties shaped solely by evolutionary pressures can reliably track truth. He also defends the objectivity of moral values, which Naturalism struggles to ground. While Nagel does not embrace Theism, he acknowledges that teleology—some intrinsic, goal-directed principle—may be necessary to explain consciousness and value.

Plantinga takes the critique further. In his Evolutionary Argument Against Naturalism (EAAN), Plantinga (2011) contends that if both evolution and Naturalism are true, we have no reason to trust our cognitive faculties, because they were shaped for survival, not truth. This renders belief in Naturalism self-defeating. Theism, by contrast, posits a rational Creator who designed human minds to know the truth. Reason, on this account, is not an accidental byproduct of natural selection but a gift intended for truth-seeking. The convergence of Nagel's secular critique and Plantinga's theistic response highlights the inadequacy of materialist explanations and strengthens the case for a theistic worldview—particularly Christian Theism—as a more coherent and comprehensive account of reality.

Christian Theism: A Coherent and Explanatory Worldview

This chapter began by exploring the profound mystery of the brain and the mind, and it ends by returning to that mystery through the lens of Theism. Christian Theism offers a more coherent explanation of consciousness, rationality, and moral realism than materialist naturalism. Human consciousness is not a mere byproduct of neural complexity but reflects the imago Dei—our creation in the image of a conscious, personal God (Genesis 1:27). Rationality is reliable not because it emerged from evolutionary necessity, but because it derives from the Logos, the rational Word through whom all things were made (John 1:1–3). Morality is not a social construct but reflects the unchanging nature of a morally perfect Creator (Psalm 19:7–9).

Naturalism, by contrast, ultimately collapses under its own premises. If our minds evolved solely for survival and not truth, belief in Naturalism itself becomes irrational. If moral norms are mere evolutionary adaptations or social conventions, they cannot explain the deep moral convictions we all share—our longing for justice, our outrage at evil, and our willingness to make sacrifices for the good.

Christian Theism resolves these dilemmas by affirming that our rational faculties, moral instincts, and longing for transcendence are not illusions or evolutionary accidents—they are the imprint of a Creator who made us for Himself. In the unity of the physical, mental, and abstract; in the convergence of reason, morality, and consciousness; in the harmony of science, philosophy, and theology—the case for Christian Theism emerges as the most coherent, comprehensive, and life-affirming explanation of reality. It is not only intellectually satisfying but existentially meaningful. It answers the riddle of the one and the many with the One who is the ground of all being, and in whom all things hold together.

Why This Matters

The significance of these arguments lies not only in their logical strength but in their philosophical origin. Thomas Nagel's critique of Naturalism is particularly compelling because it comes from a non-theist. His intellectual honesty in acknowledging the failures of materialist explanations—for consciousness, reason, and moral value—opens the door to a broader metaphysical horizon. Alvin Plantinga's theological depth and philosophical rigor help construct that doorway, demonstrating that Christian Theism uniquely accounts for the most fundamental features of human experience. In the end, if the universe is rational, moral, and intelligible, then the most reasonable explanation is that it was created by a rational, moral, and intelligent God. Christian Theism not only explains the world as we know it—it invites us into a relationship with the One who made it.

Conclusion

In this cumulative—but not exhaustive—case for Christian Theism, we have explored how this worldview offers the most coherent and comprehensive framework for understanding the transphysical dimensions of human experience, particularly as they relate to the brain metafunction framework introduced earlier. In contrast to Materialism, which reduces consciousness and morality to emergent byproducts of matter, and Naturalism, which restricts reality to the empirically observable, Christian Theism affirms a richer ontology—one that includes both the physical and the spiritual, both the measurable and the meaningful. Christian Theism situates human beings as imago Dei—created in the image of a purposeful, relational, and transcendent God. From this perspective, consciousness, free will, moral awareness, reason, and relationality are not evolutionary anomalies but essential elements of divine design. They are intelligible precisely because they reflect the nature of the Creator. Christian Theism not only accounts for the existence of these capacities but also their value and purpose. It integrates metaphysical truths, scientific discovery, and existential reality into a unified vision of human nature and destiny.

This synthesis of science, philosophy, and theology supports the claim that Christian Theism is the most coherent worldview for interpreting the brain's metafunctions as extending beyond the purely physical. It affirms what our deepest intuitions and most rigorous reasoning both suggest: that human life is marked by complexity, intention, and inherent dignity. Christian Theism provides a framework in which we are not merely biological organisms, but spiritual beings—capable of seeking truth, embodying love, discerning good, and aspiring toward the divine. In a world hungry for meaning and haunted by reductionism, Christian Theism stands as a bold and intellectually satisfying alternative. It is not merely a set of abstract doctrines, but a living invitation—to know the Creator who is the source of all truth, beauty, goodness, and being. It is a call not only to understand reality but to enter into relationship with the One who made it.

REFERENCES

Adams, R. M. (1999). *Finite and infinite goods: A framework for ethics.* Oxford University Press.

Aquinas, T. (1947). *Summa theologica* (Fathers of the English Dominican Province, Trans.). Benziger Bros. (Original work published ca. 1265–1274)

Augustine. (c. 400/1991). *The Trinity* (E. Hill, Trans.). New City Press. (Original work written c. 400).

Augustine. (1991). *Confessions.* Translated by H. Chadwick. Oxford, UK: Oxford University Press. (Original work published ca. 400 CE.)

Augustine. (2002). *The City of God* (H. Bettenson, Trans.). Penguin Classics. (Original work written ca. 426)

Ball, Philip. (2024). *How Life Works: A User's Guide to the New Biology.* London: Pan Macmillan.

Barrow, J. D., & Tipler, F. J. (1986). *The anthropic cosmological principle.* Oxford University Press.

Begley, S. (1998, July 20). Science finds God. *Newsweek.*

Behe, Michael J. (2006). *Darwin's Black Box: The Biochemical Challenge to Evolution.* 10th Anniversary ed. New York, NY: Free Press.

Behe, Michael J. (2019). *Darwin Devolves: The New Science About DNA That Challenges Evolution.* New York, NY: HarperOne.

Blackmore, S. J. (1993). *Dying to live: Near-death experiences.* Prometheus Books.
Bush, Nancy Evans. (2012). *Dancing Past the Dark: Distressing Near-Death Experiences.* Raleigh, NC: Parson's Porch.

Camus, A. (1991). *The myth of Sisyphus* (J. O'Brien, Trans.). Vintage International. (Original work published 1942)

Chalmers, David J. (1996). *The Conscious Mind: In Search of a Fundamental Theory.* New York, NY: Oxford University Press.

Chalmers, D. J. (2009). The two-dimensional argument against materialism. In B. McLaughlin, A. Beckermann, & S. Walter (Eds.), *The Oxford handbook of philosophy of mind* (pp. 313–339). Oxford University Press.

Churchland, P. M. (1986). *Neurophilosophy: Toward a unified science of the mind-brain.* MIT Press.

Collins, Robin. (2009). "The Teleological Argument: An Exploration of the Fine-Tuning of the Universe." In *The Blackwell Companion to Natural Theology,* edited by W. L. Craig and J. P. Moreland, 202–281. Malden, MA: Wiley-Blackwell.

Craig, W. L. (2016). *God over all: Divine aseity and the challenge of Platonism.* Oxford University Press.

Craig, William Lane. (2008). Reasonable Faith: Christian Truth and Apologetics. 3rd ed. Wheaton, IL: Crossway Books.

Craig, William Lane. (2010). *On Guard: Defending Your Faith with Reason and Precision.* Colorado Springs, CO: David C. Cook.

Crick, Francis. (1994). *The Astonishing Hypothesis: The Scientific Search for the Soul.* New York, NY: Scribner.

Dennett, Daniel C. (1991). *Consciousness Explained.* Boston, MA: Little, Brown and Company.

Dennett, Daniel C. (2003). *Freedom Evolves.* New York, NY: Viking Penguin.

Eliade, M. (1959). *The sacred and the profane: The nature of religion* (W. R. Trask, Trans.). Harcourt, Brace & World.

Feser, E. (2019). *Aristotle's revenge: The metaphysical foundations of physical and biological science.* Editiones Scholasticae.

Gallup, G., & Proctor, W. (1982). *Adventures in immortality: A look beyond the threshold of death.* McGraw-Hill.

Geach, P. T. (1957). *Mental acts: Their content and their objects.* Routledge & Kegan Paul.

Geisler, Norman L., and Frank Turek. (2004). *I Don't Have Enough Faith to Be an Atheist.* Wheaton, IL: Crossway.

Gleick, J. (1987). *Chaos: Making a new science.* Viking.

Greyson, B. (2003). Incidence and correlates of near-death experiences in a cardiac care unit. *General Hospital Psychiatry, 25*(4), 269–276.

Greyson, Bruce. (2010). "Implications of Near-Death Experiences for a Postmaterialist Psychology." *Psychology of Religion and Spirituality,* 2 (1): 37–45.

Guyer, Paul. (1987). *Kant and the Claims of Knowledge.* Cambridge: Cambridge University Press.

Habermas, G. R., & Moreland, J. P. (2003). *Beyond death: Exploring the evidence for immortality.* Wipf and Stock Publishers.

Harris, S. (2012). *Free will.* Free Press.

Harrison, P. (2007). *The fall of man and the foundations of science.* Cambridge University Press.

Hawking, S. (1988). *A brief history of time: From the big bang to black holes.* Bantam Books.

Heisenberg, Werner. (1927). "Über den anschaulichen Inhalt der quantentheoretischen Kinematik und Mechanik [On the Perceptual Content of Quantum Theoretical Kinematics and Mechanics]." *Zeitschrift für Physik,* 43 (3–4): 172–198. https://doi.org/10.1007/BF01397280.

Heisenberg, W. (1930). *The physical principles of the quantum theory.* University of Chicago Press.

Hick, J. (1966). *Evil and the God of love.* Harper & Row.

Hoyle, F. (1982). The universe: Past and present reflections. *Annual Review of Astronomy and Astrophysics, 20,* 1–36. https://doi.org/10.1146/annurev.aa.20.090182.000245

Jaki, S. L. (1978). *The road of science and the ways to God.* University of Chicago Press.

James, William. (1902). *The Varieties of Religious Experience: A Study in Human Nature.* New York, NY: Longmans, Green, and Co.

John Paul II. (1984, February 11). *Salvifici doloris: On the Christian meaning of human suffering.* Vatican City: Libreria Editrice Vaticana. Retrieved from https://www.vatican.va/content/john-paul-ii/en/apost_letters/1984/documents/hf_jp-ii_apl_11021984_salvifici-doloris.html

Kane, R. (2005). *A contemporary introduction to free will.* Oxford University Press.

Kant, Immanuel. (1998). *Critique of Pure Reason.* Translated by Paul Guyer and Allen W. Wood. Cambridge: Cambridge University Press. (Original work published 1781.)

Keener, Craig S. (2011). *Miracles: The Credibility of the New Testament Accounts.* Grand Rapids, MI: Baker Academic.

Krauss, L. M. (2012). *A universe from nothing: Why there is something rather than nothing.* Free Press.

Kreeft, Peter, and Ronald K. Tacelli. (1994). *Handbook of Christian Apologetics: Hundreds of Answers to Crucial Questions.* Downers Grove, IL: InterVarsity Press.

Kreeft, P. (1990). *A summa of the summa: The essential philosophical passages of St. Thomas Aquinas' Summa Theologica edited and explained for beginners.* Ignatius Press.

Kreeft, P. (2008). Because God is real: Sixteen questions, one answer. Ignatius Press.

Lewis, C. S. (1940). *The problem of pain.* HarperOne.

Lewis, C. S. (2001). *The abolition of man.* HarperOne. (Original work published 1943)

Lewis, C. S. (1947). *Miracles: A Preliminary Study.* London: Geoffrey Bles.

Lewis, C. S. (1952). *Mere Christianity.* London: Geoffrey Bles.

Mackie, J. L. (1982). *The miracle of theism: Arguments for and against the existence of God.* Oxford University Press.

Meyer, Stephen C. (2009). *Signature in the Cell: DNA and the Evidence for Intelligent Design*. New York, NY: HarperOne.

Meyer, Stephen C. (2013). *Darwin's Doubt: The Explosive Origin of Animal Life and the Case for Intelligent Design*. New York, NY: HarperOne.

Meyer, Stephen C. (2021). *The Return of the God Hypothesis: Three Scientific Discoveries That Reveal the Mind Behind the Universe*. New York, NY: HarperOne.

Moreland, J. P. (2014). *The Soul: How We Know It's Real and Why It Matters*. Chicago, IL: Moody Publishers.

Moreland, J. P., & Craig, W. L. (2003). *Philosophical foundations for a Christian worldview*. InterVarsity Press.

Moreland, J. P., & Rae, S. B. (2000). *Body and soul: Human nature & the crisis in ethics*. InterVarsity Press.

Morse, M., & Perry, P. (1990). *Closer to the light: Learning from the near-death experiences of children*. Villard Books.

Nagel, T. (1971). The absurd. *The Journal of Philosophy, 68*(20), 716–727. https://doi.org/10.2307/2024942

Nagel, T. (1974). What is it like to be a bat? *The Philosophical Review, 83*(4), 435–450. https://doi.org/10.2307/2183914

Nagel, T. (1997). *The last word*. Oxford University Press.

Nagel, T. (2012). *Mind and cosmos: Why the materialist neo-Darwinian conception of nature is almost certainly false*. Oxford University Press.

Nietzsche, F. (2006). *On the genealogy of morals* (K. Ansell-Pearson, Ed., C. Diethe, Trans.). Cambridge University Press. (Original work published 1887)

Noble, Denis. (2016). *Dance to the Tune of Life: Biological Relativity*. Cambridge: Cambridge University Press.

Noble, Denis. (2024). "It's Time to Admit That Genes Are Not the Blueprint for Life." *Nature*, Book Review Article, 05 February 2024. https://doi.org/10.1038/d41586 -024-00327-x.

Otto, Rudolf. (1923). *The Idea of the Holy*. Translated by John W. Harvey. Oxford, UK: Oxford University Press. (Original work published 1917.)

Parnia, S., Waller, D. G., Yeates, R., & Fenwick, P. (2001). A qualitative and quantitative study of the incidence, features, and aetiology of near-death experiences in cardiac arrest survivors. *Resuscitation, 48*(2), 149–156.

Penrose, R. (1994). *Shadows of the mind: A search for the missing science of consciousness*. Oxford University Press.

Penrose, Roger. (1989). *The Emperor's New Mind: Concerning Computers, Minds, and the Laws of Physics*. Oxford: Oxford University Press.

Pinker, S. (1997). *How the mind works*. W. W. Norton & Company.

Plantinga, A. (1974). *God, freedom, and evil*. Eerdmans.

Plantinga, Alvin. (1993). *Warrant and Proper Function*. New York, NY: Oxford University Press.

Plantinga, Alvin. (2000). *Warranted Christian Belief*. New York, NY: Oxford University Press.

Plantinga, Alvin. (2011). *Where the Conflict Really Lies: Science, Religion, and Naturalism*. Oxford: Oxford University Press.

Plato. (2004). *Republic* (C. D. C. Reeve, Trans.). Hackett Publishing. (Original work written ca. 380 BCE)

Polanyi, M. (1958). *Personal knowledge: Towards a post-critical philosophy*. University of Chicago Press.

Polkinghorne, John. (1998). *Belief in God in an Age of Science*. New Haven, CT: Yale University Press.

Polkinghorne, J. (2000). *Faith, science and understanding*. Yale University Press.

Rachels, J. (1990). *Created from animals: The moral implications of Darwinism*. Oxford University Press.

Ring, K., & Cooper, S. (1999). *Mindsight: Near-death and out-of-body experiences in the blind*. William James Center for Consciousness Studies.

Ring, K., & Valarino, E. E. (1998). *Lessons from the light: What we can learn from the near-death experience*. Moment Point Press.

Rorty, R. (1989). *Contingency, irony, and solidarity*. Cambridge University Press.

Ruse, Michael. (1989). *The Darwinian Paradigm: Essays on Its History, Philosophy, and Religious Implications*. London: Routledge.

Russell, B. (1935). *Religion and science*. Oxford University Press.

Scruton, R. (2012). *The soul of the world*. Princeton University Press.

Searle, J. R. (1983). *Intentionality: An essay in the philosophy of mind*. Cambridge University Press.

Spitzer, R. J. (2015). *The Soul's Upward Yearning: Clues to Our Transcendent Nature from Experience and Reason*. Ignatius Press.

Stapp, H. P. (2001). Quantum theory and the role of mind in nature. *Foundations of Physics, 31*(10), 1465–1499. https://doi.org/10.1023/A:1012682413597

Stott, John R. W. (1986). *The Cross of Christ*. Downers Grove, IL: InterVarsity Press.

Strawson, G. (1994). The impossibility of moral responsibility. *Philosophical Studies, 75*(1/2), 5–24. http://www.jstor.org/stable/4320507

Street, S. (2006). A Darwinian dilemma for realist theories of value. *Philosophical Studies, 127*(1), 109–166. https://doi.org /10.1007/s11098-005-1726-6

Swinburne, R. (1997). *The evolution of the soul* (Revised ed.). Clarendon Press.

Swinburne, R. (1998). *Providence and the problem of evil*. Oxford University Press.

Swinburne, R. (2004). *The existence of God* (2nd Ed.). Oxford University Press.

Swinburne, R. (2013). *Mind, Brain, and Free Will.* Oxford: Oxford University Press.

Tour, J. (2020, September 11). *Scientists are clueless on the origin of life | Lecture @ Andrews University* [Video]. YouTube. https://youtu.be/OYHHIBIZF8o?si=Vnrn5SBxyoMB7PM9

Tour, J. M. (2016). Animadversions of a synthetic chemist. *Inference: International Review of Science, 1*(4). https://inference-review.com/article/animadversions-of-a-synthetic-chemist

Townes, Charles H. (1966). *The Convergence of Science and Religion.* New York, NY: Columbia University Press.

Turner, J. Scott. (2017). *Purpose and Desire: What Makes Something "Alive" and Why Modern Darwinism Has Failed to Explain It.* New York, NY: HarperOne.

van Lommel, P., van Wees, R., Meyers, V., & Elfferich, I. (2001). Near-death experience in survivors of cardiac arrest: A prospective study in the Netherlands. *The Lancet, 358*(9298), 2039–2045. https://doi.org/10.1016/S0140-6736(01)07100-8

Wright, N. T. (2006). *Evil and the justice of God.* InterVarsity Press.

22

The Transphysical Seen Through Contemporary Miracles and Extraordinary Historical Events

Miracles

One of the most striking modern accounts of miraculous healing is the case of Barbara Schneider, a medically verified recovery that stunned physicians and observers alike. Diagnosed with advanced multiple sclerosis at the Mayo Clinic, Barbara's condition had steadily declined over the years. She endured numerous hospitalizations, surgeries, and complications that ultimately rendered her bedridden and dependent on full-time hospice care. According to Dr. Harold P. Adolph, a board-certified surgeon, she was "one of the most hopelessly ill patients" he had ever encountered. Her medical prognosis was grim: one lung had completely ceased to function, while the other operated at only 50% capacity. She required a tracheotomy, continuous oxygen support, and was legally blind, incontinent, and fed through a tube. Her limbs were atrophied and immobile, curled tightly against her body.

In a final act of faith, a friend requested prayer for Barbara on a Christian radio program. Over 450 individuals began interceding on her behalf. Then, on Pentecost Sunday, as friends read aloud letters of encouragement, Barbara heard a commanding voice say: "My child, get up and walk." What happened next defied all medical explanation. She removed her oxygen tube, stood up, and began walking. Her previously curled hands opened, her sight was restored, and her muscle control returned instantly. Doctors were astounded. Dr. Thomas Marshall, one of her attending physicians, later remarked, "It was medically impossible." The event was documented by both Dr. Scott J. Kolbaba and Dr. Harold P. Adolph, who emphasized the inexplicability of the healing from a medical standpoint (Kolbaba, 2016; Strobel, 2018). Today, Barbara lives an active and healthy life—her condition, once terminal, fully reversed. Her story stands as a powerful and well-documented challenge to purely naturalistic frameworks (Strobel, 2021).

Another historically significant healing took place in Lourdes, France, a well-known site of Roman Catholic pilgrimage. In 1902, Marie Bailly, suffering from advanced tuberculous peritonitis, was transported to Lourdes in critical condition. Her abdomen was distended and rigid, and her breathing labored. She was expected to die. After bathing in the spring's waters, Marie experienced a rapid and complete recovery. Her abdominal swelling subsided, her temperature normalized, and her symptoms disappeared—within hours. This recovery was not only observed by the religious community but witnessed by Nobel Prize-winning physician Dr. Alexis Carrel, who was at the time a skeptic of religious healing. Carrel's presence gave the event particular credibility. In his later reflections, published in The Voyage to Lourdes (Carrel, 1950), he wrote candidly about the limits of scientific explanation when confronted with phenomena like Marie's. While Carrel did not immediately convert to Theism, the experience profoundly altered his assumptions about reality, causality, and the role of the transphysical in healing.

Why These Accounts Matter

Both cases illustrate that not all phenomena are adequately explained by materialist assumptions. These events were witnessed by trained medical professionals, documented in medical literature, and occurred in environments that offered no naturalistic mechanism for recovery. While they do not provide "proof" in the sense of laboratory repeatability, they offer philosophical and evidential weight to the claim that transphysical causality is real and operative in the world. In complement to the philosophical and scientific arguments made in the previous chapter, these accounts provide existential and experiential support for Christian Theism. They suggest that if human beings are more than physical systems—and if reality itself includes a spiritual dimension—then miracles are not violations of natural law, but interventions by the Creator who authored those laws.

A Brief History of Miracles and Their Contemporary Relevance

Critics of miracles often appeal to the Enlightenment philosopher David Hume, whose skepticism is famously summarized in the maxim, "extraordinary claims require extraordinary evidence." Hume's critique, articulated in An Enquiry Concerning Human Understanding (1748), emerged within the intellectual climate of the Enlightenment—a period that emphasized reason, empirical verification, and the challenging of religious authority. During this time, scientific advances, such as those presented in Isaac Newton's *Philosophiæ Naturalis Principia Mathematica* (1687), contributed to a mechanistic and deterministic worldview, where natural laws were viewed as absolute and immutable. This worldview left little conceptual space for supernatural intervention. However, it's important to clarify that Newton himself did not hold to a closed naturalism. While he described the cosmos in terms of laws and mechanics, Newton also believed these laws were expressions of divine order. He maintained that God not only created the universe but actively sustained it and could intervene when necessary. Newton's theology thus reflects a nuanced relationship between science and Theism, challenging the simplistic notion that Enlightenment science necessarily excluded divine agency (Newton, 1687). Hume's empiricism—shaped by the influence of thinkers like John Locke (1689) and Pierre Bayle (1697)—led him to define knowledge as that which can be derived from sense experience. As a result, he treated miracles, by definition, as violations of "uniform experience," arguing that no amount of testimony could outweigh the improbability of a miraculous event. For Hume, the very rareness of miracles disqualified them from rational belief.

Why Hume's Argument Against Miracles Fails

Contemporary philosophers such as John Earman and William Lane Craig have offered substantial critiques of Hume's reasoning. In Hume's Abject Failure: The Argument Against Miracles, Earman (2000) argues that Hume's methodology is philosophically and mathematically flawed. Hume assumes that the improbability of a miracle automatically outweighs any supporting evidence, failing to account for the probabilistic interplay between the rarity of an event and the strength of the testimony supporting it. Drawing on Bayesian probability, Earman demonstrates that the credibility of even highly improbable events can increase dramatically if supported by robust and independent evidence. Rather than assessing the probability of a miracle in the abstract, Bayesian reasoning weighs how likely the given evidence would be if the miracle occurred, compared to if it did not. Earman shows that Hume's framework fails to account for this conditional reasoning, rendering his dismissal logically and methodologically unsound.

William Lane Craig (2010) expands this critique by reframing miracles not as inherently irrational, but as events whose improbability can be rationally overcome by cumulative, well-supported evidence—such as credible eyewitness testimony, medical documentation, and corroborated reports. Craig argues that Hume's approach confuses prior improbability with absolute impossibility and wrongly asserts that extraordinary events require an extraordinary quantity of evidence. What is needed is evidence that is highly probable given the event in question—a Bayesian framework that Craig defends with precision.

Extraordinary Events and the Nature of Evidence

The oft-repeated slogan, "extraordinary events require extraordinary evidence," has become a popular rebuttal to claims of miracles. Yet this formulation is misleading if interpreted to mean that such events demand more evidence than is logically necessary. As Earman and Craig demonstrate, the probative value of a claim does not depend on the volume of evidence but on its quality and coherence in light of the hypothesis. A single, well-documented, and independently corroborated case can outweigh a general assumption of improbability. The case of Barbara Snyder exemplifies this principle. Her healing, thoroughly documented by physicians and supported by multiple eyewitnesses, presents a challenge to naturalistic dismissal. Medical professionals including Dr. Harold P. Adolph, Dr. Thomas Marshall, and others confirmed the prior terminal state of her illness and the instantaneous, complete nature of her recovery (Kolbaba, 2016; Strobel, 2018, 2021; Adolph, 1997). These records were not speculative but based on direct clinical observation, forming a robust empirical foundation for evaluating the event.

A Firsthand Medical Testimony

The following account is presented as firsthand clinical testimony from the author, a physician who practiced in Cuba. It is offered not as anecdote, but as a medically rigorous observation grounded in well-documented diagnosis and recovery. The patient was a man in his forties diagnosed with Stage IV histiocytic lymphocytic lymphoma, an aggressive and terminal cancer. Despite comprehensive treatment efforts by a multidisciplinary team—including Professor Héctor Rodríguez Silva, a renowned clinician, and Professor Israel Borrajero Martinez, an internationally recognized pathologist—the disease progressed. He exhibited classic terminal-stage symptoms: extreme weight loss, anemia, hepatosplenomegaly, and systemic physiologic deterioration. The prognosis was short-term and fatal.

Following a visit from a Christian minister, who prayed over the patient in the name of Jesus, the man experienced a dramatic change. Within hours, vital signs began to normalize. Over several days, his strength, appetite, and energy returned. Within weeks, follow-up diagnostic tests revealed no signs of the malignancy, and the patient's organ systems had returned to normal function. The attending medical team, including specialists in oncology and internal medicine, could offer no natural explanation for the remission. The event remains unpublished in peer-reviewed literature, yet it is preserved in medical documentation and the testimony of those who observed it. This case does not prove a miracle in a strict scientific sense—such claims are rarely susceptible to replication or lab-based validation. But it does demonstrate that some accounts of extraordinary healing are supported by sufficient empirical evidence to merit philosophical and theological reflection. In the context of a broader argument for the transphysical, it offers coherent support for the possibility of divine action in the world. The historical critiques of Hume, the contemporary philosophical rebuttals, and the documented modern cases all point to the same conclusion: miracles cannot be rationally dismissed merely on the grounds of improbability. When approached with philosophical clarity and evidential openness, they constitute plausible signs of a transphysical dimension—one fully compatible with the framework of Christian Theism. Far from being superstitious relics of a pre-scientific age, miracles invite us to reconsider the boundaries of reality, causation, and human purpose.

Broaer Scientific and Medical Evidence for Miraculous Recovery

Beyond individual testimonies and theological reflection, a growing body of scientific and sociological literature supports the need for empirical investigation into miraculous healing, especially in faith-based contexts. These studies span disciplines including medical science, psychology, sociology, and religious studies, and together, they challenge the notion that claims of healing through prayer or religious practice are beyond the scope of serious scholarly inquiry. One of the most notable contributors to this field is Dr. Candy Gunther Brown, whose book Testing Prayer: Science and Healing (2012) presents findings from field-based studies on proximal intercessory prayer (PIP). Her research in Mozambique and Brazil documents significant rates of improvement, particularly in participants' vision and hearing. These findings were subject to peer-reviewed statistical analysis (Brown et al., 2010), and while Brown acknowledges the need for caution in interpretation, the outcomes challenge strictly naturalistic assumptions about healing. Her work has initiated serious dialogue in both scientific and theological circles about the legitimacy of faith-based healing as a subject of empirical study.

Complementing Brown's empirical data, Craig Keener's two-volume work Miracles: The Credibility of the New Testament Accounts (2011) offers a sweeping compilation of both historical and modern miracle claims. Keener integrates philosophical analysis, eyewitness accounts, and case studies from around the world, many of which are medically documented and defy conventional explanation. Importantly, he does not merely catalog anecdotal stories but addresses skeptical objections directly, making a robust case for the credibility and frequency of miracles in both biblical and contemporary contexts. Meredith McGuire's Ritual Healing in Suburban America (1988) adds a sociological lens to this inquiry, examining how healing rituals in Christian communities are experienced, interpreted, and socially reinforced. Her fieldwork affirms that many individuals report measurable recovery after prayer or religious rites, and that such experiences play a significant role in shaping community and individual well-being. While not all healings can be described as miraculous, McGuire's research supports the idea that religious practice contributes to health outcomes in ways that warrant further exploration.

Another important resource is Spontaneous Remission: An Annotated Bibliography by Brenda O'Regan and Caryle Hirshberg (1993). This compendium, published by the Institute of Noetic Sciences, documents hundreds of well-documented cases of unexpected recoveries, including from terminal illnesses such as cancer. Though the authors refrain from theological conclusions, their work provides critical evidence that spontaneous healing—some would say miraculous—occurs with enough frequency to merit scientific and philosophical attention. Medical literature also includes documented cases that resist reductionist explanation. Jacalyn Duffin's work in Medical Miracles offers a compelling argument against the reflexive dismissal of miracles in modern scholarship. As a physician and historian, she demonstrates that alleged miracles—particularly those scrutinized during canonization processes—are often subjected to rigorous medical evaluation. Her nuanced analysis underscores the importance of engaging with miracle claims as culturally and medically significant phenomena that challenge the boundaries of scientific certainty. By refusing to discard miracles a priori, Duffin opens space for a more honest and interdisciplinary dialogue between faith, medicine, and history.

Matthews and Clark (1999) reviewed multiple cases of recovery attributed to faith healing, urging the medical community to engage more fully with spiritual dimensions of health. Taken together, these diverse sources converge on a central claim: faith-based healing is not merely a matter of subjective belief, but an empirically observable phenomenon worthy of scientific investigation. While not all recoveries are miraculous in the strict sense, the accumulation of medically documented, statistically significant, and culturally attested cases presents a powerful case that miracles should not be dismissed a priori. Instead, they should be evaluated through a multidisciplinary lens that includes but is not limited to the physical sciences. By documenting recoveries that appear to challenge naturalistic explanations, this body of literature strengthens the broader philosophical argument for transphysical causation and supports the epistemic credibility of miracle claims within a Christian theistic framework. These findings invite not only further research but also a more integrated understanding of healing—one that includes biological, psychological, social, and spiritual dimensions of human experience.

The Life, Death, and Resurrection of Jesus

The life and teachings of Jesus Christ have had an unparalleled impact on humanity. His example and instruction provide a compelling foundation for Christian theism, reflecting the highest ideals of love, forgiveness, and humility. However, it is Jesus' death and resurrection that form the pivotal core of Christian belief. The historical evidence supporting these events—combined with Christianity's rapid growth despite severe persecution—constitutes a strong case for the truth of Christian theism. The resurrection is especially critical: it validates Jesus' divine claims and affirms the reality of theistic belief. No naturalistic explanation has sufficiently accounted for the historical phenomena surrounding these events. Gary Habermas, a leading scholar on the resurrection, developed the "minimal facts" approach, which centers on core historical details accepted by the vast majority of New Testament scholars—both skeptical and non-skeptical. These minimal facts include:

1. *Jesus' Crucifixion*: Multiple sources attest to Jesus' crucifixion under Pontius Pilate, including Roman historians (Tacitus), Jewish sources (Josephus), and all four Gospels. This fact is widely acknowledged by scholars across religious and secular perspectives and serves as a foundational point of agreement.

2. *Post-Resurrection Appearances*: Reports of Jesus' appearances after his death are central to the earliest Christian proclamations. Paul's letters, particularly 1 Corinthians 15:3–8, present an early creed listing eyewitnesses of the risen Christ: Peter, the twelve apostles, over 500 believers, and James, Jesus' previously skeptical brother. This creed is dated by most scholars to within a few years of the crucifixion, reflecting the early and consistent transmission of the resurrection belief (Habermas 2020; Dunn 2003). Paul likely received this creed from Peter and James during his visit to Jerusalem (Galatians 1:18–19), which underscores the immediacy and authenticity of the resurrection claim.

3. *Transformation of the Disciples*: The disciples' dramatic change—from fearful followers to bold proclaimers of the resurrection, despite threats and persecution—provides compelling evidence for their belief in its truth. Their willingness to endure hardship and martyrdom is difficult to reconcile with theories of fabrication or delusion (Habermas 2020).

4. *Conversions of Paul and James*: Paul and James are two particularly powerful conversion cases. Paul, a former persecutor of Christians, claimed to have encountered the risen Christ and went on to become one of Christianity's most influential apostles. James, initially skeptical of Jesus' messianic identity, later became a leader in the Jerusalem church and was martyred for his faith. These transformations are difficult to explain apart from their sincere belief that they had witnessed the resurrected Jesus (Craig 2000; Wright 2003).

5. *The Empty Tomb*: Although not part of Habermas' original minimal facts, the empty tomb is widely supported by scholarship. Multiple independent sources, including the Gospels, attest to it. The fact that women were reported as the primary witnesses adds to the credibility of the account, as their testimony held little legal value in 1st-century Jewish culture. This meets the criterion of embarrassment, suggesting that the Gospel writers prioritized accurate reporting over social expectations (Craig 1985).

The Gospels and the Early Creedal Tradition

The Gospels, our primary sources for the resurrection narrative, have undergone extensive historical scrutiny. Scholars such as O'Collins (1987), Habermas (2020), Craig (2000), Keener (2009), and Wright (2003) defend their reliability, pointing to early dating, multiple independent attestations, and alignment with the historical context of 1st-century Judea. The detail that women discovered the tomb strengthens the credibility of the narrative through the criterion of embarrassment. The early Christian creed found in 1 Corinthians 15:3–7, widely dated to within five years of Jesus' crucifixion, affirms the centrality of the resurrection from the very beginning of the Christian movement. Paul's visit to Peter and James (Galatians 1:18–19) further supports the early transmission of this belief. Together, the multiple independent sources, the transformations of skeptics, and the criterion of embarrassment provide robust historical support for the resurrection claim. These elements make a strong case that the resurrection was not a later legend, but a deeply held belief grounded in the earliest Christian witness (Wright 2003; Keener 2009).

Extrabiblical Sources

Numerous extrabiblical sources corroborate elements of Jesus' life, death, and impact. Jewish historian Flavius Josephus mentions Jesus twice in Antiquities of the Jews (93–94 CE). While the Testimonium Flavianum may contain Christian interpolations, most scholars agree that it originally affirmed Jesus as a wise man who was crucified under Pilate (Meier 1991). Josephus also mentions James, the brother of Jesus, further affirming Jesus' historical existence (Antiquities 20.9.1). Roman historian Tacitus (Annals 15.44) confirms Jesus' execution during the reign of Tiberius and describes Christianity as a "superstition" that originated in Judea. Suetonius references disturbances in Rome linked to "Chrestus," likely referring to Christ (Suetonius, Life of Claudius 25). Pliny the Younger, in a letter to Emperor Trajan (Letters 10.96–97), notes that early Christians worshiped Christ "as a god," offering insight into early Christian devotion.

Other sources also provide valuable references to Jesus and early Christian beliefs. Mara Bar-Serapion, a Stoic philosopher, compares Jesus to Socrates and Pythagoras, calling him a "wise king." The *Babylonian Talmud* (Sanhedrin 43a) refers to "Yeshu," likely Jesus, acknowledging his execution. Lucian of Samosata mocks Christians for worshiping a "crucified one," attesting to their early belief in Jesus' divinity. Celsus, a 2nd-century philosopher, criticized Jesus' miracles and birth in *The True Word* (preserved in Origen's *Contra Celsum*), yet ironically affirms elements of the Gospel tradition. Earlier writers, such as Thallus and Phlegon of Tralles, mention darkness and seismic events at the crucifixion (Julius Africanus, *Chronography*, fragment 18.1, preserved in George Syncellus, *Chronographia*; see also Eusebius, *Ecclesiastical History*, Loeb Classical Library, Kirsopp Lake, vol. 1, p. 401). Additionally, the Apostolic Fathers—including the *Didache, 1 Clement,* and *The Shepherd of Hermas*—emphasize the resurrection as central to Christian faith (Holmes, 2007). Collectively, these sources reinforce the historical existence of Jesus and the early belief in his resurrection.

Naturalistic Explanations and Their Shortcomings

Several naturalistic theories have been proposed to explain the resurrection, but each faces significant problems when assessed against the historical data:

Swoon Theory: Suggests Jesus survived the crucifixion. However, Roman execution practices—including scourging, asphyxiation, and a spear thrust—render this scenario implausible (Edwards, Gabel, and Hosmer 1986; Craig 2000).

Theft Theory: Argues that Jesus' followers stole his body. This fails to explain the disciples' willingness to die for their claims (Craig 2000).

Hallucination Theory: Suggests the resurrection appearances were grief-induced visions. But hallucinations are private, individual experiences, whereas the resurrection appearances occurred to groups, across different settings, and included skeptics like Paul and James (Keener 2009; Lüdemann 1995). Hallucinations also do not explain the empty tomb.

Vision Theory: Claims disciples experienced spiritual visions, not physical appearances. Yet the Gospels emphasize tangible, physical encounters with the risen Jesus—touching his wounds, sharing meals—which challenge this view (Wright 2003; Craig 2000).

Legend Theory: Asserts the resurrection account developed over time. However, the early dating of 1 Corinthians 15 and the Gospels undermines this claim (Keener 2009).

Wrong Tomb Theory: Proposes the women went to the wrong tomb. If so, authorities could have easily corrected the error by producing the body (Habermas and Licona 2004).

Twin or Impersonation Theory: Suggests someone impersonated Jesus. This lacks historical evidence and fails to explain the recognition of Jesus by his closest followers (Craig 2000).

Spiritual Resurrection Theory: Interprets the resurrection as symbolic, not physical. But this contradicts the New Testament accounts and the apostles' martyrdoms, which indicate belief in a physical event (Wright 2003).

Hypnosis/Trance Theory: Suggests Jesus or his followers entered a hypnotic state. This highly speculative theory cannot account for the physical details in the Gospels (Keener 2009).

Stolen Body by Enemies Theory: Suggests opponents of Jesus took the body. Yet if they had, they could have produced it to quash resurrection claims (Habermas and Licona 2004a).

Mythic Theory: Claims Jesus never existed. This view is overwhelmingly rejected by historians, given the wealth of both Christian and non-Christian sources affirming Jesus' life and death (Habermas 2020).

These naturalistic explanations fail to account for the empty tomb, the transformations of key witnesses, and the early, consistent testimony of a bodily resurrection. The resurrection, as proclaimed by early Christians, remains the best explanation for the historical data (Keener 2009; Wright 2003). The convergence of early, independent sources; the dramatic transformation of eyewitnesses; the consistent and immediate proclamation of the resurrection; and the inability of alternative naturalistic explanations to adequately account for these facts together make a strong historical case for the resurrection of Jesus. Rather than being a legend or symbolic tale, the resurrection emerges as a well-attested event that the earliest Christians sincerely believed and for which many willingly suffered and died. Their conviction was not rooted in myth or metaphor, but in what they claimed was a real, bodily encounter with the risen Christ. This foundational event not only launched the Christian movement under intense opposition but continues to serve as the cornerstone of Christian theism. The resurrection affirms the truth of Jesus' divine claims and offers a historically grounded basis for faith that speaks to both the mind and the heart.

Implications of the Resurrection

The historical evidence for Jesus' life, death, and resurrection offers compelling support for the existence of realities beyond the physical—what may be described as transphysical phenomena. As the central event in Christian theology and philosophy, the resurrection not only confirms Jesus' divine identity and mission but also serves as the foundational basis for key Christian doctrines: the forgiveness of sins, the promise of eternal life, and the transformative power of grace. The Apostle Paul emphasizes the essential nature of the resurrection in 1 Corinthians 15:14–17, asserting that if Christ has not been raised, then Christian preaching and faith are in vain. In this light, the resurrection functions as divine affirmation of Jesus as the Son of God, revealing God's power to inaugurate new creation and offering humanity hope for ultimate redemption (Craig 2000).

When examined alongside contemporary reports of miracles and rigorous historical inquiry, the resurrection provides a rational foundation for accepting theistic belief and the plausibility of transcendent realities. The well-documented miracles of Jesus' ministry—and credible accounts of miraculous events in modern contexts—suggest ongoing divine interaction with the world. In particular, the resurrection defies all naturalistic explanations, compelling serious philosophical engagement and challenging materialist assumptions about reality. By bridging historical evidence, theological significance, and metaphysical implications, the resurrection substantiates a comprehensive vision of existence that includes both physical and transphysical dimensions. This integrated perspective not only supports Christian theism but also aligns with the central thesis of this work: that the human person, including the brain's highest functions (or metafunctions), cannot be fully explained by material processes alone. The resurrection thus offers not only theological insight but also a coherent framework in which faith, reason, and human experience converge to affirm the spiritual dimension of reality and humanity's ultimate purpose.

The Shroud of Turin: A Historical Witness to the Death and Resurrection of Jesus

As we transition from a historical and theological examination of Jesus' life, death, and resurrection, we encounter a singular artifact that continues to captivate both scientific and religious communities: the Shroud of Turin. This ancient linen cloth—widely considered the most studied relic in history—is believed by many to be the burial shroud of

Jesus Christ. It bears a faint yet detailed image of a man who appears to have been crucified in a manner consistent with Roman execution, closely corresponding to the Gospel accounts of Jesus' crucifixion.

The Shroud has become a focal point of interdisciplinary inquiry, uniting rigorous scientific investigation with theological reflection. With its unique image characteristics and remarkable preservation, it stands as a silent witness to a profound historical and spiritual event, prompting deep reflection on the intersection of faith, science, and the divine. A growing body of research invites not only veneration but also serious consideration of the Shroud's authenticity. For the reasonable investigator, the convergence of scientific, historical, and forensic data may plausibly suggest a material connection between the Shroud and the resurrection—a physical artifact linked to the central claims of the Gospels.

The crucifixion and resurrection of Jesus Christ form the heart of the Christian theistic worldview. Many indicators suggest that the Shroud of Turin is indeed Jesus' burial cloth, and the extensive scientific scrutiny it has undergone continues to yield insights that support the historical reliability of these events. That the Shroud may have been providentially preserved into a modern scientific age is a striking possibility—especially as contemporary tools reveal aspects of the cloth inaccessible to earlier generations. In a time when scientific materialism often challenges belief in the divine, the Shroud may stand as a sign meant for such a time as this.

The Image on the Shroud

Modern scientific analysis of the Shroud began in 1898 when Secondo Pia's photograph revealed that the image functions as a photographic negative—a property not characteristic of known artistic techniques of the time (Vignon 1939). Subsequent studies using ultraviolet fluorescence, UV-Vis, infrared, and Raman spectroscopy found no evidence of pigments or dyes, strongly indicating that the image was not painted (Adler 1996). Microscopic and chemical analyses further confirmed the absence of pigment particles or binding agents, reinforcing the conclusion that the image was not produced by any known artistic method (Rogers 2004).

Bloodstains on the cloth have been subjected to extensive forensic testing. X-ray fluorescence (XRF) and proton-induced X-ray emission (PIXE) detected iron and other elements consistent with human blood (Jackson et al. 1977). STURP researchers identified type AB blood and hemoglobin degradation products, including elevated bilirubin levels—interpreted as consistent with trauma associated with crucifixion (Adler 1996; Bucklin 1997). Pollen samples embedded in the cloth, while contested in some academic circles, have been interpreted as evidence of origin in the Mediterranean region, including Jerusalem (Frei 1982). Textile experts note that while herringbone twill was known in medieval Europe, it was rarely used in linen, whereas the Shroud's weave is consistent with high-quality linen fabrics from the 1st-century eastern Mediterranean (Fanti & Malfi 2015; Flury-Lemberg 2000). Together, these findings provide multidisciplinary support for the Shroud's ancient provenance.

The Shroud as a Receipt

Beyond its physical characteristics, the Shroud holds deep theological significance. Russ Breault (2024, March 15) describes the Shroud as a "receipt"—a visible testament to the redemptive transaction completed at the cross. The visible wounds—pierced wrists and feet, a spear wound to the side, scourge marks, and punctures from a crown of thorns—closely correspond to the Gospel descriptions of Jesus' crucifixion (Fanti & Malfi 2015; Habermas & Flew 2004b). In the original Greek and Aramaic, Jesus' final word on the cross—tetelestai (Greek) or mashelem (Aramaic)—is translated as "It is finished," but its meaning is multifaceted. The term denotes not merely the end of a task, but its fulfillment—akin to settling a debt or completing a divine commission. When Jesus uttered this phrase, he was proclaiming the fulfillment of messianic prophecies (e.g., Psalm 22; Isaiah 53) and the completion of the redemptive plan (John 3:16). Beyond Scripture and historical testimony, the Shroud presents itself as a theological and historical artifact that powerfully aligns with the core message of the Gospels. It offers a tangible portrait of divine love, sacrifice, and hope—one that continues to resonate with believers around the world.

In addition to its theological meaning, the Shroud also presents physical characteristics that continue to provoke scientific inquiry. Some researchers have identified anatomical features—such as faint images of hand bones and teeth—suggesting that the image was not formed by direct contact, but through a process involving radiation. The hypothesis, most notably advanced by Rucker (2019, 2024), posits that the image was caused by a burst of radiation originating from within the body. While hypothetical, such a mechanism would be consistent with a transphysical event—resonating with theological descriptions of the resurrection. From this perspective, the Shroud becomes more than a historical artifact; it becomes a theological witness—a visual testimony to divine intervention. As De Caro et al. (2022) argue, the Shroud offers material data that point beyond natural processes, affirming the central Christian claims of redemption, resurrection, and eternal life.

The Shroud's Eloquent Testimony: Reassessing the 1988 Carbon-14 Dating

While the 1988 radiocarbon dating famously placed the Shroud's origin between 1260 and 1390 CE, that conclusion has since been the subject of serious critique. The sample used was taken from a single area now suspected to have undergone medieval reweaving, possibly introducing cotton fibers that skewed the results (Rogers 2005). In fact, all three laboratories—Oxford, Zurich, and Arizona—analyzed subdivided portions of what was likely a single fiber, rather than multiple samples from different locations. This raises further concerns about the representativeness of the sample and the validity of the test results. Chemical and microscopic analyses revealed inconsistencies in that section—such as dye and foreign materials not found elsewhere on the cloth—casting doubt on the reliability of the tested area (Rogers 2004, 2005).

Moreover, the test did not account for centuries of environmental contamination—exposure to fire, smoke, and frequent handling—which could have altered the cloth's molecular structure (Freer-Waters & Jull 2010). The three laboratories reported slightly differing results, and the use of a single site sample violated standard radiocarbon protocol (Damon et al. 1989). In contrast, recent analytical approaches—such as Wide-Angle X-ray Scattering (WAXS)—have yielded dating results consistent with a 1st-century origin (De Caro et al. 2022). Though still under evaluation as a dating method, WAXS represents a promising development. Additional support comes from textile analysis, forensic pathology, and botanical evidence: linen weaving styles typical of the 1st century, pollen from plants native to the Jerusalem region, and bloodstain patterns consistent with crucifixion all reinforce the plausibility of the Shroud's ancient origin (Frei 1982; Fanti & Malfi 2015).

Professor Giulio Fanti's forensic research further supports this view, identifying wound patterns consistent with Gospel narratives—such as the crown of thorns and scourging (Fanti 2024). Taken together, these findings significantly challenge the reliability of the 1988 carbon-14 test and bolster the case for the Shroud's authenticity as a 1st-century burial cloth—plausibly that of Jesus himself.

Scientism and the Shroud of Turin

Modern science arose within a worldview shaped by Christian theism. Early scientists such as Copernicus, Galileo, Kepler, and Newton understood the universe as orderly, intelligible, and governed by divine laws (Butterfield 1957; Jaki 1974; Stark 2003). However, as the Enlightenment progressed, science increasingly adopted scientism—the belief that empirical science is the only valid path to knowledge, excluding metaphysical, moral, or theological truths as either unknowable or meaningless. Philosophers such as William Lane Craig (2008) and J.P. Moreland (1987) have critiqued this stance, pointing out that science itself rests on non-empirical foundations: the laws of logic, the objectivity of moral reasoning, and the structure of mathematics. These principles point beyond the material and into the transphysical realm.

Paradoxically, while scientism dismisses supernatural possibilities, scientific research on the Shroud of Turin points to phenomena that many argue are not currently explainable by naturalistic means. Some researchers hypothesize that the image was formed by a burst of radiant energy—possibly lasting just 1/40 billionth of a second—that created a high-resolution, superficial image without damaging the linen itself (Fanti & Malfi 2015; De Caro et al. 2022; Breault 2024). No known natural mechanism has successfully replicated all of the Shroud's image features, including its chemical superficiality, three-dimensionality, and anatomical precision.

The Shroud thus presents a compelling challenge to the materialist paradigm. It invites a broader vision of reality—one that affirms the coherence of faith and reason. As Jesus declared, "I am the way, the truth, and the life" (John 14:6). As the Logos, the divine Word through whom all things were made (John 1:1–3), Jesus is both the source of rational order and the center of salvation history. Within this framework, the Shroud of Turin stands not only as a historical artifact but also as a theological symbol—calling us to explore the mystery of Christ with both scientific integrity and reverence for the divine (Wilson 2010; Antonacci 2000).

God Can Be Personally Known and Experienced

Christian theism affirms that God can be personally known and experienced—a claim supported not only by philosophical reasoning but by the lived, transformative experiences of countless believers. Philosopher J.P. Moreland emphasizes the experiential dimension of

Christianity, arguing that encounters with God bring about moral and spiritual renewal that reorients a person's values and actions toward the divine character revealed in Scripture (Moreland 1997). These transformations reflect not just psychological shifts, but a deeper moral and existential realignment rooted in relationship with a living God. Alvin Plantinga builds on this foundation, asserting in *Warranted Christian Belief* that belief in God can be *"properly basic"*—justified and rational even apart from empirical evidence. Like our intuitive knowledge of other minds or the external world, awareness of God can arise directly through personal experience. This non-inferential awareness, according to Plantinga, constitutes a legitimate ground for belief and provides a robust foundation for faith (Plantinga 2000).

Peter Kreeft similarly underscores the transformative impact of divine encounters, citing radical life changes—freedom from addiction, the triumph of love over hatred, and hope rising from despair—as signs of God's personal activity in the lives of individuals (Kreeft & Tacelli 1994; Kreeft 2002). These changes, he argues, transcend naturalistic explanations. William Lane Craig adds that personal experiences of God—marked by peace, purpose, and moral clarity—complement classical theistic arguments and offer a holistic vision of belief rooted in both reason and relationship (Craig 2008). Together, these philosophical and experiential perspectives affirm that God is not merely a concept, but a reality who can be personally encountered. The Apostle Peter captures this dynamic when he writes, *"But in your hearts revere Christ as Lord. Always be prepared to give an answer to everyone who asks you to give the reason for the hope that you have. But do this with gentleness and respect"* (1 Peter 3:15).

This chapter opened with two examples of miraculous healing, including the recovery of Orlando M. Garcia Prado, the author's father, who had been diagnosed with Stage IV histiocytic lymphocytic lymphoma—an aggressive and terminal condition. His complete and rapid recovery, following prayer in Jesus' name, defied medical expectations and was witnessed firsthand by the author and a team of seasoned physicians. I offer this account as both a physician and a witness, affirming the reality of God's power to heal and transform. While not part of the peer-reviewed literature, the case was thoroughly documented and medically validated, provoking thoughtful reflection on the possibility of divine intervention. Throughout this book, we have explored the depths of human consciousness, free will, moral knowledge, and personal identity—facets of our nature that consistently gesture beyond the boundaries of materialism. Considered alongside the resurrection of Christ and the witness of miracles, these dimensions reveal a reality that is both physical and transphysical, grounded in reason and revelation. Christian theism is not merely a metaphysical hypothesis or an abstract doctrine. It is a lived relationship with a personal God—one who knows, heals, and redeems. For those willing to see and to listen, the evidence does not merely point to God's existence; it extends a personal invitation. It invites each of us into a life of purpose, transformation, and eternal hope.

REFERENCES

Adler, A. D. (1996). Updating recent studies on the Shroud of Turin. In M. V. Orna (Ed.), *Archaeological chemistry: Organic, inorganic, and biochemical analysis* (Vol. 625, pp. 223–228). American Chemical Society.

Adolph, H. P. (1997). Personal communication.

Julius Africanus. (n.d.). *Chronography* (Fragment 18.1). In G. Syncellus, *Chronographia* (via Eusebius, *Ecclesiastical History*, Loeb Classical Library, Vol. 1, p. 401; K. Lake, Trans.).

Antonacci, M. (2000). *Resurrection of the Shroud: New scientific, medical, and archaeological evidence.* M. Evans and Co.

Babylonian Talmud. (2011). *The Babylonian Talmud* (J. Neusner, Trans.). Hendrickson Publishers.

Barcaccia, G., et al. (2015). Uncovering the sources of DNA found on the Turin Shroud. *Scientific Reports*, 5, Article 14484. https://doi.org/10.1038/srep14484

Bayle, P. (1697). *Dictionnaire historique et critique.*

Breault, R. (2024). *Shroud encounter: Explore the world's greatest unsolved mystery.* Covenant Books.

Breault, R. (2024, March 15). *Shroud of Turin: God's receipt?* [Video]. YouTube. https://www.youtube.com/watch?v=wA whc9G0EUM

Brown, C. G., Mory, S. C., Williams, R., & McClymond, M. J. (2010). Study of the therapeutic effects of proximal intercessory prayer (STEPP) on auditory and visual impairments in rural Mozambique. *Southern Medical Journal*, 103(9), 864–869.

Brown, C. G. (2012). *Testing prayer: Science and healing.* Harvard University Press.

Bucklin, R. A. (1997). An autopsy on the man of the Shroud. In M. Rinaldi (Ed.), *The Shroud of Turin: A critical summary of observations, data, and hypotheses* (pp. 125–145). E. Rizzoli.

Butterfield, H. (1957). *The origins of modern science.* Macmillan.

Carrel, A. (1950). *The voyage to Lourdes.* Harper.

Craig, W. L. (1985). The historicity of the empty tomb of Jesus. *New Testament Studies*, 31(1), 39–67.

Craig, W. L. (2000). *The Son rises: The historical evidence for the resurrection of Jesus.* Wipf and Stock.

Craig, W. L. (2008). *Reasonable faith: Christian truth and apologetics* (3rd ed.). Crossway.

Craig, W. L. (2010). *On guard: Defending your faith with reason and precision.* David C. Cook.

Damon, P. E., Donahue, D. J., Gore, B. J., et al. (1989). Radiocarbon dating of the Shroud of Turin. *Nature*, 337(6208), 611–615. https://doi.org/10.1038/337611a0

De Caro, L., et al. (2022). New radiocarbon dating of the Shroud of Turin using wide angle X-ray scattering. *Heritage Science*, 10(12), 1–11. https://doi.org/10.1186/s40494-022-00674-3

Dunn, J. D. G. (2003). *Jesus remembered.* Eerdmans.

Earman, J. (2000). *Hume's abject failure: The argument against miracles.* Oxford University Press.

Edwards, W. D., Gabel, W. J., & Hosmer, F. E. (1986). On the physical death of Jesus Christ. *Journal of the American Medical Association*, 255(11), 1455–1463. https://doi.org/10. 1001/jama.1986.03370110077025

Fanti, G. (2013). *The Shroud of Turin: First century after Christ!* World Scientific.

Fanti, G., & Malfi, P. (2015). *The Shroud of Turin: First century after Christ!* World Scientific.

Fanti, G. (2024). New insights on blood evidence from the Turin Shroud consistent with Jesus Christ's tortures. *Archives of Hematology Case Reports and Reviews*, 9(1), 001–015. https://dx.doi.org/10.17352/ahcrr.000044

Flury-Lemberg, M. (2000). Textile aspects of the Shroud of Turin. In M. Marinelli (Ed.), Proceedings of the International Scientific Symposium: The Turin Shroud: Past, Present and Future (pp. 21–43). Cantalupa (TO), Italy: Effatà Editrice.

Freer-Waters, R., & Jull, A. J. T. (2010). Investigating a dated piece of the Shroud of Turin. *Radiocarbon*, 52(4), 1521–1527. https://doi.org/10.1017/S0033822200046228

Frei, M. (1982). Pollen analysis of the Shroud of Turin. *Palynology Research Journal*, 12(3), 45–58.

Habermas, G. R. (1996). *The historical Jesus: Ancient evidence for the life of Christ.* College Press.

Habermas, G. R. (2020). *The case for the resurrection of Jesus* (Rev. Ed.). Kregel Publications.

Habermas, G. R., & Flew, A. (2004b). *Did the resurrection happen? A conversation with Gary Habermas and Antony Flew.* InterVarsity Press.

Habermas, G. R., & Licona, M. R. (2004a). *The case for the resurrection of Jesus.* Kregel Publications.

Holmes, M. W. (Ed.). (2007). The Apostolic Fathers: Greek texts and English translations (3rd Ed.). Baker Academic.

Hume, D. (1748). An enquiry concerning human understanding.

Duffin, J. (2009). *Medical miracles: Doctors, saints and healing in the modern world.* Oxford University Press.

Jackson, J. P., et al. (1977). A physical analysis of the Shroud of Turin. In *Proceedings of the 1977 Conference on the Shroud of Turin.* Albuquerque, NM.

Jaki, S. (1974). *Science and creation: From eternal cycles to an oscillating universe.* Scottish Academic Press.

Josephus, F. (93–94 CE). *Antiquities of the Jews* (W. Whiston, Trans.).

Keener, C. S. (2009). *The historical Jesus of the Gospels.* Eerdmans.

Keener, C. S. (2011). *Miracles: The credibility of the New Testament accounts* (Vols. 1–2). Baker Academic.

Kolbaba, S. J. (2016). *Physician's untold stories: Miraculous experiences doctors are hesitant to share with their patients, or anyone!.* KSB Press.

Kreeft, P. (2002). *The journey: A spiritual roadmap for modern pilgrims.* InterVarsity Press.

Kreeft, P., & Tacelli, R. K. (1994). *Handbook of Christian apologetics: Hundreds of answers to crucial questions.* InterVarsity Press.

Locke, J. (1689). *An essay concerning human understanding.*

Lucian of Samosata. (1905). *The passing of Peregrinus* (H. W. Fowler & F. G. Fowler, Trans.). Clarendon Press.

Mara Bar-Serapion. (n.d.). *Letter to his son* (ca. 73–200 CE).

Marino, J., & Benford, S. (2000). Textile evidence supports skewed radiocarbon date for Shroud of Turin. *Textile History Journal,* 18(3), 1–15.

Matthews, D. A., & Clark, C. (1999). *The faith factor: Proof of the healing power of prayer.* Viking.

McGuire, M. B. (1988). *Ritual healing in suburban America.* Rutgers University Press.

Meier, J. P. (1991). *A marginal Jew: Rethinking the historical Jesus, Volume One: The roots of the problem and the person.* Doubleday.

Moreland, J. P. (1987). *Scaling the secular city: A defense of Christianity.* Baker Academic.

Moreland, J. P. (1997). *Love your God with all your mind: The role of reason in the life of the soul.* NavPress.

Newton, I. (1687). *Philosophiæ naturalis principia mathematica.*

O'Collins, G. (1987). *Jesus risen: An historical, fundamental, and systematic examination of Christ's resurrection.* Paulist Press.

O'Regan, B., & Hirshberg, C. (1993). *Spontaneous remission: An annotated bibliography.* Institute of Noetic Sciences.

Origen. (n.d.). *Contra Celsum,* Book II, Chapter 1. Retrieved from https://www.newadvent.org/fathers/04162.htm

Phlegon of Tralles. (n.d.). *Chronicles.* Fragments cited in Julius Africanus and Eusebius.

Plantinga, A. (2000). *Warranted Christian belief.* Oxford University Press.

Pliny the Younger. (2003). *Letters* (10.96–97) (B. Radice, Trans.). Penguin Classics.

Rogers, R. N. (2004). Studies on the radiocarbon sample from the Shroud of Turin. *Thermochimica Acta*, 425(1–2), 189–194.

Rogers, R. N. (2005). *A chemist's perspective on the Shroud of Turin.* Lulu Press.

Rucker, R. A. (2019). Holistic solution to the mysteries of the Shroud of Turin (Rev. 0, December 18). Retrieved from https://0201.nccdn.net/1_2/000/000/174/1a8/how-the-image-was-formed-on-the-shroud.pdf

Rucker, R. A. (2024). *Carbon dating revisited: Why the Shroud of Turin is from the first century.* Resurrection Research Institute.

Stark, R. (2003). *For the glory of God: How monotheism led to reformations, science, witch-hunts, and the end of slavery.* Princeton University Press.

Strobel, L. (2018). *The case for miracles: A journalist investigates evidence for the supernatural.* Zondervan.

Strobel, L. (Host). (2021, February 23). *Interview with Barbara Schneider: A miraculous healing* [Video]. YouTube. https://youtu.be/TQs4zzfOAHc?si=NG1HVNqpmUbHa3gP

Suetonius, G. (2007). *The twelve Caesars* (R. Graves, Trans.). Penguin Classics.

Tacitus, C. (116 CE). *Annals* (A. J. Church & W. J. Brodribb, Trans.).

The Holy Bible. (1978). New International Version. Zondervan. (1 Peter 3:15)

Vignon, P. (1939). *The Shroud of Christ.* Houghton Mifflin.

Wilson, I. (2010). *The Shroud: The 2000-year-old mystery solved.* Bantam Press.

Wilson, K. G. (1983). Faith and healing: A psychological perspective. *American Psychologist*, 38(11), 1282–1289.

Wright, N. T. (2003). *The resurrection of the Son of God.* Fortress Press.

CONCLUSION

As we conclude this exploration of the Brain Metafunction Framework, it becomes evident that the brain is not merely an organ of cognition, but the central mediator of human experience, personhood, and relationality. It sustains thought and memory through intricate networks and dynamic processes, while supporting the core functions that shape our biological, psychological, moral, and spiritual identity. This book has examined the brain's many dimensions—from bodily regulation and integrative function to perception, memory, consciousness, and free will—through the unifying lens of the Brain Metafunction. In doing so, it has offered a comprehensive and integrative model that bridges scientific understanding with philosophical reflection and theological vision.

At its core, the Brain Metafunction Framework affirms the profound unity of the human person. No isolated structure or process can account for the fullness of our lived experience. Rather, this model brings together the tangible and intangible, the physiological and metaphysical, honoring the complexity and dignity of human existence. It reframes the individual not as a mere biological machine, but as a being whose mental, moral, and spiritual life is embodied, interconnected, and meaning-laden—a being whose experience is organized through neural architecture yet grounded in a reality that transcends neurobiology.

A central contribution of this work is the Metafunctional Theory of Perception, which reconceptualizes perception as an active and generative process. At the heart of this theory is the efferent loop—a constructive and potentially transphysical mechanism that projects perceptual coherence into the world. This challenges reductionist stimulus-response models and affirms perception as participatory and formative. Clinically, this theory proves relevant in understanding phenomena such as phantom limb pain, neurorehabilitation, and trauma recovery—contexts in which perception can be reshaped even in the absence of intact sensory input. This model maps clinical conditions along the afferent–metafunctional–efferent loop, offering a novel diagnostic and therapeutic lens. It is grounded in the integrative coherence of the brain's metafunctional organization.

The concept of Brain Metafunction Syndromes extends this framework to disruptions of personal identity—as seen in coma, dementia, and neurodegenerative disease. These syndromes deepen our understanding of the brain's role not just in sustaining biological life, but in upholding human presence and moral agency. In redefining personhood and revisiting the notion of death, the framework moves beyond biological reductionism, offering a more nuanced, ethically coherent vision of life's end.

This book also developed the concept of Metafunctional Free Will, proposing that the brain's integrative complexity provides the conditions for authentic agency. Brain Metafunction Ports—points of convergence among sensory, cognitive, and moral domains—highlight how the brain integrates diverse inputs into a unified sense of self. At the edge of this integration, the theory of Brain Metafunction Transphysicalism suggests that certain dimensions of human existence—consciousness, moral awareness, spiritual longing—are not fully explicable within materialist paradigms, but rather gesture toward a transcendent metaphysical order.

In this light, the framework resonates with the vision of Christian Theism, proposing that the brain serves not only as the seat of embodied cognition but also as a conduit between the physical and the divine. Near-death experiences, mystical perception, and documented miracles—explored throughout this work—are not dismissed as anomalies, but rather understood as events that may reveal the brain's capacity to interface with transphysical realities. This approach affirms the unity of body and soul and restores theological anthropology to its rightful place in the dialogue between science and faith. The book has also engaged critically with materialism, naturalism, and scientism. While these paradigms may offer valuable insights within their domains, they fall short of explaining the full range of human experience. In contrast, Christian Theism, rooted in metaphysical realism, offers a more coherent and comprehensive worldview—one that accounts for rationality, freedom, consciousness, moral truth, and ultimate meaning. In this vision, the Brain Metafunction Framework becomes a bridge: between biology and meaning, medicine and ethics, finitude and transcendence.

What began as a neurophilosophical inquiry into consciousness and coma has developed into a holistic vision of the human person—embodied, intelligent, morally accountable, and spiritually destined. As neuroscience, theology, and philosophy converge, the Brain Metafunction

Framework emerges as a generative model for future research and practice—one that honors scientific rigor while embracing the mystery of personhood. It calls for approaches to medicine, psychology, and care that are not merely technical, but profoundly human. In the final analysis, this framework affirms that the human being is more than the sum of neural activity. We are creatures of meaning, longing, and transcendence. To explore the brain is to approach a gateway—into consciousness, morality, relationality, and ultimately, into the mystery of God. This journey invites us to embrace the full dignity of every person and to stand in awe before the miracle of life, mind, and spirit.

Looking ahead, the Brain Metafunction Framework invites ongoing interdisciplinary exploration—bridging neuroscience, philosophy, theology, and clinical practice. It opens new avenues for inquiry into perception, spiritual experience, and consciousness, while offering a foundation for ethically and spiritually informed therapy, education, and care. It encourages a renewal of spiritual formation, grounded in a deep understanding of the person as an embodied soul, capable of reason, love, relationship, and reverence. Whether pursued through academic scholarship, clinical application, or personal faith, this model calls for a holistic vision of human flourishing—rooted in both the mysteries of the brain and the transcendent hope of the soul.

www.ingramcontent.com/pod-product-compliance
Lightning Source LLC
LaVergne TN
LVHW081943200926
848484LV00009B/2505
* 9 7 8 1 7 7 8 2 9 6 2 4 6 *